Lecture Notes of the Institute for Computer Sciences, Social Informatics and Telecommunications Engineering 370

More information about this series at http://www.springer.com/series/8197

Houbing Song · Dingde Jiang (Eds.)

Simulation Tools and Techniques

12th EAI International Conference, SIMUtools 2020
Guiyang, China, August 28–29, 2020
Proceedings, Part II

Editors
Houbing Song
Embry-Riddle Aeronautical University
Daytona Beach, FL, USA

Dingde Jiang
School of Astronautics and Aeronautic
UESTC
Chengdu, China

ISSN 1867-8211 ISSN 1867-822X (electronic)
Lecture Notes of the Institute for Computer Sciences, Social Informatics and Telecommunications Engineering
ISBN 978-3-030-72794-9 ISBN 978-3-030-72795-6 (eBook)
https://doi.org/10.1007/978-3-030-72795-6

This Springer imprint is published by the registered company Springer Nature Switzerland AG
The registered company address is: Gewerbestrasse 11, 6330 Cham, Switzerland

Preface

We are delighted to introduce the proceedings of the twelfth edition of the European Alliance for Innovation (EAI) International Conference on Simulation Tools and Techniques (SIMUTools). The conference focuses on a broad range of research challenges in the field of simulation, modeling and analysis, addressing current and future trends in simulation techniques, models, practices and software. The conference is dedicated to fostering interdisciplinary collaborative research in these areas and across a wide spectrum of application domains.

The technical program of SIMUTools 2020 consisted of 125 full papers. Coordination with the steering chair, Imrich Chlamtac, was essential for the success of the conference. We sincerely appreciate his constant support and guidance. It was also a great pleasure to work with such an excellent organizing committee team for their hard work in organizing and supporting the conference. In particular, the Technical Program Committee completed the peer-review process of technical papers and made a high-quality technical program. We are also grateful to the Conference Manager, Karolína Marcinová, for her support and to all the authors who submitted their papers to the SIMUTools 2020 conference.

We strongly believe that the SIMUTools conference provides a good forum for all researchers, developers and practitioners to discuss all scientific and technological aspects that are relevant to simulation tools and techniques. We also expect that future SIMUTools conferences will be as successful and stimulating, as indicated by the contributions presented in this volume.

March 2021

Houbing Song
Dingde Jiang

Organization

Steering Committee

Chair

Imrich Chlamtac Bruno Kessler Professor, University of Trento, Italy

Organizing Committee

General Chair

Dingde Jiang University of Electronic Science and Technology of China, China

General Co-chair

Houbing Song Embry-Riddle Aeronautical University, USA

Technical Program Committee Co-chairs

Lei Shi Institute of Technology Carlow, Ireland
Sheng Qi University of Electronic Science and Technology of China, China
Binbing Hou Louisiana State University, USA
Chenggang Li Guizhou University of Finance and Economics, China
Bangtao Zhou University of Electronic Science and Technology of China, China

Publicity and Social Media Co-chairs

Litao Zhang Zhengzhou Institute of Aeronautical Industry Management, China
Zhihao Wang University of Electronic Science and Technology of China, China
Jingyang Zhang University of Electronic Science and Technology of China, China

Workshops Co-chairs

Zhi Zhu National University of Defense Technology, China
Yihang Zhang University of Electronic Science and Technology of China, China
Zhibo Zhang University of Electronic Science and Technology of China, China
Wenbo Yan University of Electronic Science and Technology of China, China

Sponsorship and Exhibits Co-chairs

Changsheng Zhang	Northeastern University, China
Yuwen Wang	University of Electronic Science and Technology of China, China

Publications Co-chairs

Qingshan Wang	Hefei University of Technology, China
Feng Wang	University of Electronic Science and Technology of China, China

Panels Chairs

Lei Miao	Middle Tennessee State University, Murfreesboro, USA
Feilong He	University of Electronic Science and Technology of China, China

Tutorials Chairs

Peiying Zhang	China University of Petroleum, China
Li Xue	University of Electronic Science and Technology of China, China

Demos Chairs

Minhong Sun	Hangzhou Dianzi University, China
Lisha Cheng	University of Electronic Science and Technology of China, China

Posters and PhD Track Co-chairs

Lei Chen	Xuzhou University of Technology, China
JunYang Zhang	University of Electronic Science and Technology of China, China

Web Chairs

Xiongzi Ge	NetApp, Advanced Technology Group, USA
Liuwei Huo	Northeastern University, China
Jian Jin	University of Electronic Science and Technology of China, China

Local Co-chairs

Mingsen Deng	Guizhou University of Finance and Economics, China
Zhihao Wang	University of Electronic Science and Technology of China, China
Hujun Shen	Guizhou Education University, China

Technical Program Committee

Dingde Jiang	University of Electronic Science and Technology of China, China
Litao Zhang	Zhengzhou University of Aeronautics, China
Lei Shi	Institute of Technology Carlow, Ireland
Sheng Qi	University of Electronic Science and Technology of China, China
Lei Chen	Xuzhou University of Technology, China
Bangtao Zhou	University of Electronic Science and Technology of China, China
Giovanni Stea	University of Pisa, Italy
Lorenzo Donatiello	University of Bologna, Italy
Andrea D'Ambrogio	University of Rome Tor Vergata, Italy
Christian Engelmann	Oak Ridge National Laboratory, USA
Kevin Jin	Illinois Institute of Technology, USA
Gary Tan	National University of Singapore, Singapore
Jiaqi Yan	Microsoft, USA
James Byrne	Dublin City University, Ireland
Jan Himmelspach	Nordakademie University of Applied Sciences, Germany
Johannes Lüthi	University of Applied Sciences Kufstein, Austria
Linbo Luo	Xidian University, China
Stephan Eidenbenz	Los Alamos National Laboratory, USA
Yiping Yao	National University of Defense Technology, China
Francesco Quaglia	University of Rome Tor Vergata, Italy
Zongwei Luo	Southern University of Science and Technology, China

Contents – Part II

Contents – Part I

Attention Aware Deep Learning Object Detection and Simulation

Jiping Xiong(✉), Lingyun Zhu, Lingfeng Ye, and Jinhong Li

College of Physics and Electronic Information Engineering, Zhejiang Normal University, Jinhua 321300, China

xjping@zjnu.cn

Abstract. Dish recognition has certain difficulties in specific applications. Because in the actual inspection, the dishes are filled with food, and the food occupy most of the space of the dishes, and only the edges of the dishes can be seen. If you use empty dishes for training, the accuracy will be low due to insufficient feature matching during actual detection. At the same time, due to the wide variety of foods, if we collect all the food during training, the pre-processing workload will be very large. Based on the above ideas, this paper analyzes the model through three visualization methods, improves Faster R-CNN, and proposes a Cross Faster R-CNN model. This model consists of Faster R-CNN and Cross Layer, which can fuse the low-level features and high-level features of dishes. During training, the model can focus the feature extraction on the edges of the dishes, reducing the interference of food on dish recognition. This method improves the detection accuracy without significantly increasing the detection time. The experimental results show that compared with Faster R-CNN, the accuracy and recall of Cross Faster R-CNN have increased to a certain extent, and the detection speed has basically not changed significantly.

Keywords: Deep learning · Object detection · Convolutional neural network · Visualization

1 Introduction

In recent years, artificial intelligence has developed rapidly. More and more fields combined with artificial intelligence have begun to improve people's daily live. With the growth of the economy, people's requirements for living standards have gradually increased. In view of the need to solve the problems of the traditional model in the catering industry, controlling costs is one of the most important factors. Among them, labor costs account for the majority. Especially at this stage, restaurants generally use manual billing, which is not only a waste of human resources. At the same time, queueing billing is a waste of customer time. Therefore, if machines can be used for manual billing, it can reduce costs on the one hand and save time and increase passenger traffic on the other.

Artificial intelligence developed slowly before 2012, but after the advent of machine learning, there was a dawn, and the most eye-catching aspect of machine learning is deep

H. Song and D. Jiang (Eds.): SIMUtools 2020, LNICST 370, pp. 1–14, 2021.
https://doi.org/10.1007/978-3-030-72795-6_1

learning. Deep learning has broken through many areas that are difficult for artificial intelligence machine learning to break through, bringing the possibility of the further development of artificial intelligence [1, 2]. AlphaGo is a typical artificial intelligence that uses deep learning. Deep learning's effects achieved in the fields of computer vision and natural language processing have far surpassed traditional methods [3, 4]. Deep learning is a learning method that approximates humans. Through a large amount of sample data, the machine learns the inherent laws and representation levels, and finally has the ability to analyze like humans. Convolutional neural network is a successful application of deep learning in the field of image vision. It has made great breakthroughs in tasks such as image classification [5–7], object detection [8–11], semantic segmentation [12–16], and instance segmentation [17]. Since *Hinton* [18] won the championship with *AlexNet* in the ImageNet competition in 2012, convolutional neural networks have achieved great development and have been applied in daily life, such as image subtitles [19, 20, 33], visual question and answer[21, 22, 34], and autonomous driving[23, 24, 35] which are recent some research focused on, and so on.

The current research on automatic billing has two directions, one is the billing for food identification, and the other is the billing for dish recognition. The latter cannot solve the problems of lighting, occlusion, deformation, etc. with traditional image processing methods, and will be subjected to many restrictions in practical applications. Therefore, more RFID chips are embedded in the bottom of the dish to identify the dish. This method has some disadvantages, especially the RFID chips are not resistant to high temperatures. The restaurant will easily damage the chips during high temperature sterilization. The replacement of the entire dish will increase the cost and is not conducive to the long-term stable development of the restaurant.

In order to solve the above problems, this paper applies convolutional neural network to dish detection. Based on Faster R-CNN, combining the characteristics of dishes and introducing an attention mechanism, a method of dish detection based on Cross Faster R-CNN is proposed. This method uses cross layer to fuse the edge, color and texture information of dishes with deep advanced features. The experimental results show that this method can reduce the miss rate of dishes and improve the detection accuracy. The main work of this article is as follows:

(1) Visualize Faster R-CNN, analyzing the features extracted by VGG16 (a convolutional neural network) layer by layer when detecting dishes, understanding the role of the convolution kernel of each layer. At the same time, using the heat map to determine which region features of the dishes have a greater response to the detection results.
(2) According to the visual analysis, the network structure of Cross Faster R-CNN is proposed by improving Faster R-CNN and combining the idea of cross-connect. The network fuses the low-level features of the dishes with the high-level features. Then the 1×1 convolution kernel is used to increase the nonlinearity of the network and improve the network's ability to express complex features.

2 Related Work

2.1 Convolutional Neural Network

Convolutional neural network [25, 36, 37] (CNN) is a type of feed forward neural network with convolutional calculation and deep structure. It belongs to representational learning and can automatically learn high-level features from samples without artificial design features. The structure of a convolutional neural network generally consists of a convolutional layer, a pooling layer, and a full connected layer. Compared to the hidden layer of a traditional neural network, the convolutional layer can greatly reduce the amount of parameters due to the characteristics of convolutional weight sharing. Thus, CNN can train models with deeper network structures. At the same time, the convolutional layer and pooling layer in the convolutional neural network can respond to the translation invariance of the input features, and can identify similar features located in different positions in space.

In order to extract more advanced features, the current convolutional neural network generally has a deep structure and a large number of network layers. However, the weight of the convolutional neural network is updated by gradient back propagation. A deeper network structure will cause the mean and standard deviation of the data change, resulting in a covariate shift phenomenon, that is, the disappearance of the gradient. In order to solve the above problems, *Ioffe S.* [26] and others proposed batch normalization (BN) in 2015. The features will be obtained through the convolution layer, and the data will be scaled using two linear parameters to meet the variance is 1, the mean is 0, and then the activation function is used as the input of the next layer. The process of the BN layer is as follows:

$$\mu = \frac{1}{m}\sum_{i}^{m}\left(X^{(i)} - \mu\right)^2 \tag{2-1}$$

$$\sigma^2 = \frac{1}{\mathrm{m}}\sum_{i}^{m}\left(X^{(\mathrm{i})} - \mu\right)^2 \tag{2-2}$$

$$Z^{(i)} = \frac{X^{(i)} - \mu}{\sqrt{\sigma^2 + \xi}} \tag{2-3}$$

$$\hat{Z}^{(i)} = \alpha * Z^{(i)} + \beta \tag{2-4}$$

where X is the original data, μ is the mean, σ^2 is the variance, $\hat{Z}$ is the scaled data, α and β are two newly added parameters, which are trained by the convolutional neural network to replace the bias ξ. First input data $X = \{x^{(1)}, x^{(2)}, \Lambda, x^{(m)}\}$, calculate the mean μ and variance of the data σ^2, then normalize Z, train the parameters α and β, and update $\hat{Z}$ through linear transformation. In the forward propagation, the new distribution value is obtained through the learnable α and β parameters; in the back propagation, the sum α and β the related weights are obtained through the chain derivation.

2.2 Visualization

Although the convolutional neural network has excellent performance in computer vision tasks, it still hasn't the ability to use mathematical formulas to explain the deep-level feature expression of neural networks. And humans and computers have different ways of expressing features. Convolutional neural network is a type of end-to-end representation learning, and people can only understand the input and output ends. The meaning of the feature map in the middle layer still can't be clearly explained, so the convolutional neural network is also called a "black box" model.

With the in-depth research in recent years, the "black box" of convolutional neural networks has gradually been opened up, and the internal working principle of convolutional neural network can be intuitively understood in a visual way [27, 28]. Common visualization methods at this stage include the visualization of feature map, the visualization of convolution kernels, and the heat map which can be used to understand the activation characteristics of category.

The feature map is the output of the convolutional layer. The visualization of the feature map is to display the output of the convolution kernel in the form of an image [38, 39], which can help understand the role of the convolution kernel. This is commonly used in the visualization of convolutional neural networks.

The process of convolution is essentially the process of feature extraction. Each convolution kernel represents a feature. If some regions in the image respond more to a convolution kernel, the region can be considered to be similar to the convolution. Therefore, the visualization of the convolution kernel can also be regarded as an optimization problem, that is, to find an image with the largest response to a certain convolution kernel. The whole process is as follows:

First, get an image I randomly, and find the gradient of K to I for the convolution kernel to be visualized, that is:

$$Grad = \frac{\partial K}{\partial I} \tag{2-5}$$

Then update the image I by gradient ascent, that is:

$$I = I + \alpha * Grad \tag{2-6}$$

where α is the step size, which is 1 in this experiment.

In 2017, *Selvaraju R.R* [29] and others proposed a gradient weighted class activation map (Grad-CAM), which uses a heat map to display the degree of activation of the picture for each region of the output category, it is consistent with human visual characteristics. To some extent, it explains the correlation between the output of the convolutional neural network and the images. The essence of Grad-CAM is that the output category weights the feature map obtained from a certain convolution layer through gradients. The entire process is as follows:

First, calculating the probability y^c of the output category c of the convolutional neural network *softmax* layer for the gradient of the pixel value of the feature map A_{ij} obtained by the convolution layer (the original text calculates the output before the *softmax* layer, and in the experiment it was found that the heat map obtained by calculating the output probability after the *softmax* layer is more obvious), namely:

$$\frac{\partial y^c}{\partial A_{ij}} \tag{2-7}$$

where y^c is the output probability of category c, A is the convolution feature map, and k is the number of channels in the feature map.

Then the global average of the partial derivatives is obtained, that is:

$$a_k^c = \frac{1}{Z}\sum_i \sum_j \frac{\partial y^c}{\partial A_{ij}} \tag{2-8}$$

where a_k^c represents the partial linearization from the deep network in A, that is, the importance of category c relative to the k-th channel in A.

Then a_k^c and A are weighted and combined and processed by the *Relu* activation function, that is:

$$L^c_{Grad-Cam} = \mathrm{ReLU}\left(\sum_k a_k^c A^k\right) \tag{2-9}$$

where L is a class activation map for category c.

2.3 Attention Mechanism

In general, the Attention mechanism is to focus attention on important points and ignore other unimportant factors. In this article, the edge of the dishes where attention needs to be focused on, other places, including the center of the dishes, are ignored.

The general definition of attention mechanism is shown in formula (2-10) [30]:

$$Attention(Q, K, V) = soft\max\left(\frac{QK^T}{\sqrt{d_k}}\right)V \tag{2-10}$$

where $Q \in R^{m \times d_k}$, $K \in R^{m \times d_k}$, $V \in R^{m \times d_k}$.

May wish to take $q_t \in Q$, then the encoding result obtained for a single input vector q_t can be expressed as [23]:

$$Attention(q_t, K, V) = \sum_{s=1}^{m} \frac{1}{Z}\exp\left(\frac{\langle q_t, k_s\rangle}{\sqrt{d_k}}\right)v_s \tag{2-11}$$

Where Z is the normalization factor, and q, k, v are shorthand for *query, key, value.* $\sqrt{d_k}$ regulates the internal product of the activation function so that it is not too large. The $\exp\left(\frac{\langle q_t, k_s\rangle}{\sqrt{d_k}}\right)v_s$ in formula (2-11) is the core part of the attention mechanism. $\exp\left(\frac{\langle q_t, k_s\rangle}{\sqrt{d_k}}\right)v_s$ is mainly used to measure the similarity q_t with v_s. The entire formula can be understood as looking for a non-linear mapping relationship between q_t and v_s.

The general flow of formula (2-11) is the inner product of q_t and each k_s, and then the similarity of q_t and v_s is evaluated by *softmax*. The final result is a vector of d_v by weighted summation. Observing the calculation process, other parts of the input matrix besides q_t will also affect the calculation result of the vector d_v. Therefore, d_v is not only related to q_t, but also related to other parts of the input matrix, and q_t can be used as the representation of q_t in the global vector.

Based on the *Attention* mechanism[40], the *Multi-Head Attention* mechanism proposed by *Google* is used to further improve the coding ability of the model[31,32]. The *Attention* mechanism model used in this paper is mainly based on *Multi-Head Attention.* Compared with the basic model, the *Multi-Head Attention* mechanism has two differences. One is to map Q, K, V via matrix parameters, and then send them to the *Attention* model. Second, the original input is subjected to *Attention* operations without sharing parameters multiple times, and the output results are stitched. These two improvements can improve the description ability of the model. The specific model is as follows:

$$head_i = Attention\left(QW_i^Q, KW_i^K, VW_i^V\right) \tag{2-12}$$

where $W_i^Q \in R^{d_k \times \tilde{d}_k}$, $W_i^K \in R^{d_k \times \tilde{d}_k}$, $W_i^V \in R^{d_v \times \tilde{d}_v}$, the feature representation of the final output can be expressed as:

$$MultiHead(Q, K, V) = Concat(head_1, \Lambda, head_h) \tag{2-12}$$

3 Dish Detection Based on Cross Faster R-CNN

3.1 Visualizing Faster R-CNN

Visualizing Faster R-CNN can help understand its working principle in dish detection, further clarify the features that are activated when making image classification decisions. At the same time, when it involves tasks in a specific field, it can improve the network structure based on prior knowledge and the accuracy of the model for this task.

Feature Map Visualization

Visually understand the differences between the extracted features of the low-level convolution layer and the high-level convolution layer. And the visualization of the feature map obtained by Conv1_2 convolution layer, Conv2_2 convolution layer, Conv3_3 convolution layer, Conv4_3 convolution layer, Conv5_3 convolution layer and the RPN_Conv convolution layer in Faster R-CNN respectively, as shown in Fig. 1. For the convenience of analysis, each feature map shows only the first 16 channels.

It can be seen that the features extracted by the shallow convolution layer are similar to the edge and color information, and the retained image information is relatively complete. As the number of layers becomes deeper, the image feature information obtained through the convolution layer gradually decreases, and the entire feature map becomes more abstract.

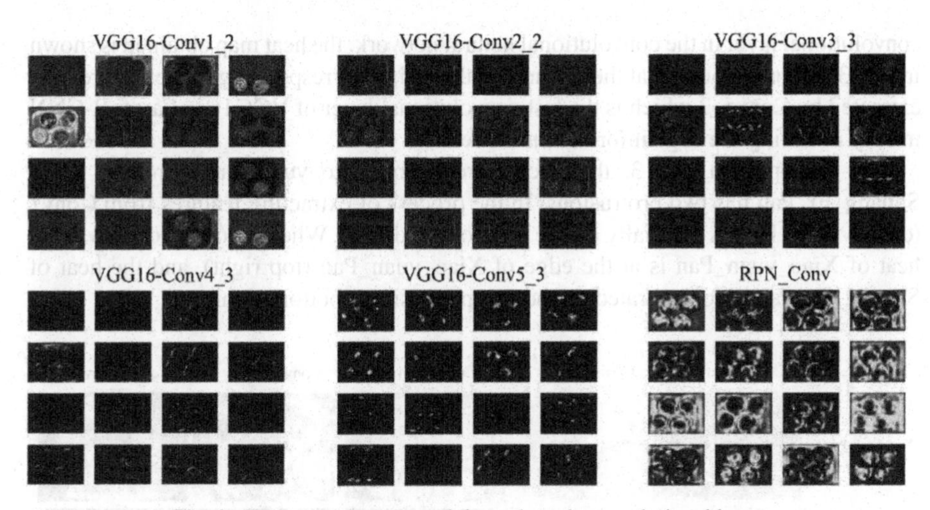

Fig. 1. Feature map extracted through each convolutional layer

RPN_Conv is obtained from the feature map obtained through the VGG16-Conv5_3 convolutional layer through a 3×3 convolutional layers and used as the input of the RPN network. From the visualization results, it can be seen that the information in this layer mainly relates to the characteristics of the dish, and also proves that the RPN network is effective for extracting foreground and background.

Convolution Kernel Visualization

The visualization of the convolution kernel can be seen from Sect. 2.2. It is necessary to randomly obtain a pair of original images. In this paper, the input image is used as the original image. Based on this, gradient rise is performed. The number of iterations is 100. The visualization results are shown in Fig. 2.

It is clear from the Fig. 2 that the low-level convolution kernel responds to features such as color, edge and texture, while the feature of interest in the high-level convolution kernel is an advanced abstracted information. These features are so complex that they are difficult to understand and explain. At the same time, it can also be found that the high-level convolution kernel is difficult to be visualized by the gradient rising method, and the obtained image has more noise points.

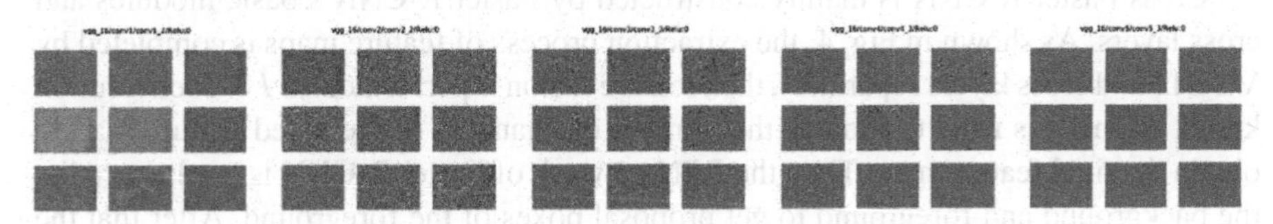

Fig. 2. Visualization of the convolution kernel at each layer

Gradient Weighted Class Activation Map

Using the Grad-CAM method to weighted and sum the feature maps obtained by the convolutional layer in the convolutional netural network, the heat map obtained is shown in Fig. 3. It can be found that the features of the dishes corresponding to the feature map extracted by Conv5_3 which is the last convolutional layer of VGG16 in Faster R-CNN mainly involving the edge information of the dish.

For example, in Fig. 3, the circumference of Xiao_yuan_Pan is round, while Shuang_Er_Pan has two protrusions. In the process of extracting features from Conv1 to Conv5, the heat is gradually on the edge of the dishes. When it comes to Conv5, the heat of Xiao_yuan_Pan is at the edge of Xiao_yuan_Pan (top right), and the heat of Shuang_Er_Pan is concentrated on the two protrusions (bottom right).

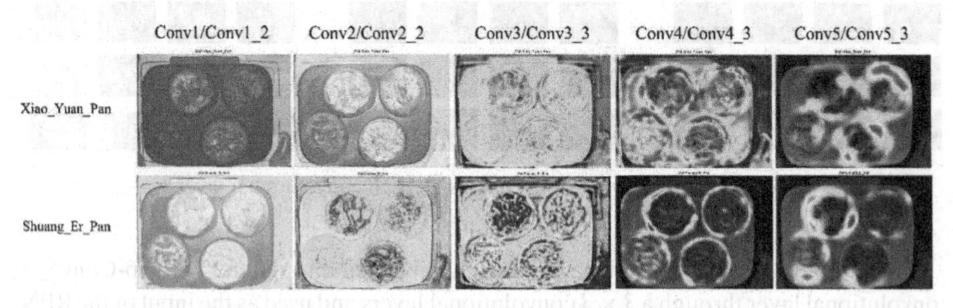

Fig. 3. Thermal map obtained by convolution of the layers corresponding to the categories Xiao_Yuan_Pan and Shuang_Er_Pan

Because transfer learning is used to train the dish detection model, VGG16′s pre-training weights come from the ImageNet dataset. At the same time, the Conv1 and Conv2 repeated convolution block weights are fixed when training the model. The heat maps of Conv1_2 and Conv2_2 are not aimed at the detection result of the dishes, and should be a general heat map for extracting shallow features.

3.2 Cross Faster R-CNN

Figure 4 is a schematic diagram of the Cross Faster R-CNN model. A Cross Layer is added to the Faster R-CNN, and the edge and texture features are combined with the deep-level advanced features extracted by VGG16 to enhance the model's feature expression ability for targets such as dishes.

Cross Faster R-CNN is mainly constructed by Faster R-CNN's basic modules and cross layers. As shown in Fig. 4, the extraction process of feature maps is completed by VGG16 and cross layers separately, then feature fusion is performed. A 1×1convolution kernel (Conv7) is used to modify the number of channels of the fused feature map to obtain the final feature map. Then the RPN network of Faster R-CNN is used to predict the background and foreground to get proposal boxes of the foreground. After that the proposal boxes and feature maps are passed through ROI Pooling (Region of Interest Pooling) to output feature maps with the same size. Finally, predicting the category and location of the object by the Full Connection network (FC).

The cross layer is composed of two convolutional layers and two pooling layers. The first convolutional layer (Conv2_6_1) which is composed of 128 3×3 convolution

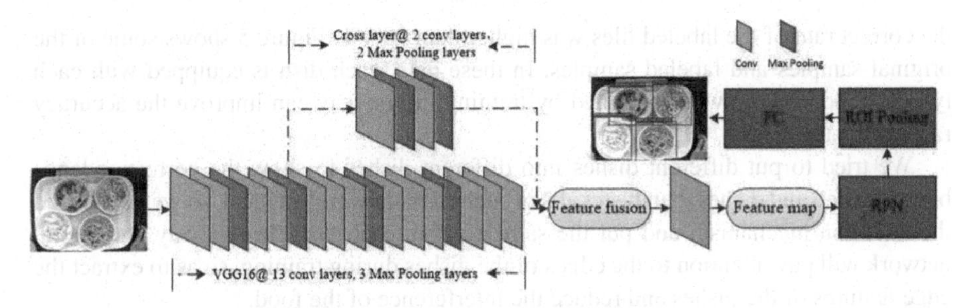

Fig. 4. Cross Faster R-CNN model

Table 1. Cross Faster R-CNN structure

Cross Faster R-	Cross Faster R-	Cross Faster R-
Cross Faster R-	Cross Faster R-	Cross Faster R-

kernels, followed by a 2×2 maximum pooling layer (Max Pooling2_6_1) reduces the height and width of the feature map, and then increases the number of channels of the feature map through a convolution layer (Conv2_6_2) composed of 256 1×1 convolution kernels, and finally a 2×2 maximum pooling layer (Max Pooling2_6_2) reduces the height and width of the feature map so that the feature dimensions obtained by cross layer match the feature dimensions obtained by VGG16. Because VGG16 speeds up training by pre-training weights, and fixed the weights of the Conv1 and Conv2 layers, the purpose of the 3×3 convolutional layer through a receptive field in the cross layer is to ensure the low-level characteristics of the shallow dishes, while the 1×1 convolutional layer is added to reduce the amount of parameters and increase the number of channels of the feature map and increase its nonlinear expression ability.

Cross Faster R-CNN does not fix the size of the input image. Table 1 shows the detailed network structure of the entire model more intuitively. VGG16 standard input is used as an example, the input image size is fixed at $224 \times 224 \times 3$. There are 11 categories and 9 anchors.

Note that the RPN_cls_score layer and the RPN_bbox_pred layer are independent of each other in the RPN structure block, and each layer is directly connected to the RPN_Conv layer. Similarly, in the FC structure block, the cls_score layer and the bbox_pred layer are also independent of each other and connected with the FC7 layer. For the convenience of explanation, the offset is not calculated in the parameters.

4 Experiment

4.1 Experimental Data

The data set used in the experiment was collected from the actual restaurant using a collection of collection and identification equipment, and are the dishes and food being sold and used in the restaurant. Samples were labeled and verified by 14 researchers and the correct rate of the labeled files was higher than 99.9%. Figure 5 shows some of the original samples and labeled samples. In these data, each dish is equipped with each type of food. The network obtained by training in this way can improve the accuracy rate during the test.

We tried to put different dishes into different dishes to show the correspondence between food and dishes, but the results of such training were not ideal, so we introduced the attention mechanism and put the same food in each dish. In this way, the neural network will pay attention to the edges of the dishes during training, so as to extract the edge features of the dishes and reduce the interference of the food.

(a) (b)

Fig. 5. Experimental dataset. (a) original sample; (b) labeled sample.

There are a total of 31,111 samples of 10 types of dishes in the entire dataset. Each sample contains 1 to 5 types of dishes and the number of each type is about 3000 to 5000, as shown in Fig. 6. The training set contains 17,421 samples, the validation set contains 7,467 samples, and the test set contains 6,223 samples.

At the same time, 150 samples were collected during the actual use of the restaurant, as shown in Fig. 7. The pictures collected in actually will have some additional background interference, such as chopsticks, mobile phones, payment cards and other items. Labelling these background interferences (labeled as background) to participate in the training, so that items such as chopsticks, payment cards, and mobile phones will be recognized as background during prediction, and will not interfere with the recognition of dishes. Cross Faster R-CNN does't fix the input size of the samples. The resolution of each sample in this dataset is $(690 \pm 10) \times (520 \pm 30)$.

4.2 Analysis of Results

The experiments were performed on a PC, the operating system was Windows 10, and the RTX 2080ti with a GPU of 11G was used to accelerate training and model testing. The models used were implemented under Python 3.6 and Tensor flow 1.8.0.

The test is performed on the test set and the data set collected in actual use, as shown in Table 2 and 3, where Table 2 is the test set result and Table 3 is the actual collected data set result. The evaluation indicators are accuracy, recall and detection speed. Among them, the accuracy rate is a comprehensive index of TOP-1 classification accuracy rate and IOU of the regression box is bigger than 0.5, the recall rate is the ratio of correctly distinguishing foreground and background, and the detection speed is the

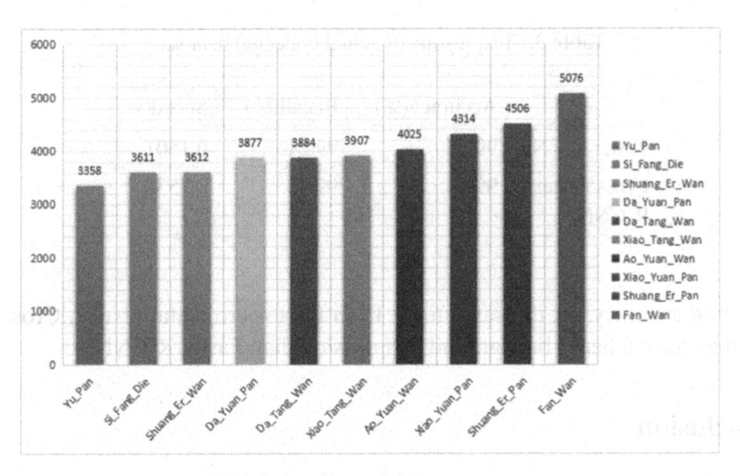

Fig. 6. Number of dishes per type

Fig. 7. The actual collection data set

time required to averagely detect an image under the GPU (excluding the first model loading time during detection). Both the test set and the actual data set are labeled, so it is only necessary to check whether the label is consistent with the actual prediction result.

Table 2. Test results

	Accuracy(%)	Recall(%)	Speed(s)
Faster R-CNN	99.87	99.51	0.1453
Cross Faster R-CNN	99.91	99.67	0.1527

Analysis of the above table shows that the detection time of Cross Faster R-CNN has little increased regardless of whether it is in the test set or the actual collected data set. The accuracy and recall of Cross Faster R-CNN are higher than Faster R-CNN, it also proves that for the detection of dishes, the fusion of low-level features can improve

Table 3. Test results of actual collected data sets

	Accuracy(%)	Recall(%)	Speed(s)
Faster R-CNN	99.52	99.21	0.1501
Cross Faster R-CNN	99.64	99.45	0.1538

the detection accuracy. At the same time, from the experimental results, Cross Faster R-CNN may have a better background suppression than Faster R-CNN.

5 Conclusion

By visualizing Faster R-CNN and analyzing the importance of features in dish detection to the detection results, this paper proposes a Cross Faster R- CNN model. The cross layer is used to combine the low-level features at the shallow level with the high-level features at the deep level. The features of the low-level layer include the features of Conv1, and the feature of the high-level layer is the feature of Conv5. The dataset has been filled with each type of food in each type of dishes to improve the detection accuracy. The experimental results show that the model improves the detection accuracy of the dishes and suppresses the background interference when the detection time does not increase significantly. In the following work, research will focus on optimizing the structure of the network model to reduce the time required for detection.

References

1. Jia, X., Li, R.: Deep learning and artificial intelligence. Neijiang Technol. **41**(06), 78–78+84 (2020)
2. Jiang, D., Wang, Y., Lv, Z., Wang, W., Wang, H.: An energy-efficient networking approach in cloud services for IIoT networks. IEEE J. Sel. Areas Commun. **38**(5), 928–941 (2020)
3. LeCun, Y., Bengio, Y., Hinton, G.: Deep learning. Nature **521**(7553), 436 (2015)
4. Qi, S., Jiang, D., Huo, L.: A prediction approach to end-to-end traffic in space information networks. Mobile Networks and Application (2019). online available
5. He, K., Zhang, X., Ren, S., et al.: Deep residual learning for image recognition. In: Proceedings of the IEEE Conference on Computer Vision and Pattern Recognition, pp. 770–778 (2016)
6. Breier, B., Onken, A.: Analysis of video feature learning in two-stream CNNs on the example of zebrafish swim bout classification. arXiv:1912.09857 (2019)
7. Jiang, D., Wang, W., Shi, L., Song, H.: A compressive sensing-based approach to end-to-end network traffic reconstruction. IEEE Trans. Network Sci. Eng. **7**(1), 507–519 (2020)
8. Levine, S., Pastor, P., Krizhevsky, A., et al.: Learning hand-eye coordination for robotic grasping with deep learning and large-scale data collection. Int. J. Robot. Res. **37**(4–5), 421–436 (2018)
9. Liang, T., Ling, H.: MFPN: a novel mixture feature pyramid network of multiple architectures for object detection. arXiv:1912.09748 (2019)
10. Liu, Y., Jin, L.: Exploring the capacity of sequential-free box discretization network for omnidirectional scene text detection. arXiv:1912.09629 (2019)

11. Jiang, D., Huo, L., Song, H.: Rethinking behaviors and activities of base stations in mobile cellular networks based on big data analysis. IEEE Trans. Network Sci. Eng. **7**(1), 80–90 (2020)
12. Zhou, Q., Yang, W., Gao, G., et al.: Multi-scale deep context convolutional neural networks for semantic segmentation. World Wide Web **22**(2), 555–570 (2019)
13. He, Y., Fritz, M.: Segmentations-leak: membership inference attacks and defenses in semantic image segmentation. arXiv:1912.09685 (2019)
14. Zhao, L., Tao, W.: JSNet: joint instance and semantic segmentation of 3D point clouds. arXiv: 1912.09654 (2019)
15. Yu, Q., Xu, D.: C2FNAS: coarse-to-fine neural architecture search for 3D medical image segmentation. arXiv:1912.09628 (2019)
16. Jiang, D., Wang, Y., Lv, Z., Qi, S., Singh, S.: Big data analysis based network behavior insight of cellular networks for industry 4.0 applications. IEEE Trans. Ind. Inform. **16**(2), 1310–1320 (2020)
17. Bai, M, Urtasun, R.: Deep watershed transform for instance segmentation. In: /Proceedings of the IEEE Conference on Computer Vision and Pattern Recognition, pp. 5221–5229 (2017)
18. Krizhevsky, A., Sutskever, I., Hinton, G.E.: Imagenet classification with deep convolutional neural networks. In: Advances in Neural Information Processing Systems, pp. 1097–1105(2012)
19. Johnson, J., Karpathy, A., Fei-Fei, L.: Densecap: Fully convolutional localization networks for dense captioning. In: Proceedings of the IEEE Conference on Computer Vision and Pattern Recognition, pp. 4565–4574 (2016.
20. Jiang, D., Huo, L., Lv, Z., Song, H., Qin, W.: A joint multi-criteria utility-based network selection approach for vehicle-to-infrastructure networking. IEEE Trans. Intell. Transp. Syst. **19**(10), 3305–3319 (2018)
21. Young, T., Hazarika, D., Poria, S., et al.: Recent trends in deep learning based natural language processing. IEEE Comput. IntelligenCe Mag **13**(3), 55–75 (2018)
22. Jiang, D., Huo, L., Li, Y.: Fine-granularity inference and estimations to network traffic for SDN. PLoS ONE **13**(5), 1–23 (2018)
23. Milz, S., Arbeiter, G., Witt, C., et al.: Visual slam for automated driving: Exploring the applications of deep learning. In: Proceedings of the IEEE Conference on Computer Vision and Pattern Recognition Workshops, pp. 247– 257(2018)
24. Wang, Y., Jiang, D., Huo, L., Zhao, Y.: A new traffic prediction algorithm to software defined networking. Mobile Networks and Applications, 2019 (2019). online available
25. Gu, J., Wang, Z., Kuen, J., et al.: Recent advances in convolutional neural networks. Pattern Recogn. **77**, 354–377 (2018)
26. Ioffe, S., Szegedy, C.: Batch normalization: Accelerating deep network training by reducing internal covariate shift. arXiv preprint arXiv:1502.03167 (2015)
27. Wagner, J., Kohler, J.M., Gindele, T., et al.: Interpretable and fine-grained visual explanations for convolutional neural networks. In: Proceedings of the IEEE Conference on Computer Vision and Pattern Recognition, pp. 9097–910 (2019)
28. Jiang, D., Zhang, P., Lv, Z., et al.: Energy-efficient multi-constraint routing algorithm with load balancing for smart city applications. IEEE Internet Things J. **3**(6), 1437–1447 (2016)
29. Selvaraju, R.R., Cogswell, M., Das, A., et al.: Grad-cam: visual explanations from deep networks via gradient-based localization. In: Proceedings of the IEEE International Conference on Computer Vision, pp. 618–626 (2017)
30. Zhong, C., Zhou, H., Wei, H.: A 3D point cloud object recognition method based on attention mechanism. Key R&D plan projects of the Ministry of Science and Technology (2019)
31. Vaswani, A., Shazeer, N., Parmar, N., et al.: Attention is all you need. In: Advances in Neural Information Processing Systems, pp. 998–6008. [S.1.], Long Beach, USA; IEEE (2017)

32. Jiang, D., Li, W., Lv, H.: An energy-efficient cooperative multicast routing in multi-hop wireless networks for smart medical applications. Neurocomputing **220**, 160–169 (2017)
33. Zhou, Y., Zhang, R.: A brief analysis of subtitle translation of documentary wild china from the perspective of eco-translatology. Theory Pract. Language Studies (TPLS) **9**(10), 1301–1308 (2019)
34. Hong, J., Park, S., Byun, H.: Selective residual learning for Visual Question Answering. Neurocomputing **402**, 366–374 (2020)
35. Liu, D., Cui, Y., Chen, Y.: Jiyong Zhang; Bin Fan, Video object detection for autonomous driving: Motion-aid feature calibration. Neurocomputing **409**, 1–1 (2020)
36. Chirra, V.R.R., Uyyala, S.R., KishoreKolli, V.K.: Deep CNN: a machine learning approach for driver drowsiness detection based on eye state. Revue d'Intelligence Artificielle, 33(6), EI (2020)
37. Vo, S.A., Scanlan, J., Turner, P., Ollington, R.: Convolutional Neural Networks for individual identification in the Southern Rock Lobster supply chain. Food Control **118**, 107419 (2020)
38. Ebani, E.J., Kaplitt, M.G., Wang, Y., Nguyen, T.D., Askin, G., Chazen, J.L.: Improved targeting of the globus pallidus interna using quantitative susceptibility mapping prior to MR-guided focused ultrasound ablation in Parkinson's disease. Clin. Imaging **68**, 94–98 (2020)
39. Madhar, S.A., Mraz, P., Mor, A.R., Ross, R.: Empirical analysis of partial discharge data and innovative visualization tools for defect identification under DC stress. Int. J. Electr. Power Energy Syst. **123**, 106270 (2020)
40. Luo, M., Wen, G., Yang, H., Dai, D., Ma, J.: Learning competitive channel-wise attention in residual network with masked regularization and signal boosting. Expert Syst. Appl. **160**, 113591 (2020)

Robust Stability of Uncertain Replicator Population Dynamics with Time Delay

Chongyi Zhong[1,2], Nengfa Wang[3(✉)], Hui Yang[2], and Wei Zhao[2]

[1] School of Mathematics and Big data, Guizhou Education University, Guiyang, China
[2] School of Mathematics and Statistics, Guizhou University, Guiyang, China
[3] School of Mathematics and Statistics, Guizhou University of Finance and Economics, Guiyang, China

Abstract. In this paper, we generalize the replicator dynamics with time delay of Tao [13] to the time-delayed uncertain population replicator dynamics. Considering the uncertainties of payoff functions, we propose three perturbed models of the traditional time-delayed replicator dynamics. The stability of the evolutionary stable strategy (ESS) is then studied. We prove the ESS of traditional replicator dynamics is also asymptotically stable under some constraints on the time delay in our first model and the result of Tao [13] is a special case of our first model. Then We further study the uniformly perturbed time-delayed replicator dynamics with two delays. In our third model, we show the stability of the ESS is irrelevant to the rate parameter. Some numerical examples are illustrated.

Keywords: Population games · Replicator dynamics · Time delay · Uncertainty · ESS · Stability

1 Introduction and Preliminaries

Evolutionary game theory is an excellent tool for analyzing strategic interactions among populations consisting of numerous individuals. The traditional static game [1] which supposes that players are usually perfectly rational or sometimes even hyper-rational. Evolutionary games however treat individuals as myopic and assume that they dynamically revise their strategies. Early in 1974, the fundamental concept of evolutionarily stable strategy (ESS) was introduced in [2], which has wide applications in the study of animal conflicts. Nevertheless, this fundamental concept neither involves the time aspect on resisting mutants, nor studies the long-term state of populations given an initial configuration. Taylor

N. Wang—This work is funded by National Nature Science Foundation of China (Nos. 61472093, 11761018), Guizhou Province Science and Technology Top Talents Project(No. KY2018047), Guizhou University of Finance and Economics (No. 2018XZD01), the Innovation Exploration and Academic New Seedling Project of Guizhou University of Finance and Economics (No. Qian Ke He Ping Tai Ren Cai[2017]5736-025).

H. Song and D. Jiang (Eds.): SIMUtools 2020, LNICST 370, pp. 15–28, 2021.
https://doi.org/10.1007/978-3-030-72795-6_2

and Jonker [3] proposed a dynamic approach to avoid this drawback. Their replicator dynamics does not rely on any assumption of rationality and describes the evolution of strategies in the way that agents always compare their payoffs to the average payoff of the population they belong to. The replicator equation is one of the most important game dynamics and it is still a hot topic in evolutionary game theory, see [4–9]. The monograph of Weibull [10] provided a thorough description to evolutionary game theory. Sandholm [11] provided a systematic and unified presentation of evolutionary game theory. He extended the evolutionary model to the context of multi-population and considered the nonlinear payoff functions. Recently, Newton [12] excellently surveyed the progress of evolutionary game theory and discussed open topics of importance to economics and the social sciences.

On one hand, classical evolutionary dynamics consider individuals instantaneously react to the system, that is to say, without time delay. It is really unreasonable. Tao and Wang [13] first investigated a two-strategy evolutionary dynamics model with time delay. They analyzed the effect of a time delay on the stability of time-delayed replicator dynamics and showed that the stability of evolutionary stable strategy will lost when the time delay is sufficiently large. The authors [14] considered two models of discrete time-delayed replicator dynamics. In their first model, they showed that large delay made the dynamics unstable just as the result in [13]. Furthermore, it was shown that the introduction of time delay had no effect on the stability in their biological-type model. Ben-Khalifa et al. [15] extended the evolutionary games by introducing two communities. In addition, they defined three different evolutionarily stable strategies with different levels of stability. Moreover, they studied the behavior of the evolutionary game dynamics with different types of time delay. On the other hand, fixed time delay is usually considered in literature. This assumption is too restrictive and demanding. Therefore, it is unrealistic in reality. For instance, it is of course impossible to assume that the eggs hatch simultaneously. Thus considering uncertain time delays is more reasonable. Zhong et al. [16] proposed the replicator dynamics with bounded continuously distributed time delay and discussed the stability conditions of the unique evolutionarily stable strategy in their model.

Our world is full of uncertainties and in some sense nothing in our life is deterministic. When individuals interact in a competition, nobody is sure of the payoffs of him or others. Therefore, the outcome is always unknown or fuzzy to everyone. Thus we should consider the payoffs to be perturbed sometimes.

We first introduce the population game and derive the traditional replicator dynamics as in [14]. Consider a large population of individuals who interact repeatedly through random match and assume that this population is haploid, which means the offspring have the identical strategies as their parents. For simplicity, there are only two different strategies denoted by C_1 and C_2. The interaction outcome is represented by the following matrix

$$G = \begin{array}{c} \\ C_1 \\ C_2 \end{array} \begin{array}{c} \begin{array}{cc} C_1 & C_2 \end{array} \\ \begin{pmatrix} a_1 & a_2 \\ a_3 & a_4 \end{pmatrix} \end{array}$$

and we assume that $a_3 > a_1$ and $a_2 > a_4$.

At time t, we denote as $x(t)$ the population share of strategy C_1. Therefore $1 - x(t)$ is the share of strategy C_2, which implies $(x(t), 1 - x(t))$ always belongs to the population state space $X = \{(x_1, x_2) \in \mathbb{R}^2_+ : x_1 + x_2 = 1\}$. Let $P_i(t)$ be the fitness or payoff function of C_i, $i = 1, 2$ at the population state $(x(t), 1 - x(t))$. Then we have

$$P_1(t) = a_1 x(t) + a_2(1 - x(t)),$$

$$P_2(t) = a_3 x(t) + a_4(1 - x(t)).$$

Definition 1 [11]. *A population state $x^\star \in X$ is a evolutionarily stable state if*

(1) *$x^\star$ is a Nash equilibrium, that is, $\langle y - x^\star, P(x^\star)\rangle \leq 0$, $\forall y \in X$,*

(2) *there is a neighborhood $U_{x^\star}$ of $x^\star$ satisfying*

$$\langle y - x^\star, F(y)\rangle < 0, \ \forall y \in U_{x^\star}.$$

Let $\alpha_1 = a_3 - a_1$, $\alpha_2 = a_2 - a_4$ and $\alpha = \alpha_1 + \alpha_2$. We can easily see that $x^\star = \alpha_2/\alpha$ is the unique ESS in the above population game G.

During a small time interval σ, assume that a σ fraction of the population participates in pairwise competitions, which is to say, they are matched randomly to play the population game G. Assume that $h_i(t)$, $i = 1, 2$, denote the the number of participants at time t using strategies C_1 and C_2 respectively. Thus $h(t) = h_1(t) + h_2(t)$ is the total number of individuals and $x(t) = \frac{h_1(t)}{h(t)}$ is the population share of strategy C_1 at time t. Accordingly, we have

$$h_i(t + \sigma) = (1 - \sigma)h_i(t) + \sigma h_i(t)P_i(t), \quad i = 1, 2. \tag{1}$$

It is not restrictive to assume that the payoff functions are non-negative, which makes sure that h_1 and h_2 are always non-negative.

The total amount of participants at time $t + \sigma$ admits

$$\begin{aligned} h(t + \sigma) &= h_1(t + \sigma) + h_2(t + \sigma) \\ &= (1 - \sigma)h(t) + \sigma\big(h_1(t)P_1(t) + h_2(t)P_2(t)\big) \\ &= (1 - \sigma)h(t) + \sigma h(t)\big(x(t)P_1(t) + (1 - x(t))P_2(t)\big) \\ &= (1 - \sigma)h(t) + \sigma h(t)\bar{P}(t), \end{aligned} \tag{2}$$

where $\bar{P}(t) = x(t)P_1(t) + (1 - x(t))P_2(t)$ represents the expected payoff at time t. Let $i = 1$ and divide (1) by (2), we have

$$\begin{aligned} x(t+\sigma) &= \frac{h_1(t+\sigma)}{h(t+\sigma)} \\ &= \frac{(1-\sigma)h_1(t) + \sigma h_1(t)P_1(t)}{(1-\sigma)h(t) + \sigma h(t)\bar{P}(t)} \\ &= \frac{(1-\sigma)x(t) + \sigma x(t)P_1(t)}{1-\sigma+\sigma\bar{P}(t)} \\ &= \frac{(1-\sigma+\sigma\bar{P}(t))x(t) + \sigma x(t)\big(P_1(t) - \bar{P}(t)\big)}{1-\sigma+\sigma\bar{P}(t)} \\ &= x(t) + \frac{\sigma x(t)\big(P_1(t) - \bar{P}(t)\big)}{1-\sigma+\sigma\bar{P}(t)}. \end{aligned} \tag{3}$$

Divide Eq. (3) by σ and let $\sigma \to 0$, then we obtain the traditional replicator dynamics proposed by Taylor and Jonker [3] and named by Schuster and Sigmund [17]

$$\frac{dx(t)}{dt} = x(t)\big(P_1(t)) - \bar{P}(t)\big) \tag{4}$$

The remaining part of the paper is organized as follows. In Sect. 2, we propose three modified models of Tao [13] considering the uncertainties and show the stability of the ESS under some constraints. In Sect. 3, we illustrate some simple numerical examples to verify our results. Discussion and conclusion follow in Sect. 4.

2 Models of Payoff Uncertainties

On one hand, in reality, it is impossible for anyone to react to the system immediately, which means there should be a time delay. For instance, a corn field needs some time to become maturity after sowing. Thus time delay should not be ignored. Therefore it is reasonable to modify model (4) to the time-delayed replicator population dynamics. On the other hand, we are living in an uncertain world. Various uncertainties really exist, which make our life colourful. For instance, the number of the offspring is not fixed and the harvest of crops is also unexpected when genes, temperature, climate, geographical position et al. are considered. Therefore in a population game model, the players' uncertainties should take into consideration and the payoffs may be disturbed.

We first assume that at time t individuals replicate according to the payoff obtained by their strategies at time $t - \nu$, $\nu > 0$, that is time delay exists. For example, at time t, only when the child animals born at time $t-\nu$ become mature after some time, they can take part in the competitions. Second, we suppose that the payoff functions are uncertain, which means there exist perturbations. We modify Eq. (1) to the following equations

$$h_i(t+\epsilon) = (1-\epsilon)h_i(t) + \epsilon h_i(t)\big(P_i(t-\nu) + d_i(t-\nu)\big), \quad i = 1, 2, \tag{5}$$

where $d_i(t)$ represents the perturbation of the payoff function $P_i(t)$.

Similar to the procedure in Sect. 1, we have

$$\begin{aligned} h(t+\sigma) &= h_1(t+\sigma) + h_2(t+\sigma) \\ &= (1-\sigma)h(t) + \sigma\big[h_1(t)\big(P_1(t-\nu)+d_1(t-\nu)\big) + h_2(t)\big(P_2(t-\nu)+d_2(t-\nu)\big)\big] \\ &= (1-\sigma)h(t) + \sigma h(t)\big[x(t)\big(P_1(t-\nu)+d_1(t-\nu)\big) + (1-x(t))\big(P_2(t-\nu)+d_2(t-\nu)\big)\big] \\ &= (1-\sigma)h(t) + \sigma h(t)\bar{P}(t-\nu), \end{aligned} \tag{6}$$

where $\bar{P}(t-\nu) = x(t)\big(P_1(t-\nu)+d_1(t-\nu)\big) + \big(1-x(t)\big)\big(P_2(t-\nu)+d_2(t-\nu)\big)$. We divide (5) by (6) as $i=1$, thus we have

$$\begin{aligned} x(t+\sigma) &= \frac{h_1(t+\sigma)}{h(t+\sigma)} \\ &= \frac{(1-\sigma)h_1(t) + \sigma h_1(t)\big(P_1(t-\nu)+d_1(t-\nu)\big)}{(1-\sigma)h(t)+\sigma h(t)\bar{P}(t-\nu)} \\ &= \frac{(1-\sigma)x(t) + \sigma x(t)\big(P_1(t-\nu)+d_1(t-\nu)\big)}{1-\sigma+\sigma\bar{P}(t-\nu)} \\ &= \frac{(1-\sigma+\sigma\bar{P}(t-\nu))x(t) + \sigma x(t)\big[P_1(t-\nu)+h_1(t-\nu)-\bar{P}(t-\nu)\big]}{1-\sigma+\sigma\bar{P}(t)} \\ &= x(t) + \frac{\sigma x(t)\big[P_1(t-\nu)+d_1(t-\nu)-\bar{P}(t-\nu)\big]}{1-\sigma+\sigma\bar{P}(t-\nu)}. \end{aligned} \tag{7}$$

Simplifying Eq. (7), we obtain the perturbed time-delayed replicator population dynamics

$$\frac{dx(t)}{dt} = x(t)\big[P_1(t-\nu)+d_1(t-\nu)-\bar{P}(t-\nu)\big] \tag{8}$$

$$= x(t)(1-x(t))\big[P_1(t-\nu)+d_1(t-\nu)-P_2(t-\nu)-d_2(t-\nu)\big]. \tag{9}$$

2.1 Uniform Uncertainty

From now on, we assume that $d_1(t-\nu) - d_2(t-\nu) = \theta\big(P_1(t-\nu) - P_2(t-\nu)\big)$, where θ is a random number satisfying $0 \le \underline{\theta} \le \theta \le \bar{\theta}$. We require that the difference between $d_1(t-\nu)$ and $d_2(t-\nu)$ proportional to that between $P_1(t-\nu)$ and $P_2(t-\nu)$ by a uncertain parameter θ. It is common in many applications. Under this assumption, we deduce the Eq. (9) to the following equation

$$\begin{aligned} \frac{dx(t)}{dt} &= x(t)(1-x(t))\big[P_1(t-\nu)+d_1(t-\nu)-P_2(t-\nu)-d_2(t-\nu)\big] \\ &= x(t)(1-x(t))\big[P_1(t-\nu)-P_2(t-\nu)+\theta\big(P_1(t-\nu)-P_2(t-\nu)\big)\big] \\ &= (1+\theta)x(t)(1-x(t))\big[P_1(t-\nu)-P_2(t-\nu)\big] \\ &= (1+\theta)x(t)(1-x(t))\big[a_1x(t-\nu)+a_2(1-x(t-\nu)) - \big(a_3x(t-\nu)+a_4(1-x(t-\nu))\big)\big] \\ &= -\alpha(1+\theta)x(t)(1-x(t))(x(t-\nu)-x^\star), \end{aligned} \tag{10}$$

where $\alpha_1 = a_3 - a_1$, $\alpha_2 = a_2 - a_4$, $\alpha = \alpha_1 + \alpha_2$ and $x^\star = \alpha_2/\alpha$.

Definition 2. *For the system (10), the stationary point $x^\star$ is uniformly robust stable with respect to θ if it is asymptotically stable for all $\theta \in [\underline{\theta}\ ,\bar{\theta}]$.*

From system (10), we know the unique ESS $x^\star$ is a stationary point of the perturbed replicator population dynamics with time delay. To examine the uniformly robust asymptotically stability of $x^\star$, we transform the rest point $x^\star$ of system (10) into a trivial zero and linearize it. We resort to the variable transformation $y(t) = x(t) - x^\star$ and linearize system (10), we obtain the following linearized system

$$\frac{dy(t)}{dt} = -(1+\theta)\alpha x^\star(1-x^\star)y(t-\nu). \tag{11}$$

It is well known from [18] that the system (10) is stable if and only if all the roots of its characteristic equation have negative real part. We assume that $y(t) = ce^\lambda t$ is a solution of the Eq. (11), where c is a real number. Thus we derive the characteristic equation of (11)

$$\lambda = -(1+\theta)\beta e^{-\lambda\nu}, \tag{12}$$

where $\beta = \alpha x^\star(1-x^\star)$.

Theorem 1. *The unique ESS $x^\star$ of the perturbed time-delayed replicator population dynamics (10) is uniformly robust stable if $0 < \nu < \frac{\pi}{2(1+\bar{\theta})\beta}$.*

Proof. Assume that $\nu < \frac{\pi}{2(1+\bar{\theta})\beta}$, we show that each root of the characteristic Eq. (12) has negative real part for any uncertain $\theta \in [\underline{\theta}\ ,\bar{\theta}]$.

Suppose that there is a root of (12) $\lambda = \zeta + i\mu$ with ζ, $\mu \in \mathbb{R}$, $i^2 = -1$ and $\zeta \geq 0$. We notice that $\lambda = 0$ is obviously not a root of (12). Furthermore, we assume that $\mu > 0$ since if $\lambda = \zeta + i\mu$ is a solution of (12), so is $\lambda = \zeta - i\mu$.

By separating the real and imaginary parts in Eq. (12), we have

$$\begin{aligned}\zeta &= -(1+\theta)\beta e^{-\zeta\nu}\cos\mu\nu,\\ \mu &= (1+\theta)\beta e^{-\zeta\nu}\sin\mu\nu.\end{aligned}$$

Since $\mu > 0$, this implies

$$\begin{aligned}0 < \mu\nu &= (1+\theta)\nu\beta e^{-\zeta\nu}\sin\mu\nu\\ &\leq (1+\theta)\nu\beta\\ &\leq (1+\bar{\theta})\nu\beta\\ &< (1+\bar{\theta})\cdot\frac{\pi}{2(1+\bar{\theta})\beta}\cdot\beta\\ &= \frac{\pi}{2}.\end{aligned}$$

The second inequality holds since $\zeta > 0$ and $\sin\mu\nu \leq 1$, the third and forth inequalities hold due to the assumption that $\theta \in [\underline{\theta}\ ,\bar{\theta}]$ and $\nu < \frac{\pi}{2(1+\bar{\theta})\beta}$. Accordingly, we obtain

$$\zeta = -(1+\theta)\beta e^{-\zeta\nu}\cos\mu\nu < 0,$$

since $\beta = \alpha x^\star(1-x^\star) > 0$. This contradicts with our assumption $\zeta \geq 0$, which prove the roots of the characteristic Eq. (12) have negative real parts.

According to Theorem 1, the stability of the perturbed time-delayed replicator population dynamics relies on the parameter θ. If θ equals to 0, we derive a special case discussed in [13], which means that we develop their consequence.

Corollary 1. *For the time-delayed replicator population dynamics of Tao [13]*

$$\frac{dx(t)}{dt} = x(t)\big(F_1(t-\nu)) - \bar{F}(t-\nu)\big),$$

the mixed ESS $x^\star$ is asymptotically stable if the delay $\nu < \frac{\alpha\pi}{2\alpha_1\alpha_2}$.

2.2 Uniform Uncertainty with Two Delays

In this section, we further study the perturbed time-delayed replicator dynamics discussed in the previous subsection. We assume that there are two time delays for each of the two strategies C_1 and C_2. Moreover, for strategy C_1, if players consider no delay with probability p_1, a time delay ν_1 with probability p_2, and a time delay ν_2 with probability p_3, where $p_1 + p_2 + p_3 = 1$, then the expected payoff of players who select strategy C_1 becomes

$$\begin{aligned}
P_1(t) &= a_1[p_1x(t) + p_2x(t-\nu_1) + p_3x(t-\nu_2)] \\
&\quad + a_2[p_1(1-x(t)) + p_2(1-x(t-\nu_1)) + p_3(1-x(t-\nu_2))] \\
&= a_1[p_1x(t) + p_2x(t-\nu_1) + p_3x(t-\nu_2)] \\
&\quad + a_2[1 - p_1x(t) - p_2x(t-\nu_1) - p_3x(t-\nu_2)].
\end{aligned}$$

In the same way, we have

$$\begin{aligned}
P_2(t) &= a_3[p_1x(t) + p_2x(t-\nu_1) + p_3x(t-\nu_2)] \\
&\quad + a_4[p_1(1-x(t)) + p_2(1-x(t-\nu_1)) + p_3(1-x(t-\nu_2))] \\
&= a_3[p_1x(t) + p_2x(t-\nu_1) + p_3x(t-\nu_2)] \\
&\quad + a_4[1 - p_1x(t) - p_2x(t-\nu_1) - p_3x(t-\nu_2)].
\end{aligned}$$

Accordingly, under the assumption of the uniform uncertainty, the perturbed replicator dynamics with two time delays is expressed as

$$\begin{aligned}
\frac{dx(t)}{dt} &= x(t)(1-x(t))\left[P_1(t) + d_1(t) - P_2(t) - d_2(t)\right] \\
&= x(t)(1-x(t))\left[P_1(t) - P_2(t) + \theta(P_1(t) - P_2(t))\right] \\
&= (1+\theta)x(t)(1-x(t))(P_1(t) - P_2(t)) \\
&= (1+\theta)x(t)(1-x(t))\left[(a_1 - a_3 + a_4 + a_2)(p_1x(t) + p_2x(t-\nu_1) + p_3x(t-\nu_2)) + a_2 - a_4\right] \\
&= -\alpha(1+\theta)x(t)(1-x(t))(p_1x(t) + p_2x(t-\nu_1) + p_3x(t-\nu_2) - x^\star),
\end{aligned} \tag{13}$$

where $\alpha_1 = a_3 - a_1$, $\alpha_2 = a_2 - a_4$, $\alpha = \alpha_1 + \alpha_2$, $x^\star = \alpha_2/\alpha$ and θ is the uncertain parameter as defined in the previous subsection.

Similar to the procedure in the previous subsection, we can derive the linearized equation of (13) around $x^\star$ as follows

$$\frac{dy(t)}{dt} = -\alpha(1+\theta)x^\star(1-x^\star)(p_1y(t)+p_2y(t-\nu_1)+p_3y(t-\nu_2)). \quad (14)$$

To investigate the stability of the unique ESS $x^\star$, we resort to the following lemma in [19].

Lemma 1. *Let $\dot{y} = -a_1y(t) - a_2y(t-h_1) - a_3y(t-h_2)$, where $h_1 > 0,\ h_2 > 0$. Suppose at least one of the following conditions holds*

(1) $a_1 > 0,\ |a_2| + |a_3| < a_1$,

(2) $0 < a_1 + a_2 + a_3,\ |a_2|h_1 + |a_3|h_2 < \dfrac{a_1+a_2+a_3}{|a_1|+|a_2|+|a_3|}$,

(3) $0 < a_1 + a_2,\ |a_2|h_1 < \dfrac{a_1+a_2-|a_3|}{|a_1|+|a_2|+|a_3|}$,

(4) $0 < a_1 + a_3,\ |a_3|h_2 < \dfrac{a_1+a_3-|a_2|}{|a_1|+|a_2|+|a_3|}$,

then the above equation is asymptotically stable.

From this lemma, we get the corresponding result for the Eq. (14).

Theorem 2. *If at least one of the following conditions are satisfied*

(1) $p_2 + p_3 < p_1$,

(2) $p_2\nu_1 + p_3\nu_2 < \dfrac{1}{\alpha(1+\theta)x^\star(1-x^\star)}$,

(3) $p_2\nu_1 < \dfrac{p_1+p_2-p_3}{\alpha(1+\theta)x^\star(1-x^\star)}$,

(4) $p_3\nu_2 < \dfrac{p_1+p_3-p_2}{\alpha(1+\theta)x^\star(1-x^\star)}$,

then the unique ESS $x^\star$ is asymptotically stable for the perturbed replicator dynamics with two time delays (13).

2.3 Exponential Uncertainty

In this part, we continue to investigate the perturbed time-delayed replicator population dynamics of form (9). We now assume that at time $t-\nu$, the difference between the two perturbations $d_1(t-\nu)$ and $d_2(t-\nu)$ satisfies

$$d_1(t-\nu) - d_2(t-\nu) = \int_{\nu}^{\infty} \xi e^{-\xi\phi}(P_1(t-\phi) - P_2(t-\phi))d\phi,$$

where ξ is called the rate parameter. Now the difference between the perturbations is relevant to the differences between the payoffs at every moment. Moreover, $e^{-\xi\phi}$ indicates the effect on the perturbation. Accordingly, the perturbed time-delayed replicator population dynamics (9) becomes

$$\begin{aligned}\frac{dx(t)}{dt} &= x(t)(1-x(t))\big[P_1(t-\nu)+d_1(t-\nu)-P_2(t-\nu)-d_2(t-\nu)\big]\\ &= x(t)(1-x(t))\big[P_1(t-\nu)-P_2(t-\nu)+\int_\nu^\infty \xi e^{-\xi\phi}(P_1(t-\phi)-P_2(t-\phi))d\phi\big]\\ &= -\alpha x(t)(1-x(t))\big[x(t-\nu)-x^\star+\int_\nu^\infty \xi e^{-\xi\phi}(x(t-\phi)-x^\star)d\phi\big]. \qquad (15)\end{aligned}$$

Compared with the normal time delay replicator dynamics in Tao[13], the system (15) is perturbed in a more complicated but realistic way since we consider every moment before time $t-\nu$. At the same time, we see that the unique ESS $x^\star$ is a also a stationary point of the exponentially perturbed time-delayed replicator population dynamics (15).

To investigate the stability of $x^\star$, we transform the stationary point $x^\star$ of the system (15) by the variable transformation $y(t)=x(t)-x^\star$ and linearize it, we obtain

$$\begin{aligned}\frac{dy(t)}{dt} &= -\alpha x^\star(1-x^\star)\big(y(t-\nu)+\int_\nu^\infty \xi e^{-\xi\phi}y(t-\phi)d\phi\big)\\ &= -\beta\big(y(t-\nu)+\int_\nu^\infty \xi e^{-\xi\phi}y(t-\phi)d\phi\big), \qquad (16)\end{aligned}$$

where $\beta=\alpha x^\star(1-x^\star)$. Same as the procedure in the last subsection, we easily derive the characteristic equation of (16) as follows

$$\lambda = -\beta\big(e^{-\lambda\nu}+\xi\int_\nu^\infty e^{-(\xi+\lambda)\phi}d\phi\big). \qquad (17)$$

The next theorem shows that the ESS $x^\star$ is still asymptotically stable under perturbation for any rate parameter, which implies $x^\star$ is nicely perfect in some sense.

Theorem 3. *The evolutionary stable strategy $x^\star$ is asymptotically stable under the exponentially perturbed time-delayed replicator population dynamics for any $\xi>0$ if $0<\nu<\frac{\pi}{4\beta}$.*

Proof. Suppose that $\nu<\frac{\pi}{4\beta}$, we prove none of the roots for the characteristic Eq. (17) have nonnegative real part for any $\xi>0$.

Let $\lambda=\zeta+i\mu$ be a root of (17) with $\zeta\geq 0$. Without loss of generality, we assume that $\mu>0$. Substituting $\lambda=\zeta+i\mu$ into (17) and separating the real and imaginary parts, we get

$$\zeta = -\beta e^{-\zeta\nu}\cos\mu\nu-\beta\xi\int_\nu^\infty e^{-(\xi+\zeta)\phi}\cos\mu\phi d\phi,$$
$$\mu = \beta e^{-\zeta\nu}\sin\mu\nu+\beta\xi\int_\nu^\infty e^{-(\xi+\zeta)\phi}\sin\mu\phi d\phi.$$

Consequently, we have

$$\begin{aligned}0 < \mu\nu &= \nu\beta e^{-\zeta\nu}\sin\mu\nu + \nu\beta\xi\int_{\nu}^{\infty} e^{-(\xi+\zeta)\phi}\sin\mu\phi d\phi \\ &\leq \nu\beta + \nu\beta\xi\int_{\nu}^{\infty} e^{-(\xi+\zeta)\phi}d\phi \\ &= \nu\beta + \nu\beta\frac{\xi}{\xi+\zeta}e^{-(\xi+\zeta)\nu} \\ &< 2\nu\beta \\ &< 2\cdot\frac{\pi}{4\beta}\cdot\beta \\ &= \frac{\pi}{2}.\end{aligned}$$

Furthermore, since $0 < \mu\nu < \frac{\pi}{2}$, we have

$$\begin{aligned}0 \leq \zeta &= -\beta e^{-\zeta\nu}\cos\mu\nu - \beta\xi\int_{\nu}^{\infty} e^{-(\xi+\zeta)\phi}\cos\mu\phi d\phi \\ &< -\beta + \beta\xi\int_{\nu}^{\infty} e^{-(\xi+\zeta)\phi}d\phi \\ &= -\beta(1 - \frac{\xi}{\xi+\zeta}e^{-(\xi+\zeta)\nu}) \\ &< 0.\end{aligned}$$

The second inequality holds since $\cos\mu\phi \geq -1$.

This contradiction prove that each of the roots of the characteristic Eq. (17) has negative real part, which implies the ESS x^* is also asymptotically stable even in the exponentially perturbed time-delayed replicator population dynamics.

3 Numerical Examples

In hawk and dove games, two individual compete for a resource. They share a same strategy set with two strategies, namely, Hawk (H) and Dove (D). The individuals choosing (H) strategy always want to monopolize the resource and never share. The strategy (D) represents a gentle attitude on the resource and never choose to fight. The payoff matrix is given as follows

$$G = \begin{array}{c} \\ H \\ D \end{array}\!\!\begin{array}{c} \begin{array}{cc} H & D \end{array} \\ \begin{pmatrix} \frac{V-C}{2} & V \\ 0 & \frac{V}{2} \end{pmatrix} \end{array},$$

where C and V are positive. V represents the resource value and C stands for the cost of fight. The cost is usually assumed to be high if agents fight, therefore we assume that $V < C$.

3.1 Uniform Uncertainty

In the case of interval perturbation model of Eq. (10), the ESS $x^\star$ is uniformly robust asymptotically if the time delay is small than $\frac{\pi}{2(1+\bar{\theta})\beta}$, according to Theorem 1. In the context of Hawk and Dove games, this value is given as $\frac{C\pi}{(1+\bar{\theta}V(V-C)}$. We depict in Fig. 1(a) the trajectory of numerical solution of the perturbed time-delayed replicator population dynamics with parameter $C = 5$, $V = 3$, the perturbation upper bound $\bar{\theta} = 0.5$ and $\nu_1 = 1.7$. We observe the convergence to the ESS. In Fig. 1(b), as we increase the value of ν to 5, the system obviously becomes to oscillate.

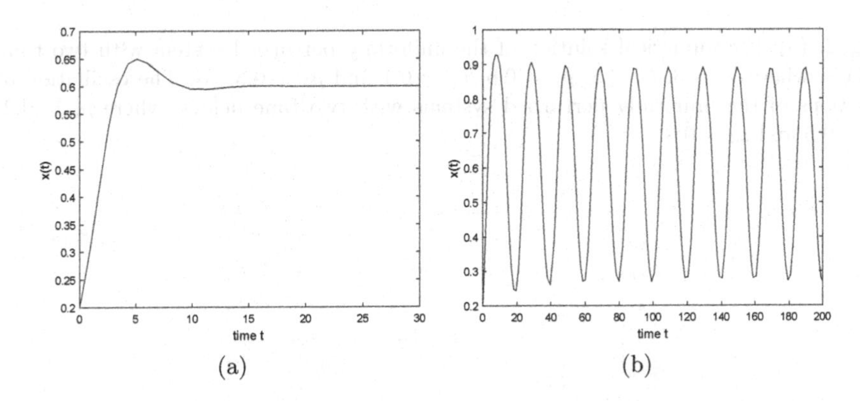

Fig. 1. (a) The numerical solution of the interval perturbed system, where $V = 3$, $C = 5$, $\bar{\theta} = 0.5$ and $\nu_1 = 1.7$. (b) The oscillation of the solution of the interval perturbed system, where $\nu_2 = 5$.

3.2 Uniform Uncertainty with Two Delays

Under the perturbed time-delayed replicator dynamics with two time delays, we derived that the unique ESS $x^\star$ is asymptotically stable as long as the probability distribution satisfies $p_2 + p_3 < p_1$, no matter how large the delays are. We see the truth from Fig. 2(a). In addition, if $p_2 + p_3 > p_1$, the solution always oscillates even if the time delays are small. From Fig. 3, we know that values of perturbation parameter θ affect only the rate of convergence of $x^\star$.

3.3 Exponential Uncertainty

When we consider the exponentially perturbed time-delayed replicator population dynamics, we proved in Theorem 3 the stability of the ESS $x^\star$ is not related to rate parameter ξ. In fact, the parameter ξ only affects the convergent rate.

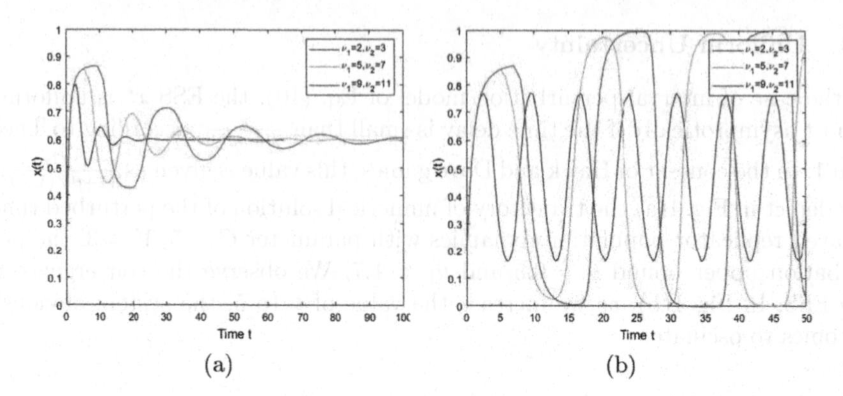

(a) (b)

Fig. 2. (a) The numerical solution of the uniformly perturbed system with two time delays, where $V = 3$, $C = 5$, $p_1 = 0.6$, $p_2 = 0.1$ and $p_3 = 0.3$. (b) The oscillation of solutions of the uniformly perturbed systems with two time delays, where $p_1 = 0.2$, $p_2 = 0.5$ and $p_3 = 0.3$.

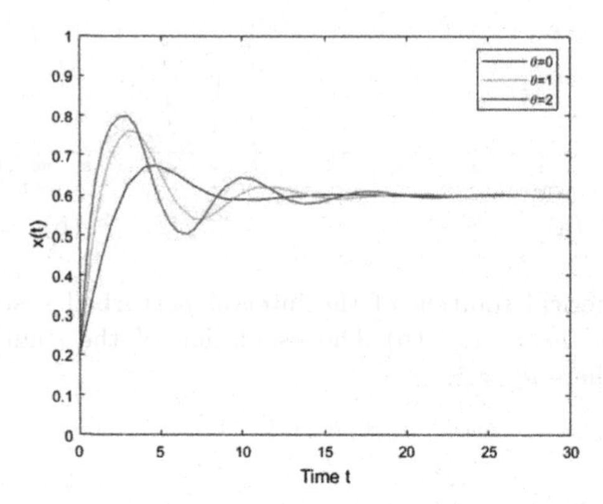

Fig. 3. The convergent rate of the numerical solutions for different perturbation parameter θ, where $V = 3$, $C = 5$.

Now The value of V and C are the same as that in the last subsection and we set $\nu = 1$ and $\xi = 0.4$. We see from Fig. 4, the ESS $x^\star = 1.7$ is asymptotically stable and becomes unstable when $\nu = 3.8$. From Fig. 5, we conclude that the rate parameter ξ only has effects on the convergent rate.

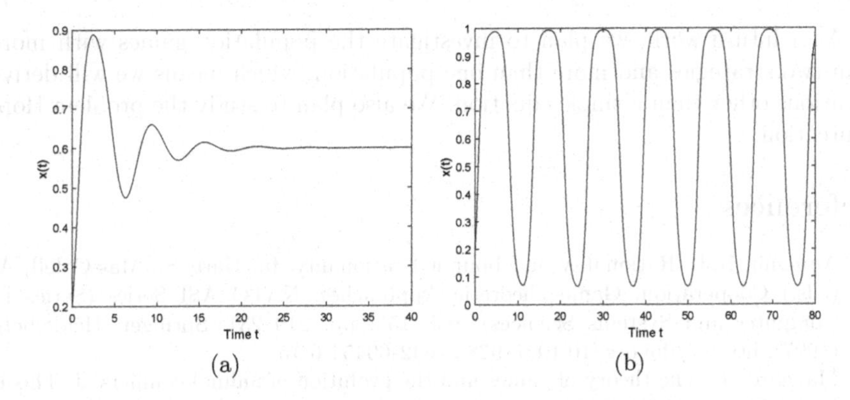

Fig. 4. (a) The numerical solution of the exponentially perturbed system, where $V = 3$, $C = 5$, $\xi = 0.4$ and $\nu_1 = 1.7$. (b) The oscillation of the solution of the exponentially perturbed system, where $\nu_2 = 3.8$.

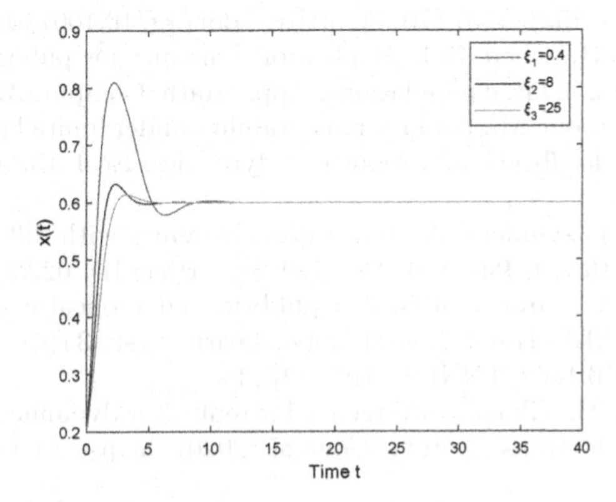

Fig. 5. The convergent rate of the numerical solutions for different values of ξ, where $V = 3$, $C = 5$.

4 Conclusion

In this paper, considering the uncertainty, we developed the traditional population replicator dynamics with time delay to time-delayed population dynamics with uncertainties. The first model treated the perturbation parameter as uncertain in a interval and we derived the uniformly perturbed replicator dynamics. The second model further investigated the uniformly perturbed replicator dynamics considering two time delays exist. The last model discussed the parameter as a stochastic variable with exponential distribution. We showed that under some constraints, the $x^\star$ of the traditional replicator dynamics is always asymptotically stable in our three models and the result in [13] is a special case of our result. We verified our results with some simple examples.

As a future work, we plan to investigate the population games with more than two strategies and more than one population, which means we will derive equations other than a single equation. We also plan to study the problem Hopf bifurcation.

References

1. Aumann, R.J.: Rationality and bounded rationality. In: Hart, S., Mas-Colell, A. (eds.) Cooperation: Game-Theoretic Approaches. NATO ASI Series (Series F: Computer and Systems Sciences), vol. 155, pp. 219–231. Springer, Heidelberg (1997). https://doi.org/10.1007/978-3-642-60454-6_15
2. Maynard, J.: The theory of games and the evolution of animal conflicts. J. Theor. Biol. **47**(1), 209–221 (1974)
3. Taylor, P., Jonker, L.: Evolutionary stable strategies and game dynamics. Math. Biosci. **40**(1), 145–156 (1978)
4. Gutierrez, S.M., Adeli, H.: Many-objective control optimization of high-rise building structures using replicator dynamics and neural dynamics model. Struct. Multidiscip. Optim. **56**(6), 1521–1537 (2017). https://doi.org/10.1007/s00158-017-1835-9
5. Wang, Q., He, N.R., Chen, X.J.: Replicator dynamics for public goods game with resource allocation in large populations. Appl. Math. Comput. **328**, 162–170 (2018)
6. Argasinski, K., Broom, M.: Evolutionary stability under limited population growth: eco-evolutionary feedbacks and replicator dynamics. Ecol. Complex. **34**, 198–212 (2018)
7. Requejo, R.J., Díaz-Guilera, A.: Replicator dynamics with diffusion on multiplex networks. Phys. Rev. E **94**(2), 022301 (2018). Article ID: 022301
8. Tan, S., Wang, Y.: Graphical Nash equilibria and replicator dynamics on complex networks. IEEE Trans. Neural Netw. Learn. Syst. **31**(6), 1831–1842 (2019). https://doi.org/10.1109/TNNLS.2019.2927233
9. Ramazi, P., Cao, M.: Global convergence for replicator dynamics of repeated snowdrift games. IEEE Trans. Autom. Control (2020). https://doi.org/10.1109/TAC.2020.2975811
10. Weibull, J.: Evolutionary Game Theory, 2nd edn. MIT Press, Cambridge (1994)
11. Sandholm, J.: Population Games and Evolutionary Dynamics. MIT Press, Cambridge (2010). Journal of Theoretical Biology
12. Newton, J.: Evolutionary game theory: a renaissance. Games **9**, 31 (2018)
13. Tao, Y., Wang, Z.: Effect of time delay and evolutionarily stable strategy. J. Theor. Biol. **187**(1), 111–116 (1997)
14. Alboszta, J., Miękisz, J.: Stability and evolutionary stable strategies in discrete replicator dynamics with delay. J. Theor. Biol. **231**(2), 175–179 (2004)
15. Ben Khalifa, N., El-Azouzi, R., Hayel, Y., et al.: Evolutionary games in interacting communities. Dyn. Games Appl. **7**(2), 131–156 (2017)
16. Zhong, C., Yang, H., Liu, Z., et al.: Stability of replicator dynamics with bounded continuously distributed time delay. Mathematics **8**(3), 431 (2020)
17. Schuster, P., Sigmund, K.: Replicator dynamics. J. Theor. Biol. **100**(3), 533–538 (1983)
18. Gopalsamy, K.: Stability and oscillations in dealy differential equations of population dynamics. Kluwer Academic Publishers, Kluwer, Dordrecht, The Netherlands (1992)
19. Berezansky, L., Braverman, E.: On stability of some linear and nonlinear delay differential equations. J. Math. Anal. Appl. **314**(2), 391–411 (2006)

Dynamics of a Single Population Model with Non-transient/Transient Impulsive Harvesting and Birth Pulse in a Polluted Environment

Yumei Zhou(✉) and Jianjun Jiao

School of Mathematics and Statistics,
Guizhou University of Finance and Economics,
Guiyang 550025, People's Republic of China

Abstract. In this paper, we present a single population model with non-transient/transient impulsive harvesting and birth pulse in a polluted environment. The sufficient conditions for system permanence is presented.

Keywords: Single population system · Non-transient/transient impulses · Birth pulse · Permanence

1 Introduction

Many investigations [1–4] devoted into impulsive equations. Clack [5] has studied the logistic equation with optimal harvesting. The environmental toxicant decreases the carrying capacity in polluted environments [6,7]. They are assumed that the inputting toxicant was continuous. Liu et al. [8] considered that the environmental toxicant is often emitted with regular pulse. In this paper, we do notation as $N = n\tau$ and $L = l\tau$.

2 The Model

In this work, we present

Supported by National Natural Science Foundation of China (11761019, 11361014).

H. Song and D. Jiang (Eds.): SIMUtools 2020, LNICST 370, pp. 29–38, 2021.
https://doi.org/10.1007/978-3-030-72795-6_3

$$\begin{cases} \frac{dx}{dt} = -(d_1 + \beta_1 c_o)x + c_1 x^2, \ t \in (N, (N+L)], \\ \left.\begin{array}{l} \Delta x = -\mu x, \\ \Delta c_0 = 0, \\ \Delta c_{e1} = 0, \\ \Delta c_{e2} = 0, \end{array}\right\} t = (N+L), \\ \frac{dx}{dt} = -(d_2 + E + \beta_2 c_o(t))x + c_2 x^2, \ t \in ((N+L), (N+1)], \\ \left.\begin{array}{l} \Delta x = bx, \\ \Delta c_0 = 0, \\ \Delta c_{e1} = D(c_{e2} - c_{e1}) + \mu_1, \\ \Delta c_{e2} = D(c_{e1} - c_{e2}) + \mu_2, \end{array}\right\} t = (N+1), \\ \left.\begin{array}{l} \frac{dc_o}{dt} = fc_{e1} - (g+m)c_o, \\ \frac{dc_{e1}}{dt} = -h_1 c_{e1}), \\ \frac{dc_{e2}}{dt} = -h_2 c_{e2}, \end{array}\right\} t \in (N, (N+1)]. \end{cases} \tag{1}$$

The biological meanings of the varies and parameters can reference [8–10].

3 The Dynamics

For (1), one subsystem of (1) is

$$\begin{cases} \frac{dx}{dt} = -d_1 x + c_1 x^2, \ t \in (N, (N+L)], \\ \Delta x = -\mu x, t = (N+L), \\ \frac{dx}{dt} = -(d_2 + E)x + c_2 x^2, \ t \in ((N+L), (N+1)], \\ \Delta x = bx, t = (N+1). \end{cases} \tag{2}$$

Integrating (2), we get

$$x(t) = \begin{cases} \frac{d_1 x(N^+) e^{-d_1(t-N)}}{d_1 + c_1 x(N^+)(1 - e^{-d_1(t-N)})}, t \in (N, (N+L)], \\ \frac{(d_2+E)x((N+l)^+) e^{-(d_2+E)(t-N)}}{(d_2+E) + c_2 x((N+l)^+)(1 - e^{-(d_2+E)(t-N)})}, \\ \qquad t \in ((N+L), (N+1)]. \end{cases} \tag{3}$$

Stroboscopic map for (2) is

$$x((n+1)\tau^+) = \frac{M_1}{M_2}, \tag{4}$$

where $M_1 = (1+b)(1-\mu)d_1(d_2+E)e^{-[d_1l+(d_2+E)(1-l)]\tau}x(N^+)$, $M_2 = d_1(d_2+E) + [(d_2+E)c_1(1-e^{-d_1l}) + (1-\mu)d_1c_2e^{-d_1l\tau}(1-e^{-(d_2+E)(1-l)\tau})]x(N^+)$.
We rewrite (4) as

$$x((N+1)^+) = \frac{A_0x(N^+)}{d_1(d_2+E) + B_0x(N^+)}. \tag{5}$$

where

$$A_0 = (1+b)(1-\mu)d_1(d_2+E)e^{-[d_1l+(d_2+E)(1-l)]\tau},$$
$$B_o = (d_2+E)c_1(1-e^{-d_1l}) + (1-\mu)d_1c_2e^{-d_1l\tau}(1-e^{-(d_2+E)(1-l)\tau}).$$

We get points $G_1(0)$ and $G_2(x^*)$ of (4), and

$$x^* = \frac{d_1(d_2+E)[(1+b)(1-\mu)e^{-(d_1l+(d_2+E)(1-l))\tau} - 1]}{(d_2+E)c_1(1-e^{-d_1l}) + (1-\mu)d_1c_2e^{-d_1l\tau}(1-e^{-(d_2+E)(1-l)\tau})}, \quad (1+b)(1-\mu)e^{-(d_1l+(d_2+E)(1-l))\tau} > 1. \tag{6}$$

Condition $(1+b)(1-\mu)e^{-(d_1l+(d_2+E)(1-l))\tau} < 1$ is made as (C_1), and condition $(1+b)(1-\mu)e^{-(d_1l+(d_2+E)(1-l))\tau} > 1$ is made as (C_2).

Theorem 1. *i*) Suppose (C_1) holds $G_1(0)$ of (4) is globally asymptotically stable;

ii) Suppose (C_2) holds, $G_2(x^*)$ is globally asymptotically stable.

Proof. Marking x^n as $x(n\tau^+)$, we rewrite (4) as

$$x_{n+1} = F(x_n) = \frac{Ax_n}{d_1(d_2+E) + Bx_n}. \tag{7}$$

i) Suppose (C_1) holds, then,

$$\frac{dF(x)}{dx}\mid_{x=0} = (1+b)(1-\mu)e^{-(d_1l+(d_2+E)(1-l))\tau} < 1. \tag{8}$$

Then, locally stable $G_1(0)$ exists, Furthermore, it is globally asymptotically stable.

ii) Suppose (C_2) holds,concerning to $G_1(0)$, we get

$$\begin{aligned}\frac{dF(x)}{dx}\mid_{x=0} &= \frac{Ad_1(d_2+E)}{[d_1(d_2+E)+Bx]^2}\mid_{x=0} \\ &= \frac{A}{d_1(d_2+E)}. \\ &= (1+b)(1-\mu)e^{-(d_1l+(d_2+E)(1-l))\tau} > 1,\end{aligned} \tag{9}$$

then, $G_1(0)$ is unstable.

Concerning $G_2(x^*)$, we also get

$$\begin{aligned}\frac{dF(x)}{dx}\mid_{x=x^*}=&\frac{Ad_1(d_2+E)}{[d_1(d_2+E)+Bx]^2}\mid_{x=x^*}\\=&\frac{1}{(1+b)(1-\mu)e^{-(d_1l+(d_2+E)(1-l))\tau}}<1.\end{aligned} \tag{10}$$

Then, locally stable $G_2(x^*)$ exists. Furthermore, it is globally asymptotically stable.

Similarly with Reference [6], the following lemma can easily be proved.

Theorem 2. *i*) Suppose (C_1) holds, 0 of (1) is globally asymptotically stable; *ii*) Suppose (C_2) holds, $\widetilde{x(t)}$ of (1) is globally asymptotically stable, and

$$\widetilde{x(t)}=\begin{cases}\dfrac{d_1x^*e^{-d_1(t-N)}}{d_1+c_1x^*(1-e^{-d_1(t-N)})}, t\in(N,(N+L)],\\ \dfrac{(d_2+E)x^{**}e^{-(d_2+E)(t-N)}}{(d_2+E)+c_2x^{**}(1-e^{-(d_2+E)(t-N)})}, t\in((N+L),(N+1)].\end{cases} \tag{11}$$

here x^* is by (6), and x^{**} is by $x^{**}=\frac{(1-\mu_1)d_1x^*e^{-d_1l\tau}}{d_1+c_1x^*(1-e^{-d_1l\tau})}$.

For (1), another subsystem of system (1) is obtained as

$$\begin{cases}\left.\begin{aligned}&\frac{dc_o}{dt}=fc_{e_1}-(g+m)c_o,\\&\frac{dc_{e1}}{dt}=-h_1c_{e1},\\&\frac{dc_{e2}}{dt}=-h_2c_{e2},\end{aligned}\right\} t\neq N,\\ \left.\begin{aligned}&\triangle c_o=0,\\&\triangle c_{e_1}=D(c_{e_2}-c_{e_1})+\mu_1,\\&\triangle c_{e_2}=D(c_{e_1}-c_{e_2})+\mu_2,\end{aligned}\right\} t=N.\end{cases} \tag{12}$$

Integrating (12), we get

$$\begin{cases}\widetilde{c_o}=c_o(N^+)e^{-(g+m)(t-N)}\\ \qquad +\dfrac{fc_{e1}(N^+)(1-e^{-(h-g-m)(t-N)})}{h-g-m}, t\in(N,(N+1)],\\ \widetilde{c_{e1}}=c_{e1}(N^+)e^{-h_1(t-N)}, t\in(N,(N+1)],\\ \widetilde{c_{e2}}=c_{e2}(N^+)e^{-h_2(t-N)}, t\in(N,(N+1)].\end{cases} \tag{13}$$

The stroboscopic map of (12) is

$$\begin{cases} c_o((N+1)^+) = c_o(N^+)e^{-(g+m)\tau} \\ \qquad\qquad + \dfrac{fc_{e1}(N^+)(1-e^{-(h-g-m)\tau})}{h-g-m}, t \in (N,(N+1)], \\ c_{e1}((N+1)^+) = (1-D)c_{e1}(N^+)e^{-h_1\tau} + Dc_{e2}(N^+)e^{-h_2\tau} + \mu_1, \\ \qquad\qquad t \in (N,(N+1)], \\ c_{e2}((N+1)N^+) = Dc_{e1}(N^+)e^{-h_1\tau} + (1-D)c_{e2}(N^+)e^{-h_2\tau} + \mu_2 \\ \qquad\qquad t \in (N,(N+1)]. \end{cases} \tag{14}$$

A unique fixed point of (14) is

$$\begin{cases} c_o^* = \dfrac{f[\mu_1(1-(1-D)e^{-h_2\tau}) + \mu_2 De^{-h_2\tau}](1-e^{-(h-g-m)\tau})}{M_3}, \\ c_{e1}^* = \dfrac{\mu_1[1-(1-D)e^{-h_2\tau}] + \mu_2 De^{-h_2\tau}}{M_4}, \\ c_{e2}^* = \dfrac{\mu_2[1-(1-D)e^{-h_1\tau}] + \mu_1 De^{-h_1\tau}}{M_4}, \end{cases} \tag{15}$$

where $M_3 = (h-g-m)(1-e^{-(h-g-m)\tau})[(1-(1-D)e^{-h_1\tau})(1-(1-D)e^{-h_2\tau}) - D^2e^{-(h_1+h_2)\tau}]$ and $M_4 = [1-(1-D)e^{-h_1\tau}][1-(1-D)e^{-h_2\tau}] - D^2e^{-(h_1+h_2)\tau}$.

Writing (14) as a map, and defining it as $F : R_+^3 \to R_+^3$

$$\begin{cases} F_1(c) = c_o(N^+)e^{-(g+m)\tau} \\ \qquad\qquad + \dfrac{fc_{e1}(N^+)(1-e^{-(h-g-m)\tau})}{h-g-m}, t \in (N,(N+1)], \\ F_2(c) = (1-D)c_{e1}(N^+)e^{-h_1\tau} + Dc_{e2}(N^+)e^{-h_2\tau} + \mu_1, \\ F_3(c) = Dc_{e1}(N^+)e^{-h_1\tau} + (1-D)c_{e2}(N^+)e^{-h_2\tau} + \mu_2. \end{cases} \tag{16}$$

Lemma 3. Suppose $D > \frac{1}{2}$ holds, $F(c_o^*, c_{e1}^*, c_{e2}^*)$ of (16) is globally asymptotically stable.

Proof. We mark $(c_o^n, c_{e1}^n, c_{e2}^n)$ as $(c_o(n\tau^+), c_{e1}(n\tau^+), c_{e2}(n\tau^+))$. Rewriting linearity of (16) as

$$\begin{pmatrix} c_o^{n+1} \\ c_{e1}^{n+1} \\ c_{e2}^{n+1} \end{pmatrix} = M \begin{pmatrix} c_o^n \\ c_{e1}^n \\ c_{e2}^n \end{pmatrix}. \tag{17}$$

Obviously, the stabilities of $F(c_o^*, c_{e1}^*, c_{e2}^*)$ is by eigenvalues $\lambda_1, \lambda_2, \lambda_3$ of M, which are less than 1.

Suppose $D > \frac{1}{2}$ holds, $0 < e^{-h_i} < 1 (i = 1, 2)$, $F(c_o^*, c_{e1}^*, c_{e2}^*)$ exists, and

$$M = \begin{pmatrix} e^{-(g+m)\tau} & \frac{f(1-e^{-(h-g-m)\tau})}{h-g-m} & 0 \\ 0 & (1-D)e^{-h_1\tau} & De^{-h_2\tau} \\ 0 & De^{-h_1\tau} & (1-D)e^{-h_2\tau} \end{pmatrix}. \tag{18}$$

For

$$\lambda_1 = e^{-(g+m)\tau} < 1,$$

$$\lambda_2 = \frac{(1-D)(e^{-h_1\tau} + e^{-h_2\tau}) + \sqrt{[(1-D)(e^{-h_1\tau} + e^{-h_2\tau})]^2 - 4(1-2D)e^{-(h_1+h_1)\tau}}}{2}$$

$$= \frac{(1-D)(e^{-h_1\tau} + e^{-h_2\tau}) + \sqrt{[D(e^{-h_1\tau} + e^{-h_2\tau})]^2 + (1-2D)(e^{-(h_1-h_1)\tau})^2}}{2}$$

$$< \frac{e^{-h_1\tau} + e^{-h_2\tau}}{2} < 1,$$

$$\lambda_3 = \frac{(1-D)(e^{-h_1\tau} + e^{-h_2\tau}) - \sqrt{[(1-D)(e^{-h_1\tau} + e^{-h_2\tau})]^2 - 4(1-2D)e^{-(h_1+h_1)\tau}}}{2}$$

$$= \frac{(1-D)(e^{-h_1\tau} + e^{-h_2\tau}) - \sqrt{[(1-D)(e^{-h_1\tau} - e^{-h_2\tau})]^2 + 4D^2e^{-(h_1+h_1)\tau}}}{2}$$

$$< \frac{(1-D)\min\{e^{-h_1\tau}, e^{-h_2\tau}\}}{2} < 1.$$

That is, $\lambda_i < 1 (i = 1, 2, 3)$, then, $F(c_o^*, c_{e1}^*, c_{e2}^*)$ is locally stable. Then, it is globally asymptotically stable.

Similar to Reference [9], we get

Lemma 4. (12) has a unique globally asymptotically stable periodic solution $(\widetilde{c_o}, \widetilde{c_{e1}}, \widetilde{c_{e2}})$ of (12), and

$$\begin{cases} \widetilde{c_o} = c_o^* e^{-(g+m)(t-N)} \\ \qquad + \dfrac{f c_{e1}^*(1 - e^{-(h-g-m)(t-N)})}{h-g-m}, t \in (N, (N+1)], \\ \widetilde{c_{e1}} = c_{e1}^* e^{-h_1(t-N)}, t \in (N, (N+1)], \\ \widetilde{c_{e2}} = c_{e2}^* e^{-h_2(t-N)}, t \in (N, (N+1)]. \end{cases} \tag{19}$$

where $c_o^*, c_{e1}^*, c_{e2}^*$ are by (15).

Remark 5. $m_o \le c_o(t) \le M_o$, $m_{e1} \le c_{e1}(t) \le M_{e1}$ and $m_{e2} \le c_{e2}(t) \le M_{e2}$ hold for t large enough, and $m_o = c_o^* e^{-(g+m)\tau} + \frac{f c_{e1}^*(1-e^{-(h-g-m)\tau})}{h-g-m} - \varepsilon > 0$, $M_o = c_o^* + \frac{f c_{e1}^*}{(h-g-m)} + \varepsilon$, $m_{e1} = c_{e1}^* e^{-h_1\tau} - \varepsilon > 0$, $M_{e1} = c_{e1}^* + \varepsilon > 0$ $m_{e2} = c_{e2}^* e^{-h_2\tau} - \varepsilon > 0$ and $M_{e2} = c_{e2}^* + \varepsilon > 0$.

Thinking (1), we get

$$\begin{cases} \dfrac{dx}{dt} < -d_1 x + c_1 x^2, \\ \qquad\qquad t \in (N, (N+L)], \\ \Delta x = -\mu x, \\ \qquad\qquad t = (N+L), \\ \dfrac{dx}{dt} < -(d_2+E)x + c_2 x^2, \\ \qquad\qquad t \in ((N+L), (N+1)], \\ \Delta x = bx, \\ \qquad\qquad t = (N+1), \end{cases} \tag{20}$$

and

$$\begin{cases} \dfrac{dx}{dt} > -(d_1 + \beta_1 M_o)x + c_1 x^2, \\ \qquad\qquad t \in (N, (N+L)], \\ \Delta x = -\mu x, \\ \qquad\qquad t = (N+L), \\ \dfrac{dx}{dt} > -(d_2 + E + \beta_2 M_o)x + c_2 x^2, \\ \qquad\qquad t \in ((N+L), (N+1)], \\ \Delta x = bx, \\ \qquad\qquad t = (N+L), \end{cases} \tag{21}$$

with their comparative impulsive systems

$$\begin{cases} \dfrac{dx_1}{dt} = -d_1 x_1 + c_1 x_1^2, \\ \qquad\qquad t \in (N, (N+L)], \\ \Delta x_1 = -\mu x_1, \\ \qquad\qquad t = (N+L), \\ \dfrac{dx_1}{dt} = -(d_2+E)x_1 + c_2 x_1^2, \\ \qquad\qquad t \in ((N+L), (N+L)], \\ \Delta x_1 = bx_1, \\ \qquad\qquad t = (N+1), \end{cases} \tag{22}$$

and

$$\begin{cases} \frac{dx_2}{dt} = -(d_1 + \beta_1 M_o)x_2 + c_1 x_2^2, \\ \qquad t \in (N, (N+L)], \\ \triangle x_2 = -\mu x_2, \\ \qquad t = (N+L), \\ \frac{dx_2}{dt} = -(d_2 + E + \beta_2 M_o)x_2 + c_2 x_2^2, \\ \qquad t \in ((N+L), (N+1)], \\ \triangle x_2 = b x_2, \\ \qquad t = (N+1)\tau. \end{cases} \tag{23}$$

Marking $(1+b)(1-\mu)e^{-[(d_1+\beta_1 M_o)l+(d_2+E+\beta_2 M_o)(1-l)]\tau} < 1$ as (C_3) and $(1+b)(1-\mu)e^{-[(d_1+\beta_1 M_o)l+(d_2+E+\beta_2 M_o)(1-l)]\tau} > 1$ as (C_4).

Similar to Theorem 2, we can get

Theorem 6. *i)* Suppose (C_3) holds, 0 of (23) is globally asymptotically stable;

ii) Suppose (C_4) holds, $\widetilde{x_2(t)}$ of (23) is globally asymptotically stable, and

$$\widetilde{x_2} = \begin{cases} \frac{d_1 x^* e^{-d_1(t-N)}}{(d_1+\beta_1 M_o)) + c_1 x^*(1 - e^{-(d_1+\beta_1 M_o))(t-N)})}, \\ \qquad t \in (N, (N+L)], \\ \frac{(d_2+E+\beta_2 M_o))x^{**} e^{-(d_2+E+\beta_2 M_o))(t-N)}}{(d_2+E+\beta_2 M_o)) + c_2 x^{**}(1 - e^{-(d_2+E+\beta_2 M_o))(t-N)})}, \\ \qquad t \in ((N+l), (N+1)]. \end{cases} \tag{24}$$

here x_2^* is by

$$x_2^* = \frac{M_4}{M_5}, (1+b)(1-\mu)e^{-[(d_1+\beta_1 M_o)l+(d_2+E+\beta_2 M_o)(1-l))\tau} > 1, \tag{25}$$

where $M_4 = (d_1 + \beta_1 M_o)(d_2 + E + \beta_2 M_o)[(1+b)(1-\mu)$ $e^{-[d_1+\beta_1 M_o)l+(d_2+E+\beta_2 M_o)(1-l)]\tau} - 1]$, $M_5 = (d_2+E+\beta_2 M_o)c_1(1-e^{-(d_1+\beta_1 M_o)l})+(1-\mu)(d_1+\beta_1 M_o)c_2 e^{-(d_1+\beta_1 M_o)l\tau}(1-e^{-(d_2+E+\beta_2 M_o)(1-l)\tau})$ and $x_2^{**} = \frac{(1-\mu_1)(d_1+\beta_1 M_o))x_2^* e^{-(d_1+\beta_1 M_o)l\tau}}{(d_1+\beta_1 M_o))+c_1 x_2^*(1-e^{-(d_1+\beta_1 M_o)l\tau})}$.

Theorem 7. *i)* Suppose (C_1) and (C_3) hold, $(0, \widetilde{c_o}, \widetilde{c_{e1}}, \widetilde{c_{e2}})$ of (1) is globally asymptotically stable;

ii) Suppose (C_2) and (C_4) hold, (1) is permanent.

Proof. *i)* From condition (C_1), (22), Theorem 2, and the theorem of the impulsive equation, we can have $x(t) \leq x_1(t) = x_1(t) \to 0$ as $t \to +\infty$. From condition (C_3) (23), Theorem 6, and the theorem of the impulsive differential equation, we can have $x(t) \geq x_2(t) \to 0$ as $t \to +\infty$. So, the globally asymptotically stable periodic solution $(0, \widetilde{c_o(t)}, \widetilde{c_{e1}(t)}, \widetilde{c_{e2}(t)})$ exists.

ii) From condition (C_2), (22), Theorem 2, we can have

$$x(t) \leq x_1(t) \leq \widetilde{x_1(t)} - \varepsilon \leq \frac{d_1 x^* e^{-d_1 l \tau}}{d_1) + c_1 x^*(1 - e^{-d_1 l \tau})}$$

$$+ \frac{(d_2 + E)x^{**} e^{-(d_2+E)(1-l)\tau}}{(d_2 + E) + c_2 x^{**}(1 - e^{-(d_2+E))(1-l)\tau}} - \varepsilon \triangleq m_1.$$

From condition (C_4), (23), Theorem 6, we can easily obtain

$$x(t) \geq x_2(t) \geq \widetilde{x_2(t)} - \varepsilon \geq \frac{d_1 x^* e^{-d_1 l \tau}}{(d_1 + \beta_1 M_o)) + c_1 x^*(1 - e^{-(d_1+\beta_1 M_o) l \tau}}$$

$$+ \frac{(d_2 + E + \beta_2 M_o))x^{**} e^{-(d_2+E+\beta_2 M_o))(1-l)\tau}}{(d_2 + E + \beta_2 M_o)) + c_2 x^{**}(1 - e^{-(d_2+E+\beta_2 M_o))(1-l)\tau}} - \varepsilon \triangleq m_2.$$

From above discussion and Remark 5, we can have $m_1 \leq x(t) \leq m_2$, $m_o \leq c_o(t) \leq M_o$, $m_{e1} \leq c_{e1}(t) \leq M_{e1}$, $m_{e2} \leq c_{e2}(t) \leq M_{e2}$. This completes the proof.

4 Discussion

In this work, we consider we construct a single population model with non-transient/transient impulsive harvesting and birth pulse in a polluted environment. From the conditions of Theorem 7, we can deduce that the transient impulsive harvesting amount μ has a threshold μ^*. If $\mu > \mu^*$, the globally asymptotically stable $(0, \widetilde{c_o(t)}, \widetilde{c_{e1}(t)}, \widetilde{c_{e2}(t)})$ exists. If $\mu < \mu^*$, system permanence. We can also obtain the threshold l^* for non-transient impulsive harvesting intervals. Our results present a biological management researching basis in a polluted environment.

References

1. Lakshmikantham, V.: Theory of Impulsive Differential Equations. World Scientific, Singapore (1989)
2. Jiao, J., et al.: An appropriate pest management SI model with biological and chemical control concern. Appl. Math. Comput. **196**, 285–293 (2008)
3. Jiao, J., Cai, S., Chen, L.: Analysis of a stage-structured predator-prey system with birth pulse and impulsive harvesting at different moments. Nonlinear Anal. Real World Appl. **12**, 2232–2244 (2011)
4. Chen, L., Meng, X., Jiao, J.: Biological Dynamics. Scientific Press, Beijing (2009)
5. Clack, C.W.: Mathematical Bioeconomics: The Optimal Management of Renewable Resources. Wiley, New York (1976)
6. Zhang, B.G.: Population's Ecological Mathematics Modeling. Publishing of Qingdao Marine University, Qingdao (1990)
7. Hallam, T.G., Clark, C.E., Lassider, R.R.: Effects of toxicant on population: a qualitative approach I. Equilibrium environmental exposure. Ecol. Modell. **18**, 291–340 (1983)

8. Liu, B., Chen, L.S., Zhang, Y.J.: The effects of impulsive toxicant input on a population in a polluted environment. J. Biol. Syst. **11**, 265–287 (2003)
9. Jiao, J.J., et al.: Dynamics of a periodic switched predator-prey system with impulsive harvesting and hibernation of prey population. J. Franklin Inst. **353**, 3818–3834 (2016)
10. Jiao, J., et al.: Threshold dynamics of a stage-structured single population model with non-transient and transient impulsive effects. Appl. Math. Lett. **97**, 88–92 (2019)

Dynamics of a Predator and Prey-Hibernation Model with Genic Mutation and Impulsive Effects

Shaohong Cai and Jianjun Jiao(✉)

School of Mathematics and Statistics, Guizhou University of Finance and Economics, Guiyang 550025, People's Republic of China

Abstract. A predator and prey-hibernation system with mutation and impulsive effects is presented in this work. The globally asymptotically stable solution $(\widetilde{x}, \widetilde{y}, 0)$ of system (1) exists. System (1) is proved to be permanent. Finally, simulations are presented to explain the results.

Keywords: Mutation · Predator and prey-hibernation system · Permanence

1 Introduction

Many authors [1,2] indicated that environmental pollutants caused many diseases. Population dynamics are studied by the theories of impulsive differential equations [3–6]. Jiao et al. [7] constructed a predator-prey system with periodic switches and impulses. Jiao and Chen [8] presented a predator-prey system with mutation and impulses. vFor convenience, we make notation $N = n\tau$, and $N + l = (n + l)\tau$.

2 The Model

In this work, we presented a predator and prey-hibernation system with mutation and impulses

Supported by National Natural Science Foundation of China (11761019, 11361014), the Science Technology Foundation of Guizhou Education Department (20175736-001, 2008038), the Project of High Level Creative Talents in Guizhou Province (No. 20164035), Major Research Projects on Innovative Groups in Guizhou Provincial Education Department (No. [2018]019).

H. Song and D. Jiang (Eds.): SIMUtools 2020, LNICST 370, pp. 39–52, 2021.
https://doi.org/10.1007/978-3-030-72795-6_4

$$
\begin{cases}
\left.\begin{aligned}
&\frac{dx}{dt} = -ax - cx^2 - \beta_1 xz,\\
&\frac{dy}{dt} = -d_1 y - \beta_2 yz,\\
&\frac{dz}{dt} = -d_2 z + k_1\beta_1 xz\\
&\qquad\qquad + k_2\beta_2 yz,
\end{aligned}\right\} t \in (N, N+l],\\
\left.\begin{aligned}
&\Delta x = 0,\\
&\Delta y = -\mu_1 y,\\
&\Delta z = -\mu_2 z,
\end{aligned}\right\} t = (N+l),\\
\left.\begin{aligned}
&\frac{dx}{dt} = -d_3 x - \beta_3 xz,\\
&\frac{dy}{dt} = -d_4 y - \beta_4 yz,\\
&\frac{dz}{dt} = -d_5 z + k_3\beta_3 xz\\
&\qquad\qquad + k_4\beta_4 yz,
\end{aligned}\right\} t \in (N+l, N+1],\\
\left.\begin{aligned}
&\Delta x = (1-\theta_1) b_1 x,\\
&\Delta y = \theta_1 b_1 x,\\
&\Delta z = 0,
\end{aligned}\right\} t = N+1,
\end{cases}
\tag{1}
$$

The biological meanings of the varies and the parameters can reference to [7,8].

3 The Lemmas

If $z = 0$, we can easily have the subsystem of system (1)

$$
\begin{cases}
\left.\begin{aligned}
&\frac{dx}{dt} = -ax - cx^2,\\
&\frac{dy}{dt} = -d_1 y,
\end{aligned}\right\} t \in (N, N+l],\\
\left.\begin{aligned}
&\Delta x = 0,\\
&\Delta y = -\mu_1 y,
\end{aligned}\right\} t = N+l,\\
\left.\begin{aligned}
&\frac{dx}{dt} = -d_3 x,\\
&\frac{dy}{dt} = -d_4 y,
\end{aligned}\right\} t \in (N+l, N+1],\\
\left.\begin{aligned}
&\Delta x = (1-\theta_1) b_1 x,\\
&\Delta y = \theta_1 b_1 x,
\end{aligned}\right\} t = N+1.
\end{cases}
\tag{2}
$$

Integrating (2), and the stroboscopic map of (2) is

$$\begin{cases} x((N+1^{+}) = [1+(1-\theta_1)b_1] \times \dfrac{ax(N^{+})e^{-[al+d_3(1-l)]\tau}}{a+cx(N^{+})(1-e^{-al\tau})}, \\ y((N+1)^{+}) = \theta_1 b_1 \times \dfrac{ax(N^{+})e^{-[al+d_3(1-l)]\tau}}{a+cx(N^{+})(1-e^{-al\tau})} \\ \qquad\qquad + (1-\mu_1)e^{-[d_1 l+d4(1-l)]\tau} y(N^{+}). \end{cases} \tag{3}$$

Points $G_1(0,0)$ and $G_2(x^*, y^*)$ are gotten as

$$\begin{cases} x^* = Ay^*, \quad \theta_1 b_1 A e^{-[al+d_3(1-l)]\tau} + B > 1. \\ y^* = \dfrac{a}{cA(1-e^{-al\tau})} \times [\dfrac{\theta_1 b_1 A e^{-[al+d_3(1-l)]\tau}}{1-B} - 1], \\ \qquad \theta_1 b_1 A e^{-[al+d_3(1-l)]\tau} + B > 1, \end{cases} \tag{4}$$

here $A = \frac{[1+(1-\theta_1)b_1](1-B)}{\theta_1 b_1} > 0$ and $B = (1-\mu_1)e^{-[d_1 l+d_4(1-l)]\tau} > 0$.

Similarly with reference [7], the following lemmas can easily be obtained.

Lemma 1. *i)* If $\theta_1 b_1 A e^{-[al+d_3(1-l)]\tau} + B < 1$, point $G_1(0,0)$ is globally asymptotically stable;

ii) If $\theta_1 b_1 A e^{-[al+d_3(1-l)]\tau} + B > 1$, point $G_2(x^*, y^*)$ is globally asymptotically stable.

Lemma 2. *i)* If $\theta_1 b_1 A e^{-[al+d_3(1-l)]\tau} + B < 1$, periodic solution $(0,0)$ of system (2) is globally asymptotically stable;

ii) If $\theta_1 b_1 A e^{-[al+d_3(1-l)]\tau} + B > 1$, periodic solution $(\widetilde{x}, \widetilde{y})$ of system (2) is globally asymptotically stable, where

$$\begin{cases} \widetilde{x} = \begin{cases} \dfrac{ax^* e^{-a(t-N)}}{a+cx^*(1-e^{-a(t-N)})}, t \in (N, (N+1)], \\ x^{**} e^{-d_3(t-(N+l))}, t \in ((N+l), (N+1)], \end{cases} \\ \widetilde{y} = \begin{cases} y^* e^{-d_1(t-N)}, t \in (N, (N+l)], \\ y^{**} e^{-d_4(t-(N+l))}, t \in ((N+l), (N+1)], \end{cases} \end{cases} \tag{5}$$

here x^*, y^* are determined as (5) and x^{**}, y^{**} are determined as

$$\begin{cases} x^{**} = \dfrac{ax^* e^{-al\tau}}{a+cx^*(1-e^{-al\tau})}, \\ y^{**} = (1-\mu_1)y^* e^{-d_1 l\tau}. \end{cases} \tag{6}$$

For (1), we know

$$
\begin{cases}
\left.\begin{aligned}
\frac{dx}{dt} &\le -ax - cx^2, \\
\frac{dy}{dt} &\le -d_1 y,
\end{aligned}\right\} t \in (N, (N+l)], \\
\left.\begin{aligned}
\Delta x &= 0, \\
\Delta y &= -\mu_1 y,
\end{aligned}\right\} t = (N+l), \\
\left.\begin{aligned}
\frac{dx}{dt} &\le -d_3 x, \\
\frac{dy}{dt} &\le -d_4 y,
\end{aligned}\right\} t \in ((N+l), (N+1)], \\
\left.\begin{aligned}
\Delta x &= (1-\theta_1) b_1 x, \\
\Delta y &= \theta_1 b_1 x,
\end{aligned}\right\} t = (N+1).
\end{cases}
\tag{7}
$$

Then, we can obtain the following remark.

Remark 3. If $\theta_1 b_1 A e^{-[al+d_3(1-l)]\tau} + B > 1$, then

$$x(t) \le x^* + x^{**},$$

and

$$y(t) \le y^* + y^{**},$$

with (x, y) of system (1).

Similar to Ref. [13], we get

Lemma 4. There exists a constant $M > 0$, which makes $x \le \frac{M}{k}$, $y \le M$ and $z \le M$ for all t large enough.

4 The Dynamics

Theorem 1. Suppose

(H_1):

$$\theta_1 b_1 A e^{-[al+d_3(1-l)]\tau} + B > 1,$$

(H_2):

$$\ln \frac{1}{1+(1-\theta_1)b_1} + al\tau + 2\ln \frac{a + cx^*(1-e^{-al\tau})}{a} + d_3(1-l)\tau > 0,$$

(H_3):

$$\ln \frac{1}{1-\mu_2} + d_2 l\tau + d_5(1-l)\tau$$

$$> \frac{k_1\beta_1}{c} \ln \frac{a + cx^*(1-e^{-al\tau})}{a} + \frac{k_2\beta_2 y^*(1-e^{-d_1 l\tau})}{d_1}$$

$$+\frac{k_3\beta_3x^{**}e^{-d_3(1-l)\tau}(1-e^{-d_1(1-l)\tau})}{d_3}+\frac{k_4\beta_4y^{**}e^{-d_5(1-l)\tau}(1-e^{-d_5(1-l)\tau})}{d_5},$$

hold, the solution $(\widetilde{x},\widetilde{y},0)$ of (1) is globally asymptotically stable, and x^*,x^{**} are by (5), y^*,y^{**} are by (7).

Proof. Defining $x_1=x-\widetilde{x}, y_1=y-\widetilde{y}, z_1=z$, for $t\in(N,(N+l)]$, the linear system of (1) are as

$$\begin{pmatrix}\frac{dx_1}{dt}\\ \frac{dy_1}{dt}\\ \frac{dz_1}{dt}\end{pmatrix}=\begin{pmatrix}-a-2c\widetilde{x} & 0 & -\beta_1\widetilde{x}\\ 0 & -d_1 & -\beta_2\widetilde{y}\\ 0 & 0 & -d_2+k_1\beta_1\widetilde{x}+k_2\beta_2\widetilde{y}\end{pmatrix}\begin{pmatrix}x_1\\ y_1\\ z_1\end{pmatrix}.$$

Then, we get the fundamental solution matrix

$$\Phi_1(t)=\begin{pmatrix}\exp(\int_N^t(-a-2c\widetilde{x})ds) & *_1 & *_2\\ 0 & \exp(-d_1(t-N)) & *_3\\ 0 & 0 & M_1\end{pmatrix}$$

where $M_1=\exp[\int_N^t(-d_2+k_1\beta_1\widetilde{x}+k_2\beta_2\widetilde{y})ds]$.

For $t\in((N+l),(N+1)]$, the linear system of (1) are also as

$$\begin{pmatrix}\frac{dx_1}{dt}\\ \frac{dy_1}{dt}\\ \frac{dz_1}{dt}\end{pmatrix}=\begin{pmatrix}-d_3 & 0 & -\beta_3\widetilde{x}\\ 0 & -d_4 & -\beta_4\widetilde{y}\\ 0 & 0 & -d_5+k_3\beta_3\widetilde{x}+k_4\beta_4\widetilde{y}\end{pmatrix}\begin{pmatrix}x_1\\ y_1\\ z_1\end{pmatrix}.$$

Then, we also get the fundamental solution matrix

$$\Phi_2(t)=\begin{pmatrix}\exp(-d_3(t-(N+l)) & *_4 & *_5\\ 0 & \exp(-d_4(t-(N+l)) & *_6\\ 0 & 0 & M_2\end{pmatrix},$$

where $M_2=\exp[\int_{(N+l)}^t(-d_5+k_3\beta_3\widetilde{x}+k_4\beta_4\widetilde{y})ds]$. There is no need to calculate the form of $*_i(i=1,2,3,4,5,6)$.

The linearization of the 4th, to 6th equations of (1) is

$$\begin{pmatrix}x_1((N+l)^+)\\ y_1((N+l)^+)\\ z_1((N+l)^+)\end{pmatrix}=\begin{pmatrix}1 & 0 & 0\\ 0 & 1-\mu_1 & 0\\ 0 & 0 & 1-\mu_2\end{pmatrix}\begin{pmatrix}x_1((N+l))\\ y_1((N+l))\\ z_1((N+l))\end{pmatrix}.$$

The linearization of the 10th, to 12th equations of (1) is

$$\begin{pmatrix} x_1((N+1)^+) \\ y_1((N+1)^+) \\ z_1((N+1)^+) \end{pmatrix} = \begin{pmatrix} 1+(1-\theta_1)b_1 & 0 & 0 \\ \theta_1 b_1 & 1 & 0 \\ 0 & 0 & 1 \end{pmatrix} \begin{pmatrix} x_1((N+1)) \\ y_1((N+1)) \\ z_1((N+1)) \end{pmatrix}.$$

The stability of $(\widetilde{x}, \widetilde{y}, 0)$ is by eigenvalues of

$$M = \begin{pmatrix} 1 & 0 & 0 \\ 0 & 1-\mu_1 & 0 \\ 0 & 0 & 1-\mu_2 \end{pmatrix} \begin{pmatrix} 1+(1-\theta_1)b_1 & 0 & 0 \\ \theta_1 b_1 & 1 & 0 \\ 0 & 0 & 1 \end{pmatrix} \Phi(\tau),$$

which are

$$\lambda_1 = [1+(1-\theta_1)b_1]\exp(\int_0^{l\tau}(-a-2c\widetilde{x})ds + d_3(1-l)\tau),$$

$$\lambda_2 = (1-\mu_1)\exp(-[d_1 l + d_4(1-l)]\tau) < 1,$$

and

$$\lambda_3 = (1-\mu_2)\exp[\int_0^{l\tau}(-d_2+k_1\beta_1\widetilde{x}+k_2\beta_2\widetilde{y})ds + \int_{l\tau}^{\tau}(-d_5+k_3\beta_3\widetilde{x}+k_4\beta_4\widetilde{y})ds].$$

According the conditions of Theorem 1, then $\lambda_1 < 1$, and $\lambda_3 < 1$. Therefore, the $(\widetilde{x}, \widetilde{y}, 0)$ is locally stable.

Choose $\varepsilon > 0$, we get

$$\rho_1 = (1-\mu_2)\exp\{\int_0^{l\tau}[-d_2+k_1\beta_1(\widetilde{x}+\varepsilon)+k_2\beta_2(\widetilde{y}+\varepsilon)]ds + \int_{l\tau}^{\tau}[-d_5+k_3\beta_3(\widetilde{x}+\varepsilon)+k_4\beta_4(\widetilde{y}+\varepsilon)]ds\} < 1.$$

From (1), we get $\frac{dx}{dt} \le -ax - cx^2, \frac{dy}{dt} \le -d_1 x$, and $\frac{dx}{dt} \le -d_3 x$, $\frac{dy}{dt} \le -d_4 x$, so

$$\begin{cases} \left.\begin{aligned} \frac{dx_1}{dt} &= -ax_1(t) - cy_1^2, \\ \frac{dy_1}{dt} &= -d_1 y_1, \end{aligned}\right\} t \ne (N, (N+l)], \\ \left.\begin{aligned} \Delta x_1 &= 0, \\ \Delta y_1 &= -\mu_1 y_1, \end{aligned}\right\} t = (N+l), \\ \left.\begin{aligned} \frac{dx_1}{dt} &= -d_3 x_1, \\ \frac{dy_1}{dt} &= -d_4 y_1, \end{aligned}\right\} t \in ((N+l), (N+1), \\ \left.\begin{aligned} \Delta x_1 &= (1-\theta_1)b_1 x_1, \\ \Delta y_1 &= \theta_1 b_1 y_1, \end{aligned}\right\} t = (N+1), \end{cases} \tag{8}$$

From Lemma 3, we have $x \leq x_1, y \leq y_1$, and

$$\begin{cases} x \leq x_1 \leq \widetilde{x} + \varepsilon, \\ y \leq y_1 \leq \widetilde{y} + \varepsilon, \end{cases} \tag{9}$$

From (1) and (14), we also get

$$\begin{cases} \dfrac{dz}{dt} \leq z[-d_2 + k_1\beta_1(\widetilde{x} + \varepsilon) + +k_2\beta_2(\widetilde{y} + \varepsilon)], \\ \qquad\qquad\qquad\qquad t \in (N, (N+l)] \\ \triangle z = -\mu_2 z, t = (N+l), \\ \dfrac{dz}{dt} \leq z[-d_5 + k_3\beta_3(\widetilde{x} + \varepsilon) + k_4\beta_4(\widetilde{y} + \varepsilon)], \\ \qquad\qquad\qquad\qquad t \in ((N+l), (N+1)], \end{cases} \tag{10}$$

Then,

$$z((N+1)) \leq (1-\mu_2)z(N^+)\exp[\int_N^{(N+l)} (-d_2 + k_1\beta_1(\widetilde{x} + \varepsilon) + k_2\beta_2(\widetilde{y} + \varepsilon)ds$$
$$+ \int_{(N+l)}^{(N+1)} (-d_5 + k_3\beta_3(\widetilde{x} + \varepsilon) + k_4\beta_4(\widetilde{y} + \varepsilon))ds].$$

and $z((N+l)) \leq z(l\tau^+)\rho_1^n$ with $z((N+l)) \to 0$ as $n \to \infty$. For $0 < z \leq z((N+l))e^{(-d_5(1-l)\tau + \frac{k_3\beta_3}{d_3}x^{**} + \frac{k_4\beta_4}{d_4}y^{**})}$ for $N < t \leq (N+l)$. Therefore $z \to 0$ as $t \to \infty$.

For (1) and $t \in (N, (N+l)]$, we have

$$\begin{cases} -(a + \beta_1\varepsilon)x - cx^2 \leq \dfrac{dx}{dt} \leq -ax - cx^2, \\ -(d_1 + \beta_2\varepsilon)y \leq \dfrac{dy}{dt} \leq -d_1 y, \end{cases} \tag{11}$$

and for $t \in ((N+l), (N+1)]$,

$$\begin{cases} -(d_3 + \beta_3\varepsilon)x \leq \dfrac{dx}{dt} \leq -d_1 x, \\ -(d_4 + \beta_4\varepsilon)y \leq \dfrac{dy}{dt} \leq -d_4 y, \end{cases} \tag{12}$$

then, $w_1 \le x \le m_1, w_2 \le y \le m_2$, with $w_1 \to \widetilde{w_1}, w_2 \to \widetilde{w_2}, m_1 \to \widetilde{x}, m_2 \to \widetilde{y}$, as $t \to \infty$. While (w_1, w_2) and (m_1, m_2) are the solutions of

$$\begin{cases} \left.\begin{array}{l} \frac{dw_1}{dt} = -(a+\beta_1\varepsilon)w_1 - cw_1^2, \\ \frac{dw_2}{dt} = -(d_1+\beta_2\varepsilon)w_2, \end{array}\right\} t \in (N, (N+l)), \\ \left.\begin{array}{l} \triangle w_1 = 0, \\ \triangle w_2 = -\mu_1 w_2, \end{array}\right\} t = (N+l), \\ \left.\begin{array}{l} \frac{dw_1}{dt} = -(d_3+\beta_3\varepsilon)w_1, \\ \frac{dw_2}{dt} = -(d_4+\beta_4\varepsilon)w_2, \end{array}\right\} t \in ((N+l), (N+1)], \\ \left.\begin{array}{l} \triangle w_1 = (1-\theta_1)b_1 w_1), \\ \triangle w_2 = \theta_1 b_1 w_1), \end{array}\right\} t = (N+1), \end{cases} \tag{13}$$

and

$$\begin{cases} \left.\begin{array}{l} \frac{dm_1}{dt} = -am_1 - cm_1^2, \\ \frac{dm_2}{dt} = -d_1 m_2, \end{array}\right\} t \in (N, (N+l)), \\ \left.\begin{array}{l} \triangle m_1 = 0, \\ \triangle m_2 = -\mu_1 m_2, \end{array}\right\} t = (N+l), \\ \left.\begin{array}{l} \frac{dm_1}{dt} = -d_3 m_1, \\ \frac{dm_2}{dt} = -d_4 m_2, \end{array}\right\} t \in ((N+l), (N+1)], \\ \left.\begin{array}{l} \triangle m_1 = (1-\theta_1)b_1 m_1, \\ \triangle m_2 = \theta_1 b_1 m_1, \end{array}\right\} t = (N+1). \end{cases} \tag{14}$$

Here $(\widetilde{w_1}, \widetilde{w_2})$ can be expressed as

$$\begin{cases} \widetilde{w_1} = \begin{cases} \frac{(a+\beta_1\varepsilon)w_1^* e^{-(a+\beta_1\varepsilon)(t-N)}}{(a+\beta_1\varepsilon) + cw_1^*(1-e^{-(a+\beta_1\varepsilon)(t-N)})}, t \in (N, (N+1)], \\ w_1^{**} e^{-(d_3+\beta_3\varepsilon)(t-(N+l))}, t \in ((N+l), (N+1)], \end{cases} \\ \widetilde{w_2} = \begin{cases} w_2^* e^{-(d_1+\beta_2\varepsilon)(t-N)}, t \in (N, (N+l)], \\ w_2^{**} e^{-(d_4+\beta_4\varepsilon)(t-(N+l))}, t \in ((N+l), (N+1)], \end{cases} \end{cases} \tag{15}$$

while w_1^*, w_2^* are determined as

$$\begin{cases} w_1^* = A_1 w_2^*, \ \theta_1 b_1 A_1 e^{-[(a+\beta_1\varepsilon)l+(d_3+\beta_3\varepsilon)(1-l)]\tau} + B_1 > 1, \\ w_2^* = \dfrac{(a+\beta_1\varepsilon)}{cA_1(1-e^{-(a+\beta_1\varepsilon)l\tau})} \times [\dfrac{\theta_1 b_1 A_1 e^{-[(a+\beta_1\varepsilon)l+(d_3+\beta_3\varepsilon)(1-l)]\tau}}{1-B_1} - 1], \\ \qquad \theta_1 b_1 A_1 e^{-[(a+\beta_1\varepsilon)l+(d_3+\beta_3\varepsilon)(1-l)]\tau} + B_1 > 1, \end{cases} \tag{16}$$

where $A_1 = \frac{[1+(1-\theta_1)b_1](1-B_1)}{\theta_1 b_1} > 0$ and $B_1 = (1-\mu)e^{-[(d_1+\beta_2\varepsilon)l+(d_4+\beta_4\varepsilon)(1-l)]\tau} > 0$, and w_1^{**}, w_2^{**} are determined as

$$\begin{cases} w_1^{**} = \dfrac{(a+\beta_1\varepsilon)w_1^* e^{-(a+\beta_1\varepsilon)l\tau}}{(a+\beta_1\varepsilon)+cw_1^*(1-e^{-(a+\beta_1\varepsilon)l\tau})}, \\ w_2^{**} = (1-\mu_1)w_2^* e^{-(d_1+\beta_2\varepsilon)l\tau}. \end{cases} \tag{17}$$

For any $\varepsilon_1 > 0$, there exists a $t_1, t > t_1$ such that

$$\widetilde{w_1} - \varepsilon_1 < x < \widetilde{x} + \varepsilon,$$

and

$$\widetilde{w_2} - \varepsilon_1 < y < \widetilde{y} + \varepsilon.$$

Let $\varepsilon \to 0$, we have

$$\widetilde{x} - \varepsilon_1 < x < \widetilde{x} + \varepsilon_1,$$

and

$$\widetilde{y} - \varepsilon_1 < y < \widetilde{y} + \varepsilon_1,$$

for t large enough, which indicates $x \to \widetilde{x}, y \to \widetilde{y}$ as $t \to \infty$.

Theorem 2. If (H_1), (H_2) and (H_4):

$$\ln\frac{1}{1-\mu_2} + d_2 l\tau + d_5(1-l)\tau$$

$$< \frac{k_1\beta_1}{c}\ln\frac{a+cx^*(1-e^{-al\tau})}{a} + \frac{k_2\beta_2 y^*(1-e^{-d_1 l\tau})}{d_1}$$

$$+\frac{k_3\beta_3 x^{**}e^{-d_3(1-l)\tau}(1-e^{-d_1(1-l)\tau})}{d_3} + \frac{k_4\beta_4 y^{**}e^{-d_5(1-l)\tau}(1-e^{-d_5(1-l)\tau})}{d_5},$$

hold, (1) is permanent, where x^*, x^{**} are by (5) and y^*, y^{**} are by (7).

Proof. By Lemma 3, we get that $x(t) \leq \frac{M}{k}, y(t) \leq \frac{M}{k}, z(t) \leq M$ for t large enough. From Theorem 1, we know $x(t) \geq \frac{ax^* e^{-al\tau}}{a+cx^*(1-e^{-al\tau})} + x^{**}e^{-d_3(1-l)\tau} - \varepsilon_2 = m_{21}$ and $y(t) \geq y^* e^{-d_1 l\tau} + y^{**}e^{-d_4(1-l)\tau} - \varepsilon_2 = m_{22}$ for t. Thus, we will seek out $m_1 > 0$ making $y(t) \geq m_1$.

By the conditions H_4, we can choose $m_3 > 0, \varepsilon_1 > 0$ small enough to have

$$\sigma = \frac{k_1\beta_1}{c} \ln \frac{(a+\beta_1 m_3) + cz_1^*(1-e^{-(a+\beta_1 m_3)l\tau})}{(a+\beta_1 m_3)} + k_1\beta_1\varepsilon_1 l\tau$$

$$+ \frac{k_2 + \beta_2 z_2^*(1-e^{-(d_1+\beta_1 m_3)l\tau})}{(d_1+\beta_1 m_3)l\tau + k_2\beta_2 m_3 l\tau} + k_2\beta_2\varepsilon_1 l\tau$$

$$+ \frac{k_3\beta_3 z_1^{**}e^{-(d_3+\beta_1 m_3)(1-l)\tau}(1-e^{-(d_1+\beta_1 m_3)(1-l)\tau})}{(d_3+\beta_1 m_3)} + k_3\beta_3\varepsilon_1(1-l)\tau$$

$$+ \frac{k_4\beta_4 z_2^{**}e^{-(d_5+\beta_5 m_3)(1-l)\tau}(1-e^{-(d_5+\beta_5 m_3)(1-l)\tau})}{(d_5+\beta_5 m_3)} + k_4\beta_4\varepsilon_1(1-l)\tau,$$

$$- \ln\frac{1}{1-\mu_2} - d_2 l\tau - d_5(1-l)\tau] > 0,$$

here z_1^*, z_2^* are by

$$\begin{cases} z_1^* = A_2 z_2^*, \quad \theta_1 b_1 A_2 e^{-[(a+\beta_1 m_3)l+(d_3+\beta_3 m_3)(1-l)]\tau} + B_2 > 1, \\ z_2^* = \dfrac{(a+\beta_1 m_3)}{cA_2(1-e^{-(a+\beta_1 m_3)l\tau})} \\ \qquad \times [\dfrac{\theta_1 b_1 A_2 e^{-[(a+\beta_1 m_3)l+(d_3+\beta_3 m_3)(1-l)]\tau}}{1-B_2} - 1], \\ \qquad \theta_1 b_1 A_2 e^{-[(a+\beta_1 m_3)l+(d_3+\beta_3 m_3)(1-l)]\tau} + B_2 > 1, \end{cases} \tag{18}$$

and z_1^{**}, z_2^{**} are by

$$\begin{cases} z_1^{**} = \dfrac{(a+\beta_1 m_3)z_1^* e^{-(a+\beta_1 m_3)l\tau}}{(a+\beta_1 m_3) + cz_1^*(1-e^{-(a+\beta_1 m_3)l\tau})}, \\ z_2^{**} = (1-\mu_1)z_2^* e^{-(d_1+\beta_2 m_3)l\tau}. \end{cases} \tag{19}$$

with $A_2 = \frac{[1+(1-\theta_1)b_1](1-B_2)}{\theta_1 b_1} > 0$ and $B_2 = (1-\mu) e^{-[(d_1+\beta_2 m_3)l+(d_4+\beta_4 m_3)(1-l)]\tau} > 0$. $y(t) < m_3$ will be proved that it can not establish. Otherwise,

$$\begin{cases} \left.\begin{aligned} &\frac{dx}{dt} \geq -(a+\beta_1 m_3)x - cx^2, \\ &\frac{dy}{dt} \geq -(d_1+\beta_2 m_3)y, \end{aligned}\right\} t \in (N,(N+l)], \\ \left.\begin{aligned} &\Delta x = 0, \\ &\Delta y = -\mu_1 y, \end{aligned}\right\} t = (N+l), \\ \left.\begin{aligned} &\frac{dx}{dt} \geq -(d_3+\beta_3 m_3)x, \\ &\frac{dy}{dt} \geq -(d_4+\beta_4 m_3)y, \end{aligned}\right\} t \in ((N+l),(N+1)], \\ \left.\begin{aligned} &\Delta x = (1-\theta_1)b_1 x, \\ &\Delta x = \theta_1 b_1 x, \end{aligned}\right\} t = (N+1). \end{cases} \tag{20}$$

By Lemma 4, we have $x \geq z_1$, $y \geq z_2$ with $z_1 \to \widetilde{z}_1, z_2 \to \widetilde{z}_2, t \to \infty$, where (z_1, z_2) is the solution of

$$\begin{cases} \left.\begin{aligned} &\frac{dz_1}{dt} = -(a+\beta_1 m_3)z_1 - cz_1^2, \\ &\frac{dz_2}{dt} = -(d_1+\beta_2 m_3)z_2 \end{aligned}\right\} t \in (N,(N+l)], \\ \left.\begin{aligned} &\Delta z_1 = 0, \\ &\Delta z_2 = -\mu_1 z_2, \end{aligned}\right\} t = (N+l), \\ \left.\begin{aligned} &\frac{dz_1}{dt} = -(d_3+\beta_3 m_3)z_1, \\ &\frac{dz_2}{dt} = -(d_4+\beta_4 m_3)z_2, \end{aligned}\right\} t \in ((N+l),(N+1)], \\ \left.\begin{aligned} &\Delta z_1 = (1-\theta_1)b_1 z_1, \\ &\Delta z_2 = \theta_1 b_1 z_2, \end{aligned}\right\} t = (N+1), \end{cases} \tag{21}$$

and

$$\begin{cases} \widetilde{z}_1 = \begin{cases} \dfrac{(a+\beta_1 m_3)z_1^* e^{-(a+\beta_1 m_3)(t-n\tau)}}{(a+\beta_1 m_3) + cz_1^*(1-e^{-(a+\beta_1 m_3)(t-N)})}, t \in (N,(N+1)], \\ z_1^{**} e^{-(d_3+\beta_3 m_3)(t-(N+l))}, t \in ((N+l),(N+1)], \end{cases} \\ \widetilde{z}_2 = \begin{cases} z_2^* e^{-(d_1+\beta_2 m_3)(t-N)}, t \in (N,(N+l)], \\ z_2^{**} e^{-(d_4+\beta_4 m_3)(t-(N+l))}, t \in ((N+l),(N+1)], \end{cases} \end{cases} \tag{22}$$

here z_1^*, z_2^* are by (18) and z_1^{**}, z_2^{**} are by (19) with $A_2 = \frac{[1+(1-\theta_1)b_1](1-B_2)}{\theta_1 b_1} > 0$ and $B_2 = (1-\mu)e^{-[(d_1+\beta_2 m_3)l+(d_4+\beta_4 m_3)(1-l)]\tau} > 0$. Therefore, there exists a $T_1 > 0$ such that

$$x \geq z_1 \geq \widetilde{z}_1 - \varepsilon_1,$$

and

$$y \geq z_2 \geq \widetilde{z}_2 - \varepsilon_1.$$

Then

$$\begin{cases} \dfrac{dz}{dt} \geq [-d_1 + k_1\beta_1(\widetilde{z}_1 - \varepsilon_1) + k_2\beta_2(\widetilde{z}_2 - \varepsilon_1)]z, \\ \qquad\qquad\qquad\qquad t \in (N, (N+l)], \\ \triangle z = -\mu_2 z, t = (N+l), \\ \dfrac{dz}{dt} \geq [-d_5 + k_3\beta_3(\widetilde{z}_1 - \varepsilon_1) + k_4\beta_4(\widetilde{z}_2 - \varepsilon_1)]z, \\ \qquad\qquad\qquad\qquad t \in ((N+l), (N+1)]. \end{cases} \tag{23}$$

For $t \geq T_1$, Let $N_1 \in N$ and $N_1\tau > T_1$, integrating (28) on $(N, (N+1)], n \geq N_1$, we have

$$y((N+1)) \geq z(N^+)(1-\mu_1)$$
$$\times e^{\int_N^{(N+l)}[-d_1+k_1\beta_1(\widetilde{z}_1-\varepsilon_1)+k_2\beta_2(\widetilde{z}_2-\varepsilon_1)]ds+\int_{(N+l)}^{(N+1)}[-d_5+k_3\beta_3(\widetilde{z}_1-\varepsilon_1)+k_4\beta_4(\widetilde{z}_2-\varepsilon_1)]ds}$$
$$= (1-\mu_1)z(N^+)e^{\sigma},$$

then $z((N_1+k)\tau) \geq (1-\mu_1)^k z(N_1\tau^+)e^{k\sigma} \to \infty$, as $k \to \infty$, which is a contradiction to the boundedness of z. Hence there exists a $t_1 > 0$ such that $z(t) \geq m_1$. The proof is complete.

5 Discussion

In this work, we consider a predator and prey-hibernation model with genic mutation and impulsive effects. If it is supposed that the variables are shown in the table below:

$x(0)$	$y(0)$	$z(0)$	a	c	d_1	d_2	d_3	d_4	d_5	β_1	β_2	β_3	β_4	k_1	k_2	k_3	k_4	θ_1	b_1	μ_1	μ_2	l	τ
1	1	1	0.3	0.1	0.3	0.3	0.3	0.1	0.1	0.4	0.6	0.3	0.4	0.1	0.1	0.1	0.1	0.6	1.5	0.4	0.1	0.5	1

system (1) is permanent (one can see Fig. 1). If it is supposed that another variables are shown in the table below:

$x(0)$	$y(0)$	$z(0)$	a	c	d_1	d_2	d_3	d_4	d_5	β_1	β_2	β_3	β_4	k_1	k_2	k_3	k_4	θ_1	b_1	μ_1	μ_2	l	τ
1	1	1	0.3	0.1	0.3	0.3	0.3	0.1	0.1	0.4	0.6	0.3	0.4	0.1	0.1	0.1	0.1	0.6	1.5	0.4	0.1	0.5	1

there exists a globally asymptotically stable solution $(\widetilde{x}, \widetilde{y}, 0)$ of system (1) (one can see Fig. 2). Our results show that the environmental pollution will reduce biological diversity of the nature world. So we must be in harmony with the environment.

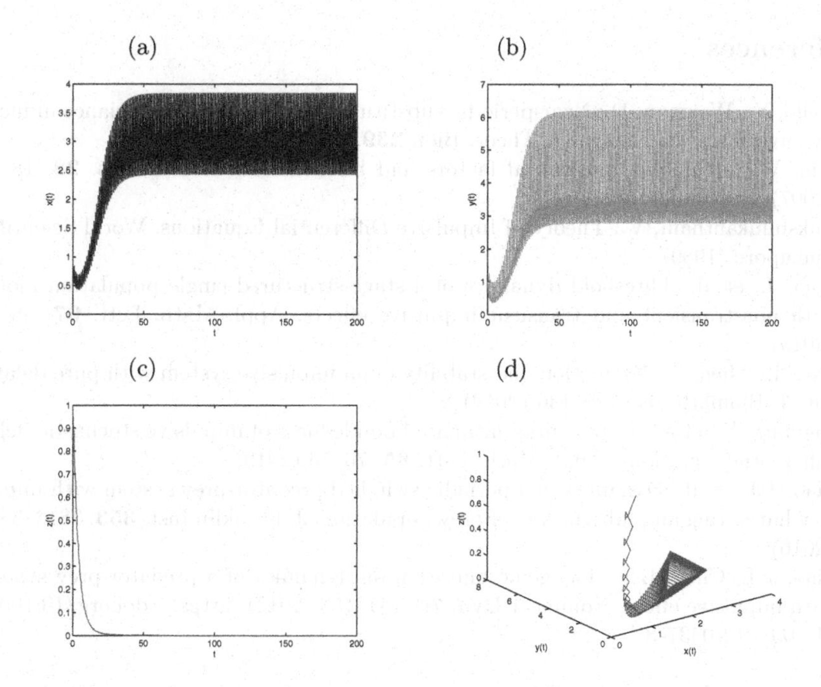

Fig. 1. The permanence of system (2.1) with parameters in the first table.

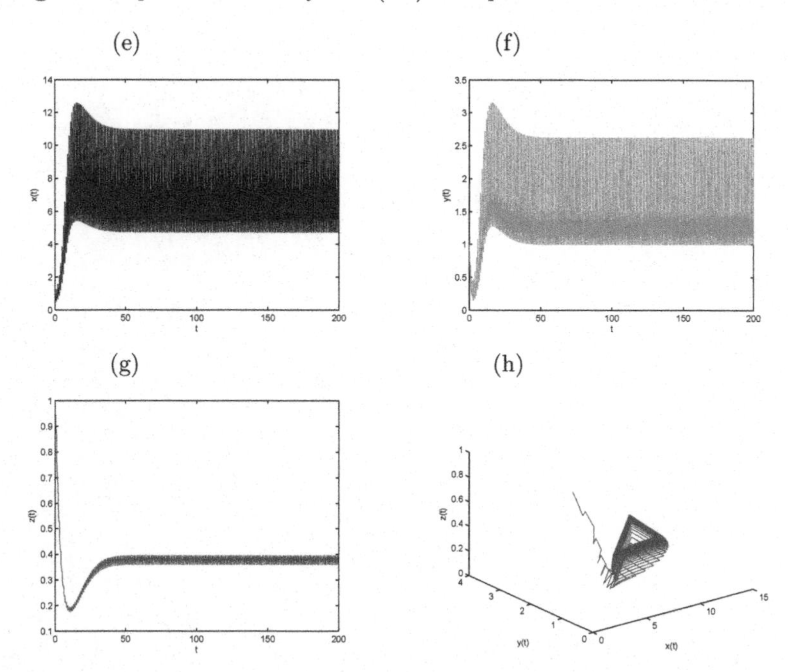

Fig. 2. The dynamics of the globally asymptotically stable microorganism-extinction with parameters in the second table.

References

1. Bello, Y., Waxman, D.: Near-periodic substitution and the genetic variance induced by environmental change. J. Theor. Biol. **239**, 152–160 (2006)
2. Xin, W., et al.: Environmental factors and birth defect. Med. Philos. **28**, 18–21 (2007). (in Chinese)
3. Lakshmikantham, V.: Theory of Impulsive Differential Equations. World Scientific, Singapore (1989)
4. Jiao, J., et al.: Threshold dynamics of a stage-structured single population model with non-transient and transient impulsive effects. Appl. Math. Lett. **97**, 88–92 (2019)
5. He, M., Chen, F.: Extinction and stability of an impulsive system with pure delays. Int. J. Biomath. **91**, 128–136 (2019)
6. Liguang, X., et al.: Exponential ultimate boundedness of impulsive stochastic delay differential equations. Appl. Math. Lett. **85**, 70–76 (2019)
7. Jiao, J.J., et al.: Dynamics of a periodic switched predator-prey system with impulsive harvesting and hibernation of prey population. J. Franklin Inst. **353**, 3818–3834 (2016)
8. Jiao, J.J., Chen, L.S.: The genic mutation on dynamics of a predator-prey system with impulsive effect. Nonlinear Dyn. **70**, 141–153 (2012). https://doi.org/10.1007/s11071-012-0437-8

Dynamics of a Population-Hibernation Model with Genic Mutation and Impulsive Effects in a Polluted Environment

Jianjun Jiao(✉)

School of Mathematics and Statistics,
Guizhou University of Finance and Economics,
Guiyang 550025, People's Republic of China
jiaojianjun05@126.com

Abstract. In this work, we present a population-hibernation system with genic mutation and impulsive effects in a polluted environment. The $(0, 0, \widetilde{c_o}, \widetilde{c_e})$ of (1) is globally asymptotically stable. (1) is also proved to be permanent.

Keywords: Population-hibernation model · Mutation · Impulses

1 Introduction

Many authors [1,2] considered that the environmental pollutants cause many diseases. Hibernation constitutes an effective strategy of animals [3,4]. Population dynamics were investigated by theories of impulsive differential equations lately [5,6]. Jiao et al. [7] investigated an impulsive predator-prey system with periodic switches and hibernation. Jiao and Chen [8] considered a predator-prey system with mutation and impulses. However, theories of impulsive differential equations introducing into system with genic mutation and hibernation have been considered very little. For convenience, we make notations as $N = n\tau$ and $N + l = (n + l)\tau$.

Supported by National Natural Science Foundation of China (11761019,11361014), the Science Technology Foundation of Guizhou Education Department (20175736-001,2008038), the Project of High Level Creative Talents in Guizhou Province (No. 20164035), Major Research Projects on Innovative Groups in Guizhou Provincial Education Department (No.[2018]019).

H. Song and D. Jiang (Eds.): SIMUtools 2020, LNICST 370, pp. 53–62, 2021.
https://doi.org/10.1007/978-3-030-72795-6_5

2 The Model

In this work, we present a single population-hibernation model with mutation and impulsive effects in a polluted environment

$$\begin{cases}
\left.\begin{aligned}
\frac{dx}{dt} &= -ax - cx^2 - \beta_1 x c_o,\\
\frac{dy}{dt} &= -d_1 y,\\
\frac{dc_o}{dt} &= fc_e - (g+m)c_o,\\
\frac{dc_e}{dt} &= -hc_e,
\end{aligned}\right\} t \in (N, (N+l)],\\
\left.\begin{aligned}
\Delta x &= (1-\theta_1)b_1 x + \theta_2 b_2 y,\\
\Delta y &= \theta_1 b_1 x + (1-\theta_2) b_2 y,\\
\Delta c_o &= 0,\\
\Delta c_e &= 0,
\end{aligned}\right\} t = (N+l),\\
\left.\begin{aligned}
\frac{dx}{dt} &= -d_2 x - \beta_2 x c_o,\\
\frac{dy}{dt} &= -d_3 y,\\
\frac{dc_o}{dt} &= fc_e - (g+m)c_o,\\
\frac{dc_e}{dt} &= -hc_e,
\end{aligned}\right\} t \in ((N+l), (N+1)],\\
\left.\begin{aligned}
\Delta x &= 0,\\
\Delta y &= -\mu_1 y,\\
\Delta c_o &= 0,\\
\Delta c_e &= \mu,
\end{aligned}\right\} t = (N+1),
\end{cases} \tag{1}$$

the biological meanings of the varies and the parameters of (1) can reference [6–8].

3 The Dynamics

For (1), the subsystem of (1) is as

$$\begin{cases}
\left.\begin{aligned}
\frac{dc_o}{dt} &= fc_e - (g+m)c_o,\\
\frac{dc_e}{dt} &= -hc_e,
\end{aligned}\right\} t \neq N,\\
\left.\begin{aligned}
\Delta c_o &= 0,\\
\Delta c_e &= \mu,
\end{aligned}\right\} t = N.
\end{cases} \tag{2}$$

Then, we give an important property of system (2) as following.

Lemma 1. [6] (2) has a globally asymptotically stable solution $(\widetilde{c_o}, \widetilde{c_e})$ as

$$\begin{cases} \widetilde{c_o} = \widetilde{c_o(0)}e^{-(g+m)(t-N)} + \dfrac{\mu f(e^{-(g+m)(t-N)} - e^{-h(t-N)})}{(h-g-m)(1-e^{-h\tau})}, \\ \widetilde{c_e} = \dfrac{\mu e^{-h(t-N)}}{1-e^{-h\tau}}, \\ \widetilde{c_o(0)} = \dfrac{\mu f(e^{-(g+m)\tau} - e^{-h\tau})}{(h-g-m)(1-e^{-(g+m)\tau})(1-e^{-h\tau})}, \\ \widetilde{c_e(0)} = \dfrac{\mu}{1-e^{-h\tau}}, \end{cases} \tag{3}$$

Remark 2. $m_o - \varepsilon \le c_o(t) \le M_o + \varepsilon$ and $m_e - \varepsilon \le c_e(t) \le M_e + \varepsilon$ hold for t large enough, where $m_o = \frac{\mu f(e^{-(g+m)\tau} - e^{-h\tau})}{(h-g-m)(e^{(g+m)\tau}-1)(1-e^{-h\tau})} > 0$, $M_o = \frac{\mu f(e^{-(g+m)\tau} - e^{-h\tau})}{(h-g-m)(1-e^{-(g+m)\tau})(1-e^{-h\tau})} + \frac{\mu f}{|h-g-m|(1-e^{-h\tau})} > 0$, $m_e = \frac{\mu e^{-h\tau}}{1-e^{-h\tau}} > 0$ and $M_e = \frac{\mu}{1-e^{-h\tau}} > 0$ for $\varepsilon > 0$.

From (1) another subsystem of (1) is also as

$$\begin{cases} \left.\begin{aligned} \frac{dx}{dt} &= -ax - cx^2, \\ \frac{dy}{dt} &= -d_1 y, \end{aligned}\right\} t \in (N, (N+l)], \\ \left.\begin{aligned} \Delta x &= (1-\theta_1)b_1 x + \theta_2 b_2 y, \\ \Delta y &= \theta_1 b_1 x + (1-\theta_2) b_2 y, \end{aligned}\right\} t = (N+l), \\ \left.\begin{aligned} \frac{dx}{dt} &= -d_2 x, \\ \frac{dy}{dt} &= -d_3 y, \end{aligned}\right\} t \in ((N+l), (N+1)], \\ \left.\begin{aligned} \Delta x &= 0, \\ \Delta y &= -\mu_1 y, \end{aligned}\right\} t = (N+1). \end{cases} \tag{4}$$

Integrating (4), we get

$$\begin{cases} x = \begin{cases} \dfrac{ax(N^+)e^{-a(t-N)}}{a + cx(N^+)(1-e^{-a(t-N)})}, t \in (N, (N+1)N], \\ x((N+l)^+)e^{-d_2(t-N+l))}, t \in ((N+l), (N+1)] \end{cases} \\ y = \begin{cases} y(N^+)e^{-d_1(t-N)}, t \in (N, (N+l)N], \\ y((N+l)^+)e^{-d_3(t-N)}, t \in ((N+l)N, (N+1)N]. \end{cases} \end{cases} \tag{5}$$

The stroboscopic map of (4) is

$$\begin{cases} x((N+1)^+) = [1+(1-\theta_1)b_1] \times \dfrac{ax(N^+)e^{-[al+d_2(1-l)]\tau}}{a+cx(N^+)(1-e^{-al\tau})} \\ \qquad\qquad\qquad +\theta_2 b_2 e^{-[d_1 l+d_2(1-l)]\tau} y(N^+), \\ y((N+1)^+) = \theta_1 b_1 (1-\mu_1) \times \dfrac{ax(N^+)e^{-[al+d_3(1-l)]\tau}}{a+cx(N^+)(1-e^{-al\tau})} \\ \qquad\qquad\qquad +[1+(1-\theta_2)b_2](1-\mu_1)e^{-[d_1 l+d_3(1-l)]\tau} y(N^+). \end{cases} \tag{6}$$

$G_1(0,0)$ and $G_2(x^*, y^*)$ are obtained and

$$\begin{cases} x^* = By^*, \ \ Be^{-al\tau} > A, \\ y^* = \dfrac{a(Be^{-al\tau} - A)}{cAB(1-e^{-al\tau})}, Be^{-al\tau} > A, \end{cases} \tag{7}$$

here $A = \frac{1-[1+(1-\theta_2)b_2](1-\mu_1)e^{-[d_1 l+d_3(1-l)]\tau}}{\theta_1 b_1 (1-\mu_1)e^{-d_3(1-l)\tau}}$, and $B = [1+(1-\theta_1)b_1]e^{-d_2(1-l)\tau}A + \theta_2 b_2 e^{-[d_1 l+d_3(1-l)]\tau}$.

Similarly with reference [6], we have the following two theorems.

Theorem 3. *i)* Suppose $Be^{-al\tau} < A$ holds, $G_1(0,0)$ of (6) is globally asymptotically stable;

ii) Suppose $Be^{-al\tau} > A$ holds, $G_2(x^*, y^*)$ of (6) is globally asymptotically stable.

Theorem 4. *i)* Suppose $Be^{-al\tau} < A$ holds, periodic solution $(0,0)$ of (2) is globally asymptotically stable;

ii) Suppose $Be^{-al\tau} > A$ holds, periodic solution $(\widetilde{x}, \widetilde{y})$ of (2) is globally asymptotically stable, and

$$\begin{cases} \widetilde{x} = \begin{cases} \dfrac{ax^* e^{-a(t-N)}}{a+cx^*(1-e^{-a(t-N)})}, t \in (N, (N+1)N], \\ x^{**} e^{-d_2(t-(N+l))}, t \in ((N+l), (N+1)], \end{cases} \\ \widetilde{y} = \begin{cases} y^* e^{-d_1(t-N)}, t \in (N, (N+l)], \\ y^{**} e^{-d_3(t-N)}, t \in ((N+l), (N+1)], \end{cases} \end{cases} \tag{8}$$

here x^*, y^* are by (7), and x^{**}, y^{**} are by

$$\begin{cases} x^{**} = [1+(1-\theta_1)b_1] \times \dfrac{ax^* e^{-al\tau}}{a+cx^*(1-e^{-al\tau})} + \theta_1 b_1 y^* e^{-d_1 l\tau}, \\ y^{**} = \theta_1 b_1 \times \dfrac{ax^* e^{-al\tau}}{a+cx^*(1-e^{-al\tau})} + [1+(1-\theta_1)b_1] y^* e^{-d_1 l\tau}. \end{cases} \tag{9}$$

From (1) and Remark 3., we get

$$-(a+\beta M_o)x - cx^2 \le \frac{dx}{dt} \le -ax - cx^2, \tag{10}$$

and

$$-(d_2+\beta M_o)x \le \frac{dx}{dt} \le -d_2 x. \tag{11}$$

Then, $(x_1(t), y_1(t))$ and $(x_2(t), y_2(t))$ are respectively the solutions of

$$\begin{cases} \left.\begin{aligned} \frac{dx_1}{dt} &= -(a+\beta_1 M_o)x_1 - cx_1^2, \\ \frac{dy_1}{dt} &= -d_1 y_1, \end{aligned}\right\} t \in (N, (N+l)], \\ \left.\begin{aligned} \triangle x_1 &= (1-\theta_1)b_1 x_1 + \theta_2 b_2 y_1, \\ \triangle y_1 &= \theta_1 b_1 x_1 + (1-\theta_2) b_2 y_1, \end{aligned}\right\} t = (N+l), \\ \left.\begin{aligned} \frac{dx_1}{dt} &= -(d_2+\beta_2 M_o)x_1, \\ \frac{dy_1}{dt} &= -d_3 y_1, \end{aligned}\right\} t \in ((N+l), (N+1)], \\ \left.\begin{aligned} \triangle x_1 &= 0, \\ \triangle y_1 &= -\mu_1 y_1, \end{aligned}\right\} t = (N+1), \end{cases} \tag{12}$$

and

$$\begin{cases} \left.\begin{aligned} \frac{dx_2}{dt} &= -ax_2 - cx_2^2, \\ \frac{dy_2}{dt} &= -d_1 y_2, \end{aligned}\right\} t \in (N, (N+l)], \\ \left.\begin{aligned} \triangle x_2 &= (1-\theta_1)b_1 x_2 + \theta_2 b_2 y_2, \\ \triangle y_2 &= \theta_1 b_1 x_2 + (1-\theta_2) b_2 y_2, \end{aligned}\right\} t = (N+l), \\ \left.\begin{aligned} \frac{dx_2}{dt} &= -d_2 x_2, \\ \frac{dy_2}{dt} &= -d_3 y_2, \end{aligned}\right\} t \in ((N+l), (N+1)], \\ \left.\begin{aligned} \triangle x_2 &= 0, \\ \triangle y_2 &= -\mu_1 y_2, \end{aligned}\right\} t = (N+1), \end{cases} \tag{13}$$

Obviously,

$$\begin{cases} x_1 \le x \le x_2, \\ y_1 \le y \le y_2. \end{cases} \tag{14}$$

Similar to theorem (4), we have

Theorem 5. *i*) Suppose $B_1e^{-(a+\beta_1 M_o)l\tau} < A_1$ holds, triviality solution $(0,0)$ of (13) is globally asymptotically stable;

ii) Suppose $B_1e^{-(a+\beta_1 M_o)l\tau} > A_1$ holds, the periodic solution $(\widetilde{x_1}, \widetilde{y_1})$ of (13) is globally asymptotically stable, and

$$\begin{cases} \widetilde{x_1} = \begin{cases} \dfrac{(a+\beta_1 M_o)x_1^* e^{-(a+\beta_1 M_o)(t-N)}}{(a+\beta_1 M_o)+cx_1^*(1-e^{-(a+\beta_1 M_o)(t-N)})}, t \in (N,(N+1)], \\ x^{**}e^{-d_2(t-(N+l))}, t \in ((N+l),(N+1)], \end{cases} \\ \widetilde{y_1} = \begin{cases} y^* e^{-d_1(t-N)}, t \in (N,(N+l)], \\ y^{**}e^{-d_3(t-N)}, t \in ((N+l),(N+1)], \end{cases} \end{cases} \tag{15}$$

here x_1^*, y_1^* are by

$$\begin{cases} x_1^* = By_1^*, \quad B_1e^{-(a+\beta_1 M_o)l\tau} > A_1, \\ y_1^* = \dfrac{(a+\beta_1 M_o)(B_1e^{-(a+\beta_1 M_o)l\tau} - A_1)}{cA_1B_1(1-e^{-(a+\beta_1 M_o)l\tau})}, B_1e^{-(a+\beta_1 M_o)l\tau} > A_1, \end{cases} \tag{16}$$

and x_1^{**}, y_1^{**} are by

$$\begin{cases} x_1^{**} = [1+(1-\theta_1)b_1] \\ \quad \times \dfrac{(a+\beta_1 M_o)x_1^* e^{-(a+\beta_1 M_o)l\tau}}{(a+\beta_1 M_o)+cx_1^*(1-e^{-(a+\beta_1 M_o)l\tau})} + \theta_1 b_1 y_1^* e^{-d_1 l\tau}, \\ y_1^{**} = \theta_1 b_1 \\ \quad \times \dfrac{(a+\beta_1 M_o)x_1^* e^{-(a+\beta_1 M_o)l\tau}}{(a+\beta_1 M_o)+cx_1^*(1-e^{-(a+\beta_1 M_o)l\tau})} + [1+(1-\theta_1)b_1]y_1^* e^{-d_1 l\tau}. \end{cases} \tag{17}$$

here $A_1 = \frac{1-[1+(1-\theta_2)b_2](1-\mu_1)e^{-[d_1 l+d_3(1-l)]\tau}}{\theta_1 b_1(1-\mu_1)e^{-d_3(1-l)\tau}}$, and $B_1 = [1+(1-\theta_1)b_1]e^{-d_2(1-l)\tau}A_1 + \theta_2 b_2 e^{-[d_1 l+d_3(1-l)]\tau}$.

Remark 6. Suppose $B_1e^{-(a+\beta_1 M_o)l\tau} > A_1$ holds, for $\varepsilon_2 > 0$, there exists a T_2 such that $x_1 \geq \widetilde{x_1} - \varepsilon_2$ and $y_1 \geq \widetilde{y_1} - \varepsilon_2$ for $t > T_2$.

From theorem (4),(4) and remark (6), we gets

Theorem 7. *i*) Suppose $Be^{-al\tau} < A$ holds, the population of (1) will go extinct.

ii) Suppose $B_1e^{-(a+\beta_1 M_o)l\tau} > A_1$ and $Be^{-al\tau} > A$ hold, (1) is permanent.

Proof. *i*) According to the condition $Be^{-al\tau} < A$, theorem (5), and (14) for any sufficient small $\varepsilon_1 > 0$, there exists a $T_1 > 0$ such that $x \leq x_2 \leq \varepsilon_1$ and $y \leq y_2 \leq \varepsilon_1$ for $t > T_1$. That is to say, for any sufficient small $\varepsilon_1 > 0$, there exists a $T_1 > 0$ such that $x \leq \varepsilon_1$ and $y \leq \varepsilon_1$ for $t > T_1$. These show that population of (1) will go extinct.

ii) According to the condition $B_1 e^{-(a+\beta_1 M_o)l\tau} > A_1$, Remark (7) and (14), for any sufficient small $\varepsilon_2 > 0$, there exists a $T_2 > 0$ such that $x \geq x_1 \geq \widetilde{x_1} - \varepsilon_2 \geq \frac{(a+\beta_1 M_o)x_1^* e^{-(a+\beta_1 M_o)l\tau}}{(a+\beta_1 M_o)+cx_1^*(1-e^{-(a+\beta_1 M_o)l\tau})} + x^{**}e^{-d_2(1-l)\tau} - \varepsilon_1 \triangleq m_1$ and $y \geq y_1 \geq \widetilde{y_1} - \varepsilon_1 \geq y_1^* e^{-d_1 l\tau} + y_1^{**}e^{-d_3(1-l)\tau} - \varepsilon_1 \triangleq m_2$ for $t > T_2$, where x_1^*, y_1^* and x_1^{**}, y_1^{**} are determined as (16) and (17) respectively. From $Be^{-al\tau} > A$, and (14), for any sufficient small $\varepsilon_2 > 0$, there exists a $T_3 > 0$ such that $x \leq \widetilde{x_2} - \varepsilon_2 \leq x_2^* + x_2^{**} + \varepsilon_2 \triangleq m_3$ and $y \leq y_2 \leq y_2^* + y_2^{**} + \varepsilon_2 \triangleq m_4$. From the above discussion, we know $m_1 < x < m_3$ and $m_2 < y(t) < m_4$ for $t > \max\{T_2, T_3\}$. That is to say, the normal population $x(t)$ and genic mutation population $y(t)$ of system (1) will be permanent.

Therefore,

Remark 8. *i*) Suppose $B_1 e^{-(a+\beta_1 M_o)l\tau} < A_1$, and $Be^{-al\tau} < A$ holds, $(0, 0, \widetilde{c_o}, \widetilde{c_e},)$ of (1) is globally asymptotically stable.

ii) Suppose $B_1 e^{-(a+\beta_1 M_o)l\tau} > A_1$, and $Be^{-al\tau} > A$ hold, (1) are permanent.

4 Discussion

In this work, we consider a single population-hibernation model with genic mutation, birth pulse and impulsive input toxin in a polluted environment. If it is supposed that the variables are shown in the table below:

$x(0)$	$y(0)$	$c_o(0)$	$c_e(0)$	a	c	d_1	d_2	d_3	β_1	β_2	f	g	m	h	θ_1	θ_2	b_1	b_2	μ_1	μ	l	τ
0.3	0.1	0.3	0.3	0.2	0.1	0.1	0.2	0.1	0.1	0.1	0.3	0.4	0.1	0.4	0.1	0.5	0.3	0.3	0.6	0.2	0.5	1

there exists a globally asymptotically stable solution $(0, 0, \widetilde{c_o}, \widetilde{c_e})$ of (1) (one can see Fig. 1). If it is supposed that another variables are shown in the table below:

$x(0)$	$y(0)$	$c_o(0)$	$c_e(0)$	a	c	d_1	d_2	d_3	β_1	β_2	f	g	m	h	θ_1	θ_2	b_1	b_2	μ_1	μ	l	τ
0.3	0.1	0.3	0.3	0.2	0.1	0.1	0.2	0.1	0.1	0.1	0.3	0.4	0.1	0.4	0.1	0.5	0.3	0.3	0.2	0.2	0.5	1

then, (1) is permanent (one can see Fig. 2). If it is supposed that variables are shown in the table below:

$x(0)$	$y(0)$	$c_o(0)$	$c_e(0)$	a	c	d_1	d_2	d_3	β_1	β_2	f	g	m	h	θ_1	θ_2	b_1	b_2	μ_1	μ	l	τ
0.3	0.1	0.3	0.3	0.2	0.1	0.1	0.2	0.1	0.1	0.1	0.3	0.4	0.1	0.4	0.1	0.5	0.3	0.3	0.9	0.4	0.5	1

there exists a globally asymptotically stable solution $(0, 0, \widetilde{c_o}, \widetilde{c_e})$ of system (1) (one can see Fig. 3).

If it is supposed that variables are shown in the table below:

$x(0)$	$y(0)$	$c_o(0)$	$c_e(0)$	a	c	d_1	d_2	d_3	β_1	β_2	f	g	m	h	θ_1	θ_2	b_1	b_2	μ_1	μ	l	τ
0.3	0.1	0.3	0.3	0.2	0.1	0.1	0.2	0.1	0.1	0.1	0.3	0.4	0.1	0.4	0.1	0.5	0.3	0.3	0.9	0.2	0.5	1

then, (1) is permanent (see Fig. 4). The results show that the amount of genic mutation rate of the normal population also plays important role for the permanence of system (1). Our results show that the impulsive environmental toxin will reduce biological diversity of the nature world.

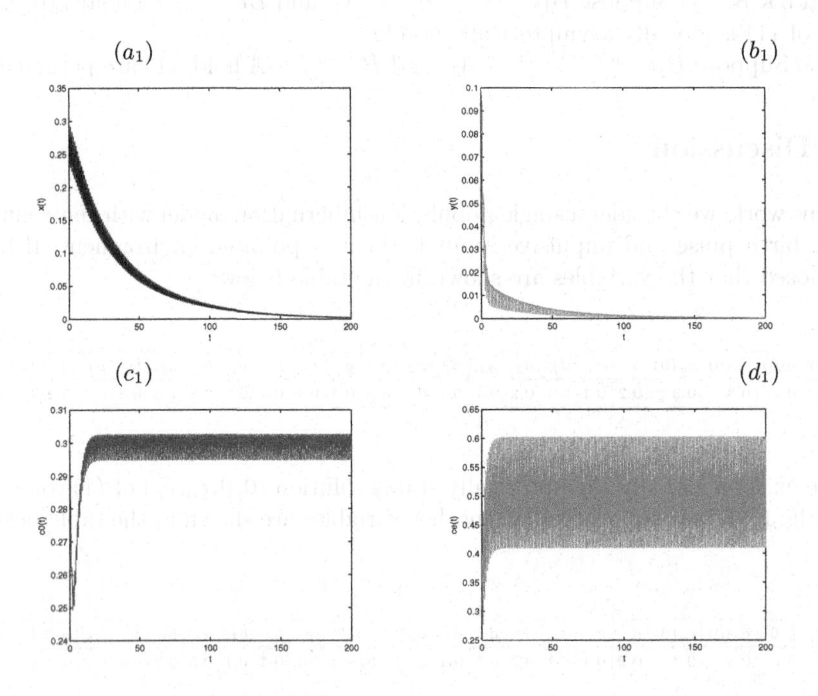

Fig. 1. The dynamics of the globally asymptotically stable microorganism-extinction with parameters in the second table.

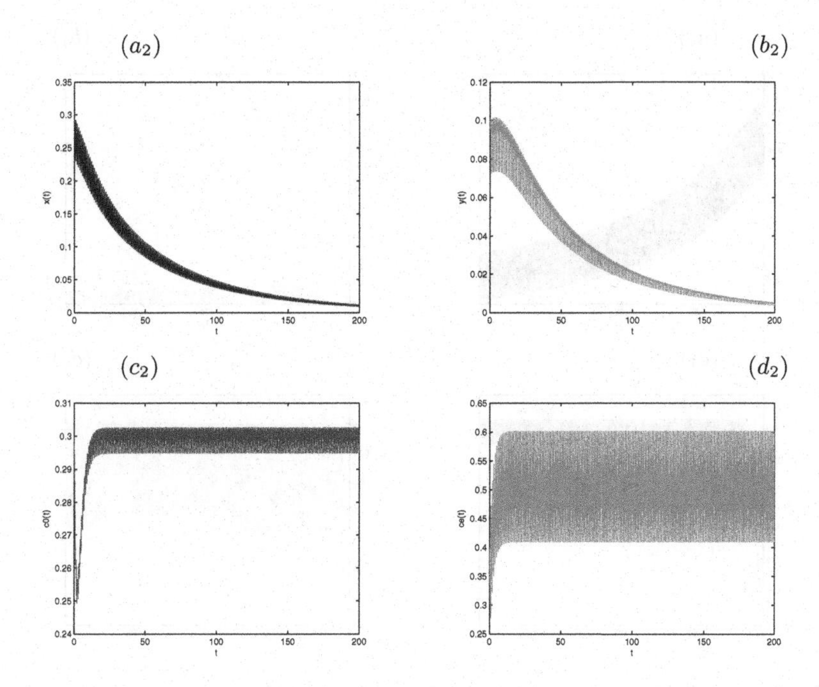

Fig. 2. The permanence of system (1) with parameters in the second table.

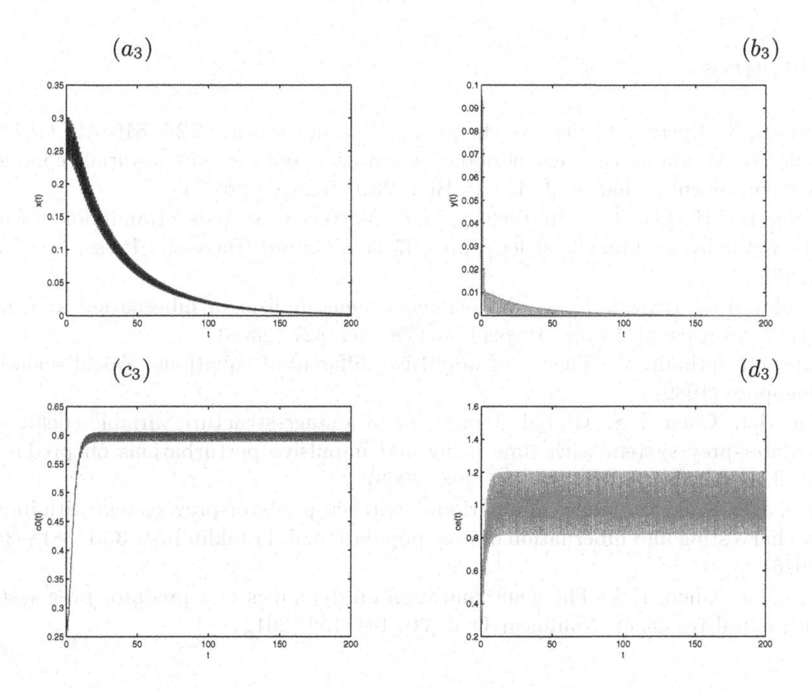

Fig. 3. The dynamics of the globally asymptotically stable microorganism-extinction with parameters in the third table.

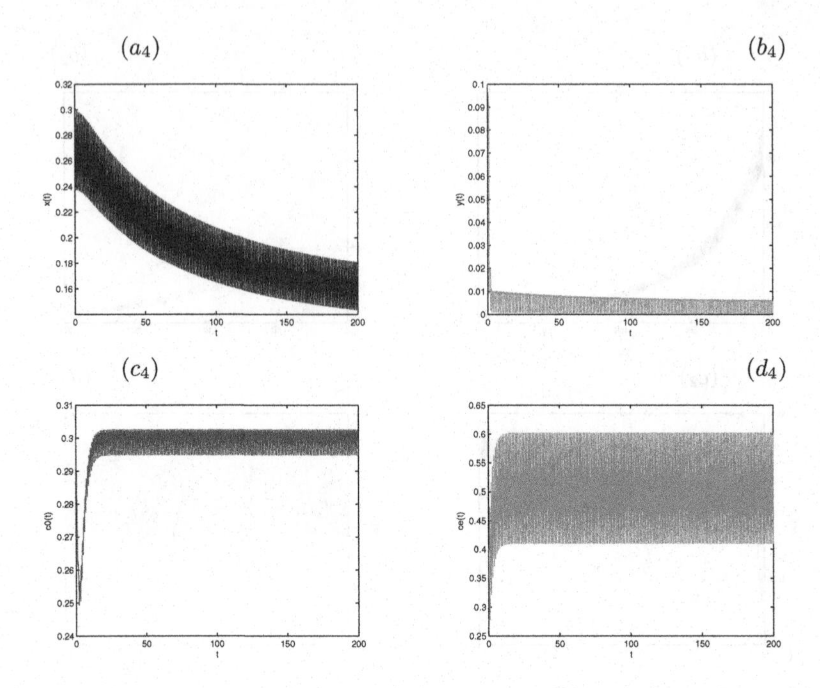

Fig. 4. The permanence of system (1) with parameters in the fourth table.

References

1. Samuel, S.: Epstein, Control of chemical pollutants. Nature **228**, 816–819 (1970)
2. Bello, Y., Waxman, D.: Near-periodic substitution and the genetic variance induced by environmental change. J. Theor. Biol. **239**, 152–160 (2006)
3. Wang, L.C.H., Lee, T.F.: In: Fregley, M.J., Blatteis, C.M. (eds.) Handbook of Physiology: Environmental Physiology, pp. 507–532. Oxford University Press, New York (1996)
4. Staples, J.F., Brown, J.C.L.: Mitochondrial metabolism in hibernation and daily torpor: a review. J. Comp. Physiol. B **178**, 811–827 (2008)
5. Lakshmikantham, V.: Theory of impulsive differential equations. World scientific, Singapore (1989)
6. Jiao, J.J., Chen, L.S.: Global attractivity of a stage-structure variable coefficients predator-prey system with time delay and impulsive perturbations on predators. Int. J. Biomathematics **1**(2), 197–208 (2008)
7. Jiao, J.J., et al.: Dynamics of a periodic switched predator-prey system with impulsive harvesting and hibernation of prey population. J. Franklin Inst. **353**, 3818–3834 (2016)
8. Jiao, J.J., Chen, L.S.: The genic mutation on dynamics of a predator-prey system with impulsive effect. Nonlinear Dyn. **70**, 141–153 (2012)

Dynamics of a Chemostat-Type Model with Impulsive Effects in a Polluted Karst Environment

Jianjun Jiao(✉)

School of Mathematics and Statistics, Guizhou University of Finance and Economics, Guiyang 550025, People's Republic of China

Abstract. In this paper, we present a chemostat-type model with impulsive effects in a polluted karst environment. The globally asymptotically stable sufficient condition are gained for a microorganism-extinction periodic solution. System permanent condition are also presented. The results are illustrated by simulations.

Keywords: Chemostat-type model · Impulsive diffusing · Pulse inputting · Polluted karst environments · Extinction

1 Introduction

There are wide underground rivers in the underground rock gaps of karst areas. It easy for pollutants and nutrient of soils to come into aquifers for the thin soil layers of karst areas [1]. Many authors [2–4] employed SDE,PDE,ODE and IDE to study chemostat systems. References [5,6] devoted themselves to the effects of toxicants on population. But there are few authors devoted themselves to karst environmental chemostat system with impulsive diffusing and pulse inputting. We mark a notation as $N = nT, N + L = (n + l)T$.

2 The Model

In this paper, we present a chemostat-type model with impulsive effects in a polluted karst environment

Supported by NNSFC (No.11791019,11361014), the Science Technology Foundation of Guizhou Education Department (No.QJK[KY]2018019).

H. Song and D. Jiang (Eds.): SIMUtools 2020, LNICST 370, pp. 63–78, 2021.
https://doi.org/10.1007/978-3-030-72795-6_6

$$
\begin{cases}
\left.\begin{aligned}
\frac{dx_1}{dt} &= D_1(x_1^0 - x_1),\\
\frac{dx_2}{dt} &= D_2(x_2^0 - x_2)\\
&\quad -\frac{1}{\delta}\times\frac{\eta x_2 Y}{\alpha + x_2},\\
\frac{dY}{dt} &= -D_2\\
&\quad +\frac{\eta x_2 Y}{\alpha + x_2} - \beta c_1 Y,\\
\frac{dc_1}{dt} &= f_1 c_2 - (g_1 + m_1)c_1,\\
\frac{dc_2}{dt} &= -h_1 c_2,
\end{aligned}\right\} t \neq (N+L), t \neq (N+1),\\
\left.\begin{aligned}
x_1^+ &= x_1,\\
x_2^+ &= x_2,\\
Y^+ &= Y,\\
c_1^+ &= c_1,\\
c_2^+ &= (1-\theta_1)c_2,
\end{aligned}\right\} t = (N+L),\\
\left.\begin{aligned}
x_1^+ &= (1-d)x_1 + dx_2 + \theta_2,\\
x_2^+ &= dx_1(1-d)x_2,\\
Y^+ &= Y,\\
c_1^+ &= (1-h_3)c_1,\\
c_2^+ &= c_2 + \theta,
\end{aligned}\right\} t = (N+1),
\end{cases} \tag{1}
$$

where system (1) is constructed by two patches, patch (1) is a non-polluted environment and patch (2) is a polluted environment. The meanings of the variables and parameters can be consulted from reference [3–5].

3 The Foundation

If $Y = 0$, there are two subsystems of system (1) as

$$
\begin{cases}
\left.\begin{aligned}
\frac{dx_1}{dt} &= D_1(x_1^0 - x_1),\\
\frac{dx_2}{dt} &= D_2(x_2^0 - x_2),
\end{aligned}\right\} t \neq N,\\
\left.\begin{aligned}
x_1^+ &= (1+d)x_1 + dx_2 + \theta_1,\\
x_2^+ &= dx_1 + (1-d)x_2,
\end{aligned}\right\} t = N,
\end{cases} \tag{2}
$$

and

$$\begin{cases} \left.\begin{aligned} \frac{dc_1}{dt} &= f_1 c_2 - (g_1 + m_1)c_1, \\ \frac{dc_2}{dt} &= -h_1 c_2, \end{aligned}\right\} t \neq N, t \neq (N+1), \\ \left.\begin{aligned} \triangle c_1 &= 0, \\ \triangle c_2 &= -\theta_1 c_2, \end{aligned}\right\} t = (N+L), \\ \left.\begin{aligned} \triangle c_1 &= -\theta_3 c_1, \\ \triangle c_2 &= \theta, \end{aligned}\right\} t = (N+1). \end{cases} \tag{3}$$

$i)$ Integrating on system (2)

$$x_i(t) = x_i^0 - [x_i^0 - x_i(nT^+)]e^{-D_i(t-nT)}. \tag{4}$$

Then,

$$\begin{cases} x_1((n+1)T^+) = (1-d)e^{-D_1T}x_1(nT^+) + de^{-D_2T}x_2(nT^+) \\ \qquad +(1-d)(1-e^{-D_1T})x_1^0 + d(1-e^{-D_2T})x_2^0 + \theta_1, \\ x_2((n+1)T^+) = de^{-D_1T}x_1(nT^+) + (1-d)e^{-D_2T}x_2(nT^+) \\ \qquad +d(1-e^{-D_1T})x_1^0 + (1-d)(1-e^{-D_2T})x_2^0. \end{cases} \tag{5}$$

From system (5), we gain

$$\begin{cases} x_1^* = \dfrac{K_1 + K_1'}{K_2 + K_2'}, \\ x_2^* = \dfrac{K_3 + K_3'}{K_4 + K_4'}, \end{cases} \tag{6}$$

where $K_1 = [(1-d)(1-e^{-D_1T}) - (1-2d)(1-e^{-D_2T})e^{-D_1T}]x_1^0 - d(1-e^{-D_2T})x_2^0 + (1-e^{-D_2T}, K_1' = de^{-D_2T})\theta_1$, $K_2 = 1-(1-d)e^{-D_1T} - (1-d)e^{-D_2T}$ $K_2' = -(1-2d)e^{-(D_1+D_2)T}$, $K_3 = -d(1-e^{-D_1T})x_1^0 + [(1-d)(1-e^{-D_2T}) - (1-2d)(1-e^{-D_1T})e^{-D_2T}]x_2^0$, $K_3' = de^{-D_1T}\theta_1$, and $K_4 = 1-(1-d)e^{-D_2T} - (1-d)e^{-D_1T}$, $K_4' = -(1-2d)e^{-(D_2+D_1)T}$.

The map $G: R_+^2 \to R_+^2$ coming from system (5) is presented as

$$\begin{cases} G_1(x) = G_{11}x_1 + G_{12}x_2 + G_{13}, \\ G_2(x) = G_{21}x_1 + G_{22}x_2, \end{cases} \tag{7}$$

where $x = (x_1, x_2) \in R_+^2$, and $G_{11} = (1-d)e^{-D_1T}, G_{12} = de^{-D_2T}, G_{13} = (1-d)(1-e^{-D_1T})x_1^0 + d(1-e^{-D_2T})x_2^0 + \theta_1, G_{21} = de^{-D_1T}, G_{22} = (1-d)e^{-D_2T}, G_{23} = +d(1-e^{-D_1T})x_1^0 + (1-d)(1-e^{-D_2T})x_2^0$.

Being similar to reference [7], two lemmas are gotten.

Lemma 1. $G^n(x) \to (x_1^*, x_2^*)$ (as $n \to \infty$) holds for (x_1, x_2) of system (7).

Lemma 2. The solution $(\widetilde{x_1}, \widetilde{x_2})$, which is defined as

$$\begin{cases} \widetilde{x_1} = x_1^0 - (x_1^0 - x_1^*)e^{-D_1(t-nT)}, N < t \leq (N+1), \\ \widetilde{x_2} = x_2^0 - (x_2^0 - x_2^*)e^{-D_2(t-nT)}, N < t \leq (N+1), \end{cases} \tag{8}$$

is globally stable.

ii) Integrating on system (3), we have

$$\begin{cases} c_1((n+1)T^+) = c_1(nT^+)e^{-(g_1+m_1)T} \\ \quad + \dfrac{c_2(nT^+)f_1(e^{-(g_1+m_1)(1-l)T} - e^{-(h_1-g_1-m_1)lT-(g_1+m_1)(1-l)T})}{(h_1-g_1-m_1)} \\ \quad + \dfrac{(1-\theta_1)c_2(nT^+)f_1(e^{-h_1lT} - e^{-(h_1-g_1-m_1)T})}{(h_1-g_1-m_1)}, \\ c_2((n+1)T^+) = (1-\theta_1)e^{-h_1T}c_2(nT^+) + \theta. \end{cases} \tag{9}$$

From (9), we have c_1^* and c_2^* with

$$\begin{cases} c_1^* = \dfrac{1-\theta_3}{1-(1-\theta_3)e^{-(g_1+m_1)T}} \\ \quad \times [\dfrac{\theta f_1(e^{-(g_1+m_1)(1-l)T} - e^{-(h_1-g_1-m_1)lT-(g_1+m_1)(1-l)T})}{(h_1-g_1-m_1)(1-(1-\theta_1)e^{-h_1T})(1-e^{-(g_1+m_1)T})} \\ \quad + \dfrac{(1-\theta_1)\theta f_1(e^{-h_1lT} - e^{-(h_1-g_1-m_1)T})}{(h_1-g_1-m_1)(1-(1-\theta_1)e^{-h_1T})(1-e^{-(g_1+m_1)T})}], \\ c_2^* = \dfrac{\theta}{1-(1-\theta_1)e^{-h_1T}}. \end{cases} \tag{10}$$

Obviously, for (9), (c_1^*, c_2^*) is globally asymptotically stable.

Lemma 3. The globally asymptotically stable $(\widetilde{c_1}, \widetilde{c_2})$ of (3) exists. and $\widetilde{c_1}, \widetilde{c_2})$ are in reference [8], c_1^*, c_2^* are as (11), and $c_1^{**} = c_1^* e^{-(g_1+m_1)lT} + \frac{f_1 c_2^*(1-e^{-(h_1-g_1-m_1)lT})}{(h_1-g_1-m_1)}$, $c_2^{**} = (1-\theta_1)e^{-h_1lT}c_2^*$.

Remark 4. There exist positive constants m_0, M_0, m_e, M_e, it is easy to get $m_0 \leq c_o(t) \leq M_0$ and $m_e \leq c_e(t) \leq M_e$.

Similar to reference [7], we can obtain

Lemma 5. There exists a positive constant $\lambda > 0$, we can easily have $x_1(t) \leq [\delta(D_1x_1^0 + D_2x_2^0) + \frac{\theta \exp(\lambda T)}{\exp(\lambda T)-1}]e^{\lambda T_1}$, $x_2(t) \leq [\delta(D_1x_1^0 + D_2x_2^0) + \frac{\theta \exp(\lambda T)}{\exp(\lambda T)-1}]e^{\lambda T_1}$, $Y(t) \leq \delta(D_1x_1^0 + D_2x_2^0) + \frac{\theta \exp(\lambda T)}{\exp(\lambda T)-1}$, $c_0(t) \leq \delta(D_1x_1^0 + D_2x_2^0) + \frac{\theta \exp(\lambda T)}{\exp(\lambda T)-1}$, and $c_e(t) \leq \delta(D_1x_1^0 + D_2x_2^0) + \frac{\theta \exp(\lambda T)}{\exp(\lambda T)-1}$.

4 Dynamical Analysis

Theorem 1. Suppose

$$d > \frac{1}{2}, \tag{11}$$

and

$$
\begin{aligned}
&(\eta - D_2)T - \frac{\alpha\eta}{D_2(\alpha + x_2^0)} \ln \frac{\alpha e^{-D_2T} + (e^{-D_2T} - 1)x_2^0 + x_2^*}{\alpha + x_2^*} \\
&+ \frac{c_1^*(1 - e^{-(g_1+m_1)lT})}{g_1 + m_1} + \frac{f_1 c_2^* lT}{h_1 - g_1 - m_1} - \frac{f_1 c_2^*(1 - e^{-(h_1-g_1-m_1)lT})}{(h_1 - g_1 - m_1)^2} \\
&+ \frac{c_1^{**}(e^{-(g_1+m_1)lT} - e^{-(g_1+m_1)T})}{g_1 + m_1} + \frac{f_1 c_2^{**}(1 - l)T}{h_1 - g_1 - m_1} \\
&- \frac{f c_2^{**}(e^{-(h_1-g_1-m_1)lT} - e^{-(h_1-g_1-m_1)T})}{(h_1 - g_1 - m_1)^2} < 0,
\end{aligned} \tag{12}
$$

hold, the globally asymptotically stable $(\widetilde{x_1}, \widetilde{x_2}, 0, \widetilde{c_1}, \widetilde{c_2})$ exists, and x_2^*, x_2^{**} are with (6), and $c_1^*, c_1^{**}, c_2^*, c_2^{**}$ are with in (10).

Proof. Doing it as $y_1 = x_1 - \widetilde{x_1}, y_2 = x_2 - \widetilde{x_2}, Y = Y, z_1 = c_1 - \widetilde{c_1}, z_2 = c_2 - \widetilde{c_2}$, then, linear system with considering for one periodic solution $(\widetilde{x_1}, \widetilde{x_2}, 0, \widetilde{c_1}, \widetilde{c_2})$ is presented as

$$
\begin{pmatrix} \frac{dy_1}{dt} \\ \frac{dy_2}{dt} \\ \frac{dY}{dt} \\ \frac{dz_1}{dt} \\ \frac{dz_2}{dt} \end{pmatrix} = \begin{pmatrix} -D_1 & 0 & 0 & 0 & 0 \\ 0 & -D_2 & K_5 & 0 & 0 \\ 0 & 0 & K_5' & 0 & 0 \\ 0 & 0 & 0 & -(g_1 + m_1) & f_1 \\ 0 & 0 & 0 & 0 & -h_1 \end{pmatrix} \begin{pmatrix} y_1 \\ y_2 \\ Y \\ z_1 \\ z_2 \end{pmatrix},
$$

where $K_5 = -\frac{1}{\delta} \times \frac{\widetilde{\eta x_2(t)}}{\alpha + \widetilde{x_2(t)}}, K_5' = -[D_2 + \beta\widetilde{c_1(t)} - \frac{\widetilde{\eta x_2(t)}}{\alpha + \widetilde{x_2(t)}}]$. Then, the fundamental solution matrix is

$$
\Phi(t) = \begin{pmatrix} e^{-D_1 t} & 0 & 0 & 0 & 0 \\ 0 & e^{-D_2 t} & \dagger & 0 & 0 \\ 0 & 0 & K_6 & 0 & 0 \\ 0 & 0 & 0 & K_6' & \ddagger \\ 0 & 0 & 0 & 0 & e^{-h_1 t} \end{pmatrix},
$$

where $K_6 = e^{\int_0^t (-D_2 + \frac{\widetilde{\eta x_2(\xi)}}{\alpha + \widetilde{x_2(\xi)}} - \beta\widetilde{c_1(\xi)})d\xi}, K_6' = e^{-(g_1+m_1)t}$. There is no need for computing $\dagger, \ddagger$.

When $t = (n+l)T$, we get

$$\begin{pmatrix} y_1(nT^+) \\ y_2(nT^+) \\ Y(nT^+) \\ z_1(nT^+) \\ z_2(nT^+) \end{pmatrix} = \begin{pmatrix} 1\,0\,0\,0 & 0 \\ 0\,1\,0\,0 & 0 \\ 0\,0\,1\,0 & 0 \\ 0\,0\,0\,1 & 0 \\ 0\,0\,0\,0 & 1-\theta_1 \end{pmatrix} \begin{pmatrix} y_1(nT) \\ y_2(nT) \\ Y(nT) \\ z_1(nT) \\ z_2(nT) \end{pmatrix}.$$

When $t = (n+1)T$, we also get

$$\begin{pmatrix} y_1(nT^+) \\ y_2(nT^+) \\ Y(nT^+) \\ z_1(nT^+) \\ z_2(nT^+) \end{pmatrix} = \begin{pmatrix} 1-d & d & 0 & 0 & 0 \\ d & 1-d & 0 & 0 & 0 \\ 0 & 0 & 1 & 0 & 0 \\ 0 & 0 & 0 & 1-\theta_3 & 0 \\ 0 & 0 & 0 & 0 & 1 \end{pmatrix} \begin{pmatrix} y_1(nT) \\ y_2(nT) \\ Y(nT) \\ z_1(nT) \\ z_2(nT) \end{pmatrix}.$$

The stability of $(\widetilde{x_1}, \widetilde{x_1}, 0, \widetilde{c_1}, \widetilde{c_2})$ is decided by eigenvalues of

$$M = \begin{pmatrix} 1-d & d & 0 & 0 & 0 \\ d & 1-d & 0 & 0 & 0 \\ 0 & 0 & 1 & 0 & 0 \\ 0 & 0 & 0 & 1-\theta_3 & 0 \\ 0 & 0 & 0 & 0 & 1-\theta_1 \end{pmatrix} \Phi(\tau).$$

From condition (12), $e^{-D_1\tau} < 1$, and $e^{-D_2\tau} < 1$, the eigenvalues of M are presented as

$$\gamma_1 = \frac{K_7 + \sqrt{K_7'}}{2} << 1,$$

$$\gamma_2 = \frac{K_7 - \sqrt{K_7'}}{2} < (1-d)e^{-D_2T} < 1,$$

$$\gamma_3 = e^{\int_0^T (-D_2 + \frac{\eta \widetilde{x_2(\xi)}}{\alpha + \widetilde{x_2(\xi)}} - \beta \widetilde{c_1(\xi)})d\xi},$$

$$\gamma_4 = (1-\theta_3)e^{-(g_1+m_1)\tau} < 1,$$

$$\gamma_5 = (1-\theta_1)e^{-h_1\tau} < 1.$$

where $K_7 = (1-d)(e^{-D_1T} + e^{-D_2T}), K_7' = (1-d)^2(e^{-D_1T} + e^{-D_2T})^2 - 4(1-2d)e^{-(D_1+D_2)T}$. For the Floquet theory [6] and condition (14), it is easily to have $\gamma_3 < 1$, then, the locally stable $(\widetilde{x_1}, \widetilde{x_2}, 0, \widetilde{c_1}, \widetilde{c_2})$ exists.

Choosing $\varepsilon > 0$, we have

$$\rho_1 = \exp[\int_0^{\tau} K_8 - \beta(\widetilde{c_1(t)} - \varepsilon)dt] < 1.$$

where $K_8 = -D_2 + \frac{\eta(\widetilde{x_2}+\varepsilon)}{\alpha+(\widetilde{x_2}+\varepsilon)}$. We get $\frac{dx_2}{dt} \leq D_2(x_2^0 - x_2)$ with considering system (1). So

$$\begin{cases} \left.\begin{aligned} \frac{dm_1}{dt} &= D_1(x_1^0 - m_1), \\ \frac{dm_2}{dt} &= D_2(x_2^0 - m_2), \end{aligned}\right\} t \neq (n+1)T, \\ \left.\begin{aligned} \Delta m_1 &= d(m_2 - m_1) + \theta_2, \\ \Delta m_2 &= d(m_1 - m_2), \end{aligned}\right\} t = (n+1)T. \end{cases} \tag{13}$$

From lemma (2), and [8], we get

$$\begin{cases} x_1 \leq m_1 \leq \widetilde{x_1} + \varepsilon, \\ x_2 \leq m_2 \leq \widetilde{x_2} + \varepsilon. \end{cases} \tag{14}$$

and

$$\begin{cases} \widetilde{c_1} - \varepsilon \leq c_1 \leq \widetilde{c_1} + \varepsilon, \\ \widetilde{c_2} - \varepsilon \leq c_2 \leq \widetilde{c_2} + \varepsilon. \end{cases} \tag{15}$$

Therefore,

$$\frac{dY}{dt} \leq Y[-D_2 + \frac{\eta(\widetilde{x_2} + \varepsilon)}{\alpha + (\widetilde{x_2} + \varepsilon)} - \beta(\widetilde{c_1} - \varepsilon)]. \tag{16}$$

Then, $Y(t) \leq Y(0^+) \exp[\int_0^t (-D_2 + \frac{\eta(\widetilde{x_2}+\varepsilon)}{\alpha+(\widetilde{x_2}+\varepsilon)} - \beta(\widetilde{c_1} - \varepsilon))ds]$, thus

$$\begin{aligned} Y((n+1)T) &\leq Y(nT^+) \\ &\times \exp[\int_{nT}^{(n+1)T} (-D_2 + \frac{\eta(\widetilde{x_2(s)} + \varepsilon)}{\alpha + (\widetilde{x_2(s)} + \varepsilon)} - \beta(\widetilde{c_1(s)} - \varepsilon))ds]. \end{aligned} \tag{17}$$

Hence $Y(n\tau) \leq Y(0^+)\rho_1^n$ and $Y(nT) \to 0$ as $n \to \infty$. So $Y(t) \to 0$ as $t \to \infty$.

For $\varepsilon > 0$, it has a $t_0 > 0$ such that $0 < Y < \varepsilon$ for all $t \geq t_0$. It is no difficulty to gain

$$\frac{\delta\alpha D_2 + \eta\varepsilon}{\delta(\alpha + M)}[\frac{\delta(\alpha + M)D_2 x_2^0}{\delta(\alpha + M)D_2 + \eta\varepsilon} - x_2] \leq \frac{dx_2}{dt} \leq D_2(x_2^0 - x_2). \tag{18}$$

(v_1, v_2), (n_1, n_2) are of

$$\begin{cases} \left.\begin{array}{l} \frac{dv_1}{dt} = D_1(x_1^0 - v_1), \\ \frac{dv_2}{dt} = \frac{\delta(\alpha+M)D_2 + \eta\varepsilon}{\delta(\alpha+M)}[\frac{\delta(\alpha+M)D_2 x_2^0}{\delta(\alpha+M)D_2 + \eta\varepsilon} - v_2], \end{array}\right\} t \neq nT, \\ \left.\begin{array}{l} \Delta v_1 = d(v_2 - v_1) + \theta_2, \\ \Delta v_2 = d(v_1 - v_2), \end{array}\right\} t = nT, \end{cases} \tag{19}$$

and

$$\begin{cases} \left.\begin{array}{l} \frac{dn_1}{dt} = D_1(x_1^0 - n_1), \\ \frac{dn_2}{dt} = D_2(x_2^0 - n_2), \end{array}\right\} t \neq nT, \\ \left.\begin{array}{l} \Delta n_1 = d(n_2 - n_1) + \theta_2, \\ \Delta n_2 = d(n_1 - n_2), \end{array}\right\} t = nT, \end{cases} \tag{20}$$

respectively, while

$$\begin{cases} \widetilde{v_1} = x_1^0 - (x_1^0 - v_1^*)e^{-D_1(t-nT)}, \\ \widetilde{v_2} = \frac{\delta(\alpha+M)D_2 x_2^0}{\delta(\alpha+M)D_2 + \eta\varepsilon} \\ \qquad -(\frac{\delta(\alpha+M)D_2 x_2^0}{\delta(\alpha+M)D_2 + \eta\varepsilon} - v_2^*)e^{-\frac{\delta(\alpha+M)D_2+\eta\varepsilon}{\delta(\alpha+M)}(t-nT)}, \end{cases} \tag{21}$$

with

$$\begin{cases} v_1^* \\ = \frac{[(1-d)(1-e^{-D_1T}) - (1-2d)(1-e^{-\frac{\delta(\alpha+M)D_2+\eta\varepsilon}{\delta(\alpha+M)}T})e^{-D_1T}]x_1^0 - d(1-e^{-\frac{\delta(\alpha+M)D_2+\eta\varepsilon}{\delta(\alpha+M)}T})\frac{\delta(\alpha+M)D_2x_2^0}{\delta(\alpha+M)D_2+\eta\varepsilon}}{1-(1-d)e^{-D_1T} - (1-d)e^{-\frac{\delta(\alpha+M)D_2+\eta\varepsilon}{\delta(\alpha+M)}\tau} - (1-2d)e^{-(D_1+\frac{\delta(\alpha+M)D_2+\eta\varepsilon}{\delta(\alpha+M)})T}} \\ \quad + \frac{(1-e^{-\frac{\delta(\alpha+M)D_2+\eta\varepsilon}{\delta(\alpha+M)}\tau} + de^{-\frac{\delta(\alpha+M)D_2+\eta\varepsilon}{\delta(\alpha+M)}\tau})\theta_2}{1-(1-d)e^{-D_1\tau} - (1-d)e^{-\frac{\delta(\alpha+M)D_2+\eta\varepsilon}{\delta(\alpha+M)}\tau} - (1-2d)e^{-(D_1+\frac{\delta(\alpha+M)D_2+\eta\varepsilon}{\delta(\alpha+M)})T}}, \\ v_2^* \\ = \frac{-d(1-e^{-D_1\tau})x_1^0 + [(1-d)(1-e^{-\frac{\delta(\alpha+M)D_2+\eta\varepsilon}{\delta(\alpha+M)}T}) - (1-2d)(1-e^{-D_1T})e^{-\frac{\delta(\alpha+M)D_2+\eta\varepsilon}{\delta(\alpha+M)}\tau}]\frac{\delta(\alpha+M)D_2x_2^0}{\delta(\alpha+M)D_2+\eta\varepsilon}}{1-(1-d)e^{-\frac{\delta(\alpha+M)D_2+\eta\varepsilon}{\delta(\alpha+M)}\tau} - (1-d)e^{-D_1T} - (1-2d)e^{-(\frac{\delta(\alpha+M)D_2+\eta\varepsilon}{\delta(\alpha+M)}+D_1)T}} \\ \quad + \frac{+de^{-D_1T}\theta_2}{1-(1-d)e^{-\frac{\delta(\alpha+M)D_2+\eta\varepsilon}{\delta(\alpha+M)}T} - (1-d)e^{-D_1T} - (1-2d)e^{-(\frac{\delta(\alpha+M)D_2+\eta\varepsilon}{\delta(\alpha+M)}+D_1)T}}, \end{cases} \tag{22}$$

For $\varepsilon_1 > 0$, existing a $t_1, t > t_1$ such that

$$\widetilde{v_1} - \varepsilon_1 < x_1 < \widetilde{n_1} + \varepsilon_1,$$

and

$$\widetilde{v_2} - \varepsilon_1 < x_2 < \widetilde{n_2} + \varepsilon_1.$$

One will gain the followings with considering $\varepsilon \to 0$,

$$\widetilde{x_1} - \varepsilon_1 < x_1 < \widetilde{x_1} + \varepsilon_1,$$

and

$$\widetilde{x_2} - \varepsilon_1 < x_2 < \widetilde{x_2} + \varepsilon_1.$$

This completes the proofs.

Theorem 2. Suppose

$$d > \frac{1}{2}, \tag{23}$$

and

$$\begin{aligned}(\eta - D_2)T &- \frac{\alpha\eta}{D_2(\alpha + x_2^0)} \ln \frac{\alpha e^{-D_2 T} + (e^{-D_2 T} - 1)x_2^0 + x_2^*}{\alpha + x_2^*} \\ &+ \frac{c_1^*(1 - e^{-(g_1+m_1)lT})}{g_1 + m_1} + \frac{f_1 c_2^* lT}{h_1 - g_1 - m_1} - \frac{f_1 c_2^*(1 - e^{-(h_1-g_1-m_1)lT})}{(h_1 - g_1 - m_1)^2} \\ &+ \frac{c_1^{**}(e^{-(g_1+m_1)lT} - e^{-(g_1+m_1)T})}{g_1 + m_1} + \frac{f_1 c_2^{**}(1 - l)T}{h_1 - g_1 - m_1} \\ &- \frac{f c_2^{**}(e^{-(h_1-g_1-m_1)lT} - e^{-(h_1-g_1-m_1)T})}{(h_1 - g_1 - m_1)^2} > 0,\end{aligned} \tag{24}$$

hold, system permanence, and x_2^* and x_2^{**} are with (6), $c_1^*, c_1^{**}, c_2^*, c_2^{**}$ are with (10).

Proof. Owing to remark (4), and lemma (5), we have obtain that (x_1, x_2, Y, c_1, c_2) is bounded. It can be easily obtained that $c_1 \geq m_o$ and $c_2 \geq m_e$ with considering with remark (4).

Therefore,

$$\begin{cases} \left.\begin{aligned} \frac{dx_1}{dt} &= D_1(x_1^0 - x_1), \\ \frac{dx_2}{dt} &\geq D_2(x_2^0 - x_2) - \frac{\eta M}{\delta\alpha} x_2, \end{aligned}\right\} t \neq nT, \\ \left.\begin{aligned} \Delta x_1 &= d(-x_1 + x_2) + \theta_2, \\ \Delta x_2 &= d(x_1 - x_2), \end{aligned}\right\} t = nT, \end{cases} \tag{25}$$

with its comparison system

$$\begin{cases} \left.\begin{aligned} \frac{dw_1}{dt} &= D_1(x_1^0 - w_1), \\ \frac{dw_2}{dt} &= \frac{\delta\alpha D_2 + \eta M}{\delta\alpha}\left[\frac{\delta\alpha D_2 x_2^0}{\delta\alpha D_2 + \eta M} - w_2\right], \end{aligned}\right\} t \neq nT, \\ \left.\begin{aligned} \Delta w_1 &= d(w_2 - w_1) + \theta_2, \\ \Delta w_2 &= d(w_1 - w_2), \end{aligned}\right\} t = nT. \end{cases} \tag{26}$$

With considering (2) and (8), $(\widetilde{w_1}, \widetilde{w_2})$ of (26) is

$$\begin{cases} \widetilde{w_1} = x_1^0 - (x_1^0 - w_1^*)e^{-D_1(t-nT)}, nT < t \leq (n+1)T, \\ \widetilde{w_2} = \dfrac{\delta\alpha D_2 x_2^0}{\delta\alpha D_2 + \eta M} - (\dfrac{\delta\alpha D_2 x_2^0}{\delta\alpha D_2 + \eta M} - w_2^*)e^{-\frac{\delta\alpha D_2 + \eta M}{\delta\alpha}(t-nT)}, \\ \qquad nT < t \leq (n+1)T, \end{cases} \tag{27}$$

with

$$\begin{cases} w_1^* = \dfrac{[(1-d)(1-e^{-D_1 T}) - (1-2d)(1-e^{-\frac{\delta\alpha D_2+\eta M}{\delta\alpha}T})e^{-D_1 T}]x_1^0 - d(1-e^{-\frac{\delta\alpha D_2+\eta M}{\delta\alpha}T})\frac{\delta\alpha D_2 x_2^0}{\delta\alpha D_2+\eta M}}{1-(1-d)e^{-D_1 T} - (1-d)e^{-\frac{\delta\alpha D_2+\eta M}{\delta\alpha}T} - (1-2d)e^{-(D_1+\frac{\delta\alpha D_2+\eta M}{\delta\alpha})T}} \\ \qquad + \dfrac{+(1-e^{-\frac{\delta\alpha D_2+\eta M}{\delta\alpha}T} + de^{-\frac{\delta\alpha D_2+\eta M}{\delta\alpha}T})\theta_2}{1-(1-d)e^{-D_1 T} - (1-d)e^{-\frac{\delta\alpha D_2+\eta M}{\delta\alpha}T} - (1-2d)e^{-(D_1+\frac{\delta\alpha D_2+\eta M}{\delta\alpha})T}}, \\ w_2^* = \dfrac{-d(1-e^{-D_1 T})x_1^0 + [(1-d)(1-e^{-\frac{\delta\alpha D_2+\eta M}{\delta\alpha}T}) - (1-2d)(1-e^{-D_1\tau})e^{-\frac{\delta\alpha D_2+\eta M}{\delta\alpha}T}]\frac{\delta\alpha D_2 x_2^0}{\delta\alpha D_2+\eta M}}{1-(1-d)e^{-\frac{\delta\alpha D_2+\eta M}{\delta\alpha}T} - (1-d)e^{-D_1 T} - (1-2d)e^{-(\frac{\delta\alpha D_2+\eta M}{\delta\alpha}+D_1)\tau}} \\ \qquad + \dfrac{+de^{-D_1\tau}\theta_2}{1-(1-d)e^{-\frac{\delta\alpha D_2+\eta M}{\delta\alpha}\tau} - (1-d)e^{-D_1 T} - (1-2d)e^{-(\frac{\delta\alpha D_2+\eta M}{\delta\alpha}+D_1)T}}. \end{cases} \tag{28}$$

Furthermore, $(\widetilde{w_1}, \widetilde{w_2})$ of (27) is globally asymptotic stable. Then, it exists a $\varepsilon > 0$ such that $x_1 \geq w_1 \geq \widetilde{w_1} - \varepsilon \geq w_1^* - \varepsilon = k_1$ and $x_2 \geq w_2 \geq \widetilde{w_2} - \varepsilon \geq w_2^* - \varepsilon = k_2$.

Since

$$\begin{aligned} &(\eta - D_2)T + \frac{\eta\alpha}{D_2(\alpha + x_2^*)} \ln \frac{\alpha e^{-D_2 T} + (e^{-D_2 T} - 1)x_2^0 + x_2^*}{\alpha + x_2^*} \\ &+ \frac{c_1^*(1 - e^{-(g_1+m_1)l\tau})}{g_1 + m_1} + \frac{fc_2^* lT}{h_1 - g_1 - m_1} - \frac{f_1 c_2^*(1 - e^{-(h_1-g_1-m_1)lT})}{(h_1 - g_1 - m_1)^2} \\ &+ \frac{c_1^{**}(e^{-(g_1+m_1)lT} - e^{-(g_1+m_1)T})}{g_1 + m_1} + \frac{f_1 c_2^{**}(1-l)T}{h_1 - g_1 - m_1} \\ &- \frac{f_1 c_2^{**}(e^{-(h_2-g_2-m_2)lT} - e^{-(h_1-g_1-m_1)T})}{(h_1 - g_1 - m_1)^2} > 0, \end{aligned} \tag{29}$$

$m_3 > 0$ and $\varepsilon_1 > 0$ can be selected to do as

$$\begin{aligned} \sigma = &(\eta - \frac{\delta\alpha D_2 + \eta m_3}{\delta\alpha} - \varepsilon)T - \frac{\eta\alpha}{D_2(\alpha - \varepsilon + \frac{\delta\alpha D_2}{\delta\alpha D_2 + \eta m_3}x_2^0)} \\ &\times \ln \frac{(\alpha - \varepsilon + \frac{\delta\alpha D_2}{\delta\alpha D_2 + \eta m_3})e^{-\frac{\delta\alpha D_2 + \eta m_3}{\delta\alpha}} - \frac{\delta\alpha D_2}{\delta\alpha D_2 + \eta m_3}x_2^0) + k_2^*}{\alpha - \varepsilon + k_2^*} \\ &+ \frac{c_1^*(1 - e^{-(g_1+m_1)lT})}{g_1 + m_1} + \frac{f_1 c_2^* l\tau}{h_1 - g_1 - m_1} - \frac{f_1 c_2^*(1 - e^{-(h_1-g_1-m_1)lT})}{(h_1 - g_1 - m_1)^2} \\ &+ \frac{c_1^{**}(e^{-(g_1+m_1)lT} - e^{-(g_1+m_1)T})}{g_1 + m_1} + \frac{f_1 c_2^{**}(1-l)T}{h_1 - g_1 - m_1} \\ &- \frac{f_1 c_2^{**}(e^{-(h_1-g_1-m_1)lT} - e^{-(h_1-g_1-m_1)T})}{(h_1 - g_1 - m_1)^2} > 0, \end{aligned} \tag{30}$$

where k_2^* is defined as (34)

$Y < m_3$ will be proved that it can not be held for $t \geq 0$. Otherwise,

$$\begin{cases} \left.\begin{array}{l} \dfrac{dx_1}{dt} = D_1(x_1^0 - x_1), \\ \dfrac{dx_2}{dt} \geq D_2(x_2^0 - x_2(t)) - \dfrac{\eta m_3}{\delta\alpha} x_2, \end{array}\right\} t \neq nT, \\ \left.\begin{array}{l} \triangle x_1 = d(-x_1 + x_2) + \theta_2, \\ \triangle x_2 = d(x_1 - x_2), \end{array}\right\} t = nT. \end{cases} \tag{31}$$

with its comparison system

$$\begin{cases} \left.\begin{array}{l} \dfrac{dk_1}{dt} = D_1(x_1^0 - k_1), \\ \dfrac{dk_2}{dt} \\ \quad = \dfrac{\delta\alpha D_2 + \eta m_3}{\delta\alpha}[\dfrac{\delta\alpha D_2}{\delta\alpha D_2 + \eta m_3} x_2^0 - k_2], \end{array}\right\} t \neq nT, \\ \left.\begin{array}{l} \triangle k_1 = d(k_2 - k_1) + \mu_2, \\ \triangle k_2 = d(k_1 {-}_2), \end{array}\right\} t = nT. \end{cases} \tag{32}$$

By lemma (2), we gain $x_1 \geq k_1, x_2 \geq k_2$ and $k_1 \to \overline{k_1}, k_2(t) \to \overline{k_2}$, as $t \to \infty$, and

$$\begin{cases} \overline{k_1} = x_1^0 - [x_1^0 - k_1^*]e^{-D_1(t-nT)}, nT < t \leq (n+1)T, \\ \overline{k_2} = \dfrac{\delta\alpha D_2}{\delta\alpha D_2 + \eta m_3} x_2^0 - [\dfrac{\delta\alpha D_2}{\delta\alpha D_2 + \eta m_3} x_2^0 - k_2^*]e^{-\frac{\delta\alpha D_2 + \eta m_3}{\delta\alpha}(t-nT)}, \\ \qquad nT < t \leq (n+1)T, \end{cases} \tag{33}$$

with

$$\begin{cases} k_1^* = \dfrac{[(1-d)(1-e^{-D_1T}) - (1-2d)(1-e^{-\frac{\delta\alpha D_2+\eta m_3}{\delta\alpha}\tau})e^{-D_1T}]x_1^0 - d(1-e^{-\frac{\delta\alpha D_2+\eta m_3}{\delta\alpha}\tau})\frac{\delta\alpha D_2}{\delta\alpha D_2+\eta m_3}x_2^0}{1-(1-d)e^{-D_1\tau} - (1-d)e^{-\frac{\delta\alpha D_2+\eta m_3}{\delta\alpha}T} - (1-2d)e^{-(D_1+\frac{\delta\alpha D_2+\eta m_3}{\delta\alpha})\tau}} \\ \qquad + \dfrac{+(1-e^{-\frac{\delta\alpha D_2+\eta m_3}{\delta\alpha}T} + de^{-\frac{\delta\alpha D_2+\eta m_3}{\delta\alpha}T})\theta_2}{1-(1-d)e^{-D_1T} - (1-d)e^{-\frac{\delta\alpha D_2+\eta m_3}{\delta\alpha}T} - (1-2d)e^{-(D_1+\frac{\delta\alpha D_2+\eta m_3}{\delta\alpha})T}}, \\ k_2^* = \dfrac{-d(1-e^{-D_1T})x_1^0 + [(1-d)(1-e^{-\frac{\delta\alpha D_2+\eta m_3}{\delta\alpha}T}) - (1-2d)(1-e^{-D_1T})e^{-\frac{\delta\alpha D_2+\eta m_3}{\delta\alpha}T}]\frac{\delta\alpha D_2}{\delta\alpha D_2+\eta m_3}x_2^0}{1-(1-d)e^{-\frac{\delta\alpha D_2+\eta m_3}{\delta\alpha}T} - (1-d)e^{-D_1T} - (1-2d)e^{-(\frac{\delta\alpha D_2+\eta m_3}{\delta\alpha}+D_1)T}} \\ \qquad + \dfrac{+de^{-D_1T}\theta_2}{1-(1-d)e^{-\frac{\delta\alpha D_2+\eta m_3}{\delta\alpha}T} - (1-d)e^{-D_1T} - (1-2d)e^{-(\frac{\delta\alpha D_2+\eta m_3}{\delta\alpha}+D_1)T}}, \end{cases} \tag{34}$$

Therefore,

$$\begin{cases} x_1 \geq k_1 \geq \overline{k_1} - \varepsilon_1, \\ x_2 \geq k_2 \geq \overline{k_2} - \varepsilon_1. \end{cases} \tag{35}$$

For $t \geq T_1$

$$\frac{dY}{dt} \geq [-D_2 + \frac{\eta(\overline{k_2} - \varepsilon_1)}{\alpha + (\overline{k_2} - \varepsilon_1)} - \beta(\widetilde{c_1} + \varepsilon_1)]Y, \tag{36}$$

Let $K_9 \in N^+$ and $K_9 \tau > T_1$, integrating (36) on $(nT, (n+1)T), n \geq K_9$, we have

$$Y((n+1)T) \geq Y(nT^+)\exp(\int_{nT}^{(n+1)T} [-D_2 + \frac{\eta(\overline{k_1} - \varepsilon_1)}{\alpha + (\overline{k_1} - \varepsilon_1)} - \beta(\overline{k_2} + \varepsilon_1)]dt)$$

$$= x(nT)e^{\sigma},$$

then $Y((K_9 + k)T) \geq Y(K_9 T^+)e^{k\sigma} \to \infty$, as $k \to \infty$, it is an illogicality with the bounded Y. Hence $Y \geq m_3$.

5 Discussion

In this work, we present a chemostat-type model with impulsive effects in a polluted karst environment. If it is supposed that the variables are shown in the table below:

$x_1(0)$	$x_2(0)$	$Y(0)$	$c_1(0)$	$c_2(0)$	D_1	D_2	x_1^0	x_2^0	δ	η	α	β	θ_1	θ_2	θ_3	θ	f_1	g_1	m_1	h_1	d	T
2	2	2	0.13	0.15	2	0.2	2	3	1	0.5	2	0.5	0.5	0.8	0.6	0.01	0.4	0.3	0.1	0.4	0.2	1

system (1) is permanent (one can see Fig. 1). If it is supposed that another variables are shown in the table below:

$x_1(0)$	$x_2(0)$	$Y(0)$	$c_1(0)$	$c_2(0)$	D_1	D_2	x_1^0	x_2^0	δ	η	α	β	θ_1	θ_2	θ_3	θ	f_1	g_1	m_1	h_1	d	T
2	2	2	0.13	0.15	2	0.2	2	3	1	0.5	2	0.5	0.5	0.8	0.1	0.01	0.4	0.3	0.1	0.4	0.2	1

there exists a globally asymptotically stable solution $(\widetilde{x_1(t)}, \widetilde{x_2(t)}, 0, \widetilde{c_1(t)}, \widetilde{c_2(t)})$ of system (1) (one can see Fig. 2). If it is supposed that another variables are shown in the table below:

$x_1(0)$	$x_2(0)$	$Y(0)$	$c_1(0)$	$c_2(0)$	D_1	D_2	x_1^0	x_2^0	δ	η	α	β	θ_1	θ_2	θ_3	θ	f_1	g_1	m_1	h_1	d	T
2	2	2	0.13	0.15	2	0.16	2	3	1	0.5	2	0.5	0.5	0.8	0.1	0.01	0.4	0.3	0.1	0.4	0.2	1

then, system (1) is permanent (see Fig. 3).

The simulations show that parameters $0 < \theta_3 < 1$ and D_2 are very important for system (1). The parameters $D_1, \theta_1, \theta_2, \theta$ and d of system (1) can also be discussed. The results will guide us how to manage the source of water management in karst areas.

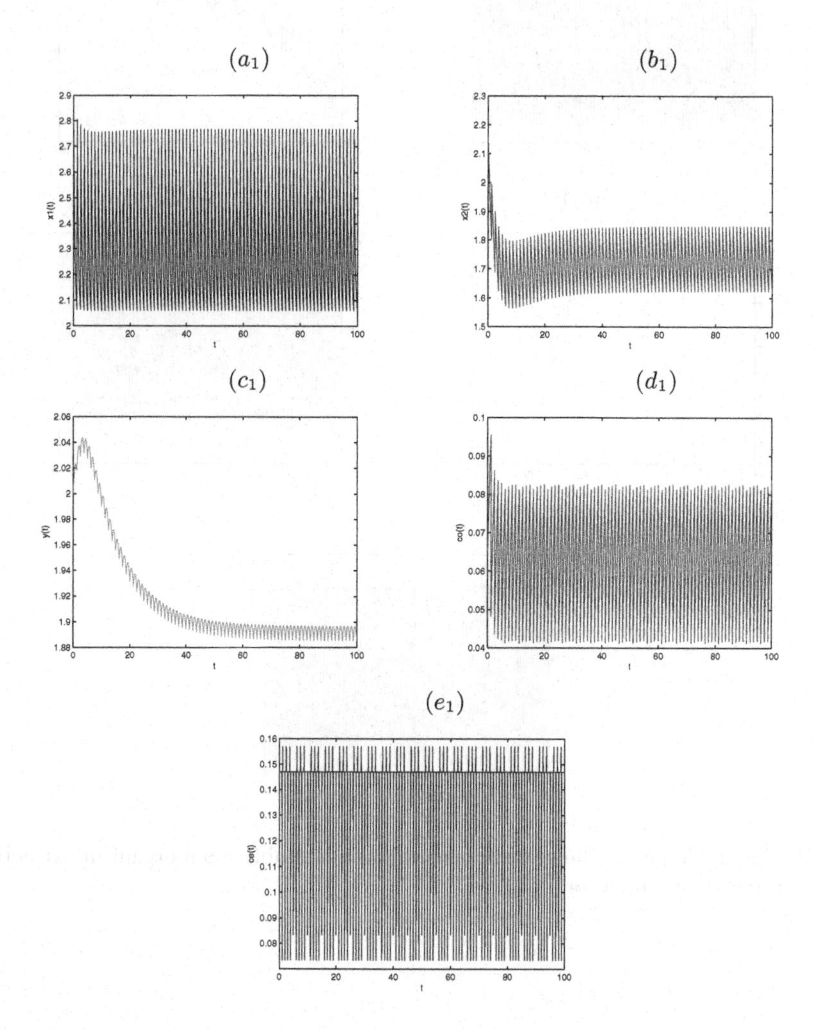

Fig. 1. The permanence of system (1) with parameters in the first table.

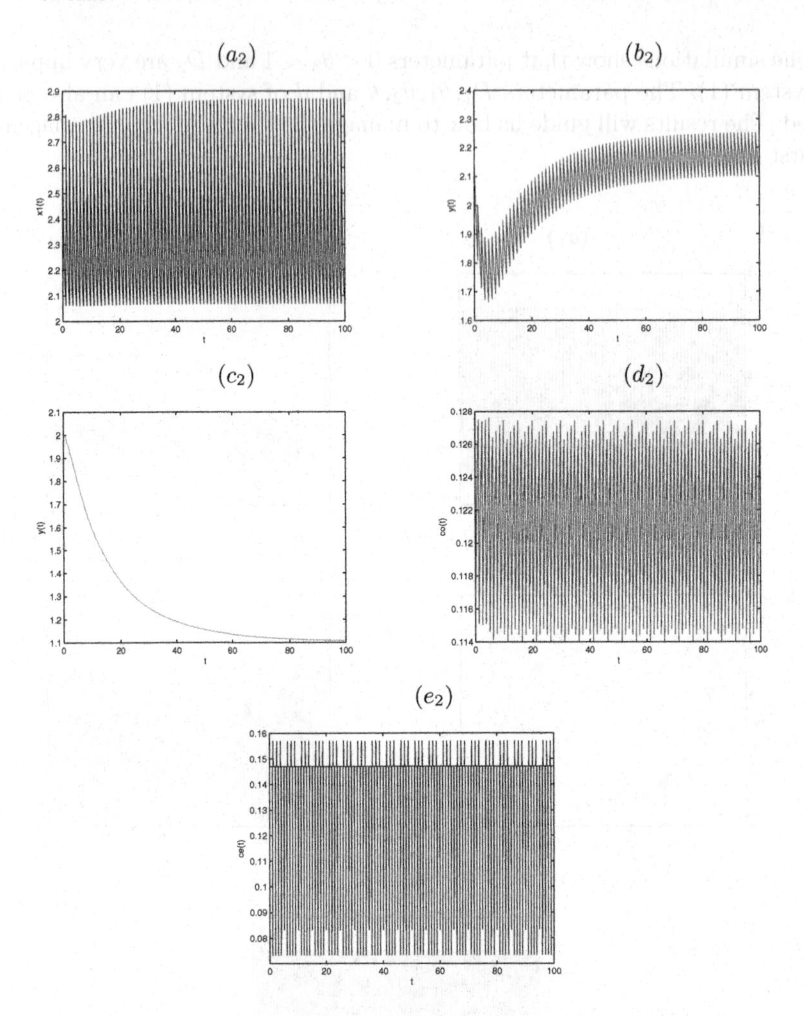

Fig. 2. The dynamics of the globally asymptotically stable microorganism-extinction with parameters in the second table.

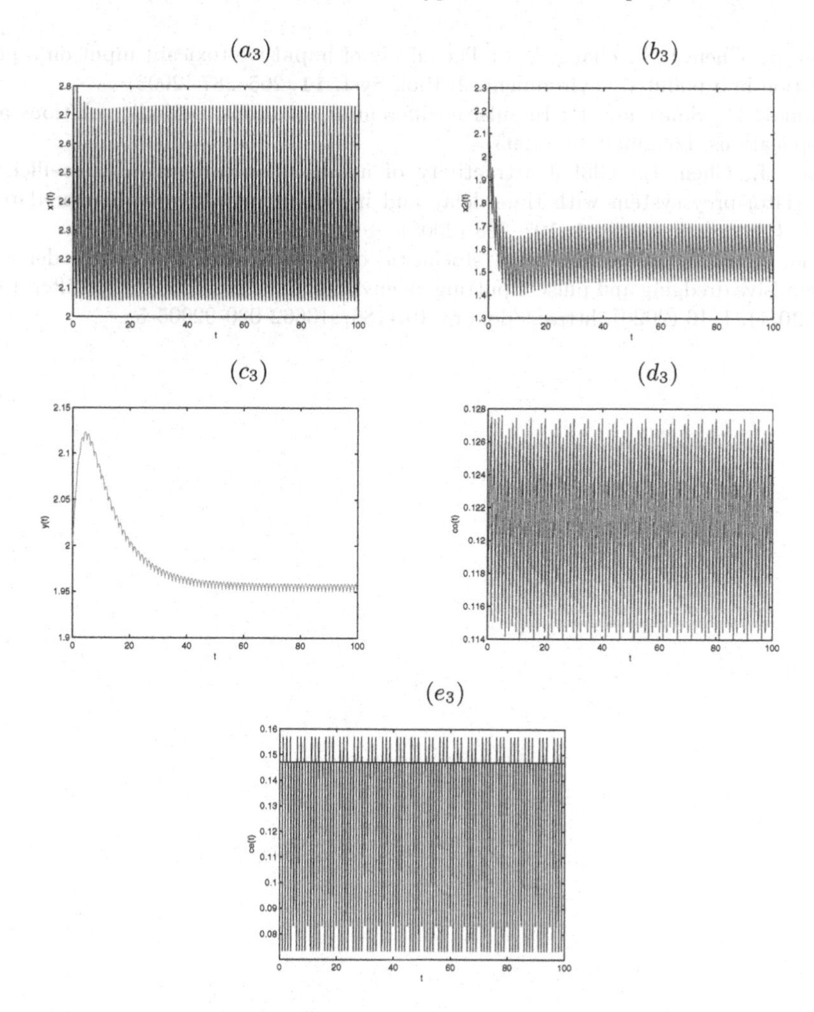

Fig. 3. The dynamics of the globally asymptotically stable microorganism-extinction with parameters in the third table.

References

1. Bo, L., Yi-Fan, Z., Bei-Bei, Z., Xian-Qing, W.: A risk evaluation model for karst groundwater pollution based on geographic information system and artificial neural network applications. Environ. Earth Sci. **77**(9), 1–14 (2018). https://doi.org/10.1007/s12665-018-7539-7
2. Zhao, D., Yuan, S.: Noise-induced bifurcations in the stochastic chemostat model with general nutrient uptake functions. Appl. Math. Lett. **103**, 106180 (2020)
3. Amster, P., et al.: Existence of -periodic solutions for a delayed chemostat with periodic inputs. Nonlinear Anal. Real World Appl. **55**, 103134 (2020)
4. Yuan, H., Zhang, C., Li, Y.: Existence and stability of coexistence states in a competition unstirred chemostat. Nonlinear Anal. Real World Appl. **35**, 441–456 (2017)

5. Liu, B., Chen, L.S., Zhang, Y.J.: The effects of impulsive toxicant input on a population in a polluted environment. J. Biol. Syst. **11**, 265–287 (2003)
6. Bainov, D., Simeonov, P.: Impulsive differential equations: periodic solutions and applications. Longman, 66 (1993)
7. Jiao, J., Chen, L.: Global attractivity of a stage-structure variable coefficients predator-prey system with time delay and impulsive perturbations on predators. Int. J. Biomathematics **1**, 197–208 (2008)
8. Jiao, J., Li, Q.: Dynamics of a stochastic eutrophication-chemostat model with impulsive dredging and pulse inputting on environmental toxicant. Adv. Differ. Equ. **2020**(1), 1–16 (2020). https://doi.org/10.1186/s13662-020-02905-5

Impact of Performance Estimation on Fast Processor Simulators

Sebastian Rachuj(✉), Dietmar Fey, and Marc Reichenbach

Friedrich-Alexander-Universität Erlangen-Nürnberg,
Martensstr. 3, 91058 Erlangen, Germany
{sebastian.rachuj,dietmar.fey,marc.reichenbach}@fau.de

Abstract. Hardware simulation is always a compromise between the speed of the simulator and the accuracy of the estimated runtime on the real hardware. Instrumenting fast simulation frameworks to estimate runtimes always results in tremendous slow-downs. In this paper, a quantization is done regarding the minimal overhead that can be expected when adding architectural models to a fast JIT enhanced emulation. Previous work is only focused on new approaches and improving available methods, but not on the unavoidable overhead that is introduced with any kind of instrumentation. Thus, the additional simulation time of calling an empty stub function instead of a fully implemented architectural model is investigated. We show relative runtimes for calling a function after executing each instruction and after executing a block of instructions. Also, a comparison against fully implemented models is done. On the test platforms, an academic and a commercial processor simulator were evaluated. The resulting average relative runtimes of the minimal introduced overhead are determined to be between 2.24 and 8.09 meaning that an emulation takes twice to eight times as long with instrumentation enabled.

1 Introduction

Processor Simulators are the means of choice when it comes to designing and evaluating new hardware systems. They not only provide the possibility to run the software before the silicon is available [3]. Additionally, design space explorations to find the best processor configuration for a specific application can be performed [15]. However, they always have to make a compromise between maximum simulation speed and accuracy [10]. Therefore, many different approaches were developed and investigated trying to optimize one dimension without sacrificing the other one. A brief overview about them is presented in Sect. 2.

The most promising method for creating a fast and precise simulator is by starting with a fast instruction set simulator (ISS) and adding additional models to make it more and more accurate [22]. Just-In-Time (JIT) compiling emulators are the fastest known ISS due to the fact that they translate the target machine code (meaning the machine code that runs on the ISS) to the host machine code. This reduces the overhead tremendously since the generated host code can be

H. Song and D. Jiang (Eds.): SIMUtools 2020, LNICST 370, pp. 79–93, 2021.
https://doi.org/10.1007/978-3-030-72795-6_7

run directly on the host processor of the emulator as soon as the translation step finishes. Prominent examples for JIT compiling emulators include the Tiny Code Generator (TCG) backend of QEMU [4], the code morphing facility of OVPsim [12], and the ARM Fast Models [2].

While these emulators are capable of executing software with very high speed, they lack means of determining nonfunctional properties like performance and energy requirements. There are plugins and extensions enabling these simulators to perform estimations of the runtime behavior. However, each activated model reduces the runtime since additional aspects of the hardware (e.g. the pipeline of the processor) have to be simulated. Even without modelling minor details, the runtime is impacted just by having to run further code next to the translated host code.

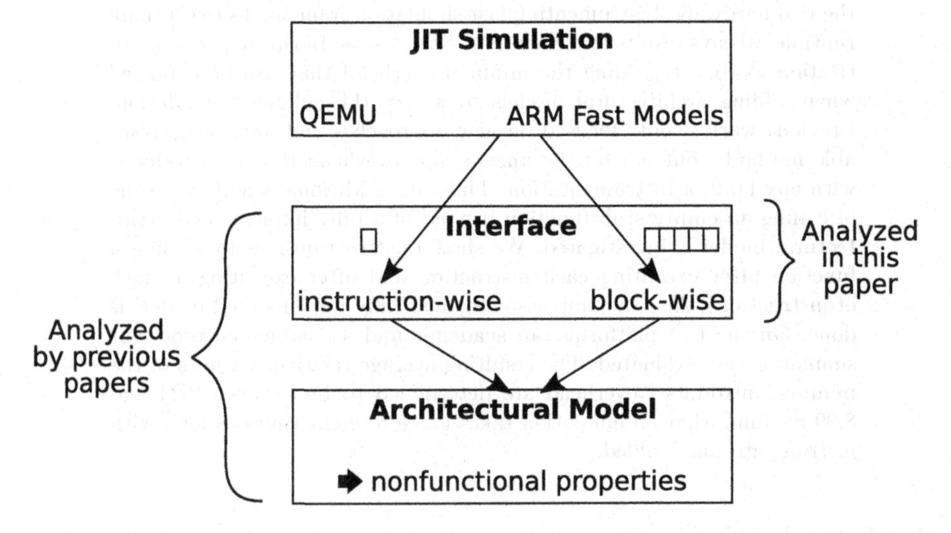

Fig. 1. The usual layout of a JIT emulator that is enhanced by architectural models. The analyzed part of this paper and parts that were previously evaluated are marked with curly braces.

The purpose of this paper is to quantize the minimal overhead that can be expected when adding functionality for estimating nonfunctional properties to a JIT compiling emulator. Figure 1 shows the usual parts of a fast simulation that is enhanced by an architectural model to estimate nonfunctional properties like time and energy. From the JIT accelerated emulation, an interface is used that transfers the information about executed instructions to the model. This is often done using callback mechanisms that are used to run a function for each instruction. While the architectural model can be adjusted with a high degree on variation depending on the requirements, the interface forwarding the information from the JIT simulation to the architectural model allows only minor modifications. Thus, the interface limits the maximum reachable instruction throughput and is therefore evaluated in this paper. Prior work only focused on a complete architectural model including the interface and did not analyze the interface separately. For this paper, QEMU as an example for an academic simulator and the ARM Fast Models as an example for a commercial simulator are investigated. While the ARM Fast Models already provide a probing infrastructure, QEMU had to be extended to support analyzing the instruction stream during the emulation process. For a faster execution, it is possible to forward multiple instructions at once in the form of translation blocks or basic blocks. This is an optimization to reduce the investigated overhead while still providing all data that is required to drive an accurate model. In Fig. 1 the two methods are called instruction-wise for running the callback for each emulated instruction and block-wise for running the callback for multiple instructions. In order to classify the additional overhead, comparisons to fully modelled architectural simulations are presented later in this paper. For QEMU, this consists of a literature review since there is no architectural model directly available for QEMU. On the other hand, the ARM Fast Models provide a pipeline implementation that was deployed and its overhead quantified.

This paper is structured as follows. At first, a brief overview is given about different methods for fast and accurate simulation. Afterwards, the simulation infrastructure for measuring the overhead is presented in Sect. 3. The findings of this paper are shown in Sect. 4. Finally, a conclusion is drawn.

2 Related Work

Many different methodologies for implementing precise and fast simulators were published. However, none of these publications investigated the minimal required overhead when fast JIT compilers have to be extended to support mechanistic or statistical models. In this section, some of these approaches are presented.

Multiple papers were written about extending QEMU with the ability to measure non-functional properties. QSim is a very promising approach combining QEMU with sim-outorder which is part of SimpleScalar [6,16]. It achieves huge speeds while still modelling the architectural aspects cycle-accurately by caching the results of the sim-outorder model of SimpleScalar for each translation block. However, small deviations in accuracy can be observed since effects spanning multiple translation blocks are not correctly modelled.

TQSIM is another way for extending QEMU with runtime estimation [13]. Here, a sampled approach has been implemented, based on an analytical model and sampled simulation, allowing a high speed simulation in comparison to a cycle-accurate simulator. On the other hand, they experience a slowdown from 3 to 14 in comparison to plain QEMU. Their accuracy is still very high with an error of only 8%.

A plugin interface for the code translator of QEMU (TCG) was recently added to the emulator[1]. It offers isolated functions for manipulating the intermediate code generation. Like implemented in this paper, it is also possible to add

[1] In git commit 68d8ef4ec540682c3538d4963e836e43a211dd17.

callback functions for each instruction or translation block and even to generate certain immediate instructions like an add on a custom virtual register. This can be used to implement an instruction counter. For this paper, the plugin system was not used as it is not as flexible as direct code changes, yet. Additionally, for interfacing with SystemC, code changes were necessary nonetheless.

In addition to the already mentioned simulator extensions that make functional simulation more accurate like statistical simulation and integration of architectural simulators, there are many more approaches for speeding up simulators while retaining a high accuracy. This can be done by parallelizing the simulation as done for example with PQEMU [8]. If the host and guest instruction set architecture is the same, even near-native speed for simulation can be achieved by exploiting parallelism and the virtualization capabilities of a platform [20]. Additionally, new simulators are also created that try to achieve new speed records when modelling a pipeline [9]. However, none of these publications contain an analysis of the minimal expected overhead with an empty model.

The peripheral system used for evaluation within this paper is implemented in SystemC which also required adjustments in QEMU. This has been done multiple times before. QBox is an example that uses a custom C-style TLM-2.0 library which is similar but not completely compatible with existing C++-style TLM-2.0 compatible models [7]. A transactor module is required to translate C-style transactions into C++-style transactions. Further implementations that are compatible to TLM-2.0 as defined in the SystemC standard include an implementation called LibSystemCTLM-SoC of Xilinx and QEMU-SystemC [18,23]. Xilinx's library uses a remote socket protocol to connect SystemC and QEMU while QEMU-SystemC is integrated into QEMU. These libraries are oftentimes more difficult to use and don't get the newest patches of QEMU. This means that new features like the RISC-V frontend is missing when using them.

Publications featuring the ARM Fast Models are sparsely found. Nonetheless, there are implementations of architectural models that enhance the Fast Models by timing behavior estimation. Available aspects that can be modelled include cache simulators and a pipeline behavior simulator. However, an analysis of the overhead introduced by the different available probes and models is not yet publicly available.

Other emulation frameworks were also extended with nearly cycle-accurate models. Examples include the works of Rosa et al. and Schreiner et al. who implemented a more accurate simulation on top of OVPsim [19,21]. But, they also did not evaluate the overhead in comparison to an OVPsim without any instrumentations.

JIT compilation techniques are also used in dynamically executed languages like Java, Python and JavaScript to implement their virtual machines. In this area, the most important features to add to the code generation are control flow analysis and execution tracking. This is done for debugging purposes or to offer additional security functionality. Kerschbaumer et al. adopted the JavaScript engine of WebKit to track the data flow within the programs and to check, if sensitive data is sent to an untrusted domain [14]. While they measured an overhead of 74%, the value is not comparable to emulation since direct changes of the JIT compilation infrastructure was done and additional control flow analysis integrated.

3 Simulation Infrastructure

For the investigation of this paper, two simulators capable of emulating the AArch64 instruction set architecture (ISA) have been investigated. Interconnecting peripherals in virtual hardware systems is commonly based on SystemC [5,11]. The abstraction level of virtual prototypes makes the TLM libraries of SystemC a sensible choice. However, the first considered simulator called QEMU, an academic and open source emulator [4], does not support interfacing with SystemC by default. Additionally, QEMU has no features for analyzing the instruction stream in more detail. The ARM Fast Models are a commercial simulator which is used within Platform Architect of Synopsys [3] for simpler usage. It supports SystemC by default and provides a probing infrastructure which is explained in the appropriate section. In the following, the adjustments that were made to the simulators to evaluate the introduced overhead are presented.

3.1 QEMU

QEMU is intended to be a fast emulation framework for running simple programs or even full operating systems. The high emulation speed is accomplished by using JIT compilation techniques meaning a translation of the guest machine code into the host machine code. While it also supports interpretation of its intermediate language and virtualization, these additional emulation modes were not of interest for this paper. Interpretation is slower than translation since more code has to be executed for each instruction. Virtualization requires the host to have the same ISA as the guest and is thus not possible if both ISAs do not match. QEMU's high speed makes it interesting for the evaluation. In the following, the execution system of the QEMU JIT backend is explained and the necessary changes to support callback functions that can be used to drive an architectural model are described.

Translation to the host machine code happens in two steps. First, the guest machine code is translated into an intermediate representation of the TCG component of QEMU. Afterwards, the resulting code can be optimized and translated again into the host machine code. A whole translation block containing multiple instructions is processed at once. It is a similar concept like basic blocks known for control flow analysis as required in optimizing compilers [1]. A basic block identifies a sequence of instructions that have only jumps to the beginning of the block and no jumps within the block with the exception of the last instruction. Translation blocks add additional constraints like a maximum length and a maximum of two successors. If a guest instruction has to be executed but the corresponding translation block is not yet within the code cache, a jump back

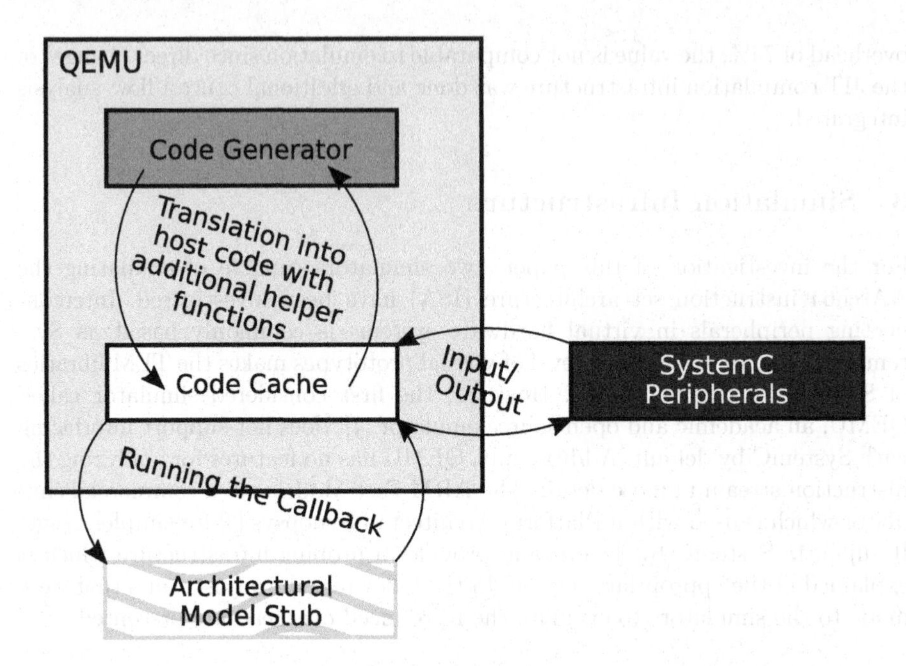

Fig. 2. Layout of the simulation framework used for this paper with QEMU as the binary translator. The code generator had to be adjusted and the peripherals and the architectural model stub were added for the evaluation.

into the code generator has to be done to translate the new translation block. This can be seen in Fig. 2.

Each guest instruction is translated into an arbitrary amount of TCG instructions. To ease implementing complex instructions, so-called helper functions can be defined. These functions can be implemented in high-level C and are called when the original guest instruction has to be emulated. While this functionality is normally used for implementing algorithmically complex guest instructions like cache maintenance or exception related operations, it is used to forward information about the execution flow to the architectural model. For this, the code generator had to be adjusted that a helper function is called for each executed guest instruction. Within this additional helper function, a detailed model can be driven. Since the aim of this paper is to evaluate the raw overhead of calling architectural functions, the model is just a stub as shown at the bottom of Fig. 2.

Since the overall simulated platform should stay the same for QEMU und the ARM Fast Models to allow an easier comparison, QEMU had to be extended with support for SystemC as also depicted in Fig. 2. For this, a custom machine type had to be implemented that is capable of interfacing with TLM2-compatible SystemC modules. It connects to plain memory by using the direct memory interface that offers a pointer referencing directly into the underlying memory. More sophisticated devices like the UART were implemented by mapping the hardware registers with the help of so-called IO ranges. When a QEMU memory access goes into an IO range, a previously defined callback is executed. For this paper, we forward the information about the memory access to the SystemC modules using the blocking transport interface of TLM2. As already mentioned in Sect. 2, there are a lot of different approaches for connecting QEMU to SystemC. Since some of these implementations change QEMU tremendously and do not profit from changes of the original code base, the previously presented connection was introduced.

3.2 ARM Fast Models

The ARM Fast Models are intended to provide a fast processor simulator for hardware system generation that uses a JIT compilation technique [2]. It does not allow radical changes like the previously introduced changes to QEMU due to the source code not being publicly available. However, a set of interfaces is provided allowing the interaction with the models. The most important interface is the master socket that allows plugging in TLM2-compatible SystemC modules by using a bridge that is supplied with the Synopsys Platform Architect which is used as a development environment. In this way, the peripherals of the simulation framework can be used by the models as depicted in the lower part of Fig. 3.

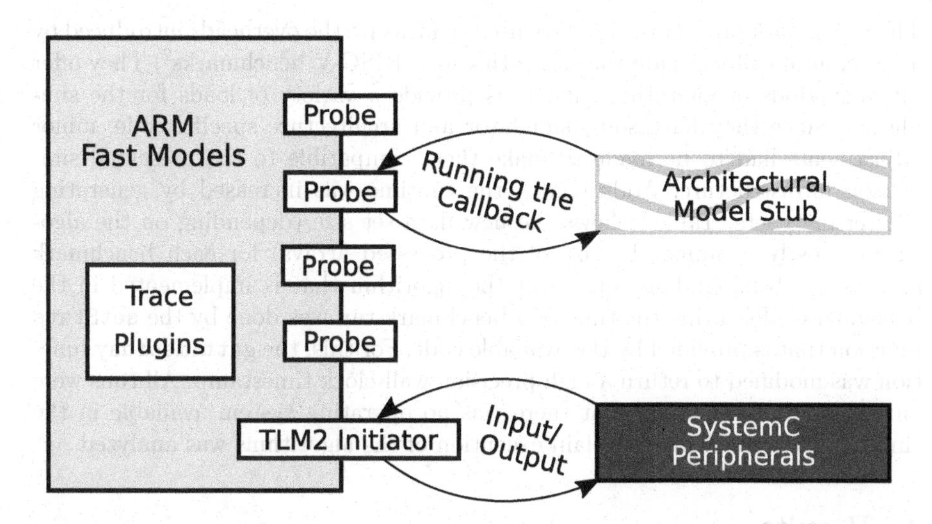

Fig. 3. Layout of the simulation framework used for this paper with the ARM Fast Models as the binary translator. The TLM2 initiator is offered by an additional bridge connected to the models. Again, the peripherals and the architectural model stub were added for the evaluation of this paper.

Additionally to the SystemC interface, the ARM Fast Models also support probes. They allow the registration of callbacks that are run on certain events during the simulation. For this paper, the `EndInstructionExecutionProbe` and

BeginBasicBlockProbe are of importance. The first one executes a callback every time the execution of an instruction finishes, while the second calls a callback as soon as a new basic block is entered. The interfaces use the observer pattern and the callbacks have to be implemented in C++. This allows driving the same architectural model stub as for QEMU. The overall layout of this simulation system is shown in Fig. 3. Usually, the implemented probes do not interact with the simulation except for gathering the information available via callbacks. This means that the information about nonfunctional properties is printed separately and is not returned to the emulation.

Another way to track the instructions is the usage of a trace plugin which can also be seen in Fig. 3. They are intended for creating custom tracing exports but can also be used for architectural models that are not implemented directly inside the ARM Fast Models. ARM delivers different plugins like a Tarmac Trace plugin and an example pipeline model. For this paper, the pipeline model was used to compare the measured callback overhead against the overhead of a full architectural simulation. All of the benchmarks were run with backdoor mode enabled meaning that the simulator uses the direct memory interface to access the memory.

3.3 Benchmarks

The benchmark programs that were used to measure the overheads introduced by architectural callbacks are the single-threaded RISC-V benchmarks[2]. They offer different kinds of algorithms and thus provide a variety of loads for the simulators. Since they have some simulator and architecture specific code, minor adjustments had to be made to make them compatible to the presented simulation infrastructure. Additionally, the runtime was increased by generating greater data sets. Table 1 shows the new data set size (depending on the algorithm, mostly meaning the size of the processed arrays) for each benchmark and gives a brief explanation about the algorithm that is implemented in the benchmark. Measuring the time of a benchmark run was done by the setStats function that is provided by the available code. For this, the gettimeofday function was modified to return a high precision wall-clock timestamp. All runs were bare-metal runs meaning that there was no operating system available in the simulated system. Only the plain execution of the algorithms was analyzed.

4 Results

Different overheads were analyzed for this paper. The baseline for comparison was measured by not having any callback executed during the simulation. This is the native speed of these simulators if just the emulation of the software is required. The first overhead that was measured is the one introduced by running a callback for all emulated instructions. As previously explained, the callback is

[2] https://github.com/riscv/riscv-tests/tree/master/benchmarks.

Table 1. Data size of benchmark programs

Benchmark	Description	Data Set Size
median	1D median filter	1048576
multiply	Inner product of two vectors	1048576
qsort	Quicksort (with insertion sort for small sizes)	1048576
rsort	Quicksort	1048576
spmv	Sparse matrix vector multiplication	R = 1024, C = 1024, NNZ = 65631
towers	Towers of Hanoi problem	20 discs
vvadd	Vector vector addition	1048576

only an architectural stub and does no actual modelling. This allows identifying the overhead of a simulator with additional instrumentation without also measuring the new functionality.

The second overhead that was investigated is based on callbacks that are only run for each translation block in case of QEMU or each basic block in case of the ARM Fast Models. This is a sensible optimization of the single-instruction callbacks since the instructions executed in all of these blocks are the same each time a certain block is emulated. A potential architectural model might profit from already having seen such a block before and might reuse older results from a previous call [16].

In the case of the ARM Fast Models, a third experiment for overhead measurement was performed. It includes a fully implemented pipeline model for a processor simulator. QEMU does not offer easily exchangeable architectural models but some related works are available as mentioned in Sect. 2. The findings of this paper are discussed and compared to the findings of these works.

All results presented in this section represent the average of at least 100 runs of the benchmark within the simulation. Before starting the simulation, the file system cache of the host operating system was warmed up with a few runs of the simulation. During the measurement no additional workload except from background processes was applied to the host system. The simulations took place on a standard desktop machine offering an Intel Core i5-7500 CPU. Results are given in the form of a relative runtime r that can be defined as seen in (1) with t_{base} being the measured time of the base simulation against which the comparison is made and t_{instr} being the measured time with additional instrumentation or pipeline models. This means that a relative runtime of one denotes no measurable overhead.

$$r = \frac{t_{instr}}{t_{base}} \tag{1}$$

4.1 QEMU

Figure 4 shows the results of the investigation in form of relative runtimes when QEMU runs the implemented callbacks. For each benchmark, two overheads, one for running a callback for each instruction and one running a callback for each basic block, are given. The relative runtime ranges between 5.88 and 1.19 depending on the callback frequency and the benchmark. On average, the relative runtime amounts to 2.76 for instruction-wise processing and 2.24 for block-wise processing.

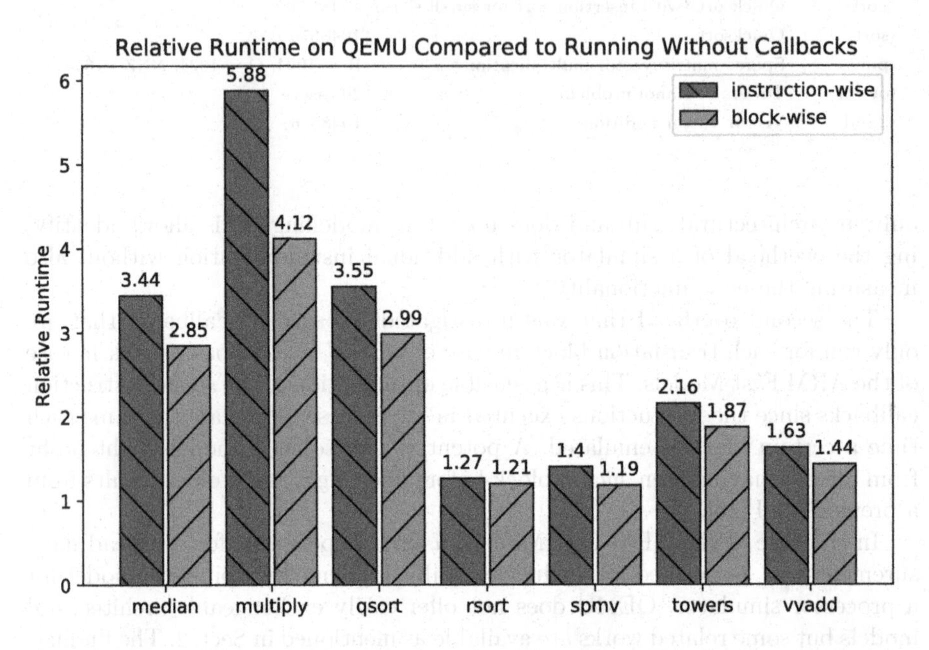

Fig. 4. The measured relative runtime for the different benchmarks when running on QEMU. Instruction-wise results were created by executing a callback for each emulated instruction, while block-wise results were created by executing a callback for each translation block.

The fluctuation between the different overheads of the benchmarks arises from the different instruction mixes of the programs. Memory-intensive applications that contain many load and store operations oftentimes have to call helper functions for performing the memory access task. Since this is already much overhead compared to arithmetic-intensive applications (like e.g. multiply), the additional overhead introduced by the callbacks to drive the architectural model is negligible. This can be seen for example by rsort and spmv which are very memory-intensive and thus have the lowest overhead. To prove this effect, two artificial benchmarks were created while one contains many more memory accesses than the other one. They showed exactly the previously described behavior.

Additionally, it can be seen in Fig. 4 that only executing one callback for each translation block reduces the overhead compared to running it for each instruction. This is due to the fact that less function calls have to be performed. Applications that have a higher overhead by instruction-wise instrumentation also benefit more from block-wise executed callbacks. If the instruction-wise callbacks are not very significant, running less of them can't result in a huge improvement. However, if the instruction-wise overhead is big, it is very noticeable when changing to block-wise callbacks.

Since there are no other architectural models available that can be directly measured, we will compare the findings against values found in publications. While for QSim no comparison against QEMU is given (only against sim-outorder), it is not suitable for giving an overhead as comparison [16]. TQSim on the other hand offers such a comparison from which we can conclude a relative runtime of 6.17 on average over all benchmarks [13]. However, since sampled instrumentation and statistical extrapolation is used, the overhead is expected to be less than having a callback executed for each instruction as evaluated in this paper. A more conventional approach is done by Miettinen et al. who extended QEMU with instruction classification, a cache and a TLB model [17]. When enabling all architectural aspects they implemented, the relative runtime in comparison to the simulation without instrumentation ranges from 27.25 to 29.33. With the findings of our paper, it can be said, that even with the fastest models, the average relative runtime will not be below 2.76 on average due to the added functions. Thus, the callback overhead is approximately 10% of the overall overhead introduced by the models of Miettinen et al. This means that one-tenth of the overhead originates from just transferring information into the architectural models. When optimizing the models, the interface will become responsible for a greater amount of overhead.

In QEMU, further optimizations to the overhead are possible. Instead of calling a helper function, the whole architectural model could be implemented using TCG intermediate code. This allows driving the model without any additional callback by integrating the code directly into the code cache. Additionally, it is possible that the created TCG code can be optimized by the optimization functionality of QEMU. However, this approach is very work intensive as it requires the model implementation to be written completely with a code that is similar to assembler languages. Additionally, it is most likely that an architectural model itself will call many further functions and thus cannot benefit much from this approach.

4.2 ARM Fast Models

The evaluation of the Fast Models showed a different behavior which can be seen in Fig. 5. Results are given as a relative runtime of executing a callback each instruction (instruction-wise) and executing a callback each basic block (block-wise) compared to running the simulation without any callback. With the exception of the SPMV benchmark, the overheads do not show significant differences ranging between 8.25 and 9.63. Additionally, the benefit from using block-wise instrumentation is nearly the same for all investigated algorithms. On average, the relative runtime amounts to 8.09 for instruction-wise and 4.45 for block-wise instrumentation.

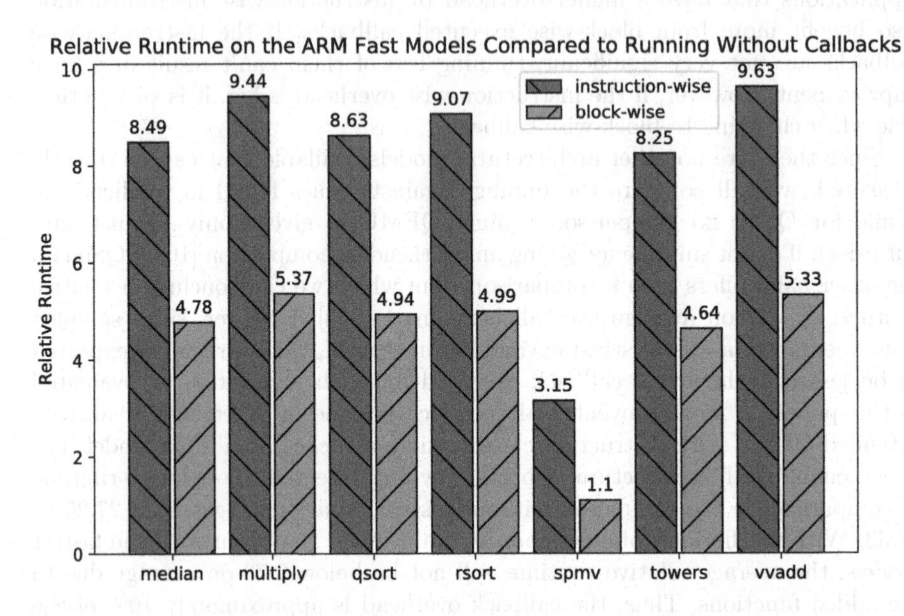

Fig. 5. The measured relative runtime for the different benchmarks when running on the ARM Fast Models. Instruction-wise results were created by executing a callback for each emulated instruction, while block-wise results were created by executing a callback for each basic block.

The particularity of the SPMV benchmarks is the result of the small overall runtime this benchmark exhibits. When increasing the problem size, the relative runtime rises near the other benchmarks. Since this behavior could not be observed for QEMU, an additional overhead at the beginning of the simulation when using the ARM Fast Models within Platform Architect is to be assumed.

Architectural plugins are also provided with the ARM Fast Models. This makes it possible to compare the previous results of the minimal expected overhead with a full pipeline model. Figure 6 shows the relative runtimes of the pipeline model in comparison with the previously measured instruction-wise runtime that can also be seen in Fig. 5. The average relative runtime of the pipeline model amounts to 65.71 meaning that the non-optimizable overhead introduced by calling additional functions amounts in average to approximately 12.31% of the simulation time.

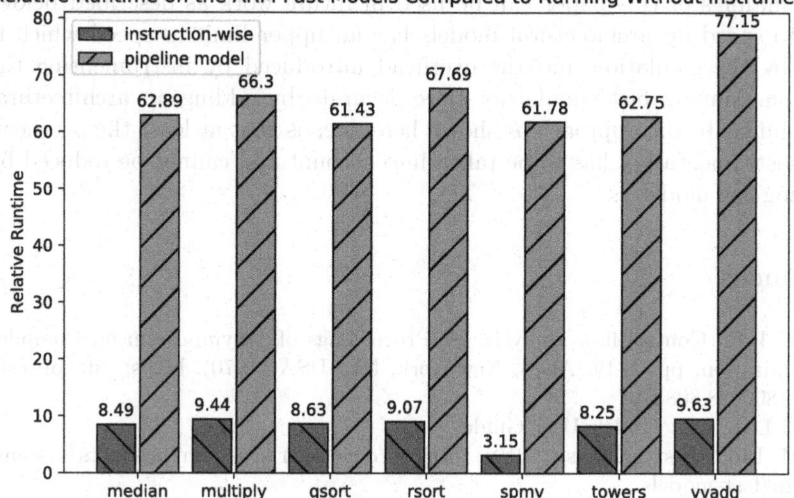

Fig. 6. The measured relative runtime for the different benchmarks when running on the ARM Fast Models. Instruction-wise results were created by executing a callback for each emulated instruction, while pipeline model results were created with the architectural pipeline model enabled.

5 Conclusion

Simulating processor cores and estimating the execution time on the final hardware is important for software development without having the final hardware available. However, since an accurate estimation requires a lot of time, a compromise between the precision and simulation speed must be found. When instrumenting a fast emulator, that uses JIT compilation techniques to speed up its execution, to drive an architectural model, additional overheads have to be taken into account that are not directly introduced by the model of the architecture. Calling anything from within the simulation without even doing anything also requires additional time. This overhead cannot be easily avoided.

In this paper, two emulators, an academic and a commercial one, were investigated regarding the introduced overhead when running callbacks during the simulation. On average, a relative runtime of 2.76 and 8.09 respectively was measured. These numbers are expected to be the minimal overhead introduced by architectural models that are driven by the emulation. Improvements can be achieved by using different approaches like driving a model only each block instead of each instruction. Here, relative runtimes of 2.24 and 4.45 could be measured which is less than the previously mentioned overhead. Thus, it could be shown that certain optimizations to this minimal overhead is still possible but require suitable methods for instrumentation. Finally, comparisons to the overhead of full architectural models were given. The investigated callbacks amount to approximately 10% to 12.31% of the overall runtime.

The findings of this paper can be used in future work as described below. Trying to speed up architectural models has an upper limit in speed which is defined by the emulation and the overhead introduced by instrumenting the simulation. Making fast simulators more accurate by adding an architectural model, similar to the approach as shown here, means that at least the overhead for the instrumentation has to be taken into account and cannot be reduced by optimizing the model.

References

1. Allen, F.E.: Control flow analysis. In: Proceedings of a Symposium on Compiler Optimization. pp. 1–19. ACM, New York, NY, USA (1970). https://doi.org/10.1145/800028.808479
2. ARM Ltd.: Fast Models User Guide
3. ARM Ltd.: Fast models (2019). https://developer.arm.com/products/system-design/fast-models
4. Bellard, F.: QEMU, a fast and portable dynamic translator. In: Proceedings of the Annual Conference on USENIX Annual Technical Conference (ATEC 2005), USENIX Association, Berkeley, CA, USA (2005)
5. Bringmann, O., et al.: The next generation of virtual prototyping: ultra-fast yet accurate simulation of hw/sw systems. In: 2015 Design, Automation Test in Europe Conference Exhibition (DATE), pp. 1698–1707 (March 2015). https://doi.org/10.7873/DATE.2015.1105
6. Burger, D., Austin, T.M.: The simplescalar tool set, version 2.0. SIGARCH Comput. Archit. News **25**(3), 13–25 (1997). https://doi.org/10.1145/268806.268810
7. Delbergue, G., Burton, M., Konrad, F., Le Gal, B., Jego, C.: QBox: an industrial solution for virtual platform simulation using QEMU and SystemC TLM-2.0. In: 8th European Congress on Embedded Real Time Software and Systems (ERTS 2016). TOULOUSE, France (Jan 2016). https://hal.archives-ouvertes.fr/hal-01292317
8. Ding, J., Chang, P., Hsu, W., Chung, Y.: PQEMU: A parallel system emulator based on QEMU. In: 2011 IEEE 17th International Conference on Parallel and Distributed Systems, pp. 276–283 (Dec 2011). https://doi.org/10.1109/ICPADS.2011.102
9. Guo, X., Mullins, R.: Accelerate cycle-level full-system simulation of multi-core risc-v systems with binary translation (2020)
10. Herglotz, C., Seiler, J., Kaup, A., Hendricks, A., Reichenbach, M., Fey, D.: Estimation of non-functional properties for embedded hardware with application to image processing. In: 2015 IEEE International Parallel and Distributed Processing Symposium Workshop, pp. 190–195 (May 2015). https://doi.org/10.1109/IPDPSW.2015.58
11. IEEE Computer Society: IEEE Standard for Standard System C Language Reference Manual, pp. 1666–2011. IEEE Standard (2012)
12. Imperas: Open virtual platforms (2017). http://www.ovpworld.org/
13. Kang, S.H., Yoo, D., Ha, S.: TQSIM: A fast cycle-approximate processor simulator based on QEMU. J. Syst. Archit. **66**, 33–47 (2016). https://doi.org/10.1016/j.sysarc.2016.04.012. http://www.sciencedirect.com/science/article/pii/S1383762116300297

14. Kerschbaumer, C., Hennigan, E., Larsen, P., Brunthaler, S., Franz, M.: Information flow tracking meets just-in-time compilation. ACM Trans. Archit. Code Optim. **10**(4), 38-1–38-25 (2013). https://doi.org/10.1145/2555289.2555295
15. Lee, J., Jang, H., Kim, J.: Rpstacks: fast and accurate processor design space exploration using representative stall-event stacks. In: 2014 47th Annual IEEE/ACM International Symposium on Microarchitecture, pp. 255–267 (Dec 2014). https://doi.org/10.1109/MICRO.2014.26
16. Luo, Y., Li, Y., Yuan, X., Yin, R.: QSIM: framework for cycle-accurate simulation on out-of-order processors based on QEMU. In: 2012 Second International Conference on Instrumentation, Measurement, Computer, Communication and Control, pp. 1010–1015 (Dec 2012). https://doi.org/10.1109/IMCCC.2012.397
17. Miettinen, A., Hirvisalo, V., Knuuttila, J.: Execution-driven simulation of nonfunctional properties of software. In: Proceedings of European Simulation and Modelling Conference (ESM) (2010)
18. Monton, M., Portero, A., Moreno, M., Martinez, B., Carrabina, J.: Mixed sw/systemc soc emulation framework. In: 2007 IEEE International Symposium on Industrial Electronics, pp. 2338–2341 (June 2007). https://doi.org/10.1109/ISIE.2007.4374971
19. Rosa, F., Ost, L., Reis, R., Sassatelli, G.: Instruction-driven timing CPU model for efficient embedded software development using OVP. In: 2013 IEEE 20th International Conference on Electronics, Circuits, and Systems (ICECS), pp. 855–858 (Dec 2013). https://doi.org/10.1109/ICECS.2013.6815549
20. Sandberg, A., Nikoleris, N., Carlson, T.E., Hagersten, E., Kaxiras, S., Black-Schaffer, D.: Full speed ahead: detailed architectural simulation at near-native speed. In: 2015 IEEE International Symposium on Workload Characterization, pp. 183–192 (Oct 2015). https://doi.org/10.1109/IISWC.2015.29
21. Schreiner, S., Görgen, R., Grüttner, K., Nebel, W.: A quasi-cycle accurate timing model for binary translation based instruction set simulators. In: 2016 International Conference on Embedded Computer Systems: Architectures, Modeling and Simulation (SAMOS), pp. 348–353 (July 2016). https://doi.org/10.1109/SAMOS.2016.7818371
22. Weaver, V.M., McKee, S.A.: Are cycle accurate simulations a waste of time. In: Proceedings of 7th Workshop on Duplicating, Deconstructing, and Debunking, pp. 40–53 (2008)
23. Xilinx: LibSystemCTLM-SoC (2019). https://github.com/Xilinx/libsystemctlm-soc

Multi-stage Network Recovery to Maximize the Observability of Smart Grids

Huibin Jia(✉), Qi Qi, Min Wang, Min Jia, and Yonghe Gai

North China Electric Power University, Baoding, China

Abstract. Large-scale natural disaster or malicious attacks could cause serious damages to communication networks in smart grids. If the damaged network cannot be recovered timely, greater threat will be brought to the secure and stable operation of smart grids. However, network recovery will take a lot of time, and the recovery process will involve multiple stages due to the limited recovery resources. In this paper, we address the problem of network recovery by selecting partial damaged links to be repaired in every stage so as to maximize the number of survival services. We formulate the problem as 0–1 programming, and then propose a heuristic algorithm to solve the problem. Extensive simulations are carried out on some topologies, the number of services, and the number of stages. The simulation results demonstrate that the heuristic algorithm is time-efficient and near-optimal.

Keywords: Smart grids · Multiple-stage · Network recovery · Large-scale Failures

1 Introduction

Communication network is the infrastructure for suring the safe and stable operation of smart grids. It is heavily depends to support the services of urgent protection and control in smart grids, especially in times of emergency. When large-scale failures are caused by nature disaster, blackout, and malicious, huge economic losses and serious social impacts will be brought, if the failures cannot be restored timely. Therefore, it is essential to recover the damaged communication network of smart grids, at least to the point where mission-critical services can be supported.

Much research on network recovery has been reported in telecommunication networks, which can be classified into single-stage [1] and multi-stage recovery [2–4]. In telecommunication networks, the objective of network recovery is to maximize the traffic of the total network [5]. The problem of efficiently restoring sufficient resources in communication networks has been addressed; An iterative split and prune algorithm, which is a polynomial-time heuristic, has been proposed [6]. The importance assessment method for damaged network components has been proposed in [6], the network components ware repaired according to their importance. A multiple stages network recovery

H. Song and D. Jiang (Eds.): SIMUtools 2020, LNICST 370, pp. 94–103, 2021.
https://doi.org/10.1007/978-3-030-72795-6_8

method was proposed in [7], and it can determine an appropriate recovery order of damaged network components and consider both prompt network service recovery and fair traffic recovery to meet traffic demand. The network recovery under uncertain knowledge of damages has been studied [8], and it was formulated as a mixed integer linear programming (MILP), which is NP-hard. An iterative stochastic recovery algorithm (ISR) has been proposed to recover the network in a progressive manner to satisfy the critical services.

However, communication networks of smart grids are different from telecommunication networks because they are mission-critical. In smart grids, the more services survive, the more components of power grids can be observed, and the more secure operation of smart grids can be guaranteed when large-scale failures occur. Therefore, the reliability requirement of communication networks is more complicated and important for smart grids. Most researches concentrate on network frameworks and multipath to improve network reliability against network failures. An end-to-end attack-resilient cyber-physical security framework for WAMPAC applications in power grids was introduced; and a defense-in-depth architecture has been described in [9]. An optimization-based restoration strategy was developed to coordinate the restoration actions between power grids and communication networks [10]. The problem was modelled as a mixed integer linear programming problem, with the objective of activating every node in both networks with the minimum number of activation/energization of branches. A self-healing phasor measurement unit (PMU) network has been investigated to exploit the features of dynamic and programmable configuration in a software-defined networking infrastructure to achieve resiliency against cyber-attacks [11].

In smart grids, much research has focused on network frameworks for large-scale failures, and the recovery method after the large-scale failure is seldom reported. Different from telecommunication networks, the traffic service is inseparable; and the much more services survive in the damaged network, the safer the operation of smart grids. Therefore, the objective of network recovery in smart grids is to maximize the surviving services so as to ensure the safe and stable operation of smart grids. In this paper, to maximize the surviving services after large-scale failures, a multiple-stage network recovery model has been established, and we decompose the multi-stage network recovery problem into multiple single-stage problems. A potential energy-based heuristic algorithm has been proposed to solve the single-stage network recovery problem. The damaged links are rationally arranged at different stages to maximize the number of damaged services in each stage, and to maximize the number of damaged services in the entire recovery process. The simulation results demonstrate that the algorithm has a strong applicability and it is also time-efficient.

The paper is organized as follows: the multi-stage network recovery in smart grids is introduced in Sect. 2; the problem of network recovery is formulated in Sect. 3; In Sect. 4, a heuristic algorithm is proposed; the simulation experiments are carried out and the time complexity is analyzed in Sect. 5; and the conclusion can be obtained in Sect. 6.

2 Multi-stage Network Recovery in Smart Grids

When a large-scale failure occurs in communication networks of smart grids, network recovery needs to be rapidly implemented. However, recovery resources are influenced by the time and geographical location, and network recovery cannot be fully carried out at the same time. During the entire recovery process, network resources need to be returned to the recovery site in batches. Therefore, network recovery can be carried out in multiple stages. In the multi-stage network recovery, the early recovery stages will affect the latter recovery work. The remaining recovery resources in the early stages can also be utilized in the latter recovery process. Also, the pre-repaired links may be used in the latter recovery process. Part of the recovery network recovered in the early stages can still provide services for the latter stages. Therefore, when the total recovery process is planned, recovery resources are first allocated at each stage, and then the damaged network components are determined to be recovered. Finally, the entire recovery process can meet the requirements of smart grids as quickly as possible.

In the multi-stage recovery process of the power communication network, different strategies will lead to different recovery results. Considering the example in Fig. 1, there are 4 damaged links and 7 damaged services. In order to illustrate the problem, assuming that the recovery resources provided in each stage can only repair one damaged link. There are four different recovery sequences to recover the network. Different recovery sequences are shown in Fig. 2. The link (1, 3), link (2, 8), link (2, 7), and link (5, 11) are listed in Fig. 2 (a). The link (2, 7), link (1, 3), link (2, 8) and link (5, 11) are given in Fig. 2 (b). The link (2, 8), link (1, 3), link (2, 7), and link (5, 11) are provided in Fig. 2 (c). The link (5, 11), link (1, 3), link (2, 7), and link (2, 8) are described in Fig. 2 (d).

In Fig. 2 (a) and Fig. 2 (c), i.e. in the first and second stages, more damaged services have been restored; in the third and fourth stages as shown in Fig. 2 (b) and (d), the damaged services increase. In order to recover more damaged services timely, the strategy is preferred. However, after the first stage, the number of restored services are 3 and 2 as shown in Fig. 2 (a) and Fig. 2 (b), respectively. The recovery sequence as shown in Fig. 2 (a) is most appropriate. Therefore, the different strategies will result in different recovery effects for the communication network of smart grids.

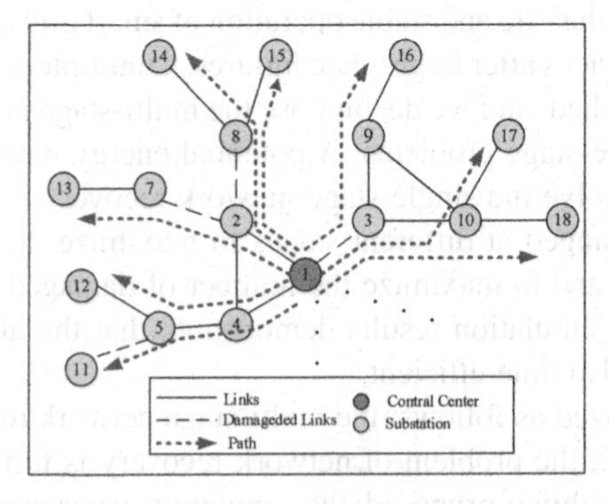

Fig. 1. The example of network recovery in smart grids

It can be seen from Fig. 2, it is an urgent problem to recover as many services as possible timely after large-scale failures of power communication networks. To solve this problem, a multi-stage mathematical model for network recovery is established to

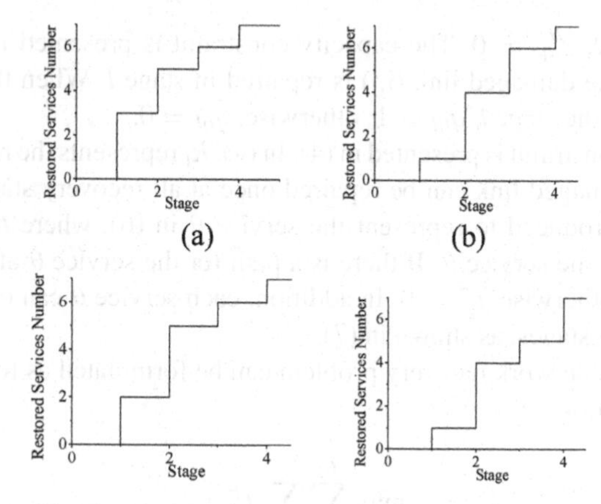

Fig. 2. The results of different recovery sequences

determine which damaged links should be recovered in each stage, so as to maximize the number of surviving services during the entire recovery process.

3 Problem Formulation

A communication network plays an important role in smart grids, and it is responsible for exchanging information between substations and regional control centers. The scale of the communication network is particularly large, sometimes there are more than 100 nodes in the communication network of smart grids. In addition, the traffic of the service cannot be split and the amount of the traffic is integer-valued. The communication services in smart grids are frequently converged to the control center.

In this paper, our assumptions under large-scale failures scenarios are as follows:

1. At least 10% of physical links are damaged;
2. The damaged links will result in the disruption of services;
3. Repairing damaged links takes resources which consist of time and human labors, and the supply of resources is staged.

We model a communication network in smart grids as an undirected graph $G(V, E)$ where V is the set of nodes and E is the set of links. The communication network in smart grids is regarded as an integer multi-commodity supply graph. Θ is the set of services. Each service can be represented by θ represents a service with the source node $s(\theta)$ and the destination node $t(\theta)$. Each link (i, j) belongs to E with the capacity C_{ij}. In order to describe the damaged network, we define E^D as the set of damaged links and E^{DR} as the set of recovered links. Let us define $f(\theta)$ as the flag, which represents if there is a path for the service θ in the graph G. If the path does not exist, $\theta = 0$; otherwise, $\theta = 1$. The objective function is to maximize the number of surviving services in the damaged network.

The flow balance constraint is presented in (2), where x_{ijk}^{θ} represents if the services θ traverses the link (i,j) at the stage k, going from the node i to the node j. If it functions,

$x_{ijk}^{\theta} = 1$; otherwise, $x_{ijk}^{\theta} = 0$. The capacity constraint is presented in (3), where q_{ijl} represents that if the damaged link (i,j) is repaired in stage l. When the damaged link (i,j) is recovered at the stage l, $q_{ijl} = 1$; otherwise, $q_{ijl} = 0$.

The recovery constraint is presented in (4). In (4), R_l represents the recovery resource of stage l. Each damaged link can be repaired once at all recovery stages, as shown in (5). Also, f_k^{θ} is introduced to represent the service θ in (6), where $t(\theta)$ refers to the destination node of the service θ. If there is a path for the service θ after k stages, and $j = t(\theta)$, $f_k^{\theta} = 1$; otherwise, $f_k^{\theta} = 0$. In addition, each service θ can only find the path once at all recovery stages, as shown in (7).

The multi-stage network recovery problem can be formulated as follows:

Objective Function:

$$\max \sum_{k=1}^{K} \sum_{\theta \in \Theta} f_k^{\theta} \tag{1}$$

S.t.:

$$\sum_{(i,j)\in E} x_{ijk}^{\theta} - \sum_{(j,i)\in E} x_{jik}^{\theta} = \begin{cases} 1 & i = s(\theta) \\ -1 & i = t(\theta) \\ 0 & others \end{cases} \tag{2}$$

$$\sum_{\theta \in \Theta} x_{ijk}^{\theta} b^{\theta} \le s_{ij} + (C_{ij} - s_{ij}) \sum_{l=1}^{k} q_{ijl} \tag{3}$$

$$\sum_{l=1}^{k} \sum_{(i,j)\in L} q_{ijl} r_{ij} \le \sum_{l=1}^{k} R_l \tag{4}$$

$$\sum_{k=1}^{K} q_{ijk} \le 1 \tag{5}$$

$$f_k^{\theta} = x_{ijk}^{\theta} \qquad j = t(\theta) \tag{6}$$

$$\sum_{k=1}^{K} f_k^{\theta} \le 1 \tag{7}$$

4 The Proposed Algorithm

The problem described above is 0–1 programming, which is NP-hard. The computation time will increase exponentially as the number of damaged links increase, especially when the extent of failures is larger. Therefore, an(a) approximation or heuristic algorithm should be proposed to solve the optimization problem.

In this paper, we proposed an algorithm based on potential energy. First, the multi-stage recovery problem is decomposed into k single-stage recovery problems with the same objective which is to maximize the number of surviving services in the damaged network. Second, the single recovery problem can be solved one by one to recover the damaged links at each stage, and a local optimum result can be obtained for the multi-stage recovery problem. Finally, all local optimum results are combined as the approximation optimum result for the multi-stage recovery problem.

Here a heuristic algorithm is proposed to solve the single recovery problem considering the characteristic of smart grids. First, in order to accurately measure the damaged links, the concept of potential energy from electromagnetic theory is adopted. The recovery resources of damaged links are taken into account to reduce or avoid the effect of only considering the distance in the algorithm. Therefore, the potential energy of a damaged link is defined in (8), where len_{ij} represents the distance from the damaged link (i,j) to the regional control center, and r_{ij} refers to the resource of repairing the damaged link (i, j). The potential energy of the damaged link can be described as the priority. The damaged links, which have the greater potential energy, will be repaired first.

$$E_{ij} = \frac{1}{len_{ij} * r_{ij}} \tag{8}$$

Algorithm:

Input: Supply graph G, the services set Θ, the set of restored service Θ_1, the set of damaged links L, the set of repaired links for each stage L_k;

1 Decompose the multi-stage recovery problem into K single-stage recovery problems
2 Calculate the potential energy E_{ij} of all damaged links;
3 Sort the damaged links in descending order by E_{ij};
4 while $k \le K$ do
5   while $L \ne \phi$ and $R_k \ge 0$ do
6   Move the damaged link (i,j) with the maximum of E_{ij} from L to L_k;
7         $R_k = R_k - r_{ij}$
8   k=k+1;
9 Repair the damaged links in L to L_K; and update G
10 for $\theta \in \Theta$ do
11   Find the path of θ by using Dijkstra' algorithm;
12   if find the path then
13   Move θ from Θ to Θ_1;
14   Update G;
15 Calculate the objective function according to Θ_1;

In the algorithm, len_{ij} can be obtained by using Dijkstra's algorithm, and the damaged links (i,j) will be sorted in descending order by E_{ij}. At each stage, the damaged links are selected one by one, and the algorithm will terminate when there is no stage or damaged links left. The number of restored services can be obtained by calculating the shortest path of θ using Dijkstra's algorithm.

5 Computational Experiments

In order to verify the performance of the proposed algorithm under the conditions of different numbers of stages and services, two experiments are designed. The optimal result is obtained according to the enumeration algorithm [12, 13]. In addition, the two experiments are implemented by using C++. In the two experiments, first, an undirected network with n nodes is designed. The capacity of each link C_{ij} randomly distributes on [1, 40]; after large-scale failures, the capacity of damaged links decreases to s_{ij}, which randomly distributes on $[0, C_{ij}]$. The resources of repairing the damaged link (i,j), r_{ij}, distributes on $[0, C_{ij} - S_{ij}]$. Moreover, the number of services in the undirected network is m, and the bandwidth requirement for each service distributes on [1, 4] unit(s) at random. The number of recovery stages is K, and resources of each stage randomly distributes on $mR/n(2*K)$, $2mR/nK$, where R represents the available resources. There are two experiments to be designed to validate the performance of the proposed heuristic algorithm under the different number of recovery stages and services in the network.

5.1 Different Network Experiment Results

In the first experiment, the communication network with 50 nodes is designed and the number of services are {50, 60, 70, 80, 90, 100}, respectively; after the large-scale failure, the number of damaged services are {28, 30, 36, 40, 44, 50}, respectively. The recovery stages, K, are {3, 5, 7}. The recovery results are shown in Fig. 3.

It can be seen from the Fig. 3 that the gap between the optimal and the recovery result of different stages is becoming smaller with the increase of the number services. Especially the higher the recovery stages, the more the number of recovered services, and the better the performance of the proposed algorithm has.

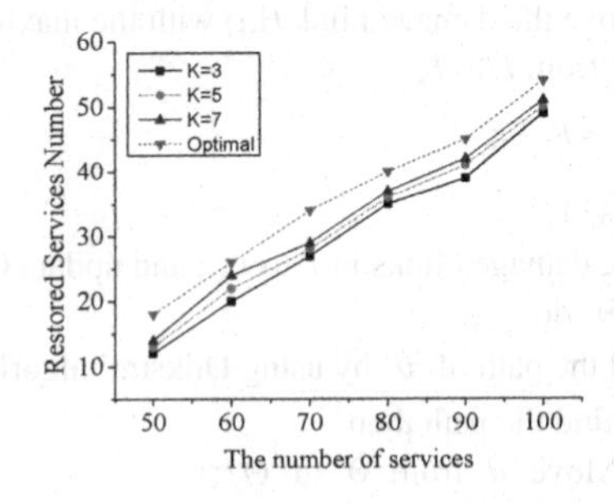

Fig. 3. The recovery results in the network with 50 nodes

In the second experiment, the communication network with 100 nodes is designed and the number of services are {100, 110, 120, 130, 140, 150}, respectively; after large-scale failures, the number of damaged services are {42, 48, 50, 56, 60, 63}, respectively. The number of stages, K, are {3, 5, 7}. The recovery results are shown in Fig. 4.

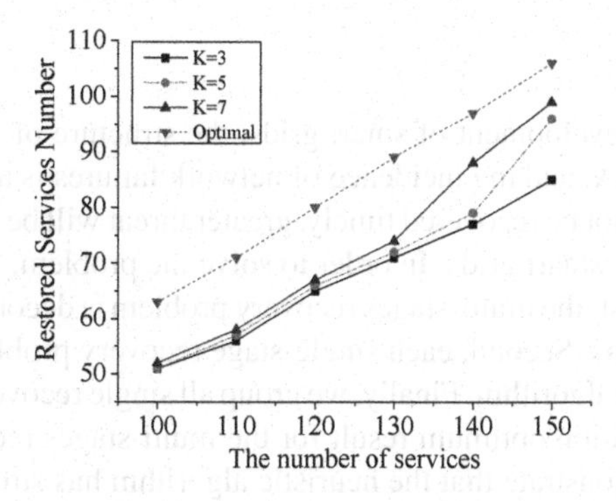

Fig. 4. The recovery results in the network with 100 nodes

It can be seen from Fig. 4 that the gap between the optimal and the proposed algorithm is becoming smaller with the increase of the number of services. At the same time, the performance is very close under the scenios of different number of recovery stages when the number of services is small.

5.2 Time Complexity

In the algorithm proposed here, the time complexity mainly depends on the step 2 and 11. In the step 2, Dijkstra's algorithm is used to calculate the potential energies of all damaged links [14]. The time complexity of the algorithm is $O(nlg(n))$, where n is the number of nodes in the network [15]. In the step 2, assume that the number of damaged links is S, so the time complexity of the step 2 is $S * O(nlg(n))$. In the step 11, the route search is still based on Dijkstra's algorithm, which is similar to the time complexity of step 2. Therefore, the time complexity of the step 11 is $M * O(nlg(n))$, where M is the number of damaged services. In summary, the time complexity of the proposed algorithm is $(2M + S) * O(nlg(n))$.

The optimal solution is obtained according to the enumeration algorithm. In the enumeration algorithm, assume that the communication network with a number of damaged components is S, the number of combinations is at most 2^S, so the time complexity is $O(2^S)$. In a word, the time complexity of the two algorithms is as follows (Table 1).

Table 1. Time complexity of algorithms

Algorithm	The Proposed Algorithm	The Enumeration algorithm
Time complexity	$(2M + S) * O(nlg(n))$	$O(2^S)$

It can be seen from the description above, that the time complexity of the proposed algorithm is linear, while the enumeration algorithm is exponential. For large-scale failures, when both S and n are large, and the enumeration algorithm has high computational complexity. Therefore, in terms of time complexity, the proposed algorithm is better.

6 Conclusion

With the continuous development of smart grids, the structure of smart grids becomes more and more complex, and the incidence of network failures is also increasing. If the damaged network cannot be recovered timely, greater threat will be brought to the secure and stable operation of smart grids. In order to solve the problem, a heuristic algorithm has been proposed. First, the multi-stages recovery problem is decomposed into k single-stage recovery problems. Second, each single-stage recovery problem is solved using a potential energy-based algorithm. Finally, we group all single recovery results and regard them as the approximation optimum result for the multi-stages recovery problem. The simulation results demonstrate that the heuristic algorithm has strong applicability and time-efficient. However, according to the experimental results, we can find that there is a gap between the heuristic algorithm and the optimal value. In the future work, a better combination optimization method should be found, so that the best recovery effect can be reached as soon as possible after the large-scale failures.

References

1. Pedreno-Manresa, J., Izquierdo-Zaragoza, J., Pavon-Marino, P.: Guaranteeing traffic survivability and latency awareness in multilayer network design. IEEE/OSA J. Opt. Commun. Networking **9**(3), B43–B53 (2017). https://doi.org/10.1364/JOCN.9.000B53
2. Bartolini, N., Ciavarella, S., La Porta, T.F., et al.: Network recovery after massive failures. Depend. Syst. Netw. 97–108 (2016)
3. Srinivasan, S.K., Hamid, S.: Multi-stage manufacturing/re-manufacturing facility location and allocation model under uncertain demand and return. Int. J. Adv. Manuf. Technol. **94**(5/8), 2847–2860 (2018)
4. De Silva, V.R.S., Ranjith, P.G.: Intermittent and multi-stage fracture stimulation to optimise fracture propagation around a single injection well for enhanced in-situ leaching applications. Eng. Fracture Mech. **220** (2019)
5. Tootaghaj, D.Z., La Porta, T., He, T.: Modeling, monitoring and scheduling techniques for network recovery from massive failures. In: 2019 IFIP/IEEE Symposium on Integrated Network and Service Management (IM), Arlington, VA, USA, pp. 695–700 (2019)
6. Bartolini, N., et al.: On critical service recovery after massive network failures. IEEE/ACM Trans. Netw. 1–15 (2017)
7. Genda. Effective network recovery with higher priority to service recovery after a large-scale failure. In: 2017 23rd Asia-Pacific Conference on Communications (APCC), Perth, WA, pp. 1–6 (2017)
8. Genda, K., Kamamura, S.: Multi-stage network recovery considering traffic demand after a large-scale failure. In: 2016 IEEE International Conference on Communications (ICC), Kuala Lumpur, pp. 1–6 (2016)
9. Zad Tootaghaj, D., Bartolini, N., Khamfroush, H., La Porta, T.: On progressive network recovery from massive failures under uncertainty. IEEE Trans. Netw. Serv. Manag. **16**(1), 113–126 (2019)
10. Jovanov, I., Pajic , M.: Relaxing integrity requirements for attack-resilient cyber-physical systems. IEEE Trans. Autom. Control **64**(12), 4843–4858 (2019). https://doi.org/10.1109/TAC.2019.2898510
11. Baidya, P.M., Sun, W.: Effective restoration strategies of interdependent power system and communication network. J. Eng. **13**(2017), 1760–1764 (2017)

12. Lin, H., et al.: Self-healing attack-resilient PMU network for power system operation. IEEE Trans. Smart Grid **9**(3), 1551–1565 (2018)
13. Amarilli, A., et al.: Enumeration on Trees with Tractable Combined Complexity and Efficient Updates (2018)
14. Gao, X., et al.: Reachability for airline networks: fast algorithm for shortest path problem with time windows. Theoret. Comput. Sci. S030439751 830063X (2018)
15. Gharib, M., Yousefizadeh, H., Movaghar, A.: Secure overlay routing for large scale networks. IEEE Trans. Netw. Sci. Eng. 1 (2018)

Investigation of Containerized-IoT Implementation Based on Microservices

Jayanshu Gundaniya and Chung-Horng Lung(✉)

Department of Systems and Computer Engineering, Carleton University, 1125 Colonel By Drive, Ottawa, ON K1S 5B6, Canada

jayanshugundaniya@carleton.ca, chlung@sce.carleton.ca

Abstract. Due to a rapid increase in potential scales of Internet of Things (IoT) systems, it becomes challenging to deploy, patch and upgrade applications effectively. Breaking these applications down into small and independent microservices and isolating them by using container techniques leads to faster development, deployment and integration as compared to the traditional monolithic development approach. This paper introduces one such experimental implementation of simulated generic IoT services deployed separately as Docker containers on different machines using an open-source container orchestration platform Docker Swarm. Such a distributed implementation enables easier deployment and expansion of the system by adding and modifying only a few services in need. The objective of the approach is to conduct rapid prototyping of container-based simulated IoT microservices and investigate the management of multiple containers using open-source Docker Swarm. We validated the concept of developing and integrating containerized IoT microservices and evaluated the effectiveness of the concept. The experimental result shows that container-based IoT applications can be efficiently developed as microservices and multiple IoT containers can be effectively managed with an orchestrator. The experience can be applied to real IoT applications using actual IoT sensors and devices.

Keywords: Internet of Things · Microservices · Containers · Docker · Docker Swarm

1 Introduction

Internet of Things (IoT), one of the most popular topics of research these days, is constantly being updated with new features. All these features lead to an increase in its scale and thus we need an architecture and design that enables us to dynamically add these new features and scale them according to the needs without undergoing a system redesign at each stage. Moreover, according to predictions conducted by Gartner, the total number of connected IoT devices would reach about 21 billion in 2020 [1]. Deployment of large IoT applications is still in the beta stage. As the number of devices is increasing, a decentralized approach is needed for solving issues regarding the communication and management of devices. To make devices and services work independently would

H. Song and D. Jiang (Eds.): SIMUtools 2020, LNICST 370, pp. 104–115, 2021.
https://doi.org/10.1007/978-3-030-72795-6_9

involve decoupling them in development and deployment. Service Oriented Architecture (SOA) is one of the widely used solutions for achieving loose coupling between services, but it lacks support for the constant patching and integration of new services.

Though the concept of microservices has been in use for years, the term "microservices" was coined in 2014 [2]. In microservices, applications consist of small and independent services that communicate using lightweight methods like HTTP API [3]. The authors in [3] have explained it as: "In short, the microservice architectural style is an approach to developing a single application as a suite of small services, each running in its own process and communicating with lightweight mechanisms, often an HTTP resource API. These services are built around business capabilities and independently deployable by fully automated deployment machinery. There is a bare minimum of centralized management of these services, which may be written in different programming languages and use different data storage technologies". This allows individual services to be tested or patched independently without disturbing the rest of the deployment. Microservices Architecture is already in use heavily by companies like Netflix and Amazon, and it is attracting more attentions in both academia and industry.

These services, if running on the same machine, need to be isolated from each other to separate the environment variables used by a service from being modified by other services. This is usually realized by containerizing those services. Application containerization is an OS-level virtualization method [14]. It can be used to run distributed applications without the overhead of launching and running a fresh operating system as an entire virtual machine (VM) for each app. Multiple isolated applications can be deployed on a single host and can utilize the API of the same OS kernel. The invention of an open-source containerizing engine called Docker [15] has made containerizing easy and popular as Docker containers are designed to run on everything from physical computers to virtual machines, bare-metal servers, and more.

Microservices and distributed applications go together with containerization, as each container operates independently of others. Microservices communicate via APIs with the container virtualization layer. Microservices could be scaled up to distribute the load and meet the rising demand for an application component, thus making the system flexible.

The main objective of this paper is to experimentally validate the effectiveness of applying microservices using containers for IoT systems which have become popular and in high demand. Specifically, this paper investigates an approach consisting of simulated IoT applications bundled as microservices designed and implemented as containers. The main evaluation criteria include the deployment of containerized-IoT applications, management of multiple containers, and extensibility of the system for adding more sensors and devices. This approach is deployed on multiple machines by using an orchestration platform Docker Swarm [16]. The experimental result demonstrates that the simulated IoT application can be effectively integrated using microservices for continuous development with container techniques.

The rest of the paper is organized as follows. Section 2 describes the background information and related work. Section 3 presents the design and implementation of containerized microservices. Section 4 describes the conclusion and future directions.

2 Background and Related Work

This section explains the research performed on the application of microservices in IoT. A great deal of research has been conducted to integrate the application design areas since microservices was coined. Key issues and design choices like orchestration, communication protocols, and microservices are the primary areas of research.

2.1 Internet of Things (IoT)

IoT is a term used for connected computing devices which may be mechanical or digital systems that communicate by transferring data over a local or global network with little or no human-to-human or human-to-computer interactions. It can be an automobile with proximity sensors, a volume measurement unit in an industry, etc.

An IoT system could consist of numerous devices that have embedded processors coupled with intelligent sensors. The sensors collect data, communicate with a gateway for data transfer and the gateway can also act on data to achieve an objective. These devices may even communicate with other related devices and act on the data they get from one another.

The deployment of IoT applications mostly is still in the development stage. Prediction released by Gartner in 2018 [1] indicates that almost half the cost of implementing IoT solutions would consist of efforts spent on integrating them with one another and the backend. Other challenges may include concurrent operations and communications of the connected devices and their security on the network.

2.2 Microservices and Containerization

The microservice architecture allows for the continuous delivery/deployment of large-scale applications having high complexity levels. It is a unique approach for developing a large single application as a collection of small independent services, each running separately and communicating with one another using lightweight mechanisms, such as a REST API [17]. A REST (Representational State Transfer) API (application program interface) defines a way in which a developer can design an application (client) to request services from a server. It usually uses HTTP GET, POST, PUT and DELETE requests for communications. A large application is broken down into small services based on their business capabilities. Their deployment could also be automated like the original monolith. Centralized management of these services is kept at a minimum to avoid tight coupling. Alshuqayran et al. [4] presented a mapping study on the research conducted for the challenges and quality requirements of microservices.

One of the ways these small independent services are deployed is to use containers [14]. A container packs the application/service under its consideration with all dependencies with it, thus making it deployable on any environment running a container engine. Although containers are memory efficient, portable, and scalable, some of its disadvantages include the inability of security monitoring tools to protect the containers as most of them are designed for hypervisors, virtual machines (VMs), and natively run OSes.

2.3 Microservices and IoT

Butzin et al. [2] introduced a microservices approach to IoT applications. They mentioned the importance of the Circuit-Breaker pattern [5] which will monitor the health of services and prevent a broken service from receiving calls. Another important concept they discussed that goes well with the Circuit-Breaker is the Load Balancer pattern [6] that would distribute workload amongst equal services and works serially with the Circuit-Breaker. Petrenko et al. [7] described the semantic workflows that are useful in IoT and discussed an orchestration approach to dynamic service-oriented systems.

Campeanu [8] published a mapping study on the progress of microservice architecture in IoT. The paper states that "the number of publications in 2016 is four times larger than to the previous year. Moreover, the number of publications in 2017 is higher (i.e., 62%) than the previous year." Datta et al. [9] provided an end-to-end skeleton architecture for deployment of microservices suitable for IoT systems both on an edge server and on cloud. Khazaei et al. [10] presented a self-managing autonomic containerized IoT platform called SAVI-IoT for various use cases, such as Big data compatibility, in-place data processing, etc. Khanda et al. [11] also investigate implementing IoT as microservices in university buildings by using Jolie, a programming language built for microservices.

2.4 Docker Containers

Docker containers have been widely used in industries these days [15]. Docker is an open source containerizing engine designed to deploy and run applications. Docker allows the developer to isolate the applications from the infrastructure or system it is being deployed on for faster delivery of the software.

Docker isolates the environment in which a container runs, thus providing an ability to deploy it on any host OS which has a Docker Engine. Each container also packs the required dependencies required by an application. Thus, multiple containers having different dependencies can separately run on a single host machine without the overhead of using a hypervisor or a VM manager. Some key features of Docker containers are highlighted as follows:

- Services: A Docker service is a bunch of containers in production. All those containers run on one image but could have different ports exposing them. A service can contain multiple replicas of a single container depending on the load required. This number could be increased or reduced, thus leading to dynamic scaling of the system.
- Swarm: Docker swarm [16] consists of a swarm manager employing various strategies to run containers on multiple nodes. Strategies such as the emptiest node, global container, etc., could be used. Some crucial features available from Docker Swarm include service discovery, load balancing, multi-host networking, etc. [2].
- Stacks: Docker stack consists of a group of interrelated services that share the same dependencies. In production, these could be orchestrated and scaled together. In this paper, only one stack is used but complex applications usually require multiple stacks.

3 Design and Implementation

In this paper, the usage of microservice architecture for a generic IoT system is analyzed. Our system architecture and design are inspired by [10] but the implementation is primarily written in JavaScript with the use of containers for isolation. We briefly discuss the design and then describe the use cases that suit such a development.

The deployment of containers in our approach was carried out using Docker on a machine running Ubuntu 16.04 with 4 GB RAM and Pentium 4 processor.

3.1 System of Containers

The applications used in this implementation are loosely based on a workshop conducted by Nearform [13]. Our system involves services like frontend, database, serializer, broker, and simulated IoT applications. The services were written in JavaScript with Nodejs runtime. The experiment carried out in this paper preserves the implementation of some of those services and adds some more to it. Moreover, we use Docker Swarm and deploy containers on multiple machines with the help of its orchestrator instead of the regular implementation as standalone Docker containers on a single machine. A generic model of the containers used in this paper and their communications is shown in Fig. 1.

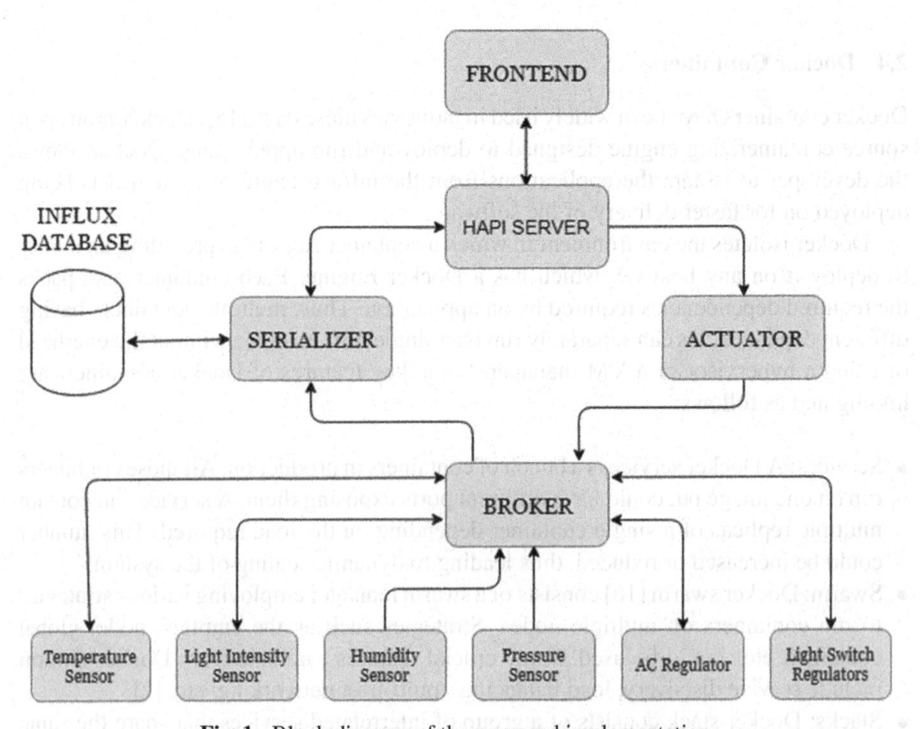

Fig. 1. Block diagram of the proposed implementation

A brief description of the containers shown in Fig. 1 is presented as follows:

- *Frontend*: This is the user interface (UI) of the system. It is a Hapi server to poll the *Database* for retrieval of values that are to be displayed.
- Serialize*r*: It writes to the *Influx Database* using the Nodejs APIs and it receives the poll from the *Frontend* and commands from the *Broker*.
- *Influx Database*: The database is a container implementation, which is a time-series database.
- *Broker*: In this implementation, the message broker is implemented using Mosca, which is an MQTT broker written in JavaScript. It receives messages on topics and publishes them to subscribers.
- *Actuator*: The component acts as an intermediary between the *Frontend* and the *Broker*. The *Broker* has a specific API used to communicate with the *Actuator*. The *Actuator* takes the overhead off the *Frontend* from calling the API.
- *Temperature, Humidity, Pressure and Light sensors*: Each of these IoT sensors randomly generates a value for its respective category every few seconds and sends it as the {topic, value} pair to the *Broker*.
- *AC and Light Regulators*: Both are subscribed to the *Broker* for receiving messages on appropriate topics and turn devices, e.g., AC or light, on or off depending on the received values and current threshold values via the *Broker.*

The *Frontend* uses the Hapi server, which uses a polling service. The polling service polls the *Influx Database* periodically through the *Serializer* using Seneca actions. The *Frontend* is also connected to the *Actuator* and calls its actions via Seneca, which are then converted to appropriate messages by the *Actuator* and passed to the *Broker* which will publish it for the subscribers to read. Every read or write or database creation done by the database is carried out by the *Serializer*. The *Broker*, on receipt of a 'write' message from any of the sensors, triggers the *Serializer* using Seneca APIs and passes the values to be written into the database. The sensors are all connected to the *Broker* and subscribed to relevant topics except humidity and pressure sensors as they neither receive offset values nor commands.

3.2 Use Cases

We adopt some of key use cases presented in [11, 12] in our evaluation of the architecture and design. There are:

- *Measurement of conditions in a smart home*: Temperature and light measurements in rooms are measured. This is a basic and yet critical requirement for many IoT applications, as new devices or sensors may need to be added or replaced. The idea can be easily applied to other similar types of sensors, e.g., humidity, fire alarm, etc.
- *Management of home appliances*: Applications for AC and Light Regulators have been tested in the experiment. Other appliances having configurable interfaces can also be more easily added in our implementation than in a usual monolithic approach, as only a small number of relevant containers may be related.
- *Service replicas*: A service can contain multiple replicas of a single container depending on configuration and the number of replicas can be changed dynamically. This feature is supported by Docker, which can improve system performance by deploying

more instances and can increase reliability in case of a failure of one instance. The failover will be handled automatically by Docker Swam.

3.3 Description of Workflow

Figure 2 shows the administrator UI of the *InfluxDB* container which is exposed on a port, i.e., 8083, (administrator interface port for the container implementation) of localhost (or any other IP of a connected swarm machine). The top right corner as depicted in Fig. 2 provides an option to select the database. In this experiment, five databases are used: temperature, light, pressure, humidity, and status. The first four databases are for storing values from their respective sensors, whereas the last one is for storing the status of AC or Lights as provided by their containers. Figure 2 displays some timed temperature values (random values in a range for illustration) obtained from sensor 1. *InfluxQL* is a SQL-like query language for interacting with *InfluxDB* and providing features specific for storing and analyzing time series data [7].

InfluxDB Write Data Documentation Database: temperature

Query: select * from temperature order by desc

Generate Query URL Query Templates

temperature

time	sensorId	temperature
2019-04-22T21:59:48.728Z	"1"	554
2019-04-22T21:59:46.734Z	"1"	514
2019-04-22T21:59:44.74Z	"1"	534
2019-04-22T21:59:42.728Z	"1"	361
2019-04-22T21:59:40.738Z	"1"	417
2019-04-22T21:59:38.726Z	"1"	534
2019-04-22T21:59:36.727Z	"1"	372
2019-04-22T21:59:34.73Z	"1"	621
2019-04-22T21:59:32.721Z	"1"	514

Fig. 2. Screenshot of an *InfluxQL* query run on administrator UI of *InfluxDB*

The top five components, as shown in Fig. 1, are designed using containers. They are all functioning in an orchestrated manner. When sensor values are to be plotted on the graph, the *Frontend* will call appropriate actions on the *Serializer* which will retrieve the values from the *InfluxDB* and provide them to the *Frontend*. A similar workflow occurs when a command from the *Frontend* needs to be sent to a sensor or regulator. The *Frontend* will first call actions from the *Actuator* which will then act on behalf of the *Frontend* to convert commands into appropriate messages for the *Broker*.

When a sensor value needs to be written to the *InfluxDB*, the sensor sends the value to the *Broker* as a message and the *Broker* will then call appropriate actions from the *Serializer* to write to *InfluxDB*.

The communications between sensors and *Actuator* are supported by message passing via the *Broker* and they work independently. On the other hand, *Frontend* needs to call appropriate actions from *Serializer* or *Actuator* to get its work done. Thus, when a new sensor is added (as a container), it is easy for it to subscribe to messages from other sensors or the *Actuator* and start working as it should.

Figure 3 shows the output of various sensors, i.e., temperature, humidity, pressure, and light being logged on the *Frontend* using the rickshaw charting framework. The *Frontend* pulls the output data values registered by the sensors in the past 20 min from the database via a Seneca [18] action by calling appropriate functions in the *Serializer*. Seneca supports microservice system by using a messaging mechanism based on pattern matching. It hides where services are located, how many of services there are, or what services do. Everything external to the microservice, e.g., databases, is hidden behind microservices. The decoupling facilitates system expandability for continuously adding new features.

The values in those graphs are the values randomly generated, for concept demonstration purpose, by the sensors based on a pre-configured offset using simulation. The offset is different for different sensors to make it easy to differentiate while testing. Temperature and Light sensors even allow the offset to be modified for better testing as the offset values affect the status of AC and lights, respectively. The offsets are not registered in the database as they are just for simulation.

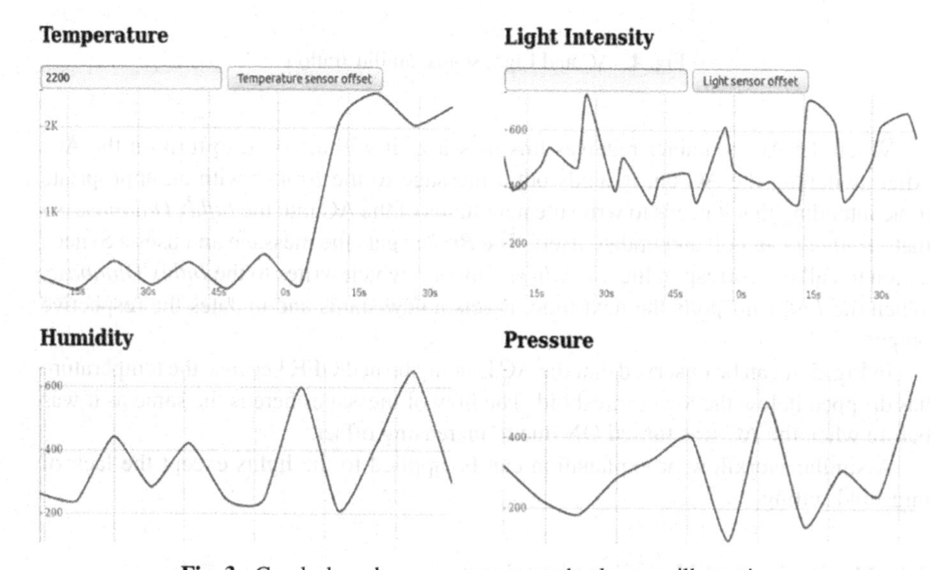

Fig. 3. Graphs based on sensor-generated values: an illustration

As an example in IoT, in Fig. 4, the upper and lower thresholds of the AC (top left) are displayed. When the Temperature offset is set, an appropriate action from the *Actuator* is invoked using a Seneca action. The *Actuator* then converts it into a message format with a topic intended to the AC Regulator and sends the message to the *Broker*. The *Broker* receives it and publishes it for the corresponding subscriber(s) to read. The temperature sensor, being subscribed to receive the offset for it, does so and starts generating values above the new offset (2200 as shown in Fig. 4). The next generated value, as before, is sent in a message with a topic to the *Broker* so that it can act on it and can also let other subscribers read it.

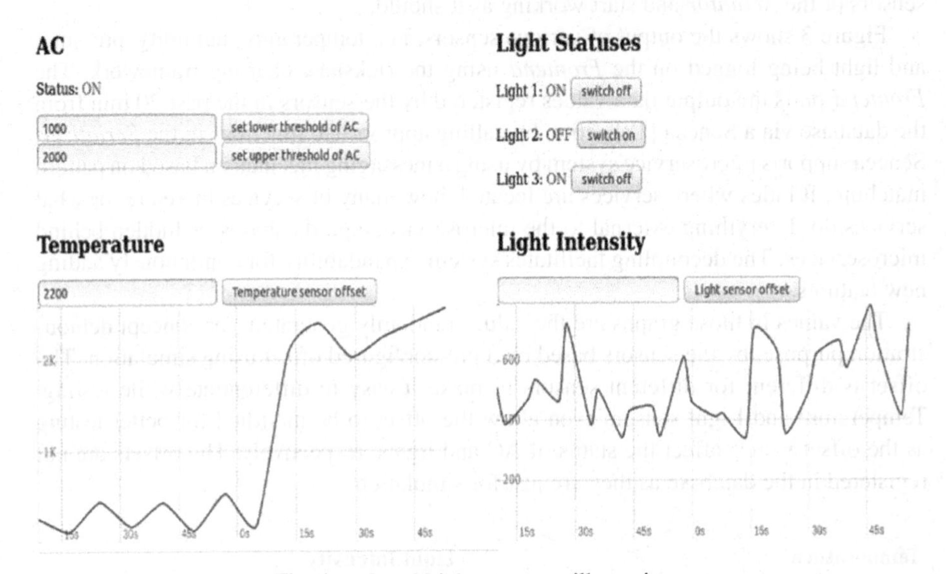

Fig. 4. AC and Light status: an illustration

When the AC container receives this message, it will act on it to turn on the AC. After switching the AC on, it sends out a message to the *Broker* with an appropriate topic intending that it needs to write the new status of the AC into the *Influx Database,* so that *Frontend* can poll and update itself. The *Broker* gets the message and uses a Seneca action to call the corresponding *Serializer* function which writes to the *Influx Database.* When the *Frontend* polls the next time, it gets a new status and updates the respective output.

In Fig. 5, it can be observed that the AC is being turned OFF, because the temperature has dropped below the lower threshold. The flow of messages here is the same as it was before when the AC was turned ON due to increasing offset.

A similar workflow or explanation can be applied to the lights except the lack of threshold setting.

3.4 Monitoring and Managing Containers

Figure 6 shows the Portainer UI which is a Docker management interface. Portainer supports monitoring and management of multiple containers by displaying the node on which the container is deployed, container ID and also the metrics of the container,

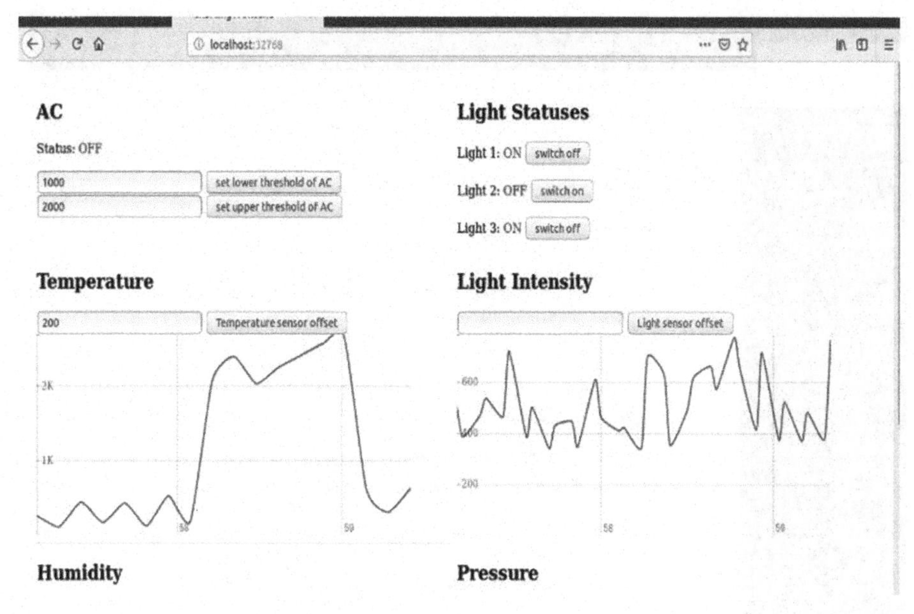

Fig. 5. Update on AC status: an illustration

such as the memory and CPU consumption of a container. Portainer also can present a network view and volume labels defined as pa,rt of a container. In addition, containers can be stopped or restarted from the management interface.

Moreover, scaling the container by creating replicas to improve reliability and possibly performance can be easily realized via the UI itself with the 'Scale' option under the 'Scheduling Mode' (the 3rd column as depicted in Fig. 6). If a containerized replication fails, the Docker management will automatically take an appropriate action, e.g., load balancing among remaining replicas or move the workload to the last running container, when the failure is detected.

The 'Swarm' option on the left can display the nodes in the swarm, as shown in Fig. 6, including the memory, total cores, hostname, and also collectively displaying this information for the entire swarm. Our deployment was on three machines. Figure 6 shows that the *Broker* is running on the node laptop-3. The 'Stacks' option shows a list of services that are deployed to a part of that stack as in the screenshot. We deployed one stack containing all the containers called 'myproject'. So, the 'Stacks' shows one entry named 'myproject'. When it is checked and expanded, it shows the list of services that are deployed to a part of that stack as in the screenshot. When a specific service is expanded, e.g., 'myproject_broker' in Fig. 6, the container (and replicas) running under the service is (are) displayed.

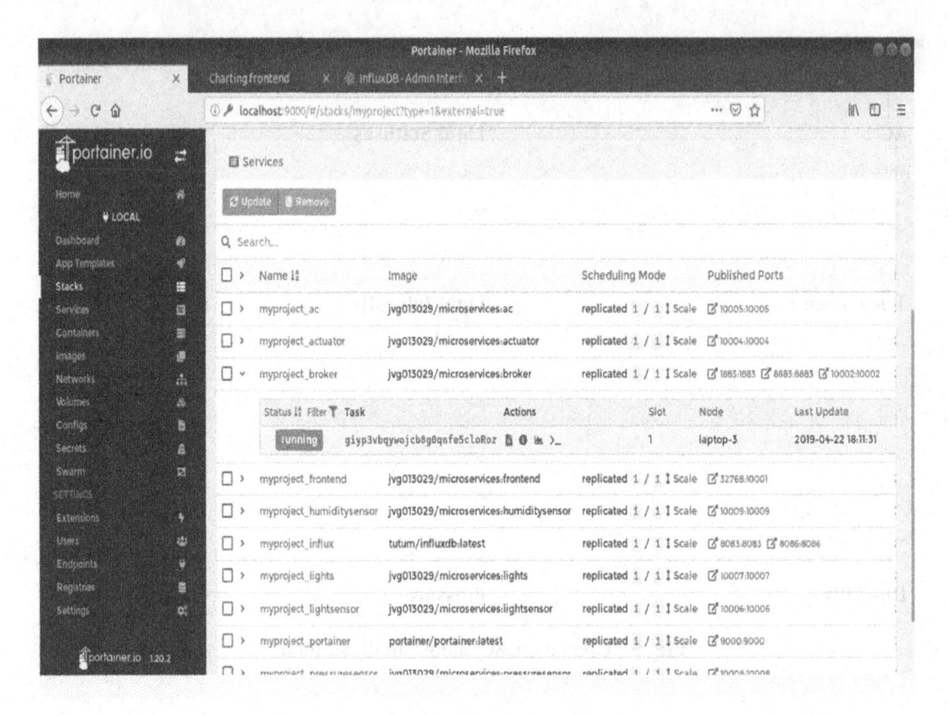

Fig. 6. Portainer UI for docker management

4 Conclusion and Future Work

IoT applications have attracted a great deal of attention and the deployment of large scale or high scalability IoT applications becomes crucial. On the other hand, containers and microservices have been popular in practice due to a number of advantages, particularly independent development, incremental deployment, high scalability. Management of multiple containers based on the microservice architecture becomes an important task, as the number of containers can be high. The objective of this paper was to experimentally validate the effectiveness of applying microservices using containers for IoT systems which has become popular.

The main benefits of such an architecture and subsequent implementation include the ease of adding new sensors or services, ease of deployment due to dependencies being packed with the applications in the containers and usage of lightweight communication protocols (Message Queueing Telemetry Transport in our case). The trade-off is the increased overhead on the system as containers are more resource heavy than individual applications combined.

In this paper, typical IoT applications have been simulated as a suite of small and independent (micro)services which were deployed as isolated containers for feasibility and effectiveness investigation. The simulation of IoT applications was realized using multiple containers and Docker Swarm for container management. The simulation demonstrated a successful rapid application prototyping experiment that preserves the benefits stated above, which reveals that the concept could be further turned into real testbed implementation using a large number of real sensors and common IoT-related controllers, e.g., Arduino or Raspberry Pi.

References

1. Gartner Report. https://www.informationweek.com/mobile/mobile-devices/gartner-21-billion-iot-devices-to-invade-by-2020/d/d-id/1323081. Accessed 16 July 2019
2. Butzin, B., Golatowski, F., Timmermann, D.: Microservices approach for the Internet of Things. In: Proceedings of the IEEE 21st International Conference on Emerging Technologies and Factory Automation (ETFA), pp. 1–6 (2016)
3. Lewis, J., Fowler, M.: Microservices: a definition of this new architectural term. https://martinfowler.com/articles/microservices.html. Accessed 16 July 2020
4. Alshuqayran, N., Ali, N., Evans, R.: A systematic mapping study in microservice architecture. In: Proc. of the 9th IEEE International Conference on Service-Oriented Computing and Applications (SOCA), pp. 44–51 (2016)
5. Circuit Breaker pattern. https://msdn.microsoft.com/de-de/library/dn589784.aspx. Accessed 16 July 2020
6. Cloud Computing Patterns Load Balancer. https://patterns.arcitura.com/cloud-computing-patterns/mechanisms/load_balancer. Accessed 16 July 2020
7. Petrenko, A., and Bulakh, B.: Intelligent service discovery and orchestration. In: Proceedings of IEEE 1st International Conference on System Analysis & Intelligent Computing (SAIC), pp. 1–5 (2018)
8. Campeanu, G.: A mapping study on microservice architectures of internet of things and cloud computing solutions. In: Proceedings of the 7th Mediterranean Conference on Embedded Computing (MECO), pp. 1–4 (2018)
9. Datta, S.K., Bonnet, C.: Next-generation, data centric and End-to-End IoT architecture based on microservices. In: Proceedings of IEEE International Conference on Consumer Electronics - Asia (ICCE-Asia), pp. 206–212 (2018)
10. Khazaei, H., Bannazadeh, H., Leon-Garcia, A.: SAVI-IoT: a self-managing containerized IoT platform. In: Proceedings of the IEEE 5th International Conference on Future Internet of Things and Cloud, pp. 227–234 (2017)
11. Khanda, K., et al.: Microservice-based IoT for smart buildings. In: Proceedings of the 31st International Conference on Advanced Information Networking and Applications Workshops (WAINA), pp. 302–308 (2017)
12. Soliman, M., Abiodun, T., Hamouda, T., Zhou, J., Lung, C.-H.: Smart home: integrating Internet of Things with web services and cloud computing. In: Proceedings of IEEE 5th International Conference on Cloud Computing Technology and Science, pp. 317–330 (2013)
13. Nearform. https://github.com/nearform/micro-services-tutorial-iot. Accessed 16 July 2020
14. Container. https://www.docker.com/resources/what-container. Accessed 16 July 2020
15. Docker. https://www.docker.com/why-docker. Accessed 16 July 2020
16. Docker Swarm. https://docs.docker.com/engine/swarm/swarm-tutorial/. Accessed 16 July 2020
17. REST. https://restfulapi.net/. Accessed 16 July 2020
18. Senaca. https://senecajs.org/getting-started/. Accessed 16 July 2020

Detecting Dark Spot Eggs Based on CNN GoogLeNet Model

Min-lan Jiang(✉), Pei-lun Wu, and Fei Li

College of Physics and Electronic Information Engineering, Zhejiang Normal University, Jinhua 321004, China

xx99@zjnu.cn

Abstract. Aiming at the problems of high labor intensity and low efficiency in detecting dark spot eggs, a method of detecting dark spot eggs based on GoogLeNet model is proposed. This method uses Inception convolution module in GoogLeNet model to automatically extract dark spot eggs features and realize the detection. A device for collecting transparent images of eggs was set up in the experiment, and the sample collection experiments were designed to acquire samples. A total of 1200 dark spot eggs images and 8850 normal eggs images were obtained. Selecting 1200 samples of these two kinds for network modeling. The experimental results show that the detection accuracy of dark spotted eggs based on CNN GoogLeNet model is 98.19%. In order to further verify the GoogLeNet model, this paper repeats the above experiments using the VGG16 and VGG19 models of CNN model, and compares the accuracy. To further validate the GoogLeNet model, this paper repeats the above experiments using VGG16 and VGG19 models, and compares the accuracy. The results show that the three CNN models together have high detection accuracy, and the GoogLeNet model is highest, which provides a new method for egg quality detection.

Keywords: Dark spot eggs · Convolutional Neural Network · GoogLeNet model · HOG-SVM model

1 Introduction

Chicken egg's dark spots are dark spots formed when water in the egg contents permeates and accumulates on the eggshell. The appearance of dark spots is a manifestation of the decline in egg quality. Dark spot eggs lose water quickly, decrease freshness quickly and are more susceptible to microbial contamination, which have adverse effects on egg storage performance. Eggs are the national preferred nutritional food. It is an inevitable trend to accurately detect and remove dark spots of eggs, and to detect and grade the quality of eggs before they are put on shelves [2], which can ensure the health and safety of food.

At present, the detection of dark spot eggs is mainly completed by humans. It is of practical significance to find a high-efficiency detection method for the reasons such as the great difference in the coverage rate of the dark spots on the eggshell, the unstable

H. Song and D. Jiang (Eds.): SIMUtools 2020, LNICST 370, pp. 116–126, 2021.
https://doi.org/10.1007/978-3-030-72795-6_10

position, the insignificant color of some dark spots, and the high intensity of labor. The detection method of machine vision combined with machine learning has been widely studied in egg quality detection due to its advantages of low cost and high convenience. For example, wang qiaohua [3–6] collected the light-transmitting image of eggs, extracted the color information inside the eggs and detected the freshness of the eggs by using the morphological features such as the yolk area ratio and the air chamber area ratio. Tu kang et al. [7] marked egg surface stains with threshold segmentation method to realize nondestructive detection of egg stains. Li xincheng et al. [8] use four characteristic parameters of egg freshness including ratio of egg yolk area, air chamber area, chamber height and chamber diameter and established single-element regression model with Haff values. The above research has made some progress in egg quality detection, however there are also some limitations. For example, only tens or hundreds of samples of egg translucent images were collected in the experimental which are easy to lead to over-fitting. Also, the distribution of small samples is uneven and the distribution of features is unbalanced which lead to in the freshness test, the samples are concentrated in grade b eggs and the number of super and grade a sample is far less than grade b. In addition, traditional machine learning methods such as BP neural network and support vector machine (SVM) are mostly used in the above studies. BP neural network needs to update network parameters repeatedly and SVM is excessively dependent on parameter adjustment. What's more, there are many pretreatment steps and large interference of the extraction methods based on color and morphological characteristics, so the accuracy of the existing egg quality detection methods is not high and there is a big gap with the actual production.

The Convolutional Neural Network (CNN) model [9] of deep learning uses the common operation of "Convolutional" and "descending sampling" to process and classify the multi-dimensional sample features, which has the advantage of processing multiple data. It has been widely used in the field of image recognition and detection, however, there are only a few studies use the CNN model to detect the quality of eggs. In this paper, normal eggs and dark spot eggs were taken as classified samples, and the method of combining machine vision with CNN was applied to dark spot eggs detection. The GoogLeNet [10] model was put forward for learning the characteristics of dark spot eggs to establish the dark spotted egg detection model, and to verify the effectiveness and accuracy of the model. The detection method proposed in this paper solves the problems existing in previous studies and provides a reference for the detection of dark spot eggs.

2 Materials and Methods

2.1 Experimental Materials

The test materials were 500 fresh JingBai 939 powdered-shell eggs provided by lanxi poultry farm in JinHua, China. In this experiment, the eggs were numbered, placed horizontally in the egg tray, and stored in indoor environment with temperature of 26–28 C and relative humidity of 70%–80%.

2.2 Instruments and Equipment

In order to obtain the transparent image of eggs, an egg transparent image acquisition system as shown in Fig. 1. The light source is OPT-RID-150 sphere integrated light source that is used in the process of image acquisition to supplement the light of the collected objects. In this experiment, a plane mirror was placed at the bottom of the light source, and the light from the original light source convergent at the bottom of the device. In this experiment, a plane mirror is placed at the bottom of the light source, so that the light originally collected at the bottom of the device pass through the top light transmission hole after being reflected through the plane mirror. The light obtained in the experiment is uniform, the light intensity is better and transmission effect are far better than those of incandescent light sources used in previous studies. The camera uses an industrial color CMOS camera (effective pixels: 2592 × 1944) with a 5-megapixel 12 mm lens. The illumination intensity in the image acquisition environment is 2–10 lx, and the collected images are RGB color images with a resolution of 1920 × 1440. Firstly, the brightness of light source was adjusted to highlight the dark spots of eggs. The process of collecting experimental images is as follows: firstly, adjust the brightness of the light source to the maximum; secondly, adjust the height of the camera by the iron frame, so that the whole egg is just of a moderate size and can be clearly photographed; then, the egg is placed in the light hole at the top of the light source in sequence according to its' number; finally, the image is collected by using the computer acquisition software.

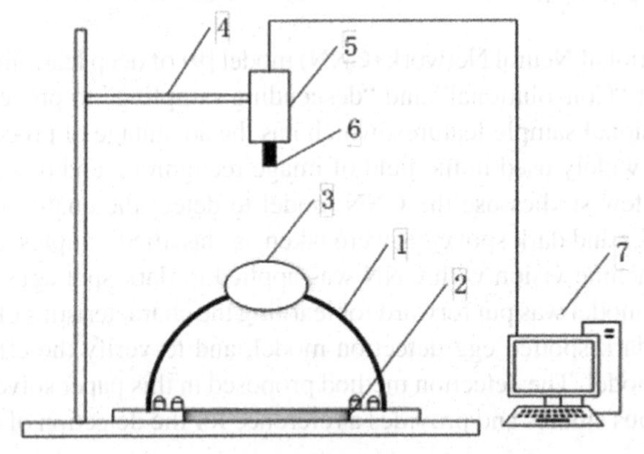

Fig. 1. Egg picture collecting device. 1. Spherical integral light source; 2. Plane mirror; 3. Egg; 4. Iron frame; 5. Industrial color CMOS camera; 6. Camera lens; 7. Computer image acquisition

2.3 Sample Collection

In the actual detection, eggs pass through the camera at a random Angle. If an image is collected from a single egg, dark spots on the egg shell cannot be completely collected, also the sample size cannot meet the modeling requirements. In order to get close to the practical application scene, this experiment adopts the multi-angle collection method for dark spotted eggs. After the egg image collection, it is turned over to 90° and repeated collection. If there is no dark spot on the eggshell at a certain angle, it is not collected. Normal eggs are collected one image sample individually. Studies have shown that [11] with the extension of storage time, dark spots will gradually increase and expand and appear on normal eggs. In order to learn the dark spot characteristics, the experiment repeated the above collection every day with a sampling period of 20 days. A total of 1200 images of dark speckled eggs and 8850 images of normal eggs were obtained.

2.4 Image Processing

The color of transmitted light image of eggs collected in this paper is yellow to red, and the color of dark spots of eggs is dark red. The color contrast between the two is not high. Previous studies have suggested that [12] color enhancement of RGB images can enhance color contrast. After color enhancement experiments, it was found that the G component in RGB space was easier to recognize the dark spots of eggs, so the G component in the sample picture was enhanced four times. Then, an interpolation algorithm is used to reduce the size of the enhanced image to 1/8 of the original size to meet the fast training and testing of the CNN network model. The resulting image is shown in Fig. 2.

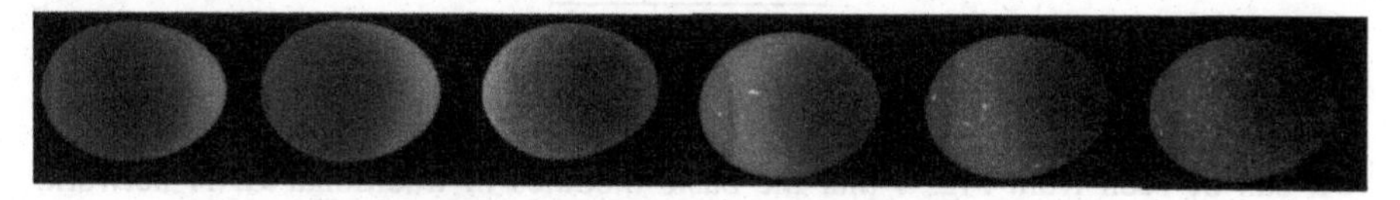

Fig. 2. Sample images of partial dark spot eggs

2.5 Model Description

CNN network, as a deep neural network model [13], can automatically learn and extract features from data, avoiding complex image preprocessing in the early stage, and has been widely used in such fields as pattern classification, object detection and object recognition. CNN network composed of input layer, feature extraction, the fully connection layer and output layer. The feature extraction layer is the core of the network, the more layers there are, the stronger the feature extraction capability is. It mainly includes convolution layer and lower sampling layer. The convolution layer is used to extract features of input images. Features extracted from different convolution cores are different. The more the number of cores in the convolution layer, the more the features extracted. The lower sampling layer can reduce the amount of data processing and ensure the computing speed.

As a CNN network model, GoogLeNet uses Inception module as a convolutional layer to introduce multi-scale convolution to extract multi-scale local features. Its structure is shown in Fig. 3. The module contains several 1×1, 3×3 and 5×5 convolution kernel branches, the multi-core structure can extract and learn the features of different forms of egg spots, which is suitable for the dispersed, multi-morphological and multi-scale characteristics of egg spots [14]. Moreover, an additional 1×1 convolution kernel is added to the structure, which not only increases the network depth, but also improves the nonlinear degree of the network. It also reduces the dimensions of convolution objects in other convolution kernel and reduces the computation. The CNN GoogLeNet model structure used in this article is shown in Fig. 4.

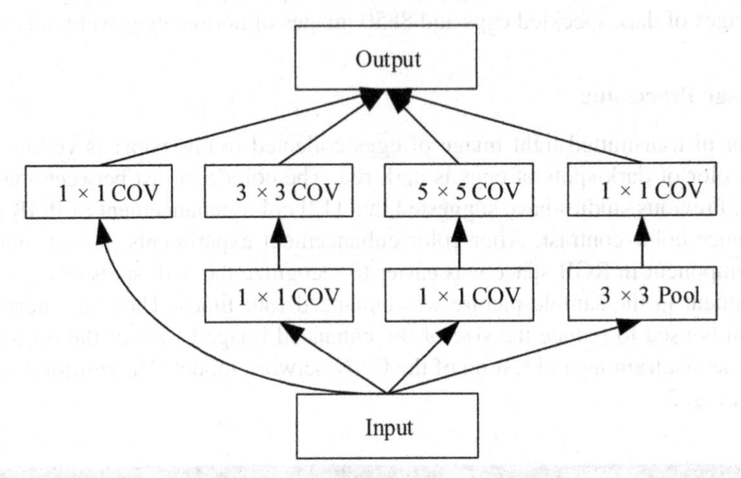

Fig. 3. Inception module structure

It can be seen from Fig. 4 that the basic modules of traditional CNN network, namely convolution layer and pooling layer, are adopted near the image input layer. Considering that the characteristics of the middle layer have been able to identify to a certain extent, and considering that the gradient vanishing in the optimization process of random gradient descent algorithm is easily caused by the excessively deep network layer, GoogLeNet added two additional full-connected SoftMax classifiers to the side of the trunk network.In the process of model optimization, network model parameters are updated by adding the gradient of loss function of trunk and branch classifier. In the test process, the branch classifier was removed and only the trunk classifier was used for dark spot egg detection and grading.

2.6 Testing Process

A total of 1200 dark spot egg images and 8850 normal egg images were obtained. To balance the number of two types of samples, 1200 samples were randomly selected from the normal egg samples, and the two types of samples were processed with the above image processing. Secondly, the number of training sets and test sets is selected, according to the ratio of 1:3. This paper randomly takes 900 samples from each category as model training samples and 300 as test samples. The labels of dark spot eggs and normal eggs are coded as 0001 and 0010 by one-hot code. Then, the training samples and labels are substituted into the input and output of the CNN GoogLeNet model for training, and the random gradient descent algorithm (SGD) is adopted for weight update. When the error or iteration number reaches the threshold value, the training stops. Finally,

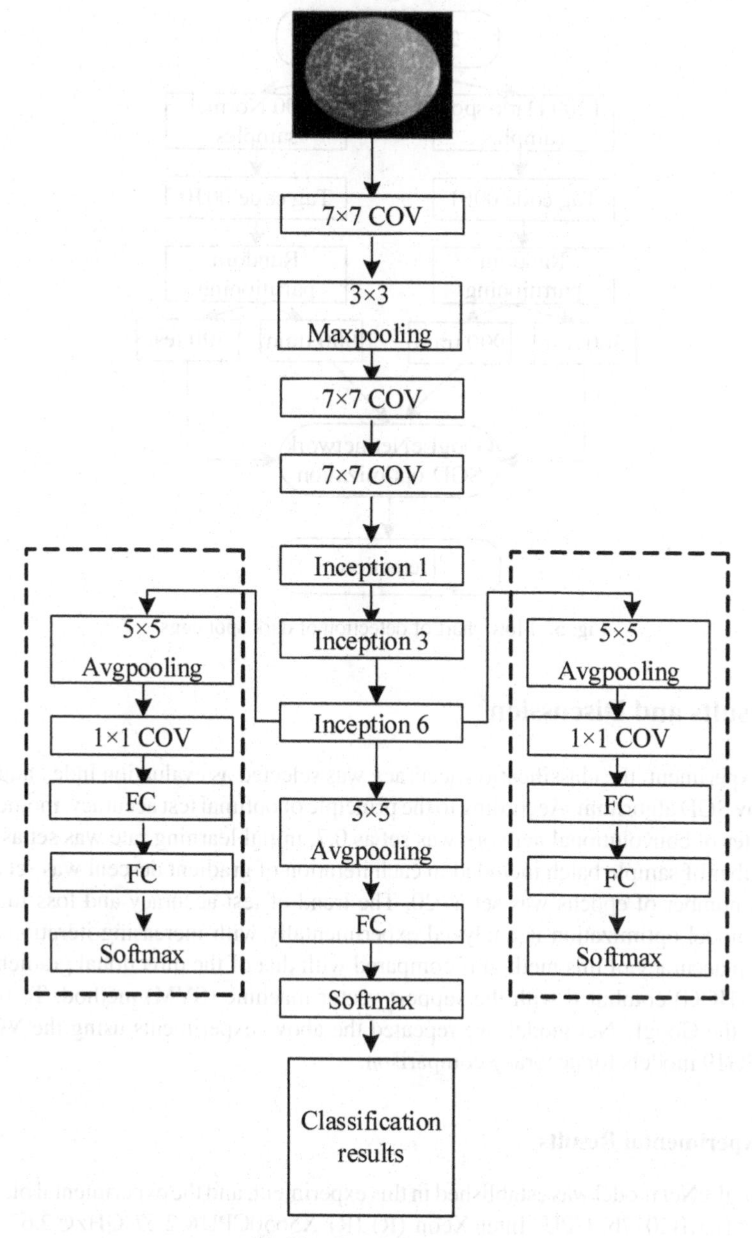

Fig. 4. GoogLeNet model

test samples are substituted into the trained network and test results are obtained. The flow chart is shown in Fig. 5.

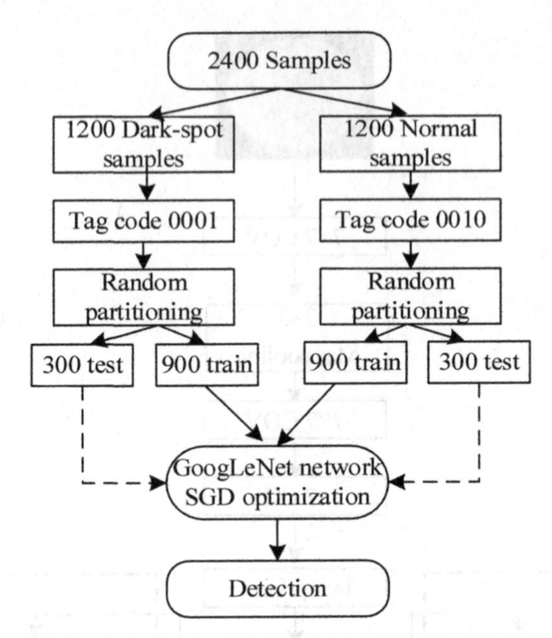

Fig. 5. Flow chart of detection of dark spot eggs

3 Results and Discussion

In the experiment, the classification accuracy was selected as evaluation index and optimized by SGD algorithm. According to the principle of optimal test accuracy, momentum parameter of convolutional network was set as 0.7, initial learning rate was set as 1e-4, the number of sample batch included in each iteration of gradient descent was set as 32, and the number of epochs was set as 20. The trend of test accuracy and loss function during model optimization is analyzed experimentally with increasing iterations. The detection accuracy of this method is compared with that of the directional gradient histogram (HOG) combined with the support vector machine (SVM) method. To further validate the GoogLeNet model, we repeated the above experiments using the VGG16 and VGG19 models for accuracy comparison.

3.1 Experimental Results

The GoogLeNet model was established in this experiment, and the experimental platform was MATLAB2017b, CPU: Intel Xeon (R) (R) X5650CPU@2.67 GHz@2.67 GHz; Memory size: 48 GB. When training the model in this experiment, the training situation of the model is evaluated according to the loss and accuracy curves, and the network parameters are adjusted accordingly, as shown in Fig. 6.

It can be seen from Fig. 6 that the training loss function presents a downward trend in the training process, and the prediction loss deviation of the response model gradually decreases in the optimization process by updating the loss function gradient. Meanwhile, as the number of iterations increases, the prediction accuracy of the model on the test set

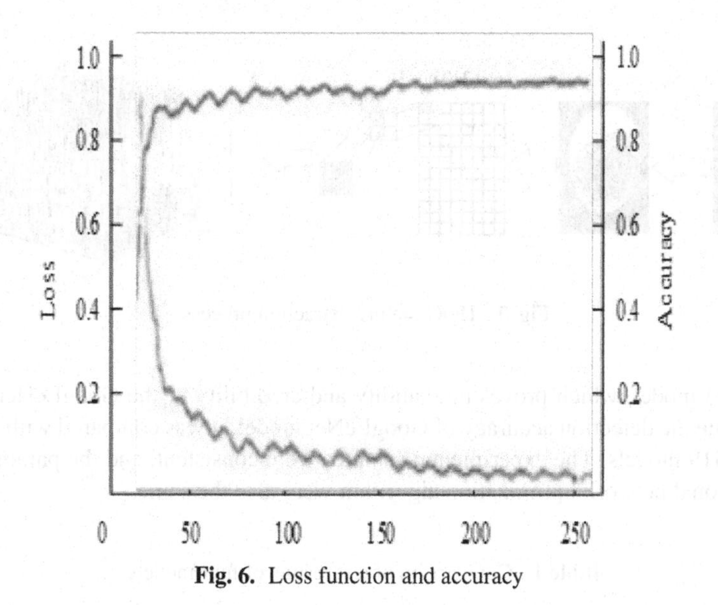

Fig. 6. Loss function and accuracy

increases as a whole. The performance of the response model can be optimized in the process of constantly updating the parameters. When the number of iterations reaches 260, it basically converges. The detection accuracy of dark spot eggs was 97.04%, that of normal eggs was 98.89%, and the total detection accuracy of test samples was 98.19%.

3.2 Model Validation

In order to verify the effectiveness and credibility of this method, the GoogLeNet model and HOG-SVM method are compared. As a widely used image feature extraction method, HOG feature [16] can describe image morphology by using the density distribution of gradient and edge direction in the sample to be measured. To extract HOG feature, firstly, color normalization of the image is required. In this paper, the optimal threshold value is obtained by combining with the traditional large law, and the optimal adaptive threshold value is 0. 294. Then the color sample image is binarized. After that, the sample of 240×180 pixels is divided into several blocks of 2×2 cells, each cell is 8×8 pixels in size. Histogram statistics of the gradient directions facing all pixels in each cell is conducted to obtain a multi-dimensional feature vector. Finally, the feature vectors of adjacent units are connected to get higher-dimensional feature vectors, and the HOG feature of the entire sample is obtained. The process is shown in Fig. 7.

SVM [15] model, as a classical machine learning method, can produce better classification results under high-dimensional data. This paper combines SVM with HOG and applies HOG-SVM to dark spot egg detection. The SVM toolbox is the least squares support vector machine (LS-SVM) toolbox and the kernel function is RBF kernel function. The combination of grid search and crossover verification is used to find the best kernel parameter. The experimental results are shown in Table 1. It can be seen from Table 1 that GoogLeNet detection accuracy is more than 10% higher than that of the

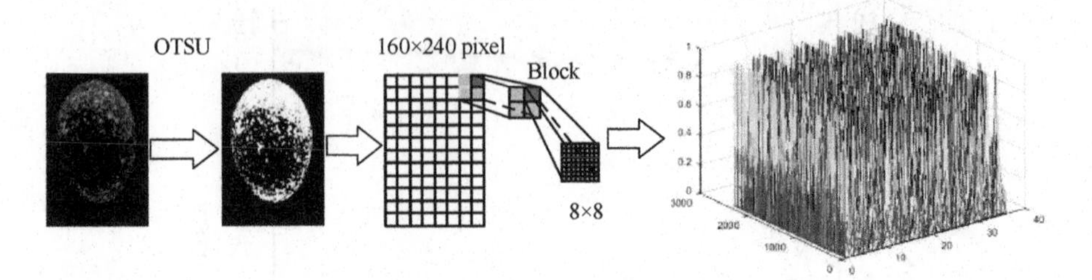

Fig. 7. HOG feature extraction process

Hog-SVM model, which proves the validity and credibility of the GoogLeNet model. To evaluate the detection accuracy of GoogLeNet model, it was compared with VGG16 and VGG19 models. The experimental samples were consistent, and the parameters of convolutional network optimization algorithm were also the same.

Table 1. Comparison of test results of four models

Network model	Feature extraction	Test set accuracy/%	Dark-spots eggs accuracy/%	Normal eggs accuracy/%
GoogLeNet	Inception	98.19	97.04	98.89
SVM	HOG	86.12	85.34	88.73
VGG16	Conv	96.99	95.83	97.64
VGG19	Conv	96.76	95.53	97.52

As can be seen from Table 1, the CNN based dark spot egg detection method proposed in this paper is feasible and reliable. The accuracy of the three CNN models can reach over 96%, among which the GoogLeNet model has a better detection accuracy of 98.19%. It can be concluded from the experiment that the detection method based on GoogLeNet model is feasible and has highest detection accuracy.

4 Conclusion

This paper presents a method of dark spot eggs detection based on the convolutional neural network GoogLeNet model. Aiming at the problems of too few samples, too many pretreatment steps and low precision of model in the previous research on egg quality detection. This paper improves the egg image acquisition experiment which greatly expanding the number of samples and balancing the number of samples of two kinds of egg images. The GoogLeNet model, which can automatically learn and extract features, is applied to the detection of egg dark spots. The experiment dividing training and testing samples were used in GoogLeNet model training. The result shows that the detection accuracy of dark spot eggs is 98.19%. This proves that the method based on CNN network can detect dark spot eggs with high detection accuracy without too many pre-processing steps. In order to verify the effectiveness and credibility of this method, the GoogLeNet model is compared with the HOG-SVM model and the other two CNN models. The results show that the detection results of the three CNN models are far superior to the traditional image classification algorithm HOG-SVM model, while the GoogLeNet model is more accurate than other CNN models.

In this paper, although machine vision combined with CNN model has been applied to the detection of dark spotted eggs and good results have been obtained, there are still researches to be further carried out: firstly, In order to highlight the characteristics of samples, the illumination intensity of the collected environment in this paper is low. In order to enhance the adaptability of the detection method to the illumination changes, the diversity of samples under different illumination conditions should be increased; Secondly, This paper proposes that the detection method is still in the laboratory stage, and further research on egg transfer and sorting device should be carried out if it is to be put into production. Thirdly, the detection method proposed in this paper is not only limited to the detection of dark spots, but also applicable to the detection of egg freshness in a wide range of studies, and the detection of other poultry eggs can also be tried, which is worthy of further research.

References

1. Zhang, M., Ye, J., et al.: Research on effects of chicken eggshell dark spots on chicken egg storage. Food Mach. **32**(06), 118–122 (2016)
2. Liu, Y., Zhong, N.: Research on prediction model of egg freshness based on image processing. Food Mach. **33**(12), 103–109 (2017)
3. Wang, Q., Wang, C., Ma, M.: Duck eggs' freshness detection based on machine vision technology. J. Chin. Inst. Food Sci. Technol. **17**(08), 268–274 (2017)
4. Zheng, L., Yang, X., et al.: Nondestructive detection of egg freshness based on computer vision. Trans. CSAE **25**(S2), 335–339 (2009)
5. Wang, Q., Wen, Y., et al.: Correlation between egg freshness and morphological characteristics of light transmission image of eggs. Trans. CSAE **24**(03), 179–183 (2008)
6. Wang, Q., Reng, Y., Wen, Y.: Study on non-destructive detection method for fresh degree of eggs based on BP neural network. Trans. Chin. Soc. Agric. Mach. **37**(01), 104–106 (2006)
7. Tu, K., Pan, L., et al.: Dirt detection on brown eggs based on computer vision. J. Jiangsu Univ. (Nat. Sci. Edn.) **28**(03), 189–192 (2007)
8. Li, X., Zhao, D., Shi, H.: Non-destructive testing method of egg quality based on machine vision. J. Food Saf. Qual. **10**(2), 489–493 (2019)
9. Lecun, Y., Bottou, L., Bengio, Y., et al.: Gradient-based learning applied to document recognition. Proc. IEEE **86**(11), 2278–2324 (1998)
10. Szegedy, C., Liu, W., Jia, Y.Q., et al.: Going deeper with convolutions. In: IEEE Conference on Computer Vision and Pattern Recognition, Boston, MA, USA (2015)
11. Wang, D.: Mechanism exploration for translucent egg formation. China Agricultural University Doctoral dissertation (2017)
12. Liu, Y., Li, Q., Huang, X., et al.: Egg characteristics extraction from light transmission image and egg freshness model training. Sci. Technol. Eng. **15**(25), 72–77 (2015)
13. Huang, S., Sun, C., et al.: Rice panicle blast identification method based on deep convolution neural network. Trans. Chin. Soc. Agric. Eng. **33**(20), 169–176 (2017)
14. Yu, C., Zhou, L., Wang, X., et al.: Hyperspectral detection of unsound kernels of wheat based on convolutional neural network. Food Sci. **38**(24), 283–287 (2017)

15. Xu, Y., Xu, X., et al.: Pedestrian detection combining with SVM classifier and HOG feature extraction. Comput. Eng. **42**(01), 56–60+65 (2016)
16. Cortes, C., Vapnik, V.: Support-vector networks. Mach. Learn. **20**(3), 273–297 (1995)

An Edge Server Placement Method with Cyber-Physical-Social Systems in 5G

Xing Zhang[1], Jielin Jiang[1], Lianyong Qi[2], and Xiaolong Xu[1(✉)]

[1] School of Computer and Software, Nanjing University of Information Science and Technology, Nanjing, China

[2] School of Information Science and Engineering, Qufu Normal University, Qufu, China

Abstract. Recently, cyber-physical-social systems (CPSS), an advanced information system, have been introduced to promote the development of the smart society. It makes the control of machines in all walks of life more intelligent in an efficient, convenient, and stable way. Besides, with the maturity of edge computing, the task requests emitted by users in CPSS are tent to be transmitted edge servers for immediate processing. Nevertheless, some problems, i.e., high latency and low utilization, exist in current network placement. It leads to the problem that users in the CPSS are unable to enjoy instant and efficient processing. Given these problems, this paper designs an edge server placement method (ESPM) to alleviate this situation. To be specific, a system model designed according to this scenario is presented firstly. Then, the multi-objective evolutionary algorithm, i.e., improving the strength pareto evolutionary algorithm (SPEA2), is applied in this paper to optimize the access delay and load balance variance with the propose of enhancing the service experience of users. Furthermore, the normalization methods, i.e., the technique for order preference by similarity to an ideal solution (TOPSIS) and multi-criteria decision-making (MCDM) are selected to produce the standard data and optimal strategy. Finally, the experimental results show the effectiveness of ESPM.

Keywords: CPSS · Server placement · Evolutionary algorithm · Edge computing

1 Introduction

Business innovation and industrial intelligence pave the way for intelligent society, smart machines and networking in the future. Also, the persistent improvement of big data, cloud computing, Internet of things (IoT) and so on in recent years has promoted the further integration of traditional physical systems and advanced information technologies as well as the maturity of cyber-physical-social systems (CPSS) [1]. CPSS cover embedded environment perception, dynamic analysis of personnel organization behavior, network communication and network control, which makes the physical system have the functions of

H. Song and D. Jiang (Eds.): SIMUtools 2020, LNICST 370, pp. 127–139, 2021.
https://doi.org/10.1007/978-3-030-72795-6_11

calculation, communication, precise control and remote cooperation [2]. CPSS will be applied in many fields, such as intelligent enterprise, intelligent transportation, intelligent home, intelligent medical treatment and so on [3]. CPSS enable the physical system to have efficient computing, instant communication and other functions. This makes machine system more evolved and human organizations operate machines more stably and immediately through cyberspace. Besides, CPSS also lead to the rapid development of IoT technology and the more widespread utilization in a variety of intelligent scenarios [4].

However, it is precise because handheld devices, IoT devices and various intelligent scenarios are becoming more and more intelligent, more users have higher standards and stricter requirements for these application scenarios [5]. Besides, when the physical system combines network information and social information, the mass data transmission also puts forward higher requirements for network communication (i.e., bandwidth and speed) [6]. This also reflects people's inability to enjoy high-quality and low latency network services. Therefore, how to deal with these massive data in CPSS timely and effectively is an urgent problem to be solved [7]. There is no doubt that it is meaningless to consider the quality of service outside the network performance which signifies the basis of the quality of service [8]. The introduction of 5G network to accelerate the evolution of intelligent application scenarios can not only improve the data transmission rate and reduce the delay but also make the intelligent scenarios more advanced [9].

To provide timely and efficient feedback for users in CPSS, it is undoubtedly necessary to make full use of edge computing, so that users are able to experience high-quality service applications in real-time [10]. Technically, edge computing has rich computing resources. It could effectively reduce the task offloading delay by placing the execution task on the computing node close to the terminal device. It enables users to be geographically close to the servers that are processing resources, thus reducing the delay of offloading tasks and obtaining a higher quality of service [11]. Specifically, in CPSS, base stations are processed to evolve into edge computing nodes for serving users covered by them. In addition to the advantages of edge computing in offloading tasks, the proximity of distance also reduces the possibility of the harm to users caused by traditional information interception, thus increasing the security of users' privacy.

In general, placing edge server is an infrastructure requirement for edge computing. Unfortunately, most of today's researches on edge computing are only for theory, and there are few studies on practical applications such as edge server placement. In this paper, we focus on the edge server placement in a mobile edge computing environment that provides wireless internet coverage for mobile users.

In the edge server (ES) placement strategy, each ES node has limited computing resources. Most of them cannot handle the massive tasks at the same time [12]. This will lead to produce some services delay in the calculation process, affect service execution and reduce service efficiency. It is unacceptable for users in the 5G scenarios. Therefore, we need to significantly reduce the overall delay. Besides, we need to consider how to improve the resource utilization as much as possible owing to the limited computing resources of each node. In addition, we need to trade-off all nodes to ensure the stability of each node for performance maximization.

Given these facts, achieving a reasonable ES placement strategy by reducing transmission latency and ensuring load balance to improve overall performance is a huge challenge. In this paper, an edge server placement method, namely ESPM, is designed for the placement of edge devices in 5G network.

Specifically, the pivotal motivations and contributions of this paper are shown below:

- Few studies research on the edge server placement method while pursuing the minimum access delay and load balance variance in CPSS scenario. So a unique task offload method in CPSS based on edge computing is designed in this paper.
- The evolutionary algorithm and normalization method are collectively deployed in this paper to obtain the feasible offloading strategies and select the optimal strategy.

2 Related Work

The characteristics of edge computing, such as large storage capacity and strong computing power, are particularly prominent in CPSS. In the previous literature, CPSS and its outstanding advantages have been studied in depth.

CPSS have the advantages of high immediacy, efficiency, reliability and so on. It is widely used in various aspects. Wang et al. fully described how CPSS is transformed into CPSS, and also introduced the definition, classification and application of CPSS, the contribution and significance of CPSS [13]. Han et al. proposed to introduce dynamic and diverse human behaviors into the vehicle network to make it become a CPSS system and evolve it into a parallel vehicle network to achieve efficient traffic state and low data communication delay between vehicles [2]. Wang et al. proposed a new unified method of CPSS framework based on cloud parallel driving with the purpose of realizing collaborative online automatic driving and carried out parallel testing, learning and reinforcement learning for the framework [14].

Given the massive data in the CPSS scene, it is difficult for traditional methods to deal with these data effectively and timely. Therefore, we apply edge computing to CPSS to solve the above problems. Edge computing introduces the advantages of cloud computing to the network edge cloud in various scenes close to CPSS to provide efficient services [15].

Wang et al. studied the edge server layout in the mobile edge computing environment of smart cities, which is described as a multi-objective (i.e., minimizing the access delay and achieving the load balance) constrained optimization problem [16]. Li et al. studied the energy-aware edge server layout problem, tried to find a more efficient and low-energy layout scheme, and designed an energy-aware edge server layout algorithm based on particle swarm optimization to find the optimal solution [17]. Zeng et al. studied how to efficient and economic deployment in the wireless metropolitan area network edge server problem, and put forward a kind of based on greedy algorithm as well as a simulated annealing based global optimization method to solve the problem [18].

It is undeniable that there are indeed many achievements and developments in edge computing. However, as far as we know that there is little research have investigated the ES placement strategies. Most studies of this respect have focus on the traffic prediction, load balancing and minimum latency under the conditions of determining certain placement strategy. Compared with the previous works, this paper first presents a task offloading framework based on CPSS. Then server placement method based on CPSS, whose purpose is offering better experience for users in the condition of minimum time consumption and load balance variance, has been designed accordingly. At last, the proof experiments are given in the paper to prove the effectiveness of our method.

3 System Model

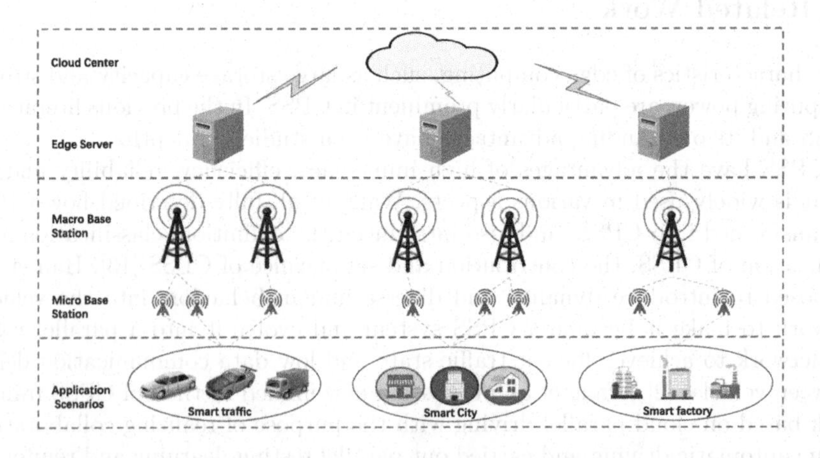

Fig. 1. An edge server placement framework based on CPSS with edge.

The framework in Fig. 1 is composed of user layer, base station layer and the service layer. Three typical scenes are presented in user layer, include smart factory, city and traffic. These scenes tend to generate lots of data which require low latency and high reliability. Such high requirements are difficult to realize in traditional 4G environments. For example, automatic driving in IoV scenario requires time response to reach millisecond level. Thus, the gradual maturation of 5G technology provides a good opportunity for these requirements. All task requests are defined as different tasks which are denoted as $T = \{t_1, t_2, ..., t_{Num}\}$.

Base station layer is divided into micro base station layer and macro base station layer. The former usually receive the data of users and send them to macro base station in real time. The collection of macro base stations is denoted as $BS = \{bs_1, bs_2, ..., bs_N\}$, and micro base stations collection is defined as $BS^i = \{bs_1^i, bs_2^i, ..., bs_O^i\}$. In 5G scene, the tasks produced from users will transmit to the nearest micro base station, then to macro base station. At last, the tasks will be coped with edge server. Besides, we define that an edge server e_i covers K macro base stations in any location. All the stations are denoted as a collection $\mathrm{B}_{cb}^m = \{b_1^{e^m}, b_2^{e^m}, ..., b_K^{e^m}\}$.

The ES collection is denoted as $E = \{e_1, e_2, ..., e_M\}$. This paper is studying the placement strategies of ES, so a unique collection $PS = \{ps_1, ps_2, ..., ps_U\}$ is defined to store different placement methods. In edge servers, the computing resources are measured by virtual machines (VMs) which are denoted as $V = \{V_m^1, V_m^2, ..., V_m^D\}$. The collection means that the m-th edge server has D VMs.

3.1 Delay Model

This paper assumes that the access delay includes three segments in all. AB_m^n is defined to judge whether the n-th ($n = 1, 2, ..., N$) base station bs_n is combined with the m-th ($i = 1, 2, ..., M$) edge server es_m. N and M represent the largest number of base stations placed and edge servers placed respectively.

$$AB_m^n = \begin{cases} 1, \text{ if } bs_n & \text{combines with } es_m, \\ 0, & \text{otherwise.} \end{cases} \tag{1}$$

The first segment is transmission delay for transmitting task from the base station to the target edge server, which is defined by

$$TT_n(X) = (1 - AB_m^n) \cdot \frac{ts_n}{tr} \cdot ap_n, \tag{2}$$

where ts_n represents the size of t_n coming from bs_n in the coverage of es_m. Besides, ap_n signifies the amount of the passing base stations in the process of task propagation, and tr on behalf of the propagation rate between base stations.

The second segment is processing time for executing ts_n coming from bs_n, which is calculated by

$$TP_n(X) = \frac{ts_n}{NV_n \cdot \lambda}, \tag{3}$$

where NV_n represents the amount of resource units demanded by the task of ts_n, and λ represents the executing power of the unit in VM.

The third segment is the feedback time of processed result coming from es_m, which is calculated by

$$TF_n(X) = \frac{ts_n'}{tr}, \tag{4}$$

where ts_n' represents the task size of the results unloaded from bs_n.

The total access delay for answering t_n is calculated by

$$OD_n(X) = TT_n(X) + TP_n(X) + TF_n(X). \tag{5}$$

The average access delay for overall ES placement strategy is calculated by

$$AAD = \frac{1}{Num} \cdot \sum_{n=1}^{Num} OD_n(X). \tag{6}$$

3.2 Load Balance Model

In the load balance model, we define the data forwarded by the base station as the tasks that the edge server needs to process [19]. In this section, we assume that each edge server covers the same number of base stations. The task collection in one certain edge server is defined by

$$T_e^i = \sum_{k=1}^{K} t_k, \tag{7}$$

where t_k represent the task number of k-th station covered by i-th server.

The capacity of ES and the computing resource requirement of data are measured by the VMs number. In this scene, the number of VMs required for a certain server is calculated by

$$VM_i = \frac{T_e^i}{R}, \tag{8}$$

where R represents a coefficient calculated jointly by N and M.

The usage of the computing resources in the m-th edge server e_m is measured by

$$U_{cal}^m = \frac{\sum_{r=1}^{7} VM_r}{C}, \tag{9}$$

where C on behalf of the capacity of each edge server.

The average usage of the computing resources in ES is calculated by

$$\mathrm{U_{ave}} = \frac{\sum_{i=1}^{M} U_{cal}^i}{M}. \tag{10}$$

The average load balance variance for the whole ES is calculated by

$$U_{bal} = \frac{\sum_{i=1}^{M} \left(U_{cal}^i - U_{ave}\right)^2}{M}. \tag{11}$$

3.3 Problems Formulation

The aim in this paper is to achieve the minimum delay and to realize the load balance when placing ES around base stations. The multi-objective optimization problem is given as

$$\begin{cases} \min D_{ave}, \\ \min U_{bal}. \end{cases} \tag{12}$$

A few constrains are formalized as follows for acquiring better results by

$$s.t.\ M \leq N, \tag{13}$$

$$\sum_{i=1}^{N} t_{bs_i} \leq C \times M. \tag{14}$$

The constraint (13) means that the number of the ES is less than the number of edge servers. The constraint (14) represents that each active base station has its own server which responsible for handling forwarded tasks, and the sum of the processing power of all servers is greater than the tasks need to be processed.

4 The Design of ESPM

In this section, the elaborative process of designing ESPM is fully presented. Firstly, the process of selecting placement strategies with SPEA2 is designed. Then, the process of identifying placement strategy by TOPSIS and MCDM is presented. Finally, the overview of how to design ESPM is summarized in the last.

4.1 Generating Optimal Placement Strategy

There exists a fact that evolutionary algorithms are widely applied in multi-objective problem. And different algorithms are suitable for different multi-objective scenes. SPEA2 adopts a reasonable fitness allocation strategy and integrates the nearest neighbor density estimation technology, which makes the search process more accurate. In view of this, SPEA2 is used to solve the above problems. This process consists of four stages, i.e., encoding the relevant objective function, transforming the objective function into fitness function and adding constraints, using selection, crossover, mutation and other operators to mutate the algorithm, and finally, generating all the solutions.

Encoding. The objective functions, i.e., to minimize the access delay and load balance variance, to be optimized needs to be mapped into a mathematical problem. Coding is to transform the objective functions into different forms. The purpose of coding is that genetic algorithm can operate the functions easily. There are many ways of coding, different ways determine the efficiency of genetic evolution. In this method, binary coding method is proposed, which is easy to code and decode, and its crossover, mutation and other genetic operations are easy to realize.

Fitness Functions and Constraints. The fitness function is also called the evaluation function, which is a criterion for distinguishing the good or bad of individuals in a group according to the objective function. The specific process includes three processes, i.e., obtain the individual's phenotype according to the processing of the coding part, calculate the objective function value of the individual through the individual's phenotype, obtain the individual's fitness from the objective function value according to the principle of minimization. Besides, the constraint functions are set to constrain the algorithm to a reasonable range. Specific constraints have been shown in (13) and (14).

Selection Operator. Selection operation is utilized to determine how to select those individuals from the parent population in a certain way so as to inherit them to the next generation population. The selection operation is used to determine the recombination or cross individuals, and how many offspring will be generated by the selected individuals. In this method, the expectation selection operator is selected, and the random selection operation is performed according to the survival expectation of each individual in the next generation group.

Crossover Operator. Cross operation refers to the exchange of some genes between two paired chromosomes in a certain way, so as to form two new individuals. In this method, we choose the uniform crossover mode, which can make the genes on each locus of two matched individuals exchange with the same crossover probability, thus forming two new individuals.

Mutation Operator. Mutation operation refers to the replacement of gene values at some points in the coding string of individual chromosomes with other alleles at that point, so as to form a new individual. In this method, the mutation operation is performed on a bit or bits in the individual coding string which are randomly assigned by the mutation probability only because of the value on the seat.

Selecting Optimal Strategy by TOPSIS and MCAM. After generating placement strategies, the TOPSIS and MCAM owing to be selected to generate the optimal strategy. By detecting the distance between the evaluation object and the optimal solution as well as the worst solution, if the evaluation object is the closest to the optimal solution while the farthest from the worst solution, it is the best. Otherwise, it is not the best. Each index value of the optimal solution reaches the optimal value of each evaluation index. Each index value of the worst solution reaches the worst value of each evaluation index.

4.2 The Overview of ESPM

ESPM aims to realize the optimization of the objective functions (6) and (11). The overview of ESPM has showed in Algorithm 1. In this algorithm, the number of the population is A, the maximum amount of inheritance is B, and the exportation of ESPM is the optimal strategy OS.

Algorithm 1. Obtaining the best strategy by utilizing ESPM

Require: R
Ensure: OS
1: **for** $a = 1$ to A **do**
2: $b = 1$
3: **while** $b <= B$ **do**
4: **for** individuals in population **do**
5: Calculate access dalay time by (6)
6: Calculate load balance variance by (11)
7: Execute the selection, crossover and mutation operators to gengrate better offspring
8: **end for**
9: $b = b + 1$
10: **end while**
11: Obtain the optimal solution by TOPSIS and MCDM
12: **end for**
13: **return** OS

---

5 . Experimental Analysis

In this section, we will compare Benchmark, FFD, BFD and ESPM in the same experimental environment to verify the effectiveness of our method.

5.1 Performance Evaluations on ESPM

Experimental Results on the Average Access Delay. Correspondingly, the average access delay of tasks is calculated. As is shown in the Fig. 2, the average access delay keeps growing as the number of tasks increases. The picture describes that the delay calculated by ESPM is lower than Benchmark, FFD and BFD visually. The average access delay of ESPM are 0.15, 0.25, 0.43, 0.55 and 0.70 (s) when the scale of the tasks are set to 50, 100, 150, 200 and 250.

Experimental Results on the Average Resource Utilization. The average resource utilization, an important indicator, reflects the overall utilization of the system. This indicator directly represents the degree of usage of cells in the virtual machine, and expects to be a high degree. We can see that the Fig. 3 shows the average resource utilization calculated by ESPM and the other three classical methods intuitively. The average resource utilization are 0.74, 0.80, 0.85, 0.88 and 0.92 (s) when the scale of the tasks equal 50, 100, 150, 200 and 250.

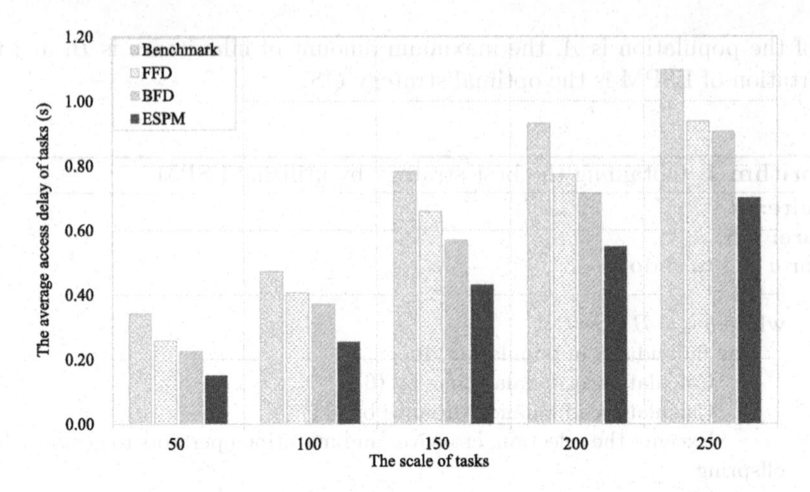

Fig. 2. Experimental results on the average access delay.

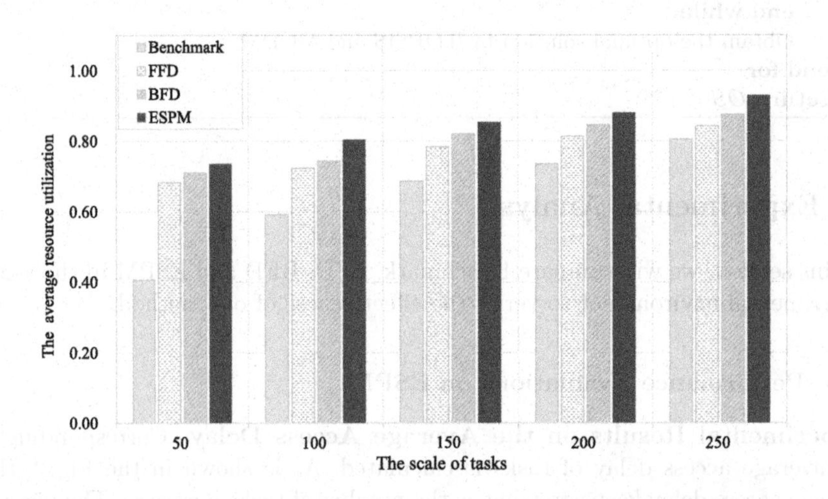

Fig. 3. Experimental results on the average resource utilization.

Experimental Results on the Load Balance Variance. Load balance variance is our main objective function in our paper. Figure 4 shows that the variance calculated by ESPM is lower than the other methods and the variance keeps growing as the number of tasks increases. Besides, the performance of ESPM keeps better than the other methods no matter the number of the tasks. After the detailed statistics, the load balance variance of ESPM are 0.09, 0.31, 0.43, 0.55, and 0.69 when the amount of the tasks are 50, 100, 150, 200 and 250.

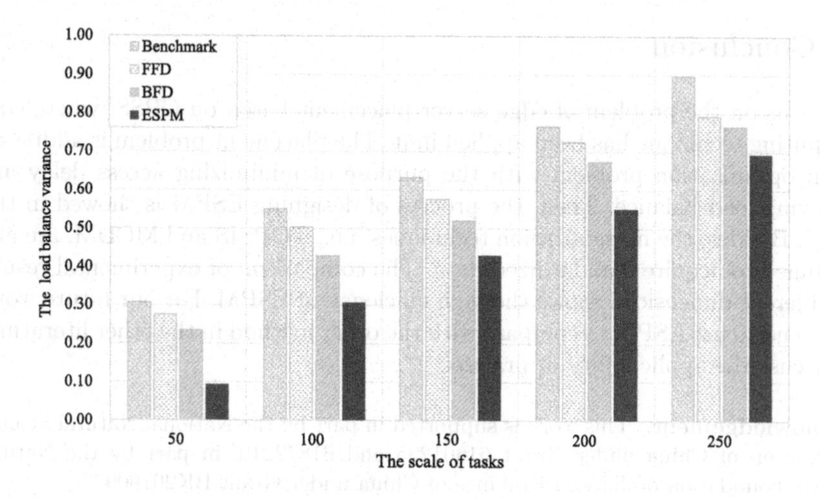

Fig. 4. Experimental results on the load balance variance.

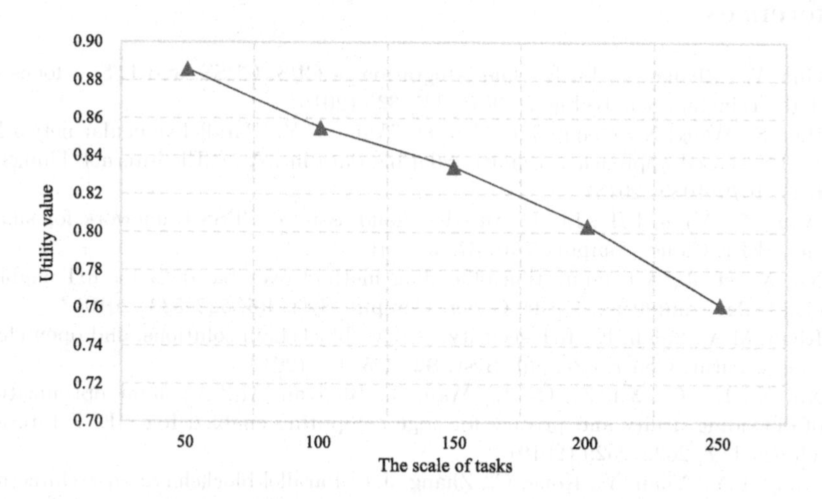

Fig. 5. Experimental results on the utility value.

Experimental Results on the Utility Value. Among all the unload strategies, the one with the highest utility value is the optimal strategy we want to obtain. Figure 5 shows that the highest utility value in different scale will decrease with the increasing of the task scale. Besides, when the scale of the tasks equals 50, the utility value achieves the highest value in all the solutions. The utility value of ESPM are 0.89, 0.85, 0.83, 0.80 and 0.79 when the amount of the tasks are set to 50, 100, 150, 200 and 250.

6 Conclusion

We focus on the problem of edge server placement based on CPSS, where edge computing technique has been applied in it. The placement problem is addressed as an optimization problem with the purpose of minimizing access delay and achieving load balance. Then, the process of designing ESPM is showed in this paper. Besides, the normalization techniques, i.e., TOPSIS and MCDM, are also combined to acquire standardized data. The comparison of experimental results on different dimensions shows the high efficiency of ESPM. For our future work, we intend to use ESPM to compare with the other method in the other literatures to discuss the applicability in practice.

Acknowledgement. This work is supported in part by the National Natural Science Foundation of China under Grant 61601235 and 61872219, in part by the Natural Science Foundation of Jiangsu Province of China under Grant BK20160972.

References

1. Guo, W., Zhang, Y., Li, L.: The integration of CPS, CPSS, and ITS: a focus on data. Tsinghua Sci. Technol. **20**(4), 327–335 (2015)
2. Han, S., Wang, X., Zhang, J.J., Cao, D., Wang, F.Y.: Parallel vehicular networks: a CPSS-based approach via multimodal big data in IoV. IEEE Internet Things J. **6**(1), 1079–1089 (2018)
3. Wang, P., Yang, L.T., Li, J.: An edge cloud-assisted CPSS framework for smart city. IEEE Cloud Comput. **5**(5), 37–46 (2018)
4. Xu, X., et al.: A computation offloading method over big data for IoT-enabled cloud-edge computing. Futur. Gener. Comput. Syst. **95**, 522–533 (2019)
5. Khan, M.A., Salah, K.: IoT security: review, blockchain solutions, and open challenges. Futur. Gener. Comput. Syst. **82**, 395–411 (2018)
6. Xu, X., He, C., Xu, Z., Qi, L., Wan, S., Bhuiyan, M.Z.A.: Joint optimization of offloading utility and privacy for edge computing enabled IoT. IEEE Internet Things J. **7**, 2622–2629 (2019)
7. Wang, F.Y., Yuan, Y., Rong, C., Zhang, J.J.: Parallel blockchain: an architecture for CPSS-based smart societies. IEEE Trans. Comput. Soc. Syst. **5**(2), 303–310 (2018)
8. Karakus, M., Durresi, A.: Quality of service (QoS) in software defined networking (SDN): a survey. J. Netw. Comput. Appl. **80**, 200–218 (2017)
9. Xu, X., et al.: An edge computing-enabled computation offloading method with privacy preservation for internet of connected vehicles. Futur. Gener. Comput. Syst. **96**, 89–100 (2019)
10. Mach, P., Becvar, Z.: Mobile edge computing: a survey on architecture and computation offloading. IEEE Commun. Surv. Tutor. **19**(3), 1628–1656 (2017)
11. Hu, Y.C., Patel, M., Sabella, D., Sprecher, N., Young, V.: Mobile edge computing-a key technology towards 5G. ETSI white paper 11(11):1–16 (2015)
12. Mao, Y., You, C., Zhang, J., Huang, K., Letaief, K.B.: A survey on mobile edge computing: the communication perspective. IEEE Commun. Surv. Tutor. **19**(4), 2322–2358 (2017)

13. Wang, F.: The emergence of intelligent enterprises: from CPS to CPSS. IEEE Intell. Syst. **25**(4), 85–88 (2010)
14. Wang, F.Y., Zheng, N.N., Cao, D., Martinez, C.M., Li, L., Liu, T.: Parallel driving in CPSS: a unified approach for transport automation and vehicle intelligence. IEEE/CAA J. Automatica Sinica **4**(4), 577–587 (2017)
15. Xu, X., et al.: An IoT-oriented data placement method with privacy preservation in cloud environment. J. Netw. Comput. Appl. **124**, 148–157 (2018)
16. Wang, S., Zhao, Y., Xu, J., Yuan, J., Hsu, C.H.: Edge server placement in mobile edge computing. J. Parallel Distrib. Comput. **127**, 160–168 (2019)
17. Li, Y., Wang, S.: An energy-aware edge server placement algorithm in mobile edge computing. In: 2018 IEEE International Conference on Edge Computing (EDGE), pp. 66–73. IEEE (2018)
18. Zeng, F., Ren, Y., Deng, X., Li, W.: Cost-effective edge server placement in wireless metropolitan area networks. Sensors **19**(1), 32 (2019)
19. Xu, Z., Liu, X., Jiang, G., Tang, B.: A time-efficient data offloading method with privacy preservation for intelligent sensors in edge computing. EURASIP J. Wirel. Commun. Netw. **2019**(1), 1–12 (2019). https://doi.org/10.1186/s13638-019-1560-8

Simulation of Software Reliability Growth Model Based on Fault Severity and Imperfect Debugging

Xuejie Sun[✉] and Jiwei Li

School of Information Science and Engineering, Ocean University of China, Songling Road No. 238, Qingdao, China
sunxuejie@stu.ouc.edu.cn

Abstract. The existing software reliability growth model (SRGMs) usually assumes that the detected faults can be eliminated well when considering different types of software faults, to simplify the problem. Therefore, given these existing defects, we propose a new non-homogeneous Poisson process (NHPP) SRGM based on considering different fault severity. According to the complexity of the fault, we define the software fault as three levels: Level I is a simple fault, Level II is a general fault, and Level III is a severe fault. In the process of fault detection, the model comprehensively considers the tester's ability to find problems and the number of remaining issues. In the process of debugging, the problems of imperfection and new fault introduction are considered. Two kinds of real data sets, fault classification and non-classification, were selected and we made simulation for the proposed model and other traditional SRGMs on the PyCharm platform. The experimental results show that the software reliability model considering fault severity has excellent performance of fault fitting and prediction on both types of data sets.

Keywords: Fault severity · Non-homogeneous Poisson process · Software reliability growth models

1 Introduction

With the comprehensive application of computer software technology in various systems such as daily life and safety-critical applications, software quality is an essential guarantee for software survival, and software reliability is an important index to measure software quality [1–3]. In the process of software development, it is necessary to consider the time of software release, the number of failures after software release, and the severity of the failures. Therefore, as an important means of quantitative evaluation and prediction of software reliability, in recent years, many software reliability growth models (SRGMs) based on time-domain have been proposed and successfully applied to the development process of various types of security-critical software. Among all SRGMs, the non-homogeneous Poisson process (NHPP) class SRGM is recognized as

H. Song and D. Jiang (Eds.): SIMUtools 2020, LNICST 370, pp. 140–152, 2021.
https://doi.org/10.1007/978-3-030-72795-6_12

the most effective and widely used model because of its excellent characteristics such as easy to understand and easy to use.

In 1979, Goel and Okumoto [4] first used NHPP to describe the SRGM, known as the G-O model. The model assumes that the failure detection rate function is constant. Many models are based on the G-O model to improve some assumptions and modify the G-O model to make the newly established model achieve a better fitting effect. Yamada [5] proposed the Delay S-Shaped model, which believed that the failure detection rate was a non-decreasing function changing with time. Meanwhile, Ohba [6, 7] considers that there is more than one fault in the software. Kapur et al. [8] introduced the concept of Fault Severity Factor (FSF). They proposed a SRGM with two types of fault. The first type is the model proposed by Goel and Okumoto. The second type introduced the logistic rate during the removal process. Later many materials indicate that there is more than one level of software failure in the software system. However, in the models that have been proposed, it is usually assumed that the faults detected are immediately eliminated to simplify the calculations. In other words, they assume that the troubleshooting process is perfect. This assumption ignores the fact that all detected faults cannot be eliminated due to resource constraints and the introduction of new faults in the troubleshooting process. Therefore, the models can not be well applied to a practical application environment. The fitting ability and prediction ability of the model needs to be improved.

In this paper, we show how to classify software failures into three categories: severe, general, and simple. According to the severity of the software fault, we proposed a new SRGM to quantify the software level. In the process of fault detection, for simple failure, in different time intervals, the probability of testers finding problems is only related to the number of residual failures. Therefore, unlike Kapur et al., we assume that the fault detection rate of this fault level is a decreasing function over time. In the other two more complex fault levels, we will consider the ability of testers to detect problems and the number of remaining issues. Similarly, during the troubleshooting process, we introduced the non-removable software failure rate into consideration of the possibility of imperfect debugging. We entered the failure lead-in rate parameter into account of the option of adding new errors. Based on the above analysis, we carried out experiments on two types of failure data and compared the experimental results with the existing SRGMs. From the results, we can see that the new model has a better fitting and prediction effect.

The rest of the paper is arranged as follows. In the second part, three models of different severity fault of software reliability are given. This section also describes the proposed new model. The third part presents the evaluation criteria of the model. Part four discusses estimating model parameters using the Least Square Method (LSE) [4, 25] and applying these models to two failure data sets. Finally, in the fifth part, the experimental results are compared with other classical SRGMs.

2 The Fault Levels Model

We improved the assumptions of the earliest NHPP class model and obtained the following assumptions [9–11].

1. The software test runs in the same way as the actual running profile.
2. The different types of software faults are mutually independent.
3. Assume that $m(t)$ is the mean value function (MVF) of the expected number of problems detected in time $(0, t)$. The cumulative errors to time t follow the Poisson process where the MVF is $m(t)$. We can get the predicted function $m(t)$ of the increasing error number is a bounded non-subtractive function that satisfies the requirement that $m(0) = 0$.
4. The expected number of errors at any time interval $(t, t + \Delta t)$ is proportional to the number of errors remaining at time t. The ratio is the failure detection rate $b(t)$. The function of the failure detection rate over time for different problem levels is assumed as follows:

   1) For Level I, in different time intervals, the probability of testers finding minor problems is only related to the number of residual failures, and the failure detection rate function is $b_1(t)$.
   2) For Level II, the probability of the tester finding non-minor problems is related not only to the number of residual failures, but also to the tester's learning ability. The failure detection rate function is $b_2(t)$.
   3) For Level III, similarly, we can assume the fault detection rate function of Level III is the same as that of Level II, which is $b_3(t)$.

5. We assume that the original fault content in the software is N. N_1, N_2 and N_3 represent the initial fault number of simple, general, and severe levels respectively. Eliminating errors isn't all perfect for problems of different grades. Thus, we introduce the failure introduction rate a.

   1) For Level I and Level II, when the fault level is low, the developer does not introduce new errors in the debugging process, the failure introduction rate are a_1 and a_2, and $a_1 = a_2 = 0$.
   2) For Level III, at this point, developers may introduce new problems when solving the problem, so assume that the failure introduction rate is a_3.

6. In practice, due to limited test resources, the skill and experience of the tester, and different severity of fault, it is not possible to eliminate all detected faults in the test phase. Therefore, we introduce the non-removable failure rate c.

   1) For Level I and Level II, when the fault is relatively simple, we assume that the fault can be completely removed, so the non-removable failure rate is $c_1 = c_2 = 0$.
   2) For Level III, there are some software glitches that the software development team can not eliminate, so we assume that the non-removable failure rate is c_3.

Based on the above assumptions, we can construct the model as follows, where, the $m(t)$ of Level I, II, III are respectively expressed as $m_1(t), m_2(t), m_3(t)$.

2.1 Level I SGRM

According to the above assumptions, and from [12, 13], the model of the simple problem can be expressed as

$$\frac{dm(t)}{dt} = b_1(t) \times [N - m_1(t)] \tag{1}$$

Since the failure detection rate of minor problems by testers is only related to the number of remaining faults, with the continuous correction of the failures, the number of residual failures in the software become less and less, and the probability of detection becomes lower and lower. Therefore, it is assumed that the function of the failure detection rate over time of the tester for minor problems satisfies the following equation.

$$b_1(t) = \frac{b_1}{1+t}(0 \le b_1 \le 1) \tag{2}$$

In the formula, b_1 denotes a fault detection rate of simple faults found by the tester at the initial time.

Substitute (2) into (1), and solving (1) under the condition $m_1(0) = 0$, we can get the MVF of Level I as follows

$$m_1(t) = N[1 - (1+t)^{-b_1}] \tag{3}$$

2.2 Level II SGRM

Similarly, according to the above assumptions, the model of the middle problem can be formulated as

$$\frac{dm(t)}{dt} = b_2(t) \times [N - m_2(t)] \tag{4}$$

The logistic Testing-Effort Function (TEF) [14–18] can well describe extensive test work, we use the ratio of test coverage growth rate and uncovered code to express the failure detection rate of the Level II fault.

The logistic TEF formula for the period $(0, t]$ is

$$W(t) = \frac{W_{max}}{1 + A\exp(-\alpha t)} \tag{5}$$

Where A is a constant, α is the consumption rate of testing effort, and W_{max} is the total testing effort that can be consumed finally. The current TEF rate at test time t can be shown as

$$W'(t) = \frac{W_{max} A\alpha \exp(-\alpha t)}{[1 + A\exp(-\alpha t)]^2} \tag{6}$$

For simplicity of calculation, we assume $W_{max} = 1$. The function of the failure detection rate over time can be represented as

$$b_2(t) = \frac{W'(t)}{1 - W(t)} = \frac{\alpha}{1 + A\exp(-\alpha t)} \tag{7}$$

With the tester's continuous understanding of the software under test, the tester can write better test cases, and the failure detection rate will increase; at the same time, as the failure is constantly corrected, the remaining failure in the software is less and less, and the probability of detection is lower and lower. Therefore, the fault detection rate at this time is affected by these two aspects.

Substitute (7) into (4), and solving (4) under the boundary condition $m_2(0) = 0$, we can obtain the MVF of Level II as follow

$$m_2(t) = \frac{N[\exp(\alpha t) - 1]}{A + \exp(\alpha t)} \tag{8}$$

2.3 Level III SGRM

According to the above assumptions, the model of the complex problem can be shown as

$$\frac{dm_3(t)}{dt} = b_3(t) \times [N(1 + a_3 t) - m_3(t)] - c_3 m_3(t) \tag{9}$$

According to assumptions 2, the parameters of $b_2(t)$ and $b_3(t)$ should be different. Therefore, the function of the failure detection rate over time of Level III can be formulated as

$$b_3(t) = \frac{\alpha_1}{1 + A_1 \exp(-\alpha_1 t)} \tag{10}$$

Substitute (10) into (9), and solving (9) with the condition $m_3(0) = 0$, the MVF of Level III is

$$m_3(t) = \frac{N\alpha_1}{[\exp(\alpha_1 t) + A_1](\alpha_1 + c_3)^2 \exp(c_3 t)} \times [(1 + a_3 t)(\alpha_1 + c_3)\exp(\alpha_1 t + c_3 t) - a_3 \exp(\alpha_1 t + c_3 t) + a_3 - \alpha_1 - c_3] \tag{11}$$

Therefore, we can assume different parameters in front of the formulas of varying Level and get the new SGRM as follow

$$m(t) = \sum_{i=1}^{k=3} p_i m_i(t) \tag{12}$$

We call it a fault levels model. In Eq. (12), p_1, p_2 and p_3 need to satisfy $\sum_{i=1}^{k=3} p_i = 1$. And the initial fault number satisfies $p_i N = N_i (i \in 1, 2, 3)$.

3 Model Comparison Criteria

We analyze models based on the ability to fit software failures and the ability to predict future software behavior based on observed failure data sets. The four standards of model comparison are:

3.1 The Fitting Effect Criterion

To quantitatively compare the effects of model fitting data, we use the Sum of Squared Errors (SSE), the Mean Square of Fitting Errors (MSE), and the R-square (R) [19–22].

MSE. The MSE formula is shown below

$$MSE = \frac{\sum_{i=1}^{k} (m(t_i) - m_i)^2}{k} \tag{13}$$

The smaller the value of MSE, the lower the fitting error, and the better the performance.

SSE. The calculation formula of SSE is as follows

$$SSE = \sum_{i=1}^{k} (m(t_i) - m_i)^2 \tag{14}$$

Similarly, the smaller the value of SSE is, the lower the fitting error is, that is, the better the performance is.

R. The formula for R is

$$R = 1 - \sum_{i=1}^{k} (m(t_i) - m_i)^2 \Big/ \sum_{i=1}^{k} (m_i - m_{ave})^2 \tag{15}$$

Unlike the above, the closer the value of R is to one, the better the fitting effect will be.

3.2 The Predictive Goodness Criterion

The ability of a model to predict failure behavior based on the current number of failures is called predictive validity. Musa [8, 23, 24] proposed a method that could be used to calculate the Relative Error (RE) of the data set to represent the predictive validity.

$$RE = \frac{\hat{m}(t_q) - q}{q} \tag{16}$$

First, assuming that q faults are found at the end of the test time t_q, we use the failure data before $t_e (t_e \leq t_q)$ to predict the parameters of $m(t)$. By substituting the values of these prediction parameters into MVF, we can obtain the number of failures $m(t_q)$ over time t_q. The second step is to compare the predicted value with the actual amount q. Third, repeat the process for different t_e values. The validity of the prediction can be verified by drawing the relative errors of different t_q values. The closer the number is to zero, the better the prediction. Where the positive error represents an overestimate; A negative error indicates underestimation.

4 Model Simulation and Result Analysis

In this section, the proposed model has been tested on two real data sets to evaluate its validity. At the same time, the model with better performance on each data set and the classical models are used as the comparison model. In this paper, we use PyCharm software as a simulation platform. The Least Square Estimation (LSE) method is used to estimate the model parameters [4, 25, 26], and the estimation results generated by LSE are unbiased.

4.1 Data Set I

The data is from Misra, which is the failure data of software developed in the contract between IBM's Federal Systems Division and NASA's Johnson [27, 28]. The software was tested for 38 weeks, during which 2456.4 computer hours were used, and 231 faults were removed. It can be seen that faults are classified when failure data is recorded. The proposed model has been compared with the model proposed by Kapur et al. [29, 30], who also used the data set for experiments.

Analysis of Fitting Results. The Parameter Estimation result and the goodness of fit results for the proposed SRGM are given in Table 1. It is observed that the proposed model has the smallest value of SSE and MSE when compared with the SDE model. The two models have the same value of R. From the weight coefficient values of the proposed model, we found that Level I and Level II faults account for a significant proportion of the DS-I, while Level III accounts for a small percentage. The SDE model also reflects this phenomenon. The fitting results of the two models are close to the original data set, which further proves the validity of the model. Compared with the SDE model, the total number of faults fitted by the proposed model is closer to the total number of faults in the original data set. Figure 1 describes the comparison between the fitting value $m(t_i)$ of each failure data in the DS-I by the two models and the actual observed failure data m_i. It can be seen from Fig. 1 that the fitting results of the two models basically coincide with the real data. Combined with Table 1 and Fig. 1, the proposed model performs better in DS-I fitting.

Analysis of Prediction Results. We train with the failure data of the first 22 weeks, and compare the predicted value with the real cost to get the RE curves in Fig. 2. It can find that the RE value of the proposed model is the closest to zero as a whole when compared with the SDE model. It means that the proposed model predicts more accurately.

Table 1. Fitting parameters for DS-I.

Models under comparisons	Parameter estimation	MSE	SSE	R
Stochastic differential equation-based (SDE) model	$b_1 = .059, b_2 = .104, b_3 = .378, \beta = 66.593$ $p_1 = .64, p_2 = .342, p_3 = .018, a = 420$ $\sigma_1 = .048, \sigma_2 = .185, \sigma_3 = .599$	7.22	274.35	.999
Proposed SRGM (fault level (FL) model)	$p_1 = .711, p_2 = .268, p_3 = .021$ $N = 382, A = 12.01, A_1 = -8.14$ $\alpha = 1.98, \alpha_1 = 2.91, a_3 = .006, c_3 = 2.89$	6.32	240.15	.999

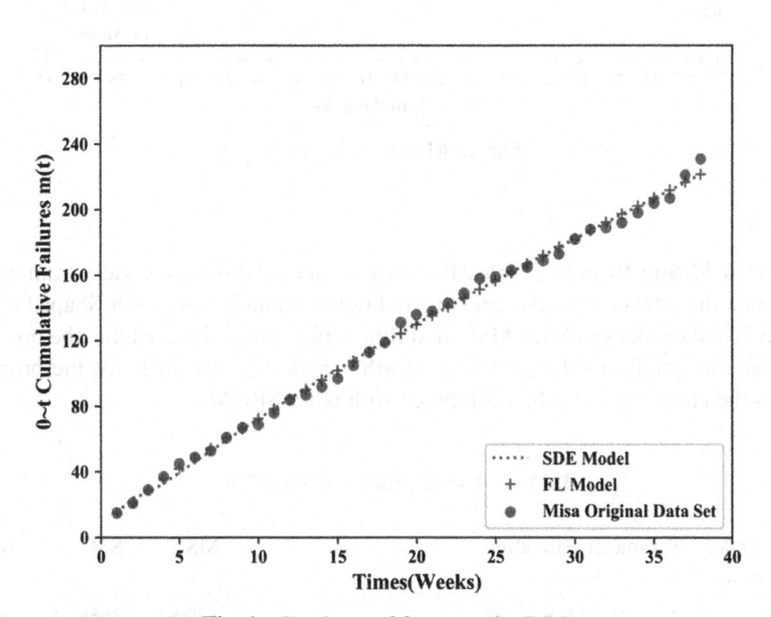

Fig. 1. Goodness of fit curves for DS-I.

4.2 Data Set II

The Ohba data set is mentioned in a paper written by Ohba for a database software system that contains approximately 1317,000 lines of code [12, 31, 32]. The software was tested for 19 weeks, during which 47.65 computer hours were used, and 328 faults were removed. Different from the DS-I, DS-II does not classify the failures when recording the failure data. Therefore, we choose three models that are tested with DS-II to compare with the proposed and prove that the proposed model has an excellent performance in both classified and unclassified data sets.

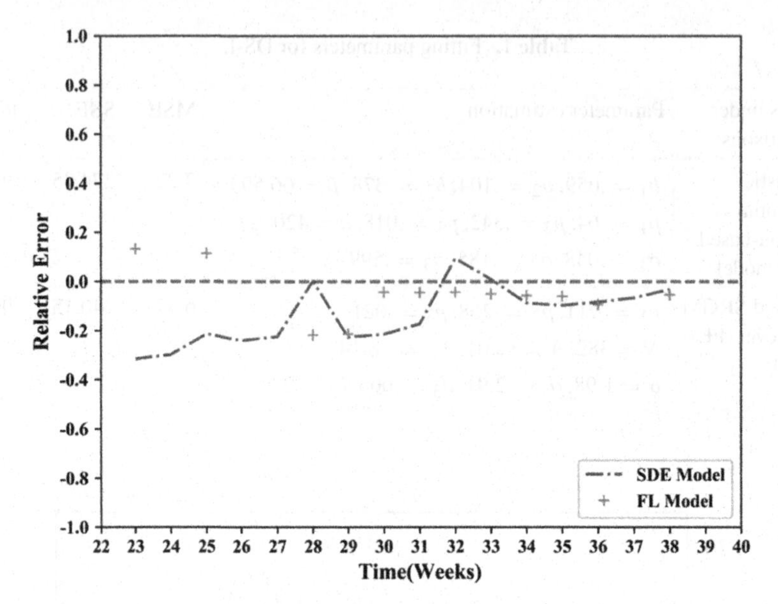

Fig. 2. RE curves on DS-I.

Analysis of Fitting Results. Table 2 lists the estimates of different model parameters on DS-II, including the G-O model, and the traditional Yamada Delayed S-Shaped model. We also give the values of SSE, MSE, and R in Table 2. It is observed that the proposed model has the smallest value of SSE and MSE, and the value of R for the proposed model is the closest to one when compared with other SRGMs.

Table 2. Fitting parameters for DS-II.

Models under comparisons	Parameter estimation	MSE	SSE	R
G-O model	$N = 760.53, r = .03$	139.82	2656.48	.986
Delay S-shaped model	$N = 374.05, r = .21$	168.67	3204.79	.984
Improved G-O model	$N = 451.32, b = 13.03, A = .04$	89.81	1706.43	.991
Proposed SRGM (fault level (FL) model)	$p_1 = .073, p_2 = .755, p_3 = .172$ $N = 506, A = 19.28, A_1 = 61.88$ $\alpha = .301, \alpha_1 = 1.64, a_3 = -0.095, c_3 = .47$	32.47	616.95	.997

Different from the DS-I, we found that Level II and Level III faults account for a significant proportion of the DS-II, while Level I accounts for a small percentage. This is because different software development environments and application scenarios have different fault level distribution.

To clearly show the fitting effect diagram, we chose the two models with the best fitting effect for comparison. The two models with better fitting effect are the proposed model, and the Improved G-O model. Figure 3 describes the contrast between the appropriate value ($m(t_i)$) of each failure data in the DS-II by the above two models and the actual observed failure data (m_i). On average, the proposed model performs better in data set fitting.

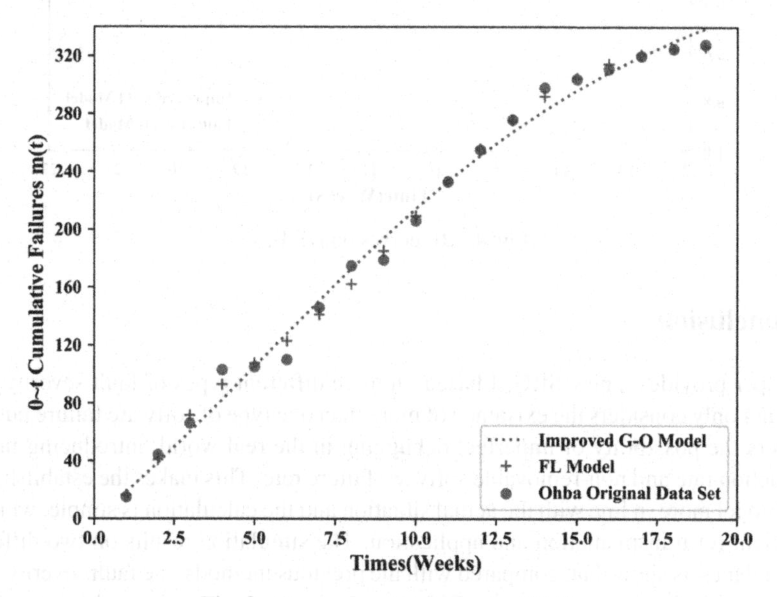

Fig. 3. Goodness of fit curves for DS-II.

Analysis of Prediction Results. In DS-II, we train with the failure data of the first 12 weeks, and compare the predicted value with the real value to get the RE curves in Fig. 4. For the convenience of observation, the RE curves of two models with a better fitting effect on the DS-II are depicted in Fig. 4. It is worth noting the curve of the Improved G-O model deviates from zero by a large margin. The RE value of the proposed model is the closest to zero as a whole, and the speed of the curve approaching zero is the fastest after eighteen weeks, which indicates that the proposed model has excellent predicted results on the DS-II.

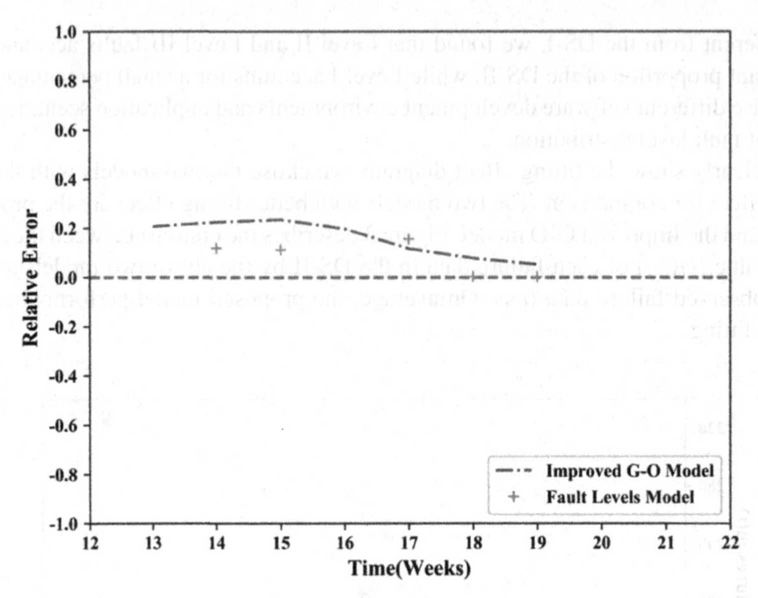

Fig. 4. RE curves on DS-II.

5 Conclusion

This paper provides a new SRGM based on three different types of fault severity. The model not only considers the existence of more than one type of software failure but also considers the possibility of imperfect debugging in the real world, introducing failure introduction rate and non-removable software failure rate. This makes the establishment of the model more in line with the actual situation and the calculation is simple, which is convenient for transplantation and application. The simulation results on two different types of data sets show that, compared with the previous methods, the fault severity classification can effectively improve the fitting effect and prediction effect of the traditional software reliability model, which plays an essential role in the theoretical research and engineering application of the software reliability model.

References

1. Mengmeng, Z., Hoang, P.: A software reliability model incorporating martingale process with gamma-distributed environmental factors. Ann. Oper. Res., 1–22 (2018)
2. Jaiswal, A., Malhotra, R.: Software reliability prediction using machine learning techniques. Int. J. Syst. Assur. Eng. Manag. **9**(1), 230–244 (2018)
3. Chatterjee, S., Shukla, A.: A unified approach of testing coverage based software reliability growth modelling with fault detection probability, imperfect debugging, and change point. J. Softw. Evol. Process **31**(3), e2150 (2019)
4. Aihua, G., Chunyang, Z., Qingxin, H.: Improvement of G-O model of software reliability growth. J. Inner Mongolia Univ. Nat. Sci. Edn. **45**(2), 84–87 (2014)
5. Kumar, R., Kumar, S., Tiwari, S.K.: A study of software reliability on big data open source software. Int. J. Syst. Assur. Eng. Manage. **10**(2), 1–9 (2019)

6. Mengmeng, Z., Hoang, P.: A two-phase software reliability modeling involving with software fault dependency and imperfect fault removal. Comput. Lang. Syst. Struct. **53**, 27–42 (2017)
7. Hwang, S., Pham, H.: Quasi-renewal time-delay fault-removal consideration in software reliability modeling. IEEE Trans. Syst. Man Cybern. **39**(1), 200–209 (2009)
8. Garmabaki, A.H., Aggarwal, A.G., Kapur, P.K.: Multi up-gradation software reliability growth model with faults of different severity. In: Industrial Engineering and Engineering Management, pp. 1539–1543 (2011)
9. Goseva-Popstojanova, K., Trivedi, K.S.: Failure correlation in software reliability models. IEEE Trans. Reliab. **49**(1), 37–48 (2000)
10. Singh, V.B., Sharma, M., Pham, H.: Entropy based software reliability analysis of multi-version open source software. IEEE Trans. Softw. Eng. **44**(12), 1207–1223 (2018)
11. Yaghoobi, T.: Parameter optimization of software reliability models using improved differential evolution algorithm. Math. Comput. Simul. **17**, 46–62 (2020)
12. Ohba, M.: Software reliability analysis models. IBM J. Res. Dev. **28**(4), 428–443 (1984)
13. Nagaraju, V., Fiondella, L., Zeephongsekul, P., Jayasinghe, C.L., Wandji, T.: Performance optimized expectation conditional maximization algorithms for nonhomogeneous poisson process software reliability models. IEEE Trans. Reliab. **66**(3), 722–734 (2017)
14. Vizarreta, P.: Assessing the maturity of SDN controllers with software reliability growth models. IEEE Trans. Netw. Serv. Manage. **15**(3), 1090–1104 (2018)
15. Chatterjee, S., Singh, J.B., Roy, A.: NHPP-Based software reliability growth modeling and optimal release policy for N-Version programming system with increasing fault detection rate under imperfect debugging. Proc. Natl. Acad. Sci. India Sect. A Phys. Sci. **90**, 11–26 (2020)
16. Peng, R., Ma, X., Zhai, Q.: Software reliability growth model considering first-step and second-step fault dependency. J. Shanghai Jiaotong Univ. (Sci.) **24**(4), 477–479 (2019)
17. Li, Q., Pham, H.: A generalized software reliability growth model with consideration of the uncertainty of operating environments. IEEE Access **7**, 84253–84267 (2019)
18. Utkin, L.V., Coolen, F.P.A.: A robust weighted SVR-based software reliability growth model. Reliab. Eng. Syst. Saf. **176**(8), 93–101 (2018)
19. Briones, B.A.: Wiley encyclopedia of electrical and electronics engineering. Charleston Adv. **21**(2), 51–54 (2019)
20. Dai, Y., Xie, M., Poh, K.: Modeling and analysis of correlated software failures of multiple types. IEEE Trans. Reliab. **54**(1), 100–106 (2005)
21. Rani, P., Mahapatra, G.S.: A novel approach of NPSO on dynamic weighted NHPP model for software reliability analysis with additional fault introduction parameter. Heliyon **5**(7) (2019)
22. Lin, C., Huang, C.: Enhancing and measuring the predictive capabilities of testing-effort dependent software reliability models. J. Syst. Softw. **81**(6), 1025–1038 (2008)
23. Li, Q., Pham, H.: A testing-coverage software reliability model considering fault removal efficiency and error generation. PLOS ONE **12**(7), e0181524 (2017)
24. Huang, C., Kuo, S., Lyu, M.R.: An assessment of testing-effort dependent software reliability growth models. IEEE Trans. Reliab. **56**(2), 198–211 (2007)
25. Jing, Z., Hongwei, L., Gang, C., Xiaozong, Y.: A software reliability growth model considering differences between testing and operation. J. Comput. Res. Dev. **43**(3), 503 (2006)
26. Misra, P.N.: Software reliability analysis. IBM Syst. J. **22**(3), 262–270 (1983)
27. Hui, Z., Liu, X.: Research on software reliability growth model based on gaussian new distribution. Procedia Comput. Sci. **166**, 73–77 (2020)
28. Kapur, P.K., Anand, S., Yamada, S., Yadavalli, V.S.: Stochastic differential equation-based flexible software reliability growth model. Math. Probl. Eng. **2009**, 1–15 (2009)
29. Xie, J., Jinxia, A.N., Zhu, J.: NHPP software reliability growth model considering imperfect debugging. J. Softw. **21**(5), 942–949 (2010)

30. Chatterjee, S., Chaudhuri, B., Bhar, C.: Optimal release time determination using FMOCCP involving randomized cost budget for FSDE-based software reliability growth model. Int. J. Reliab. Qual. Saf. Eng. **27**(1), 257–279 (2020)
31. Yamada, S., Ohba, M., Osaki, S.: S-shaped reliability growth modeling for software error detection. IEEE Trans. Reliab. **32**(5), 475–484 (1983)
32. Laura, P.: Software fault tolerance techniques and implementation. Artech (2001)

Research on Intelligent Recommendation Method of e-commerce Hot Information Based on User Interest Recommendation

Huang Jingxian(✉)

Guizhou Minzu University, Guiyang 550025, Guizhou, China

Abstract. In order to improve the effect of information intelligent recommendation, this paper puts forward the optimization research of hot information intelligent recommendation method based on user interest recommendation. The quality of user collecting modeling directly determines the quality of personalized recommendation by collecting user interest and making intelligent recommendation based on information classification model. The basic process, principle and algorithm of personalized recommendation system are studied. Based on the algorithm principle of user data collection, user model representation, user model learning and user model updating, this paper optimizes the recommendation methods, completes the design and implementation of the intelligent recommendation engine in the special system. The recommendation system simulates the store salesperson to provide product to customers and help users find the required information. Finally, through the experiment, it is proved that the intelligent recommendation method of e-commerce hot spot information based on user interest recommendation has improved the customer satisfaction and the effectiveness of recommendation information in the practical application process, which fully meets the research requirements.

Keywords: User interest · Interest recommendation · E-commerce · Hot information · Intelligent recommendation

1 Introduction

With the development of Internet, website provides more and more information for users, and its structure becomes more and more complex. It is more and more difficult to find the required information in the massive information on the network in time. The recommendation system simulates the store salesperson to provide product recommendation to customers and help users find the required information. It can effectively retain users and improve the click through rate and user loyalty of the website. Because recommendation system can help enterprises to achieve the purpose of personalized marketing, and then improve sales, and create the biggest profit for enterprises, coupled with the rise of the concept of personalized service, many e-commerce enterprises begin to pay attention to the application of recommendation system. The good development

H. Song and D. Jiang (Eds.): SIMUtools 2020, LNICST 370, pp. 153–166, 2021.
https://doi.org/10.1007/978-3-030-72795-6_13

and application prospect of intelligent recommendation system has gradually become an important research content of Web Intelligent Technology, which has been widely concerned by many researchers. In recent years, recommendation system has been developed rapidly in theory and practice, but with the further expansion of the scale of the applied system, intelligent recommendation system is also facing a series of challenges.

In reference [1], a recommendation algorithm based on non-negative matrix decomposition (NMD) based on user clustering is proposed. Based on the original NMD model, users are clustered based on clustering idea and combined with users' rating data to fully mine the correlation among users, on this basis, the intelligent recommendation method of hot information of e-commerce is carried out, but the method has the problem of low user satisfaction, which is difficult to be widely used in practice. Reference [2] proposed an information personalized recommendation algorithm for e-commerce shopping guide platform based on big data and artificial intelligence. The task is decomposed into multiple tasks by Map, and the final processing result is obtained by combining the decomposed multi task processing results with Reduce. Two MapReduce and one Map are used to parallelize the user preference acquisition algorithm in the platform, However, this method has the problem of low effectiveness of recommendation information. Reference [3] proposed a new e-commerce hot information group recommendation method, aiming at the problems of low accuracy of e-commerce hot information group recommendation and the difficulty of fusion of preference conflicts among group members. This method combines the influence factor of group leader and the influence factor of project heat. Based on K-nearest neighbor as the target group, this paper designs a group recommendation algorithm based on preference fusion, which is used for intelligent recommendation of hot e-commerce information. However, this method has the problem of low user satisfaction, and the actual application effect is not ideal.

In this paper, the design of recommendation algorithm and the architecture of recommendation system in intelligent recommendation system are studied. The main content of this paper is to apply neural network and fuzzy logic to the recommendation system, mainly involving the real-time performance of the recommendation system, the recommendation system architecture and the application research of the recommendation system based on Web mining. The overall design scheme of e-commerce hot information intelligent recommendation method based on user interest recommendation is as follows:

(1) The user interest information is collected and the information classification model is constructed.
(2) Design the basic process, principle and algorithm of personalized intelligent recommendation of hot e-commerce information.
(3) Experiments are designed to verify the advantages of the proposed method.
(4) Draw a conclusion and look forward to the future.

2 Intelligent Recommendation Method of Hot Information in E-commerce

2.1 Hot Spot Information Collection Method of e-commerce

Recommendation system is an important way for e-commerce websites and content-based app users, and also an important means for them to achieve profits. It is an artificial intelligence recommendation system. Intelligent recommendation is an important example of the application of recommendation system in the field of video. Recommendation system products have different forms in specific application scenarios, but they are ultimately to solve the long tail problem of content and goods, and increase the exposure of content and goods. According to the collected user preferences and user's historical behavior, the recommendation system finds out what users may like and what products they buy, and then recommends the results to users. It makes e-commerce websites and content industries reduce a lot of manual editing links in the work of recommending users [4]. Using some algorithm models to recommend, it can surpass manual editing recommendation and achieve better recommendation effect. Recommendation engine can recommend content and items of interest to all users, and can let different users get different recommendation results because of the use of information filtering technology in recommendation engine. The purpose of recommender system is to let users get satisfactory recommender items, so the only index to evaluate recommender system is user satisfaction [5]. If the recommendation set contains too many similar items, the redundancy is very high, which has little effect on users. On the contrary, if the number of recommended items is small, users can make better choices. Classification and regression methods are generated from machine learning technology and can be used in UCI machine learning database. They generally evaluate some algorithm by error rate. But it is not very good to use this method only in the recommendation system. Sometimes, considering the performance of the algorithm, the accuracy has to be affected. Sometimes even if the accuracy is affected a little, the algorithm performance is good, and most users can accept it. Therefore, one of the principles that must be paid attention to in evaluation is the balance between accuracy and performance [6]. Offline evaluation is more suitable for the evaluation of recommendation system, mainly because online evaluation requires building a comprehensive engineering system, which needs many users, so it is generally considered that offline evaluation is easier to implement, especially in the test of experimental algorithm. Figure 1 shows how the recommendation engine works.

Analysis of Fig. 1 shows that the e-commerce hot information recommendation engine will first collect a variety of information about items, keywords, user gender, age and other information, as well as the user's purchase records. Through the integration and analysis of these information, it recommends different e-commerce hot information for different users.

The recommendation engine will adopt different recommendation mechanisms. One recommendation mechanism is to analyze the data in the data source, get certain rules, and recommend items to users according to these rules. Another recommendation mechanism is to predict and calculate the preference degree of the target user for the items, and recommend the items to the user according to the order of preference degree [7].

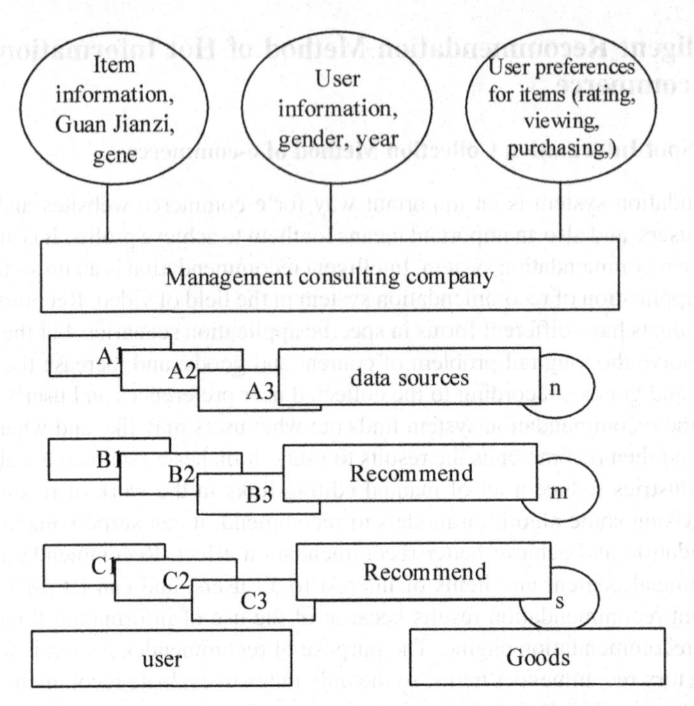

Fig. 1. Working principle of hot information recommendation engine of e-commerce

It's very simple to use only one recommendation engine in the recommendation system. In order to make the recommendation system achieve better recommendation effect, we often adopt a recommendation strategy with a combination of multiple recommendation mechanisms in line with business scenarios. For example, in the recommendation system, the recommendation mechanism considering fusion includes similar video recommendation based on users' historical behavior browsing or comments, popular video recommendation based on video popularity, etc. The purpose is to recommend the interested video information to the user more accurately and give the reason of recommendation to the user better.

2.2 User Interest Recommendation Algorithm

The recommendation algorithm has many branches. This paper focuses on the collaborative filtering based recommendation algorithm [8]. Collaborative filtering can be divided into neighborhood based and model-based collaborative filtering. Neighborhood based collaborative filtering, also known as memory based collaborative filtering, uses the method of calculating similarity to get neighborhood users with similar interests. Neighborhood collaborative filtering is divided into user based and item based collaborative filtering [9].

Table 1. User interest model collaborative filtering information

Algorithm level	Attribute data	Attribute example
Collaborative filtering recommendation algorithm	Demographic attributes	Collaborative filtering recommendation on population attribute
Content based recommendation	Geographical attributes	Content recommendation on geographical attributes
Method hybrid push algorithm	Asset attribute	Make mixed recommendation on asset attributes
Popularity recommendation algorithm	Interest attribute	Make popularity recommendation on Xing attribute
	Demographic attributes	Kmeans
	Geographical attributes	naive bayesian classification linear regression
	Asset attribute	SvD
	Interest attribute	A model of implicit meaning
Model based collaborative filtering	Demographic attributes	LFM-Apriori
clustering algorithm	Geographical attributes	confidence
Classification algorithm	Asset attribute	Wide and De
Regression algorithm	Interest attribute	Iterative processing
Matrix decomposition	Demographic attributes	Scoring vector
association algorithm	Geographical attributes	similarity analysis
neural network	Asset attribute	correlation coefficient
Feature recognition	Interest attribute	Interest information trend
Collaborative filtering recommendation algorithm	Demographic attributes	Collaborative filtering recommendation on population attribute
Content based recommendation	Geographical attributes	Content recommendation on geographical attributes

Model based collaborative filtering uses historical data and algorithms to build prediction models. The Table 1 shows the types of model-based collaborative filtering.

According to the idea of data mining, there are three ways to recommend offline evaluation:

(1) Consider recommendation as information retrieval, that is, from the selection of a subset of user related data. From this point of view, generally from the accuracy and return of two aspects.

(2) Think of recommendation as a regression problem.
(3) Think of recommendation as a classification problem.

At present, the academic circles mainly use the first way of thinking to regard recommendation as an information retrieval problem.

The method of small sample gradient descent is used to extract the user's interest information at random and mine the received feature information. If r is the interest information, u is the independent feature value of information search, i and j respectively represent the scoring vector and similarity of data features, then the collected information feature correlation numerical algorithm can be recorded as follows:

$$simr_{ij} = \frac{\sum (r_{ui} - \hat{r}_i)(r_{uj} - \hat{r}_j)}{2\left\|\hat{r}_i\right\|_j * \left\|\hat{r}_j\right\|_i - 1} \tag{1}$$

Combined with the information feature value, the evaluation standard of interest data feature similarity is further standardized. Euclidean distance can be used to evaluate the relationship between two users [10]. The smaller the distance is, the more relevant the user is, and vice versa. If Q is all the data sets searched by users, a and b are the pre-selected score value of similarity and the mean value of similarity score respectively, then the correlation coefficient algorithm can be recorded as:

$$V = \frac{|Q_a + Q_b|^{i-j} - (i-j)|Q_a * Q_b|}{\sum \lim\limits_{1 \to \infty} 2simr_{ij}} \tag{2}$$

According to the data feature coefficients, the different data similarity results are further classified. If $V > 1$, the data can be classified into similar feature sets. If $V \leq 1$, the data features are further classified and mined to calculate their feature similarity categories. According to different information feature similarity values, the feature levels are divided and integrated into feature sets, so as to evaluate the difference feature values in different levels and select the optimal evaluation scale for recording [4]. Combined with the fuzzy control algorithm, the network traffic information features are collected, the user retrieval information trends are further mined, and the reference feature values are standardized. The restriction condition can be recorded as Y, then the interest degree algorithm of information feature data mining is:

$$I(n) = [\sup(simr_{ij} * \Delta V), \mathit{congfidence}(Y)] \tag{3}$$

In the actual operation of data feature collection, the interest information patterns with the same characteristics need to be classified into a set, and the value of the additional attribute of the same density in the set is 1 or 0, which is recorded as $I_e(n) = \begin{cases} 1 \\ 0 \end{cases}$, By calculating the additional attributes of the user's interested information, the additional attributes of the feature data required by the user are found out, and the support degree is calculated [11]. If the calculation result is greater than the preset support degree feature value, the depth feature mining of the data is carried out according to the horizontal division principle. Under the same data feature structure, the global analysis of user interest data is carried out, and the feature subset of any data is recorded as $sim(x, y)$, In the case of different feature structures, deep mining is carried out according to the vertical division principle. If each feature and other categories will contain a group column, it is recorded as h, and the common feature attribute between each feature subset is recorded as u, and the difference feature value is recorded as e. The number of times of mining implicit information in the data is t, and the number of data feature categories is n. Then we can get the optimal value of multi distributed information feature data mining based on the principle of cryptic meaning, standardize the association rules of the current personalized information number recommendation system, and input the standard into the system to ensure that the information recommendation always follows the rule instructions for statistical processing [12].

set up $A_i = \left(y_1, y_2, \ldots, y_j\right)$ a feature data set representing the book selection direction of the target user. The combined values of multiple weakly classified linear feature data are recorded as $A_m(x)(\mathrm{m} = 1, 2, \cdots, \mathrm{M})$.

If there are m users in the system at the same time, if the target user is recorded as D_i, the similarity feature threshold $L(x)$ between multiple users is calculated in the process of business hotspot information category retrieval by multiple users. The specific algorithm is as follows:

$$L(x) = \sum\sum \lim_{0\to\infty} \frac{I(n)\prod m - D_i}{2\|A_m(x) - A_i\|} \tag{4}$$

The feature values of multiple user retrieval are calculated, and the feature values of user retrieval belong to the range of $L(x)$ calculation results. Then the similar data are further classified and processed [13]. If there are n similarity retrieval users, then based on the above algorithm, the information of the detected nearest neighbor users is further output, and the reference value z is selected as the feature similarity prediction data set. Furthermore, the user similarity level is judged by Pearson system. The specific algorithm is as follows:

$$sim(A, B, C, D) = L(x) \sum\sum \frac{(p - f)(m - n)}{2\sqrt{L(x) + z(a + b)}} - 1 \tag{5}$$

In the above algorithm, p and f respectively represent the mean value of the feature score of the book grade.

$$E = \sum sim(x, y) - h\{ue_{ij}^{-\frac{1}{2}tn}\Big|I_e(n)_{ij}^{n-1}\} \tag{6}$$

Based on the above algorithm, we can effectively check and detect the non-important keywords of users, and classify the feature categories according to the feature values, so as to better help users to filter their favorite interest information. In different recommendation scenarios, the calculation methods of the selected similarity are also different. Let the range of A_n point eigenvector is $(a_1, a_2, a_3, ..., a_n)$, The value range of feature vector of B_m point is $(b_1, b_2, b_3, ..., b_n)$. Then the expression formula of the common characteristics of A_n point and B_m point is:

$$sim(A_n, B_m) = \ln E * \sqrt{\sum\sum \lim_{1\to\infty} (x - y)I(n)^{\frac{1}{2}}} \tag{7}$$

Feedback and detection are carried out according to the browsing and collection of user's historical information, the highest retrieval and data click rate of users are judged, data search results are judged, effective information is filtered, and personalized recommendation list is provided [14]. Through multi-channel analysis of data feature association rules and personalized recommendation, according to the information detected and recommended for comprehensive evaluation, and based on the evaluation results to adjust the content and order of recommendation, and finally realize the humanized service, effectively meet the needs of customers. In order to ensure the recommendation effect, the evaluation standard algorithm is optimized, and the relevant data association model is set according to the mining results, so as to ensure the steps of personalized recommendation are simple, convenient, efficient and accurate.

2.3 Hot Information can only be Recommended

Based on the previous research steps, we can only recommend methods to improve the hot information of e-commerce. There are many ways for e-commerce recommendation system to recommend to customers. It can be predicted by calculation, and it can also be other customers' personal evaluation and comments on products. Which way to choose depends on how the e-commerce website wants customers to use recommendation. According to the interface forms of the recommendation system, it is mainly divided into the following types:

1) Browse: customers put forward the query requirements for specific products, and the recommendation system returns high-quality recommendations according to the query requirements.
2) Similarity: the recommendation system recommends similar products according to the products in the customer's shopping basket or the products that the customer is interested in, and provides personalized recommendation for customers.
3) E-mail: the recommendation system informs customers of the product information they may be interested in by e-mail, so that the website can keep in touch with customers, improve customers' trust in the website, and thus increase the number of visits to the website.
4) Comment information: the recommendation system provides customers with other customers' comments on the corresponding products, and the customer root

Make your own judgment according to others' evaluation of the product.

5) Rating evaluation: the recommendation system provides customers with rating evaluation of corresponding products from other customers, rather than product comment information. Through the corresponding statistics and analysis of rating evaluation, it can intuitively show other customers' views or opinions on products, so that customers can easily accept the recommendation.
6) Top-N: the recommendation system recommends to customers n products that are most likely to attract them according to their preferences. On the one hand, it can transform the visitors of the website into customers, and on the other hand, it can help customers decide whether to buy the products they initially feel hesitant about.

7) Ranking of search results: the recommendation system lists all search results and arranges the search results in descending order of customer interest.

Based on this, the intelligent recommendation mode of hot e-commerce information is further optimized, as shown in Fig. 2.

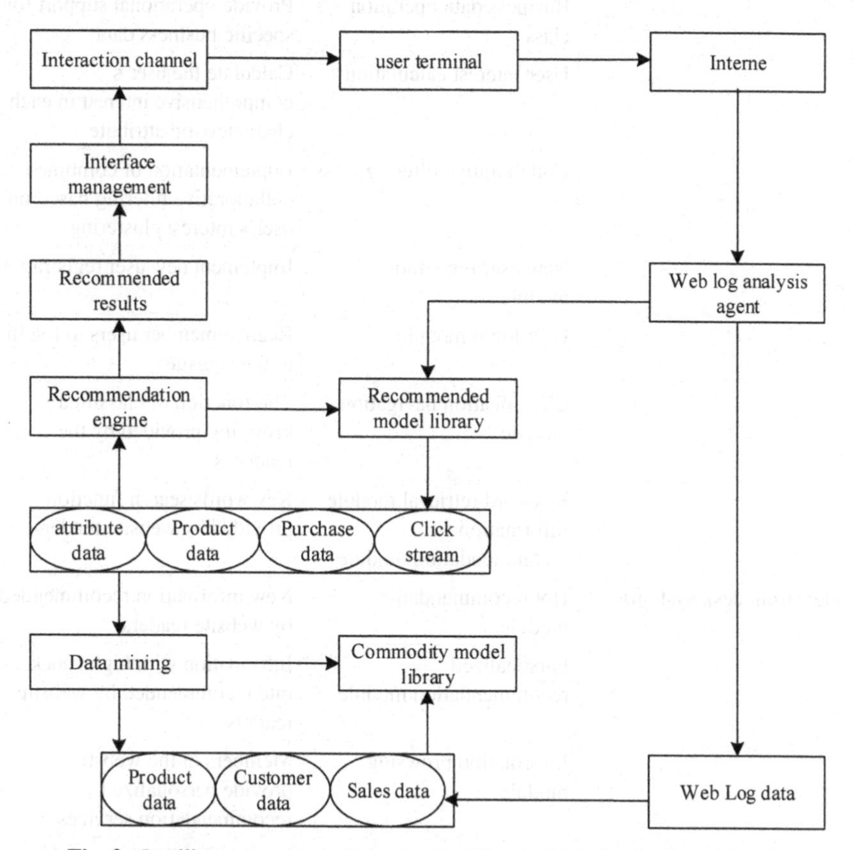

Fig. 2. Intelligent recommendation mode of hot information in E-commesrce

Based on the above steps, users' interests can be effectively collected, and targeted hot spot information of e-commerce can only be recommended.

3 Analysis of Experimental Results

In order to verify the effectiveness of the intelligent recommendation method based on user interest recommendation for hot e-commerce information, the method in reference [3] is used as the experimental comparison method for experimental test.

Table 2. Table captions should be placed above the tables.

Category	Functional module	Remarks
System class (encapsulation of common functions)	Basic data operation class	Provide database connection and data operation support
	Business data operation class	Provide operational support for specific business data
	User interest calculation	Calculate the user's comprehensive interest in each characteristic attribute
	Collaborative filtering class	Implementation of combined collaborative filtering based on user's interest clustering
	New user registration module	Implement new user registration
	User login module	Realize member users to log in to the website
	Classification navigation module	The function of classified browsing provided by the reader is
	Keyword retrieval module Information recommendation module	Key words search function provided by website readers
Reader (front desk system)	Hot recommendation module	New information recommended by website readers
	Personalized recommendation module	Information with high shock rate recommended by website readers
	Information browsing module	Members of the website provide personalized recommendation services
	Relevant information recommendation module	Provide detailed information for website visitors
	User interest feature information	Recommendation and information under survey belong to the same category
	Information acquisition module	Get the user's browsing behavior information and calculate the degree of interest
	Information filtering module	Collaborative filtering of information

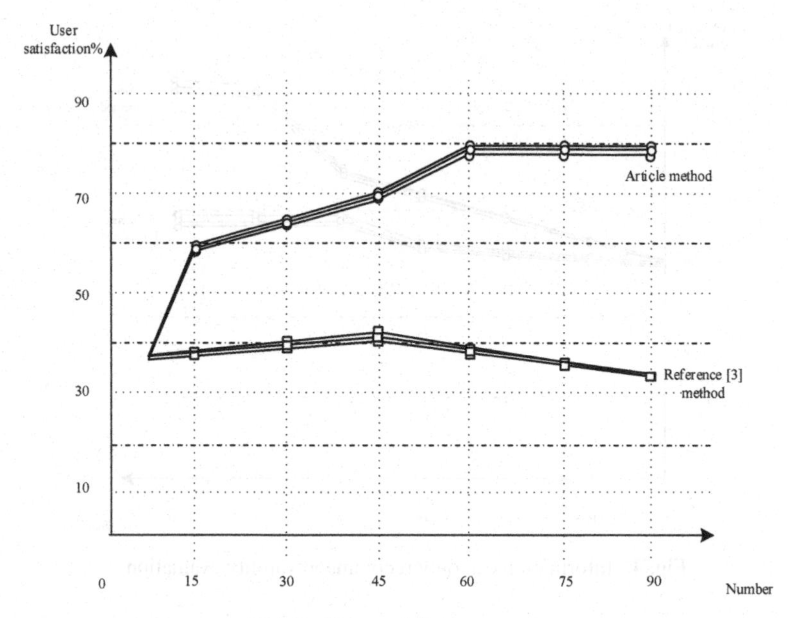

Fig. 3. User satisfaction survey

The format of the data file downloaded from the movie lens site is transformed, and it is imported into the SQL Server 2000 database as the experimental data after being sorted out. We randomly selected the experimental data set: including 60000 scoring data, which is the scoring data of 410 users for 3910 movies. The data set was converted into a user item scoring matrix A (mxn). In the experiment, we also consider the sparse level of the data set, which is defined as the percentage of items not scored in the user item scoring matrix. The sparse level of data set is 1 – 60000/(410 * 3910) = 0.96257. Experiments can be tested on different datasets. The hardware configuration of the experiment machine is: Intel Pentium core 2 processor, 1G memory, 120g hard disk. The operating environment is: the operating system Windows 2003 development platform is Microsoft Visual Studio. Net 2003 programming language is C, and the database system is SQL Server 20000. Based on this, the function of intelligent recommendation system module is standardized, as shown in the following Table 2:

According to the data obtained during the trial operation of this experiment for a period of time, the number of registered users in the first stage is 4–5, 168 after the second stage and 329 after the third stage. The more registered users, the higher the satisfaction of users. After the second stage, the satisfaction of users with personalized recommendation results has been higher than that of traditional recommendation results.

Moreover, the satisfaction of users in traditional recommendation mode is decreasing with the increase of registered users, while the satisfaction of users in personalized recommendation mode is increasing. The data collected online must be cleaned, integrated and transformed before entering the data warehouse, which requires offline data processing.

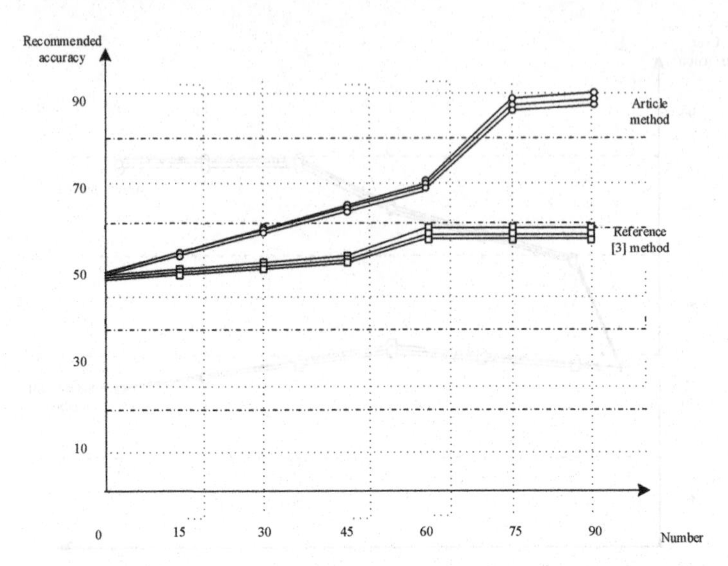

Fig. 4. Information can only recommend validity evaluation

There are two parts in the data processing phase. One is web log data processing, that is, the original logs left by users visiting the website are sorted into transaction databases. It includes data cleaning, user identification, session identification, improvement of access path and transaction identification, etc.; second, user evaluation data standardization, that is, the evaluation of recommended goods by the users to be collected will be standardized. The result of data processing in offline work is to generate the data warehouse of commodity sales website, which is the basic data needed for commodity recommendation. The recommendation engine needs to use the user interest feature information table, the product feature information table and the user rating table in the data warehouse. The running result shows that the user model proposed in this paper can correctly show the real interest of users in a certain program, and realize the dynamic tracking and updating of user interest. This result also shows that the user model proposed in this paper improves the problem of "data sparsity" faced by combined collaborative filtering algorithm to a certain extent, and improves the accuracy of recommendation results. Further carry out comparative investigation and summary on user satisfaction and recommendation effect, and record the results into a grap. The comparison of user satisfaction and recommendation effect is shown in Fig. 3 and Fig. 4.

Analysis of Fig. 3 shows that the user satisfaction rate of this method is between 38% and 80%, while that of the traditional method is between 38% and 43%, indicating that the user satisfaction of the method in this paper is always higher than that of the traditional method. Analysis of Fig. 4 shows that the effectiveness of information intelligent recommendation of this method is between 50%–90%, and that of reference [3] method is between 50%–60%, indicating that the effectiveness of information intelligent recommendation of this method is always higher than that of reference [3] method.

In conclusion, in the process of information intelligent recommendation, the proposed e-commerce hot information intelligent recommendation method based on user interest recommendation has higher customer satisfaction and information recommendation effectiveness than reference [3] method. The reason is that the method collects user interest information and constructs an information classification model. On this basis, the basic process, principle and algorithm of personalized intelligent recommendation of e-commerce hot information are designed to improve the customer satisfaction and the effectiveness of information recommendation.

4 Concluding Remarks

The existing e-commerce recommendation system cannot meet the needs of large-scale e-commerce recommendation under the network conditions. The combination of knowledge grid technology, semantic ontology and e-commerce recommendation technology can meet the requirements of effective acquisition, aggregation and intelligent recommendation of commodity knowledge, user demand knowledge and recommendation knowledge under the grid conditions. Through the research on the theory and method of e-commerce intelligent recommendation based on knowledge grid, it provides the theoretical basis for the research and development of large-scale, high-quality and strong real-time requirements of Distributed E-Commerce intelligent recommendation system. The theory and method of e-commerce intelligent recommendation based on knowledge grid are divided into three levels: knowledge-based e-commerce intelligent recommendation system, e-commerce intelligent recommendation knowledge grid, and e-commerce intelligent recommendation knowledge grid service community; This paper studies and designs the knowledge grid model of e-commerce intelligent recommendation, and studies and designs the structure, generation, organization mechanism and self-organization optimization algorithm of e-commerce intelligent recommendation knowledge grid service community.

References

1. Zi, L., Hua, L., Yubin, S., et al.: Clustering-based non-negative matrix factorization recommendation algorithm. Commun. Technol. **51**(11), 2675–2679 (2018)
2. Jiahua, L.: Information personalized recommendation algorithm of artificial intelligence cross border e-commerce shopping guide platform based on big data. Sci. Technol. Eng. **19**(14), 280–285 (2019)
3. Xiangshun, W.: Group recommendation based on preference fusion. J. Nanjing Univ. Inf. Technol. (Natural science edition) **15**(2), 1 (2019)
4. Lim, K.H., Chan, J., Leckie, C., Karunasekera, S.: Personalized trip recommendation for tourists based on user interests, points of interest visit durations and visit recency. Knowl. Inf. Syst. **54**(2), 375–406 (2017). https://doi.org/10.1007/s10115-017-1056-y
5. Klatte, J.M., Kopcza, K., Knee, A., et al.: Implementation and Impact of an Antimicrobial Stewardship Program at a Non-freestanding Children's Hospital. J. Pediatr. Pharmacol. Ther. **23**(2), 84–91 (2018)
6. Kushwaha, N., Sun, X., Singh, B., Vyas, O.P.: A lesson learned from PMF based approach for semantic recommender system. J. Intell. Inf. Syst. **50**(3), 441–453 (2017). https://doi.org/10.1007/s10844-017-0467-2

7. Venugopal, S.: A proficient web recommender system using hybrid possiblistic fuzzy clustering and bayesian model approach. Int. J. Intell. Eng. Syst. **11**(6), 190–198 (2018)
8. Mijović, V., Tomašević, N., Janev, V., Stanojević, M., Vraneš, S.: Emergency management in critical infrastructures: a complex-event-processing paradigm. J. Syst. Sci. Syst. Eng. **28**(1), 37–62 (2018). https://doi.org/10.1007/s11518-018-5393-5
9. Curry, S.J., Krist, A.H., Owens, D.K., et al.: Screening for cardiovascular disease risk with electrocardiography: us preventive services task force recommendation statement. JAMA, J. Am. Med. Assoc. **319**(22), 2308 (2018)
10. Aleid, H., Alkhalaf, A.A., Taemees, A.H., et al.: Framework to classify and analyze social media content. Soc. Network. **07**(2), 79–88 (2018)
11. Lian, D., Zheng, K., Ge, Y., et al.: GeoMF++: scalable location recommendation via joint geographical modeling and matrix factorization. ACM Trans. Inf. Syst. **36**(3), 1–29 (2018)
12. Qian, T.-Y., Liu, B., Hong, L., You, Z.-N.: Time and location aware points of interest recommendation in location-based social networks. J. Comput. Sci. Technol. **33**(6), 1219–1230 (2018). https://doi.org/10.1007/s11390-018-1883-7
13. Vilakone, P., Park, D.-S., Xinchang, K., et al.: An Efficient movie recommendation algorithm based on improved k-clique. Human Centric Comput. Inf. Sci. **8**(1), 38 (2018)
14. Dewi, R.K., Ananta, M.T., Fanani, L., et al.: The development of mobile culinary recommendation system based on group decision support system. Int. J. Interactive Mob. Technol. **12**(3), 209 (2018)

Research on the Optimization of E-commerce Logistics Model with User Interest Tracking: A Case Study of Japan

Huang Jingxian(✉)

Guizhou Minzu University, Guiyang 550025, Guizhou, China

Abstract. The common e-commerce logistics mode is analyzed by SWOT analysis, including self-built logistics system mode, sharing logistics system mode with traditional business, outsourcing to third-party logistics mode, and logistics alliance mode which appears relatively late. This paper mainly analyzes the advantages and disadvantages of these four modes, and briefly analyzes the opportunities and threats faced by each mode. Secondly, the evaluation index of logistics mode selection is studied. Based on the idea of balanced scoring method, this paper analyzes and judges the logistics infrastructure, logistics service ability and logistics operation ability from multiple perspectives, and determines the index of logistics mode selection. Finally, the index weight is determined by fuzzy evaluation method and AHP.

Keywords: Traceability · User interest · E-commerce · Business logistics

1 Introduction

E-commerce is an indispensable means of modern computer processing. It is a new way of data information processing. It combines e-commerce with logistics and manages each other. It can greatly improve the efficiency of logistics services, meet the market demand of higher requirements, and also meet the needs of consumers [1]. In the process of entering the new social development of knowledge economy in the twenty-first century, it is necessary to improve the level of cross-border logistics management, and become the key plan for colleagues to implement. The cross-border e-commerce is closely related to the logistics industry in the designated economic circulation zone, and the two of them influence each other and cannot be separated. Based on the background of e-commerce development, this paper introduces the economic development status of logistics industry to professionals, and introduces some current trends of cross-border e-commerce logistics industry in Japan, providing basis for the future mainstream development of society [2].

In the mainstream B2C e-commerce model at this stage, logistics has become the key to competition [3]. However, the development of Japanese logistics industry is lagging behind, the logistics capacity is insufficient, the service level is not high, and the

H. Song and D. Jiang (Eds.): SIMUtools 2020, LNICST 370, pp. 167–179, 2021.
https://doi.org/10.1007/978-3-030-72795-6_14

degree of informatization is low. There is a big gap between the logistics requirements of e-commerce. By analyzing Japan's existing e-commerce logistics model, e-commerce companies can better understand the advantages and problems of various logistics models, and they can be scientific and reasonable when choosing logistics models, and use lower costs. Provide better service to customers.

2 Investigation on the Development of E-Commerce Logistics in Japan

Japan is an island country with long and narrow land. In 1950's development strategy, Japan put forward the idea of developing logistics and "maritime power", which is committed to developing a fast controlled and flexible logistics model. The Japanese government attaches great importance to the modernization of the transportation industry. At the same time, the storage is always regarded as the central link of the logistics [4]. It has supervised the establishment of a large-scale storage group, which has become the center of the logistics in Japan. In the process of economic recovery, Japan attaches great importance to learning the advanced skills and governance experience of the United States [5]. They inspected the transportation status of American factories, and formally introduced the concept of "logistics" into Japan after knowing the status of the overall factory plan and handling skills related to the transportation in the factory, such as the handling equipment, handling methods, stacking methods of inventory materials, etc. At the same time, with the strong support of Japan's economic development and the promotion of logistics demand, the application of logistics technology has received widespread attention. In addition, Japan also attaches great importance to the study and organization of logistics management [6]. Therefore, after entering the 1970s, Japan has been in the forefront of the world in the exploration of logistics skills and governance. Since the general development of e-commerce, Japan's good logistics foundation has formed strong support for its e-commerce business. The development of many large Japanese e-commerce companies is closely related to their good logistics foundation.

Compared with the traditional retail price, e-commerce has advantages. Most consumers are attracted by the high discount rate of e-commerce and choose to shop online. Therefore, e-commerce enterprises must complete the economic layout of logistics: on the one hand, reduce the circulation cost as far as possible, so that the total cost of online shopping is lower than the cost of shopping in stores, to attract consumers to patronize; On the other hand, it adds value to the distribution [7]. E-commerce logistics still has some problems, such as receiving goods, querying, returning goods and so on. There are two satisfactory ways to deliver the goods that consumers need: one is to send the goods to designated places, such as chain stores and other similar places. In Japan, goods are usually delivered to nearby convenience stores. The other is to send the goods to the user's mailbox or the unit and home [8]. This is the most convenient way for consumers to distribute, and it is also a common way adopted by many developed countries at present. E-commerce is usually delivered at a designated place, and the convenience of consumers to pick up goods is limited.

One of the characteristics of C2C mode of e-commerce is the "scatter" of "C". Not only are goods and consumers scattered, but also suppliers. Because for individual consumers, generally only one product needs to be mailed [9]. Even some small batches of goods will be delivered to different destinations. Therefore, compared with the traditional business model, there is a huge waste in goods packaging and transportation. Logistics refers to the use of modern information technology and equipment, the goods from the place of supply to the place of receipt of accurate, timely, safe, quality and quantity, door-to-door reasonable service mode and advanced service process [10]. Logistics appears with the emergence of commodity production and develops with the development of commodity production, so logistics is an ancient traditional economic activity. Modern logistics refers to a new type of integrated management activity that integrates information, transportation, storage, inventory, loading and unloading, and packaging. Its task is to reduce the total cost of logistics as much as possible and provide the best service for customers [11]. Many Japanese experts and scholars believe that "modern" logistics is a process of transferring logistics from the supply place to the demand place with the most economic cost according to the needs of customers. It mainly includes transportation, storage, processing, packaging, loading and unloading, distribution and information processing. The basic activities included are shown in the figure.

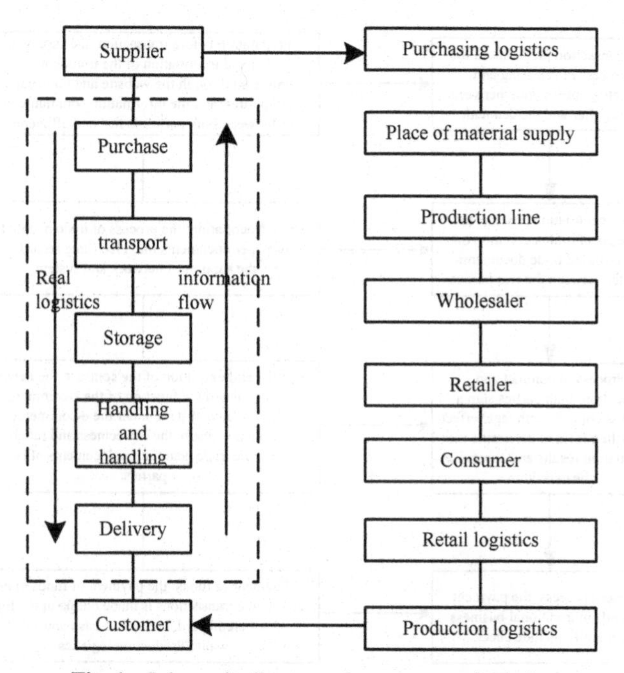

Fig. 1. Schematic diagram of e-commerce logistics

Through a large number of literature retrieval and analysis of retail e-commerce and logistics, on the basis of previous studies, this paper analyzes and studies the current logistics mode in Japan, and explores the selection of logistics distribution mode of retail e-commerce [12]. Firstly, the research scope of e-commerce logistics is defined. This paper analyzes the mode and advantages and disadvantages of e-commerce logistics in Japan, puts forward the evaluation method of e-commerce logistics mode selection, and finally puts forward the development suggestions for our country's e-commerce logistics. In the era of electronic service, the way of online business has changed a lot [13]. The traditional way is to find customers from the products, while the e-commerce business mode is to create different services to attract customers. After gathering a group of customers, we can develop the market from inside, see what the customers need, find sales opportunities here, and finally sell the products customers need. This process can be divided into the following three parts according to the internal management activities of the organization:

1. Transaction flow: refers to all documents and practical operation process of trade agreement.
2. Logistics: refers to the flow process of goods.
3. Capital flow: the flow of capital units (including banks) in a transaction.

Optimize the operation process of e-commerce logistics management, as shown in the figure:

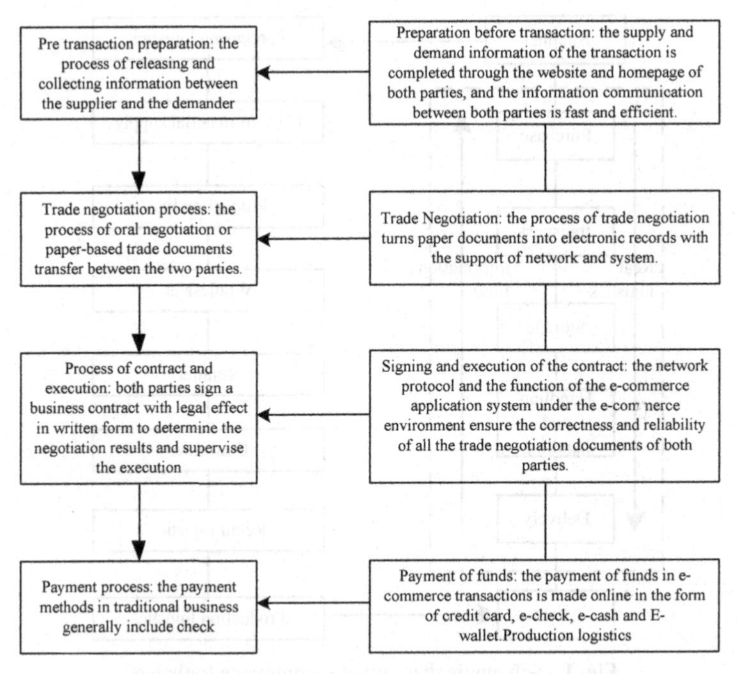

Fig. 2. E-commerce logistics management process

In the process of e-commerce transaction, although the operation process of business practice is also divided into pre-transaction preparation, trade negotiation, contract signing and execution, capital payment and other links, the force and method used in the specific operation of the transaction is totally different from the traditional business force [14]. Compared with the traditional business model, e-commerce model has the following advantages:

(1) Reduce transaction cost, transaction preparation time and negotiation time;
(2) Can reduce enterprise inventory. Through the Internet market demand information can be transmitted among enterprises faster, so that the production enterprises can make faster decision-making and production, as well as timely supply of suppliers, so as to achieve the goal of reducing inventory.
(3) Shorten the production cycle. E-commerce has changed the past self trust and closed phased cooperation mode into collaborative cooperation of information sharing, reducing the waiting time due to information closure.
(4) Increased labor productivity. The cooperation between suppliers and distributors can automatically process the workflow through the network, which improves the efficiency.
(5) Increase business opportunities and expand market scope. Online business can be carried out to traditional marketers and one! Advertising promotion can not reach the market scope, not limited by time and space.
(6) Provide personalized service for customers. Customers can customize business, enterprises can provide recommendations, and provide customers with personalized information services.

3 E-commerce Logistics User Interest Tracking Model Optimization

Completely independent logistics operation mode refers to the establishment of e-commerce enterprises' own logistics system in order to meet their own logistics needs. E-commerce enterprises invest in purchasing logistics facilities and equipment and allocating logistics services [15]. The independent organization manages the logistics business and selects the specific logistics operation and management methods. For the purpose of winning in the fierce e-commerce competition, many large websites in Japan try to use a third-party logistics mode, and then start to try the self-supporting logistics mode [16]. This is because most of the existing third-party logistics enterprises in Japan are unable to meet the needs of e-commerce in terms of internal management, information level and network level. Many large electronic enterprises have to start to build their own logistics system to strengthen their competitiveness [17]. Although self-built logistics needs sufficient funds as support, in order to gain competitive advantage, self-built logistics has become the first choice of many e-commerce websites. Based on this, users' interest information is tracked and recorded as follows:

As a new business model, e-commerce is the result of the development and application of Internet [18]. However, as a business activity, e-commerce transaction is inextricably linked with traditional transaction. Even in essence, there is no difference between e-commerce and traditional commerce. Through the realization of product sales, we can create profits for enterprises and create value for customers. Only in the way of transaction, the Internet as a medium for communication inquiry negotiations and other content. Therefore, e-commerce enterprises can also use the traditional commercial logistics system to carry out their own e-commerce logistics business [19]. While many shopping malls invest hundreds of millions of funds to expand their logistics systems, many large-scale e-commerce platforms are also actively following up and making efforts in logistics. According to the customer requirements, the scheduling objectives include the completion time to meet the customer delivery time and the supplier resource yield. The specific calculation formula is as follows:

Table 1. User interest information of logistics distribution

	Traditional logistics	Electronic logistics
Business drivers	Material wealth	IT Technology
Scope of service	Logistics services (transportation, storage, packaging, loading and unloading, distribution, etc.)	Comprehensive logistics services, while providing a wider range of business, such as online front-end services
Means of communication	Fax, telephone, etc	Massive application of Internet and EDI technology
Storage	Concentrated distribution	Distributed and distributed centers are closer to customers
Packing	Batch packaging	Individual package, small package
Transport frequency	Low	High
Delivery speed	Slow	Fast
Delivery speed	Slow	Fast
Order	Less	Many

$$\begin{cases} MAX\sum_{g=1}^{m} z_g t_g^r, g = 1, 2, ..., m \\ t_g^r \subseteq |t_{m+1} - t_0| \end{cases} \tag{1}$$

Where z_g represents the delivery deadline of the scheduling task; t_g^r represents the start time of the scheduling task; g represents the end time of the scheduling task; t_0 represents the end time of the task; t_{m+1} represents the appointment time of the task on the resource. The specific algorithm is as follows:

$$\begin{cases} MIN|q_n - q_f| \\ n, f = 1, 2, ..., m \\ n \neq f \end{cases} \tag{2}$$

In the formula: f represents the unit revenue coefficient of resource n when executing the scheduling task; q_n represents the execution time of scheduling task m; e-commerce logistics requires that the logistics supply subject provide the required comprehensive logistics services to the internal logistics demand subject by effectively and reasonably organizing and using various resources within its logistics system in a certain period of time. In order to measure and compare the capabilities of different logistics systems scientifically and objectively, this paper uses the idea of balanced scoring method for reference, and constructs the evaluation index system of e-commerce logistics from four dimensions: logistics infrastructure support capability, information system support capability, operation management capability and logistics service capability, as shown in the table.

Table 2. Evaluation indexes of business logistics

	Index category	Index explanation	Index name and code
E-commerce logistics index evaluation system	Logistics infrastructure support capacity	It reflects the basic ability of logistics service of the supplier	Utilization rate of storage equipment Utilization rate of transportation equipment Popularization rate of information equipment dot density Utilization rate of logistics infrastructure Per capita freight volume
	Information system capability	Reflect the application ability of modern information technology	Support capability of credit system Direct economic benefit evaluation Indirect economic benefit evaluation
	Operation and management capability	Reflect the operation ability of logistics management	Ratio of professional and technical personnel Logistics cost and profit rate Training cost per employee Market share
	Logistics service capacity	It reflects the demand of the demand side for the response ability of the supply side	Order response time Delivery delay rate Damage rate Delivery flexibility Delivery error rate Customer satisfaction

Further, starting from the logistics bottleneck problem in the development of e-commerce, this paper studies the effective method to solve the problem, which is to use modern information technology to realize the logistics electronation. On the basis of understanding the meaning and characteristics of e-Logistics, this paper analyzes the current e-logistics mode and its development trend [20]. The emergence of e-commerce has changed the traditional business model and created value. The value source of e-commerce refers to the elements that can increase the total value created by e-commerce. The operation of e-commerce mainly includes four interrelated value sources: efficiency, complementarity, barrier and innovation. Based on the research of the value source of e-commerce, this paper describes the unique advantages of e-logistics system, including the realization of seamless links among systems, enterprises, capital flow, logistics and information flow, the realization of online tracking of goods sent out, and the real-time supervision and control of goods; Provide system integration service solutions for customers, combine the front-end service of customers with the back-end logistics business closely, evaluate the value of e-Logistics, analyze that e-logistics can not only realize the cost advantages of enterprises, but also meet the interests of consumers, so as to achieve win–win economic benefits, and finally put forward suggestions for the development and improvement of e-Logistics.

4 Suggestions on the Optimization of E-commerce Logistics Mode

Logistics information technology refers to the application of modern information technology in all aspects of logistics. It is one of the most important fields of logistics modernization, especially the application of rapid development of computer network technology makes logistics information technology reach a new level. Logistics information technology is an important symbol of logistics modernization, and also the fastest developing field in logistics technology. From the bar code system of data collection to the computer and Internet in the office automation system, all kinds of hardware and computer software such as terminal equipment are developing rapidly. At the same time, with the continuous development of logistics information technology, a series of new logistics concepts and new logistics management methods have emerged, which has promoted the reform of logistics. A complete logistics process includes the whole process of production of products by the manufacturer, transportation, storage, processing, distribution to users and consumers. It can be divided into the following aspects: first, the manufacturer packs the single products and concentrates multiple products in large packing boxes; second, through the transportation, wholesale and other bad links, larger packaging is usually needed in this link; finally, the products are circulated to consumers through the zero-sale link, and the products are usually restored to a single product in this link. People call the management of the above process supply chain logistics management. In the process of trade, the logistics process of goods from manufacturer to end-user is objective. For a long time, people have never taken the initiative, systematically and as a whole to consider, so they fail to play the overall advantages of the system. The supply chain logistics system is connected with many production enterprises, transportation industry, distribution industry and users, and changes with the change of demand and supply, so the system management must have enough flexibility and variability; The supply chain logistics system from production, distribution, sales to users is not an isolated behavior, but a link by link, mutual restriction and complementary. Therefore, it must be coordinated to play its maximum economic and social benefits. The EDI framework structure of e-commerce logistics is shown in Fig. 3.

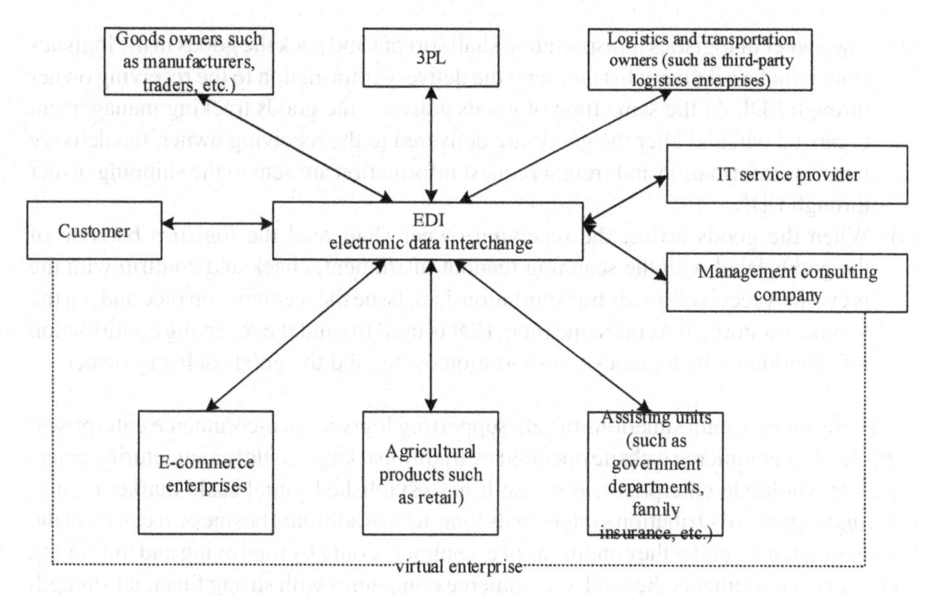

Fig. 3. EDI framework structure of e-commerce logistics

Now let's look at an example of applying logistics EDI system. This model is a logistics model which is composed of goods sending owners, logistics transportation owners and goods receiving owners. The operation steps of this logistics model are as follows:

(1) After receiving the order, the delivery owner shall make the delivery plan, and send the list and delivery schedule of the delivered goods to the logistics transportation owner and the receiving owner through EDI, so that the logistics transportation owner can make the vehicle deployment plan in advance and the receiving owner can make the goods acceptance plan.
(2) According to the requirements of customer ordering and goods delivery plan, the owner of the goods shall issue the delivery order, sort and distribute the goods, print out the goods label of logistics barcode, and paste it on the goods packing box. Meanwhile, the owner of the goods shall send the type, quantity, packaging and other information of the goods to the owner of logistics transportation and receive the goods property allocation order through EDI.
(3) When the owner of logistics transportation takes the goods from the owner of sending goods, the owner uses the on-board scanning readout to read the goods.

The logistics barcode of the label shall be checked with the previously received goods transportation data to confirm the delivery of goods.

(4) The owner of logistics transportation shall sort out and pack the goods in the logistics center, make a delivery list and send the delivery information to the receiving owner through EDI. At the same time of goods delivery, the goods tracking management is carried out, and after the goods are delivered to the receiving owner, the delivery business information and freight request information are sent to the shipping owner through EDI.
(5) When the goods arrive, the receiving owner shall read the logistics barcode of the goods label with the scanning readout instrument, check and confirm with the previously received goods transportation data, issue the receiving invoice and put the goods into storage. At the same time, EDI is used to send the receiving confirmation information to the logistics transportation owner and the goods delivery owner.

There are two main situations of self-supporting logistics in e-commerce enterprises: one is BtoB e-commerce website operated by traditional large-scale manufacturing enterprises or wholesale enterprises. Because it has established initial scale marketing network and logistics distribution system in its long-term traditional business, it can meet the logistics matching under the conditions of e-commerce only by improving and improving it Delivery requirements; Second, e-commerce companies with strong financial strength and large business scale can establish a smooth and efficient logistics system to meet business needs and provide comprehensive logistics services to other logistics service demanders (such as other e-commerce companies) when the third-party logistics cannot meet their cost control objectives and customer service requirements We should make full use of its logistics resources to achieve scale efficiency.

5 Empirical Analysis

In order to verify the effectiveness of the proposed e-commerce logistics model optimization research method, a verification experiment is carried out. The experimental data comes from the MySQL database, and the total amount of data is 10 GB. In the model of logistics cost, there is a law of two rate reverse, that is, there is a relationship between the items of logistics cost. The decrease of one item cost will lead to the increase of another item cost. Considering that transportation and warehousing are two main parts of logistics cost, we can use a simplified model to represent logistics cost, as shown in the figure.

C * in the figure represents the lowest total cost. In consideration of reducing the number of warehouses, although the storage cost can be reduced, the transportation distance will become longer and the number of transportations will increase, resulting in the increase of transportation cost. If the increase of transportation cost exceeds the decrease of storage cost, the total logistics cost will increase instead, so the measures to reduce the number of warehouses is meaningless. When choosing and designing the logistics system, we must test the total cost of the system. Firstly, we should consider the strategic position of logistics in the enterprise, secondly, we should consider the ability level of logistics, finally, we should evaluate the total cost of logistics and choose the logistics system with the lowest cost. The third-party logistics has changed the traditional enterprise logistics mode, freed the enterprise from the shackles of "big and all" and "small and all", and enabled the enterprise to enhance its core competitiveness and maintain the advantages of market competition from the strategic height. However, any enterprise must consider its operation cost and opportunity cost when deciding. What is the economic value of 3PL operation. Let's take inventory as an example to analyze 3PL's economic value. For general production enterprises, inventory related cost is an important part of logistics cost, which consists of ordering cost and storage cost, and these two kinds of costs are also a kind of relationship between the two, as shown in the figure.

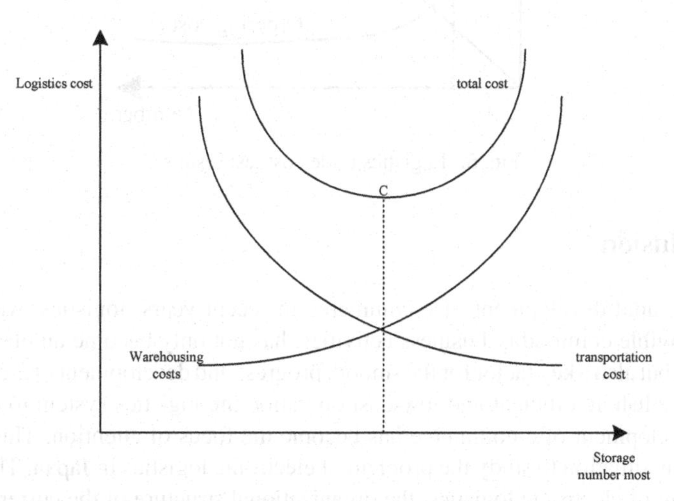

Fig. 4. Test results of simplified logistics cost model

In Fig. 5, represents the lowest inventory cost. According to the theory of consumer surplus value, we can know the relationship between total value B, price corpse and cost C. Under the condition of market economy, the demanders of logistics activities can automatically control the same kind of services with different prices and quality, and hope to obtain the maximum profit surplus (B-P). From the perspective of service providers, they should expand their profit space (P–C) as much as possible, and do not want the demanders to get more surplus. Therefore, under the effect of competition, the surplus of demander will become a constant, that is, B-P = s (constant). It has been proved that the most remarkable feature of using e-commerce and Internet technology to complete the coordination, control and management of the whole logistics process is the integration of various software technologies and logistics services. It can make seamless links among systems, enterprises, capital flow, logistics and information flow, and this link has the function of foresight at the same time. It can provide a transparent visibility function between upstream and downstream enterprises, and help enterprises to control and manage the inventory to the maximum extent.

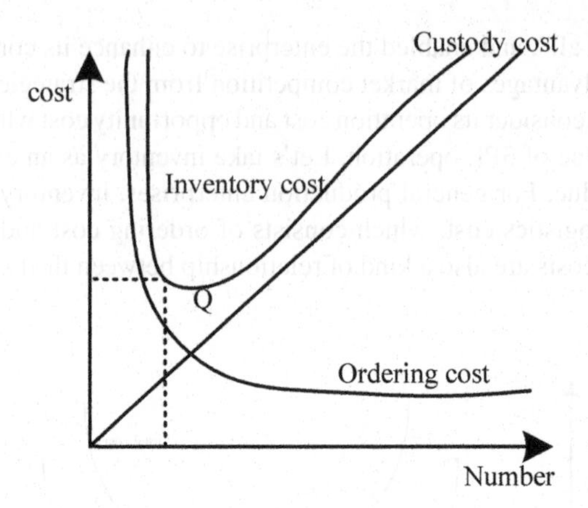

Fig. 5. Logistics trade cost test results

6 Conclusion

Due to the rapid development of e-commerce in recent years, logistics, which is the basis of tangible commodity business activities, has not only become an obstacle to e-commerce, but also a key factor for the smooth progress and development of e-commerce. How to establish an efficient and low-cost operation the logistics system to ensure the smooth development of e-commerce has become the focus of attention. This article is based on this situation to study the problem of electronic logistics in Japan. Through the development of electronic logistics, the organizational structure of the current logistics system can be reformed; through standardized and orderly electronic logistics procedures, logistics can enter a benign track that makes full use of existing resources, reduces logistics costs, and improves logistics operation efficiency. Therefore, the proposed optimization method can improve e-commerce logistics transportation performance in all aspects.

References

1. Woarawichai, C., Naenna, T.: Solving inventory lot-sizing with supplier selection under alternative quantity discounts and vehicle capacity. Int. J. Logist. Syst. Manag. **30**(2), 179 (2018)
2. Fanizzi, F.P., Lanfranchi, M., Natile, G., et al.: Platinum (II) Complexes with Monocoordinated 2,9-Dimethyl-1,10-phenanthroline and Phosphine Ligands. Exchange of the donor nitrogen and rotation about the Pt-P and P-C bonds studied by NMR spectroscopy: Arene Stacking as an Intramolecular Brake. Inorg. Chem. **33**(15), 253–268 (2018)
3. Petraška, A., Čižiūnienė, K., Prentkovskis, O., et al.: Methodology of selection of heavy and oversized freight transportation system. Trans. Telecommun. J. **19**(1), 45–58 (2018)
4. Moosavi, S.A., Rahimi, V., Ebrahimnejad, S.: A proposed grey fuzzy multi-objective programming model in supplier selection: a case study in the automotive parts industry. Int. J. Logist. Syst. Manag. **29**(4), 409 (2018)

5. Kohnehrouz, B.B., Talischian, A., Dehnad, A., et al.: Novel recombinant traceable c-Met antagonist-Avimer antibody mimetic obtained by bacterial expression analysis. Avicenna J. Med. Biotechnol. **10**(1), 9–14 (2018)
6. Shayanfar, E., Paul, M.S.: Selecting and scheduling interrelated projects: application in urban road network investment. Int. J. Logist. Syst. Manag. **29**(4), 436 (2018)
7. Jauhari, W.A.: A collaborative inventory model for vendor-buyer system with stochastic demand, defective items and carbon emission cost. Int. J. Logist. Syst. Manag. **29**(2), 241 (2018)
8. Tansakul, N., Suanmali, S., Ammarapala, V.: Perception of logistics service provider regarding trade facilitation for cross border transportation: a case study of east-west economic corridor. Int. J. Logist. Syst. Manag. **29**(2), 131 (2018)
9. Ashcroft, L., Coll, J.R., Gilabert, A., et al.: A rescued dataset of sub-daily meteorological observations for Europe and the southern Mediterranean region, 1877–2012. Earth Syst. Sci. Data. **10**(3), 1613–1635 (2018)
10. Byrd, D., Christopfel, R., Arabasz, G., et al.: Measuring temporal stability of positron emission tomography standardized uptake value bias using long-lived sources in a multicenter network. J. Med. Imaging. **5**(1), 11–16 (2018)
11. Baek, S.I., Shin, H.R., Park, H.M., et al.: A study on the management of drug logistics using beacon technology. J. Comput. Theoret. Nanosci. **24**(3), 1979–1985 (2018)
12. Chen, X., Mark, L.G., Matthew, R.R., et al.: Logistics of air medical transport: when and where does helicopter transport reduce prehospital time for trauma. J. Trauma Acute Care Surg. **85**(1), 1 (2018)
13. Perotti, S., Marchet, G., Sassi, C., et al.: Types of logistics outsourcing and related impact on the 3PL buying process: empirical evidence. Int. J. Logist. Syst. Manag. **30**(2), 139 (2018)
14. Wu, Y., Chen, J.: Collaborative logistics information service framework and reference model: based on the perspective of service ecosystem. J. Serv. Sci. Manag. **11**(1), 1–12 (2018)
15. Avdasheva, S., Golovanova, S., Yusupova, G.: Advance freight rate announcements (GRI) in liner shipping: European and Russian regulatory settlements compared. Marit. Econ. Logist. **2018**(4), 1–15 (2018)
16. Benedyk, I.V., Peeta, S.: A binary probit model to analyze freight transportation decision-maker perspectives for container shipping on the Northern Sea Route. Marit. Econ. Logist. **20**(3), 358–374 (2018)
17. Lee, T.-Z., Wang, H.-Z., Hsu, Y.-H.: The feasibility study of promotion activities in farmers' markets with regional agricultural products. Int. J. Logist. Econ. Glob. **7**(3), 248 (2018)
18. Xue, F., Dong, T., Qi, Z.: An improving clustering algorithm for order batching of e-commerce warehouse system based on logistics robots. Int. J. Wireless Mob. Comput. **15**(1), 10–15 (2018)
19. Wu, P.J., Lin, K.C.: Unstructured big data analytics for retrieving e-commerce logistics knowledge. Telematics Inform. **35**(1), 237–244 (2018)
20. Gajewska, T., Zimon, D.: Study of the logistics factors that influence the development of e-commerce services in the customer's opinion. Arch. Trans. **45**(1), 25–34 (2018)

A Novel Multi-objective Squirrel Search Algorithm: MOSSA

Xinyuan Wang, Fanhao Zhang, Zhuoran Liu(✉), Changsheng Zhang, Qidong Zhao, and Bin Zhang

Northeastern University, Shenyang 110819, People's Republic of China
paper820@sohu.com

Abstract. This paper suggests a non-dominated sorting genetic algorithm II (NSGA-II) as a multi-objective framework to construct a multi-objective optimization algorithm and uses the squirrel search algorithm (SSA) as the core evolution strategy. And a multi-objective improved squirrel search algorithm (MOSSA) is proposed. MOSSA establishes an external archive of the population to maintain the elitists in the population. The probability density is applied to limit the size of the merged population to maintain population diversity, based on roulette wheel selection. Also, this paper designs a fitness mapping evaluation according to the individual fitness value of each object. Compared with the original SSA, the generational gap is introduced to make the seasonal condition suitable for multi-objective optimization, which could keep the solution from the local convergence. This paper simulates MOSSA and other algorithms on multi-objective test functions to analyze the convergence and diversity of PF. It is concluded that MOSSA has a good performance in solving multi-objective problems.

Keywords: Multi-objective optimization · Non-dominated sorting · Squirrel search algorithm · Mapping fitness evaluation · Roulette wheel selection

1 Introduction

SSA is a novel natural heuristic optimization paradigm called squirrel search algorithm [1, 2] to simulate the flying squirrel seek for food on trees. It was proposed by Jain M, Singh V, Rani A, and others after studying the foraging behavior of southern flying squirrels in 2019. The algorithm was tested with several classic and modern unconstrained benchmark functions. Compared with other optimization algorithms reported in the literature, the SSA algorithm has significant convergence. Besides, for advanced highly complex CEC 2014 benchmark functions [3], SSA has the same good convergence. SSA provides quite competitive results on both numerical optimization and real-time problems.

However, many optimization problems encountered in reality are multi-objective problems (MOPs), such as communication engineering [3], transportation problems [5, 6], power systems [7] and other many fields [8]. Many complicated problems could be simulated to mathematical questions and solved efficiently [9, 10]. All the scenarios can be

H. Song and D. Jiang (Eds.): SIMUtools 2020, LNICST 370, pp. 180–195, 2021.
https://doi.org/10.1007/978-3-030-72795-6_15

simulated with computer methods to be multi-objective optimization problems mathematically [11, 12]. Usually, the sub-goals of multi-objective optimization problems are contradictory to each other. Improvements in one sub-goal may lead to degraded performance in another or other several sub-goals. In other words, it is impossible to make all the multiple sub-objectives reach the optimal value at the same time. You can only coordinate and compromise between them to optimize each sub-goal as much as possible. The essential difference between it and the single-objective optimization problem is that its solution is not unique, that is, there exists a set of optimal solution sets composed of many Pareto optimal solutions. Each element in the set is called a Pareto optimal solution or a non-suboptimal optimal solution.

There is no unique global optimal solution for multi-objective optimization problems. Too many non-inferior solutions cannot be directly applied, so it is necessary to find a final solution when solving. There are three main methods to find the final solution at this stage [11, 12]:

1. Decomposition method: Convert to a single-objective problem decomposition method based on the relative importance between objectives given by the decision-maker in advance.
2. Interaction method: The final solution is gradually obtained through the interaction between the analyst and the decision-maker.
3. Generating method: Find a large number of non-inferior solutions and then obtain the final solution according to the decision maker's intention.

Many experts and scholars have applied different algorithms to solve multi-objective optimization problems, such as multi-objective evolutionary algorithms [11], multi-objective particle swarm optimization (MOPSO) [13], multi-objective evolutionary algorithm based on decomposition (MOEA/D) [14], a nondominated sorting genetic algorithm II (NSGA-II) [15] and many other algorithms.

Because the SSA algorithm has an efficient search capability, it is beneficial to obtain the optimal solution in the sense of multiple objectives. The algorithm searches for many non-inferior solutions by representing the total number of solution sets, that is, searching for many Pareto optimal solutions. At the same time, SSA is more versatile, suitable for processing various types of objective functions and constraints, and easy to combine with traditional optimization methods, thereby improving its limitations and solving problems more effectively. Therefore, the application of SSA to multi-objective optimization problems has excellent advantages.

Therefore, how to design a single-objective SSA for optimizing multi-objective problems has become a research hotspot. This paper proposes a multi-objective squirrel search algorithm (MOSSA). In the process of expanding a single target to multiple targets, the following challenges are encountered: In the single target optimization process, due to the characteristics of the single target, the global optimal solution and the suboptimal solution can be selected relatively easily. In MOSSA, there are multiple mutually restricted objects. Individuals cannot simply determine a learning sample by comparing a single target. Therefore, how to judge the fitness value of an individual is a key step of the MOSSA algorithm. Besides, since multi-objective optimization problems usually have a set of non-inferior solutions, how to choose the optimal solutions from a variety of non-inferior solutions becomes a huge challenge.

This paper applies the following methods from different perspectives to design an efficient and novel multi-objective algorithm, MOSSA. The contributions of this paper are outlined as follows:

- An external archive of the population is established to reserve the elites in the population. After merging two continuous populations, the probability density is applied to limit the population size and improve the extension of distribution.
- The original operation of calculating the crowded distance in NSGA-II may cause all the dense solutions to be filtered out at one time, and meanwhile some solutions which can be used to maintain diversity are accidentally deleted. Therefore, roulette wheel selection is introduced to make the dense grids have a higher probability of deletion, rather than necessarily deletion.
- Because multiple objects are restricted to each other in multi-objective optimization, individuals cannot be simply sorted by the fitness values. This paper uses the Pareto level and grid density to construct a mapping strategy to calculate the fitness values of individuals.
- The original solution to avoid the local convergence in SSA, seasonal condition, is no longer applicable in multi-objective optimization problems. The seasonal condition in MOSSA is improved by generational gap. It prevents MOSSA from the local convergence in multi-objective optimization problems and enhances the spread of PF.
- MOSSA is simulated on the Zitzler Deb Thiele series test functions [16] in order to compare with NSGA-II, MOPSO, MOEA/D. This paper analyzes the convergence, uniformity, and spread of all the algorithms through several indicators and PF. Experimental results reveals that MOSSA has excellent performance and is an efficient multi-objective optimization algorithm.

The rest of this paper is organized as follows. Section 2 shows the related work. And Sect. 3 presents the concept of the algorithm design. Experimental results analysis is concluded in Sect. 4. At last, Sect. 5 shows the conclusion of this paper.

2 Related Work

From the current research of multi-objective optimization by experts [17], the main task of solving multi-objective optimization problems is to find the Pareto optimal solution set. This solution set can weigh multiple objective functions and achieve 3 goals [18–21]. In general, some evaluation indexes can be used to reflect it.

1. Convergence of the solution set is used to evaluate the distance between the solution obtained by the algorithm and the real Pareto front is minimized. Generally, the obtained solution set is required to make the convergence as small as possible.
2. Uniformity of the solution set is used to evaluate the uniformity and evenness of the individual distribution. Generally, each solution in the obtained solution set should be distributed as uniformly as necessary.

3. Spread of the solution set is used to evaluate the level of the entire obtained solution set distributed in the target space. Generally, the solution set should be as wide as necessary to show the Pareto Front as completely as necessary.

These goals could be achieved through different algorithms. Many various methods are applied to solve multi-objective optimization problems more efficiently. Multi-objective evolutionary algorithm is a kind of global probabilistic optimization search method formed by simulating biological evolution mechanisms [11]. And a multi evolutionary algorithm based on decomposition (MOEA/D) was proposed [14], so multi-objective optimization problem is decomposed into multiple scalar optimization sub-problems. Multiple objective particle swarm optimization (MOPSO) was presented to use Pareto dominance to decide the next direction of swarm [13]. There are two main methods [22], namely, methods that are not based on Pareto optimization and methods that are based on Pareto optimization. On this basis, some scholars have proposed the concept of external archive [23]. The external archive sets save all the non-dominated individuals of the current generation so that the solution set maintains a good distribution. A multi-objective evolutionary algorithm with an external set is put more emphasis on the efficiency and effectiveness of the algorithm [11]. The more typical multi-objective evolutionary algorithms are NSGA-II [15], PESA2 [24], and SPEA2 [25]. NSGA-II took the nondominated sorting into a multi-objective optimization algorithm. The advantage of PESA2 is that its solution converges very well, and it is easier to approach the optimal surface, especially in the case of high-dimensional problems; but its disadvantage is that the selection operation can only select one individual at a time, which consumes much time and has a class The diversity is poor. The advantage of SPEA2 is that it can obtain a well-distributed solution set, especially for solving high-dimensional problems, but its clustering process takes a long time to maintain diversity and is not efficient.

At present, the demand for multi-objective optimization algorithms has become more extensive, not only in real life but in solving the processes of many algorithms, many multi-objective optimization problems are waiting to be solved. There are more multi-objective optimization algorithms that have also been successfully used in function optimization [26], neural network training [27], pattern classification [28], fuzzy system control [29], and other application areas.

3 Background

3.1 Squirrel Search Algorithm

When the squirrel begins to forage, the search process begins. During this time, they began to migrate and explore various forested areas. Squirrels form their own migration routes based on the fitness of their companions. As the weather changes, they adjust their foraging strategies to increase the likelihood of survival. This foraging strategy runs through the entire life of each squirrel (Fig. 1).

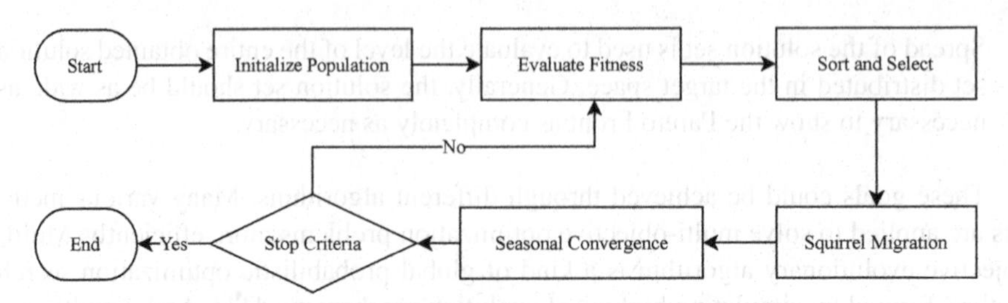

Fig. 1. A procedure of Squirrel Search Algorithm (SSA).

3.2 Basic Concept of NSGA-II

In the process of NSGA-II, fast nondominated sorting is a vital method to take advantage of sorting solutions with various Pareto dominance levels. Before the description of our design, some concepts of NSGA-II which are also applied in this paper are listed here.

Pareto dominance: if and only if $SF_{i,k} \leq SF_{j,k}$, and $SF_{i,k} < SF_{j,k}$, $k = 1, 2, \ldots, m$.

In this case, we say that S_i (i^{th} squirrel) dominates S_j or S_i being dominated by S_i, is written as $S_i \leq S_j$.

Non-dominated individuals: Individuals are non-dominated individuals in the population, if and only if there is no individual S_j, satisfying S_j dominates S_i.

Pareto Front (PF): The Pareto front is a hyperplane in the target space that is fitted by the optimal solution set of a theoretical optimization problem. In experimental research, the Pareto front is often used to represent the problem by a set of known non-dominated solutions. D_i a set of solutions that the solution S_i dominates.

4 Design of MOSSA

4.1 Basic Concept of MOSSA

To improve the performance of the multi-objective evolutionary algorithm NSGA-II in solving multi-objective optimization problems, this paper uses NSGA-II as a multi-objective optimization algorithm, combined with the efficient search strategy of the squirrel search algorithm (SSA), a pseudo-code of this multi-objective improved squirrel search algorithm (MOSSA) as shown in Algorithm 1.

First, based on the original SSA, a random initialization method suitable for multi-objective problems is designed. During the iteration process, an external archive of the population reserves the elites in the population. Then, MOSSA sorts non-dominantly the population and the generational gap is calculated for the seasonal condition. Multiple objectives are restricted to each other in multi-objective optimization, individuals cannot be simply sorted by the fitness values. In that case, MOSSA uses the Pareto level and grid density to construct a mapping function to calculate the fitness values of individuals. Squirrels then migrate according to the strategy in SSA. MOSSA uses novel seasonal condition which could avoid the local convergence. After ensuring that it meets the seasonal change conditions which means it is winter, the normal squirrels fly following the Levy flight. Two continuous populations should be merged into the archive as the current population, the probability density is applied to limit the population size and improve the distribution of PF. Finally, the feasible solution set is obtained.

Algorithm 1. Pseudocode for MOSSA

```
Input:
  Population number, test problem, parameter definition
Output
  Feasible solution set
Begin
  1. initialized population randomly;
  2. while stop criteria==false do
  3.     archive current population as set R;
  4.     non-dominated sort and calculate grid density (DG_i);
  5.     obtain mapping fitness based on R and grid density (DG_i);
  6.     squirrels migrate to form new population set T;
  7.     calculate SC and determine season;
  8.     if season==winter then
  9.         normal squirrels Levy flight;
  10.    else
  11.        continue;
  12.    end if
  13.    archive current population as set T';
  14.    merge two populations R and T';
  15. end while
  16. return feasible solution set;
End
```

4.2 Random Initialization

The position of all squirrels (SP) can be represented by the following matrix [1], where $SP_{i,j}$ represents the jth dimension (d1 in total) of ith squirrel. The objective function value of all squirrels (SF) can also be represented by the following matrix [2], where $SF_{i,k}$ represents the kth objective function value (if dimension is d2 in total) of ith squirrel.

$$SP_{i,j} = \begin{bmatrix} SP_{1,1} & SP_{1,2} & \cdots & \cdots & SP_{1,d1} \\ SP_{2,1} & SP_{2,2} & \cdots & \cdots & SP_{2,d1} \\ \vdots & \vdots & \vdots & \vdots & \vdots \\ \vdots & \vdots & \vdots & \vdots & \vdots \\ SP_{n,1} & SP_{n,2} & \cdots & \cdots & SP_{n,d1} \end{bmatrix}$$

Matrix [1]: decision space

$$SF_{i,k} = \begin{bmatrix} SF_{1,1} & SF_{1,2} & \cdots & \cdots & SF_{1,d2} \\ SF_{2,1} & SF_{2,2} & \cdots & \cdots & SF_{2,d2} \\ \vdots & \vdots & \vdots & \vdots & \vdots \\ \vdots & \vdots & \vdots & \vdots & \vdots \\ SF_{n,1} & SF_{n,2} & \cdots & \cdots & SF_{n,d2} \end{bmatrix}$$

Matrix [2]: target space

A uniform distribution is used to assign the initial position of each squirrel.

$$SP_i = SP_{min} + U(0,1) \times (SP_{max} - SP_{min}) \tag{1}$$

Where U (0, 1) is a uniformly distributed random number in the range [0, 1] and SP_{min}, SP_{max} are lower and upper limits of ith squirrel in j^{th} dimension.

4.3 Non-dominated Sorting

The pseudo-code of the non-dominated sorting part of MOSSA which the concept of the fast-nondominated-sorting [15] in NSGA-II is adapted to is presented in Algorithm 2. The current unsorted population P is sorted by Pareto levels. After processing of this section, the pareto rank levels of solutions are obtained. Each candidate solution is detected whether it is dominated by other individuals in the population. The solutions which are not dominated by any others are marked as nondominated solutions with the least level.

Algorithm 2. Fast-nondominated-sorting

Input: Unsorted population P
Output: All pareto rank levels L_i (i^{th} level)
Begin
1. for each $p \in P$
2. for each $q \in P$
3. **if** $p \preccurlyeq q$ **then**
4. $D_p = \{p\} \cup D_p$;
5. $DS_p = DS_p + 1$; //DS_p means how many solutions dominated p
6. **else if** $q \preccurlyeq p$ **then**
7. $n_p = n_p + 1$; //n_p means the number of solutions dominating p
8. **end if**
9. **if** $n_p = 0$ **then** //if no solution dominates p then
10. $L_1 = \{p\} \cup L_1$;
11. **end if**
12. $i = 1$;
13. **while** $L_i \neq \emptyset$
14. $TEM = \emptyset$;
15. **for each** $p \in L_i$
16. **for each** $q \in D_p$
17. $n_p = n_p - 1$;
18. **if** $n_p = 0$ **then** $TEM = \{p\} \cup TEM$
19. **end while**
20. $i = i + 1$;
21. $L_i = TEM$;
End

4.4 Grid Division

If a multi-objective problem has m objective functions, then it constitutes an m-dimensional target space. In order to make the population more diverse, we mesh the target space. That is, this target space is divided into $K1 \times K2 \times \ldots \times Ki \times \ldots \times Km$ grids, and the grid width of the ith target(GW_i) of each grid is:

$$GWi = (SFmaxi - SFmini)/Ki \quad (2)$$

Where Ki is the number of grids which the ith dimension objective function is divided by and SF^i_{max} and SF^i_{min} are the maximum and minimum value of objective function on

i^{th} dimension. Now we can calculate the density of each grid (DG_i) which means the number of individual in this certain grid and the grid coordinates of the ith $target(GC_i)$.

$$GC_i = (SF_i - SF_{min})/GW_i \tag{3}$$

4.5 Fitness Mapping Function Design

After the target space is divided into several grids, the fitness value mapping function of each squirrel needs to be determined according to two indicators:

Grid density (DG_i): The number of squirrels corresponding to the grid which contains the ith squirrel in the target space.

Dominant strength (DS_i): The number of other squirrels dominated by the ith squirrel.

Therefore, the ith squirrel's mapping fitness (MF_i) function can be defined as the kth evolution process as:

$$MF_i^k = DG_i^k/(DS_i^k + 1) \tag{4}$$

Where DG_i^k is obtained by counting the number of particles with the same coordinate, DS_i^k represents the number of other particles dominated by the particle i in the kth iteration, which can be obtained by the definition of domination. Adding 1 is to prevent the denominator from being zero. It can be seen that the more a particle controls other particles, the better the fitness, and the better the position of the particle, the healthier the particle is, the smaller the fitness obtained. The purpose of this definition is to obtain particles that are close to the real Pareto front end, with uniform distribution and good scalability.

The population is sorted according to the mapped fitness values, and the squirrels at the 3 optimal positions (SP_{best}) with the smallest fitness values are selected, and the squirrels at the 9 sub-optimal positions (SP_{sub}) with the smaller fitness values are selected. The default population in the algorithm is 100, so there are 88 ordinary solutions (SP_{other}) left.

4.6 Squirrel Migration

According to squirrel habits, we think that squirrels will begin to migrate when their natural enemies are not present. Suppose that the probability of the appearance of natural enemies is P, so that squirrels whose random numbers fall between [P, 1] in the interval [0,1] can migrate so that random numbers that fall between [0, P] should be randomly hidden.

In the first case, sub-best squirrels will move towards best squirrels.

$$SP_{sub,j}^{new} = SP_{sub,j} + const \times \left(SP_{best,j} - SP_{sub,j}\right) \tag{5}$$

In the second case, normal squirrels will move towards sub-best squirrels.

$$SP_{other,j}^{new} = SP_{other,j} + const \times \left(SP_{sub,j} - SP_{other,j}\right) \tag{6}$$

In the third case, some normal squirrels have already been sub-best squirrels, so they will move towards best squirrels.

$$SP_{other,j}^{new} = SP_{other,j} + const \times \left(SP_{best,j} - SP_{other,j}\right) \tag{7}$$

4.7 Seasonal Condition

In the single-objective squirrel algorithm, seasonal conditions are used to measure the degree of population aggregation during each iteration, but in the multi-objective squirrel search algorithm, we need to make some changes to apply to multi-objective optimization.

Generation *gap(G)* is one of the classic convergence indicators. It is mainly used to describe the distance between the non-dominated solution obtained by the algorithm During two consecutive iterations. The smaller *G* is, the more likely the population is to fall into local convergence.

$$G = \frac{\sqrt{\sum_{i=1}^{n} gap_i}}{n} \tag{8}$$

Where n is the number of non-dominated solutions obtained by the algorithm, gap_i is the shortest Euclidean distance between the non-inferior solution and all the solutions in the new generation.

When the two populations reach seasonal constant (SC), which means winter is coming(G ≥ SC). On the contrary(G < SC), it means that summer is coming.

$$SC = \frac{10E^{-6}}{(365)^{t/(t_m/2.5)}} \tag{9}$$

Where t means the current iteration number, t_m means maximum iteration number.

4.8 Levy Flight

Levy distribution can help the algorithm perform a global search in a better and more efficient way, and Levy flight helps the algorithm find new locations far from the current best location. Levy flight is a method of randomly changing the step size, where the step size is derived from the Levy distribution.

$$Levy = \frac{0.01 \times r_a \times \sigma}{|r_b|^{\frac{1}{\beta}}} \tag{10}$$

$$\sigma = \left(\frac{\Gamma(1+\beta) \times \sin(\frac{\pi\beta}{2})}{\Gamma\left(\frac{1+\beta}{2}\right) \times \beta \times 2^{\left(\frac{\beta-1}{2}\right)}} \right)^{\frac{1}{\beta}} \tag{11}$$

$$\Gamma(x) = (x-1)! \tag{12}$$

Where r_a and r_b are two normally distributed random numbers on the interval [0,1], β = 1.5. When the population reaches seasonal constant (SC), the ordinary individuals follow Levi's flight and update the squirrel's position.

$$SPi = SPmin + Levy \times (SPmax - SPmin) \tag{13}$$

4.9 Merge Population

After merging the two populations, because the size of new population exceeds the preset number of members of the population, corresponding screening is required. After non-dominated sorting and calculation of the grid density, the individual with the lowest Pareto rank is not what we required. The distribution of the obtained solutions is usually sparse and uneven. It is generally considered that those dense solutions are relatively poor in distribution, which need to be eliminated to ensure uniform solution distribution on the entire front. It is a problem to eliminate multiple individuals with high grid density at one time, which will probably cause one original dense grid empty. In that case, it makes the distribution of the Pareto solution worse. In response to this deficiency, this paper proposes the use of probability selection based on roulette wheel selection. Individuals with worst Pareto level are eliminated first. When the Pareto levels are the same, the relative density of the grids where these squirrels are located is regarded as the eliminating probability. First calculate the relative density (RD_j) of the jth squirrel.

$$RD_i = DG_i / \sum_{i=1}^{m} DG_i \tag{14}$$

Where m means the number of squirrels which located on worst level. Squirrels of the same Pareto level divides a disc into m parts, in which the center angle of the jth fan is $2\pi \cdot RD_j$. When making a selection, you can imagine turning the dial, and if the pointer falls into the jth sector, delete the individual j. The implementation process is as follows: First generate a random number R in [0,1]. If $\sum_{j=1}^{j-1} RD_j \le R < \sum_{j=1}^{j} RD_j$, then delete individual j. It can be seen that larger the central angle means more. More probably the individual will be eliminated to maintain population diversity.

5 Simulation Results Analysis

In this experimental simulation studies, this paper uses classic multi-objective optimization algorithms such as NSGA-II, MOPSO, and MOEA/D to compare with this algorithm. From the current research literature on multi-objective optimization problems, the evaluation indicators for multi-objective algorithms are mainly designed around convergence and diversity [11, 12, 18–21]. We use three indicators to evaluate the algorithm in this section, which are generational distance, spacing metric and diversity metric.

In the implementation process, the number of test instances was uniformly set to 100 and the number of iterations was 100 generations. Besides, non-dominated solutions obtained by these four algorithms are simulated on multi-objective optimization test functions (ZDT6, ZDTi, i = 1 − 4) [16], each experiment was executed 10 times, and the average values of evaluation parameters are represented to be compared.

5.1 Pareto Front Analysis

These 25 figures represent dominated solutions obtained using these four algorithms and also the ideal PF on a series of Zitzler-Deb-Thiele test functions in Fig. 2.

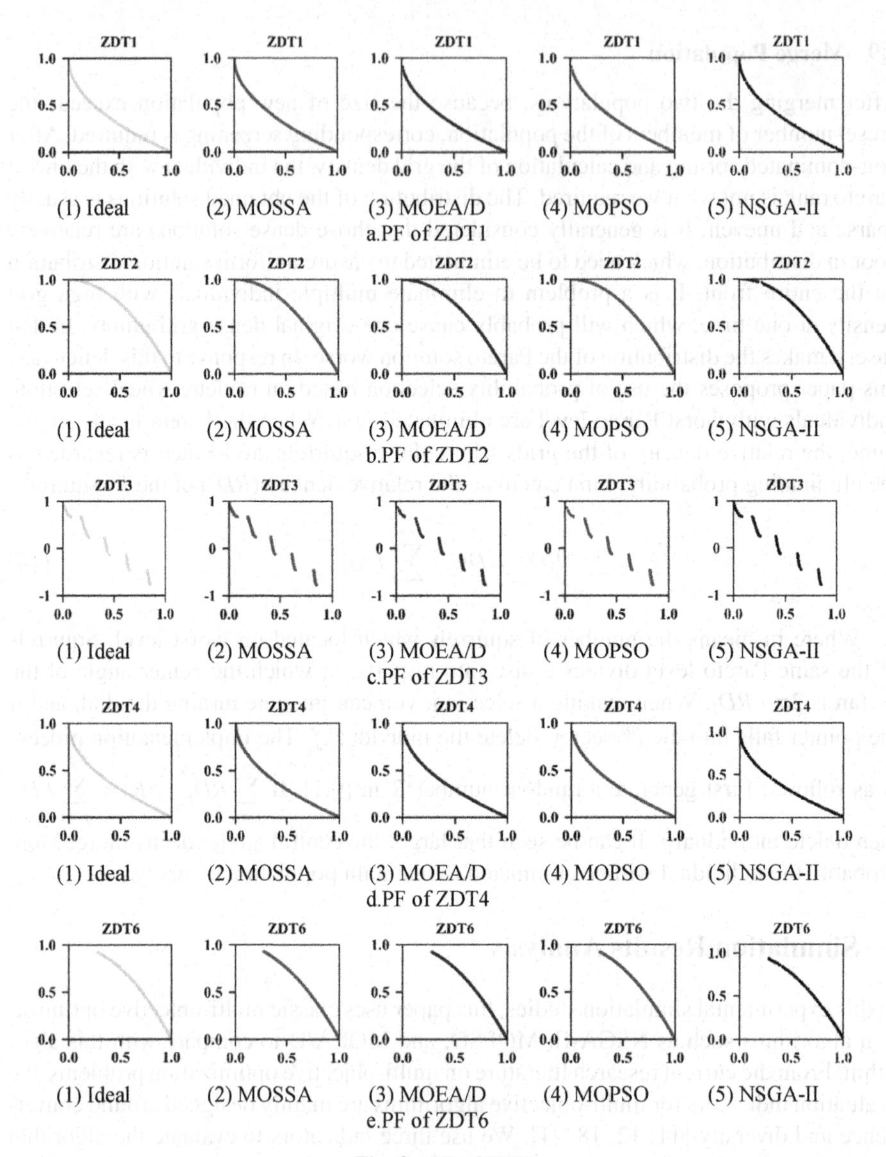

Fig. 2. PF of ZDT.

By comparing the ideal Pareto front and the experimentally obtained solution set, the solution set obtained by MOSSA almost coincides with the real Pareto front, which proves that MOSSA has good convergence. Because the differences in convergence from other algorithms is not obvious now, later in this paper we will also quantitatively compare the convergence of all algorithms.

It can be seen that the distribution of MOEA/D and MOPSO in extreme solutions is not as ideal as MOSSA, resulting in an uneven solution set. Although NSGA-II is relatively uniformly distributed, the distance between two consecutive solutions is larger than MOSSA. It can be seen that MOSSA is superior to the other three algorithms in terms of convergence, uniformity, and diversity on the ZDT test function.

Therefore, MOSSA solves the problem of the unsatisfactory distribution of MOEA/D and MOPSO at extreme solutions and solves the problem that NSGA-II has uniform and extensive solutions at the Pareto front, but the distance of continuous solutions is relatively longer. The uniformity and diversity of MOSSA in the non-dominated solution set are better than other algorithms.

5.2 Analysis with Generational Distance (GD)

Generational Distance (GD) is one of the classic convergence indicators [30]. It is mainly used to describe the distance between the non-dominated solution obtained by the algorithm and the real Pareto front end. The smaller the GD, the better the convergence.

Where n is the number of non-dominated solutions obtained by the algorithm, $dist_i$ is the shortest Euclidean distance between the non-inferior solution and all the solutions in the real Pareto front end.

$$GD = \frac{\sqrt{\sum_{i=1}^{n} dist_i}}{n} \tag{15}$$

For indicator GD, the smaller its value, the better the convergence of the algorithm. As shown in Table 1, every minimum value is bold for each column. It's observed from Table 1 that MOSSA is significantly better than the other three algorithms on ZDT1, ZDT4, and ZDT6 functions. However, for relatively poor performance on ZDT2, ZDT3 functions, we can also clearly see that MOSSA is the second best algorithm. The gap compared to the best algorithm is not obvious enough, so we can analyze from this indicator that the MOSSA algorithm has a good performance in the convergence of the multi-objective optimization algorithm.

Table 1. GD values of the solutions found by all the algorithms on all the test problems.

Algorithm	ZDT1	ZDT2	ZDT3	ZDT4	ZDT6
MOSSA	**7.19E–04**	8.77E–4	0.004136	**0.001723**	**3.185E–4**
MOEA/D	9.62E–04	0.001494	**0.004013**	0.002115	8.30E–4
NSGAII	0.002306	0.00135	0.005052	0.006056	0.001584
MOPSO	9.07E–04	**7.85E–04**	0.004813	0.005592	7.24E–4

5.3 Analysis with Spacing Metric (SP)

Spacing Metric (SP) measures the standard deviation of the minimum distance from each solution to other solutions [31]. The smaller the Spacing value, the more uniform the solution set.

$$SP = \frac{\sqrt{\frac{1}{n}\sum_{i=1}^{n}\left(\bar{d} - d_i\right)^2}}{\bar{d}} \tag{16}$$

Where n is the number of non-dominated solutions obtained by the algorithm; di is the shortest Euclidean distance between the ith solution and all solutions in the real Pareto front end. The smaller the SP, the better the distribution, and the better the diversity. When SP = 0, all the solutions in the Pareto solution set are uniformly distributed.

For indicator SP, the smaller value represents the better uniformity of the algorithm. As shown in Table 2, every minimum value is bold for each column. The data reveals that MOSSA has the smallest SP value on all of the ZDT series functions, which shows that MOSSA has a superior uniform performance.

Table 2. SP values of the solutions found by all the algorithms on all the test problems.

Algorithm	ZDT1	ZDT2	ZDT3	ZDT4	ZDT6
MOSSA	**0.009245**	**0.001546**	**0.003498**	**0.002913**	**0.001289**
MOEA/D	0.016064	0.009078	0.03842	0.01566	0.008262
NSGAII	0.009456	0.013029	0.01517	0.01247	0.01048
MOPSO	0.015692	0.002360	0.01976	0.009101	0.001835

5.4 Analysis with Diversity Metric (DM)

Diversity Metric (DM) measures how extensive the solution set is [15].

$$DM = \frac{d_f + d_l + \sum_{i=1}^{n-1} |d_i - \bar{d}|}{d_f + d_l + (n-1)\bar{d}} \tag{17}$$

Where d_f and d_l are the Euclidean distances between two extreme solutions of the true Pareto front and two boundary solutions of the non-dominated set obtained by experiments. And d_i is the Euclidean distance between two successive solutions in the non-dominated solution set. Assuming that the optimal non-dominated front has n solutions, then there are n − 1 consecutive distances, so $\bar{d}$ is the average of di ($\bar{d} = \frac{1}{n-1}\sum_{j=1}^{n-1} d_j$).

A uniform distribution makes all di approach $\bar{d}$, and when the distribution is extensive enough, $d_f = d_l = 0$ (there are extreme solutions in non-dominated sets). Therefore, for the most extensively and uniformly expanded non-dominated solution set, the numerator of DM will approach zero, making DM zero. For any other distribution, the value of the metric will be greater than zero. In other words, this evaluation parameter can measure both uniformity and breadth to achieve diversity. For two distributions with the same d_f and d_l values, DM has a higher value and a worse solution distribution in the extreme solution.

For the DM indicator, the smaller its value, the higher the diversity of the solution set generated by the multi-objective optimization algorithm. Every minimum value is bold for each column. At the same time, by analyzing data in Table 3, we can figure out that in all ZDT series functions MOSSA has excellent diversity. Although some gaps may not be obvious, the excellent diversity of MOSSA still can be observed.

Table 3. D values of the solutions found by all the algorithms on all the test problems.

Algorithm	ZDT1	ZDT2	ZDT3	ZDT4	ZDT6
MOSSA	**0.6768**	**0.6841**	**0.7808**	**0.7072**	**0.6963**
MOEA/D	0.7524	0.8008	0.9317	0.7517	0.7495
NSGAII	0.7773	0.7951	0.7956	0.7877	0.8079
MOPSO	0.7529	0.7033	0.9549	0.7552	0.7084

The analysis of non-dominated solutions reveals that MOSSA can improve the extreme solutions and large gaps between consecutive solutions, resulting in a solution set which is uneven and diversity. Besides, three indicators which reflect the convergence, uniformity and diversity of multi-objective optimization algorithms also perform better in MOSSA then NSGA-II, MOPSO and MOEA/D.

6 Conclusion

This paper takes NSGA-II as the multi-objective framework and SSA as the main evolution strategy to construct a new improved multi-objective squirrel search algorithm (MOSSA). On their basis, this article has made several improvements. First, MOSSA established an external archive of the population to retain the elite individuals in the population. Moreover, after two consequent populations are merged, this paper suggests the grid density as an eliminated probability to limit population size, which guarantees the diversity of PF. Therefore, the concept of the grid density is introduced to further maintain the uniformity and the spread of the solution set. Based on SSA, the generational gap is introduced to improve the seasonal condition, which solves the problem that solutions may be trapped locally in multi-objective optimization. In addition, this paper designs a mapping function according to Pareto sorting level and grid density to calculate individual fitness values.

Finally, these algorithms MOSSA, MOEA/D, NSGA-II, and MOPSO are used to perform simulation experiments on a multi-objective test function set. This analysis presents the Pareto front and some indicators obtained by each algorithm on the test functions. Through quantitative analysis of GD, SP and DM indicators, it is observed that MOSSA performs well on the convergence of the solution set, the uniformity of solution distribution, and the spread of the distribution. It can be concluded that this novel multi-objective optimization algorithm, MOSSA provides a framework of SSA extended to multi-objective optimization problems and also has a satisfactory performance. The proposed algorithm, MOSSA, could solve MOPs in better performance, so that it could be applied to many fields in real life, such as the transportation, the finance, the engineering, and the biology. For example, vehicle routing programming problems, shopping tradeoff, network routing optimization problems, and protein-ligand docking problems.

References

1. Jain, M., Singh, V., Rani, A.: A novel nature-inspired algorithm for optimization: squirrel search algorithm. Swarm Evol. Comput. S2210650217305229 (2018)
2. Wang, Y., Du, T.: A multi-objective improved squirrel search algorithm based on decomposition with external population and adaptive weight vectors adjustment. Physica A: Stat. Mech. Appl. **542**, (2020)
3. Liang, J.J., Qu, B.Y., Suganthan, P.N.: Problem definitions and evaluation criteria for the CEC 2014 special session and competition on single objective real-parameter numerical optimization. Computational Intelligence Laboratory, Zhengzhou University, Zhengzhou China and Technical Report, Nanyang Technological University, Singapore, 635 (2013)
4. Gunantara, N.. A review of multi-objective optimization: methods and its applications. Cogent Eng. **5**(1), 1502242 (2018)
5. Guo, Z., Liu, L., Yang, J.: A multi-objective memetic optimization approach for green transportation scheduling. In: 2015 International Conference on Intelligent Informatics and Biomedical Sciences (ICIIBMS), IEEE (2015)
6. Dai, M., Tang, D., Giret, A., Salido, M.A.: Multi-objective optimization for energy-efficient flexible job shop scheduling problem with transportation constraints. Robot. Comput. Integrated Manuf. **59**, 143–157 (2019)
7. Zaro, F.R., Abido, M.A.: Multi-objective particle swarm optimization for optimal power flow in a deregulated environment of power systems. In: 2011 11th International Conference on Intelligent Systems Design and Applications, IEEE (2019)
8. Liu, Z., Jiang, D., Zhang, C., et al.: A novel fireworks algorithm for the protein-ligand docking on the AutoDock. Mob. Netw. Appl. 1–12, 53 (2019)
9. de Villiers, D.I., Couckuyt, I., Dhaene, T.: Multi-objective optimization of reflector antennas using kriging and probability of improvement. In: 2017 IEEE International Symposium on Antennas and Propagation & USNC/URSI National Radio Science Meeting, pp. 985–986. IEEE, July 2017 (2007)
10. Delgarm, N., Sajadi, B., Kowsary, F., Delgarm, S.: Multi-objective optimization of the building energy performance: a simulation-based approach by means of particle swarm optimization (PSO). Appl. Energy **170**, 293–303 (2016)
11. von Lücken, C., Barán, B., Brizuela, C.: A survey on multi-objective evolutionary algorithms for many-objective problems. Comput. Optimization Appl. 1–50 (2014)
12. Cho, J.H., Wang, Y., Chen, R., et al.: A survey on modeling and optimizing multi-objective systems. IEEE Commun. Surv. Tutorials **19**(3), 1867–1901 (2017)
13. Xue, B., Zhang, M., Browne, W.N.: Particle swarm optimization for feature selection in classification: a multi-objective approach. Cybern. Trans. IEEE **43**(6), 1656–1671 (2013)
14. Zhang, Q., Li, H.: Moea/d: a multiobjective evolutionary algorithm based on decomposition. IEEE Trans. Evol. Comput. **11**(6), 712–731 (2008)
15. Deb, K., Pratap, A., Agarwal, S., et al.: A fast and elitist multiobjective genetic algorithm: NSGA-II. IEEE Trans. Evol. Comput. **6**(2), 182–197 (2002)
16. Mashwani, W.K., Salhi, A., Yeniay, O., et al.: Hybrid non-dominated sorting genetic algorithm with adaptive operators selection. Appl. Soft Comput. **56**, 1–18 (2017)
17. Li, K., Wang, R., Zhang, T., et al.: Evolutionary many-objective optimization: a comparative study of the state-of-the-art. IEEE Access **6**, 26194–26214 (2018)
18. Liu, Z., Zhang, C., Zhao, Q., et al.: Comparative study of evolutionary algorithms for protein-ligand docking problem on the AutoDock. International Conference on Simulation Tools and Techniques, pp. 598–607. Springer, Cham (2019)
19. Azzouz, R., Bechikh, S., Said, L.B.: Dynamic Multi-objective Optimization using Evolutionary Algorithms: A Survey. Recent Advances in Evolutionary Multi-objective Optimization, pp. 31–70. Springer, Cham (2017)

20. Bechikh, S., Elarbi, M., Said, L.B.: Many-objective Optimization using Evolutionary Algorithms: A Survey. Recent Advances in Evolutionary Multi-objective Optimization, pp. 105–137. Springer, Cham (2017)
21. Falcón-Cardona, J.G., Coello, C.A.C.: Indicator-based multi-objective evolutionary algorithms: a comprehensive survey. ACM Comput. Surv. (CSUR) **53**(2), 1–35 (2020)
22. Shamshirband, S., Shojafar, M., Hosseinabadi, A.A.R., Abraham, A.: A solution for multi-objective commodity vehicle routing problem by NSGA-II. International Conference on Hybrid Intelligent Systems. IEEE (2015)
23. Luo, G., Wen, X., Li, H., et al.: An effective multi-objective genetic algorithm based on immune principle and external archive for multi-objective integrated process planning and scheduling. The Int. J. Adv. Manuf. Technol. **91**(9–12), 3145–3158 (2017)
24. Gadhvi, B., Savsani, V., Patel, V.: Multi-objective optimization of vehicle passive suspension system using NSGA-II, SPEA2 and PESA-II. Procedia Technol. **2016**(23), 361–368 (2016)
25. Zitzler, E., Laumanns, M., Thiele, L.: SPEA2: improving the strength Pareto evolutionary algorithm. TIK-report, 103 (2001)
26. Wang, Y., Han, M.: Research on multi-objective multidisciplinary design optimization based on particle swarm optimization. In: 2017 Second International Conference on Reliability Systems Engineering (ICRSE). IEEE (2017)
27. Kaoutar, S., Mohamed, E.: Multi-criteria optimization of neural networks using multi-objective genetic algorithm. International Conference on Inteligent Systems & Computer Vision ISCV (2017)
28. Rosales-Perez, A., Garcia, S., Gonzalez, J.A., Coello, C.A.C., Herrera, F.: An evolutionary multiobjective model and instance selection for support vector machines with pareto-based ensembles. IEEE Trans. Evol. Comput. **21**(6), 863–877 (2017)
29. Juang, Chia-Feng., Jeng, Tian-Lu, Chang, Yu-Cheng: An interpretable fuzzy system learned through online rule generation and multiobjective ACO with a mobile robot control application. IEEE Trans. Cybern. **46**(12), 2706–2718 (2017)
30. Sheikholeslami, F., Navimipour, N.J.: Service allocation in the cloud environments using multi-objective particle swarm optimization algorithm based on crowding distance. Swarm and Evol. Comput. **35**, 53–64 (2017)
31. Tian, Y., Cheng, R., Zhang, X., et al.: Diversity assessment of multi-objective evolutionary algorithms: performance metric and benchmark problems [research frontier]. IEEE Comput. Intell. Magazine **14**(3), 61–74 (2019)

A GTAP Simulation-Based Analysis Method for Impact of the US Exit from WTO and China's Strategic Choice

Zheng Kou[1](✉), Qun Gao[1], and Man Zhang[2]

[1] College of Business, Yancheng Teachers University, Yancheng 224007, Jiangsu, China
[2] Economic and Trade Department, Yancheng Polytechnic College, Yancheng 224005, Jiangsu, China

Abstract. In this paper, the GTAP model is used to simulate the impact of the withdrawal of U.S. from the WTO on China and even the global economy. It is found that the withdrawal of the United States from WTO has both positive and negative effects on China's economy and these effects expand with the rise of tariff level between the United States and WTO members. The positive influence comes from the fact that after the United States imposes tariffs on other WTO members, it will worsen the terms of trade between the United States and other WTO members, reduce import and export, and bring favorable effects to China. The negative effects are mainly due to the escalating trade war between China and the United States. The withdrawal of the United States from the WTO and the imposition of tariffs on China will lead to the deterioration of trade between China and the United States, which is not conducive to the export of Chinese products to the United States. We have obtained the import and export data of the world in recent three years. Based on the GTAP model and correlation analysis method, we analyze and model the data. Through the simulation experiment based on the GTAP model, we derive a decision-making model with great reference value.

Keywords: Sino-US trade · Tariff · Non-trade barriers · GTAP model · Grey relational analysis

1 Introduction

As emerging economies have joined the WTO and are continuously integrated into the global trading system, developed countries have begun to change their attitudes towards the WTO, and this major international trade organization. The more difficult it is to "knife tofu-two sides light" (Yunkun Zhang 2020). The international multilateral system is facing unprecedented challenges (Runting Jin 2019). Therefore, it is urgent for us to study in depth what challenges will be brought to Global trade by "the withdrawal of the United States from the WTO"? How to affect Chinese interests? Take precautions to avoid risks and seize opportunities. A small amount of studies have focused on two aspects: the motivation of the U.S. to withdraw from the WTO and the possible impact.

H. Song and D. Jiang (Eds.): SIMUtools 2020, LNICST 370, pp. 196–205, 2021.
https://doi.org/10.1007/978-3-030-72795-6_16

Judging from the reasons for the withdrawal of the United States from the WTO, non-reciprocal or unequal trade is the root cause of the withdrawal of the United States from the WTO (Zhongying Pang 2018). In particular, the United States believes that China, as a "developing country" among WTO members, has taken advantage of tax rates, quotas and many other aspects, resulting in damage to the interests of the United States (Xinquan Tu and Xiaojing Shi 2019). It is an important reason for the United States to withdraw from the WTO to put pressure on the multilateral mechanism with the method of "retreat for progress" (Longyue Zhao 2018). The withdrawal of the United States from the WTO is the result of various reasons.

The reasons for Sino-US trade frictions are complex, including trade, investment, intellectual property rights and other economic and trade factors, but also contain US worries and fears about China's rise (Di Zhao 2020). The underlying reason is the collective decline of Western countries headed by the United States, and the collective rise of developing economies headed by China (Changbin Zhang 2019). The Sino-U.S. Trade war is essentially a manufacturing struggle, and curbing the development of China's manufacturing is an important reason for the U.S. to launch a trade war (Junjie Hong and Zhihao Yang 2019). The trade war cannot fundamentally improve the huge US trade deficit. In the export industry, China's transportation equipment, machinery and other manufacturing industries have a greater impact on exports, while the US agriculture, oil and other industries have a greater impact (Mengran Li 2020). It is expected that my country's GDP will fall by 0.06 to 0.07% points from 2018 to 2020, and the three industries of electrical and electronic products, mechanical equipment and transportation equipment manufacturing will have a greater impact (Yaxiong Zhang and Jifeng Li 2018). But overall, the impact of trade friction itself on China's macroeconomic economy is limited and controllable (Qing Guo and Weiguang Chen 2019).

Effective measures must be taken to reduce the factors of Sino-US economic and trade frictions (Chen et al. 2020). China will accelerate its mastery of core technologies, continue to maintain the strategic strength of comprehensive reform and opening-up in the "Belt and Road" initiative, and calmly deal with the various risks of the US initiating a trade war (Han Y.H. 2020). Due to the strong economic strength and trade volume of the United States, the comprehensive multilateral trade mechanism will be in disorder (Jingwei Zhang 2018), which will have a big shock to the global political and economic governance system, and the influence of the WTO on the global trade market will be completely overturned (Qingjiang Kong 2019). The United States may face huge economic costs, such as the loss of important global trade access arrangements, huge losses in service trade, trade-related intellectual property rights, trade-related investment issues and so on (Huifang Qiu 2017).

In summary, although the academic community has conducted some research on the reasons and impact of the US withdrawal from the WTO, the current research results mainly use qualitative research, and there are few articles using data model to analyze the impact of the U.S. withdrawal from the WTO, especially those of "Analysis of China's Economic Impact". In view of this, this paper uses GTAP model to simulate and analyze the tariff level changes between the United States and other WTO members caused by the withdrawal of the United States from the WTO, studying how the major "black swan" timing of the United States' withdrawal from the WTO will affect Chinese economic interests.

2 Simulation Scheme and Modeling Analysis

"Withdrawing from the WTO" means that the United States does not have to abide by the principles and agreements of WTO, and is no longer subject to the requirements of the most-favored-nation and binding tax rate in the WTO agreements, but will increase the tariff rate of foreign trade for its own interests. In the face of the changes of US tariff policy, other WTO member states will also impose retaliatory tariffs on US. For example, after the United States imposed a 25% tariff on steel exported to the United States by Canada, Mexico and the European Union, The European Union and Canada also impose tariffs to varying degrees on US imports. Therefore, the first scheme assumes that the tariffs between the United States and developed countries among WTO members will increase after the United States withdraws from the WTO. Compared with tariff, non-tariff measures have the characteristics of diversity, flexibility and concealment. According to the TBT agreement of WTO, once the United States withdraws from WTO, serious non-tariff barriers will be produced. Therefore, the second scheme assumes that the United States and WTO member states will mutually enhance their non-tariff barriers after withdrawing from the WTO. When the WTO was established, it provided preferential policies to China and India.

With the rise of developing countries, the United States thought that the trade with developing countries is very disadvantageous. As the largest trade deficit country of the United States, China was bound to be imposed punitive tariffs by the United States once the United States withdraws from the WTO. Therefore, the third scheme assumes that the United States will impose punitive tariffs on China after withdrawing from the WTO. Based on the above analysis, three schemes are set up to simulate impact variables of tariff and non-tariff barriers in this paper.

The classification of production factors in this paper adopts five basic element classification methods of land, unskilled labor, skilled labor, capital and natural resources. This article divides the product sector into the following three groups based on the GTAP industry sector database combined with the characteristics of US tariffs: agriculture, manufacturing, and service industries. The specif ic division is shown. This article also divides the 121 country data and 20 regional collections collected in the GTAP (10th Edition) database into the following 14 groups: China, Japan, United Kingdom, Germany, South Korea, Canada, United States, Mexico, France, Belgium, Singapore, India, other countries and regions in the world. Through grouping, we analyze the impact of the US withdrawal from the WTO and the corresponding strategies of our country.

According to the simulation results (Table 1), under the three simulation schemes, the withdrawal of the US from the WTO will not only have an impact on its own GDP, terms of trade and welfare, but also on other member countries of the WTO. In scheme 1, most countries and regions show negative changes in.

GDP, terms of trade and welfare, while the U.S. GDP, terms of trade and welfare are still in positive changes, which is inseparable from the strong economic and national strength of the United States. Among the countries and regions with negative changes,

Table 1. Changes in GDP, terms of trade and welfare.

Country or region	GDP change (unit: %)			Change in terms of trade (in %)			Changes in benefits (in millions of US dollars)		
	Scheme 1	Scheme 2	Scheme 3	Scheme 1	Scheme 2	Scheme 3	Scheme 1	Scheme 2	Scheme 3
CHN	2.27	- 0.30	- 8.75	1.69	- 0.75	- 4.96	28833.52	- 14676.44	- 123046.55
JPN	- 4.99	0.03	0.18	- 4.35	- 0.70	0.85	- 34475.79	- 5784.22	4115.87
GBR	- 2.54	0.37	0.35	- 1.99	- 0.10	0.44	- 17791.07	- 891.98	3385.86
DEU	- 2.00	0.32	0.10	- 1.53	- 0.14	0.23	- 22774.59	- 2153.50	1231.32
KOR	- 4.10	- 0.24	- 0.19	- 2.57	- 0.58	0.33	- 13326.42	- 3534.75	1130.03
CAN	- 19.53	- 0.46	3.63	- 15.02	- 0.97	1.86	- 70950.16	- 4577.76	8420.90
USA	4.30	0.12	2.29	4.85	- 0.49	2.30	112983.46	- 11474.19	40003.02
FRA	- 1.64	0.32	0.06	- 1.13	- 0.16	0.16	- 10983.38	- 1415.45	471.97
BEL	- 2.00	0.20	0.04	- 0.75	- 0.13	0.04	- 3873.89	- 889.22	146.54
SGP	- 4.48	0.56	- 0.27	- 2.28	0.01	0.15	- 4931.77	43.47	- 85.92
IND	1.33	- 0.20	0.29	0.83	- 0.80	0.52	2780.11	- 3602.22	733.14
MEX	6.28	- 0.37	3.94	3.69	- 0.99	2.48	8912.59	- 2848.60	6211.73
NLD	- 1.65	0.25	0.10	- 0.95	- 0.19	0.21	- 3897.14	- 765.70	119.18
Other	0.81	- 0.11	0.18	0.64	- 0.63	0.34	35475.57	- 47743.43	15449.92

Canada has the largest absolute value of negative changes in GDP, terms of trade and welfare. In scheme three, if the United States withdraws from the WTO and imposes punitive tariffs on China, the US GDP growth rate will reach 2.29%, the terms of trade improvement will be 2.30%, and the welfare will increase by 40003.02 million US dollars, while China's GDP, terms of trade, and welfare will show negative changes, which shows that the United States, as China's largest trading partner, can levy punitive tariffs on China and benefit from it.

The probability of item set A and B occurring simultaneously is the support degree (relative support degree) of association rules is as follow.

$$Support(A \Rightarrow B) = P(A \cup B) \tag{1}$$

The probability of occurrence of item set B is the confidence degree of association rules is as follow.

$$Confidence\ (A \Rightarrow B) = P\ (B|A) \tag{2}$$

For the data in Table 1, a can represent GDP change (unit: %), and B represents change in terms of trade (in %) or changes in benefits (in millions of US dollars). Then (1.1) the formula can be changed into the following form.

$$Support\ (\mathrm{A} \Rightarrow B) = \frac{Support\_count\ (A \cap \mathrm{B})}{Total\_count\ (\mathrm{A})} \tag{3}$$

$$Confidence\ (A \Rightarrow B) = \frac{Support\_count\ (A \cap B)}{Support\_count\ (A)} \tag{4}$$

where $Support\_count(A \cap \mathrm{B})$ represents the number of times A and B grow simultaneously. Through the Apriori algorithm, we can calculate the relationship between the change of GDP and the change of trade, and the relationship between the change of GDP and the changes in benefits.

Based on the above data, we can use the grey correlation method to analyze the change relationship between different schemes in different countries and regions. Grey relational analysis is a quantitative description and comparison method for the development and change trend of a system.

The specific calculation steps of grey correlation analysis are as follows:

Step 1: determine the analysis sequence. The reference sequence as follow.

$$Y = \{Y(k)|k = 1, 2, ..., n\} \tag{5}$$

The comparison sequence as follow.

$$X_i = \{X_i(k)|k = 1, 2, \Lambda, n\}, i = 0, 1, 2, \Lambda, m \tag{6}$$

Step 2: dimensionless variables.

$$x_i(k) = \frac{X_i(k)}{X_i(l)}, k = 1, 2, \Lambda, n; i = 0, 1, 2, \Lambda, m \tag{7}$$

Step 3: calculate the correlation coefficient.

The correlation coefficient of $x_0(k)$ and $x_i(k)$ is as follow.

$$\xi_i(k) = \frac{\min_i \min_i |y(k) - x_i(k)| + \rho \max_i \max_k |y(k) - x_i(k)|}{|y(k) - x_i(k)| + \rho \max_i \max_x |y(k) - x_i(k)|} \tag{8}$$

Step 4: calculate the correlation degree.

$$r_i = \frac{1}{n} \sum_{k=1}^{n} \xi_i(k), k = 1, 2, \Lambda, n \tag{9}$$

Step 5: ranking of relevance degree.

Through the grey relational analysis model, we can calculate the correlation degree between different schemes in different regions and countries. According to the simulation results (Table 2), if the United States withdraws from the WTO, it will have a significant impact on various industrial sectors such as agriculture, manufacturing, and service industries in China and the United States.

Judging from the export situation of various industries in China, in Option 1, if the United States withdraws from the WTO and raises the tariff rate on developed countries, the developed countries will also impose retaliatory tariffs on the United States, thus triggering an increase in tariffs among developed countries.

In view of the impact of different strategies on different schemes, we can use the method of association analysis to analyze. As before, we use the Apriori algorithm for analysis and modeling. First, we define the Minimum support as follow. Minimum support: a user or expert defined threshold to measure support, which represents the lowest statistical importance of a project set.

Table 2. Changes in import and export of various industries in China and the United States (unit: %).

Industry	US export/China Import			US import/China export		
	Scheme 1	Scheme 2	Scheme 3	Scheme 1	Scheme 2	Scheme 3
Agriculture	– 13.28	– 9.36	– 24.16	5.93	– 5.99	– 136.31
manufacturing industry	– 24.45	– 1.78	– 34.09	48.76	– 2.13	– 150.73
Service industry	– 8.19	– 0.72	– 23.14	4.50	0.72	– 62.09

Then we can calculate according to the data in the table the Minimum support between agriculture and manufacturing industry is:

$$Support\ (agric \Rightarrow manuf) = \frac{Incease\_count\ (agric \cap manuf)}{Total\_incearse\_count\ (agric)} \tag{10}$$

Minimum support between manufacturing industry and service industry as follow:

$$Support\ (manuf \Rightarrow serv) = \frac{Incease\_count\ (manuf \cap serv)}{Total\_incearse\_count\ (manuf)} \tag{11}$$

Minimum support between agriculture and service industry as follow:

$$Support\ (agric \Rightarrow serv) = \frac{Incease\_count\ (agric \cap serv)}{Total\_incearse\_count\ (serv)} \tag{12}$$

Based on the above, we can analyze the correlation support among agriculture, industry and service industry. Furthermore, we can analyze the degree of their correlation in time.

3 Simulation Results and Analysis

By calculating the degree of association between changes in GDP and change in terms of trade, we calculate the proportion of the number of times changes in GDP and change in terms of trade increase at the same time in the total number of changes in GDP growth in three schemes. Similarly, in order to calculate the correlation between changes in GDP and changes in benefits, we also calculate the proportion of the number of times changes in GDP and changes in benefits increase at the same time in the total number of changes in GDP growth. Using our association analysis algorithm, we can calculate the degree of association between changes in GDP and change in terms of trade, and between changes in GDP and changes in benefits in different countries and regions. As shown in Fig. 1, we calculate the association between changes in GDP and change in terms of trade and changes in benefits in different countries.

We calculated the correlation degree between import and export transportation. We used grey correlation analysis method to calculate the correlation degree of every two schemes located in import and export transportation respectively. As shown in Fig. 2, we can see that the direct correlation degree of schemes in the same period tends to be one, while the correlation degree between schemes in different periods is relatively low. For schemes of different periods, we can see that their correlation is at a low level, which reflects that the fluctuation of import and export does not have a strong time correlation, that is, the impact of import and export fluctuations on future import and export fluctuations cannot be predicted. Based on this, we can conclude that there is a great correlation between import and export transportation in terms of time. For China US trade, a big import and export country, it is more important to create new opportunities to increase the volume of import and export trade than to observe the past. This is also an important conclusion of our analysis of time correlation.

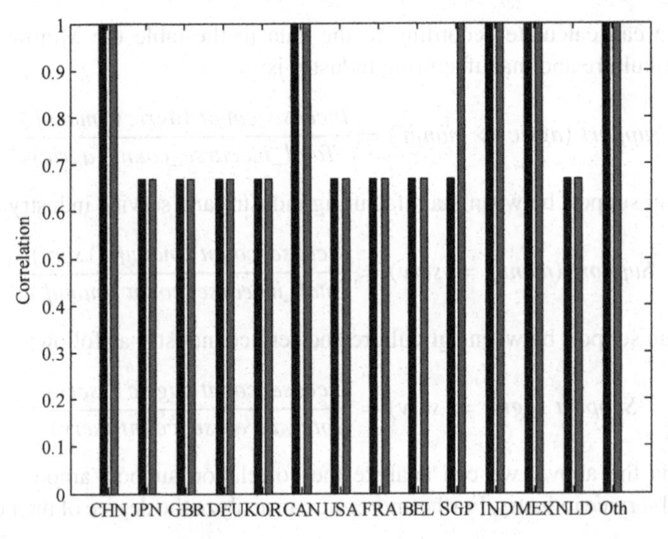

Fig. 1. The correlation between GDP and trade or benefits

We use the correlation analysis model to calculate the support between different industries. According to different schemes of import and export, we can plot their changes in different industries. To facilitate display, we use a three-dimensional diagram to show our information, as shown in Fig. 3. From the figure, we can see that the overall trend of the three industries is consistent in different schemes, which indicates that the time correlation between agriculture, industry and service industry is high. From the figure, we can also see that the fluctuation of the three major industries of scheme in different periods is quite different, which reflects a trend of economic change. Although it is difficult to predict this trend, it can be predicted according to the time correlation of different industries. Through past increases and fluctuations, we can predict the probability of future changes in agricultural, industrial, and service output values. Based on the above

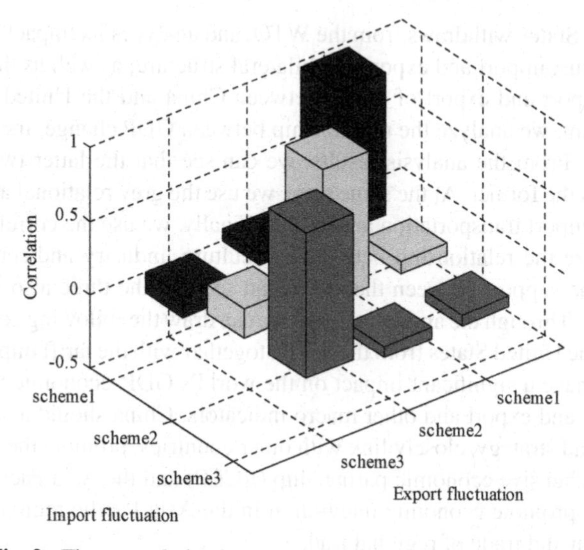

Fig. 2. Time correlation between import and export transportation

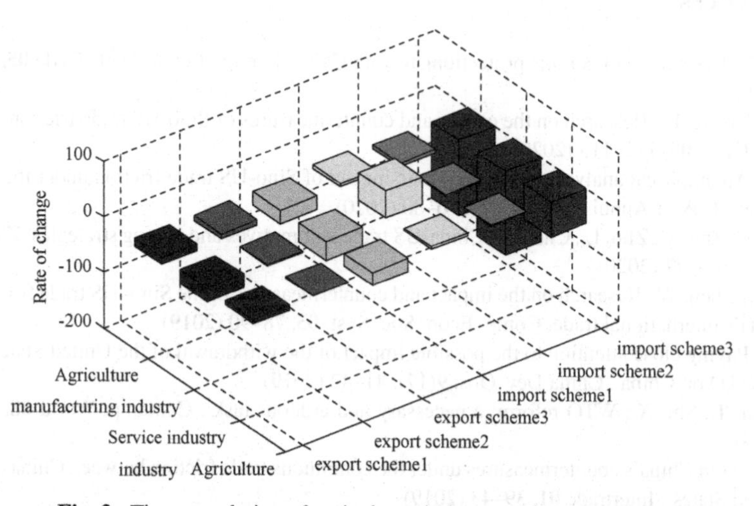

Fig. 3. Time correlation of agriculture, industry and service industry

analysis, we can conclude that the growth of Sino US trade changes is extremely time-dependent, which has sufficient reference value for predicting the changes of output value of different industries and import and export trade.

4 Conclusions

This paper uses the GTAP model to simulate the economic impact of tariff barriers and non-tariff barriers re-established between the United States and other WTO members after the United States withdraws from the WTO, and analyzes its impact on GDP, trade and welfare status, import and export and bilateral structure, as well as the relationship between the import and export of goods between China and the United States. Using Apriori algorithm, we analyze the relationship between GDP change, trade change and welfare change. From the analysis results, we can see that the latter two have strong correlation with the former. At the same time, we use the grey relational analysis model to analyze the import transportation and export. Finally, we use the correlation analysis model to analyze the relationship between agriculture, industry and service industry, and calculate the support between them. We can see that the three also show a strong time correlation. Through the above analysis, we can draw the following conclusions: the withdrawal of the United States from the WTO, together with the tariff imposed by WTO members, will have a significant impact on the world's GDP, economic welfare, terms of trade, import and export and other macro indicators. China should actively promote one belt, one road strategy, closely link with other countries, promote the negotiation of regional comprehensive economic partnership (RCEP) and the Asia Pacific Free Trade Area (FTAAP), promote economic integration in the Asia Pacific region and facilitate the liberalization and trade of regional trade.

References

Zhao, D.: Research on US trade protectionism with China. Foreign Econ. Relat. Trade **05**, 16–20 (2020)

Cui, J., Zhang, R.: Research on the causes and countermeasures of Sino-US trade friction . Econ. Res. Guid. **08**, 112–113 (2020)

Li, M.: An empirical analysis of the economic impact of Sino-US trade friction under the GTAP model . J. West Anhui Univ. **36**(01), 76–80 (2020)

Chen, W., Zhu, Y., Zhu, L., Chen, P.: China-US trade war motives and coping strategies. Zhejiang Econ. (04), **77** (2020)

Guo, Q., Chen, W.: Research on the impact and countermeasures of the Sino–US trade friction on China's international trade. Comp. Econ. Soc. Syst. **05**, 78–90 (2019)

Jin, R.: Paying close attention to the possible impact of the withdrawal of the United States from the WTO on China . China Dev. Obs. **9**(17), 41–42 (2019)

Xinquan, T., Shi, X.: WTO reform: a necessary and arduous task . Contemp. World **08**, 30–36 (2019)

Kong, Q.: On China's countermeasures under the continuous trade friction between China and the United States . Intertrade **01**, 39–43 (2019)

Donghui, F.: On defeating Trump's trade war from WTO law . Intertrade **01**, 59–67 (2019)

Zhang, C.: Analysis of the deep-seated reasons why the United States provoked a trade war against China. In: 70th Anniversary of the Founding of the People's Republic of China and Human Development Economics-China·Human Development in 2019 Proceedings of Academic Conference of Development Economics Society, pp. 102–103 (2019)

Hong, J., Yang, Z.: The impact of Sino-US trade friction on China's manufacturing and China's countermeasure. Intertrade **08**, 21–27 (2019)

Zhang, Y., Li, J.: Model calculation of the impact of Sino-US trade friction on my country's economy . China Econ. Trade Guide **13**, 14–17 (2018)

Qiu, H.: The possible impact of the U.S. withdrawal from the World Trade Organization. Econ. Trade Update (21), 25–29 (2017)

Chinese Ministry of Commerce. Research report on China-US economic and trade relations [EB/OL]. Accessed 12 Sep 2018
Zhang, J.: U.S. withdrawal from WTO is a dangerous game. China Business Times. https://epaper.cbt.com.cn/epaper/uniflows/html/2018/09/06/03/03_52.htm. Accessed 9 June 2018
Pang, Z.: Is the dream of the U.S. exit from WTO coming true?. China Times. https://www.chinatimes.cc/. Accessed 9 Oct 2018
Song, G.: Making two-step preparations for the threat of US withdrawal from WTO.Global Times. https://opinion.huanqiu.com/article/9CaKrnKc5C0. Accessed 9 Jan 2018
Zhang, Y.: Guarding the WTO "Guardian of Free Trade". Global Times

A Simulation Environment for Autonomous Robot Swarms with Limited Communication Skills

Alexander Puzicha(✉) and Peter Buchholz

Informatik 4, TU Dortmund, Dortmund, Germany
{alexander.puzicha,peter.buchholz}@cs.tu-dortmund.de

Abstract. We present a novel real-time simulation tool for modeling and analyzing a swarm of distributed autonomous mobile robots communicating over an unreliable and capacity restricted communication network. The robots are setup as ground vehicles and use C-PBP [8] as model predictive closed loop controller. This tool offers the ability to simulate rural as well as completely urban scenarios with static obstacles, dynamic obstacles with scripted movement, soil condition, noise floor, active jammers and static and dynamic obstacles for the links with adjustable damping. The goal of this simulation is the analysis of swarm behavior of robots for given missions such as terrain exploration, convoy escorting or creation of a mobile ad hoc network in disaster areas under realistic environmental conditions.

Keywords: Autonomous robots · Model predictive control · Real-time simulation · Swarm behavior

1 Introduction

Swarms of autonomous robots are used for a wide variety of missions during disasters. Typical tasks are the exploration of an area, the escorting of convoys or the set up of MANETs if infrastructure breaks down. The autonomous control of robots is a challenging task in particular under tough environmental conditions that limit the possibilities of the robots to communicate. This implies that each robot has only a local view with limited information about the rest of the swarm and of its environment. Before the robot swarm can be deployed in disaster areas, the control software has to be tested carefully. Unfortunately, field tests are expensive and sometimes almost impossible. Thus, the only viable alternative are simulations. However, the test of software with real-time requirements has to be done in an real-time environment which puts very strict demands on the simulation software. The control software has to be part of the simulator, the dynamics of the robots has to be simulated realistically, environmental conditions have to be described and the mobile communication has to be modeled including limited communication ranges and packet losses. All this should be done in user-friendly way and implemented efficiently to allow the real-time simulation of larger swarms as they are used in practice.

H. Song and D. Jiang (Eds.): SIMUtools 2020, LNICST 370, pp. 206–226, 2021.
https://doi.org/10.1007/978-3-030-72795-6_17

Related Work: The presented approach uses model predictive control which has been applied for autonomous robots in several papers [4,18]. However, the control algorithms often work under unrealistic assumptions like the availability of complete information for every robot or the existence of a central instance which computes the trajectories for every member of the swarm [3,16]. Only recently, feedback control algorithms over wireless networks have been published [2,14]. Several other simulators of autonomous robots exist, but only those related to our work are presented in this paper. [12] describe a simulation environment for swarms of robots without a detailed communication model. In [19], a co-simulation approach for communicating autonomous vehicles is introduced. The approach combines the robotic simulator Gazebo and the network simulator OMNeT++. In contrast to our tool, the software combines several tools and is mainly tailored to autonomous driving.

Contribution of this Paper: In this paper we present a new modular structured simulation environment for swarms of autonomous robots. The software allows one to describe complex environments in which the robots have to perform their mission. The environment may contain different types of obstacles which hinder the movement of robots and possibly also the communication between robots. A graphical interface shows the movement of the robots in their environment and allows the user to interact with the robots, e.g. by modifying the mission during operation. The simulator performs a detailed simulation of the swarm by exploiting the parallel features of contemporary processors. Some examples show that with the simulation software swarms of a realistic size can be simulated in real time on a modern workstation. The whole approach has been developed in [15].

Structure of the Paper: In the following section we describe the basic model of an autonomous robot and the model predictive control algorithm. Furthermore, modeling of the environment using appropriate cost functions and the communication model are introduced. Section 3 describes the simulation software. In Sect. 4 some example runs of the simulator are presented. The paper ends with a short summary and some topics for future research.

2 A Model of Autonomous Robots

At first we have to model the autonomous swarm robots. Their key behavior is adapted from people's strategic planning techniques. First, a robot gathers information from the environment, then a plan is created based on the information, mission target, communication possibilities and the prediction of states of the other robots and dynamic objects in world. Due to optimization, this results in a local plan that forms a trajectory in space and time. It is sent to other robots so their plan will be created with respect to this one. The plan leads to an action that causes a change of the environment which in turn produces new information for the sensors. This closed loop can be transformed to a model predictive controller. Therefore, it has to be split into parallel working threads. One thread tackles the problem of sensor evaluation and object detection, another one serves as communication module to handle and filter messages and data packages from other robots. To maintain the connections and to be aware of signal losses it is important that this module is able to monitor and predict the connection quality (see Sect. 2.3). Furthermore, the actuator has to be represented by a separate thread as well as the planning algorithm.

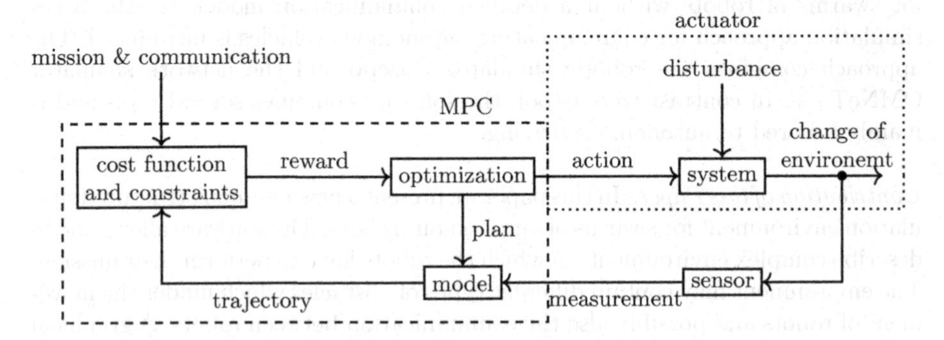

Fig. 1. Four components of the robot: communication module, model predictive controller (MPC), actuator and sensor

2.1 Autonomous Robots and Their Control

As shown in Fig. 1 the model predictive control (MPC, see [7]) algorithm needs a kinematic model of the system and due to the focus on ground based mobile robots and tracked vehicles, we use a simple discrete, nonlinear model for non-holonomic systems that have the ability to turn on the spot. Ground vehicles have less degrees of freedom than flying drones, but there are many more obstacles on the ground with limited possibilities to avoid those. Based on this model, a trajectory is created by the predicted control values and rated through cost functions and constraints which are presented in the next section. The basic model is time discrete and will be briefly described first.

$$\mathbf{x}_k = f_k(\mathbf{x}_{k-1}, \mathbf{u}_k) + \boldsymbol{\xi} = \underbrace{\begin{bmatrix} x_k \\ y_k \\ \theta_k \end{bmatrix}}_{\mathbf{x}_k} = \underbrace{\begin{bmatrix} x_{k-1} \\ y_{k-1} \\ \theta_{k-1} \end{bmatrix}}_{\mathbf{x}_{k-1}} + \begin{bmatrix} v_k \cdot \sin(\theta_{k-1}) \\ v_k \cdot \cos(\theta_{k-1}) \\ \dot{\theta}_k \end{bmatrix} \cdot \Delta k + \boldsymbol{\xi} \quad (1)$$

$$\text{with } \mathbf{u}_k = \begin{bmatrix} v_k \\ \dot{\theta}_k \end{bmatrix} \text{ and } \boldsymbol{\xi} \sim \mathcal{N}(0, \mathbf{Q}) \quad (2)$$

The state space vector $\mathbf{x}_k$ for discrete time k consists of the $x \in \mathbb{R}$ and $y \in \mathbb{R}$ coordinates and the orientation angle $\theta \in [0, 2\pi]$. The discrete transfer function

$f_k : \mathbb{X} \times \mathbb{U} \rightarrow \mathbb{X}$ describes the transfer from $\mathbf{x}_{k-1} \in \mathbb{X}$ to $\mathbf{x}_k$ by using $\mathbf{u}_k \in \mathbb{U}$. The control vector $\mathbf{u}_k$ for the actuator describes velocity $v \in \mathbb{R}$ into current direction and rotation speed $\dot{\theta} \in \mathbb{R}$. Since knowledge about the current state is always faced with some uncertainty, let $\boldsymbol{\xi}$ be zero mean Gaussian noise with covariance matrix $\mathbf{Q}$. With this function it is possible to predict a trajectory $\mathbf{x}_1, \mathbf{x}_2, \ldots, \mathbf{x}_k$ from a starting point $\mathbf{x}_0$ in state space by using a control vector sequence $\mathbf{u}_1, \mathbf{u}_2, \ldots, \mathbf{u_k}$ (see [7]).

As *Model Predictive Control* (MPC) we used the *Control Particle Belief Propagation* (C-PBP) algorithm [8]. It combines a Markov Random Field factorization and multimodal, gradient-free sampling to perform simultaneous path finding and smoothing in high-dimensional spaces. A drawback of this sample based MPC is some noise in the output signal that causes a slightly modified trajectory after each calculation. In general, the algorithm can be divided into three functional steps. C-PBP starts with an initial point in state space. Then it samples the control space with additional information like upper and lower bounds for each control dimension and a normal distribution at the mean control value. The drawn control value samples are applied to the system model to predict the following states. This generates a tree of trajectories, known as guided random walkers (see Fig. 2). These are evaluated by cost functions. If the costs are *normal*, then the algorithm will proceed in the same way, otherwise it performs a *resampling step* (see blue lines in Fig. 2). This step prunes all *expensive* trajectories and copies the *cheap* ones up to the current step. Then the algorithm can proceed as before. After the maximum number of time steps, known as control horizon, is exceeded, the best trajectory selected by total minimal costs is chosen (see violet trajectory in Fig. 3). Then a backward local refinement algorithm smooths it to gain the optimal trajectory (see red trajectory in Fig. 3). The resulting trajectory does not need to be globally optimal, but it is the optimum of all sampled trajectories. After this calculation, the first control vector is applied to the system and the sensors update the real physical state. The former optimal trajectory will be shifted by one time step and used as initial guess for the next trajectory planning call (see Fig. 4).

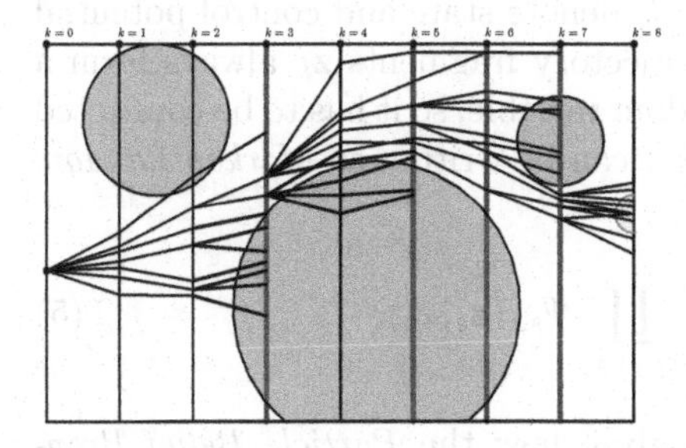

Fig. 2. Step 1 and 2 (see [2,8]) (Color figure online)

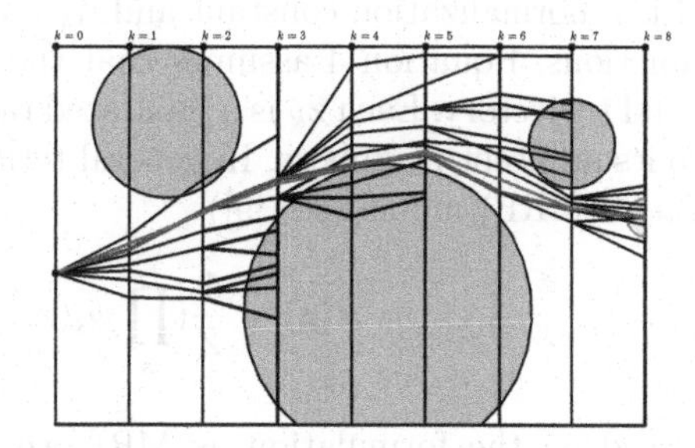

Fig. 3. Step 3 (see [2,8]) (Color figure online)

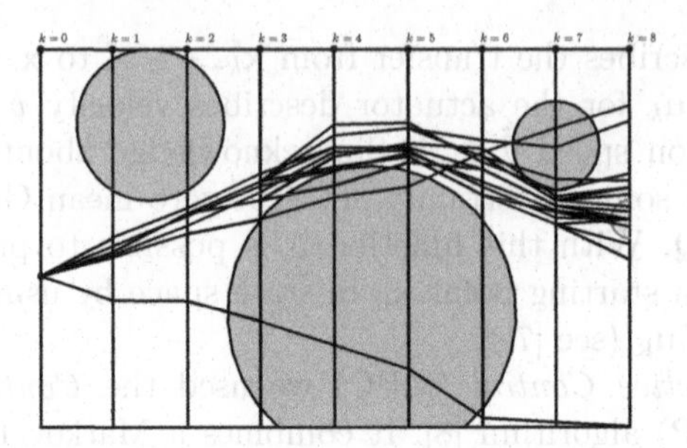

Fig. 4. Knowledge transfer (see [2,8])

A cost function based on the robot state and control vector has to be created to evaluate optimality of the control. However, no goal state or reference trajectory is available to evaluate the distance to the optimum. This is different from classic control tasks where the optimum is known. Since the only assumption of C-PBP is that the cost function $l(\mathbf{x}_k, \mathbf{u}_k, k)$ can be split into a function for the state space and one for the control space, problems because of discontinuities do not appear. Moreover, the functions have to be defined for the whole state, control and time space:

$$l(\mathbf{x}_k, \mathbf{u}_k, k) = s(\mathbf{x}_k, k) + c(\mathbf{u}_k, k) \tag{3}$$

Let $\mathbf{z}_k = [\mathbf{x}_k \; \mathbf{u}_k]^T$ denote the optimization vector consisting of state vector $\mathbf{x}_k$ and control vector $\mathbf{u}_k$ at time step k. For minimization, the cost function is transformed to maximize the probability density $\mathcal{P}(\mathbf{z})$ of the complete trajectory $\mathbf{z} = [\mathbf{z}_0, \mathbf{z}_1, ..., \mathbf{z}_K]$. Let K be the planning horizon. The transformation is done through exponentiation of the cost function to approximate sampling from $\mathcal{P}(\mathbf{z})$.

$$\begin{aligned} \mathcal{P}(\mathbf{z}) &= \tfrac{1}{Z} \prod_k \exp\left[-\tfrac{1}{2}(s(\mathbf{x}_k, k) + c(\mathbf{u}_k, k))\right] \\ &= \prod_k \alpha_k(\mathbf{x}_k)\beta_k(\mathbf{u}_k) \qquad = \prod_k \psi_k(\mathbf{z}_k) \end{aligned} \tag{4}$$

Z is a normalization constant and α_k and β_k denote state and control potential functions. Equation 4 assumes that the trajectory fragments $\mathbf{z}_k$ always form a valid trajectory. Each $\mathbf{z}_k$ is a separated random variable, so it has to be connected to its adjacent neighbors. In general terms it can be written as *Markov Random Field* (MRF) model (see [9]):

$$\mathcal{P}(\mathbf{z}) = \frac{1}{Z}\Big(\prod_s \psi_s(\mathbf{z}_s)\Big)\Big(\prod_{\{s,t\}\in\mathbb{E}} \Psi_{s,t}(\mathbf{z}_s, \mathbf{z}_t)\Big) \tag{5}$$

Based on the formulation as MRF we analyze how the *Particle Belief Propagation* (PBP), as a general sample based belief propagation method, can be applied. The main reason for the application of this method is the combination of global sampling to explore multimodal optimization landscapes "and dynamic programming to fight the curse of dimensionality" [8]. The method evaluates

$\mathcal{P}_k(\mathbf{z}_k)$ instead of much higher-dimensional $\mathcal{P}(\mathbf{z})$, so it processes trajectory segments and not the whole trajectory at once. The segments themselves are not interesting, but they offer the possibility to gain a set of complete trajectories due to the help of belief propagation. Equivalent to other belief propagation methods, PBP tries to calculate the beliefs $\mathcal{B}_k(\mathbf{z}_k)$ of the graph variables. If the graph is free of cycles, then the beliefs are proportional to the marginal probability density $\mathcal{P}_k(\mathbf{z}_k)$. Thus, messages $m_{s\to k}(\mathbf{z}_k)$ from other nodes form the belief of node k:

$$\mathcal{B}_k(\mathbf{z}_k) = \psi_k(\mathbf{z}_k) \prod_{s\in\Gamma_k} m_{s\to k}(\mathbf{z}_k) \tag{6}$$

$$m_{s\to k}(\mathbf{z}_k) = \sum_{\mathbf{z}_s\in\mathbb{Z}_s} \Psi_{s,k}(\mathbf{z}_s, \mathbf{z}_k)\psi_s(\mathbf{z}_s) \prod_{u\in\Gamma_s\setminus k} m_{u\to s}(\mathbf{z}_s) \tag{7}$$

"Here Γ_s denotes the set of neighbors of node s, and $\mathbb{Z}_s$ is the domain of $\mathbf{z}_s$. The messages $m_{s\to k}$ from node s to node k can be considered as (unnormalized) probability density functions of the target node variables $\mathbf{z}_k$. The potentials $\psi_k(\mathbf{z}_k)$ represent the *evidence* for $\mathbf{z}_k$ (here based on the state and control costs), which propagate through the graphical model via the messages. The belief $\mathcal{B}_k(\mathbf{z}_k)$ equals the product of the direct and propagated evidence. In Particle Belief Propagation, the messages and beliefs of Eqs. 6 and 7 are estimated by samples $\mathbf{z}_k^{(i)} \sim q_k(\mathbf{z}_k^{(i)})$, where i denotes the sample index and $q_k(\mathbf{z}_k^{(i)})$ is an arbitrary proposal distribution. Division by the proposal then gives the importance-weighted sample messages and beliefs [8,9]:

$$\hat{m}_{s\to k}(\mathbf{z}_k^{(i)}) = \frac{1}{N}\sum_{j=1}^{N} \Psi_{s,k}(\mathbf{z}_k^{(j)}, \mathbf{z}_k^{(i)}) \frac{\psi_s(\mathbf{z}_s^{(j)})}{q_s(\mathbf{z}_s^{(j)})} \prod_{u\in\Gamma_s\setminus k} \hat{m}_{u\to s}(\mathbf{z}_k^{(i)}) \tag{8}$$

$$\hat{\mathcal{B}}_k(\mathbf{z}_k^{(i)}) = \frac{\psi_k(\mathbf{z}_k^{(i)})}{q_k(\mathbf{z}_k^{(i)})} \prod_{s\in\Gamma_k} \hat{m}_{s\to k}(\mathbf{z}_k^{(i)}) \tag{9}$$

Let N be the sample size. The sample belief $\hat{\mathcal{B}}_k(\mathbf{z}_k^{(i)})$ represents the marginal probability density of the trajectory segment that is generated by a simulation step ending in $\mathbf{x}_k$ using $\mathbf{u}_k$. PBP has already the ability to approximate the multimodal marginal distributions corresponding to different paths to pass obstacles.

CPBP extends the basic PBP algorithm in several aspects for control [8]:

1. Selection of proposals to sample feasible trajectories.
2. Adaptive resampling to adjust between local and global search.
3. A local refinement backward pass.
4. Information propagation between function calls, to use the old optimal trajectory as optimizer warm start.

2.2 Cost Functions

After the model and optimization has been introduced, we tackle the last element of the model predictive controller. In classic control theory exists a working state, a goal state or a reference trajectory, thus an error value can be calculated by metric distances. In a completely unknown environment there exists in principle a global minimum as desired working state and often an optimal path towards it, but both are unknown. Therefore, a relative rating is impossible. Consequently, the environment will be modeled with absolute cost referring to time and space. The C-PBP method does not require continuity nor differentiability of cost functions because it works gradient free, so the only assumption is that the functions are defined on the whole space created by the product of state space, control space and time. Here, a high positive value represents repulsive unwanted locations in the state space and low values or high negative values represent recommended locations of the cost function l.

$$l(\mathbf{x}_k, \mathbf{u}_k, k) = s(\mathbf{x}_k, k) + c_t(\mathbf{u}_k, k) \quad \text{with } l : \mathbb{X} \times \mathbb{U} \times \mathbb{R}_0^+ \to \mathbb{R} \tag{10}$$

For most of the objects in the environment as well as for collision avoidance, a linear or quadratic function based on the distance is suitable to prevent collisions.

$$d = \|\mathbf{p}_k - \mathbf{x}_k\|_{2D} = \sqrt{(\Delta x)^2 + (\Delta y)^2}$$
$$\text{with } \mathbf{p}_k = \begin{bmatrix} x \\ y \\ \theta \end{bmatrix}, \mathbf{x}_k = \begin{bmatrix} x \\ y \\ \theta \end{bmatrix} \text{ and } a, b \in \mathbb{R}, d, r, \gamma \in \mathbb{R}_0^+ \tag{11}$$

$$l_{\text{lin}}, l_{\text{quad}} : \mathbb{R}_0^+ \to \mathbb{R}$$

$$l_{\text{lin}}(d) = \begin{cases} \frac{-a}{r} \cdot d + a & , \text{if } d \leq r \\ 0 & , \text{else} \end{cases} \tag{12}$$

$$l_{\text{quad}}(d) = \begin{cases} \frac{-a}{r^2} \cdot d^2 + a & , \text{if } d \leq r \\ 0 & , \text{else} \end{cases} \tag{13}$$

To form a global attractive point, for example as desired goal position, a rational function is more appropriate. However, the function should always be bounded to a maximum value.

$$\lim_{d \to \infty} l_{\text{unbounded}}(d) = 0 \tag{14}$$

$$\lim_{d \to 0} l_{\text{unbounded}}(d) = \pm\infty \tag{15}$$

$$l_{\text{unbounded}} : \mathbb{R}_0^+ \to \mathbb{R}$$

$$l_{\text{unbounded}}(d) = \begin{cases} \frac{b}{d} & , \text{if } |\frac{b}{d}| \leq |a| \\ a & , \text{else} \end{cases} \tag{16}$$

If entities like convoy vehicles have a repulsive area to prevent collision and an attractive area so that the robots surround vehicles, then the robot social

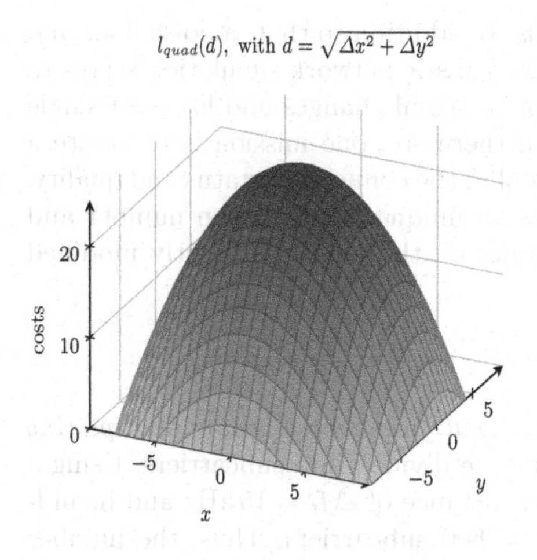

Fig. 5. Quadratic bounded cost function width radius $r = 10$ and maximum $a = 25$

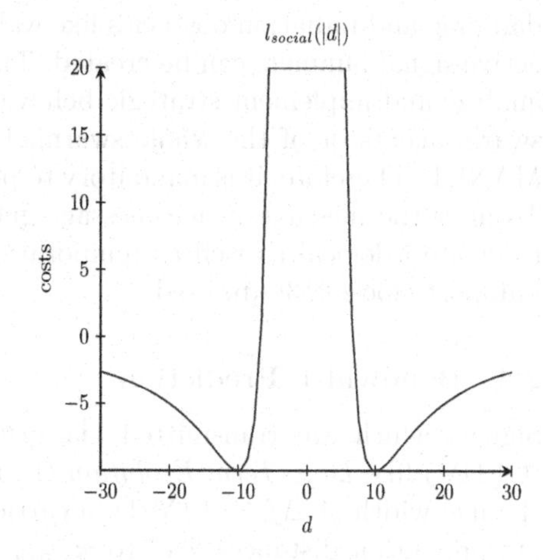

Fig. 6. Robot social function with desired minimum costs $L_{\min} = -10$, maximum costs $L_{\max} = 20$ and a desired distance $d_{\text{desired}} = 10$ for the minimum

function can be used. This is a rational function too, but with some additional constraints (Figs. 5 and 6):

$$l_{\text{social}}(d) = \begin{cases} \frac{b_1}{d^{\gamma_1}} - \frac{b_2}{d^{\gamma_2}} & , \text{ if } |\frac{b_1}{d^{\gamma_1}} - \frac{b_2}{d^{\gamma_2}}| \leq |a| \\ a & , \text{ else} \end{cases}$$

$$\text{with } b_1, b_2 \geq 0, \gamma_1 > \gamma_2 > 0 \tag{17}$$

Multidimensional functions for complex polygons and obstacles like sectors for towers, or complex missions like setting up a MANET or connection awareness are implemented as well, but they are often aggregations of the presented functions.

Furthermore, we developed a cost function considering energy consumption and accessibility of the terrain to cover different types of ground like rocks, mud, streets, fields, etc.

$$l_{\text{movement}}(\mathbf{x}_k, \mathbf{u}_k, k) = v^2 \cdot G(\mathbf{x}_k) \cdot \rho_T + \dot{\theta}^2 \cdot G(\mathbf{x}_k) \cdot \rho_R \tag{18}$$

$G(\mathbf{x}_k)$ denotes a position depending influence factor of the ground. ρ_T and ρ_R describe the effect of this influence factor on the translation and rotation movement. The velocity values are squared values to prefer slower actions.

2.3 Communication Model

A necessary condition for entities to form a swarm is the information transfer via communication. Therefore, we created a network simulation model based on signal power dissemination depending on the used frequency including free space damping and radiation obstacles like walls. In addition to that, a noise floor and active signal jammers can be created. This realistic network simulation serves to analyze and implement strategic behavior to signal changes and losses of single swarm agents or of the whole swarm. Furthermore, one mission is to create a MANET. Therefore, it is mandatory to predict the connection status and quality. To filter the messages, each message gets an unique identification number and to create a logical as well as temporal order on the messages, lightly modified Lamport clocks [13] are used.

2.4 Bandwidth Prediction

Signals which are transmitted via *orthogonal frequency-division multiplexing* (OFDM) like *Long-Term-Evolution* (LTE) are divided into subcarriers. Using a channel width of $\Delta f_c = 10\,\text{MHz}$, a carrier distance of $\Delta f_t = 15\,\text{kHz}$ and including protection distances lead to $\#_{\text{carrier}} = 600$ subcarriers. Thus, the number of subcarriers is proportional to $60\,\frac{1}{\text{MHz}}$. The carrier distance determines the symbol time:

$$t_s = \frac{1}{\Delta f_t} \approx 66,7\,\mu\text{s} \tag{19}$$

Once per symbol time each subcarrier is modulated with one symbol. The symbol width w_{symbol} depends on the modulation method. In the simulation environment *Quadrature Phase-Shift Keying* (QPSK), 16 *Quadrature Amplitude Modulation* (QAM), 64 QAM and 256 QAM are available. The maximum bandwidth for a channel width 10 MHz, carrier distance 15 kHz and 256 QAM modulation method is for example given by:

$$\Delta f_c \div \Delta f_t = 10\,\text{MHz} \div 15\,\text{kHz} \rightarrow 600 = \#_{\text{carrier}} \tag{20}$$

$$\frac{\#_{\text{carrier}} \cdot w_{\text{symbol}}}{t_s} = \#_{\text{carrier}} \cdot w_{\text{symbol}} \cdot \Delta f_t = 600 \cdot 8\,\text{bit} \cdot 15\,\text{kHz} = 72\,\text{Mbit/s} \tag{21}$$

To increase this bandwidth Multiple Input Multiple Output [1] methods are supported. Based on the maximum, we use the received signal strength index (RSSI) and the signal to noise ration (SNR) to estimate the real bandwidth.

$$\text{RSSI} = \underbrace{P_t + G_t + G_r}_{\text{transmission power}} + \underbrace{20 \cdot \log_{10}\left(\frac{c}{f \cdot 4\pi \cdot d}\right)}_{\text{free space damping}} - \underbrace{\sum^{i} P^{i}_{\text{obstacle}}}_{\text{obstacle damping in line of sight}} \tag{22}$$

The extended Friis Eq. 22 [5] describes the RSSI of the receiver in decibel based on one milliwatt (dBm). P_t denotes the sending power, G_t and G_R denote the antenna gain of the sender and receiver. The free space damping depends on the speed of light c, the sending frequency f and the distance d between the modules. For calculating the SNR, a noise floor power P_{noise} has to be given.

$$\text{SNR} = \text{RSSI} - P_{\text{noise}} \tag{23}$$

The software module can be easily extended by more complex reflection models for electromagnetic waves or interface OMNeT++ [11] for other models.

3 Structure of the Simulator

The simulation software is structured in a graphical visualization with an interface and the simulation core because this core should be usable for other simulation tools and visualization engines. Core functions can be used on real world robots to validate the simulation and to create digital twins of them in the visualization.

3.1 Structure of the Simulation Software

The software is implemented in C++ because of its efficiency, the available libraries and OpenMP support. We used an existing implementation of C-PBP as the optimization part for the model predictive controller (see [8]). In addition to OpenMP, *Eigen* [10] is included as a fast numeric library. Moreover, a *yaml-cpp* library offers the possibility to read YAML configuration files for the simulated scenarios. These files configure the whole scenario and all parameters of each entity. Furthermore, the appearance of the interface and images can be configured as well. Consequently, no programming knowledge is necessary to use the simulator.

The application is divided into two domains (see Fig. 7). The first domain is visualization of simulation results. It consists of one thread for rendering and a few callbacks for interaction with the user. The second domain is the simulation core. It has one thread to calculate the physics of the objects and one thread to calculate the network states like bandwidth. Furthermore, each robot creates four threads on his own. These are needed to realize a realistic simulation implementation of the hardware and software components in real world. A robot consists of a communication module, an actuator, a sensor unit and a planning component (see Fig. 1).

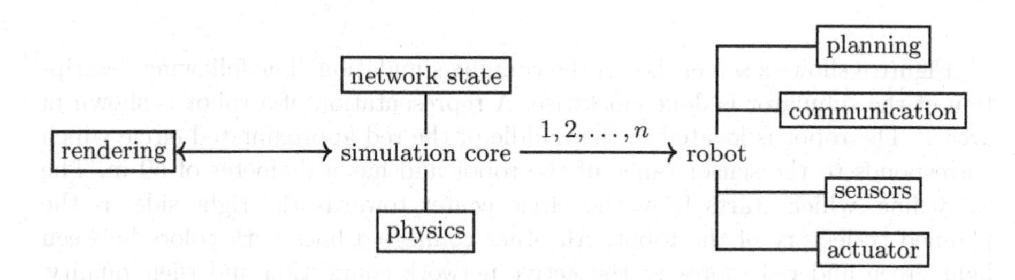

Fig. 7. Thread structure of the simulator

For analysis purposes, the software offers the possibility to display the value of cost functions in vicinity of a selected robot by using Gnuplot. In addition to this function, the scenario data such as noise floor and terrain accessibility values and all robot states including planned trajectories, network states, cost function values of the surrounding area and calculation times can be exported as CSV and text files with a special format for easy processing in statistic tools, e.g. Matlab®.

3.2 Graphical User Interface

The graphical user interface uses the Cocos2d-x game engine. It is a lightweight cross-platform game engine written in C++ and optimized for 2D applications. The application can only be controlled via keyboard at the moment.

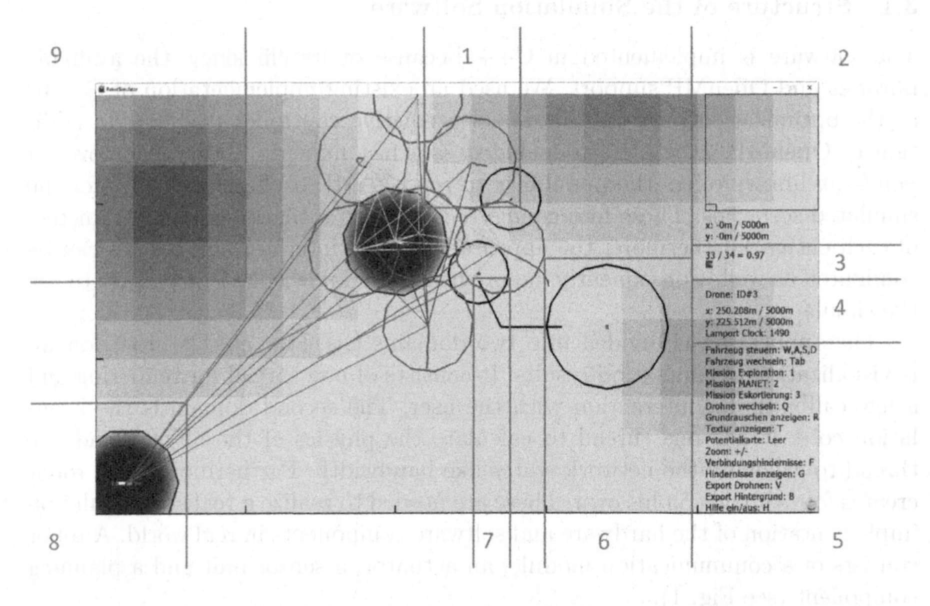

Fig. 8. Screenshot of the simulator (Color figure online)

Figure 8 shows a screenshot of the running simulation. The following description of the simulator is done clockwise. A representation of a robot is shown in area 1. The robot is located in the middle of the red approximated circle which corresponds to the sensor range of the robot and has a diameter of 60 m. The wavy line, which starts from the circle center towards the right side, is the planned trajectory of the robot. All other connected lines with colors between light green and red represent the active network connection and their quality. The green color denotes maximum bandwidth, and completely red the lowest possible bandwidth. The size of the robot images are so small because its real world dimension is less than 1 m^2 and the displayed area is 600 m by 275 m. The area number 2 shows a minimap and the current rendered extract of the simulated region which is 25 km^2 in total. The next section displays at first the simulated time in seconds, then the actual running time of the application. Thus, the division result describes the real-time factor. Area number 4 presents an identification number of the currently selected robot and its coordinates as well as its logical Lamport clock state for messages. At the lower right corner, control information is given to the user. The 6th section highlights a radial obstacle with a bounded effect radius which is shown as a black approximated circle. The obstacle uses one of the distance based cost functions. Then a line based obstacle follows that represents a wall which shall not be passed. A guidance vehicle to send data or missions to the robots is presented in the lower left corner denoted by 8. This vehicle demonstrates the different connection qualities. Normally, the quality decreases with distance so it is dyeing from green to red. But some of the shorter connection lines intersect one of the orange lines. They represent obstacles which causes a damping on the transmission power. So, the bandwidth is significantly reduced. The last area number 9 shows the accessibility of the terrain. The darker a ground tile is, the harder is the access and movement, so it increases the use of energy and cause the robots to try to move slower. Additional features are zooming, hiding entities and displaying the noise floor instead of the terrain accessibility (see Fig. 9). Furthermore, the obstacles can move with scripted behavior and the guidance vehicles can be controlled by a script or the user and send missions to the robots.

Fig. 9. Screenshot of the simulator (Color figure online)

Figure 9 shows the noise floor of the environment. Equal to the terrain accessibility, the noise level increases, the darker the cyan tiles are. The dark circular discoloration at almost the center of the screenshot indicates the position of an active jammer. Moreover, the qualities of the connections of the robot next to the jammer is significantly reduced, although the connection module distances are as short as the light green connections with higher bandwidth between the other robots.

3.3 Simulation Internals

To solve the given missions each robot creates for each other known robot a model to maintain the data it receives from that robot and to do predictions concerning its states in the future.

3.4 Exploration Mission

The goal of the exploration mission is to discover a dynamic environment. Since the dynamic information becomes obsolete, it has to be renewed from time to time. The area covered by a robot sensor is deemed to be explored. Here, the precision of exploration decreases with the sensor distance, so peripheral areas are less discovered. In addition to that, the robot should create a vanishing trace of the discovered area. This specification leads to the use of radial cost functions with a limited effect radius, which represent the sensor range. Each robot model as well as the robot itself gets a FIFO list with 18 elements per default. After a specified time, the default is 10 s because of the given maximum speed, such an radial cost function object is added to this list. If the list is already filled, the oldest object gets deleted. Moreover, the cost of each object decreases during simulation time.

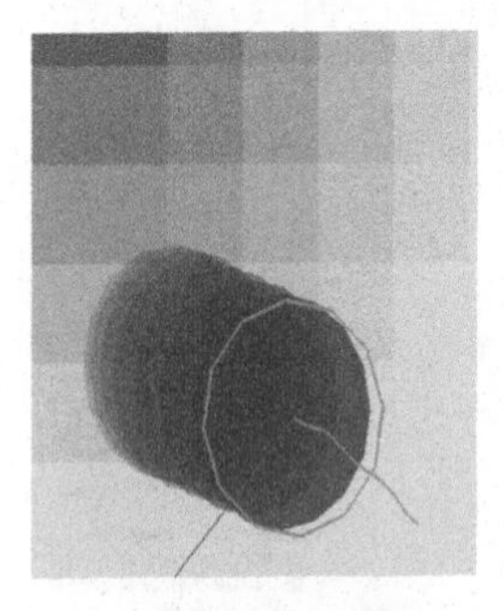

Fig. 10. Visualization of the trace. Gray scale represents terrain accessibility. (Color figure online)

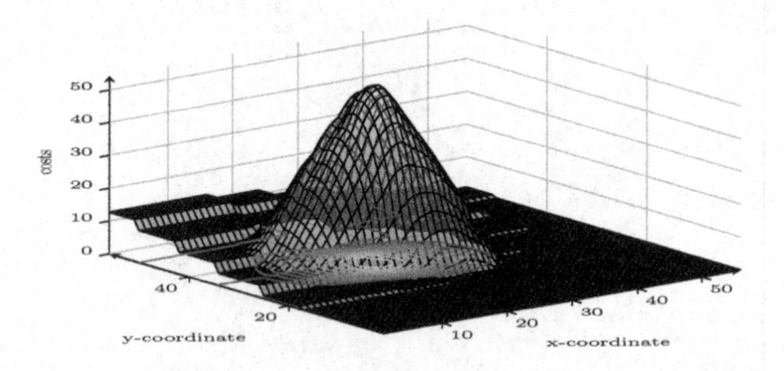

Fig. 11. Visualization of the cost function for exploration at the same time. The robot is located in the center.

In Fig. 10 the blue filled circles visualize the created cost function objects. The lighter the color is, the older is the object. The color as well as the height represent the costs in Fig. 11. The rectangular surfaces visualize the costs which are generated through movement on terrain with different accessibility.

3.5 Convoy Escorting Mission

During the escorting of a convoy the robots should place themselves around the specified target vehicle in an optimal distributed way. To maintain the desired distance, the robot social function is used. Therefore, the target vehicle sends its position and velocity vector to the robots Thus, they are able to predict the movement, even when the connection gets lost. To gain an optimal distribution around the target, each robot calculates the intersecting sensor area with other robots and tries to minimize it by using the robot model. This leads to a maximized covered area at the desired distance. Figure 12 shows a target vehicle in the center and an optimal distribution of the robots around. The blue circle denotes the desired distance and the red circles the sensor range of each robot. The green lines are the connections like before. The corresponding cost function is shown in Fig. 13.

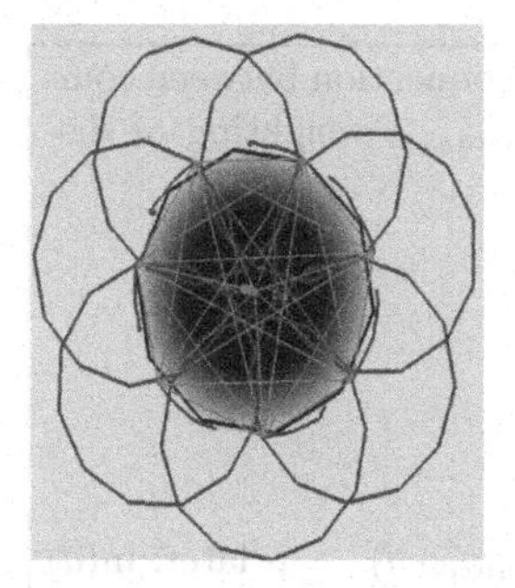

Fig. 12. Visualization of the distribution of the robots around the escorted vehicle. (Color figure online)

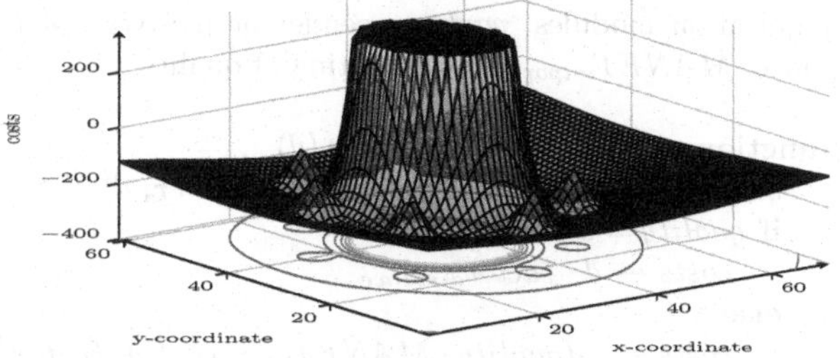

Fig. 13. Visualization of the convoy mission cost function at the same time.

3.6 MANET Mission

The MANET cost function cannot be created by an aggregation of the presented functions, because as long as the connection quality is constant, a large distance is desired to cover a larger area. If the quality decreases it has to be weighed out whether a larger distance or a higher bandwidth is prioritized. However, if the connection gets lost, the distance is not a positive aspect anymore and has to be reduced. There are two reasons for a connection loss; it could be the distance or it could be a change in the environment, which cannot always be undone by the robot itself. Thus, a distance reduction is necessary. This is the reason why the active jammer in Fig. 14 causes smaller distances between some robots while they have at most the connection quality of the other robots. Algorithm 1 shows the implementation of the MANET cost function (Fig. 15).

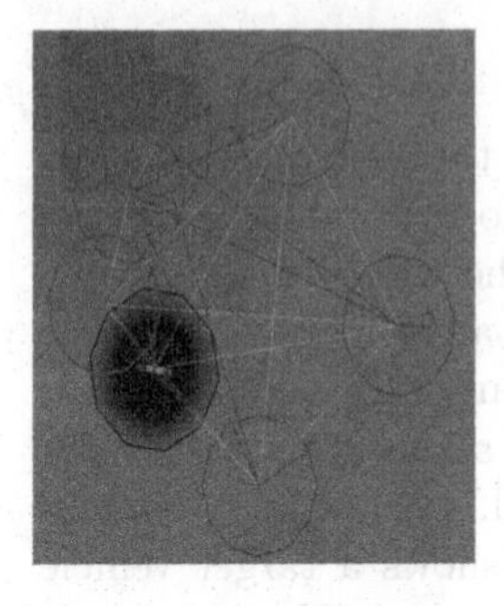

Fig. 14. Visualization of the creation of a mobile ad hoc network.

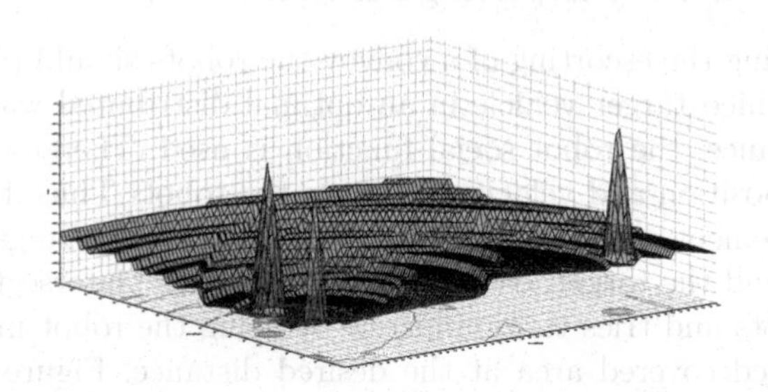

Fig. 15. Visualization of the mobile ad hoc network cost functions for nearest robot to the top of the image. Each cone visualizes the collision avoidance area of other robots.

Require: d: distance between communication modules, $link$: connection between communication modules, $partner$: sender or receiver, $factor_{distance}$: weighting of distance, $MANET_{importance}$: weighting of quality

function GETMANETPOTENTIAL(d)
  $quality \leftarrow link$.predictConnection($partner$, d)
  if $quality = 0$ **then**
    $costs \leftarrow d \cdot factor_{distance}$
  else
    $costs \leftarrow -(quality \cdot MANET_{importance} + factor_{distance} \cdot d)$ ▷ later: $\ln(d)$
  return $costs$

Algorithm 1: MANET cost function calculation

4 Experiments

We begin with some performance experiments for a reference scenario. It consists of two static radial obstacles, two static wall obstacles, one dynamic radial obstacle, one static and one dynamic radiation obstacle with damping as well as terrain accessibility between 0 and 36. Moreover, the noise floor is equally distributed between −120 dBm and −60 dBm. So, most of the simulation features are present in this scenario. Based on the maximum robot speed we figured out 300 ms as the upper bound for trajectory planning time. Table 1 shows that we can simulate up to 680 simulation steps of a trajectory for 8 robots which uses $N = 20$ parallel paths for model predicted controller optimization each. A simulation step has a time delta of 0.5 s. Thus, each robot predicts the next 340 s each 300 ms. This is a waste of resources in a highly dynamic environment.

Instead, we increase the number of parallel plans per robot to $N = 24$ to increase the sampled area to get better trajectory results. Furthermore, we use the available resources to simulate a larger number of swarm agents. Figure 17

Table 1. Comparison of the calculation times per trajectory in ms with different amount of simulation steps

Robots	Steps	Plans	Min.	Max.	Mean	Standard deviation
8	15	20	4.728	7.381	6.005	0.567
8	40	20	8.55	16.886	10.999	1.685
8	200	20	29.161	71.702	37.739	4.476
8	680	20	111.631	242.886	124.861	8.926
8	840	20	141.346	342.455	196.722	42.165

shows that up to thirty robots, which correspond to $3 + 30 \cdot 4 = 123$ threads, can be simulated on a gaming PC in real time. However, 200 equidistant simulation steps do not perform well. The majority of steps are in a later uncertain future and unknown environment which causes the robots to ignore obstacles in vicinity. Thus, we reduce the number of predicted steps to $K = 60$ and switch to an exponential distribution of simulation steps while maintaining the visual range (Fig. 16).

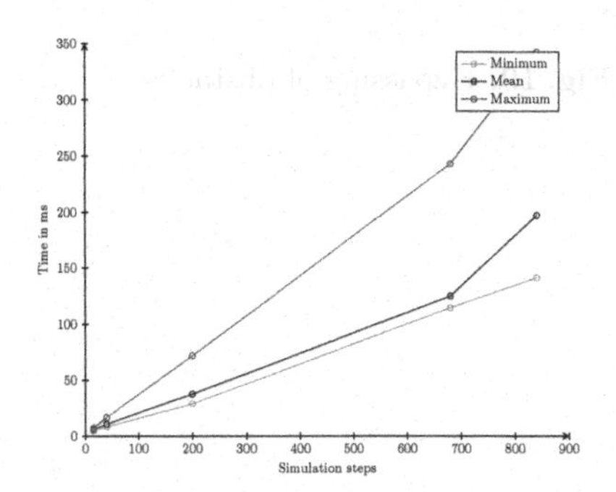

Fig. 16. Statistic of plan calculation time with 8 robots and $N = 20$ parallel plans per robot

Fig. 17. Statistic of the plan calculation time with $K = 200$ steps and $N = 24$ parallel plans per robots

$$\Delta t_i = t_{step} \cdot t^i_{factor}$$

$$\text{with } i = 0, 1, \ldots, K-1; t_{step} = 0,075 \text{ and } t_{factor} = 1,055 \tag{24}$$

$$\sum_{i=0}^{59} \Delta t_i \approx 32,5\,\mathrm{s} \approx 60 \cdot 0,5\,\mathrm{s}; \sum_{i=0}^{39} \Delta t_i \approx 10,2\,\mathrm{s} \tag{25}$$

This has the advantage that obstacles in vicinity are not ignored. Equation 25 shows that relatively more sampled points are in near future, as well as that a coarse outlook on the target is still given. Next, we compare the performance of the simulation on different hardware. First, a mobile PC with an quad core Intel® Core™ i7-8650U CPU, 16 GB RAM and a desktop PC with an eight core Intel® Core™ i9-9900K CPU, 32 GB RAM and a NVIDIA® GeForce® RTX 2080 Ti GPU. Both systems use SSD storage. Figure 18 indicates that a real-time simulation with up to 18 robots is possible on the mobile device, if additional robots are added it suffers from missing cores to handle the threads as well as a GPU to render so many objects.

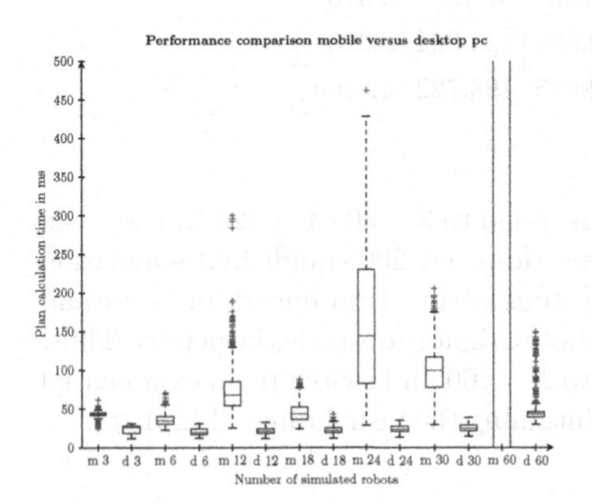

Fig. 18. Performance comparison between mobile and desktop PC for different swarm size. "m" denotes mobile PC and "d" desktop PC. The number denotes the amount of swarm agents.

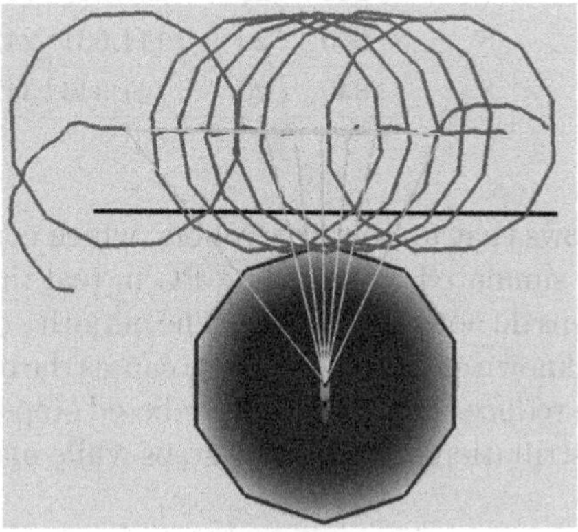

Fig. 19. Bypassing of obstacles

4.1 Convoy Escorting

In disaster areas or in rural environments a lot of situations exist where trucks or the convoy can pass an obstacle, e.g. because of a higher wading depth. The robots have to find a way around the obstacle. This situation is presented in Fig. 19. The goal is to analyze the maximum dimensions of the obstacles that do not lead to a disintegration of the convoy structure.

Table 2. Measurement series to analyze the escorting behavior at obstacles

Length	120 m	150 m	180 m	210 m	240 m	360 m	390 m	420 m
Following robots	7	6	5	5	4	3	0	0

Table 2 contains the relevant obstacle lengths. It should be noted that the obstacles at each end also have an effective radius of 20 m, so that the actual range of influence is 40 m larger than the ranges given in the table. This effect prevents the robots from cutting corners which would lead to a collision. The sensor range, which is the radius of the red circle, is set to 30 m, so only many times the amount of the range is analyzed. Up to four times the range all robots can follow the vehicle and up to 12 times the range the outer robots can follow. Crucial for bypassing of the obstacle is the swarm size and the cost function that penalizes the intersecting area. Therefore, the robots are distributed along the line. With seven robots, the minimum distance for overlap-free distribution is 420 m. If the influence of the obstacle exceeds this distance (see Table 2 at 390 m + 40 m), no robot can detect an alternative path. Furthermore, the increased distance to the edge of the obstacle reduces the attractive potential of the mission. Here, the activation of the exploration mission can support to further distribute the robots.

4.2 MANET

The functionality of the MANET is evaluated by the distribution of the robots. A circular or star-shaped arrangement is desired in order to achieve a high and

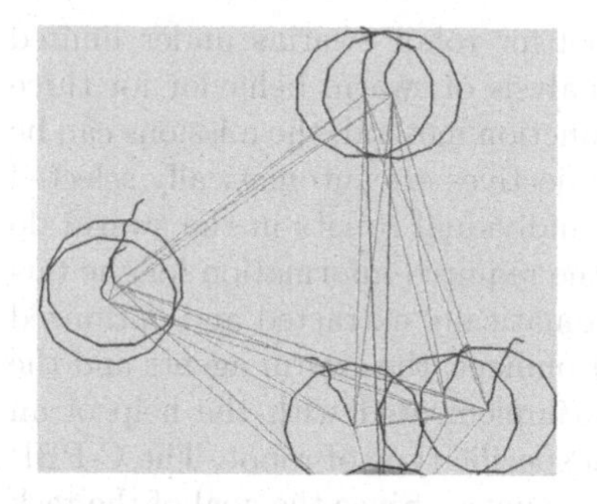

Fig. 20. Triangle structure with old MANET cost function

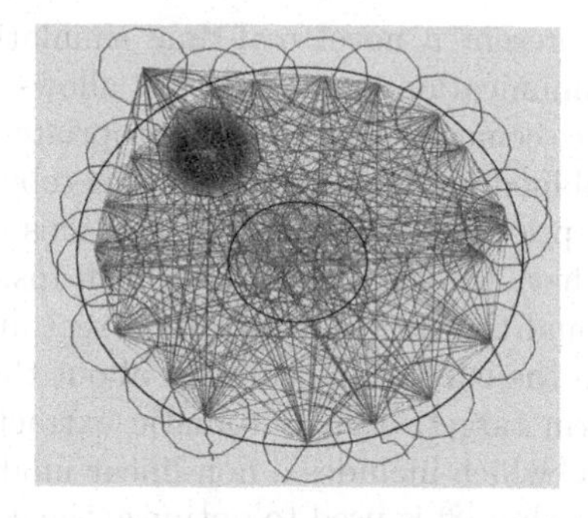

Fig. 21. Creation of ring structure with 30 robots, version 1 (Color figure online)

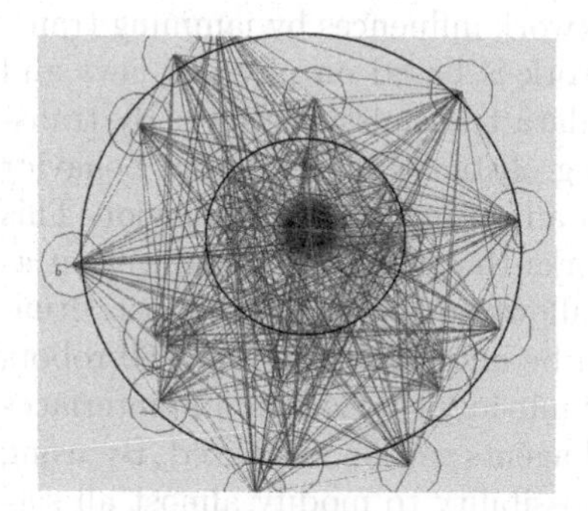

Fig. 22. Creation of ring structure with 30 robots, version 2 (Color figure online)

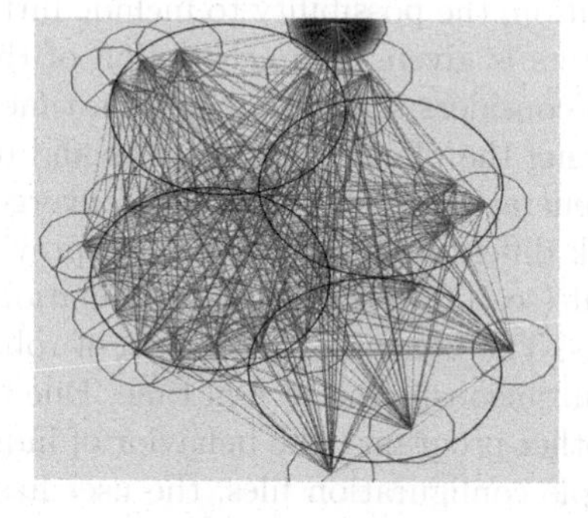

Fig. 23. Creation of ring and star structure with 30 robots (Color figure online)

slightly redundant network coverage of an area. However, experiments using the current cost function produce approximately equilateral triangles (see Fig. 20). This is because an equilateral triangle has optimum properties regarding to distances. As a result, the cost function has to be modified. The idea of always maximizing the minimum distance is not appropriate because the connection with the minimum distance does not have to be the optimal connection due to obstacles and sources of interference. As a solution, a logarithmic function may be used. However, the strictly monotonous and concave logarithmic function for the distance prefers small equal distances between the robots, instead of one large and many smaller distances. Figures 21, 22 and 23 demonstrate the functionality of the adjusted cost function because the desired ring or star structures are always formed. The position of the blue marked guidance vehicle is irrelevant and has no influence on the cost functions. Black circles are only for visualization purpose.

5 Conclusion

We present a novel real-time simulation tool for robot swarms under limited communication capabilities. It allows the analysis of swarm behavior for three basic scenarios. Due to an appropriate cost function method, the missions can be combined and parameterized, whereby the objectives are automatically selected and prioritized. The target positions of the individual robots in the swarm do not have to be explicitly specified. Instead, the required information for the target and the trajectory to the target are automatically extracted and optimized from the current information about the environment, the swarm agents and the swarm target. This automatic extraction is implemented with the help of an MPC which includes a non-linear model of a specific type of robot. The C-PBP algorithm [8] is used to optimize the control sequence. Since the goal of the task emphasizes a limited communication capability, the simulation of a complete wireless network is implemented. This takes into account fundamental properties of propagation, shadowing and attenuation of electromagnetic signals. In addition, the possibility to include further network influences by jamming transmitters is given. The calculation of the network is based on physical laws and also considers signal propagation times and data transmission times. By transmitting the planned trajectories and detecting obstacles, an emergent behavior is generated which favors the convergence to an optimal swarm behavior. This work differs fundamentally from previous works that accept communication as ideal (see [3,6,16,17]). The simulation tool allows to reproduce realistic situations. The behavior of a swarm of robots can be analyzed with up to 60 robots simultaneously and in real time. Due to the built-in visualization and interfaces to other programs, the behavior of individual agents can be visualized. By using simple configuration files, the user has the possibility to modify almost all settings and parameters. [Gefördert durch die Deutsche Forschungsgemeinschaft (DFG) – 276879186/GRK2193]

References

1. Ahlers, E.: Funk-übersicht: WLAN-Wissen für Gerätewahl und Fehlerbeseitigung. c't **2015**(15), 178–181 (2015). https://www.heise.de/ct/ausgabe/2015-15-WLAN-Wissen-fuer-Geraetewahl-und-Fehlerbeseitigung-2717917.html
2. Baumann, D., Mager, F., Zimmerling, M., Trimpe, S.: Control-guided communication: efficient resource arbitration and allocation in multi-hop wireless control systems. IEEE Control Syst. Lett. **4**(1), 127–132 (2020)
3. Euler, J.: Optimal cooperative control of UAVs for dynamic data-driven monitoring tasks. Dissertation, Technische Universität Darmstadt, Darmstadt (2018)
4. Euler, J., von Stryk, O.: Optimized vehicle-specific trajectories for cooperative process estimation by sensor-equipped UAVs. In: 2017 IEEE International Conference on Robotics and Automation, ICRA 2017, Singapore, Singapore, 29 May–3 June 2017, pp. 3397–3403. IEEE (2017)
5. Friis, H.T.: A note on a simple transmission formula. Proc. IRE **34**(5), 254–256 (1946)
6. Fukushima, H., Kon, K., Matsuno, F.: Distributed model predictive control for multi-vehicle formation with collision avoidance constraints. In: Proceedings of the 44th IEEE Conference on Decision and Control, pp. 5480–5485. IEEE (2005)
7. Grüne, L., Pannek, J.: Nonlinear Model Predictive Control: Theory and Algorithms. CCE. Springer, Cham (2017). https://doi.org/10.1007/978-3-319-46024-6
8. Hämäläinen, P., Rajamäki, J., Liu, C.K.: Online control of simulated humanoids using particle belief propagation. In: Proceedings of SIGGRAPH 2015. ACM, New York (2015)
9. Ihler, A., McAllester, D.: Particle belief propagation. In: International Conference on Artificial Intelligence and Statistics, vol. 5, pp. 256–263 (2009)
10. Jacob, B.: Eigen. http://eigen.tuxfamily.org/index.php?title=Main_Page
11. Kuntz, A., Schmidt-Eisenlohr, F., Graute, O., Hartenstein, H., Zitterbart, M.: Introducing probabilistic radio propagation models in OMNeT++ mobility framework and cross validation check with NS-2. In: Molnár, S., Heath, J.R., Dalle, O., Wainer, G.A. (eds.) Proceedings of the 1st International Conference on Simulation Tools and Techniques for Communications, Networks and Systems & Workshops, SimuTools 2008, Marseille, France, 3–7 March 2008, p. 72. ICST/ACM (2008)
12. Lächele, J., Franchi, A., Bülthoff, H.H., Robuffo Giordano, P.: SwarmSimX: real-time simulation environment for multi-robot systems. In: Noda, I., Ando, N., Brugali, D., Kuffner, J.J. (eds.) SIMPAR 2012. LNCS (LNAI), vol. 7628, pp. 375–387. Springer, Heidelberg (2012). https://doi.org/10.1007/978-3-642-34327-8_34
13. Lamport, L.: Time, clocks, and the ordering of events in a distributed system. Commun. ACM **21**(7), 558–565 (1978)
14. Mager, F., Baumann, D., Jacob, R., Thiele, L., Trimpe, S., Zimmerling, M.: Feedback control goes wireless: guaranteed stability over low-power multi-hop networks. In: Liu, X., Tabuada, P., Pajic, M., Bushnell, L. (eds.) Proceedings of the 10th ACM/IEEE International Conference on Cyber-Physical Systems, ICCPS 2019, Montreal, QC, Canada, 16–18 April 2019, pp. 97–108. ACM (2019)
15. Puzicha, A.: Modeling and analysis of a distributed non-linear model-predictive control for swarms of autonomous robots with limited communication skills (in German). Master's thesis, Department of Computer Science, TU Dortmund (2019)
16. Ritter, T.: PDE-based dynamic data-driven monitoring of atmospheric dispersion processes. Dissertation, Technische Universität Darmstadt, Darmstadt (2017)

17. Strobel, A.: Verteilte nichtlineare modellprädiktive Regelung von unbemannten Luftfahrzeug-Schwärmen. Dissertation, Technische Universität Darmstadt, Darmstadt (2016)
18. Tricaud, C., Chen, Y.: Optimal Mobile Sensing and Actuatin Policies in Cyber-physical Systems. Springer, London (2011). https://doi.org/10.1007/978-1-4471-2262-3
19. Vieira, B., Severino, R., Filho, E.V., Koubaa, A., Tovar, E.: Copadrive - a realistic simulation framework for cooperative autonomous driving applications. In: Proc. 8th IEEE International Conference on Connected Vehicles and Expo (ICCVE 2019). IEEE (2019)

Strategic Provision of Trade Credit in a Dual-Channel Supply Chain

Wan Qin[1(✉)], Huang Yu[1], and Lu Meili[2]

[1] School of Economics and Management, Southwest Petroleum University, Chengdu 610500, Sichuan, People's Republic of China

[2] Business Administration College, Shanxi University of Finance and Economics, Taiyuan 030006, People's Republic of China

Abstract. This paper focuses on a dual-channel supply chain composed of a capital-constraint bricks-and-mortar retailer and a manufacturer, where a manufacturer can sell products through traditional retail channel and direct online channel simultaneously. Supplementary pricing strategy and competitive pricing strategy are simulated in our model, and we find that the former one is the better choice for manufacture when the retailer suffers capital constraint. In our analysis, the capital constraint on retailer could mitigate the price competition between two channels, and it may be beneficial to the manufacturer under certain conditions. Our findings show that the manufacturer should provide trade credit to retailers strategically rather than provide it unconditionally. We present two trade-credit strategies (trade credit with positive interest rate and trade-credit with zero interest rate) and suggest manufacture choose appropriate trade-credit strategies according to the initial capital of retailer. To guide the manufacturer when and how to provide trade credit, we conduct several numerical simulations based on our results, and further plot out the feasible region for each trade credit strategy based on corresponding conditions.

Keywords: Dual-channel supply chain · Capital constraint · Trade credit · Direct sales cost · Acceptance level of the direct online channel

1 Introduction

Nowadays, most manufacturers prefer to join the dual-channel supply chain, which combines direct online channel with a physical retail channel. Manufacturers such as P&G, IBM, Nike, Sony, Huawei, Granz, Gree, etc. have succeeded in selling products through establishing official shopping websites or founding flagship stores on some B2C platforms. Some retail giants such as Wal-Mart and Gome have owned mature physical stores and often play a dominated role in a supply chain through intervening manufacturers' pricing strategies such as wholesale price or retail price etc. So some manufacturers may break up the cooperation with retail giants if they cannot endure the disadvantage in the supply chain. Like the Chinese manufacturer, Gree Inc had withdrawn all kinds of products from the retail giant Gome's physical stores in 2004,

H. Song and D. Jiang (Eds.): SIMUtools 2020, LNICST 370, pp. 227–245, 2021.
https://doi.org/10.1007/978-3-030-72795-6_18

and turn to cooperate with other small or medium retailers. However, although small or medium retailers will not intervene in manufacturers' pricing strategies, they usually have some capital constraint problems and require the manufacturer to provide financing support. It is common that small and medium firms are difficult to obtain financial supports from financial intermediaries such as banks [1]. So it is a very practical issue to help manufacturers explore how to offer financial service to small or medium retailers strategically.

Trade credit is the common economic phenomenon in a supply chain, it is used by the supplier for encouraging a capital-constraint retailer to order more quantity. Delay in payment is the most prominent feature of trade credit, and it is classified into several categories by Piasecki [2], Molamohamadi et al. [3, 4], which includes (1) pay as sold; (2) pay as sold during a predefined period; (3) pay as sold after a and (4) pay for the prior order at the time of establishing the new replenishment. Trade credit usually means the third type of delay in payment, which means that the retailer is permitted predefined period to delay payment until the end of the credit period and be charged interest by supplier. With trade credit, a capital-constraint retailer could get full order quantity by paying an initial capital and paying off the unpaid amount based on an interest rate after the sales season ends (Cai et al. [5]). Trade credit is widely studied in the single-channel supply chain to demonstrate it is beneficial to supply chain members, but few studies address trade credit in a dual-channel supply chain. In the single-channel supply chain, the retailer's capital constraint traditionally been delivered to have adverse side effects on manufacturer's profit because it causes the retailer cannot afford full order quantity. However, the capital constraint on retailer could mitigate the price competition between channels in dual-channel supply chain; thus manufacturer may benefit form it. Different from the traditional view, this paper puts forward that retailer's capital constraint is a double-edged sword for the manufacturer, and it is important to manufacturer to make a strategic provision of trade credit. In this research, we try to investigate when and how a manufacturer should adopt trade credit in a dual-channel supply chain.

In dual-channel supply chain, manufacturer should afford the cost of selling products through direct online channel, including building cost, advertising cost and operating cost, etc. Therefore, there is a need to incorporate direct channel sales cost. But most of the existing papers assume the sales cost in a direct channel is zero for simplicity of calculation, only a few literatures consider it. Such like Chen et al. [6], Xu et al. [7], Xiong et al. [8], and Yan et al. [9] respectively discuss different kinds of strategies for dual-channel supply chain considering direct channel sales cost. Extending to these papers, we probe into the manufacturer's pricing and trade credit strategy considering the sales cost in the direct channel.

Furthermore, an important difference between two channels is that consumers can really inspect and immediately possess product through the traditional retail channel, but consumers not only face some uncertainty due to the lack of detailed physical inspection but also need the patience to wait for the delivery lead time through direct online channel. Balasubramanian [10], Liang and Huang [11], and Kacen et al. [12] present that consumers' acceptance level of direct online channel is generally lower than that of traditional retail channel. Then many studies are conducted based on their researches, such as Chiang et al. [13], Zhang et al. [14], Pei and Yan [15], Hua et al. [16] and Rofin and Mahanty [17] study dual-channel supply chain considering consumers' acceptance level of direct channel. Therefore, besides the sales cost in direct channel, the consumers' acceptance level of the direct channel also has nonnegligible effect to manufacturer's profit.

Our model focus on the Stackelberg game between manufacturer and retailer, by incorporating sales cost in the direct channel and consumers' acceptance of direct channel simultaneously, there are three key questions should be taken into consideration by manufacture:

(1) Is the capital constraint on retailer is always bad for itself?
(2) If the capital constraint on retailer is harmful to manufacture's benefit, is providing trade credit to retailer the better choice?
(3) If manufacture decides to provide trade credit to retailer, how should manufacture make appropriate pricing and trade credit strategy?

In this paper, the first contribution is demonstrating that manufacture may benefit from the capital constraint on retailer in a dual-channel supply chain, especially when consumers' acceptance level of the direct channel is relatively high and the unit sales cost in the direct channel is relatively low. The second contribution is that we present the equilibrium pricing and trade credit strategy in different situations. Lastly, we plot out the feasible region for trade credit strategies, which help the manufacturer provide trade credit strategically.

We organized the remainder of this paper as follows. In Sect. 2, we review the relevant literature. In Sect. 3, we introduce the benchmark model which without trade credit. In Sect. 4, we expand the benchmark model with trade credit setting. In Sect. 5, we comprehensively condect serval numerical simulations. In Sect. 6, we finally make conclusions of our results.

2 Relevant Literature

There have been many researches on trade credit. Haley and Higgins [18] firstly discuss trade credit policy by jointly considering the optimal order quantity and payment time. Goyal [19] develops economic order quantity (EOQ) model under trade credit and present the optimal replenishment policy. Based on Goyal's work, many researchers study inventory models under trade credit and serval scholars have reviewed existing literature. Chang et al. [20] present a review of inventory literature about trade credit by dividing related articles into five categories. According to different model characteristics, Soni et al. [21] classify inventory studies into three categories. Extending to Chang et al. [20]'s work, Molamohamadi [4] reviews the related literature exhaustively through classifying trade credit contract into six categories. Since our emphasis is trade credit provision to a capital-constraint retailer in dual-channel supply chain, this research is primarily relevant to three literature streams, including trade credit in a capital-constraint supply chain, and the sales cost in direct channel, and consumers' acceptance of direct online channel.

The first stream of literature addresses trade credit in a capital-constraint supply chain. Xu and Birge [22] discuss the optimal decisions depend on a firm's financial constraint under trade credit. Daripa and Nilsen [23] investigate the theory of inter-firm credit which consists of prepayment and delayed payment (trade credit) and find that most of the credit is provided at a zero interest rate. Chen and Wang [24] prove supply chain members can share the risk from the market in a trade credit contract and points that not only the order quantity but also the trade credit contract terms are sufficiently effected by retailer's initial budget. Some literature make comparisons between trade credit financing (TCF) and bank credit financing (BCF) for capital-constraint supply chain. Such as considering supplier and retailer are capital-constraint simultaneously, Kouvelis and Zhao [25] investigate Stackelberg games between supply chain members, and find that retailer is more likely to choose TCF rather than BCF if supplier could provide an optimal trade credit contract. By only considering the retailer suffers capital constraint problem, Jing et al. [26] find that the equilibrium region of TCF will shrink if the initial buget of retailer is improved to high enough. Gupta [27] compares TCF with BCF considering both financing tools are not competitively priced. Different from Gupta [27]'s work, Chen [28] assumes there exists competition between TCF and BCF, and demonstrates both supply chain members are worse off under BCF than TCF, but TCF is more difficult to implement in practice than BCF since the retailer may cheat on manufacturer about the true initial budget. Cai et al. [5] present TCF and BCF can be complementary or substitutable if the retailer's internal capital satisfy corresponding conditions. Most literature in this stream reveal that the optimal decision of supply chain members and the optimal trade credit contract terms are significantly influenced by the retailer's initial capital budget level. However, few of them incorperate operational considerations within a capital-constraint and dual-channel supply chain setting. To fill the literature gap, this paper explores how does the retailer's initial capital level influences the decesions of supply chain members and the terms of trade credit contract under a capital-constraint and dual-channel supply chain.

Next, we review the literatures related to the sales cost in direct channel. Most of the existing papers assume the sales cost in direct channel is zero for simplicity of calculation in dual-channel supply chain, only few literature consider it. Such as Chen et al. [6] suggest manufacturer adopt a share-profit channel strategy when the sales cost in direct channel is relatively high as well as the inconvenience cost of retailer is relatively low. Xu et al. [7] investigate the optimal price and delivery lead time decisions in a dual-channel supply chain and find that the manufacturer should choose appropriate channel structure according to the value of direct channel cost and consumer acceptance of the online channel. Xiong et al. [8] point out that the manufacturer would like to adopt direct channel only when its sales cost is low enough. Yan et al. [9] present that expansion of e-channel could bring Pareto gains to dual-channel supply chain members if direct sales cost and product durability satisy certain conditions. Extending to above literature, this paper aims to study the influence of sales cost in direct channel on dual-channel supply chain members' decision-making, and demonstrates that the direct sales cost should not be ignored in trade credit strategy.

The last stream of studies is related to consumers' acceptance of direct online channel. Based on building normative model or conducting empirical research, Balasubramanian [10] and Liang and Huang [11] prove that consumers' acceptance level of traditional retail channel is generally higher that that of direct online channel. Considering consumers' acceptance level of direct online channel, Chiang et al. [13] explore the pricing strategy under dual-channel supply chain. Extending Chiang et al. [13]'s work, Yan et al. [29], Xu et al. [7] and Zhang et al.[14] respectively investigate the influence of consumers' acceptance on advertisement strategy, service strategy and channel selection strategy under dual-channel supply chain. Kacen et al. [12] find that both the categories and attributes of goods affect consumers' acceptance level of online channel. Consumers' acceptance level is also called "channel substitutability" or "channel preference" etc. by some scholars. Such as Pei and Yan [15] present that higher channel substitutability is more beneficial to channel members if the manufacturer invests in an added national advertising in dual-channel supply chain. Hua et al. [16] find that the customers' preference level of direct channel has a positive effect on direct sales price. Rofin and Mahanty [17] present that the direct channel preference level is positively related to direct channel demand, and a high direct channel preference level will lead to a high online sale price and a low offline sale price. Existing literature show that the consumers' acceptance of direct online channel significantly affect all kinds of strategies in a dual-channel supply chain, so this paper also accommodates this factor into our model.

Above all, in this research, we consider the sales cost in direct channel and consumers' acceptance of direct channel simultaneously, and try to explore a capital-constraint and dual-channel supply chain fundamental for the efficiency of trade credit strategy.

3 Dual-Channel Supply Chain Without Trade Credit Service

Considering a dual-channel supply chain composed of a manufacturer and a capital-constraint physical retailer, the manufacturer produces a kind of product at a unit production cost c and sells the product through both direct online channel (hereafter called direct channel) and traditional retail channel (hereafter called retail channel). In direct channel, the manufacturer directly sells the product to consumers at the price p_{do} and endures a unit sales cost in direct channel z_d; in retail channel, manufacturer supplies retailer at a wholesale price w, and the retailer sells the product at the price p_{ro} in physical store and occurs a unit sales cost in retail channel z_r. Since the wholesale price of product is the result of competition among multiple manufacturers in the market, and it changes infrequently relative to the retail price, in line with Gupta [27], Chen and Wang [24] and Cai et al. [5], we assume the wholesale price w is an exogenous parameter. All notations are summarized in Table 1. Three scenarios, denoted by the subscript $i = o, c, tc$, are identified. They include (1) the retailer has no capital constraint ($i = o$); (2) the retailer suffers capital constraint but without trade credit provision ($i = c$); (3) the retailer suffers capital constraint and is financed by trade credit from manufacturer ($i = tc$).

Table 1. Notation summary

c	Unit production cost
z_d	Unit sales cost in the direct channel
z_r	Unit sales cost in the retail channel
w	Wholesale price
v	Consumer's valuation of the product
t_r	Unit transportation cost, $t_r \geq 0$
t_d	Logistic delivery cost, $t_d \geq 0$
θ	Consumers' acceptance level of direct channel,$0 < \theta \leq 1$
x	Consumer's location, $xU[0, 1]$
B	Retailer's initial capital budget
q_i	Retailer's order quantity of under scenario i, $i = o, c, tc$
r	The interest rate for delay payment in trade credit, $r \geq 0$
p_{di}	The sales price in the direct channel under scenario i, $i = o, c, tc$
p_{ri}	The retail price under scenario i, $i = o, c, tc$
L	Delay payment of retailer, $L \geq 0$

3.1 Consumer Utility and Purchase Decision

Similar to Chiang et al. [13], Yan et al. [29], Xu et al. [7] and Zhang et al. [14], we respectively denote a consumer's valuation of the product by v in retail channel and θv in direct channel, where the parameter θ is called the consumers' acceptance level of direct channel. According to the research results of Liang and Huang [11] and Kacen et al. [12], most of the consumers are prefer buying goods from bricks and mortar to web-based channel since there are more uncertainty and risk in online buying. So, we develop model in this paper with$0 < \theta \leq 1$. Assume each consumer's location x ($xU[0, 1]$) is uniformly distributed along a unit line and the retailer is at the location 0. If a consumer located at x ($xU[0, 1]$) travels to retailer's store will incur a transportation cost$t_r x$, where t_r) is the unit transportation cost. Then the utility function of one consumer who buy the product through traveling to retailer's store is: ($i = o, c, tc$)

$$U_{ri} = v - p_{ri} - t_r x \quad (1)$$

As for the consumer who to buy from the direct channel, besides the product's sales price, he needs to afford logistic delivery cost rather than transportation cost. Form any manufacturers' official shopping websites (such like Huawei, Gree Inc), it can be easy to find that the difference of logistic delivery cost among different regions in a country is not significant, and the delivery cost is even free if the consumption amount attains to some threshold at most cases. Therefore, we assume a consumer needs to pay a constant logistic delivery cost t_d ($t_d \geq 0$) if he buy a product through the manufacturer's website directly, and his utility function is: ($i = o, c, tc$)

$$U_{di} = \theta v - p_{di} - t_d \quad (2)$$

To assure that the manufacturer has an incentive to exploit direct channel, it is appropriate to assume consumer's valuation of the product in retail channel satisfies $v < w + z_r + 2t_r$. Condition $v < w + z_r + 2t_r$ indicates the retailer cannot completely cover the whole market, so the manufacturer could improve market demand efficiently through building direct channel. And we further assume $v > w+z_r$ to ensure the demand in retail channel is non-negative. In addition, we normalize consumers' amount to 1, and suppose any individual consumer can buy (at most) one product from one channel. A consumer will buy from retail channel (or direct channel) if his utility generated by the channel is non-negative, but he would like to choose the channel which can bring bigger utility to him if both channels can generate positive consumer utility simultaneously. Then we can find the threshold of consumer's location is $\frac{v-p_r-\theta v+p_d+t_d}{t_r}$, consumers located in $0 \le x \le \frac{v-p_r-\theta v+p_d+t_d}{t_r}$ will buy from the retail channel, and whom located in $\frac{v-p_r-\theta v+p_d+t_d}{t_r} < x \le 1$ will buy from the direct channel.

3.2 Pricing Strategy of the Manufacturer When Retailer Has no Capital Constraint ($i = o$)

When retailer has no capital constraint, we respectively denote the online direct selling price and retail price of the product by p_{do} and p_{ro}. According to the consumers' purchase decisions in Sect. 3.1, the order quantity of retailer q_o equals to $\frac{v-p_{ro}-\theta v+p_{do}+t_d}{t_r}$, then the retailer's profit function $\pi_{ro}(p_{ro})$ is:

$$\pi_{ro}(p_{ro}) = (p_{ro} - w)q_o = (p_{ro} - w - z_r)\frac{v - p_{ro} - \theta v + p_{do} + t_d}{t_r} \tag{3}$$

The market share obtained by manufacturer through building the direct channel is $\left(1 - \frac{v-p_{ro}-\theta v+p_{do}+t_d}{t_r}\right)$, and the profit function $\pi_{mo}(p_{do})$ is:

$$\begin{aligned}\pi_{mo}(p_{do}) =& (w - c)q_o + (p_{do} - c - z_d)(1 - q_o)\\ =& (w - c)\frac{v - p_{ro} - \theta v + p_{do} + t_d}{t_r} + (p_{do} - c - z_d)\left(1 - \frac{v - p_{ro} - \theta v + p_{do} + t_d}{t_r}\right)\end{aligned} \tag{4}$$

The first part of Eq. (4) is the wholesale revenue in retail channel, and the latter part is the profit generated by direct channel. Suppose both manufacturer and the retailer are independent of each other and respectively maximize own profit. We utilize a manufacturer-leader Stackelberg game to explore the optimal prices of supply chain members. The manufacturer first decides the direct sales price p_{do}, then retailer makes the optimal response and decide the retail price p_{ro} in the physical store. To ensure that both channels' market demand and the marginal profit are non-negative, we can reasonably infer the constraint conditions $w + z_r \le p_{ro} \le v$ and $c + z_d \le p_{do} \le \theta v - t_d$ about the direct sales price p_{do} and the retail price p_{ro}. Thus the retailer and the manufacturers' optimization problems are respectively as follows:

$$Max_{p_{ro}}\pi_{ro}(p_{ro}),\ st.w + z_r \le p_{ro} \le v \tag{5}$$

$$Max_{p_{do}}\pi_{mo}(p_{do}),\ st.c + z_d \le p_{do} \le \theta v - t_d \tag{6}$$

Base on backward induction method, we present the equilibrium prices for both channels in Table 2, where $z_{d1} = \theta v - t_d - 2t_r + v - 2w - z_r, z_{d2} = \theta v - c - t_d$. It can be found that the equilibrium prices of both channels are simultaneously affected by the consumers' acceptance level of direct channel θ and the unit sales cost in direct channel z_d.

Table 2 shows that if the value of consumers' acceptance level of direct channel θ is no more than the threthold $\frac{t_d+c}{v}$, the manufacturer doesn't need to build direct channel. If the value of θ is higher than $\frac{t_d+c}{v}$, the direct channel is also unattractive to manufacturer as long as the unit sales cost in direct channel z_d is relatively high, i.e., exceeds threshold z_{d2}. Otherwise, both players in dual-channel supply chain should decide the optimal price based on the value of θ and z_d. In addition, the equilibrium prices p_{d0} and p_{r0} are monotonically increasing and decreasing with respect to θ respectively, it shows that a higher consumers' acceptance of the direct channel will bring a higher direct sales price and lower retail price. These findings are in line with the results presented by Hua et al. [16] and Rofin and Mahanty [17].

Table 2. Equilibrium prices when the retailer has no capital constraint

When $\theta > \frac{t_d+2t_r-v+2w+z_r}{v}$	$0 \le z_d < z_{d1}$	$z_{d1} \le z_d < z_{d2}$	$z_d \ge z_{d2}$
p_{r0}	$\frac{1}{4}(1-\theta)v + w + \frac{1}{4}z_d + \frac{1}{4}t_d + \frac{1}{2}t_r + \frac{3}{4}z_r$	$\frac{1}{2}v + \frac{1}{2}w + \frac{1}{2}z_r$	$\frac{1}{2}v + \frac{1}{2}w + \frac{1}{2}z_r$
p_{d0}	$w + \frac{1}{2}z_d - \frac{1}{2}t_d + t_r - \frac{1}{2}(1-\theta)v + \frac{1}{2}z_r$	$\theta v - t_d$	N/A
When $\frac{t_d+c}{v} < \theta \le \frac{t_d+2t_r-v+2w+z_r}{v}$	$0 \le z_d < z_{d2}$		$z_d \ge z_{d2}$
p_{r0}	$\frac{1}{2}v + \frac{1}{2}w + \frac{1}{2}z_r$		$\frac{1}{2}v + \frac{1}{2}w + \frac{1}{2}z_r$
p_{d0}	$\theta v - t_d$		N/A
When $0 < \theta \le \frac{t_d+c}{v}$	irrelevant to z_d		
p_{r0}	$\frac{1}{2}v + \frac{1}{2}w + \frac{1}{2}z_r$		
p_{d0}	N/A		

There are two pricing strategies for manufacturer in direct channel. The one is the supplementary pricing strategy which sets a relatively high direct sales price to just absorbing the market demand which cannot be satisfied by the retailer, obviously, there is no competition between p_{d0} and p_{r0} in supplementary pricing strategy. Another one is the competitive pricing strategy. Besides absorbing the market demand which cannot be satisfied by the retailer, the manufacturer could poach the retailer's market share by setting a relatively low direct sales price. We point the optimal pricing strategy for manufacturer in Proposition 1.

Proposition 1. If the unit online direct selling cost z_d does not exceed the threshold z_{d2}, the manufacturer should build a direct channel, and choose appropriate pricing strategy according to the value of the consumers' acceptance level of direct channel θ and the unit sales cost in direct channel z_d:

(1) when $\theta > \frac{t_d+2t_r-v+2w+z_r}{v}$, the manufacturer should adopt the competitive pricing strategy with setting $p_{do} = w + \frac{1}{2}z_d - \frac{1}{2}t_d + t_r - \frac{1}{2}(1-\theta)v + \frac{1}{2}z_r$ if $0 \le z < z_{d1}$, and supplementary price strategy with setting $p_{do} = \theta v - t_d$ if $z_{d1} \le z < z_{d2}$.
(2) when $\frac{t_d+c}{v} < \theta \le \frac{t_d+2t_r-v+2w+z_r}{v}$, the manufacturer should adopt the supplementary pricing strategy with setting direct sales piece $p_{do} = \theta v - t_d$;
(3) when $0\,\theta \le \frac{t_d+c}{v}$, manufacturer doesn't need to build direct channel.

Proposition 1 shows that the manufacturer should adopt the competitive pricing strategy in direct channel only when consumers' acceptance level of direct channel is relatively high as well as the unit sales cost in direct channel is relatively low. Otherwise, the supplementary pricing strategy should be the better choice.

3.3 Pricing Strategy of the Manufacturer When Retailer Suffers Capital Constraint ($i = c$)

According to the result presented by Proposition1, the manufacturer doesn't need to build a direct channel when the unit sales cost in direct channel z_d exceeds the threshold z_{d2}, so this paper only pays attention to the situation that z_d satisfies $0 \le z_d < z_{d2}$. In this section, We consider the retailer has an initial capital B which is insufficient to afford order quantity q_o, i.e., $B < w \cdot q_o$ (where $q_o = \frac{v-p_{ro}-\theta v+p_{do}+t_d}{t_r}$). According to Table 1, the equilibrium prices of both channel (p_{ro}, p_{do}) depend on the value of parameter θ and z_d, sequentially the optimal order quantity of retailer q_o is also determined by the value of the two parameters. So retailer will face a capital constraint problem under the following situations presented in Lemma 1.

Lemma 1. The retailer will face a capital constraint problem,

(1) when $\theta \le \frac{t_d+2t_r-v+2w+z_r}{v}$, the retailer's initial capital satisfies $B \le \frac{w}{2}\frac{(v-w-z_r)}{t_r}$;
(2) when $\theta > \frac{t_d+2t_r-v+2w+z_r}{v}$, the retailer's initial capital satisfies $B < \frac{w}{4}\frac{((1-\theta)v+z_d+t_d+2t_r-z_r)}{t_r}$ if $0 \le z < z_{d1}$, and $B < \frac{w}{2}\frac{(v-w-z_r)}{t_r}$ if $z_{d1} \le z_d < z_{d2}$.

With a capital constraint problem, the retailer can only make order quantity $\frac{B}{w}$ based on his initial capital, then the equilibrium prices in two channels should be further explored. The product's direct sales price and retail price are respectively denoted by p_{dc} and p_{rc}. The retailer's profit function $\pi_{rc}(p_{rc})$ and the manufacturer's profit function $\pi_{mc}(p_{dc})$ are as follows:

$$\pi_{rc}(p_{rc}) = (p_{rc} - w - z_r)\frac{B}{w} \tag{7}$$

$$\pi_{mc}(p_{dc}) = (w-c)\frac{B}{w} + (p_{dc} - c - z)\left(1 - \frac{B}{w}\right) \quad (8)$$

The manufacturer should set $c + z \leq p_{dc} \leq \theta v - t_d$ to ensure that the market demand and the marginal profit of the product in the direct channel are non-negative. As for the retailer, besides positive marginal profit, it is very important to control the market demand not less than the order quantity $\frac{B}{w}$ in the retail channel through setting the retail price satisfying $w + z_r \leq p_{rc} \leq \frac{((1-\theta)v + p_{dc} + t_d)w - Bt_r}{w}$. So the optimization problems for dual-channel supply chain players are as follows:

$$Max_{p_r}\pi_{rc}(p_{rc}), st.w + z_r \leq p_{rc} \leq \frac{((1-\theta)v + p_{dc} + t_d)w - Bt_r}{w} \quad (9)$$

$$Max_{p_d}\pi_{mc}(p_{dc}), st.c + z_d \leq p_{dc} \leq \theta v - t_d \quad (10)$$

According to the backward induction method, we try to explore the equilibrium solutions to the manufacturer-leader Stackelberg game. Since the retailer's profit function $\pi_{rc}(p_{rc})$ is monotonically increasing with retail price p_{rc}, then the response function of retail price can be firstly obtained:

$$p_{rc}(p_{dc}) = \frac{((1-\theta)v + p_{dc} + t_d)w - Bt_r}{w} \quad (11)$$

Because the profit function of manufacturer is monotonically increasing with the direct sale price, we can obtain

$$p_{dc} = \theta v - t_d \quad (12)$$

After substituting (12) into (11), it can be found that the optimal retail price is:

$$p_{rc} = \frac{vw - Bt_r}{w} \quad (13)$$

In Proposition 2, we present the manufacturer's pricing strategy and the influence from the retailer's capital constraint on manufacturer's profit.

Proposition 2. If a retailer has a capital constraint and owns initial capital B,

(1) manufacturer should always adopt the supplementary pricing strategy, and the equilibrium prices in two channels are $p_{dc} = \theta v - t_d$ and $p_{rc} = \frac{vw - Bt_r}{w}$ respectively;
(2) compare to without capital constraint, the retailer has a capital constraint will increase the manufacturer's profit if $\theta > \frac{t_d + w}{v}$ and $0 \leq z_d < z_{d3}$; Otherwise, it will decrease the manufacturer's profit, where $z_{d3} = \theta v - t_d - w$.

When the retailer has a capital constraint problem, Proposition 2 firstly shows that the supplementary pricing strategy is the better choice for manufacturer rather than the competitive pricing strategy. There is no need to compete with the retailer on price because the maximum profit can be realized through setting the highest direct sales price

$p_{dc} = \theta v - t_d$. Secondly, the equilibrium retail price p_{rc} is positively related with the retailer's initial capital B and irrelevant to the consumers' acceptance level of the direct channel θ. While the equilibrium direct sales price p_{dc} is positively correlated with θ.

Furthermore, unlike the traditional view that the capital constraint on retalier is only bad for manufacturer's profit, Proposition 2 reveals that the retailer's capital constraint has a two-way impact on manufacturer's profit. On the one hand, the capital constraint on retailer could ease price competition between two channels, thus it has a positive effect on improving the marginal profit of manufacturer. On the other hand, it reduces the retailer's order quantity, which has a negative impact on manufacturer's wholesale income. When consumers' acceptance level of direct channel is relatively high and the unit sales cost in direct channel is relatively low, the positive impact from the retailer's capital constraint is bigger than negative impact, so the capital constraint on retailer is good for manufacturer. In other cases, the retailer's capital constraint is harmful to manufacturer's profit, so it is reasonable that there exists an incentive for manufacturer to provide financial service to retailer. To clearly describing when is retailer's capital constraint beneficial to manufacturer, we further plot out the benifical region for manufacturer according to the corresponding conditions presented in Proposition 2 (see Fig. 1). Figure 1 shows that the capital constraint on retailer is good to manufacturer's profit in *Area II*, but the opposite occurs in *Area I*.

4 A Dual-Channel Supply Chain with Trade Credit Service ($i = tc$)

We accommodate a specific financial service which is trade credit in this section, and further discuss trade credit strategy in a dual-channel supply chain. To avoid default risk, suppliers often encourage retailers to pay off unpaid amount early by designing the appropriate trade credit contract. In a single-channel supply chain, Kouvelis and Zhao [25] incorporate interest rate and discounted wholesales price into the trade credit contract, and Cai et al. [5] discuss credit limit and interest rate in the trade credit contract. Different from them, we jointly optimize the pricing strategy and trade credit strategy by just considering the interest rate into the contract under a dual-channel supply chain.

Based on Proposition 2, it can be found that the manufacturer has an incentive to provide trade credit only when its profit is pulled down by retailer's capital constraint. The corresponding conditions are identified in Lemma 2.

Lemma 2. Manufacturer will have incentive to provide trade credit in following two cases:

Case (1): consumers' acceptance level of the direct channel is too low, i.e., $\frac{t_d+c}{v} < \theta \leq \frac{t_d+w}{v}$ and $z_{d3} \leq z_d < z_{d2}$.
Case (2): both consumers' acceptance level of the direct channel and the unit sales cost in direct channel are relatively high, i.e.,$\theta > \frac{t_d+w}{v}$ and $z_{d3} \leq z_d < z_{d2}$.

Above cases are included in the *Area I* of Fig. 1, if financed by trade credit, the retailer who owns a capital constraint could realize larger order quantity than that can be afforded by his initial capital *B*. Served by trade credit, the retailer only needs to make a partial payment with his initial capital B firstly, then realize full order quantity $\frac{B+L}{w}$ right away, where $L(L \geq 0)$ is the delay payment. And the retailer must pay off the delay payment L to the manufacturer at an interest rate $r(r \geq 0)$ after the sales season ends. In the manufacturer-leader Stackelberg game, manufacturer firstly decides the terms in trade credit contract which including online direct selling price p_{dtc} and the interest rate r, then the retailer makes the optimal response to manufacturer's strategy and decide the delay payment L and retail price p_{rtc}.The retailer's profit function $\pi_{rtc}(p_{rtc}, L)$ and the manufacturer's profit function $\pi_{mtc}(p_{dtc}, r)$ can be described as follows:

$$\pi_{rtc}(p_{rtc}, L) = (p_{rtc} - w - z_r)\frac{B+L}{w} - L(1+r) \tag{14}$$

$$\pi_{mtc}(p_{dtc}, r) = (w - c)\frac{B+L}{w} + (p_{dtc} - c - z)\left(1 - \frac{B+L}{w}\right) + Lr \tag{15}$$

As the same as Sect. 3.3, the manufacturer should set $c + z_d \leq p_{dtc} \leq \theta v - t_d$ to ensure that the market demand and the marginal profit of the product in direct channel are non-negative. Besides positive marginal profit, it is important for the retailer to control the market demand not less than the order quantity $\frac{B+L}{w}$ in retail channel through setting the retail price satisfying $w + z_r \leq p_{rtc} \leq \frac{((1-\theta)v + p_{dtc} + t_d)w - (B+L)t_r}{w}$. So the optimization problems for dual-channel supply chain players are as follows:

$$Max_{p_{rtc},L}\pi_{rtc}(p_{rtc}, L), st.\begin{cases} w + z_r \leq p_{rtc} \leq \frac{((1-\theta)v + p_{dtc} + t_d)w - (B+L)t_r}{w} \\ L \geq 0 \end{cases} \tag{16}$$

$$Max_{p_{dtc},r}\pi_{stc}(p_{dtc}, r), st.\begin{cases} c + z_d \leq p_{dtc} \leq \theta v - t_d \\ r \geq 0 \end{cases} \tag{17}$$

The equilibrium solutions and are presented in Table 3, where $B_1 = \frac{w(v-2w-z_r)}{2t_r}$, $B_2 = \frac{((1+\theta)v - z_d - z_r - t_d - 3w)w}{2t_r}$.

Table 3. Equilibrium outcomes with trade credit service

	$0 \leq B < B_2$	$B_2 \leq B < B_1$	$B_1 \leq B < \frac{w}{2}\frac{(v-w-z_r)}{t_r}$
p_{rtc}	$\frac{1}{4}\frac{((3+\theta)v - z_d + z_r - t_d)w - 2Bt_r + w^2}{w}$	$\frac{1}{2}v + w + \frac{1}{2}z_r$	$\frac{vw - Bt_r}{w}$
L	$\frac{1}{4}\frac{((1-\theta)v + z_d - z_r + t_d)w - 2Bt_r - w^2}{t_r}$	$\frac{1}{2}\frac{vw - 2Bt_r - 2w^2 - wz_r}{t_r}$	N/A(no trade credit)
p_{dtc}	$\theta v - t_d$	$\theta v - t_d$	$\theta v - t_d$
r	$\frac{1}{2}\frac{((1+\theta)v - z_d - z_r - t_d)w - 2Bt_r - 3w^2}{w^2}$	0	N/A(no trade credit)

After making a comparison between the value of p_{rtc} in Table 3 and p_{rc} in Proposition 2, it can be found that $p_{rc} \geq p_{rtc}$ is always holding. It indicates that the retailer will decrease the retail price of the product under trade credit, so trade credit intensifies the price competition between two channels. We further investigate whether the provision of trade credit is beneficial to the manufacturer, and present the pricing strategy and trade credit strategy in Proposition 3.

Proposition 3. When manufacturer's profit is pulled down by retailer's capital constraint, the manufacturer should adopt the supplementary pricing strategy with $p_{dtc} = \theta v - t_d$, and strategically provide trade credit based on the value of the retailer's initial capital B and the value of unit sales cost z_r in retail channel:

(1) When the value of unit sales cost in retail channel z_r is low enough to satisfy $z_r < (1+\theta)v - t_d - z_d - 3w$, (1 * roman) the manufacturer should provide trade credit to the retailer with a positive interest rate $r = \frac{1}{2}\frac{((1+\theta)v-z_d-z_r-t_d)w-2Bt_r-3w^2}{w^2}$ if $0 \leq B < B_2$,; (2 * roman) the manufacturer should provide trade credit to the retailer for free with $r = 0$ if $B_2 \leq B < B_1$; (3 * roman) the manufacturer does not need to provide trade credit if $B_1 \leq B < \frac{w}{2}\frac{(v-w-z_r)}{t_r}$.
(2) When the value of unit sales cost in retail channel z_r satisfies $(1+\theta)v - t_d - z_d - 3w \leq z_r < v - 2w$, (1 * roman) the manufacturer should provide trade credit to the retailer for free with $r = 0$ if $0 \leq B < B_1$; (2 * roman) the manufacturer does not need to provide trade credit if $B_1 \leq B < \frac{w}{2}\frac{(v-w-z_r)}{t_r}$.
(3) When the value of unit sales cost in retail channel z_r is high enough to satisfy $v - 2w \leq z_r < v - w$, the manufacturer does not need to provide trade credit.

Proposition 3 puts forward two trade credit strategies: trade credit with positive interest rate and trade credit with zero interest rate. The results guide the manufacturer when and how to provide trade credit. Although trade credit could increase the manufacturer's wholesale revenue and bring income by interest rate, it also will lead the manufacturer to suffer profit loss since the price competition between two channels is intensified. On one hand, when the value of unit sales cost in retail channel is relatively low, the increment of manufacturer's profit is lower than the profit loss under the trade credit if the value of the retailer's initial capital is relatively high, i.e., $B_1 \leq B < \frac{w}{2}\frac{(v-w-z_r)}{t_r}$, so there is no need to provide trade credit at all. If the value of the retailer's initial capital is relatively low, i.e., $0 \leq B < B_2$, the opposite is true and the manufacturer should provide trade credit with a positive interest rate. And note that when the value of retailer's initial capital is neither relatively high nor relatively low, i.e., $B_2 \leq B < B_1$, the manufacturer would like to choose a zero interest rate to expand the wholesale income. This result is consistent with the conclusion of Daripa and Nilsen [23] that much of the trade credit is supplied at zero interest. On the other hand, when the value of unit sales cost in retail channel z_r is extremely high (i.e., $v - 2w \leq z_r < v - w$), the retailer would not chose trade credit service even if manufacturer provides it, since low retail price and high sales cost lead to low marginal profit under trade credit. So there is no need to provide trade credit under this situation.

To guide the manufacturer when and how to provide trade credit, we further plot out the feasible region for each trade credit strategy based on the corresponding condition (see Fig. 5). Our results implicate that making trade credit strategy according to the retailer's initial capital and the unit sales cost in retail channel can effectively avoid the cheating problem which is presented by Chen [28] that the retailer may cheat on the manufacturer about the true initial budget.

5 Numerical Simulation

Base on several numerical simulation examples, we comprehensively explore more managerial insights of pricing and trade credit strategies in a dual-channel supply chain.

5.1 Without Trade Credit

We first test the two-way impact from the retailer's capital constraint on the manufacturer's profit. With the following base parameter set $w = 0.3$, $t_d = 0.2$, $t_r = 0.3$, $c = 0.1$, $v = 0.8$, $B = 0.2$, we obtain Fig. 1 based on the results presented in Proposition 2.

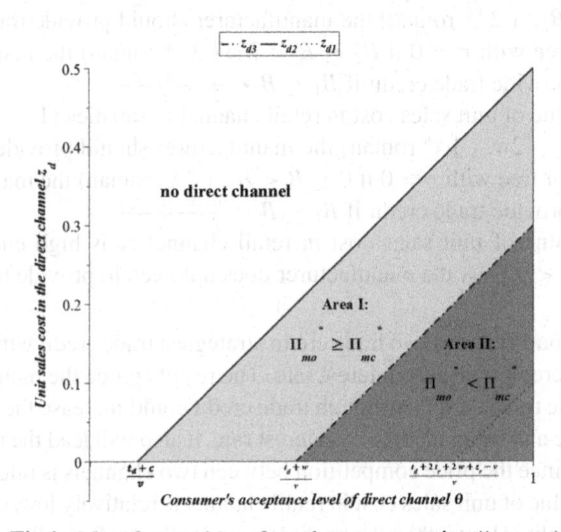

Fig.1. Manufacturer's profit under two scenarios ($i = o, c$)

Figure 1 shows that when consumers' acceptance level of the direct channel is relatively high and the unit sales cost in direct channel is relatively low (i.e., in *Area II*), manufacture can benefit from the capital constraint on retailer, while the opposite occurs in *Area I*.

5.2 With Trade Credit

Respectively explore the impact of trade credit on retailer's order quantity, retail price, manufacturer's profit, and further describe the feasible region for each trade credit strategy based on corresponding conditions. According to Lemma 2, the next four numerical studies have been developed under conditions which both consumers' acceptance level of the direct channel and unit sales cost in direct channel are relatively high (i.e., $\theta > \frac{t_d+w}{v}$ and $z_{d3} \leq z < z_{d2}$). With the following base parameter set $\theta = 0.65$, $w = 0.2$, $t_d = 0.2$, $t_r = 0.3$, $c = 0.1$, $v = 0.8$, $z_r = 0.1$, $z_d = 0.8$, we first respectively plot the order quantity of retailer Q_o, $Q_c(B)$, and $Q_{tc}(B)$ under three scenarios: (1) the retailer has no capital constraint; (2) the retailer suffers capital constraint but without trade credit

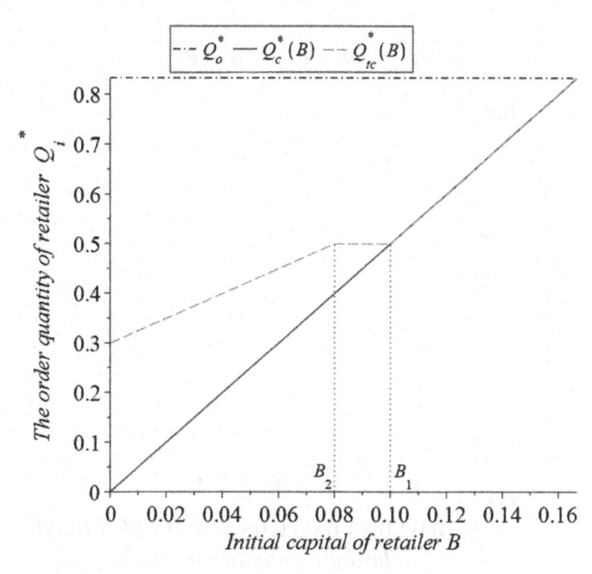

Fig. 2. Retailer's optimal order quantity under three scenarios ($i = o, c, tc$)

provision; (3) the retailer suffers capital constraint and is financed by trade credit from the manufacturer.

Figure 2 shows that under the above three scenarios, the retailer's order quantity always meets $Q_c(B) \leq Q_{tc}(B) \leq Q_o$. It indicates that capital constraint compel the retailer to reduce his order quantity from the Q_o to $Q_c(B)$, while the trade credit could improve his order quantity from $Q_c(B)$ to $Q_{tc}(B)$. It's worth noting that the retailer will not improve his order quantity $Q_{tc}(B)$ up to Q_o even if the manufacturer serves him with trade credit for free. Since the trade credit with zero interest rate puts down the retail price although it improves the order quantity, therefore blindly pursuing the order level without capital constraint may not bring the maximum profit.

Then we describe the retail price p_{ro}, $p_{rc}(B)$, and $p_{rtc}(B)$ under the above three scenarios in Fig. 3.

Figure 3 shows that $p_{ro} \leq p_{rtc}(B) \leq p_{rc}(B)$ is holding. It can be interpreted that capital constraint make the retailer rise the retail price from p_{ro} to $p_{rc}(B)$, while trade credit could decrease the retail price from $p_{rc}(B)$ to $p_{rtc}(B)$. Note that even financed by trade credit for free, the retailer with capital constraint is reluctant to drop the retail price $p_{rc}(B)$ to p_{ro} since a small marginal profit is not beneficial to him. It's interesting that if the trade credit interest rate in a single-channel supply chain is zero, the order quantity and retail price decided by the retailer with capital-constraint can be the same as that in the case of no capital constraint. However, in the dual-channel supply chain of this paper, the trade credit with zero interest rate doesn't make the retailer's order quantity and retail price reach the level of no capital constraint.

Next, we make a comparison among the profit of manufacturer under the above three scenarios in Fig. 4.

If the retailer's capital constraint is harmful to manufacturer (such as in the Cases described in Lemma 2), Fig. 4 exhibits that providing trade credit can greatly improve

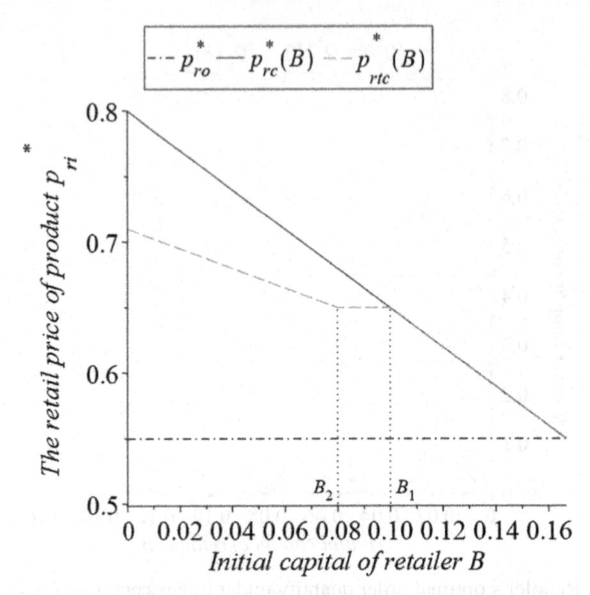

Fig. 3. The equilibrium retail price under three scenarios ($i = o, c, tc$)

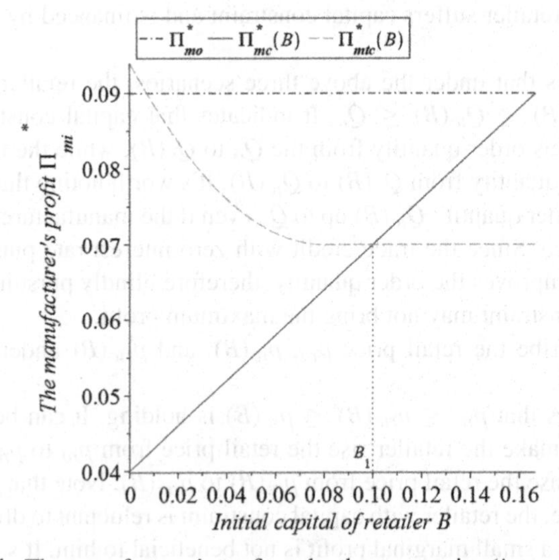

Fig. 4. The manufacturer's profit under three scenarios ($i = o, c, tc$)

manufacturer's profit to far exceed that without capital constraint when retailer B's initial capital is extremely low. But for the manufacturer, benefits brought by trade credit provision gradually shrinks to zero as B increases up to B_1, so no provision of trade credit is the better choice for the manufacturer when $B \geq B_1$.

Last, according to the results presented Proposition 3, the conditions about z_r (the unit sales cost in retail channel) and the conditions about B (the initial capital of retailer) could form a closed area in a two-dimensional plane. With the parameter set $\theta = 0.65$, $w = 0.2$, $t_d = 0.2$, $t_r = 0.3$, $c = 0.05$, $z_d = 0.25$ and $v = 0.8$, we plot out the feasible region of each trade credit strategy for manufacturer in Fig. 5.

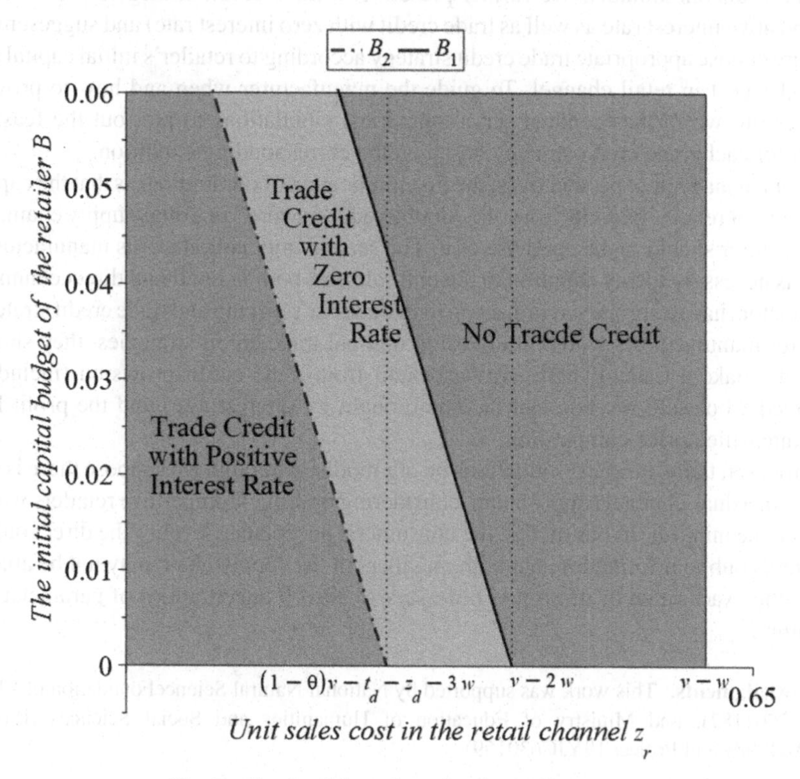

Fig. 5. The feasible region of trade credit strategy

Figure 5 is a two-dimensional plan that takes z_r as the X-axis and B as the Y-axis. The boundaries of each region respectively correspond to the thresholds of B presented in Proposition 3. We can clearly observe the feasible regions for manufacturer to adopt each trade credit strategy.

6 Conclusions

It is important for a manufacturer to effectively provide trade credit to a retailer who suffer capital constraint in a dual-channel supply chain. We carefully study pricing and trade credit strategies for manufacture in this research.

Through building Stackelberg game model with a manufacturer-leader and a retailer-follower, we first reveal the closed-form equilibrium prices for both supply chain members and find that the equilibrium prices are simultaneously affected by unit direct sale cost and consumers' acceptance level of the direct channel. Supplementary pricing strategy and competitive pricing strategy are presented in our results, we conclude that the former one is the better choice for manufacture when retailer suffers capital constraint. Moreover, when manufacturer's profit is threatened by retailer's capital constraint, we verify the manufacturer should provide trade credit to retailer strategically rather than provide it unconditionally. We further present two trade credit strategies (trade credit with positive interest rate as well as trade credit with zero interest rate) and suggest manufacture choose appropriate trade credit strategy according to retailer's initial capital and unit sales cost in retail channel. To guide the manufacturer when and how to provide trade credit, we further conduct serval numerical simulations to plot out the feasible region for each trade credit strategy based on the corresponding condition.

From management perspectives, the first implication of our findings is that the capital constraint of retailer plays the role of a "double-edged sword" in a dual-supply chain, the manufacturer should make good use of it. The second implication warns manufacturers that it is necessary to pay attention to the unit sale cost both in retail and direct channels, these factors have a significant impact on manufacturer's pricing and trade credit strategy. Last, for manufacturers who aim to design optimal trade credit strategies, they should carefully make a tradeoff between the benefit from trade credit provision (including expanded wholesale revenue and income brought by interest rate) and the profit loss from intensified price competition.

However, there are some limitations on our modle. Although we suppose there is one retailer in a dual-channel supply chain, considering multiple-competitive retailers would be more meaningful. In our model, the consumers' acceptance level of the direct online channel is public information, but both members of the supply chain may not be able to obtain the exact value of it. So it is necessary to absorb uncertainties of parameters in the future.

Acknowledgments. This work was supported by National Natural ScienceFoundation of China (No. 72001182), and Ministry of Education of Humanities and Social Sciences (Project 18YJA630071 and Project 19YJC630159).

References

1. Petersen, M.A., Rajan, R.G.: Trade credit: theories and evidence. Rev. Finan. Stud. **10**(3), 661–691 (1997)
2. Piasecki, D.: Consignment inventory: what is it and when does it make sense to use it? White Paper, Inventory Operations Consulting (2004)
3. Molamohamadi, Z., Rezaeiahari, M., Ismail, N.: Consignment inventory: review and critique of literature. J. Basic Appl. Sci. Res. **3**(6), 707–714 (2013)
4. Molamohamadi, Z., Ismail, N., Leman, Z., Zulkifli, N.: Reviewing the literature of inventory models under trade credit contact. Discret. Dyn. Nat. Soc. **2014**, 1–9 (2014)
5. Cai, G.G., Chen, X., Xiao, Z.: The roles of bank and trade credits: theoretical analysis and empirical evidence. Prod. Oper. Manag. **23**(4), 583–598 (2014)
6. Chen, K.-Y., Kaya, M., Özer, Ö.: Dual sales channel management with service competition. Manuf. Serv. Oper. Manag. **10**(4), 654–675 (2008)
7. Xu, H., Liu, Z.Z., Zhang, S.H.: A strategic analysis of dual-channel supply chain design with price and delivery lead time considerations. Int. J. Prod. Econ. **139**(2), 654–663 (2012)

8. Xiong, Y., Yan, W., Fernandes, K., Xiong, Z.-K., Guo, N.: "Bricks vs. clicks": the impact of manufacturer encroachment with a dealer leasing and selling of durable goods. Eur. J. Oper. Res. **217**(1), 75–83 (2012)
9. Yan, W., Li, Y., Wu, Y., Palmer, M.: A rising e-channel tide lifts all boats? The impact of manufacturer multichannel encroachment on traditional selling and leasing. Discret. Dyn. Nat. Soc. **2016**, 1–8 (2016)
10. Balasubramanian, S.: Mail versus mall: a strategic analysis of competition between direct marketers and conventional retailers. Mark. Sci. **17**(3), 181–195 (1998)
11. Liang, T.-P., Huang, J.-S.: An empirical study on consumer acceptance of products in electronic markets a transaction cost model. Decis. Support Syst. **24**, 29–43 (1998)
12. Kacen, J.J., Hess, J.D., Kevin Chiang, W.-Y.: Bricks or clicks? Consumer attitudes toward traditional stores and online stores. Glob. Econ. Manag. Rev. **18**(1), 12–21 (2013)
13. Chiang, W., Chhajed, D., Hess, J.: Direct marketing, indirect profits: a strategic analysis of dual-channel supply-chain design. Manag. Sci. **49**(1), 1–20 (2003)
14. Zhang, P., He, Y., Shi, C.: Retailer's channel structure choice: online channel, offline channel, or dual channels? Int. J. Prod. Econ. **191**, 37–50 (2017)
15. Pei, Z., Yan, R.: National advertising, dual-channel coordination and firm performance. J. Retail. Consum. Serv. **20**(2), 218–224 (2013)
16. Hua, G., Wang, S., Cheng, T.C.E.: Price and lead time decisions in dual-channel supply chains. Eur. J. Oper. Res. **205**(1), 113–126 (2010)
17. Rofin, T.M., Mahanty, B.: Optimal dual-channel supply chain configuration for product categories with different customer preference of online channel. Electron. Commer. Res. **18**(3), 507–536 (2017). https://doi.org/10.1007/s10660-017-9269-4
18. Haley, C.W., Higgins, R.C.: Inventory policy and trade credit financing. Manag. Sci. **20**(4-part-i), 464–471 (1973)
19. Goyal, S.K.: Economic order quantity under conditions of permissible delay in payments. J. Oper. Res. Soc. **36**(4), 335–338 (1985)
20. Chang, C.-T., Teng, J.-T., Goyal, S.K.: Inventory lot-size models under trade credits: a review. Asia-Pacific J. Oper. Res. **25**(1), 89–112 (2008)
21. Soni, H., Shah, N.H., Jaggi, C.K.: Inventory models and trade credit: a review. Control. Cybern. **39**(3), 867–880 (2010)
22. Xu, X., Birge, J.R.: Joint-production-and-financing-decisions-modeling-and-analysis. Working paper (2004). https://ssrn.com/abstract=652562
23. Daripa, A., Nilsen, J.: Ensuring sales: a theory of inter-firm credit. Am. Econ. J. Microecon. **3**(1), 245–279 (2011)
24. Chen, X., Wang, A.: Trade credit contract with limited liability in the supply chain with budget constraints. Ann. Oper. Res. **196**(1), 153–165 (2012)
25. Kouvelis, P., Zhao, W.: Financing the newsvendor: supplier vs. bank, and the structure of optimal trade credit contracts. Oper. Res. **60**(3), 566–580 (2012)
26. Jing, B., Chen, X., Cai, G.G.: Equilibrium financing in a distribution channel with capital constraint. Prod. Oper. Manag. **21**(6), 1090–1101 (2012)
27. Gupta, D.: Technical note: financing the newsvendor. Working paper, University of Minnesota, Twin Cities, Minneapolis (2008)
28. Chen, X.: A model of trade credit in a capital-constraint distribution channel. Int. J. Prod. Econ. **159**, 347–357 (2015)
29. Yan, R., Ghose, S., Bhatnagar, A.: Cooperative advertising in a dual channel supply chain. Int. J. Electron. Mark. Retail. **1**(2), 99–114 (2006)

Remote Vehicular Control Network Test Platform

Lei Chen, Ping Cui(✉), Yun Chen, Kailiang Zhang, and Yuan An

Jiangsu Province Key Laboratory of Intelligent Industry Control Technology, Xuzhou University of Technology, Xuzhou 221018, China
xzcuiping@xzit.edu.cn

Abstract. In this paper, architecture is proposed to test remote control network for low speed vehicle remote driving. By using 4G cellular network access to the cloud service platform, the platform is easy to deployed in common commercial networks. A control signal transmit experiment is executed in the commercial network crossing one more thousand miles; the performance show that the common 4G cellular and backbone networks can support the real-time signal transmit for low speed vehicle. Video latency is tested using different cameras, and the MOS is defined to measure how difficult to drive a remote vehicle under certain video latency.

Keywords: Remote drive · Vehicular video · Cloud computing · UDP · TCP · User experience

1 Introduction

Some new approaches are proposed to improve network performance in new wireless communication services [1–3]. These new approaches are used to improve the quality of service (QoS) for end user [4–7]. Through analyzing the end user behavior, the traffic flow model is able to be built [8–10]. Based on these models, the traffic can be described accurately and be used to build test scenarios [11, 12]. On the other hand, some traffic reconstructions methods are proposed to improve network efficiency [13–15], such as effective capacity [16–18], network utilization [19, 20], spectral-efficiency [21–23] and energy-efficiency [24–26]. Besides the object index, the subject user experience is critical index for commercial network [27–30]. Traffic prediction helps to improve the quality of end user experience [31–33]. However, the remote driving poses challenges to QoS request and objective quality measurement. At present, the self-driving technology cannot handle the vehicle operation under all working conditions. Considering complex scenarios and critical safe requirement, it is unlikely that self-driving will be deployed in common commercial scenarios. Combined with mobile communication technology, a few remote drivers can handle the complex situation of a large number of vehicles, which is expected to achieve man-machine cooperation driving. Many low-speed and medium-speed unmanned vehicle with fixed working scope and simple environment may take the lead because of their small risks.

H. Song and D. Jiang (Eds.): SIMUtools 2020, LNICST 370, pp. 246–255, 2021.
https://doi.org/10.1007/978-3-030-72795-6_19

C-V2X Communication is an emerging 5G scenario. 3GPP requires the 5G remote driving communication performance obtain 99.999% reliability, 25 Mb/s upload bandwidth, and 1 Mb/s download, and 20 ms delay. 5G-based self driving standards are rapidly moving towards commercial deployment [34, 35]. With the large-scale commercial deployment of mobile interconnection services based on cloud platform, IMT-2020 develops a small-scale field environment testing method of "Cellular + Vehicle Networking" (C-V2X). Although 5G cellular network and edge computing greatly reduce end-to-end responding time [36], most of remote driving system only can be deployed in LAN scenarios [37]. There is a lack of research on remote driving test for common commercial network. The cost of developing remote driving parallel management and control system is very high, the test in commercial network is necessary [38, 39].

Aiming at the scenario of remote parallel management and control low-speed vehicle, this paper proposes the test method in cellular scenario. The paper is divided into five sections. The related work is given in section two. Test system and experiment are illustrated in section tree and section four respectively. We conclude in section five.

2 Related Work

In January 21, 2019, Ericsson, Jiangsu Mobile, Qingdao Hui Tuo, and Intel test remote driving accessing to 5G cellular network in Intelligent Vehicle Research and Development Center located in Suzhou Changshu. Current researches focus on simplifying complex physical systems and building artificial systems, analyzing the behavior of physical systems, effectively managing the operation of complex systems by parallel execution between physical systems and artificial system [40]. By keeping the consistency between the real system and the artificial system, not only accurate control can be carried out, but also the whole life cycle of the equipment can be analyzed and predicted [41]. On the basis of collecting a large amount of historical data, a simulation environment can be constructed in the artificial system to predict and correct the consequences of operation, which is helpful to improve the capability of large-scale remote control system [42, 43]. To reduce the cost, it is expected that a common cloud platform and common commercial cellular access network will be employed into remote control system. The artificial system will be deployed in cloud platform. The physical system will linked to artificial system via cellular network. The latency constraint between physical system and artificial system is critical for remote driving. Different from conventional user experience measurement, the drivers' experience is related to its task. Therefore, the remote driving experience should measure how difficult to control the real vehicle.

3 Architecture of Remote Driving

3.1 Network Architecture

In order to control remote vehicles in real-time, we design architecture as in Fig. 1:

(1) The vehicles collect video and send to the control center through a common cloud platform. If the vehicle fails for its task, the human driver requires the real-time video from cloud platform.

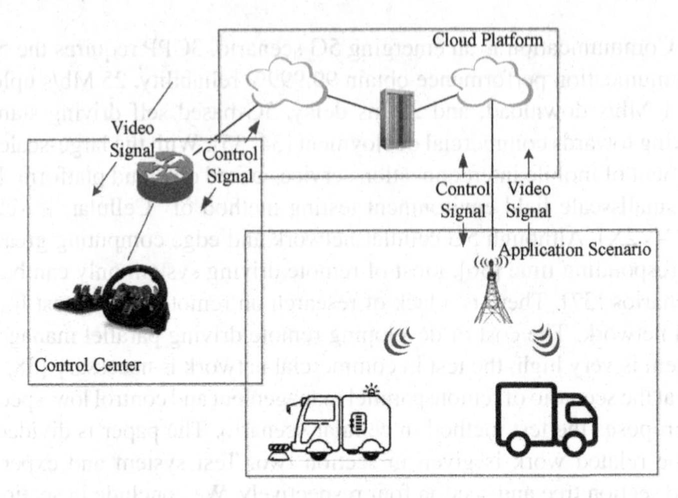

Fig. 1. Architecture diagram of the vehicle network.

(2) The drivers' operations on G29 are collected by machines; the machine encodes the operations into control signal within packet. The cloud server sends signals and some necessary parameters between control center and vehicles.
(3) The system transmits the control signal defined frequency; therefore it is easy to check whether the signal arrives in time. Since the vehicles response the control center with an acknowledge packet, the test system can know the round trip latency.

3.2 Instruction Collection Design

We adopt the G29 wheel equipment to collect the human driver operations. The Table 1 presents the control signal transmitting to remote vehicle.

Table 1. The operation parameters.

Operation	Type	Length	Upper bound	Down bound
Wheel-turn	float	64 bit	100	0
Shifter-gear	float	64 bit	– 1, 1, 2, 3, 4, 5, 6	
Pedals-gas	float	64 bit	1	0
Pedals-brake	float	64 bit	1	0
Pedals-clutch	float	64 bit	1	0

As a real-time system, the overdue control signals are useless; therefore we adopt UDP protocol. The control center sends control signals to the cloud server; the server sends the signal to the vehicle according to identification number. The vehicle response with an acknowledgement signal; the round trip latency is recorded.

3.3 Video Latency Test

In order to reduce the video transmission time, we set up the following camera network topology to test the video transmission time of multiple cameras. The architecture is illustrated in Fig. 2.

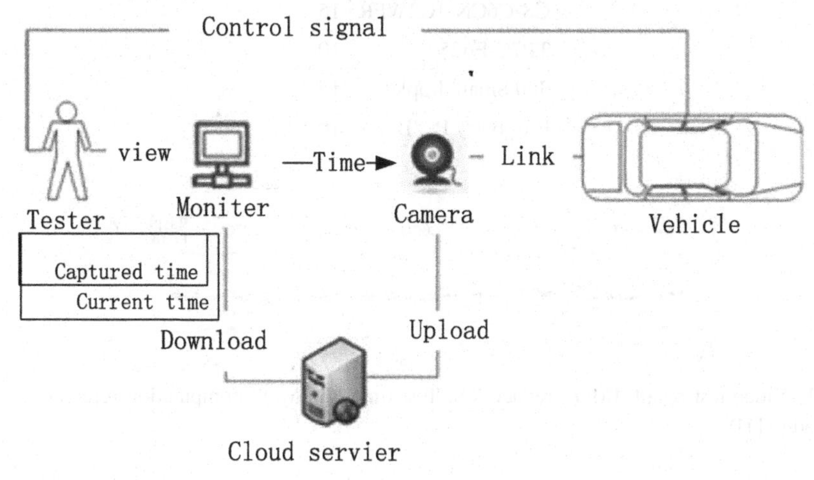

Fig. 2. Video test diagram.

(1) The camera, router, terminal server, and client are connected as in Fig. 2. For accurate test, the camera captures the picture of monitor and sends the captured picture to the terminal server. Note that the picture carries the time showing in the monitor.
(2) Both current time and captured time show on the monitor. The whole latency including encode latency, decode latency, and transmission latency is obtained by comparing the two time.

4 Experiment

4.1 Remote Video Test

We tested a total of 6 cameras and collected certain data at the same time. The following Table 2 shows the models and parameters of the 6 cameras:

During the test, the respective private protocols were used. In the LAN environment, the video delay was calculated. After testing, we get the following picture (Fig. 3).

In this experiment, the Raspberry Pi 3B used RTMP protocol, and the rest used its own private protocol. In order to obtain a more accurate delay, we used the method of comparing the recorded time on the video with the real time to obtain the delay difference. According to experiment results, we can conclude that under 4G conditions, the optimal delay time is Raspberry Pi 3B. The maximum value is 372 ms and the minimum value

Table 2. Video test parameters.

Model	Frame
CS-C6TC-32WFR	15
TL-IPC42EW-4	15
CS-C6CN-1C2WFR	15
TP7C-E625	10
360 Small droplets	15
Raspberry Pi 3B	10

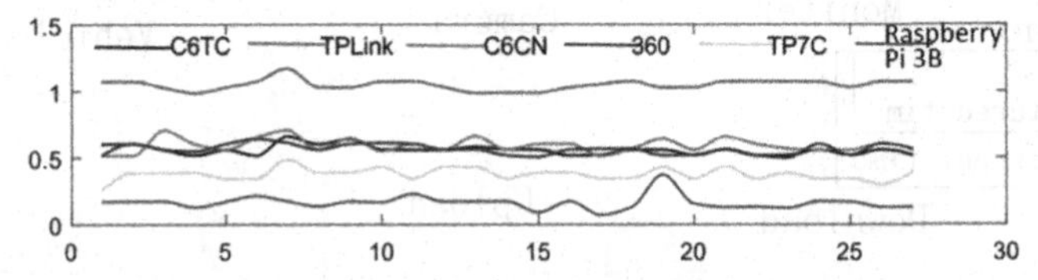

Fig. 3. Video test result. (MOS reduce VS. Test time, the MOS computation reference to the method in [1]).

is only 71 ms. The stability of the delay is generally within the acceptable range. For the Raspberry Pi 3B, the delay of the TP7C is also in the acceptable range, with an average delay of 375 ms and a maximum of 486 ms. Therefore, according to the MOS value of the experimental results, under the conditions of 4G and Raspberry Pi 3B video delay, it can basically meet the requests of driving video transmission of low-speed remote vehicles.

However, the movement effect cannot be well simulated under the conditions of static indoor conditions. Therefore, another data transmission experiment is executed in the moving vehicle. We chose the following different sites for testing, corridors, roads, campuses and suburb.

Table 3. Video test results.

Condition	Average(s)	Maximum(s)	Minimum(s)	Standard deviation
Corridor	0.267	0.364	0.216	0.04126
Way	0.275	0.651	0.16	0.11048
Campus	0.255	0.304	0.175	0.03082
Suburb	0.262	0.304	0.217	0.02374

Experimental results are presented in Table 3. According to the experimental results, under the condition of good network conditions, the transmission of delay is basically stable, which can basically meet the video conditions of low-speed remote driving.

The camera used in this experiment has a resolution of 720 * 1280. From an average point of view, the delay on campus is lower than that tested elsewhere; the transmission delay on roads is too high and at the same time the lowest. In comparison, the stability of roads is worse, and the stability in remote areas is the highest.

Experimental summary: through this mobile test, the effect of a moving object on a 4G network is realistically simulated. In different local environments, the data is not very different. Generally speaking, the overall situation of delay is in an acceptable range. At the same time, it can be compared with the static delay data, observe the differences, and imagine the experimental conclusions. Not only that, using mobile delay data as reference data for driverless driving is more contrastive and convincing.

4.2 Control Experience Test

Since the transmission of video is bound to a certain delay under the condition of 4G, in order to investigate the difficulty level of driving a remote vehicle with different speeds and certain delay, we propose experiment to measure the driver experience using driving game with certain video latency and speed.

In this experiment, for the sake of safety, we use the video game instead of the actual vehicle. The game picture by camera, to the server, and then by the monitor get to video data (data is compared with the real vehicle at this time there is a delay). At this time, the experimenter controlled the vehicle according to the acquired experimental data. A total of five test people were selected for this experiment, and data acquisition was performed multiple times. During the time delay control, we use multiple cameras to increase the delay. Note that this method introduces some fluctuations.

In the experiment environment, we adopt Learn Che Bao as simulation software. Cameras are Raspberry Pi 3B and Tencent Class, For Raspberry Pi, we still choose 10 frames with a resolution of 720 * 1280, and Tencent Class uses default parameters. During the experimental process, we firstly shoot through the Raspberry Pi camera to get the minimum average delay, then we adopt Tencent class camera, which has higher transmit delay than Raspberry, to increase latency, and then increase the latency by using the Tencent Class recording screen that show the video recorded by Raspberry Pi 3B. Through this method, we can increase certain latency by increasing cameras before the end monitor. And the drivers give their subject feeling about how difficult to control the car as the latency increases. The lower MOS means more difficulty.

The experimental results are shown in Fig. 4. The results obtained in the figure show that under the condition of low delay, the vehicle can be basically controlled under the condition of 4G under the condition of 40 km/h. With the latency higher than 500 ms, according to the feedback of the driver, the vehicle barely is controlled. With latency lower than 500 ms and speed higher than 40 km/h, it is also in an uncontrollable state. According to the results in the game, after the MOS value is roughly below 3, the car accident rate in the test data reaches 81.58%. With the latency less than 40 km/h, the MOS is high. Therefore, we conclude that in 4G conditions, low-speed long-distance driving can be satisfied.

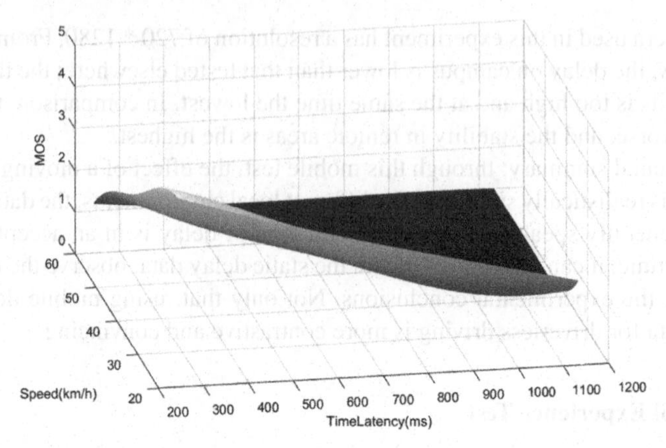

Fig. 4. Video test result.

4.3 Remote Control Signal

We deploy the vehicle and control center in the city of Xuzhou of China. The cloud platform is in Guangzhou locating one more thousand miles apart from Xuzhou. The vehicles access to backbone network via wired link and cellular network respectively. The round-trip latency is presented in Fig. 5 and Fig. 6.

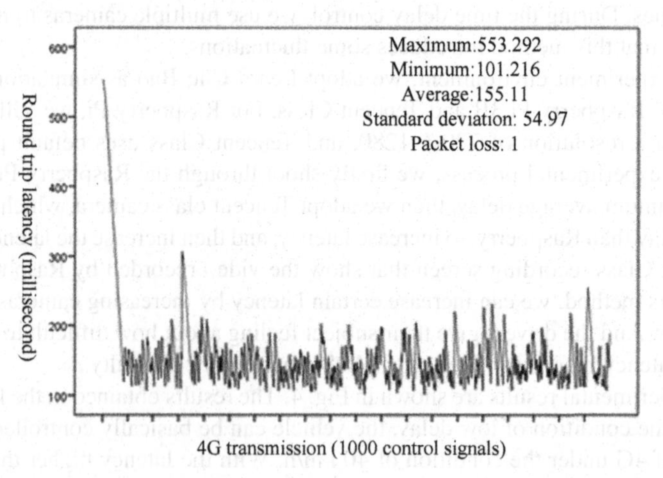

Fig. 5. 4G transmission.

Figure 5 and 6 shows the data transmission time of two access models. The number of tests was 1000. Analyzing the chart, the bottleneck of data transmission time is still air interface. We can expect a significant improvement in 5G network which provide the low latency and high reliability air interface communication.

The 155 ms average round trip latency can support remote control of some low speed special vehicles.

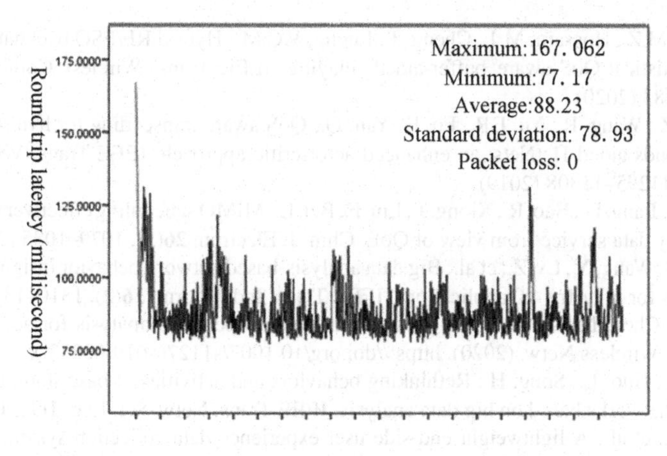

Fig. 6. Wired transmission.

5 Summary

Based on the characteristics of remote control vehicle, we propose test method based on common cloud platform and commercial cellular networks. The UDP is adopted to transmit control signal. With well designed test platform, the driver's subjective experience is able to be measured. The experiment results show that the 4G networks can support remote control for low speed vehicle. And the relationship between video latency and controllable speed is presented.

Acknowledgment. This work is partly supported by Jiangsu technology project of Housing and Urban-Rural Development (No. 2018ZD265) and Jiangsu major natural science research project of College and University (No. 19KJA470002).

References

1. Zhang, K., Chen, L., An, Y., Cui, P.: A QoE test system for vehicular voice cloud services. Mob. Netw. Appl. (2019). https://doi.org/10.1007/s11036-019-01415-3
2. Wang, F., Jiang, D., Qi, S.: An adaptive routing algorithm for integrated information networks. China Commun. **7**(1), 196–207 (2019)
3. Huo, L., Jiang, D., Lv, Z., et al.: An intelligent optimization-based traffic information acquirement approach to software-defined networking. Comput. Intell. **36**, 1–21 (2019)
4. Tan, J., Xiao, S., Han, S., Liang, Y., Leung, V.C.M.: QoS-aware user association and resource allocation in LAA-LTE/WiFi coexistence systems. IEEE Trans. Wireless Commun. **18**(4), 2415–2430 (2019)
5. Wang, Y., Tang, X., Wang, T.: A unified QoS and security provisioning framework for wiretap cognitive radio networks: a statistical Queueing analysis approach. IEEE Trans. Wireless Commun. **18**(3), 1548–1565 (2019)

6. Hassan, M.Z., Hossain, M.J., Cheng, J., Leung, V.C.M.: Hybrid RF/FSO backhaul networks with statistical-QoS-aware buffer-aided relaying. IEEE Trans. Wireless Commun. **19**(3), 1464–1483 (2020)
7. Zhang, Z., Wang, R., Yu, F.R., Fu, F., Yan, Q.: QoS aware transcoding for live streaming in edge-clouds aided HetNets: an enhanced actor-critic approach. IEEE Trans. Veh. Technol. **68**(11), 11295–11308 (2019)
8. Chen, L., Jiang, D., Bao, R., Xiong, J., Liu, F., Bei, L.: MIMO scheduling effectiveness analysis for bursty data service from view of QoE. Chin. J. Electron. **26**(5), 1079–1085 (2017)
9. Jiang, D., Wang, Y., Lv, Z., et al.: Big data analysis-based network behavior insight of cellular networks for industry 4.0 applications. IEEE Trans. Ind. Inform. **16**(2), 1310–1320 (2020)
10. Bao, R., Chen, L., Cui, P.: User behavior and user experience analysis for social network services. Wireless Netw. (2020). https://doi.org/10.1007/s11276-019-02233-x
11. Jiang, D., Huo, L., Song, H.: Rethinking behaviors and activities of base stations in mobile cellular networks based on big data analysis. IEEE Trans. Netw. Sci. Eng. **1**(1), 1–12 (2018)
12. Chen, L., et al.: A lightweight end-side user experience data collection system for quality evaluation of multimedia communications. IEEE Access **6**(1), 15408–15419 (2018)
13. Jiang, D., Wang, W., Shi, L., et al.: A compressive sensing-based approach to end-to-end network traffic reconstruction. IEEE Trans. Netw. Sci. Eng. **5**(3), 1–2 (2018)
14. Huo, L., Jiang, D., Zhu, X., et al.: An SDN-based fine-grained measurement and modeling approach to vehicular communication network traffic. Int. J. Commun. Syst. 1–12 (2019)
15. Huo, L., Jiang, D., Qi, S., Song, H., Miao, L.: An AI-based adaptive cognitive modeling and measurement method of network traffic for EIS. Mob. Netw. Appl. (2019). https://doi.org/10.1007/s11036-019-01419-z
16. Chen, L., Zhang, L.: Spectral efficiency analysis for massive MIMO system under QoS constraint: an effective capacity perspective. Mob. Netw. Appl. (2020). https://doi.org/10.1007/s11036-019-01414-4
17. Guo, C., Liang, L., Li, G.Y.: Resource allocation for low-latency vehicular communications: an effective capacity perspective. IEEE J. Sel. Areas Commun. **37**(4), 905–917 (2019)
18. Shehab, M., Alves, H., Latva-aho, M.: Effective capacity and power allocation for machine-type communication. IEEE Trans. Veh. Technol. **68**(4), 4098–4102 (2019)
19. Wang, F., Jiang, D., Qi, S., Qiao, C., Shi, L.: A dynamic resource scheduling scheme in edge computing satellite networks. Mob. Netw. Appl. (2020). https://doi.org/10.1007/s11036-019-01421-5
20. Jiang, D., Huo, L., Lv, Z., et al.: A joint multi-criteria utility-based network selection approach for vehicle-to-infrastructure networking. IEEE Trans. Intell. Transp. Syst. **19**(10), 3305–3319 (2018)
21. You, L., Xiong, J., Zappone, A., Wang, W., Gao, X.: Spectral efficiency and energy efficiency tradeoff in massive MIMO downlink transmission with statistical CSIT. IEEE Trans. Signal Process. **68**, 2645–2659 (2020)
22. Ji, H., Sun, C., Shieh, W.: Spectral efficiency comparison between analog and digital RoF for mobile fronthaul transmission link. J. Lightwave Technol. (2020)
23. Hayati, M., Kalbkhani, H., Shayesteh, M.G.: Relay selection for spectral-efficient network-coded multi-source D2D communications. In: 2019 27th Iranian Conference on Electrical Engineering (ICEE), Yazd, Iran, pp. 1377–1381 (2019)
24. Jiang, D., Zhang, P., Lv, Z., et al.: Energy-efficient multi-constraint routing algorithm with load balancing for smart city applications. IEEE Internet Things J. **3**(6), 1437–1447 (2016)
25. Jiang, D., Li, W., Lv, H.: An energy-efficient cooperative multicast routing in multi-hop wireless networks for smart medical applications. Neurocomputing **220**(2017), 160–169 (2017)
26. Jiang, D., Wang, Y., Lv, Z., et al.: Intelligent optimization-based reliable energy-efficient networking in cloud services for IIoT networks. IEEE J. Sel. Areas Commun. (2019)

27. Barakabitze, A.A., et al.: QoE management of multimedia streaming services in future networks: a tutorial and survey. IEEE Commun. Surv. Tutor. **22**(1), 526–565 (2020)
28. Orsolic, I., Skorin-Kapov, L.: A framework for in-network QoE monitoring of encrypted video streaming. IEEE Access **8**, 74691–74706 (2020)
29. Song, E., et al.: Threshold-oblivious on-line web QoE assessment using neural network-based regression model. IET Commun. **14**(12), 2018–2026 (2020)
30. Seufert, M., Wassermann, S., Casas, P.: Considering user behavior in the quality of experience cycle: towards proactive QoE-aware traffic management. IEEE Commun. Lett. **23**(7), 1145–1148 (2019)
31. Jiang, D., Huo, L., Li, Y.: Fine-granularity inference and estimations to network traffic for SDN. PLoS ONE **13**(5), 1–23 (2018)
32. Wang, Y., Jiang, D., Huo, L., Zhao, Y.: A new traffic prediction algorithm to software defined networking. Mob. Netw. Appl. (2019). https://doi.org/10.1007/s11036-019-01423-3
33. Qi, S., Jiang, D., Huo, L.: A prediction approach to end-to-end traffic in space information networks. Mob. Netw. Appl. (2019). https://doi.org/10.1007/s11036-019-01424-2
34. Nakagawa, T., et al.: A human machine interface framework for autonomous vehicle control. In: 2017 IEEE 6th Global Conference on Consumer Electronics (GCCE), Nagoya, pp. 1–3 (2017)
35. Verma, B., et al.: Framework for dynamic hand gesture recognition using Grassmann manifold for intelligent vehicles. IET Intel. Transport Syst. **12**(7), 721–729 (2018)
36. Emara, M., Filippou, M.C., Sabella, D.: MEC-assisted end-to-end latency evaluations for C-V2X communications. In: 2018 European Conference on Networks and Communications (EuCNC), Ljubljana, Slovenia, pp. 1–9 (2018)
37. Bazzi, A., Cecchini, G., Zanella, A., Masini, B.M.: Study of the impact of PHY and MAC parameters in 3GPP C-V2V mode 4. IEEE Access **6**, 71685–71698 (2018)
38. Wang, Y., Wang, C., Shi, C., Xiao, B.: A selection criterion for the optimal resolution of ground-based remote sensing cloud images for cloud classification. IEEE Trans. Geosci. Remote Sens. **57**(3), 1358–1367 (2019)
39. Tsokalo, I.A., Wu, H., Nguyen, G.T., Salah, H., Fitzek, F.H.P.: Mobile edge cloud for robot control services in industry automation. In: 2019 16th IEEE Annual Consumer Communications & Networking Conference (CCNC), Las Vegas, NV, USA, pp. 1–2 (2019)
40. Xiong, G., Shen, D., Dong, X., Hu, B., Fan, D., Zhu, F.: Parallel transportation management and control system for subways. IEEE Trans. Intell. Transp. Syst. **18**(7), 1974–1979 (2017)
41. Qi, Q., Tao, F.: Digital twin and big data towards smart manufacturing and industry 4.0: 360 degree comparison. IEEE Access **6**, 3585–3593 (2018)
42. Tao, F., Zhang, H., Liu, A., Nee, A.Y.C.: Digital twin in industry: state-of-the-art. IEEE Trans. Industr. Inf. **15**(4), 2405–2415 (2019)
43. Laaki, H., Miche, Y., Tammi, K.: Prototyping a digital twin for real time remote control over mobile networks: application of remote surgery. IEEE Access **7**, 20325–20336 (2019)

How to Deploy Timer for Retransmission

Jiamin Cheng, Lei Chen(✉), Ping Cui, Yuan An, and Kailiang Zhang

Jiangsu Province Key Laboratory of Intelligent Industry Control Technology, Xuzhou University of Technology, Xuzhou 221018, China
chenlei@xzit.edu.cn

Abstract. In some emerging mobile applications, the UDP has advantage for service latency. However, retransmission mechanism is essential to deal with the losing packets while the service is based on UDP protocol. The timer for retransmission could be deployed on sender side or receiver side. To compare these two approaches, we investigate the relationship between service latency and losing packet rate in simulation. The results show that deploying the timer on receiver side obtains higher performance than the traditional method does within unstable channel environment. A new retransmission algorithm is proposed to emerging mobile applications.

Keywords: Retransmission · Timer · Mobile computing · IOT · UDP · TCP

1 Introduction

A large number of mobile applications are emerging as 5G networks are being deployed. To satisfy the traffic demand, some new approaches are proposed to improve network routing and measurement [1–3]. Based on user behavior [4–6] and traffic flow analysis methods [7, 8], we can improve the transmission process. Some new scheduling strategies are proposed to improve user scheduling [9–12], to raise spectral resources utilization [13–16], save energy [17–19], guarantee quality of service [20–23]. Based on end side network traffic analysis [24–26], we can manage and improve the end user experience [27–30]. Through traffic reconstruction [31–33], we can characterize the traffic flow and the transport protocol can be redesigned [34–36]. However, the high latency of edge wireless networks deteriorates the user experience of mobile applications. In some cloud based services, especially vehicular service, the TCP protocol causes extra delay [37].

For reliable transportation, the timer is very important [38, 39]. In the network protocol layer, the commonly used transport layer protocols are transmission control protocol (TCP) and user datagram protocol (UDP). TCP provides the high reliability. UDP takes lower delay. TCP is a connection-oriented transport layer protocol, which requires the establishment of a connection by "three handshakes" before carrying out transmission work. It is also a kind of packet loss detection and message order guarantee. It can guarantee the correctness and correct order of the data, but the relative transmission speed is slow. UDP is a kind of connection-oriented transport layer protocol, that is, the sending and receiving parties can communicate without establishing a connection, and

H. Song and D. Jiang (Eds.): SIMUtools 2020, LNICST 370, pp. 256–265, 2021.
https://doi.org/10.1007/978-3-030-72795-6_20

provides transaction-oriented simple and unreliable information delivery services. The transmission speed of UDP is fast and the system overhead is small.

To satisfy the high reliability and low latency demand, we need to combine transmission and UDP protocols. TCP is a connection-oriented service, resulting in the disadvantages of slow transmission rate, low efficiency, and high overhead and easy to attack. UDP is a simple datagram-oriented, unconnected transport layer protocol. Without TCP's handshake, confirmation, Windows, retransmission, congestion control and other mechanisms, user data transmission rate has been improved to some extent. UDP is more efficient than TCP and is suitable for communications or broadcast communications that require high speed transmission and real-time. The internet of vehicles requires the transmission of a large amount of voice information and road information. The speed of data feedback is required to be high. If the speed is relatively slow, it cannot guarantee the access to real-time road information, resulting in inaccurate information acquisition and user experience will decline. A lot of research shows that the UDP protocol is superior to TCP protocol in networks performance. Since the unstable channel of mobile communication, however, UDP needs an additional retransmission mechanism.

As a common approach, some applications adopt UDP in network layer, and take retransmission mechanism in application layer. In this paper, we reconsider the retransmission mechanism. We compare the approach of deploying retransmission timer on sender side and the approach of deploying retransmission timer on receiver side. In [40], a timer is deployed in client to accelerate the retransmission. We further let the receiver totally trigger the retransmission.

This paper is divided into four sections. In the second section, we describe the two approaches and give the algorithms. In the third section, we show and analyze the simulation results. We conclude in the fourth section.

2 Alternative Retransmission Mechanism and Algorithm

2.1 Deploy Timer on Sender Side

Like TPC, we can set timer on sender side. The retransmission mechanism is illustrated in Fig. 1.

1) The sender sends packets to receiver, and start timer.
2) If receiver receives packets, it will send acknowledgement packet to sender to confirm the received packets.
3) When timer is timeout, the sender will check ACK message and resend unconfirmed packets. The timer is restarted. The loss of ACK message and data packets will cause resending.
4) Loop the process until all packets are confirmed by ACK message.

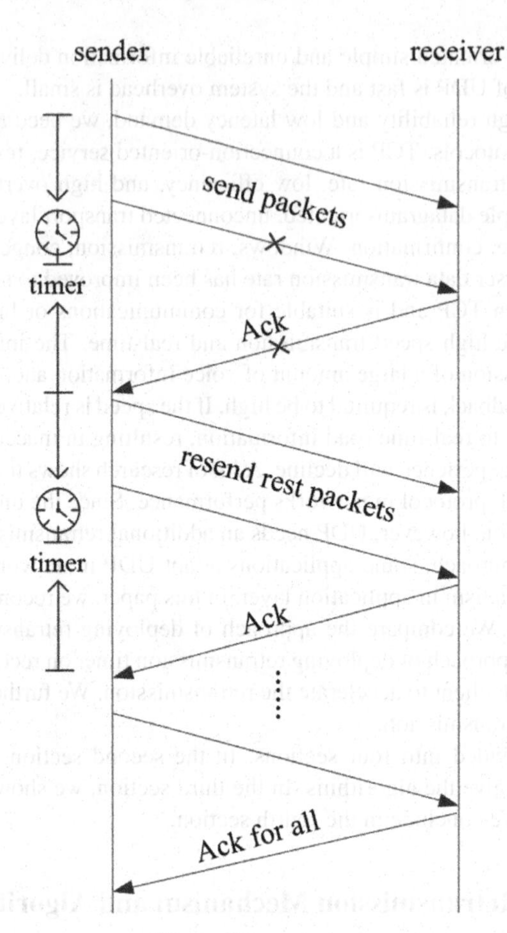

Fig. 1. Timer on sender.

2.2 Deploy Timer on Receiver Side

Unlike TPC, we deploy timer in receiver. The sender firstly tells receiver the number of packets that will be transmitted. The receiver requests for these packet and gives the sequence number it has not received. The timer is deployed on receiver side in this approach. If timeout, the receiver will send a new request message, namely NACK. The retransmission mechanism is illustrated in Fig. 2.

1) The receiver sends request message to sender, and start timer.
2) If receiver receives packets, it will record the sequence numbers of the packets.
3) When timer is timeout, the receiver will check received packets. If it has not get all wanted packets, it will send a new request message to sender and restart the timer.
4) Loop the process until all packets are received by receiver.

In this model, the request message can also be called NACK message which includes packet numbers the receiver want and has not received. The second model can reduce duplicate packets caused by ACK message losing in crowed channel.

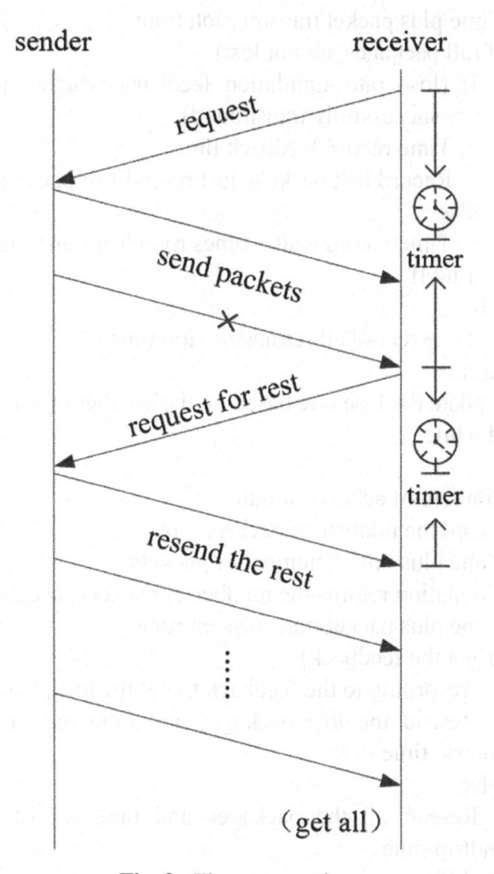

Fig. 2. Timer on receiver.

3 Simulation and Analysis

In the simulation, we arbitrarily set the transport period is 1 time slot, and set that the machine would cost 0.01 time slot to transmit one packet to channel. The timeout period is 3 round trip periods, namely 6 transport periods.

The simulation algorithm of ACK model is listed in Fig. 3.

```
Step1   While (NumPackets not null)
Step2     Record the number of packets sent
Step3       Time plus 0.01 * number of packets
Step4       simulation returns the number of packets successfully sent
Step5       Time plus packet transmission time
Step6       If (all packages are not lost)
Step7         If (lose rate simulation feedback whether the ACK was
                  successfully transmitted)
Step8           Time record feedback time
Step9           Record lost packets, just resend lost packets next time
Step10        else
Step11          Time record wait 3 times roundtrip time time
Step12        End If
Step13      Else
Step14        Time record all retransmission time
Step15      End If
Step16      Update the lose rate based on the number of users
Step17  End while
```

```
Step1   While (NumPackets not null)
Step2     Record the number of packets sent
Step3       Time plus 0.01 * number of packets
Step4       simulation returns the number of packets successfully sent
Step5       Time plus packet transmission time
Step6       If(get the feedback)
Step7         According to the feedback to get the lost packages
Step8         Resend the lost packages and time record wait 3 times
        roundtrip time
Step9       Else
Step10        Resend all the packages and time record wait 3 times
        roundtrip time
Step11      End If
Step12      Update the lose rate based on the number of users
Step13  End while
```

Fig. 3. Timer on sender.

The simulation algorithm of NACK model is listed in Fig. 4.

```
Step1     Do
Step2        Record the number of packets sent
Step3        Time plus 0.01 * number of packets
Step4        Record the number of lost packets
Step5        If (Packets are missing)
Step6           Record the Time to wait for feedback packet
Step7           While (feedback NACK is not received)
Step8                Record resend NACK time
Step9           End while
Step10          Record the wait time after the last successful NACK
Step11       End If
Step12       Update the lose rate based on the number of users
Step13    End while(all packets are received)
```

```
Step1     Do
Step2        Record the number of packets sent
Step3        Time plus 0.01 * number of packets
Step4        Record the number of lost packets
Step5        If (Packets are missing)
Step6           Record the Time to wait for feedback packet
Step7           While (feedback NACK is not received)
Step8                Resend the NACK and time 3 times roundtrip time plus
          sending NACK's time
Step9           End while
Step10          Record the wait time after the last successful NACK
Step11       End If
Step12       Update the lose rate based on the number of users
Step13    End while(all packets are received)
```

Fig. 4. Timer on receiver.

The simulation results are showed in Fig. 5 and Fig. 6.

Because the lose rate would increase in crowed channel, the NACK model that reduces non-effective duplicate retransmission obtain more stable performance.

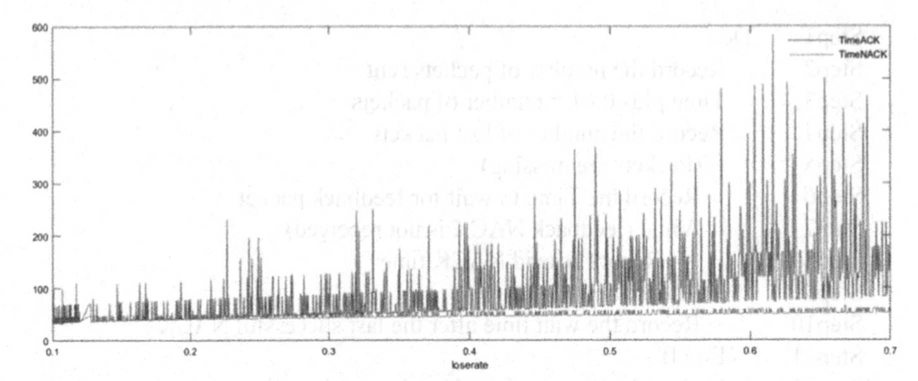

Fig. 5. Cost time for send 100 packets. (Horizontal axis is initial lose rate, the lose rate would update in the process; vertical axis is cost time).

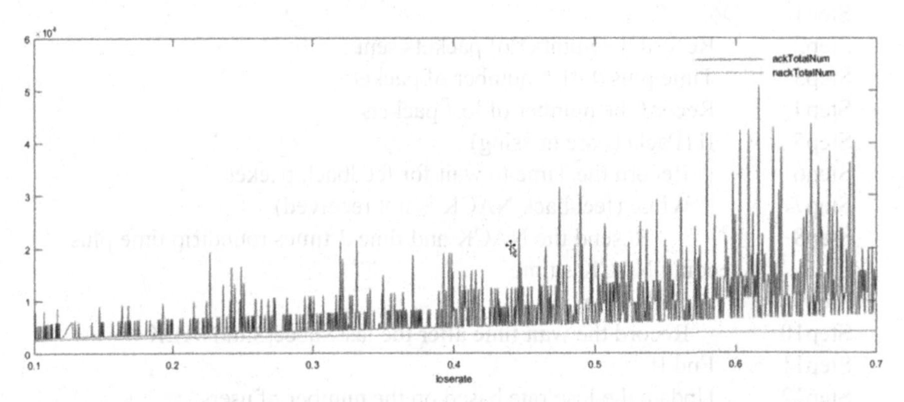

Fig. 6. Amount of sending for send 100 packets. (Horizontal axis is initial lose rate, the lose rate would update in the process; vertical axis is cost time).

4 Conclusion

In this paper, we combine the design principles and advantages of TCP and UDP protocols to design the UDP based retransmission mechanism. The timer is deployed in receiver; the retransmission process is triggered by receiver. We compare two retransmission models. The first approach sets timer on sender side like TCP, and the second one sets timer on receiver side. The simulation results show that the second model could reduce duplicate retransmission and obtain a stable performance.

Acknowledgements. This work is partly supported by Jiangsu technology project of Housing and Urban-Rural Development (No.2018ZD265) and Jiangsu major natural science research project of College and University (No. 19KJA470002).

References

1. Zhang, K., Chen, L., An, Y., Cui, P.: A QoE test system for vehicular voice cloud services. Mob. Netw. Appl. (2019). https://doi.org/10.1007/s11036-019-01415-3
2. Wang, F., Jiang, D., Qi, S.: An adaptive routing algorithm for integrated information networks. China Commun. **7**(1), 196–207 (2019)
3. Huo, L., Jiang, D., Lv, Z., Singh, S.: An intelligent optimization-based traffic information acquirement approach to software-defined networking. Comput. Intell. **36**(1), 151–171 (2019)
4. Chen, L., Jiang, D., Bao, R., Xiong, J., Liu, F., Bei, L.: MIMO scheduling effectiveness analysis for bursty data service from view of QoE. Chin. J. Electron. **26**(5), 1079–1085 (2017)
5. Jiang, D., Wang, Y., Lv, Z., et al.: Big data analysis-based network behavior insight of cellular networks for industry 4.0 applications. IEEE Trans. Ind. Inform. **16**(2), 1310–1320 (2020)
6. Bao, R., Chen, L., Cui, P.: User behavior and user experience analysis for social network services. Wireless Netw. (2019). https://doi.org/10.1007/s11276-019-02233-x,(
7. Jiang, D., Huo, L., Song, H.: Rethinking behaviors and activities of base stations in mobile cellular networks based on big data analysis. IEEE Trans. Netw. Sci. Eng. **1**(1), 1–12 (2018)
8. Chen, L., et al.: A lightweight end-side user experience data collection system for quality evaluation of multimedia communications. IEEE Access **6**(1), 15408–15419 (2018)
9. Chen, L., Zhang, L.: Spectral efficiency analysis for massive MIMO system under QoS constraint: an effective capacity perspective. Mob. Netw. Appl. (2020). https://doi.org/10.1007/s11036-019-01414-4
10. Wang, F., Jiang, D., Qi, S., Qiao, C., Shi, L.: A dynamic resource scheduling scheme in edge computing satellite networks. Mob. Netw. Appl. (2020). https://doi.org/10.1007/s11036-019-01421-5
11. Jiang, D., Huo, L., Lv, Z., et al.: A joint multi-criteria utility-based network selection approach for vehicle-to-infrastructure networking. IEEE Trans. Intell. Transp. Syst. **19**(10), 3305–3319 (2018)
12. Jiang, D., Zhang, P., Lv, Z., et al.: Energy-efficient multi-constraint routing algorithm with load balancing for smart city applications. IEEE Internet Things J. **3**(6), 1437–1447 (2016)
13. Lee, Y., Kim, Y., Park, S.: A machine learning approach that meets axiomatic properties in probabilistic analysis of LTE spectral efficiency. In: 2019 International Conference on Information and Communication Technology Convergence (ICTC), Jeju Island, Korea (South), pp. 1451–1453 (2019)
14. Ji, H., Sun, C., Shieh, W.: Spectral efficiency comparison between analog and digital RoF for mobile fronthaul transmission link. J. Lightwave Technol. **38**(20), 5617–5623 (2020)
15. Hayati, M., Kalbkhani, H., Shayesteh, M.G.: Relay selection for spectral-efficient network-coded multi-source D2D communications. In: 2019 27th Iranian Conference on Electrical Engineering (ICEE), Yazd, Iran, pp. 1377–1381 (2019)
16. You, L., Xiong, J., Zappone, A., Wang, W., Gao, X.: Spectral efficiency and energy efficiency tradeoff in massive MIMO downlink transmission with statistical CSIT. IEEE Trans. Signal Process. **68**, 2645–2659 (2020)
17. Jiang, D., Li, W., Lv, H.: An energy-efficient cooperative multicast routing in multi-hop wireless networks for smart medical applications. Neurocomputing **220**(2017), 160–169 (2017)
18. Jiang, D., Wang, Y., Lv, Z., et al.: Intelligent optimization-based reliable energy-efficient networking in cloud services for IIoT networks. IEEE J. Sel. Areas Commun. **38**(5), 928–941 (2019)
19. Wiatr, P., Chen, J., Monti, P., Wosinska, L.: Energy efficiency versus reliability performance in optical backbone networks [invited]. IEEE/OSA J. Opt. Commun. Netw. **7**(3), A482–A491 (2015)

20. Tan, J., Xiao, S., Han, S., Liang, Y., Leung, V.C.M.: QoS-aware user association and resource allocation in LAA-LTE/WiFi coexistence systems. IEEE Trans. Wireless Commun. **18**(4), 2415–2430 (2019)
21. Wang, Y., Tang, X., Wang, T.: A unified QoS and security provisioning framework for wiretap cognitive radio networks: a statistical queueing analysis approach. IEEE Trans. Wireless Commun. **18**(3), 1548–1565 (2019)
22. Hassan, M.Z., Hossain, M.J., Cheng, J., Leung, V.C.M.: Hybrid RF/FSO backhaul networks with statistical-QoS-aware buffer-aided relaying. IEEE Trans. Wireless Commun. **19**(3), 1464–1483 (2020)
23. Zhang, Z., Wang, R., Yu, F.R., Fu, F., Yan, Q.: QoS Aware transcoding for live streaming in edge-clouds aided HetNets: an enhanced actor-critic approach. IEEE Trans. Veh. Technol. **68**(11), 11295–11308 (2019)
24. Jiang, D., Wang, W., Shi, L., et al.: A compressive sensing-based approach to end-to-end network traffic reconstruction. IEEE Trans. Netw. Sci. Eng. **5**(3), 1–2 (2018)
25. Qi, S., Jiang, D., Huo, L.: A prediction approach to end-to-end traffic in space information networks. Mob. Netw. Appl. (2019). https://doi.org/10.1007/s11036-019-01424-2
26. Jiang, D., Huo, L., Li, Y.: Fine-granularity inference and estimations to network traffic for SDN. PLoS ONE **13**(5), 1–23 (2018)
27. Barakabitze, A.A., et al.: QoE management of multimedia streaming services in future networks: a tutorial and survey. IEEE Commun. Surv. Tutor. **22**(1), 526–565 (2020)
28. Orsolic, I., Skorin-Kapov, L.: A framework for in-network QoE monitoring of encrypted video streaming. IEEE Access **8**, 74691–74706 (2020)
29. Song, E., et al.: Threshold-oblivious on-line web QoE assessment using neural network-based regression model. IET Commun. **14**(12), 2018–2026 (2020)
30. Seufert, M., Wassermann, S., Casas, P.: Considering user behavior in the quality of experience cycle: towards proactive QoE-aware traffic management. IEEE Commun. Lett. **23**(7), 1145–1148 (2019)
31. Wang, Y., Jiang, D., Huo, L., Zhao, Y.: A new traffic prediction algorithm to software defined networking. Mob. Netw. Appl. (2019). https://doi.org/10.1007/s11036-019-01423-3
32. Huo, L., Jiang, D., Qi, S., Song, H., Miao, L.: An AI-based adaptive cognitive modeling and measurement method of network traffic for EIS. Mob. Netw. Appl. (2019). https://doi.org/10.1007/s11036-019-01419-z
33. Huo, L., Jiang, D., Zhu, X., et al.: An SDN-based fine-grained measurement and modeling approach to vehicular communication network traffic. Int. J. Commun. Syst. 1–12 (2019)
34. Polese, M., Chiariotti, F., Bonetto, E., Rigotto, F., Zanella, A., Zorzi, M.: A survey on recent advances in transport layer protocols. IEEE Commun. Surv. Tutor. **21**(4), 3584–3608 (2019)
35. Morelli, A., Provosty, M., Fronteddu, R., Suri, N.: Performance evaluation of transport protocols in tactical network environments. In: MILCOM 2019 - 2019 IEEE Military Communications Conference (MILCOM), Norfolk, VA, USA, pp. 30–36 (2019)
36. Dwivedi, U., Agarwal, A., Roy, A., Chakraborty, A., Mukherjee, A.: A survey on wireless sensor network protocols for transport layer. In: 2018 Second International Conference on Electronics, Communication and Aerospace Technology (ICECA), Coimbatore, pp. 390–394 (2018)
37. Assefi, M., Wittie, M., Knight, A.: Impact of network performance on cloud speech recognition. In: International Conference on Computer Communication and Networks, pp. 1–6. IEEE (2015)
38. Gomez, C., Crowcroft, J.: Multimodal retransmission timer for LPWAN. IEEE Internet Things J. **7**(6), 4827–4838 (2020)

39. Albashear, A.M.M., Ali, H.A., Ali, A.M.: Detection of man-in-the-middle attacks by using the TCP retransmission timeout: key compromise impersonation attack as study case. In: 2018 International Conference on Computer, Control, Electrical, and Electronics Engineering (ICCCEEE), Khartoum, pp. 1–8 (2018)
40. Kishimoto, S., Osada, S., Tarutani, Y., Fukushima, Y., Yokohira, T.: A TCP Incast avoidance method based on retransmission requests from a client. In: 2019 International Conference on Information and Communication Technology Convergence (ICTC), Jeju Island, Korea (South), pp. 153–158 (2019)

Benefit Optimization of SDN Honeynet System Based on Mimic Defense

Desheng Zhang and Lei Chen(✉)

Jiangsu Province Key Laboratory of Intelligent Industry Control Technology, Xuzhou University of Technology, Xuzhou 221018, China
chenlei@xzit.edu.cn

Abstract. The role of Internet technology applications in the overall economic and social development is becoming more and more obvious and the risks and challenges it brings are also increasing, and cyberspace threats and risks are increasing. In recent years, software-defined networks have provided new solutions for the field of cyberspace security with the characteristics of simplicity, rapid deployment and maintenance, flexible expansion, and openness. Mimic defense is based on the dynamic heterogeneous redundant structure of the endogenous security mechanism in cyberspace, and provides a brand new defense idea in the face of various threats. Combining SDN and mimic defense technology to form a more powerful intelligent honeynet has become a new research direction in network security. Based on the predecessors, this paper constructs a SDN-based mimic defense honeynet. Through theoretical calculations, the benefits of both offense and defense are affected by various data, and find the optimal solution in the mimic defense honeynet, and verify each This kind of data reasoning finally achieves the optimal benefit of the defense system.

Keywords: Mimic defense · Software defined network · Honeynet · Intrusion detection system

1 Introduction

In recent years, the large-scale development of new technology applications such as cloud computing [1, 2], artificial intelligence [3, 4], big data [5, 6], and the Internet of Things [7] in developing countries has been applied to many previous Some people have never had a preliminary experience, such as applications in smart cities [8, 9], smart medical [10, 11], Internet of Vehicles [12] and so on. These technologies require a more powerful basic network, especially many rely on the mobile Internet [13, 14]. With the rapid development of modern networks, cyberspace security is also facing new problems and challenges [15]. Many network attack methods have begun to merge, interchange and quickly development of. However, traditional network security defense technologies, such as firewall [16], intrusion detection [17, 18], etc., face many limitations in the face of evolving network attack technologies, among which existing defenses have been discovered by some Vulnerabilities, viruses, etc., require a certain accumulation of

H. Song and D. Jiang (Eds.): SIMUtools 2020, LNICST 370, pp. 266–277, 2021.
https://doi.org/10.1007/978-3-030-72795-6_21

technology. Most of the attacks used by the attackers are hidden and unknown threats and breakthroughs.

Honeynet [19] is an active defense [20] technology, which can deceive the attacker to turn into what he wants to attack. At the same time, after capturing the attack behavior, it can analyze the attacker's attack strategy and attack. Methods, targets, etc. These data can be repaired and expanded to enhance the defense capabilities of their own systems. Furthermore, digital forensics [21] can be used to outline the characteristics of the attacker's portrait, etc., which can be used as the basis for reverse investigations. However, traditional honeynet deployment is very complicated, costly, dynamic adjustment is complicated, and flow control is difficult. Faced with the modern network environment, neither can it be adjusted in time according to changes in network traffic, nor can it obtain effective information from the attacker.

SDN [22] is a technology developed in the face of increasingly large, complex, and diverse network environments. SDN realizes the separation of data control and data transmission, in order to achieve extremely excellent traffic analysis [23, 24] control Ability, a more flexible dynamic adjustment [25] capability, is just suitable for the deployment of honeynets, and it can also make certain predictions [26], and defend the network structure of honeynets according to the predicted results [27, 28]. Mimic defense [29, 30] is proposed by Academician Wu Jiangxing to solve the problem of inequality in defense and defense. Through a large number of overlapping dynamic scheduling, the attacker can become unknown, change, and the effect of both time and space. Mimic defense Can better conceal or camouflage the defensive scene and defensive behavior of the target object, so that the target object can obtain a more reliable advantage in the continuous, extremely concealed, high-intensity man-machine attack and defense game [31], especially In the face of the current biggest security threat-unknown breakthrough backdoors, virus Trojan horses and other infinite threats, it has significant effects and overcomes many problems of traditional security methods. The honeynet of mimic defense is realized through SDN, and end-to-end network control [32, 33] is realized, which realizes a honeynet with convenient deployment, flexible dynamic scheduling and powerful functions. A large part of the effect of the honeynet is that the IDS server can accurately identify attacks, and can accurately handle the geographical balance to deal with normal user access and abnormal operations. IDS servers can use different identification strategies when working. Here we analyze different Security strategy has a defensive impact on honeynets and how to choose the best strategy.

2 Problem Introduction

First, construct the basic honeypot network structure as shown in Fig. 1. Use multi-interaction honeypots to connect to each other through the network to simulate the simulation working network, which can be simulated as a We b server, file server, etc., when the attacker enters, it will be mistaken I thought I had successfully entered the target's work network. Before the target enters, it must pass through the firewall and IDS server. After IDS analysis, it is determined whether the target traffic goes to the real server or the honeypot server. For the attacker's traffic, network managers can analyze the attack process, attack vulnerabilities, tools used, and so on. In order to achieve the effect of making the system more secure.

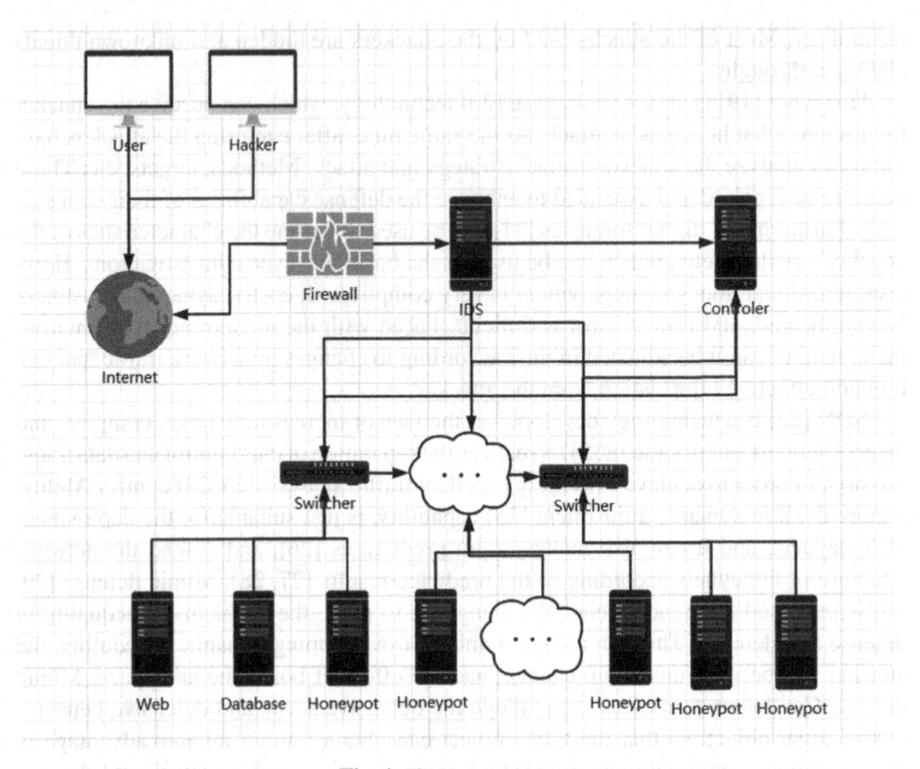

Fig. 1. System structure.

All switches in the honeynet are connected with the SDN controller, and exchange link status and flow table information with the controller through the Openflow protocol [34]. SDN controller can be programmed with its highly scalable, careful process flow for different attacks, regardless of the attacker what kind of attacks, the system will provide a highly realistic simulated network.

2.1 Theoretical Analysis of Mimic Honeynet

From previous studies on SDN mimic defense [35], it can be understood that the mimic defense honeynet can achieve the defense effect when q satisfies the following formula.

$$\begin{cases} p < 1 \\ \frac{\mu\alpha+\beta-\gamma}{\mu\alpha+\beta+\lambda} < q < \frac{1}{2} \end{cases} \tag{1}$$

Where the attack on behalf of the attacker's probability p, q represents the probability that the system uses honeypots, α indicates income, which $0 < \mu < 1$ indicates the degree of influence by the service itself, $\beta > 0$, indicates in addition to their other properties have been serving extent of damage degrees. Then the real honeypot services and services while providing service access side common with users normally access but do not attack the attacker can achieve this ideal combination of strategy Bayesian Nash equilibrium, only this time the game equilibrium condition with honeypot presence probability q is related but not related to the attacker's attack probability p, so it can be concluded that the mimic SDN virtual honeynet defense can achieve the purpose of active defense.

Because the user's revenue EVn meets the following formula:

$$\begin{aligned} EVn &= P(Sr|S1) \times \alpha + P(Sh|S1) \times (-\alpha) \\ &= (1-q) \times \alpha + q \times (-\alpha) \\ &= (1-2q)\alpha \end{aligned} \tag{2}$$

So from the user's point of view, q should be as small as possible.

But the system revenue EVa meets the following formula:

$$\begin{aligned} EVa &= P(Sr|S1) \times (\mu\alpha + \beta - \gamma) + P(Sh|S1) \times (-\lambda - \gamma) \\ &= (1-q) \times (\mu\alpha + \beta - \gamma) + q \times (-\lambda - \gamma) \\ &= \mu\alpha + \beta - \gamma - q(\mu\alpha + \beta + \lambda) \end{aligned} \tag{3}$$

From the system point of view, q should be as small as possible while λ should be as large as possible.

The greatest impact on the probability of q should be the false alarm rate of the IDS server system. IDS has two cases for false alarms of data:

$$\begin{aligned} IDSe &= \{En, Ea\} \\ &= \{\textit{Normal users falsely report as attackers}, \\ &\quad \textit{attacker was falsely reported as a normal user}\} \end{aligned} \tag{4}$$

The corresponding probabilities are respectively $\{Pen, Pea\}$. The relationship between the false alarm rate and q should be $q = (1-w) * Pen + w(1-Pea)$.Where w represents the proportion of malicious visits by attackers in all visits.

At this time, the attacker's income EVa becomes:

$$\begin{aligned} EVa &= P(Sr|S1) \times (\mu\alpha + \beta - \gamma) + P(Sh|S1) \times (-\lambda - \gamma) \\ &= Pea \times (\mu\alpha + \beta - \gamma) + (1 - Pea) \times (-\lambda - \gamma) \\ &= Peax(\mu\alpha + \beta + \lambda) - \lambda - \gamma \end{aligned} \tag{5}$$

The income EVn of ordinary users becomes:

$$\begin{aligned} EVn &= P(Sr|S1) \times \alpha + P(Sh|S1) \times (-\alpha) \\ &= (1 - Pen) \times \alpha + Pen \times (-\alpha) \\ &= (1 - 2Pen) \times \alpha \end{aligned} \tag{6}$$

The income ESr of the real server becomes:

$$\begin{aligned} ESr &= P(Vn|V1) \times \alpha + P(Va|V1) \times (-\mu\alpha - \beta) \\ &= (1-w) \times (1 - Pen) \times \alpha + w \times Pea \times (-\mu\alpha - \beta) \end{aligned} \tag{7}$$

The income of the honeypot becomes ESh as:

$$\begin{aligned} ESh =& P(Vn|V1) \times (-\alpha) + P(Va|V1) \times (\lambda) \\ =& (1 - w) \times Pen \times (-\alpha) + w \times (1 - Pea) \times (\lambda) \end{aligned} \tag{8}$$

For the entire server, the actual revenue ES becomes:

$$\begin{aligned} ES(S1, S2) =& ESr(S1) + ESh(S1) \\ =& (1 - w) \times (1 - Pen) \times \alpha + w \times Pea \\ & \times (-\mu\alpha - \beta) + (1 - w) \times Pen \times (-\alpha) \\ & + w \times (1 - Pea) \times (\lambda) \end{aligned} \tag{9}$$

Compared with the IDS server's false alarm probability, the user's single-access revenue α, once the server is set up, its business process and service objects are determined, α there will be basically no major changes, and the degree of impact on the service itself μ and other performance outside of its own service The extent of damage β depends largely on the attacker. Under the premise of a certain honeypot revenue, the judgment strategy of the IDS server greatly affects the entire system. Among them, the judgment performance of the IDS server, when the machine performance and detection technology cannot be improved, whether it is the Pen or Pea tends to zero, it will affect the other. The relationship between the two should be $Pen * Pea = k$, where k is a fixed constant, which represents the performance of the IDS server, that is, Pen and Pea are positively correlated. We can find the optimal strategy of the system in the following experiments.

3 Simulation Results and Analysis

In order to better practice before the projected offensive and defensive earnings results, we build a defense based on mimicry of SDN to the Honeynet testing, need to use Mininet [36] create multiple switches and host used to form SDN network, one of the hosts as a real server, you can use Python -m SimpleHTTPServer 80 & to open the command Mininet in http service. Then use VM ware to create a few new servers to install SDN controllers, firewalls, and IDS servers.

3.1 Basic Data Simulation

The intrusion detection system is divided into two modes according to the behavior of intrusion detection:

$$T = \{Ty, Tw\} = \{\text{Abnormal detection}, \ \text{Misuse detection}\} \tag{20}$$

The order is based on the income α of the user's single visit $\alpha = 1$. The proportion of attackers in all access locations w is 0.1%, that is, normal commercial servers face more ordinary users, and the degree of impact on the service itself $\mu = 0.1$ means that the attacker does not care about the service provided, and the degree of damage to performance $\beta = 100$ is The damage to the system is more serious. Honeypot revenue $\lambda = 10000$ is the ability to analyze attackers' attack methods, fix loopholes, digital forensics, etc. The revenue of honeypots needs to be far greater than the revenue of the attacker and the loss of real services.

3.2 Anomaly Detection Mode

In the anomaly detection mode, we must first establish a model of system access to normal behavior. Any visitor's behavior that does not conform to this model will be judged as an intrusion. Therefore, Pea can be very small and regarded as zero. The matching method is too strict except in Decreasing the abnormal rate of attackers being falsely reported as normal users, Pea, will inevitably increase the detection rate of normal users being falsely reported as attackers, Pen. Different Pen effects when pea is equal to 0 The real server revenue ESr is shown in Fig. 2. The honeypot server revenue ESh is shown in Fig. 3. The total income is shown in Fig. 4.

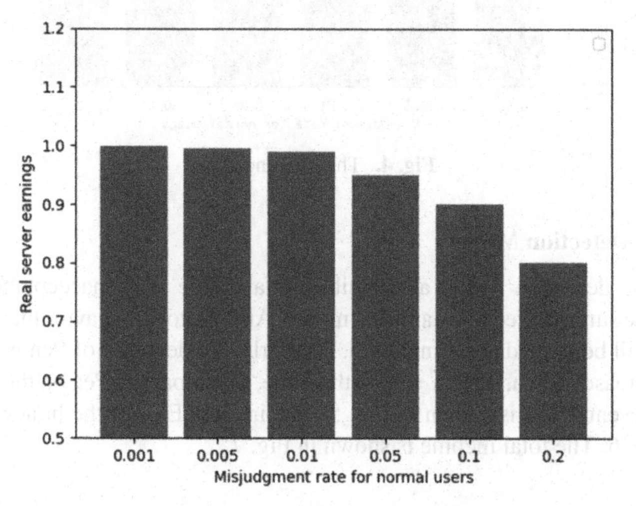

Fig. 2. The real server revenue

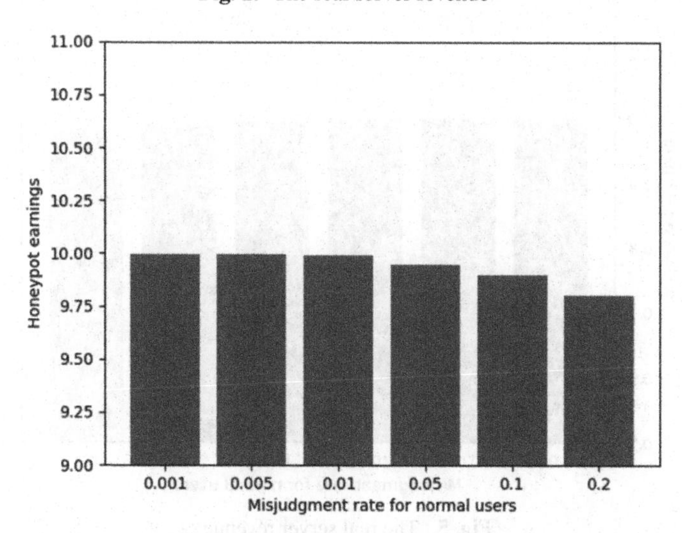

Fig. 3. The honeypot server revenue

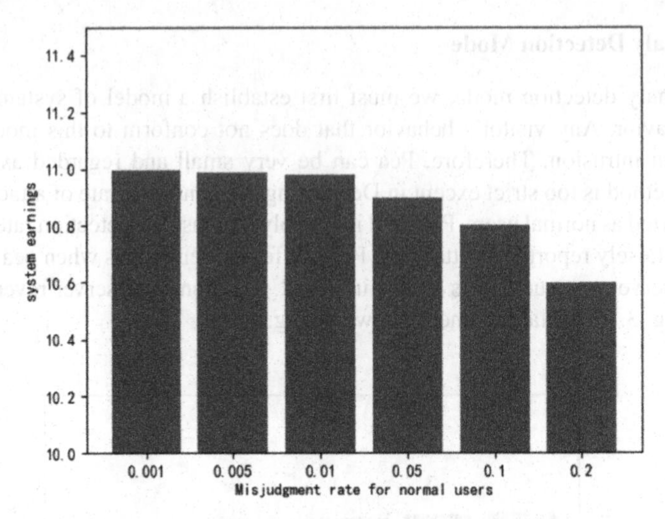

Fig. 4. The total income

3.3 Misuse Detection Mode

In the misuse detection mode, all possible unfavorable and unacceptable behaviors should first be summarized to establish a model. Any visitor's behavior that conforms to this model will be judged as an intrusion. Similarly, the decrease of Pen will also bring about the increase of Pea, so Pen = 0, at this time, the impact of Pea on the system. The real server revenue ESr is shown in Fig. 5. The income ESh of the honeypot server is shown in Fig. 6. The total income is shown in Fig. 7.

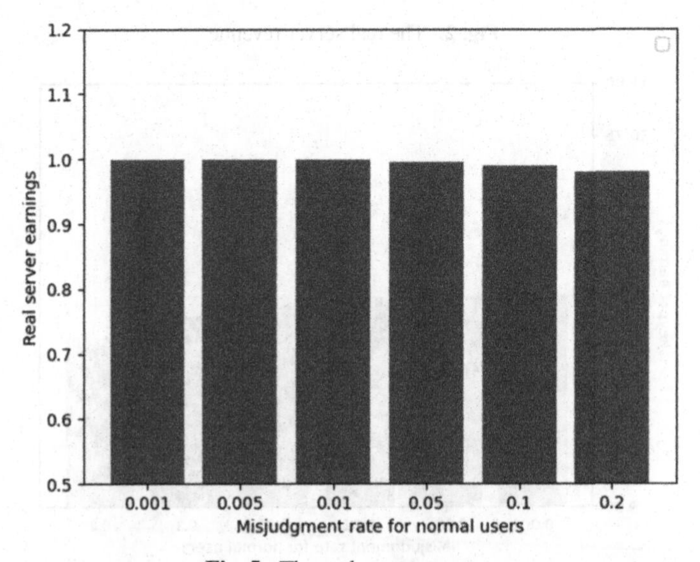

Fig. 5. The real server revenue

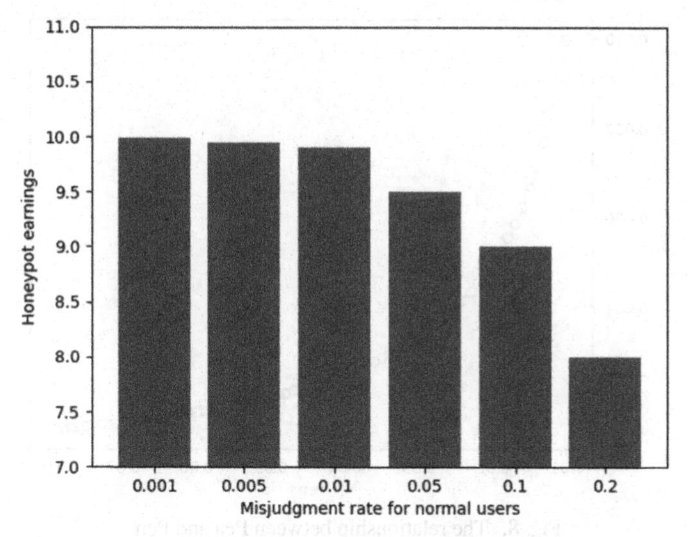

Fig. 6. The honeypot server revenue

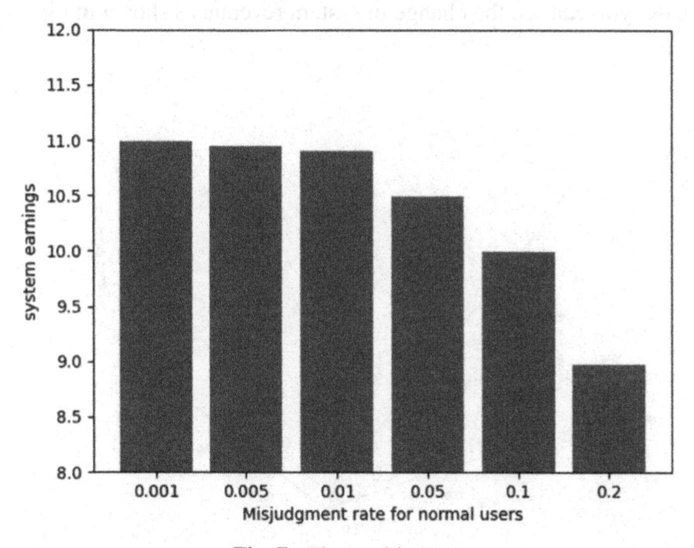

Fig. 7. The total income

3.4 System Best Profit

From the data results, the loss caused by the increase of Pea of the same magnitude will be higher than the loss caused by Pen. In the case that machine performance and detection technology cannot be improved, whether it is to turn Pen or Pea towards Zero will affect the other. In the case of $k = 0.001$, the mutual influence between the two is shown in Fig. 8.

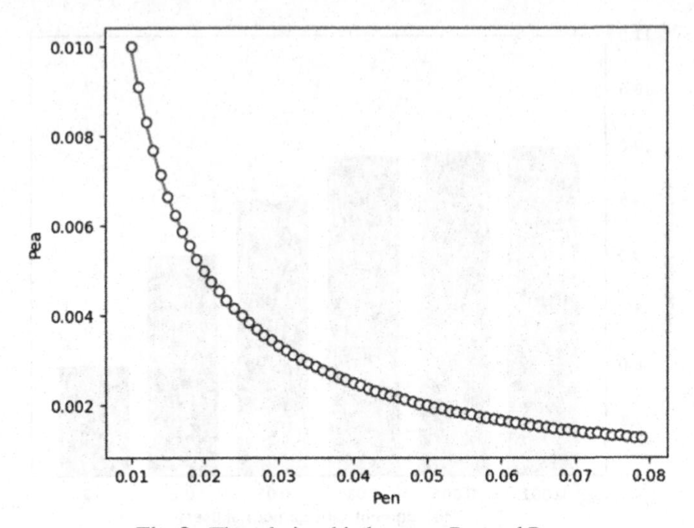

Fig. 8. The relationship between Pea and Pen

In this case, you can see the change in system revenue as shown in Fig. 9 below:

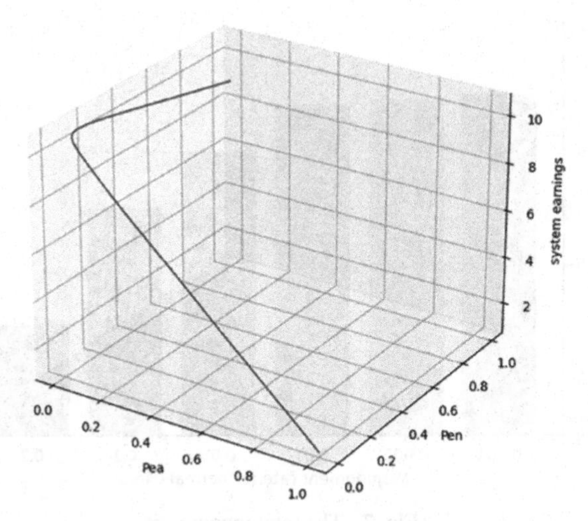

Fig. 9. Overall system benefits

We can be seen that the unilateral misjudgment rate reduction cannot be blindly pursued. Under the limit of the constant k, when $k = 0.001$, the overall profit of the system achieves the maximum boundary value $Pea = 0.014$ $Peh = 0.0714$ $E = 10.715$.

4 Conclusion

Through SDN-related technologies, the honeynet that realizes mimic defense has the characteristics of convenient deployment and convenient flow control compared with traditional honeynets. It can easily control multiple network devices and add and delete multiple services. This article also uses related data reasoning Calculated, demonstrated the relevant influence of various service strategies of IDS server on the final defense effect, and demonstrated through simulation experiments, how to obtain the best defense effect, give full play to the role of real servers and honeypot servers, deceive attackers, and collect The attacker's information and attack methods are ultimately used to analyze the characteristics of the network, and in turn repair and strengthen the entire defense system based on relevant information, forming a virtuous circle, and ultimately achieving the effect of active defense.

References

1. Bahrami, M.: Cloud computing for emerging mobile cloud apps. In: 2015 3rd IEEE International Conference on Mobile Cloud Computing, Services, and Engineering, San Francisco, CA, pp. 4–5 (2015)
2. Jiang, D., Wang, Y., Lv, Z., Wang, W., Wang, H.: An energy-efficient networking approach in cloud services for IIoT networks. IEEE J. Sel. Areas Commun. **38**(5), 928–941 (2020)
3. Arsenijevic, U., Jovic, M.: Artificial intelligence marketing: chatbots. In: 2019 International Conference on Artificial Intelligence: Applications and Innovations (IC-AIAI), Belgrade, Serbia, p. 193 (2019). https://doi.org/10.1109/IC-AIAI48757.2019.00010
4. Huo, L., Jiang, D., Qi, S., Song, H., Miao, L.: An AI-based adaptive cognitive modeling and measurement method of network traffic for EIS. Mob. Netw. Appl. (2019). https://doi.org/10.1007/s11036-019-01419-z
5. Mian, M., Teredesai, A., Hazel, D., Pokuri, S., Uppala, K.: Work in progress - in-memory analysis for healthcare big data. In: 2014 IEEE International Congress on Big Data, Anchorage, AK, pp. 778–779 (2014). https://doi.org/10.1109/BigData.Congress.2014.119
6. Jiang, D., Wang, Y., Lv, Z., Qi, S., Singh, S.: Big data analysis based network behavior insight of cellular networks for industry 4.0 applications. IEEE Trans. Ind. Inform. **16**(2), 1310–1320 (2020)
7. Singh, S., Singh, N.: Internet of things (IoT): security challenges, business opportunities & reference architecture for E-commerce. In: 2015 International Conference on Green Computing and Internet of Things (ICGCIoT), Noida, pp. 1577–1581 (2015). https://doi.org/10.1109/ICGCIoT.2015.7380718
8. Ceballos, G.R., Larios, V.M.: A model to promote citizen driven government in a smart city: use case at GDL smart city. In: 2016 IEEE International Smart Cities Conference (ISC2), Trento, pp. 1–6 (2016). https://doi.org/10.1109/ISC2.2016.7580873
9. Jiang, D., Zhang, P., Lv, Z., et al.: Energy-efficient multi-constraint routing algorithm with load balancing for smart city applications. IEEE Internet Things J. **3**(6), 1437–1447 (2016)
10. Lu, S., et al.: A study on service-oriented smart medical systems combined with key algorithms in the IoT environment. China Commun. **16**(9), 235–249 (2019). https://doi.org/10.23919/JCC.2019.09.018
11. Jiang, D., Li, W., Lv, H.: An energy-efficient cooperative multicast routing in multi-hop wireless networks for smart medical applications. Neurocomputing **2017**(220), 160–169 (2017)
12. Jiang, D., Huo, L., Lv, Z., Song, H., Qin, W.: A joint multi-criteria utility-based network selection approach for vehicle-to-infrastructure networking. IEEE Trans. Intell. Transp. Syst. **19**(10), 3305–3319 (2018)

13. Raghunandan, G.H., Chaithanya, G.H., Hajare, R.: Independent robust mesh for mobile adhoc networks. In: 2017 4th International Conference on Electronics and Communication Systems (ICECS), Coimbatore, pp. 125–128 (2017). https://doi.org/10.1109/ECS.2017.8067852
14. Jiang, D., Huo, L., Song, H.: Rethinking behaviors and activities of base stations in mobile cellular networks based on big data analysis. IEEE Trans. Netw. Sci. Eng. **7**(1), 80–90 (2020)
15. Zainudin, Z.S., Nuha Abdul Molok, N.: Advanced persistent threats awareness and readiness: a case study in Malaysian financial institutions. In: 2018 Cyber Resilience Conference (CRC), Putrajaya, Malaysia, pp. 1–3 (2018). https://doi.org/10.1109/CR.2018.8626835
16. SenthilKumar, P., Muthukumar, M.: A study on firewall system, scheduling and routing using pfsense scheme. In: 2018 International Conference on Intelligent Computing and Communication for Smart World (I2C2SW), Erode, India, pp. 14–17 (2018). https://doi.org/10.1109/I2C2SW45816.2018.8997167
17. Penya, Y.K., Bringas, P.G.: Experiences on designing an integral intrusion detection system. In: 2008 19th International Workshop on Database and Expert Systems Applications, Turin, pp. 675–679 (2008). https://doi.org/10.1109/DEXA.2008.54
18. Borkar, A., Donode, A., Kumari, A.: A survey on Intrusion detection system (IDS) and internal intrusion detection and protection system (IIDPS). In: 2017 International Conference on Inventive Computing and Informatics (ICICI), Coimbatore, pp. 949–953 (2017). https://doi.org/10.1109/ICICI.2017.8365277
19. Watson, D., Riden, J.: The honeynet project: data collection tools, infrastructure, archives and analysis. In: 2008 WOMBAT Workshop on Information Security Threats Data Collection and Sharing, Amsterdam, pp. 24–30 (2008). https://doi.org/10.1109/WISTDCS.2008.11
20. Zhang, H., Wang, J., Yu, D., Han, J., Li, T.: Active defense strategy selection based on static Bayesian game. In: Third International Conference on Cyberspace Technology (CCT 2015), Beijing, pp. 1–7 (2015). https://doi.org/10.1049/cp.2015.0806
21. Akremi, A., Sallay, H., Rouached, M., Sriti, M., Abid, M., Bouaziz, R.: Towards a built-in digital forensics-aware framework for web services. In: 2015 11th International Conference on Computational Intelligence and Security (CIS), Shenzhen, pp. 367–370 (2015). https://doi.org/10.1109/CIS.2015.95
22. Kim, D., Gil, J.-M., Wang, G., Kim, S.-H.: Integrated SDN and non-SDN network management approaches for future internet environment. In: Park, J.J.H., Ng, J.-Y., Jeong, H.Y., Waluyo, B. (eds.) Multimedia and Ubiquitous Engineering. LNEE, vol. 240, pp. 529–536. Springer, Dordrecht (2013). https://doi.org/10.1007/978-94-007-6738-6_64
23. Jiang, D., Huo, L., Li, Y.: Fine-granularity inference and estimations to network traffic for SDN. PLoS ONE **13**(5), 1–23 (2018)
24. Akyildiz, H.A., Hökelek, İ., Saygun, E., Çırpan, H.A.: Flow re-routing based traffic engineering for SDN networks. In: 2017 25th Signal Processing and Communications Applications Conference (SIU), Antalya, pp. 1–4 (2017). https://doi.org/10.1109/SIU.2017.7960321
25. Huo, L., Jiang, D., Lv, Z., Singh, S.: An intelligent optimization-based traffic information acquirement approach to software-defined networking. Comput. Intell. **36**(1), 151–171 (2019)
26. Wang, Y., Jiang, D., Huo, L., Zhao, Y.: A new traffic prediction algorithm to software defined networking. Mob. Netw. Appl. (2019). https://doi.org/10.1007/s11036-019-01423-3
27. Perepelkin, D., Tsyganov, I.: SDN cluster constructor: software toolkit for structures segmentation of software defined networks. In: 2019 XVI International Symposium Problems of Redundancy in Information and Control Systems (REDUNDANCY), Moscow, Russia, pp. 195–198 (2019). https://doi.org/10.1109/REDUNDANCY48165.2019.9003334
28. Wang, F., Jiang, D., Qi, S.: An adaptive routing algorithm for integrated information networks. China Commun. **7**(1), 196–207 (2019)
29. Jiangxing, W.: The original intent and prospect of mimic computing and mimic security defense. Telecommun. Sci. **30**(07), 2–7 (2014). (in Chinese)

30. Ma, B., Zhang, Z.: Security research of redundancy in mimic defense system. In: 2017 3rd IEEE International Conference on Computer and Communications (ICCC), Chengdu, pp. 2910–2914 (2017). https://doi.org/10.1109/CompComm.2017.8323064
31. Sun, Y., Liu, Z., Jiang, Z., Meng, X., Hu, W.: Conceptual model of situational awareness for advanced persistent threats. Inf. Secur. Res. **6**(06), 482–490 (2020). (in Chinese)
32. Jiang, D., Wang, W., Shi, L., Song, H.: A compressive sensing-based approach to end-to-end network traffic reconstruction. IEEE Trans. Netw. Sci. Eng. **7**(1), 507–519 (2020)
33. Qi, S., Jiang, D., Huo, L.: A prediction approach to end-to-end traffic in space information networks. Mob. Netw. Appl. (2019). https://doi.org/10.1007/s11036-019-01424-2
34. McKeown, N., Anderson, T., Balakrishnan, H., et al.: Openflow: enabling innovation in campus networks. ACM SIGCOMM Comput. Commun. Rev. **38**(2), 69–74 (2008)
35. Lian, Z., Yin, X., Xi, Q., Tan, R.: A SDN virtual honeynet based on mimic defense mechanism. Comput. Eng. Appl. **55**(01), 109–114 (2019)
36. Qureshi, S., Braun, R.: Mininet topology: mirror of the optical switch fabric. In: 2019 29th International Telecommunication Networks and Applications Conference (ITNAC), Auckland, New Zealand, pp. 1–6 (2019). https://doi.org/10.1109/ITNAC46935.2019.9078014

SDN Controller Scheduling Decision Mechanism Based on Mimic Defense Affects Safety

Desheng Zhang and Lei Chen(✉)

Jiangsu Province Key Laboratory of Intelligent Industry Control Technology, Xuzhou University of Technology, Xuzhou 221018, China
chenlei@xzit.edu.cn

Abstract. Nowadays, with complex network structures, diversified network equipment, and diversified network applications, more and more complex functions and requirements have led to traditional network management methods that are no longer competent for current networks. The emergence of SDN relies on the characteristics of separation of data forwarding and control, and has a huge advantage in network management in the new era, but SDN also faces security threats that have not been experienced by traditional networks. Since the control of the SDN network is centralized, the security of the SDN controller is particularly important. This article uses mimic defense to ensure the security of the controller, focusing on the security impact of different decisions of the decision maker in the mimic defense system, and optimizes the decision-making method, Found the decision-making method that can maintain the safe operation of the system for the longest time, and verified it through simulation experiments.

Keywords: Mimic defense · Software defined network · Controller

1 First Section

With the rapid development of computer technology, many emerging technologies have emerged, such as some big data [1, 2], Internet of Things [3, 4], industrial Internet [5, 6], cloud computing [7, 8] and other excellent Technology, these technologies have even changed our cities [9, 10], they also put forward higher requirements for the basic network system of the information highway, and through the development of mobile Internet [11, 12], network With the continuous expansion of scale, the variety of network equipment, and the application of new network technologies to more environments [13, 14], it is only a method of manual configuration by network practitioners to achieve each end-to-end stable link [15, 16] It is difficult to guarantee. Faced with these challenges, software-defined networking SDN [17, 18] ushered in rapid development.

SDN defines and controls the network in the form of software programming, and realizes the separation of its control plane and forwarding plane. SDN has mature applications in different places [19, 20]. Due to the characteristics of SDN, SDN can be combined with many technologies. Similar artificial intelligence technologies [21, 22], the

H. Song and D. Jiang (Eds.): SIMUtools 2020, LNICST 370, pp. 278–288, 2021.
https://doi.org/10.1007/978-3-030-72795-6_22

key to achieving many functions lies in the control of traffic [23, 24] With traffic status, routing can be better processed Forward [25, 26] to achieve traffic prediction [27, 28]. SDN is the trend of future network development and will bring qualitative changes to the entire Internet.

Mimic defense technology [29] is a dynamic discrete superposition [30] structure based on the endogenous security mechanism of cyberspace. The mimic defense technology integrates multiple active defense [31, 32] elements as a whole. The realization of mimic defense needs to be established the dynamic redundancy architecture realizes the dynamics, randomness, and diversity of the system while ensuring the replacement of system functions. Mimic technology can promote the improvement of the security of SDN controllers [33, 34]. The actual operation can be through multiple alternative controllers, while receiving the entire winding state and updating it in real time, but when there is a routing request that needs to be issued flow rules, It is necessary to go through the decider first, and according to a certain decision rule, send the theoretically safest flow rule to the network device. One more single controller is just the basis for mimic defense to improve the system's defense capabilities. The core of the real feasible and reasonable play of multiple alternate controllers lies in whether the strategy of the decider is reasonable.

2 Problem Introduction

The arbiter used in mimic defense has a variety of decision-making strategies, such as majority selection, rotation selection, random selection and so on. No matter what the strategy is adopted, to ensure the normal function of the premise, as far as possible to enhance the whole system security body, will be the same pursuit. We build a basic mimic defense structure at the SDN control layer. There are n heterogeneous controllers $\{C_1, C_1, C_2, C_3 \cdots C_n\}(n \geq 2)$, Run on k different operating systems $\{OS_1, OS_2, OS_3 \cdots OS_n\}$, all the controllers are connected to a same arbitrator A of 0. The arbiter N0 through Openflow [35] controls complex network 〖Net〗 _0 protocols below, There are m switches in the network N0 $\{S_1, S_2, S_3 \cdots S_n\}(m \geq 1)$, shown in Fig. 1.

In this case For Net_0 each switch S_m, it is logically linked with a single controller, meridians Openflow protocol controller normal exchange of link information, wherein mimicry defense portion of the link switch, either by The dedicated link can link the entire system to the Net_0 inside of the network. All n controllers and operating systems running on the n controllers may not necessarily be all different, but it is necessary to ensure that the combinations of controllers and operating systems are not the same. For example: the same controller C_a can be placed in the operating system, respectively $\{0S_a, 0S_b\}$, form two combinations $\{(0S_a, C_a), (0S_b, C_a)\}$, with corresponding operating systems can run different controller to form a new combination.

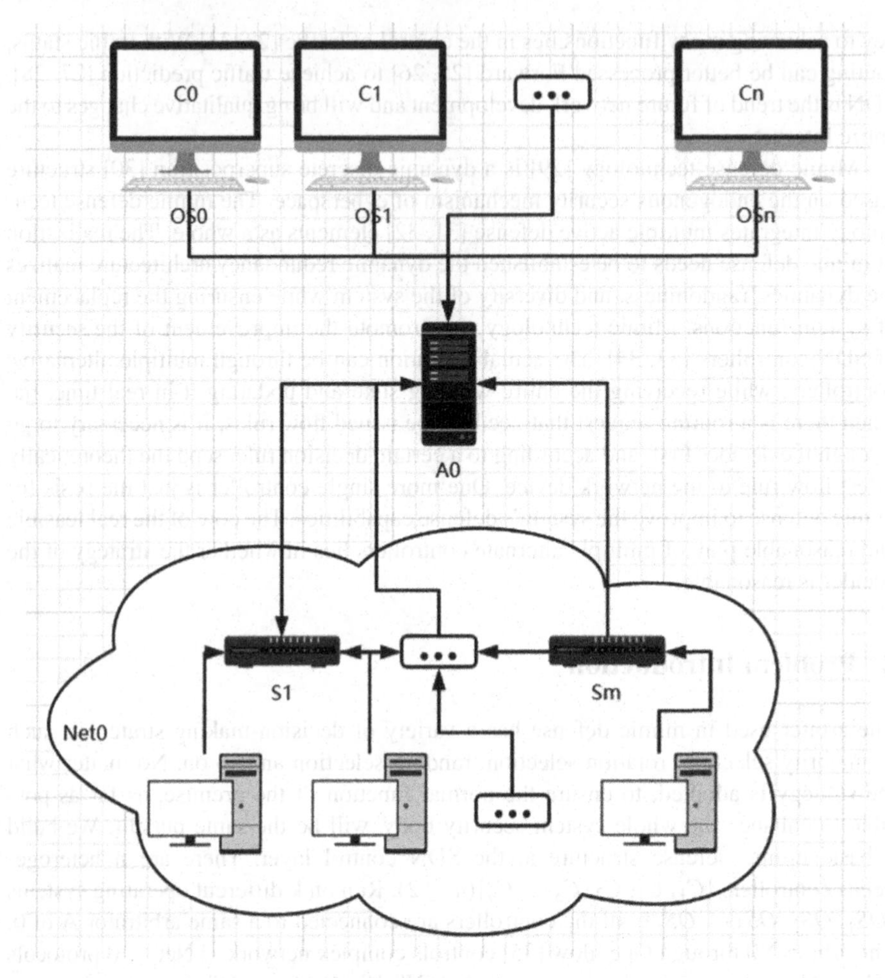

Fig. 1. Based defense mimic the structure.

2.1 Safety of a Single Controller

First of all, we must understand that the general security of the controller can be ensured by reasonable security configuration and timely security patches. However, when faced with advanced sustainable threats, the security of a single controller is difficult to ensure. During the attack, the probability P of a single controller being compromised will increase with the passage of time, and will eventually approach 1 infinitely. The probability of the controller being broken can be approximated to an exponential distribution, and the distribution function of the probability P of being broken over time T is as follows:

$$P = F(t) = \begin{cases} 1 - e^{-\lambda t}, t \geq 0 \\ 0, t < 0 \end{cases} \tag{1}$$

Among them, λ is affected by the product quality q and the attacker's ability x:

$$\lambda = K\frac{x}{q} \tag{2}$$

2.2 Security Policy Decision Maker

There are many selection methods that the selector can use, such as majority selection, alternate selection, random selection, and so on. Under the premise that a single controller must be breached within a certain period of time, no matter whether you choose to execute in turn or execute randomly, there must be a time set $T = \{T_1, T_2, T_3 \cdots\}$, and the same controller is running. Within these time periods, the entire system A pure SDN network is no different, and an attacker can continue to attack a single controller after it is taken. The entire system will be controlled by the attacker within time T. In this case, the attacker can use or destroy the entire link system through the compromised controller even in a short period of time. Single controller design break time T_{single}, the entire system is compromised time T_{system} is:

$$T_{system} = T_n(n = \min(\{n| \sum\nolimits_{i=1}^{i=n} len(T_i \geq T_{single}\})) \tag{3}$$

And a single controller is compromised as a probability P_{single}, the probability of the entire system is compromised P_{system} is:

$$P_{system} = P_{single} \tag{4}$$

We can be seen that this is not a particularly reliable choice.

At present, the most mainstream choice of mimic defense is the majority option, that is, each time the flow table is issued, the arbiter compares all the results, and finally selects the result of the majority controller. This allows a single controller to be compromised and does not affect the overall operation of the system. Even if an attacker takes down a single controller, it does not affect the overall operation of the system. If the attacker wants to control the entire system, in a total of n controllers under the premise, the attacker would need to take $k = \lceil \frac{n}{2} \rceil$ controllers. The time when the entire system was breached T_{system} is:

$$T_{system} = k \times T_{single} = \lceil \frac{n}{2} \rceil \times T_{single} \tag{5}$$

In most cases, the probability of the system being compromised P_{system} is:

$$P_{system} = \sum\nolimits_{k=\lceil \frac{n}{2} \rceil}^{n} C_n^k P_{single}{}^k (1 - P_{single})^{n-k} \tag{6}$$

Can be seen from the equation that, when the safety factor is high enough based controller, with the safety factor of the overall system increases the number of controllers will rise quickly when the controller itself is excellent enough mass, to be a plurality of states defense Redundancy will have a very limited improvement in the entire system. When the probability of the controller being compromised is higher than 0.5, the entire multiple redundant components will have a reverse effect and reduce the safety factor of the entire system.

2.3 Handling of Abnormal Controller

For most selected cases, when a single controller sends out wrong results multiple times, we can judge that this controller is abnormal and may have been controlled by an attacker. You can choose to save the scene for digital forensics for the controller first, and then restore it by mirroring. Time attacks and handling exceptions at different times will bring different security effects. Because some controllers will be shut down during operation, the safety factor of the entire system changes with time. In order to improve the overall safety of the system as much as possible, the abnormal controller is shut down as late as possible. The safety factor changes with time as shown in the figure below. The time to break a single controller is, the time to T_{single}, the time to close the i-th controller should not be less than $\lfloor \frac{n}{2}t \rfloor$. This can significantly increase the time the system lasts to defend against attacks. However, if it is just a simple mirror restoration, the time cost for an attacker to use the original vulnerability to compromise the controller again will be greatly reduced. The time cost of the first attack is t_0, the time cost of the second attack is t_1, and the time cost of the nth attack is t_n. The relationship between them should be:

$$t_n = \sqrt[m]{t_{n-1}} \tag{7}$$

Among them, m is a constant, which means that the cost of reattacking the controller after being attacked by a vulnerability will quickly drop to a constant level. Therefore, the number of restarts should be limited and at an appropriate time.

The cost of closing the controller directly or saving the image will be high. There is also a more cost-effective control method, that is, the selector is not a simple choice for the majority, and the controller can be weighted. Depending on the controller settings for different controller associated weights W, issued in all controllers m result set {R1, R2, R3···}, wherein the controller {nx, ny, nz···} The issued result is R1, and its corresponding weight is {wx, wy, wz···}. The weight of a single result, that is, W_i is:

$$W_i = \sum w_k,\ w_k \text{ is the weight of the controller } n_k \in \{\text{Output is } R_i\} \tag{8}$$

The final result R selected by the selector is:

$$R = R_{\max(w1,w2,w3\cdots wm)} \tag{9}$$

As a result, the controller updates the weight of the controller according to the result after each command:

$$\begin{cases} wi = wi + 1, n_i \ result = finall \ result \\ wi = \sqrt{wi}, n_i result \ \neq \ finall \ result \end{cases} \tag{10}$$

It is also possible to achieve results equivalent to shutdown.

3 Simulation Results and Points Analysis

3.1 Experimental Environment

In order to simply test the previous theories and solutions, we build a simple SDN test structure, try VMware to install multiple operating systems, such as Windows server,

Ubuntu, Debian, etc. Is not mounted in different virtual machines to make the same control device, such as RYU, Opendaylight, the Floodlight etc., using Mininet [36] analog SDN switches and Python write arbiter and process the data. The SDN network has no special requirements, as long as it is not too simple, the process of the controller being compromised can indicate the abnormality of the controller by actively asking the controller to send an error flow table.

3.2 Experimental Evaluation

Among the most choices, the probability of each controller being compromised by the attacker is p. The influence of probability p and the number of controllers on the entire system is shown in Fig. 2:

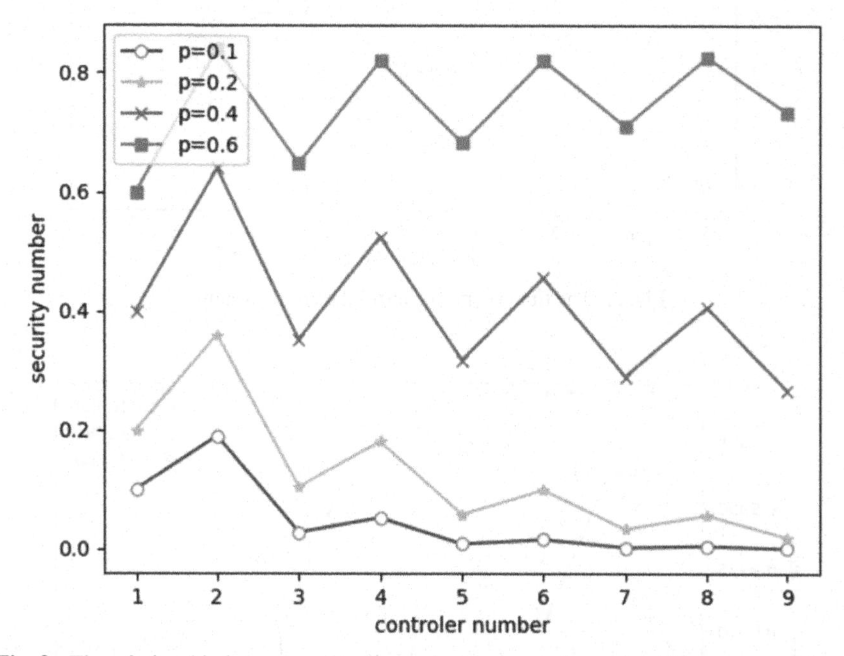

Fig. 2. The relationship between controller and system security under different p conditions.

It can be seen that when the basic safety factor of the controller is high enough, the safety factor of the entire system will increase rapidly as the number of controllers increases, and when the quality of the controller itself is not good enough, multiple redundancy of mimic defense The improvement of the entire system is very limited. When the probability of the controller being compromised is higher than 0.5, the entire multiple redundant components will have a reverse effect, reducing the safety factor of the entire system.

The average time it takes for the attacker to obtain the control of a single controller is T, and the time spent to control the entire system with the authority is shown in Fig. 3, which shows the non-processing of exceptions and the previous strategy. Shut down multiple controllers. At the same time, Fig. 4 shows the trend of the controller safety factor over time.

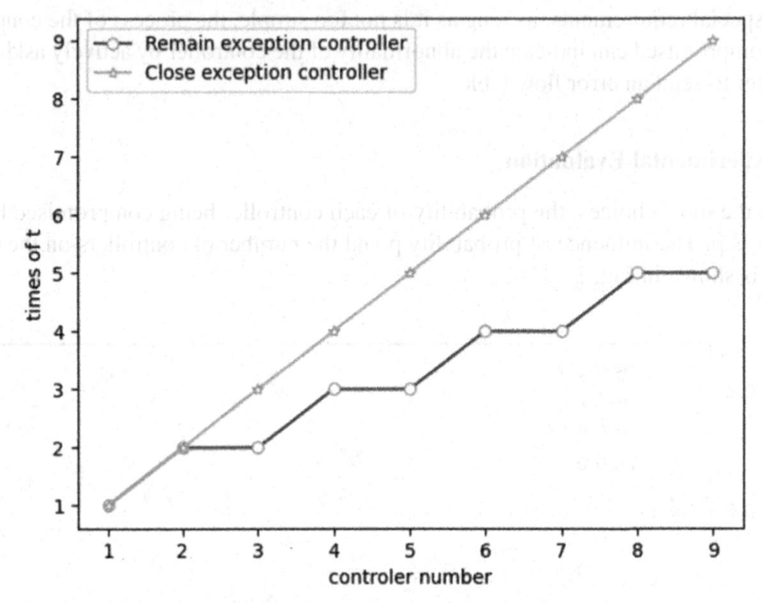

Fig. 3. The time spent to control the entire system.

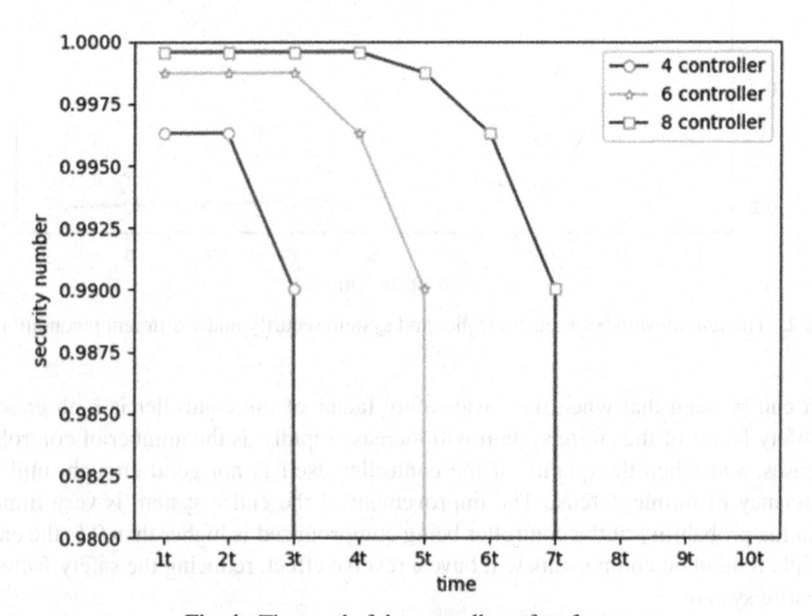

Fig. 4. The trend of the controller safety factor.

It can be seen that the method of gradually turning off the controller can not only effectively extend the time for the system to resist attacks, but also because it is turned off accordingly, it can also maintain the overall safety factor of the system as much as possible.

Regarding the weight processing ruling method, we send erroneous messages at regular intervals to indicate that the system is abnormal. At the initial weight of 5000, there are four controllers for mimic defense. After each abnormal message is sent three times, we can see the result as follows Fig. 5 weight changes. And the more intuitive weight ratio is shown in Fig. 6.

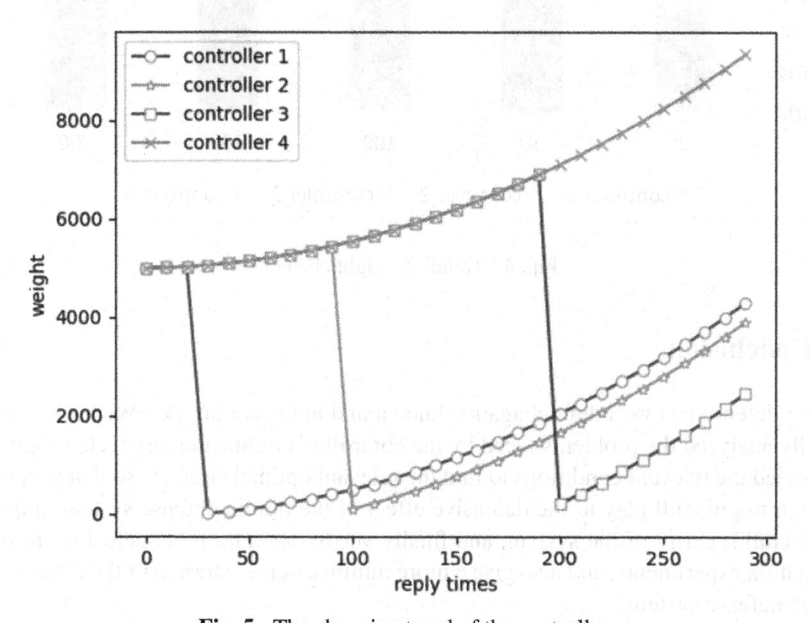

Fig. 5. The changing trend of the controller.

We can be seen that only after 30 times, 100 times and 200 times, the controller only performs three abnormal operations and its weight drops rapidly, which is equivalent to "temporarily shutting down" the controller.

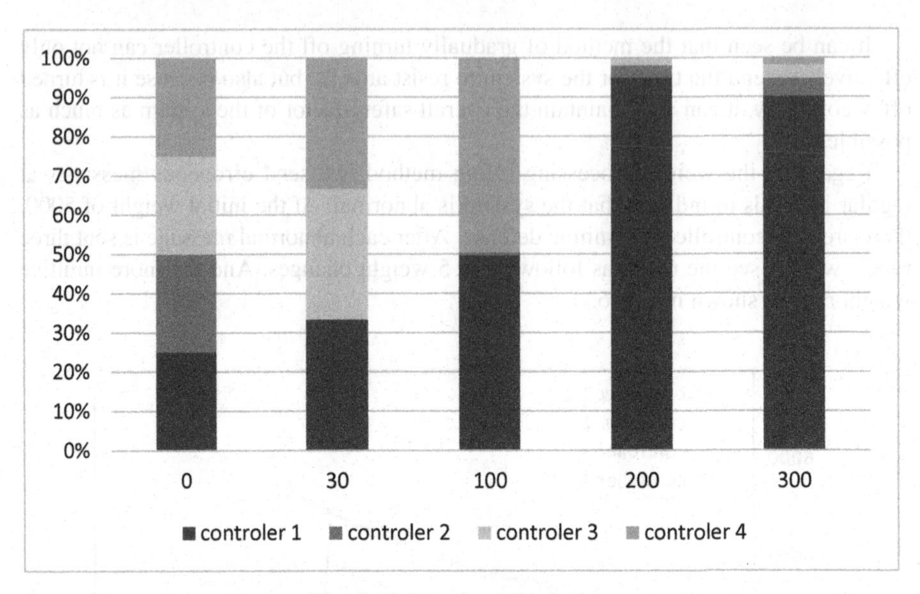

Fig. 6. Trend of weight change.

4 Conclusion

Mimic defense can well defend against known and unknown attacks. We have systematically analyzed the problems faced by the controller's mimic defense technology, and quantified the relevant conditions to find the relevant optimal strategy, so that we are all Enough to give full play to the defensive effect of the mimic defense system, improve the overall security of the system, and finally verify our relevant conclusions through simulation experiments, and also give a more intuitive demonstration of the effect of the mimic defense system.

References

1. Malhotra, S., Doja, M.N., Alam, B., Alam, M.: Bigdata analysis and comparison of bigdata analytic approches. In: 2017 International Conference on Computing, Communication and Automation (ICCCA), Greater Noida, pp. 309–314 (2017). https://doi.org/10.1109/CCAA.2017.8229821
2. Jiang, D., Wang, Y., Lv, Z., Qi, S., Singh, S.: Big data analysis based network behavior insight of cellular networks for industry 4.0 applications. IEEE Trans. Ind. Inf. **16**(2), 1310–1320 (2020)
3. Gupta, A.K., Johari, R.: IOT based electrical device surveillance and control system. In: 2019 4th International Conference on Internet of Things: Smart Innovation and Usages (IoT-SIU), Ghaziabad, India, pp. 1–5 (2019). https://doi.org/10.1109/IoT-SIU.2019.8777342
4. Park, D., Bang, H., Pyo, C.S., Kang, S.: Semantic open IoT service platform technology. In: 2014 IEEE World Forum on Internet of Things (WF-IoT), Seoul, pp. 85–88 (2014). https://doi.org/10.1109/WF-IoT.2014.6803125

5. Nagpal, C., Upadhyay, P.K., Shahzeb Hussain, S., Bimal, A.C., Jain, S.: IIoT based smart factory 4.0 over the cloud. In: 2019 International Conference on Computational Intelligence and Knowledge Economy (ICCIKE), Dubai, United Arab Emirates, pp. 668–673 (2019). https://doi.org/10.1109/ICCIKE47802.2019.9004413
6. Jiang, D., Wang, Y., Lv, Z., Wang, W., Wang, H.: An energy-efficient networking approach in cloud services for IIoT networks. IEEE J. Sel. Areas Commun. **38**(5), 928–941 (2020)
7. Bahrami, M.: Cloud computing for emerging mobile cloud apps. In: 2015 3rd IEEE International Conference on Mobile Cloud Computing, Services, and Engineering, San Francisco, CA, pp. 4–5 (2015). https://doi.org/10.1109/MobileCloud.2015.40
8. Linthicum, D.S.: Connecting fog and cloud computing. IEEE Cloud Comput. **4**(2), 18–20 (2017). https://doi.org/10.1109/MCC.2017.37
9. Abu-Matar, M., Mizouni, R.: Variability modeling for smart city reference architectures. In: 2018 IEEE International Smart Cities Conference (ISC2), Kansas City, MO, USA, pp. 1—8 (2018). https://doi.org/10.1109/ISC2.2018.8656967
10. Jiang, D., Zhang, P., Lv, Z., et al.: Energy-efficient multi-constraint routing algorithm with load balancing for smart city applications. IEEE Internet Things J. **3**(6), 1437–1447 (2016)
11. Bai, B., Guo, Z.: Dynamic complexity of mobile internet business ecosystem. In: 2017 4th International Conference on Systems and Informatics (ICSAI), Hangzhou, pp. 502–506 (2017). https://doi.org/10.1109/ICSAI.2017.8248344
12. Jiang, D., Huo, L., Song, H.: Rethinking behaviors and activities of base stations in mobile cellular networks based on big data analysis. IEEE Trans. Netw. Sci. Eng. **7**(1), 80–90 (2020)
13. Jiang, D., Huo, L., Lv, Z., Song, H., Qin, W.: A joint multi-criteria utility-based network selection approach for vehicle-to-infrastructure networking. IEEE Trans. Intell. Transp. Syst. **19**(10), 3305–3319 (2018)
14. Jiang, D., Li, W., Lv, H.: An energy-efficient cooperative multicast routing in multi-hop wireless networks for smart medical applications. Neurocomputing **220**, 160–169 (2017)
15. Jiang, D., Wang, W., Shi, L., Song, H.: A compressive sensing-based approach to end-to-end network traffic reconstruction. IEEE Trans. Netw. Sci. Eng. **7**(1), 507–519 (2020)
16. Qi, S., Jiang, D., Huo, L.: A prediction approach to end-to-end traffic in space information networks. Mobile Netw. Appl. (2019)
17. Prajapati, A., Sakadasariya, A., Patel, J.: Software defined network: future of networking. In: 2018 2nd International Conference on Inventive Systems and Control (ICISC), Coimbatore, pp. 1351–1354 (2018). https://doi.org/10.1109/ICISC.2018.8399028
18. Raychev, J., Hristov, G., Kinaneva, D., Zahariev, P.: Modelling and evaluation of software defined network architecture based on queueing theory. In:, 2018 28th EAEEIE Annual Conference (EAEEIE), Hafnarfjordur, pp. 1–5https://doi.org/10.1109/EAEEIE.2018.8534289
19. Theodorou, T., Mamatas, L.: CORAL-SDN: a software-defined networking solution for the internet of things. In: 2017 IEEE Conference on Network Function Virtualization and Software Defined Networks (NFV-SDN), Berlin, pp. 1–2 (2017). https://doi.org/10.1109/NFV-SDN.2017.8169870
20. Tantayakul, K., Dhaou, R., Paillassa, B.: Mobility management with caching policy over SDN architecture. In: 2017 IEEE Conference on Network Function Virtualization and Software Defined Networks (NFV-SDN), Berlin, pp. 1–7 (2017). https://doi.org/10.1109/NFV-SDN.2017.8169830
21. Huo, L., Jiang, D., Qi, S., et al.: An AI-based adaptive cognitive modeling and measurement method of network traffic for EIS. Mobile Netw. Appl. (2019)
22. An AI based Approach to Secure SDN Enabled Future Avionics Communications Network Against DDoS Attacks
23. Jiang, D., Huo, L., Li, Y.: Fine-granularity inference and estimations to network traffic for SDN. PLoS ONE **13**(5), 1–23 (2018)

24. Huo, L., Jiang, D., Lv, Z., et al.: An intelligent optimization-based traffic information acquirement approach to software-defined networking. Comput. Intell. 1–21 (2019)
25. Wang, F., Jiang, D., Qi, S.: An adaptive routing algorithm for integrated information networks. China Commun. **7**(1), 196–207 (2019)
26. Raza, S.M., Thorat, P., Challa, R., Choo, H., Kim, D.S.: SDN based inter-domain mobility for PMIPv6 with route optimization. In: 2016 IEEE NetSoft Conference and Workshops (NetSoft), Seoul, pp. 24–27 (2016). https://doi.org/10.1109/NETSOFT.2016.7502436
27. Abadi, A., Rajabioun, T., Ioannou, P.A.: Traffic flow prediction for road transportation networks with limited traffic data. IEEE Trans. Intell. Transp. Syst. **16**(2), 653–662 (2015). https://doi.org/10.1109/TITS.2014.2337238
28. Wang, Y., Jiang, D., Huo, L., Zhao, Y.: A new traffic prediction algorithm to software defined networking. Mobile Netw. Appl. (2019)
29. Ma, B., Zhang, Z.: Security research of redundancy in mimic defense system. In: 2017 3rd IEEE International Conference on Computer and Communications (ICCC), Chengdu, pp. 2910–2914 (2017). https://doi.org/10.1109/CompComm.2017.8323064
30. Yoon, G., Lee, S., Kwon, D., Kwon, S., Park, Y.: RAPIEnet based redundancy control system. In: 2011 11th International Conference on Control, Automation and Systems, Gyeonggi-do, pp. 140–145 (2011)
31. Research on the active defense security system based on cloud computing of wisdom campus network
32. Xia, J., Cai, Z., Hu, G., Xu, M.: An active defense solution for ARP spoofing in OpenFlow network. Chinese J. Electron. **28**(1), 172–178 (2019). https://doi.org/10.1049/cje.2017.12.002
33. Tatang, D., Quinkert, F., Frank, J., Röpke, C., Holz, T.: SDN-guard: protecting SDN controllers against SDN rootkits. In: 2017 IEEE Conference on Network Function Virtualization and Software Defined Networks (NFV-SDN), Berlin, pp. 297–302 (2017). https://doi.org/10.1109/NFV-SDN.2017.8169856
34. Boukria, S., Guerroumi, M., Romdhani, I.: BCFR: blockchain-based controller against false flow rule injection in SDN. In: 2019 IEEE Symposium on Computers and Communications (ISCC), Barcelona, Spain, pp. 1034–1039 (2019). https://doi.org/10.1109/ISCC47284.2019.8969780
35. Qureshi, S., Braun, R.: Mininet topology: mirror of the optical switch fabric. In: 2019 29th International Telecommunication Networks and Applications Conference (ITNAC), Auckland, New Zealand, pp. 1–6https://doi.org/10.1109/ITNAC46935.2019.9078014
36. Pereíni, P., Kuzniar, M., Kostic, D.: OpenFlow needs you! A call for a discussion about a cleaner OpenFlow API. In: 2013 Second European Workshop on Software Defined Networks, Berlin, pp. 44–49 (2013).https://doi.org/10.1109/EWSDN.2013.14

Use Cases for QoE Test in Heterogeneous Networks Scenarios

Yun Chen, Lei Chen(✉), Ping Cui, Kailiang Zhang, and Yuan An

Jiangsu Province Key Laboratory of Intelligent Industry Control Technology, Xuzhou University of Technology, Xuzhou 221018, China
chenlei@xzit.edu.cn

Abstract. In heterogeneous environment, different services have different QoE requirements. The purpose of this paper is to identify the characteristics of emerging typical mobile multimedia applications including video services, audio services and burst data services. For each use case, we analyze the QoE test requirement. This analysis will offer the important guideline for design of the system architecture for QoE test. Furthermore, the bottlenecks of existing wireless access technologies are discussed.

Keywords: Heterogeneous networks · Quality of experience · Wireless communication

1 Introduction

To improve the performance of mobile network, many new approaches of network routing and measurement [1–3] are proposed. To internet service provider, the most important performance index is experience of user (QoE). Based on researches about user behavior and traffic analysis methods [4–7], new scheduling strategies are designed to raise resources utilization [8, 9] and improve QoE. With the rapid technology improvement and infrastructure deployment in wireless communication technologies including 5G (and later LTE-Advanced) and high-speed Wi-Fi, there have been a dramatic growth of mobile multimedia applications, for example, mobile video, 3D video stream, VoIP (video conferences), etc. [10]. These diverse content-rich multimedia applications lead to high complex traffic patterns and face user high requirements on QoE [11]. To measure the performance of these new scheduling strategies, flow level traffic reconstruction becomes an important topic [12, 13]. However, the inherent features of wireless communications, such as scarce bandwidth, interference, fading, error-prone channels, diverse access technologies and mobility, lead to a high level of dynamics of available communication resources that can deteriorate severely the quality of mobile multimedia applications with QoE constraints. There are bottlenecks for applying existed wireless techniques for ensuring wireless multimedia QoE. Such mismatch between the multimedia quality requirements and the service offered by the underlying communication infrastructure makes it a great challenge to develop mobile multimedia applications over

H. Song and D. Jiang (Eds.): SIMUtools 2020, LNICST 370, pp. 289–304, 2021.
https://doi.org/10.1007/978-3-030-72795-6_23

wireless networks. Although the network operators and service providers make huge investments to improve the system availability, security and performance, mobile multimedia users still suffer from poor QoE frequently. Thus, new and efficient technologies are needed to improve the QoE for wireless multimedia applications.

As an efficient way to improve the wireless transmission efficiency, compared with traditional cellular-based homogeneous networks, heterogeneous wireless networks have attracted tremendous research efforts from both academia and industry over the past decade. However, most of the existing studies have mainly contributed to the performance enhancement of communication networks and paid less attention to the user-perceived QoE that desires the joint adaption of multimedia contents and the Quality-of-Service (QoS) enhancement of the underlying communication infrastructure. The optimal interaction between adaptive multimedia processing and heterogeneous wireless networking through a cross-layer design plays an important role in the efficient utilization of scarce wireless communication resources to balance the mismatch between the QoE requirements of mobile multimedia users and the QoS provisioning of underlying heterogeneous wireless networks with the aim of creating a new era where multimedia service providers, network operators and end users all benefit.

Due to the fast deploying mobile devices and surging amount of data exchange, the demands on better mobile applications and service qualities are always increasing. The conventional homogeneous framework could not be able to meet such requirements in a long run, especially when facing limited wireless resources and continuously emerging multimedia applications. Although several approaches, such as the densification of base stations, adding small cells and dynamic radio spectrum management, somewhat relieve the bandwidth pressure, they cannot effectively or efficiently solve the problem when more and more various communication demands emerge, from the perspective of service quality and costs.

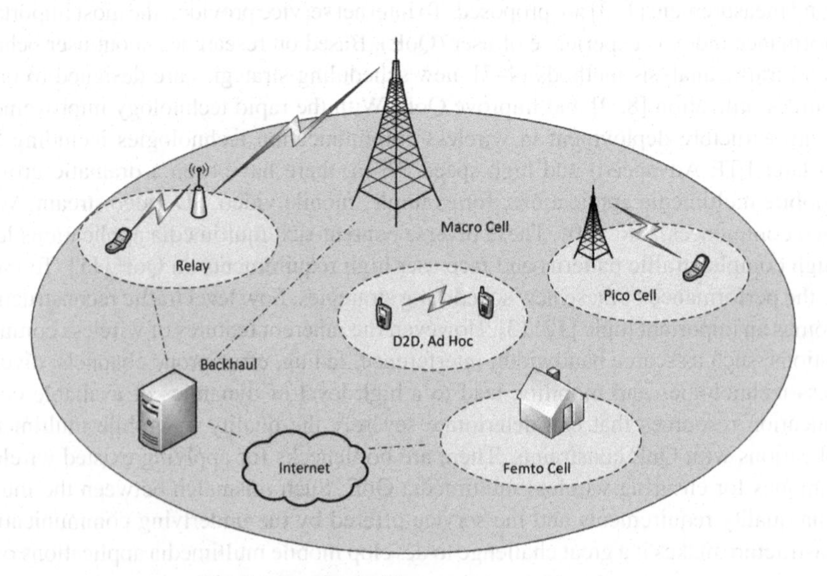

Fig. 1. Topology of heterogeneous networks.

On the other hand, heterogeneous networks provide a natural solution to boost the network capacity, to increase the network traffic and to enlarge the coverage. In contrast to homogeneous networks, heterogeneous networks consist of various types of transmission methods, radio access networks, and nodes with different power levels in the same network. The topology of heterogeneous networks can be illustrated in Fig. 1.

From the perspective of base station deployment, heterogeneous networks can be decomposed into two parts: one is the macro cells, which consist of high power base stations; another is low-power cells, such as Pico, Femto or Micro cells. Therefore, a heterogeneous network can also be treated as a combination of a macro cell and multiple small cells. More details on differences between the traditional cellular and heterogeneous networks can be found in [14]. Moreover, from the side of radio access technologies (RATs), the components in heterogeneous networks can be categorized into two classes [15]: one is single-RAT multitier components, and another is multi-RAT components. The former components are used in macro, pico, femto and micro cells, as well as the relays or client relays. The application situations for the latter group include WiFi offload, virtual carrier and mobile hotspots and personal area networks (PANs).

2 Challenging for QoE in Heterogeneous Networks

Lots of research efforts have been devoted to exploit the benefits of heterogeneous networks. Firstly, heterogeneous networks have the capability of addressing the problems induced by the rapid increase of network traffic. Secondly, heterogeneous networks are more energy-efficient since they allow communications with low power consumption. Thirdly, heterogeneous networks split the entire communication area into multiple cells with smaller size, which guarantees better QoS for user terminals. Finally, heterogeneous networks support a variety of radio access technologies such that the network performance has opportunity to be optimized with limited communication resources. A more detailed comparison can be found in Table 1.

Table 1. Comparison between traditional cellular networks and heterogeneous networks [14].

Aspect	Traditional cellular	Heterogeneous networks
Performance Metric	Outage/coverage probability distribution (in terms of SINR) or spectral efficiency (bps/Hz)	Outage/coverage probability distribution (in terms of rate) or area spectral efficiency (bps/Hz/m^2)
Topology	BSs spaced out, have distinct coverage areas. Hexagonal grid is an ubiquitous model for BS locations	Nested cells (pico/femto) inside macrocells. BSs are placed opportunistically and their locations are better modeled as a random process

(continued)

Table 1. *(continued)*

Aspect	Traditional cellular	Heterogeneous networks
Cell Association	Usually connect to the strongest BS, or perhaps two strongest during soft handover	Connect to BS(s) able to provide the highest data rate, rather than signal strength. Use biasing for small BSs
Downlink vs. Uplink	Downlink and uplink to a given BS have approximately the same SINR. The best DL BS is usually the best in UL too	Downlink and uplink can have very different SINRs; should not necessarily use the same BS in each direction
Mobility	Handoff to a stronger BS when entering its coverage area, involves signaling over wired core network	Handoffs and dropped calls may be too frequent if use small cells when highly mobile, overhead a major concern
Backhaul	BSs have heavy-duty wired backhaul, are connected into the core network. BS to MS connection is the bottleneck	BSs often will not have high speed wired connections. BS to core network (backhaul) link is often the bottleneck in terms of performance and cost
Interference Management	Employ (fractional) frequency reuse and/or simply tolerate very poor cell edge rates. All BSs are available for connection, i.e. "open access"	Manage closed access interference through resource allocation; users may be "in" one cell while communicating with a different BS; interference management is hard due to irregular backhaul and sheer number of BSs

Despite lots of merits by applying heterogeneous networks, there are still several challenges for using heterogeneous networks [15, 16]:

(1) The cooperation is a crucial issue for heterogeneous networks. It is evident that, in heterogeneous networks, the compatibility of subsystems and interface standards determines their seamless interoperation.
(2) From the technical point of view, since heterogeneous networks allow various types of low-power nodes, there will be sever imbalance in the utilization of limited resources, especially.
(3) The heterogeneous networks can support far more user terminals than homogeneous networks and the designs of each subsystem can be compatibly different, therefore, the network infrastructure and radio link management will be in a very high complexity.

Despite high designing complexity, the heterogeneous networks exhibit great potentials in supporting rapidly emerging mobile applications, and satisfying the demands and expectations toward their service requirements from both service provider and end-user perspectives. This is especially true for an important form of mobile multimedia services, which will be introduced in the following subsection.

Benefiting from existing research efforts, QoE can be partly quantified, and its value can be predicted in some level by using objective methodology and subjective projection models. The heterogeneous networks scheduling strategies pose challenge to evaluate how scheduling affects the user experience.

3 Use Cases Analysis

Video, Audio and bursty data services are main scenarios in heterogeneous networks.

3.1 User Case 1, Video Streaming

Motivations

Currently, video is a dominant application in multi-media services. A recent report from Cisco shows that [17], without counting video exchanged through peer-to-peer (P2P) file sharing, the global consumer Internet video traffic will rise from 64% in 2014 to 80% of all consumer Internet traffic in 2019. It also forecasts that, the sum of all forms of video (TV, video on demand (VoD), Internet, and P2P) will be in the range of 80%–90% of global Internet traffic by 2019. For the mobile data traffic, it is predicted that [17], by 2019, global mobile data traffic will surpass 24.3 exabytes per month, and roughly 75% of the world's mobile data traffic will be video by 2019. The results indicate that the mobile video service is acting a more and more important role globally. Up to now, the QoE assessment of video services has experienced four stages [18]: QoS monitoring, subjective test, objective quality model and data-driven analysis, and the comparison among those four assessment methods are given in Table 2.

Table 2. Comparison of video quality assessment methods [18].

	Direct measure of QoE	Objective or subjective	Real-time	Wide application	Cost
QoS monitoring	No	Objective	Yes	Wide	Not sure
Subjective test	Yes	Subjective	No	Limited	High
Objective quality model	No	Objective	Yes/No	Limited	Low
Data-driven analysis	Yes	Objective	Yes	Wide	Not sure

The video flow cost lot energy in network devices, the energy-efficiency is also important. Some new energy-efficiency strategies are proposed to save energy [19, 20]. It is necessary to evaluate how these strategies affect the QoE. The purpose of the evolution of assessment methodologies enables better QoE, and it also provides a motivation to investigate the mobile video service as well as its QoE in a systematic way.

Case Description
For streaming video in the Internet, one of most prevailing technologies is the HTTP adaptive streaming (HAS), which meets the growing consumer demands for mobile video services and enhances the QoE. Comparing with conventional streaming technologies, HAS has following four appealing and significant advantages [21]. Firstly, multiple rates of videos are provided, such that the delivered video can be adapted to the required standards by users. Secondly, different service levels and/or pricing schemes are available. Thirdly, the flexible service is offered to meet users demand on different kinds of streaming videos. Finally, the videos can be adapted to their best rate according to the current states of network and facilities in real-time. The last property, which can be interpreted as the reduced interruptions of the video playback and higher bandwidth utilization, is the most important one among all advantages over classic HTTP video streaming, and it contributes to the improvement of QoE of video streaming.

In HAS, a video is segmented into intervals that have durations between two and ten seconds, and each segment is encoded in multiple quality versions, where higher qualities correspond to higher rates, thus the number of segments equals to the number of rate versions. For HAS in wireless scenarios such as LTE network [22, 23], the progress is almost the same.Generally speaking, taking the QoE metric into account, the video transmission consists of four parts: encoder, networks, decoder and end users. In each part, there are factors that may finally degrade user experience. The typical visual distortions resulted by distorted videos are usually categorized by blocking effect, blurring, edginess and motion jerkiness, and those impacts seriously influence the QoE of video. As a subset of video service, HAS is also facing those factors that may deteriorate the QoE.

QoE of Use Case
For managing the point to multipoint multimedia communications in cellular networks, the Multimedia Broadcast Multicast Service (MBMS) is defined as a solution to deliver multicast and broadcast services over cellular networks by 3GPP [24]. The terminal devices can also provide feedback to the eNodeB using MBMS. From 3GPP Release 8, MBMS has been extended to the Long Term Evolution (LTE) standard. This extension is called evolved MBMS (eMBMS) [25]. To enhance QoE of users with degraded channels, the D2D resolution has been proposed for MBMS in [26]. In our opinion, MBMS should make full use of all available access networks to guarantee user experience. The most terminal devices using MBMS of cellular network support more access modes such D2D, Wi-Fi and blue teeth. We expect that the heterogonous network could improve QoE in the changing channel environment and save cellular communication resource.

In terms of performane measures, buffering time (startup delay), average PSNR and interruption percentage (rebuffering percentage) are proposed as three main QoE indices for MBMS in [27]. However, buffering time and interruption time/percentage are not strict QoE indices in psychological sense. We have to project this actual time to psychological time. Because no truly rigorous model can accurately describe the visual system, PSNR is still main index to evaluate user experience. Therefore, we have to check more indices for accurately describing user experience. For 2D video, the resolution, color rendition, motion portrayal, overall quality, and sharpness should be checked as important QoE indices. For 3D video QoE evaluation, the image quality, depth perception, naturalness, presence, and comfort degree are important indices closely related to user experience [28]. Although researchers have already achieved progress in the area of video QoE evaluation, we still meet some limitations and challenges for improving QoE of MBMS. The first problem is how to measure the effect of these perceptual indices. Although some standards have given the main human perceived features, it is still lacks the synthesis model to integrate these features. On the other hand, the user perceptual feature can also help efficiently use transport resource. For instance, based on the video resolution asymmetry between the left and right eye views, the main and mobile hybrid delivery for 3-D video services is proposed to maximize the channel efficiency in [29]. This means that the adaptive multimedia taking into account the user perceptual features, networking environments and media content could be efficient resolution for improving QoE of mobile multimedia services.

System Requirement of Use Case

We have to build more detailed index and explore the relationship between perceptual attributes of 2D/3D video and network performance index according to the objective and subjective evaluation standardizations recommended by Video Quality Expert Group (VQEG), ITU and European Broadcasting Union (EBU). In spite of network environment and service quality requirement, the adaptive multimedia systems should consider the user perceptual features of mobile video services. The locations of servers and clients should be also considered for access mode selection. This requires that the terminal devices should be able to gather and report the information related to QoE performance. This information is essential for the system adopting the adaptive scheduling scheme according to the current available channel resource, traffic state and QoE requirements.

To enhance the QoE, more and more Content Delivery Network servers are deployed on the core network border, or even in access network devices such as BSs. The proxy servers are deployed to reduce the handover frequency and support seamless handover for fast moving clients. Therefore, the test case should be able to set the moving mode of end users.

3.2 User Case 2, Wireless Audio Services

Motivations

Communications by voice is always a popular way for connecting people, since it is efficient and convenient. With the rapid development of wireless communication and significant proliferation of mobile devices, more powerful speech processing functions have been integrated into user facilities, such that a large variety of voice service, i.e. voice inquiry, voice remote control and audio conferencing, can be fulfilled to meet the ever-increasing needs in wireless and mobile scenarios. Those enhancement and effort make wireless voice service more attractive and popular.

As pointed out by, from the scenario point of view, the voice service can be classified into five categories: the PSTN voice service, voice services over IP (VoIP), hybrid voice service, and two other services that cross IP, interworking function and PSTN. Among all application scenarios, as Internet is playing an increasingly important role in modern society, the VoIP will see a more significant growth than its counterparts. Recent study and survey show that, over one third of enterprises now use VoIP and Session Initiation Protocol (SIP) trunking, and VoIP will take 66.51% of North America market in a few years [30]. For the sake of simplicity to discuss the topic, we narrow our scope to VoIP as a typical user case of wireless audio service.

For wireless VoIP service, clients care more about the network delay jitter [31], which can be regarded as a crucial QoE index. Furthermore, for some emerging intelligent interactive speech service, the edge computation strategies are also important for end user experience [32, 33]. The general metrics of QoE, instead of QoS, are also available for wireless VoIP service. Thus, there is a strong motivation to investigate its QoE performance.

Case Description

Generally, the variation of network determines the QoE of the VoIP service. The QoE of a VoIP service heavily relies on the network variation, such as the delay, jitter, and packet-loss [34]. Usually, the voice over Internet adopts the user datagram protocol (UDP), which can tolerate the dropping of several packets that cannot arrival at the receiver side before the deadline. It is known that, the transport protocol UDP is not reliable, and this kind of unreliability may yield severe degeneration of QoE of VoIP service. Regarding the mentioned problem, there are two classifications on the mechanisms to enhance the QoE of VoIP service, namely network-centric strategies and application-centric strategies [31]. The former utilizes a collection of compatible QoS mechanisms within the entire network to meet the requirements of services, such that the QoE of the VoIP service can be enhanced. For the latter strategies, the QoE is improved by optimizing the control mechanisms for end-users, such that the transmitter or receiver can adaptively cope with the voice data according to the network state.

By following the mechanisms and access technologies for wireless networks, it is easy to extend the conventional wired VoIP service to the wireless scenarios. An extensive application of such case can be referred to [35], where the SIP-based VoIP service in wireless mesh networks is investigated.

There have been lots of research efforts devoted to the investigation on network-centric and application-metric strategies, and their combination has gradually become a more popular tendency for improving the QoE of VoIP service. Such an approach is expected to be applied in a cross-layer design fashion in heterogeneous networks [36]. The merit of the combination is significant. On the one hand, all intermediate nodes in the network have the responsibility to monitor the variation of channel and conduct the appropriate actions to satisfy the requirements of QoS. On the other hand, applications behave adaptively to optimize the VoIP service. Both efforts improve the QoE of VoIP service.

QoE of Use Case

For the case of voice conversations, there are methods for QoE evaluation: subjective and objective speech quality assessment methodologies [37]. The subjective methodology gathers the perceptual feedbacks from users by using a collection of human subjects, to evaluate the QoE. Several subjective approaches are proposed to measure the quality of the degraded test speech with various objectives. However, many researcher found out that, the subjective methodology is time consuming and expensive, and they prefer to follow the objective methodology to overcome the drawbacks of its subjective counterpart. As stated in [37], one suitable objective speech quality assessment algorithm, which is closely related to intended goals and measurement context [38], can automatically, efficiently and reliably estimate the QoE. Furthermore, a game-theoretical method to manage the QoE of VoIP services in wireless scenarios is presented in [39], which explores a new vision to treat the QoE. From the user aspects, the QoE in wireless VoIP service is regarded as a function of the amount of efforts that the user has to put to preserve the conversation. These approaches need to be validated in well-designed test case.

System Requirement

To test QoE of audio service, the test case design must consider the characters of audio service in heterogeneous networks.

(1) The test case must have different access mode with different capacity.
(2) The capacity of different access mode should be dynamic, and the capacity can be observed.
(3) The system should have the capability of detecting the wireless environment of client, since the environment might change frequently with time.
(4) The QoE metrics should include latency and subject sense. The subject algorithm, such as P.862 and P.563, should be employed.

An Special Case: Mobile Voice Cloud

In addition to VoIP, we will briefly introduce another use case fitting the mobile audio services. This is called mobile voice cloud, which has recently attracted the attentions of both multimedia service providers and wireless network operators. The motivation behind mobile voice cloud services can be briefly described as follows. Due to the fact that normal mobile devices are small in size, the conventional way of inputting data to them (i.e., through typing buttons and/or pointing tools), requires sufficient attention and skills, and thus becomes inconvenient for many end users (e.g., those with poor eyesight, or attention drawn by other things such as driving). There is thus a need for intelligent user interfaces for supporting hand-free operation. Speech input (to computers) provides an obvious and promising solution for improved data entry/retrieval flexibility. There has also been a recent trend in interactive business and information applications such as call centers to move from a touch-tone based solution to a voice-touch driven approach. The voice input function can potentially be enhanced by the mobile voice cloud services. Specifically, upon receiving a voice request, the mobile client device sends the original voice data or speech features of the voice to the cloud service system. The cloud system recognizes the request and sends the response multimedia data to the users. Benefiting from this intelligent interaction system, users can therefore pay less attention to operate their mobile devices.

Architecture of the mobile voice cloud service can be illustrated in Fig. 2. The core technologies are voice and multimedia data transmission, distributed speech recognition, natural language processing, and information retrieval. The voice cloud service is a kind of request/response service. Therefore, the key performance indices for the user experience are response latency, success rate and the quality of response information. The adaptive transmission technology significantly affects response latency and the speech recognition accuracy, which determines the successful service rates. For improving the user experience in a changing environment, a cross-layer QoE-aware scheduling can be provided, by using network-aware adaptive multimedia processing at the application layer and multimedia-driven heterogeneous wireless networking at the network layer.

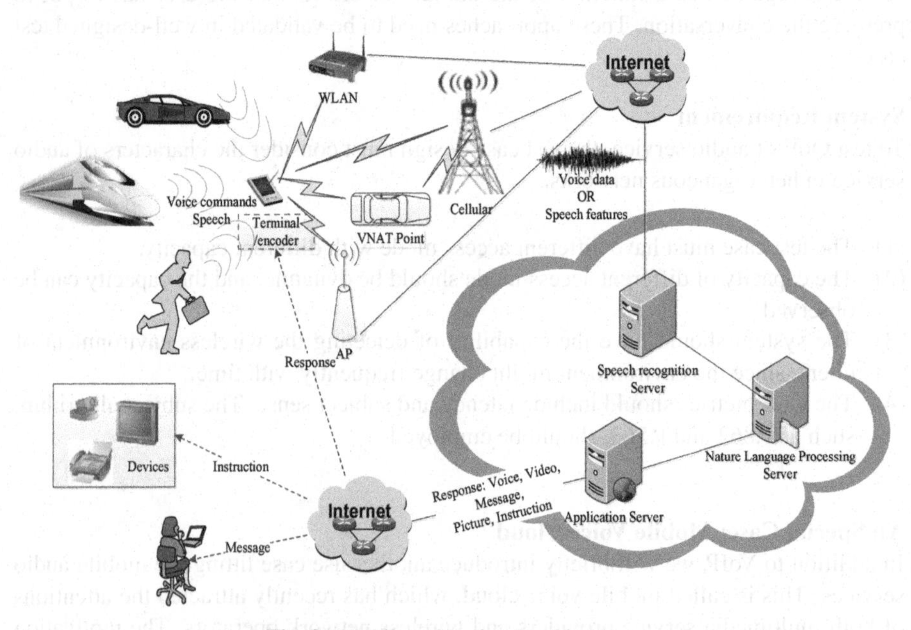

Fig. 2. Mobile voice cloud architecture.

For these use cases, because the end users are moving fast. It is difficult to predict the network traffic flow. The traffic prediction method in end side is key factor affecting the user experience [40, 41].

After the multimedia-driven heterogeneous wireless network selection at the network layer, the network-aware adaptive multimedia processing will find a balance among different QoE indices such as response time and speech recognition success rate. This adaptive processing works in both of request phase and response phase. When the client assesses the available transmission capacity at the borderline of the request phase, the adaptive speech feature encoding scheme must find a balance between the accuracy of speech recognition and the transmission latency. At the response phase, multimedia content adaption methods also need to cope with the bandwidth-fluctuant radio channels and time-varying wireless communication conditions to achieve a better content quality and service profile that are suitable for the network environment. Multimedia coding transforms input media content into output content in a form that adapts to the channel conditions and meets the user's needs. In the following, we will give some potential system design requirements that can actually improve the QoE performance of mobile voice cloud service:

In order to evaluate the QoE performance of very different services, one should build QoE index set for each type of service according to the way they affect the user experience.

The QoE evaluation model should consider both of the network parameters and service parameters. This model must be easy to guide the multimedia-driven heterogeneous wireless network selection and the network-aware adaptive multimedia processing crossing the application layer and the network layer.

A proxy for seamless handoff of voice cloud service in the heterogonous wireless network environment should also be used.

3.3 User Case 3, Mobile Bursty Data Service

Motivations
The mobile bursty data services, such as SNS, IM, OA, e-business, and so on, have been playing important roles in modern information exchanges. Yet, information exchanges among so many people generate the complex network traffic patterns. For instance, burst traffics often happen in access networks. Thus, single access mode cannot provide enough bandwidth and connection resource to prevent QoE deteriorating caused by traffic burst for all users. Since the smart terminal devices support simultaneously multiple access modes (e.g., cellular, WIFI), we need to study heterogonous network approaches, which may guarantee the user experience for mobile bursty data services.

Case Description
Currently, many people established and maintained real-time social connections with each other using bursty data services such as SNS, IM, Mobile OA, Mobile finance, in their mobile terminal devices. The characters of the transmission mechanism and devices determine how these services affect the user experience. On one hand, in general, these services adopt a round-trip transmission protocol in the application layer or the transmission layer. Therefore, the response time and the success rate of the whole round-trip transit process will affect the user experience in various ways. These services generally need to be constant online for use. Yet the high dropping rate on the move will certainly deteriorate the user experience. On the other hand, the short battery life has become a problem for mobile terminals for long time. From the point of user experience, the power consumption has to be considered in the QoE-driven heterogonous networks. The bursty data services usually have a relative long online time, even throughout the night, but they have a much shorter active time. Most traffic bursts on the networks between devices and

access points are caused by a series of user actions, and the crowded people often caused traffic congestion in the backhaul of local access networks. User behavior analysis in application layer is therefore essential for the QoE improvement in the heterogonous networks.

Thus, the user actions and events trigger data stream in bursty data services. The data transmission qualities of some actions affect the user experience. According to a rough classification in reference [42], bursty data applications belong to transactions-oriented applications characterized by request/response data flow corresponding to bidirectional data transfers. Thus the user experience is mainly related to the delay of the answer to a request. ITU-T has proposed the criterion for web-based service QoE evaluation [43]. Although the response latency is still the most important QoE index for bursty data services, most bursty data services cannot be referred simply as web browsing. Usually, the user can only perceive the quality of whole transmission process triggered by one action. This requires that the scheduling scheme must be able to find the scope of the human-perceived user actions in a number of transmission layer packets. In spite of waiting time, the long delay might causes high dropping rates that also deteriorate the user experience and leads to a disconnection, which may confuse to the users using SNS or IM, even if users cannot directly perceive the delay.

QoE of Use Case

The latency and success rate of human perceived round-trip processes in the services are closely related to user experience. We split the round-trip process of bursty data services into two classes. (1) The process can be perceived by users. The delay and success probabilities of some actions can be perceived by users, e.g., the delay of publishing a short video in web site, and that of submitting a comment on blog. (2) The process cannot be perceived by users. Although the delay of some actions cannot be directly perceived by users, a too long delay will cause confusion to users. For instance, at the chat window of IM services, users cannot perceive the latency of sending messages. However, long delay might cause the messages to arrive at the receivers in wrong orders and thus lead to a terrible logical mess to the users. Note that since the high dropping rate might lead to negative experience, the round-trip communication for keeping constant online also might influence the user experience. Moreover, the energy consumption of service might influence the user experience. There are two key principles for identifying these human-perceived round-trip processes. (1) The whole human-perceived round-trip process should be identified as a basic element in QoE-driven scheduling. This process usually includes several round-trip communication processes in the application layer and network layer, which cannot be perceived by users. (2) Some processes deteriorate QoE with long response time, and some cause negative experience with failed communications. There are different concerns for different types of processes. Moreover, the different type of action affects the user behavior in different way. For instance, some actions are delay-sensitive and some actions can tolerate relatively longer delay. Therefore, the projection models from actual time to psychological time should be very different. Since the high discharge rate is also influence the user experience, the well-designed relationship model among battery life, service energy consumption and user expectation is also necessary for QoE-driven heterogonous network design. Because most bursty data services have their own proprietary application layer protocols between server and client, understanding those proprietary protocols is another key problem to identify the type of action and to adopt adaptive KPI-QoE projection model in the heterogonous network design.

Test Requirement of Cases and the Concerns

The key of QoE-driven heterogonous network design is to identify user perceived index. Although some index sets have been build for web-based services [44–46], they cannot cover the burst data services. Only a few researches focused on the relationship model between these indices and QoE.

The user behavior is important for design bursty data service test case. Because of the complexity of user behaviors for bursty data services, it is not appropriate to validate the QoE-driven heterogonous network design in single user simple behavior scenarios. The test bed should use real data to validate the ability of scheme to identify service type and adopt corresponding strategy. Conventional testbeds usually activate data flow and build scenarios according to the traffic model. The traffic model usually characterizes the fluctuation of overall data flow [47, 48]. It is difficult to analyze the relationship between user behavior and user experience. The most traditional test systems send random data so that they cannot test the ability of service type identification of scheduling scheme.

According to the challenges for QoE improvement, heterogonous network design and system validation, the system requirements are described as follows:

(1) We have to build the user perceived index sets for each popular burst data service type, and investigate the connections between the indices and user experience.
(2) A test-bed for burst data services should be able to assemble the different scenarios based on understanding the user behavior in real world, and evaluate the performance index according to different service type. The testing process is illustrated in Fig. 3.
(3) The test-bed must be able to replay and analyze real data packet.
(4) We have to build a typical user behavior warehouse for assemble different scenarios based on user behavior analysis in real world.
(5) The test system has to measure the traffic via intelligent algorithms, and build test cases based on the traffic scenarios [49, 50].

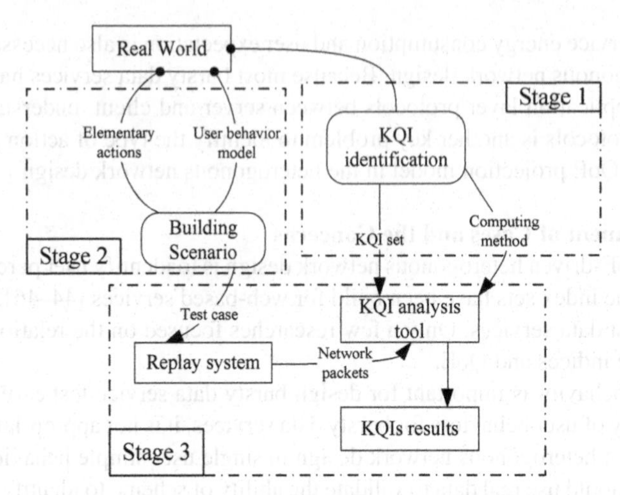

Fig. 3. Structure of the test-bed for burst data services.

4 Conclusion

In the document, we summarize the recent development on the multimedia application in heterogeneous networks. Especially, we discuss the recent development on Quality of experience of wireless multimedia applications. A brief introduction on heterogeneous networks, mobile service, QoE and their analysis method is given. For different kinds of services, the requirements of test cases are proposed based on the use case analysis.

Acknowledgements. This work is partly supported by Jiangsu technology project of Housing and Urban-Rural Development (No.2018ZD265) and Jiangsu major natural science research project of College and University (No. 19KJA470002).

References

1. Wang, F., Jiang, D., Qi, S.: An adaptive routing algorithm for integrated information networks. China Commun. **7**(1), 196–207 (2019)
2. Huo, L., Jiang, D., Lv, Z., et al.: An intelligent optimization-based traffic information acquirement approach to software-defined networking. Comput. Intell. **36**, 151–171 (2019)
3. Zhang, K., Chen, L., An, Y., Cui, P.: A QoE test system for vehicular voice cloud services. Mobile Netw. Appl. **1**, 6 (2019). https://doi.org/10.1007/s11036-019-01415-3
4. Chen, L., Jiang, D., Bao, R., Xiong, J., Liu, F., Bei, L.: MIMO Scheduling effectiveness analysis for bursty data service from view of QoE. Chinese J. Electron. **26**(5), 1079–1085 (2017)
5. Jiang, D., Wang, Y., Lv, Z., et al.: Big data analysis-based network behavior insight of cellular networks for industry 4.0 applications. IEEE Trans. Ind. Inf. **16**(2), 1310–1320 (2020)
6. Jiang, D., Huo, L., Song, H.: Rethinking behaviors and activities of base stations in mobile cellular networks based on big data analysis. IEEE Trans. Netw. Sci. Eng. **1**(1), 1–12 (2018)
7. Chen, L., et al.: A replay approach for remote testing user experience of mobile bursty data application. Int. J. Online Eng. **9**, 18–23 (2013)

8. Jiang, D., Huo, L., Lv, Z., et al.: A joint multi-criteria utility-based network selection approach for vehicle-to-infrastructure networking. IEEE Trans. Intell. Transp. Syst. **19**(10), 3305–3319 (2018)
9. Jiang, D., Zhang, P., Lv, Z., et al.: Energy-efficient multi-constraint routing algorithm with load balancing for smart city applications. IEEE Internet Things J. **3**(6), 1437–1447 (2016)
10. Andrews, J., et al.: What Will 5G Be? IEEE J. Sel. Areas Commun. (JSAC) **32**, 1065–1082 (2014)
11. Martini, M.G., Chen, C., Chen, Z., Dagiuklas, T., Sun, L., Zhu, X.: Guest editorial QoE-aware wireless multimedia systems. IEEE J. Sel. Areas Commun. (JSAC) **30**, 1153–1156 (2012)
12. Jiang, D., Wang, W., Shi, L., et al.: A compressive sensing-based approach to end-to-end network traffic reconstruction. IEEE Trans. Netw. Sci. Eng. **5**(3), 1–2 (2018)
13. Jiang, D., Huo, L., Li, Y.: Fine-granularity inference and estimations to network traffic for SDN. PLoS ONE **13**(5), 1–23 (2018)
14. Andrews, J.G.: Seven ways that HetNets are a cellular paradigm shift. IEEE Commun. Mag. **51**(3), 136–144 (2013)
15. Yeh, S., et al.: Capacity and coverage enhancement in heterogeneous networks. IEEE Wirel. Commun. **18**(3), 32–38 (2011)
16. Damnjanovic, A., et al.: A survey on 3GPP heterogeneous networks. IEEE Wirel. Commun. **18**(3), 10–21 (2011)
17. Cisco Systems, Inc.: Cisco Visual Networking Index: Global Mobile Data Traffic Forecast Update 2014–2019 White Paper. Whitepaper (2015)
18. Chen, Y., Wu, K., Zhang, Q.: From QoS to QoE: a tutorial on video quality assessment. IEEE Commun. Surv. Tutor. **17**(2), 1126–1165 (2015)
19. Jiang, D., Li, W., Lv, H.: An energy-efficient cooperative multicast routing in multi-hop wireless networks for smart medical applications. Neurocomputing **220**(2017), 160–169 (2017)
20. Jiang, D., Wang, Y., Lv, Z., et al.: Intelligent optimization-based reliable energy-efficient networking in cloud services for IIoT networks. IEEE J. Sel. Areas Commun. (2019)
21. Seufert, M., Egger, S., Slanina, M., Zinner, T., Hobfeld, T., Tran-Gia, P.: A survey on quality of experience of HTTP adaptive streaming. IEEE Commun. Surv. Tutor. **17**(1), 469–492 (2015)
22. De Vriendt, J., De Vleeschauwer, D., Robinson, D.C.: QoE model for video delivered over an LTE network using HTTP adaptive streaming. Bell Labs Tech. J. **18**(4), 45–62 (2014)
23. Staelens, N., et al.: Subjective quality assessment of longer duration video sequences delivered over HTTP adaptive streaming to tablet devices. IEEE Trans. Broadcast. **60**(4), 707–714 (2014)
24. 3GPP: General aspects and principles for interfaces supporting multimedia broadcast multicast service (MBMS) within E-UTRAN, Rel. 11: Eur. Telecommun. Stand. Inst., Route des Lucioles, France. Technical report 36.440 (2012)
25. Carla, L., Chiti, F., Fantacci, R., Khirallah, C., Tassi, A.: Power efficient resource allocation strategies for layered video delivery over eMBMS networks. In: Proceedings IEEE International Conference on Communication (ICC), Sydney, NSW, Australia, pp. 3505–3510 (2014)
26. Militano, L., Condoluci, M., Araniti, G., Molinaro, A., Iera, A., Muntean, G.-M.: Single frequency-based device-to-device-enhanced video delivery for evolved multimedia broadcast and multicast services. IEEE Trans. Broadcast. **61**(2), 263–278 (2015)
27. Kumar, U., Oyman, O.: QoE evaluation for video streaming over eMBMS. In: International Conference on Computing, Networking and Communications (ICNC), pp. 555–559 (2013)
28. Hewage, C.T.E.R., Martini, M.G.: Quality of experience for 3D video streaming. IEEE Commun. Mag. **51**(5), 101–107 (2013)

29. Jooyoung, L., et al.: A stereoscopic 3-D broadcasting system using fixed and mobile hybrid delivery and the quality assessment of the mixed resolution stereoscopic video. IEEE Trans. Broadcast. **61**(2), 222–237 (2015)
30. ipsmarx. https://www.ipsmarx.com/blog/voip-market-trends-one-third-of-enterprises-now-using-voip/
31. Jelassi, S., et al.: Quality of experience of VoIP service: a survey of assessment approaches and open issues. IEEE Commun. Surv. Tutor. **14**(2), 491–513 (2012)
32. Chen, L., Zhang, L.: Spectral efficiency analysis for massive MIMO system under QoS constraint: an effective capacity perspective. Mobile Netw. Appl. **1**, 9 (2020). https://doi.org/10.1007/s11036-019-01414-4
33. Wang, F., Jiang, D., Qi, S., et al.: A dynamic resource scheduling scheme in edge computing satellite networks. Mobile Netw. Appl. (2019)
34. Scheets, G., Parperis, M., Singh, R.: Voice over the internet: a tutorial discussing problems and solutions associated with alternative transport. IEEE Commun. Surv. Tutor. **6**(2), 1–0 (2004)
35. Rong, B., Yi, Q.: An enhanced SIP proxy server for wireless VoIP in wireless mesh networks. IEEE Commun. Mag. **46**(1), 108–113 (2008)
36. Srivastava, V., Mehul, M.: Cross-layer design: a survey and the road ahead. IEEE Commun. Mag. **43**(12), 112–119 (2005)
37. Rix, A.W., Beerends, J.G., Kim, D.-S., Kroon, P., Ghitza, O.: Objective assessment of speech and audio quality – technology and applications. IEEE Trans. Audio Speech Lang. Process. **14**(6), 1890–1901 (2006)
38. Kitawaki, N.: Perspectives on multimedia quality prediction methodologies for advanced mobile and ip-based telephony. In: Workshop Wideband Speech Quality in Terminals and Networks: Assessment and Prediction (2004)
39. Hassan, J., et al.: Managing quality of experience for wireless VOIP using noncooperative games. IEEE J. Sel. Areas Commun. **30**(7), 1193–1204 (2012)
40. Wang, Y., Jiang, D., Huo, L., et al.: A new traffic prediction algorithm to software defined networking. Mobile Netw. Appl. (2019)
41. Qi, S., Jiang, D., Huo, L.: A prediction approach to end-to-end traffic in space information networks. Mobile Netw. Appl. (2019)
42. Gand, F., Sergio, B.: Towards real-time anomalies monitoring for QoE indicators. Aannals of Telecommunications - annales des télécommunications **65**, 59–71 (2010)
43. ITU. ITU-T Recommendation G.107, The E-model, a computational model for use in transmission planning (2005)
44. TM Forum. GB923, Wireless Service Measurements Handbook, 3rd ed (2004)
45. TM Forum. GB917-2, SLA Management Handbook., 2.5 ed. (2005)
46. Kim, D., Lim, H., Yoo, J., et al.: Analysis of service transaction flow based on user's actions to develop KQIs for WiBro service. In: Tthe Fourth Advanced International Conference on Telecommunications. Athens, Greece, pp. 77–84 (2008)
47. Vishwanath, K.V., Vahdat, A.S.: Realistic and responsive network traffic generation. IEEE/ACM Trans. Networking **17**(3), 712–725 (2009)
48. Klemm, A.: Traffic modeling and characterization for UMTS networks. In: 2001 IEEE Global Telecommunications Conference, San Antonio, Texas, vol. 3, pp. 1741–1746 (2001)
49. Huo, L., Jiang, D., Qi, S., et al.: An AI-based adaptive cognitive modeling and measurement method of network traffic for EIS. Mobile Netw. Appl. (2019)
50. Huo, L., Jiang, D., Zhu, X., et al.: An SDN-based fine-grained measurement and modeling approach to vehicular communication network traffic. Int. J. Commun. Syst. 1–12 (2019)

A Massive MIMO User Selection Algorithm Based on Effective Capacity

Jing Huang, Lei Chen(✉), Ping Cui, Kailiang Zhang, and Yuan An

Jiangsu Province Key Laboratory of Intelligent Industry Control Technology, Xuzhou University of Technology, Xuzhou 221018, China
chenlei@xzit.edu.cn

Abstract. In massive multiple-input multiple-output (MIMO) systems, to maximize spectral efficiency (SE), optimal scheduled user number for a time slot has been investigated. One of the important QoS index is the probability of delay violation, which depends on transmission stableness over a long period rather than the momentary transmission rate. We formulate the relationship between effective SE, QoS constraint, and scheduled user number as a continuous function through employing the effective capacity (EC) theory of wireless channels. Nonetheless, obtaining the definitive solution for optimal scheduled user number is still problematic and unfathomed. Therefore, in this paper, an approach to obtain the suboptimal scheduled user number is presented. Simulation results show that the number of selected users is close to the theoretical value.

Keywords: Multiple-input multiple-output systems · Vehicular networks · Quality of experience · Wireless communication · Spectral efficiency

1 Introduction

With the expeditious development of next generation networks, many researchers have presented numerous novel network routing algorithm and measurement approaches [1–3]. For optimize the core and edge network, researches about user behavior and traffic analysis methods [4–7] have been conducted. Aiming at enhance user experience [8–11], resources utilization [12–15] and energy-efficiency [16–18], novel scheduling strategies are designed to optimize network efficiency. Meanwhile, the performance evaluation of these scheduling strategies is getting more important. In order to evaluate the performance of these scheduling strategies, traffic reconstruction on flow level has drawn a lot of attention of research [19–21]. To enhance the quality of user service in 5G edge network, massive multiple-input multiple-output (MIMO) systems are proposed. For dealing with massive users and devices, spectral efficiency is critical [22–25]. To guarantee service quality and utilize spectral efficiency of massive MIMO, it is essential to obtain the explicit traffic prediction [2, 26, 27] and resource scheduling according to the traffic model [28–31].

It is indicated by a lot of researches that the spectral efficiency and the experience of end user can be both improved through massive antennas effectively. However, most

H. Song and D. Jiang (Eds.): SIMUtools 2020, LNICST 370, pp. 305–314, 2021.
https://doi.org/10.1007/978-3-030-72795-6_24

topics of these researches usually focus on the way that the QoS affected by network traffic. Therefore, we put emphasis on how the delay violations caused by the fluctuation of momentary rates in the long term, in this paper. The relationship between the scheduled user number and QoS requirement has been investigated, based on the effective capacity model of wireless communication theory the theory of spectral efficiency model in massive MIMO systems [12]. From our simulation results, it can be drawn that better stableness need to be met to obtained higher QoS requirements. Hence, in each time slot, more users should be scheduled and served transmit signals to by the base station to achieve higher stableness. The momentary transmission rates fluctuation can be diminished by the increment of user scheduling frequency. While it is possible that the SE is reduced with the increasing of the scheduled user number. Hence, in this paper, we mainly formulate a function of effective spectral capacity to indicate the relationship between scheduled user number, SE, and the QoS requirement. Nevertheless, the optimal or explicit solution of relationship between them can not be obtained. Approaches to access the suboptimal solution effectively for the scheduling scheme problem is proposed.

The following content of this paper will be organized in five parts. The system model of the approach will be introduced in the second section. And the third section presents the user selection algorithm for massive MIMO systems. Simulation methodology and results are detailed in the fourth section. Finally, in the fifth section concludes the paper.

2 System Model

2.1 Spectral Efficiency of Massive MIMO

As proposed in [33], in a typical massive MIMO system model, the BS has an array of M antennas serving it cell. And we consider the mobile terminals have single antenna for receiving signal. The number of mobile terminals is N. The sequence number l is assigned to each cell. We use the notation $z_{lk} \in R^2$ to represent the geographical position of each terminal, where the $k \in \{1, \ldots, K\}$ is the sequence number of mobile terminal and $l \in L$ is the cell number of the mobile terminal. The model of the massive MIMO system is described in Fig. 1, where the inverted triangles represent the antennas of the base station and terminals. The arrows between base station and terminal denote the service-providing relationship between them. In Fig. 1, each base station with M antennas serves and communicates with several terminals at the same time. Therefore, the spectral efficiency may be influenced by the scheduled user number. In this paper, the key problem is how to find an optimal or nearly optimal solution of the number of scheduled user number.

The channel response between BS j and UE k in cell l can be denoted by $h_{jlk} \in R^N$. And it can be drawn as realizations from zero-mean circularly symmetric complex Gaussian distributions [32]:

$$h_{jlk} \sim CN\big(0, d_j(z_{lk}\mathrm{I}_M)\big) \tag{1}$$

Where $\mathbf{I}_M$ is the $M \times M$ identity matrix.

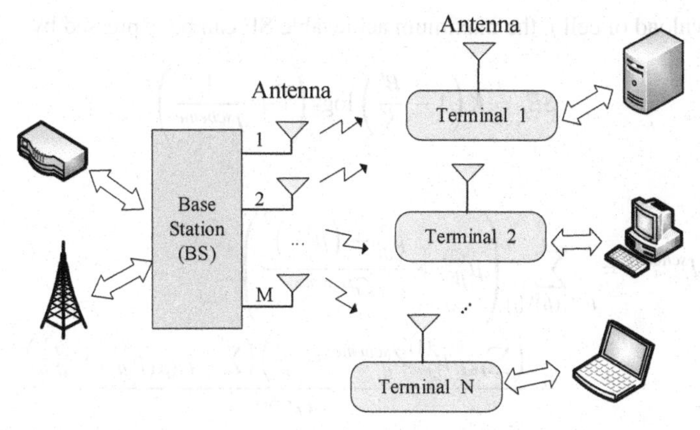

Fig. 1. The massive MIMO system

The variance of the channel attenuation from BS j to the positions of any **UE**, z can be obtained by the function $d_j(z)$. We describe the amount of symbols transmitted in each frame as S, and reserve B' out of the S symbols for pilot signaling, allocating the remaining $S - B'$ symbols for payload data. $p_{lk} = \frac{\rho}{d_{j(z_{lk})}}$ indicates the transmit power of one symbol, in which ρ is a design parameter for the channel attenuation inversion policy, ensuring that the average effective channel gain remains the same for all UEs:$E\{p_{lk}\|\boldsymbol{h}_{llk}\|^2\} = M_\rho$. And at UE k in cell j, the received download signal in a frame can be expressed as:

$$y_{jk} = \sum_{l \in L} \sum_{m=1}^{K} \mathrm{h}_{ljk}^{T} \mathbf{w}_{lm} x_{lm} + n_{jk} \tag{2}$$

where $(\cdot)^T$ indicates the transpose operation, the symbol transmitted to UE m in cell l is x_{lm}, $w_{lm} \in C^M$ is the vector of corresponding precoding, and $||w_{lm}||^2$ is the power of allocated download transmit, which can be expressed as:

$$w_{lm} = \sqrt{\frac{p_{jk}}{\mathrm{E_h}\left\{||g_{jk}||^2\right\}}} g_{jk}. \tag{3}$$

In Eq. (3), the average power of transmitting $p_{jk} \geq 0$ is a function of the UE position but not of the momentary channel realizations. The spatial directivity of the transmission is defined by vector $g_{jk} \in C^M$ which is based on the acquired CSI. The SNIR is given in [33]:

$$SINR_{jk} = \frac{p_{jk} \frac{\mathrm{E_h}\left\{||\mathbf{g}_{jk}\mathbf{h}_{jjk}||^2\right\}}{\mathrm{E_h}\left\{||\mathbf{g}_{jk}||^2\right\}}}{\sum_{l \in L} \sum_{m=1}^{K} P_{lm} \frac{\mathrm{E_h}\left\{||\mathbf{g}_{lm}\mathbf{h}_{ljk}||^2\right\}}{\mathrm{E_h}\left\{||\mathbf{g}_{lm}||^2\right\}} - p_{jk} \frac{\mathrm{E_h}\left\{||\mathbf{g}_{jk}\mathbf{h}_{jjk}||^2\right\}}{\mathrm{E_h}\left\{||\mathbf{g}_{jk}||^2\right\}} + \sigma^2} \tag{4}$$

In the download of cell j, the maximum achievable SE can be expressed by:

$$SE_j = K\left(1 - \frac{B'}{S}\right)\log_2\left(1 + \frac{1}{I_j^{scheme}}\right) \tag{5}$$

where

$$I_j^{scheme} = \sum_{l \in L_|(\beta)\backslash\{j\}} \left(\mu_{jl}^{(2)} + \frac{\mu_{jl}^2 + \left(\mu_{jl}^{(1)}\right)^2}{G^{scheme}}\right) + \frac{\left(\sum_{l \in L} \mu_{jl}^{(1)} Z_{jl}^{scheme} + \frac{\sigma^2}{\rho}\right)\left(\sum_{\ell \in L_j(\beta)} \mu_{jl}^{(1)} + \frac{\sigma^2}{B'\rho}\right)}{G^{scheme}}. \tag{6}$$

Here the different receive combining schemes employed determined by the G^{scheme} and Z_{jl}^{scheme}. In previous works, the compassion between these two techniques has been conducted in massive MIMO system. As the current research results, $G^{MR} = M$ and $Z_{jl}^{scheme} = K$ with MR combining, whereas $G^{ZF} = M - k$ and

$$Z_{jl}^{ZF} \begin{cases} K\left(1 - \frac{\mu_{jl}^{(1)}}{\sum_{\ell \in L_j(\beta)} \mu_{jl}^{(1)} + \frac{\sigma^2}{B'\rho}}\right) & if\ l \in L_j(\beta), \\ K & if\ l \notin L_j(\beta). \end{cases} \tag{7}$$

In Eq. (7):

$$\mu_{jl}^{(w)} = \mathrm{E}_{Z_{lm}}\left\{\left(\frac{d_j(z_{lm})}{d_l(z_{lm})}\right)^w\right\} for\ w = 1, 2. \tag{8}$$

2.2 Effective Capacity Model

Generally, we regard the QoS problem for a stable channel as a relationship between the covering of arrival rate and the stable transmission capacity of the system. In order to model a QoE/QoS problem for a stable channel in massive MIMO systems, in each time slot, a threshold of SINR can be set for the small burst data services. Regrettably, the future 5G networks will be consist of many emerging big buffering services, such as speech cloud and remote control based on real-time video. During the transmissions of them, the time-varying capacity, fast-moving users, and low reliability of wireless networks must be considered. How to model the time-varying channel is the key consideration for QoS guaranteeing. Inefficient user scheduling and resource allocation will be caused in traditional QoS/QoE models. Therefore, the EC concept is proposed, which is a function of the nonempty buffer probability and the QoS connection exponent. The EC concepts is more applicable for the description of the delay violation probability of time-varying channels, in many scenarios [34–38].

We denote the QoS requirement as.

$$P_r\left(\max_{1 \le i \le N} Q_i(0) > B\right) \le \varepsilon. \tag{9}$$

For a large buffer size B, a QoS constraint ϵ is set and $\theta = -\log(\epsilon)/B$, the QoS requirement can be expressed as an EC problem:

$$\lambda \le \min_{1 \le j \le N} C_k(\theta) \tag{10}$$

where

$$C_k(\theta) = \frac{1}{\theta} \lim_{n\to\infty} \frac{-1}{n} \ln \mathrm{E}\left(e^{-\theta \sum_{t=1}^{n} r_k(t)} \right) \tag{11}$$

and $r_k(t)$ is the rate allocated to user k in cell j at time t.

It is assumed that the K users out of a set of N active users is picked by the scheduling scheme at the base station for stochastic transmission with the same probability. Thus, we can denote the r_k as

$$\mathrm{r}_k(t) = \begin{cases} \frac{N_f}{K}\left(1 - \frac{B}{S}\right) \log_2\left(1 + \frac{1}{I_j^{scheme}}\right), & w.p. \frac{K}{N}, \\ 0, & w.p. 1 - \frac{K}{N}. \end{cases} \tag{12}$$

2.3 Spectral Effective Capacity of Massive MIMO

Higher fluctuation of EC model will account for lower EC, causing higher probability of delay violation. The spectral efficiency model and EC model are combined into spectral effective capacity model for massive MIMO in [33]. In this paper, we compute the outage probability of the QoS constraint due to the difficulty of the predication of a certain momentary CSI and the end users' number who access the cell. For describing the outage probability, one of the effective tools is EC theory. We denote the $\nu = N_f\left(1 - \frac{B}{S}\right) \log_2\left(1 + \frac{1}{I_j^{scheme}}\right)$, $\mathrm{P}\left\{\left(\sum_{t=1}^{n} r_k(t)\right) = \tau \frac{\nu}{K}\right\}$ as the probability of the total transmitted capacity with τ times scheduled in the future n time slots. Massive MIMO system's effective capacity can be expressed by:

$$\mathrm{E}\left(e^{-\theta \sum_{t=1}^{n} r_k(t)} \right) = \sum_{\tau=0}^{n} \left(e^{-\theta\tau \frac{\nu}{K}} P\left\{ \left(\sum_{t=1}^{n} r_k(t) \right) = \tau \frac{\nu}{K} \right\} \right). \tag{13}$$

We assume that the end devices have same probability to be scheduled and K users for each time slot are scheduled by BS. Therefore, we rewrite the expected transmission capacity of a user in the future n time slots as:

$$\mathrm{E}\left(e^{-\theta \sum_{t=1}^{n} r_k(t)} \right) = \sum_{\tau=0}^{n} \left(\binom{n}{\tau} \left(e^{-\theta \frac{\nu}{K}} \frac{K}{N} \right)^{\tau} \left(1 - \frac{K}{N} \right)^{(n-\tau)} \right). \tag{14}$$

And we rewrite the effective capacity of user k as.

$$C_k(\theta) = \frac{1}{\theta} \lim_{n\to\infty} \frac{-1}{n} \ln \sum_{\tau=0}^{n} \left(e^{-\theta\tau \frac{\nu}{K}} \binom{n}{\tau} \left(\frac{K}{N} \right)^{\tau} \left(1 - \frac{K}{N} \right)^{(n-\tau)} \right)$$

$$=\frac{1}{\theta}\lim_{n\to\infty}\frac{-1}{n}\ln\sum_{\tau=0}^{n}\left(\binom{n}{\tau}\left(e^{-\theta\frac{\nu}{K}}\frac{K}{N}\right)^{\tau}\left(1-\frac{K}{N}\right)^{(n-\tau)}\right)$$
$$=\frac{1}{\theta}\lim_{n\to\infty}\frac{-1}{n}\ln\left[1-\frac{K}{N}\left(1-e^{-\theta\frac{\nu}{K}}\right)\right]^{n}. \quad (15)$$

According to L'Hospital's rule, the Eq. (15) can be expressed as [33]:

$$C_k(\theta)=\frac{1}{\theta}\lim_{n\to\infty}\frac{-1}{n}\ln\left[1-\frac{K}{N}\left(1-e^{-\theta\frac{\nu}{K}}\right)\right]^{n}$$
$$=\frac{-1}{\theta}\lim_{n\to\infty}\frac{\frac{\partial\ln\left[1-\frac{K}{N}\left(1-e^{-\theta\frac{\nu}{K}}\right)\right]^{n}}{\partial n}}{\frac{\partial n}{\partial n}}$$
$$=\frac{-1}{\theta}\ln\left[1-\frac{K}{N}\left(1-e^{-\frac{\theta\nu}{K}}\right)\right]. \quad (16)$$

If the $\theta\to\infty$, the EC is zero. While $\theta\to 0$, the EC is $\frac{K}{N}\frac{N_f}{K}\left(1-\frac{B}{S}\right)\log_2\left(1+\frac{1}{I_j^{scheme}}\right)$
It is assumed that all users have the same QoS requirements. Therefore, for a fixed θ, at each time slot, we consider the $f(K)=C_k(\theta)$ as function of the scheduled users number K. The derivative of this EC function with respect to K is

$$\nabla f(K)=\frac{Ke^{\frac{\nu\theta}{K}}-K-\theta\nu}{(NK\theta)\left(Ne^{\frac{\nu\theta}{K}}-Ke^{\frac{\nu\theta}{K}}+K\right)}. \quad (17)$$

However, it is impractical to obtain the solution of $\nabla f(K)=0$. Therefore, the optimal users scheduled number is impossible to calculate.

3 User Selection Algorithm

Causing the impossibility of obtaining the explicit solution, a fast user selection algorithm is proposed in order to solve the problem by finding a suboptimal solution, which is shown in Table 1.

Table 1. User selection algorithm

Step1	Del = $\frac{N}{20}$, N is the number of candidate users
Step2	n = 1, ΔESC = $\nabla f(1)+\nabla f(1+Del)$
Step3	While ΔESC > 0 and $n<N-Del$
Step4	n = n + Del
Step5	ENDWHILE
Step6	n = n + Del/2
Step7	RETURN n

Figure 2 illustrates the calculation process of the proposed fast user selection algorithm. To find a suboptimal solution, we first define a *Del* value, which can be calculated by *N/20*, where *N* denotes the number of candidate users. At the beginning of the algorithm, we assign the *n* as 1 and $\Delta \text{ESC} = \nabla f(1) + \nabla f(1 + Del)$. And the *n* self-increases progressively by *Del*, until the $\Delta \text{ESC} > 0$ and $n < N - Del$. At last, we take $n = n + Del/2$ as the suboptimal solution of the scheduled user number, which will be the approximate solution of the problem.

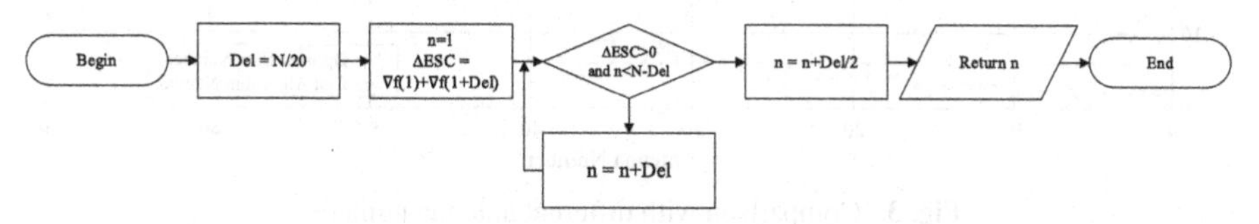

Fig. 2. The fast user selection approach.

It can be proved that the algorithm can converge in a few loops. The interval of optimal solution can be found by using Eq. (17), which is the derivative of the EC function. The symbol of the derivative of EC should be different during the two ending intervals. And the suboptimal solution can be obtained as intermediate point after several loops of the interval.

4 Simulation and Results

A simple scenario with one cell is simulated, for investigating the relationship between the QoS requirement and the scheduled user number. We set the simulation according the parameters in urban environment. In the simulation experiment, one base station equipped with two hundred antennas servers two hundred users with the same QoS requirement. We set the pilot reuse factor as 1 and the coherence block length to 400. The SNR is set to 5 dB and pathloss factor is set to 3.7. The QoS parameter θ is in a range from e^{-10} to e^{10}, in which low θ indicates a non-strict demand for real-time transmission, whereas high value of θ infers that a strict real-time request and high stableness must be satisfied. The unit of EC is bits/S/Hz.

Figure 3 shows the simulation results between theoretical optimal user number and user number obtained from proposed fast algorithm. It can be drawn from the figure that the suboptimal selected user number obtained by the proposed scheduling scheme is very close to the theoretical optimal number.

Furthermore, we conduct the simulation of the comparison between theoretical optimal user number and user number obtained from proposed fast algorithm under different QoS parameter. Figure 4 describes the comparison result of the scheduled user number. It can be seen that with the QoS parameter increasing, the user number will be decreased, which means one base station serve less terminals. With different QoS parameter, the calculation results of proposed fast scheduling algorithm are close to theoretical user number, showing the practicability of our approach.

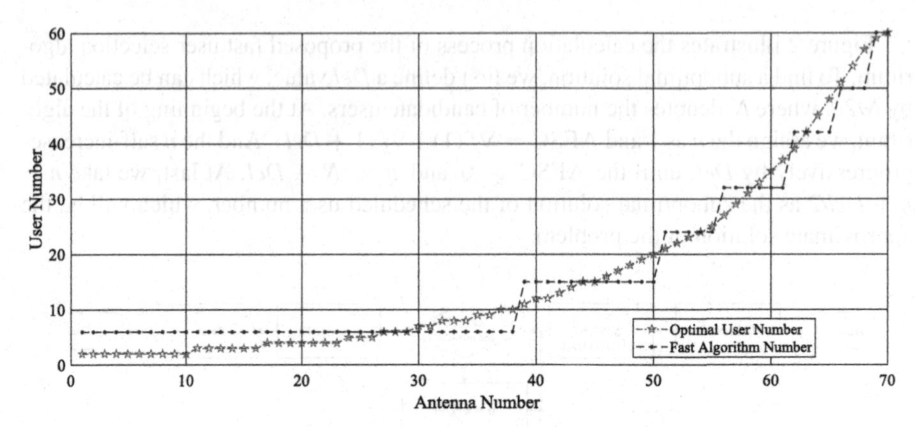

Fig. 3. Comparison with different antenna number.

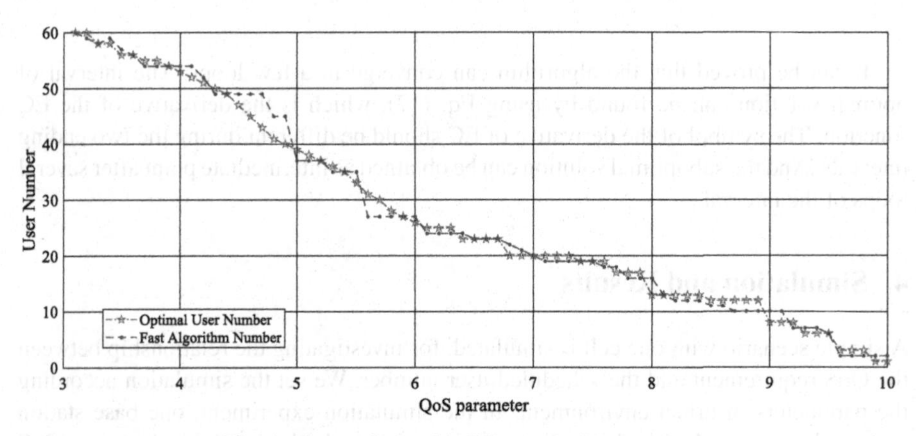

Fig. 4. Comparison with different QoS parameter.

5 Conclusion

In this paper, the optimal scheduling user number with fixed QoS constraint problem is investigated to obtain the maximum of the effective spectral capacity. Due to the difficulty to seek the explicit user number, a fast algorithm to obtain a suboptimal solution is proposed in this paper. The appropriate and suboptimal user number can be calculated with a low computation cost. The simulation on the massive MIMO system shows that the selected user number from our algorithm is near to the theoretical value, which means our approach is effective and practical.

Acknowledgements. This work is partly supported by Jiangsu technology project of Housing and Urban-Rural Development (No. 2018ZD265) and Jiangsu major natural science research project of College and University (No. 19KJA470002).

References

1. Wang, F., Jiang, D., Qi, S.: An adaptive routing algorithm for integrated information networks. China Commun. **7**(1), 196–207 (2019)
2. Huo, L., Jiang, D., Lv, Z., et al.: An intelligent optimization-based traffic information acquirement approach to software-defined networking. Comput. Intell. **36**, 1–21 (2019)
3. Zhang, K., Chen, L., An, Y., et al.: A QoE test system for vehicular voice cloud services. Mob. Netw. Appl. (2019). https://doi.org/10.1007/s11036-019-01415-3
4. Chen, L., Jiang, D., Bao, R., Xiong, J., Liu, F., Bei, L.: MIMO Scheduling effectiveness analysis for bursty data service from view of QoE. Chinese J. Electron. **26**(5), 1079–1085 (2017)
5. Jiang, D., Wang, Y., Lv, Z., et al.: Big data analysis-based network behavior insight of cellular networks for industry 4.0 applications. IEEE Trans. Ind. Inform. **16**(2), 1310–1320 (2020)
6. Jiang, D., Huo, L., Song, H.: Rethinking behaviors and activities of base stations in mobile cellular networks based on big data analysis. IEEE Trans. Netw. Sci. Eng. **1**(1), 1–12 (2018)
7. Chen, L., et al.: A lightweight end-side user experience data collection system for quality evaluation of multimedia communications. IEEE Access **6**(1), 15408–15419 (2018)
8. Barakabitze, A.A., et al.: QoE management of multimedia streaming services in future networks: a tutorial and survey. IEEE Commun. Surv. Tutorials **22**(1), 526–565 (2020)
9. Orsolic, I., Skorin-Kapov, L.: A framework for in-network QoE monitoring of encrypted video streaming. IEEE Access **8**, 74691–74706 (2020)
10. Song, E., et al.: Threshold-oblivious online web QoE assessment using neural network-based regression model. IET Commun. **14**(12), 2018–2026 (2020)
11. Seufert, M., Wassermann, S., Casas, P.: Considering user behavior in the quality of experience cycle: towards proactive QoE-aware traffic management. Commun. Lett. **23**(7), 1145–1148 (2019)
12. Chen, L., Zhang, L.: Spectral efficiency analysis for massive MIMO system under QoS constraint: an effective capacity perspective. Mob. Netw. Appl. (2020). https://doi.org/10.1007/s11036-019-01414-4
13. Wang, F., Jiang, D., Qi, S., Qiao, C., Shi, L.: A dynamic resource scheduling scheme in edge computing satellite networks. Mob. Netw. Appl. (2020). https://doi.org/10.1007/s11036-019-01421-5
14. Jiang, D., Huo, L., Lv, Z., et al.: A joint multi-criteria utility-based network selection approach for vehicle-to-infrastructure networking. IEEE Trans. Intell. Transp. Syst. **19**(10), 3305–3319 (2018)
15. Jiang, D., Zhang, P., Lv, Z., et al.: Energy-efficient multi-constraint routing algorithm with load balancing for smart city applications. IEEE Internet Things J. **3**(6), 1437–1447 (2016)
16. Wiatr, P., Chen, J., Monti, P., Wosinska, L.: Energy efficiency versus reliability performance in optical backbone networks [invited]. IEEE/OSA J. Opt. Commun. Networking **7**(3), A482–A491 (2015)
17. Jiang, D., Li, W., Lv, H.: An energy-efficient cooperative multicast routing in multi-hop wireless networks for smart medical applications. Neurocomputing **220**, 160–169 (2017)
18. Jiang, D., Wang, Y., Lv, Z., et al.: Intelligent optimization-based reliable energy-efficient networking in cloud services for IoT networks. IEEE J. Sel. Areas Commun. (2019)
19. Jiang, D., Wang, W., Shi, L., et al.: A compressive sensing-based approach to end-to-end network traffic reconstruction. IEEE Trans. Netw. Sci. Eng. **5**(3), 1–12 (2018)
20. Jiang, D., Huo, L., Li, Y.: Fine-granularity inference and estimations to network traffic for SDN. Plos One **13**(5), 1–23 (2018)
21. Wang, Y., Jiang, D., Huo, L., et al.: A new traffic prediction algorithm to software defined networking. Mob. Netw. Appl. (2019). https://doi.org/10.1007/s11036-019-01423-3

22. Lee, Y., Kim, Y., Park, S.: A machine learning approach that meets axiomatic properties in probabilistic analysis of LTE spectral efficiency. In: 2019 International Conference on Information and Communication Technology Convergence (ICTC), Jeju Island, Korea (South), pp. 1451–1453 (2019)
23. Ji, H., Sun, C., Shieh, W.: Spectral efficiency comparison between analog and digital RoF for mobile Fronthaul transmission link. J. Lightwave Technol. **38**, 5617–5623 (2020)
24. Hayati, M., Kalbkhani, H., Shayesteh, M.G.: Relay selection for spectral-efficient network-coded multi-source D2D communications. In: 2019 27th Iranian Conference on Electrical Engineering (ICEE), Yazd, Iran, pp. 1377–1381 (2019)
25. You, L., Xiong, J., Zappone, A., Wang, W., Gao, X.: spectral efficiency and energy efficiency tradeoff in massive MIMO downlink transmission with statistical CSIT. IEEE Trans. Signal Process. **68**, 2645–2659 (2020)
26. Qi, S., Jiang, D., Huo, L.: A prediction approach to end-to-end traffic in space information networks. Mob. Netw. Appl. (2019). https://doi.org/10.1007/s11036-019-01424-2
27. Huo, L., Jiang, D., Zhu, X., et al.: An SDN-based fine-grained measurement and modeling approach to vehicular communication network traffic. Int. J. Commun. Syst. 1–12 (2019). https://doi.org/10.1002/dac.4092
28. Tan, J., Xiao, S., Han, S., Liang, Y., Leung, V.C.M.: QoS-aware user association and resource allocation in LAA-LTE/WiFi coexistence systems. IEEE Trans. Wireless Commun. **18**(4), 2415–2430 (2019)
29. Wang, Y., Tang, X., Wang, T.: A unified QoS and security provisioning framework for wiretap cognitive radio networks: a statistical queueing analysis approach. IEEE Trans. Wireless Commun. **18**(3), 1548–1565 (2019)
30. Hassan, M.Z., Hossain, M.J., Cheng, J., Leung, V.C.M.: Hybrid RF/FSO Backhaul networks with statistical-QoS-aware buffer-aided relaying. IEEE Trans. Wireless Commun. **19**(3), 1464–1483 (2020)
31. Zhang, Z., Wang, R., Yu, F.R., Fu, F., Yan, Q.: QoS aware transcoding for live streaming in edge-clouds aided hetnets: an enhanced Actor-Critic approach. IEEE Trans. Veh. Technol. **68**(11), 11295–11308 (2019)
32. Gao, X., Edfors, O., Rusek, F., Tufvesson, F.: Massive MIMO performance evaluation based on measured propagation data. IEEE Trans. Wireless Commun. **14**(7), 3899–3911 (2015)
33. Björnson, E., Larsson, E., Debbah, M.: Massive MIMO for maximal spectral efficiency: how many users and pilots should be allocated? IEEE Trans. Wireless Commun. **15**(2), 1293–1308 (2016)
34. Guo, C., Liang, L., Li, G.Y.: Resource allocation for low-latency vehicular communications: an effective capacity perspective. IEEE J. Sel. Areas Commun. **37**(4), 905–917 (2019)
35. Shehab, M., Alves, H., Latva-aho, M.: Effective capacity and power allocation for machine-type communication. IEEE Trans. Veh. Technol. **68**(4), 4098–4102 (2019)
36. Cui, Q., Gu, Y., Ni, W., Liu, R.P.: Effective capacity of licensed-assisted access in unlicensed spectrum for 5G: from theory to application. IEEE J. Sel. Areas Commun. **35**(8), 1754–1767 (2017)
37. Xiao, C., Zeng, J., Ni, W., Liu, R.P., Su, X., Wang, J.: Delay guarantee and effective capacity of downlink NOMA fading channels. IEEE J. Sel. Top. Signal Process. **13**(3), 508–523 (2019)
38. Björnson, E., Larsson, E.G., Debbah, M.: Massive MIMO for maximal spectral efficiency: how many users and pilots should be allocated? IEEE Trans. Wireless Commun. **15**(2), 1293–1308 (2016)

Effective Capacity Analysis on Communication of Blockchain Synchronization Software-Defined Industrial Internet of Things

Yun Chen, Lei Chen(✉), Ping Cui, Kailiang Zhang, and Yuan An

Jiangsu Province Key Laboratory of Intelligent Industry Control Technology, Xuzhou University of Technology, Xuzhou 221018, China
chenlei@xzit.edu.cn

Abstract. For the blockchain network, the status of the account is saved via the checkpoint nodes, and updated. These knots are decentralized connected to each other. To get the current status, the IOT device must synchronize with the blockchain replica stored in the test bench. Delay prediction is an important indicator, to define the performance of software block chain Internet of things to knives. Based on the transport model generated by the synchronization protocol and the effective capacity (EC) theory of the wireless channel, the connections of QoS requests and key parameters of the synchronization protocol are studied. Simulation results show the effect of the new block creation rate on QoS level.

Keywords: Software Defined Networks · Quality of Service · Internet of Things · Blockchain · Effective Capacity Analysis

1 Introduction

The network of blockchains is one important point in the information system [1, 2]. Thanks to the decentralized communication mechanism, which is used only for Internet of Things (IoT) network [3–5]. In recent years, the new research point of network architecture has changed to be algorithms of network routing [6, 7]. Many people focus on analysis of network and traffic behavior, which contributes to make improvements of Quality of Service (QoS), which is the main objective of optimizing traffic forecasts [13–16]; Through traffic forecasting, several important indicators (KPI) can be optimized [17–19]. These optimized KPIs include frequency efficiency [20–23] and energy efficiency [24–26]. Several traffic flow forecasting methods are proposed [27–29]. On the IoT internet, with limited resources, these optimizations are more important. Under some restrictions, a channel Limited capacity and an effective capability (EC) associated with the service [30–34] are analyzed. The focus of the survey in analysis is to measure network flow and rebuild river flow [35–37]. The IoT network communication efficiency defined by the software is analyzed.

Researchers have discovered that many antennas improve Spectrum efficiency and user experience. In this area, researcher studies that the arrival of traffic may affect the

H. Song and D. Jiang (Eds.): SIMUtools 2020, LNICST 370, pp. 315–322, 2021.
https://doi.org/10.1007/978-3-030-72795-6_25

quality. The service researchers study the number of scheduled users and QoS requests [38], while [39] studies some synchronization protocols between single network devices and blockchain networks, and assesses the quality impact of the communication link and blocks chain parameters in the synchronization process. Result show that we should also consider communication parameters in the design of the chain service cycle and block device. This paper is divided into five sections. In the second section, system model is introduced. Simulation parameters and results have been showed in the third section. Finally, our conclusions come to the fourth section.

2 System Model

According to reference [39], exchange of model data expected by protocol P1 depends on the fact that device *u* is static or inactive to the implementation of protocol. Especially, when *u* wakes up, run the next sequence: firstly, *u* require new blocks from each of the *BN* nodes through the message of l block queue. After that, choose *BN* node to cast its new-block. The synchronization process can also inform *u* of the selection of BN to receive blocks. The Fig. 1 shows such a procession of requests and casts of blocks.

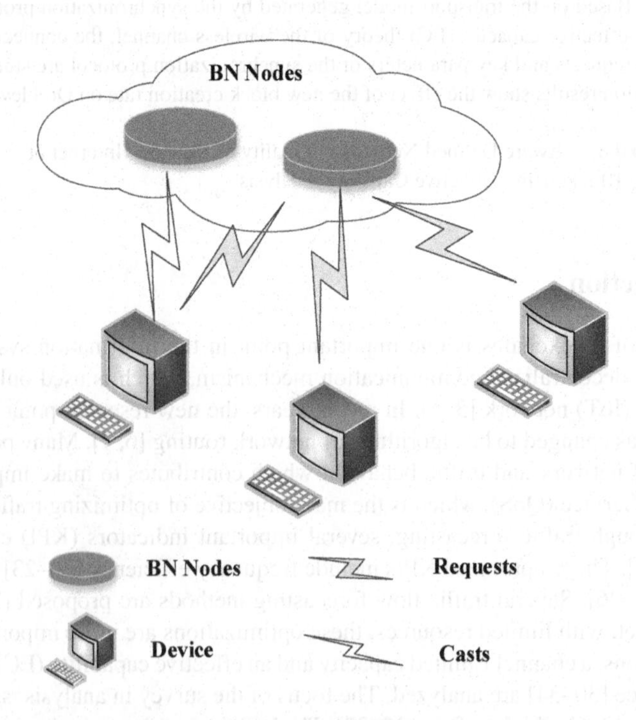

Fig. 1. Devices require new blocks from each of the BN nodes through the message of l block queue. After that, choose BN node to cast its new-block

The number of new blocks X_k is observed as an irreducible positive recursive Markov process. In order to analyze its likelihood of transition, we analyze the samples that depend on the protocol used, derived from the following paper probability of the transition from state $X_k = n$ to $X_{k+1} = m$, where the state equals to amount of missing blocks in blockchain local copy, depending on whether u is inactive between X_k and X_{k+1} [39]:

$$P\left[X_{k+1} = m|X_k = n\right] = (1 - p_s) \cdot p_{s0}(m|n) + p_s \cdot p_{s1}(m|n) \tag{1}$$

Where $p_{s0}(m|n)$ is the probability of the case that u stays active when implementing two protocols, which is the probability that blocks arrive during the protocol execution time t_p:

$$p_{s0}(m|n) = \int_0^{\infty} p_{s0}\left(t_p|n\right) \frac{\left(\lambda_B t_p\right)^m}{m!} e^{-\lambda_B t_p} dt_p \tag{2}$$

The λ_B is the parameter of P arriving, t_s is sleeping period. In other cases, u sleeps between X_k and X_{k+1}, the protocol execution time, depending on n and the number of blocks accumulated during inactive time, expressed by q.

$$\begin{aligned} p_{s1}(m|n) &= \sum_{q=0}^{\infty} p(m|n, q)p(q) \\ &= \sum_{q=0}^{\infty} \left(\int_0^{\infty} p_{s1}\left(t_p|n, q\right) \frac{\left(\lambda_B t_p\right)^m}{m!} e^{-\lambda_B t_p} dt_p \right) \frac{(\lambda_B t_s)^q}{q!} e^{-\lambda_B t_s} \end{aligned} \tag{3}$$

The device send message by Δt_x. In case that u is awake and synchronized, the protocol execution time is

$$\begin{aligned} t_{s,1} =& \Delta t_{s,1} + \Delta t_{s,2} + \Delta t_{s,3} \\ =& g_{DL}^{-1} l_r + t_w + g_{UL}^{-1} l_r + g_{DL}^{-1} l_b. \end{aligned} \tag{4}$$

If u is awake and not synchronized, the duration of the execution is a function of the number of missing blocks n:

$$\begin{aligned} t_{w,1}(n) =& \Delta t_{w1,1} + \Delta t_{w1,2} + \Delta t_{w1,3} + \Delta t_{w1,4} \\ =& g_{UL}^{-1} N l_n + g_{DL}^{-1} N l_r + g_{UL}^{-1} l_n + g_{DL}^{-1} n l_b. \end{aligned} \tag{5}$$

The distributions of the protocol execution time t_p are then

$$p_{s0}\left(t_p|n\right) = \begin{cases} \delta\left(t_{s,1}\right) & n = 0 \\ \delta\left(t_{w,1}(n)\right) & otherwise, \end{cases} \tag{6}$$

$$p_{s1}\left(t_p|n, q\right) = \delta\left(t_c + t_{w,1}(n+q)\right), \tag{7}$$

where $\delta(\cdot)$ is the Dirac. If u inactive for protocol execution, the connection time t_c should be included as well:

Define $t_a(k) = t_c + t_{w,1}(k)$, and the Eq. (2) and (3) can be rewritten as [39]:

$$p_{s0}(m|n) = \begin{cases} \frac{\left(\lambda_B t_{s,1}\right)^m}{m!} e^{-\lambda_B t_{s,1}} p_M, \ n = 0 \\ \frac{\left(\lambda_B t_{w,1}(n)\right)^m}{m!} e^{-\lambda_B t_{w,1}(n)}, \ otherwise \end{cases} \tag{8}$$

$$p_{s_1}(m|n) = \sum_{q=0}^{\infty}\left(\frac{(\lambda_B t_a(n+q))^m}{m!} e^{-\lambda_B t_a(n+q)}\right)\frac{(\lambda_B t_s)^q}{q!} e^{-\lambda_B t_s} \tag{9}$$

In wire networks, the switches and routing devices often use the dampers when the arrival speed of the service source is greater than the EM transmission speed in a short period of time. Suppose these networks channels are stable, so the most important problem is that the characteristics of the arrival column. Therefore, the key problem is that the characteristics of the arrival columns indicate that the peak and medium speed analysis may change the rate of arrival. It described that QoS problem is generally considered as the relationship between the system envelope and the transmission capacity. In IoT blockchain networks, the arrival rate is time-varying, which is depended by the new blocks.

The QoS requirements are showed as [38]

$$P_r\left(\max_{1\le i\le N} Q_i(0) > B\right) \le \epsilon. \tag{10}$$

For a large buffer size B, given a QoS constraint ϵ and by choosing $\theta = -\log(\epsilon)/B$, the QoS requirement can be expressed as an EC problem:

$$\lambda \le \min_{1\le j\le N} C_k(\theta) \tag{11}$$

Where

$$C_k(\theta) = \frac{1}{\theta}\lim_{n\to\infty}\frac{-1}{n}\ln \mathrm{E}\left(e^{-\theta\sum_{t=1}^{n} r_k(t)}\right) \tag{12}$$

and $r_k(t)$ is the rate allocated to user k in cell j at time t. Assuming these devices share a common channel, the number of devices receiving new blocks increases, and the opportunity for devices to acquire blocks decreases.

$$r_k(t) = \begin{cases} \mathrm{r\ W.P.1 - P(X_{k+1} = 0|X_k = 0)} \\ \mathrm{0\ W.P.\ \ P(X_{k+1} = 0|X_k = 0)} \end{cases} \tag{13}$$

In this case, the effective capacity can be written as:

$$\mathrm{C}(\theta) = \frac{1}{\theta}\lim_{n\to\infty}\frac{-1}{n}\sum_{\tau=0}^{n} e^{-\theta\tau}\mathrm{P}\left\{\left(\sum_{t=1}^{n} r_k(t)\right) = \tau r\right\} \tag{14}$$

Finally, the EC of device can be formulated as:

$$\mathrm{C}(\theta) = \frac{-1}{\theta}\ln\left(1 - \frac{1 - e^{-\theta r}}{\mathrm{K}(1 - \mathrm{P}(X_{k+1} = 0|X_k = 0))}\right) \tag{15}$$

3 Simulation and Results

This article studies the relationship between QoS requirements, as well as the sleep time of blockchain devices in IoT networks. We use the BN for simulating simple scenes. The sleep interval is [10, 2000], and the QoS parameters vary from 0.1 to 0.9. The channel capacity is stable with a value of 100.

The results of the simulation are shown in the Fig. 2. Increasing sleep time is a powerful first step in reducing EC experience, which means building new ones blocks collections. The high value of θ means QoS strict of request. EC get reduced as increasing the quality of QoS requests (Fig. 3).

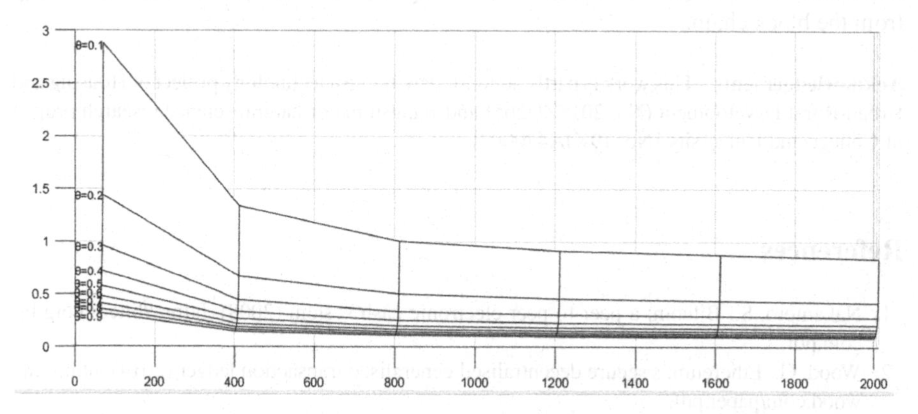

Fig. 2. Effective capacity Vs. sleeping period with fixed QoS request.

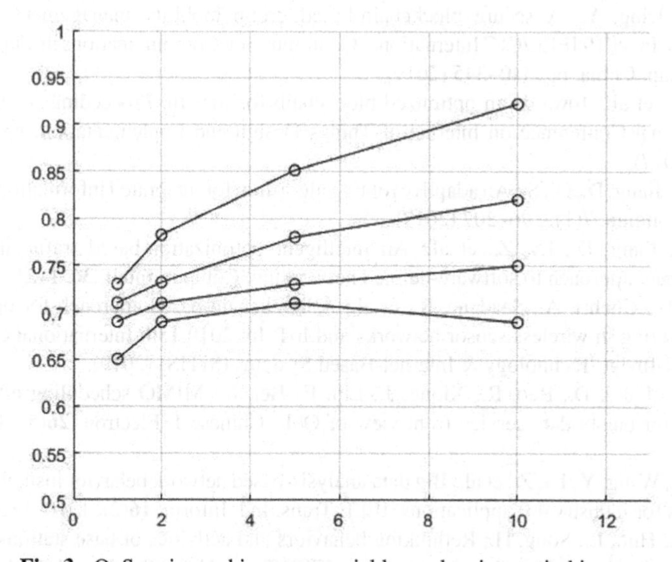

Fig. 3. QoS varies and increases quickly as sleeping period increases.

4 Conclusion

In recent years, the blockchain protocol has sparked great interest from researchers and industry. It seems a promising technology, but it is immature. This protocol family of bitcoin specification proposes a password-based solution that tries to replicate several parameters in a block chain synchronization protocol consistent database distributed by a network not licensed. The device sleep time, energy efficiency and reaction time of the new equipment are affected, either. Using our effective capability model on the street, the results of the summary sleep time simulation and the study of the QoS requirement ratio show that the emergence of the TEM sleep time has a significant impact on the requirements of the QoS. This is useful for designing network synchronization strategy from the block chain.

Acknowledgements. This work is partly supported by Jiangsu technology project of Housing and Urban-Rural Development (No. 2018ZD265) and Jiangsu major natural science research project of College and University (No. 19KJA470002).

References

1. Nakamoto, S.: Bitcoin: a peer-to-peer electronic cash system (2008). https://bitcoin.org/bitcoin.pdf
2. Wood, G.: Ethereum: s secure decentralised generalised transaction ledger (2014). http://gavwood.com/paper.pdf
3. Qiu, C., Yu, F.R., Xu, F., et al.: Permissioned blockchain-based distributed software-defined industrial Internet of Things. In: IEEE GLOBECOM 2018. IEEE (2018)
4. Lai, C., Ding, Y.: A secure blockchain-based group mobility management scheme in VANETs. In: 2019 IEEE/CIC International Conference on Communications in China (ICCC), Changchun, China, pp. 340–345 (2019)
5. Dorri, A., et al.: Towards an optimized blockchain for IoT. In: Proceedings of the Second International Conference on Internet-of-Things Design and Implementation, pp. 173–178. ACM (2017)
6. Wang, F., Jiang, D., Qi, S.: An adaptive routing algorithm for integrated information networks. China Commun. **7**(1), 196–207 (2019)
7. Huo, L., Jiang, D., Lv, Z., et al.: An intelligent optimization-based traffic information acquirement approach to software-defined networking. Comput. Intell. **36**, 1–21 (2019)
8. Lazrag, H., Chehri, A., Saadane, R., et al.: A blockchain-based approach for optimal and secure routing in wireless sensor networks and IoT. In: 2019 15th International Conference on Signal-Image Technology & Internet-Based Systems (SITIS) (2019)
9. Chen, L., Jiang, D., Bao, R., Xiong, J., Liu, F., Bei, L.: MIMO scheduling effectiveness analysis for bursty data service from view of QoE. Chinese J. Electron. **26**(5), 1079–1085 (2017)
10. Jiang, D., Wang, Y., Lv, Z., et al.: Big data analysis-based network behavior insight of cellular networks for industry 4.0 applications. IEEE Trans. Ind. Inform. **16**(2), 1310–1320 (2020)
11. Jiang, D., Huo, L., Song, H.: Rethinking behaviors and activities of base stations in mobile cellular networks based on big data analysis. IEEE Trans. Netw. Sci. Eng. **1**(1), 1–12 (2018)
12. Chen, L., et al.: A lightweight end-side user experience data collection system for quality evaluation of multimedia communications. IEEE Access **6**(1), 15408–15419 (2018)

13. Tan, J., Xiao, S., Han, S., Liang, Y., Leung, V.C.M.: QoS-aware user association and resource allocation in LAA-LTE/WiFi coexistence systems. IEEE Trans. Wireless Commun. **18**(4), 2415–2430 (2019)
14. Wang, Y., Tang, X., Wang, T.: A unified QoS and security provisioning framework for wiretap cognitive radio networks: a statistical queueing analysis approach. IEEE Trans. Wireless Commun. **18**(3), 1548–1565 (2019)
15. Hassan, M.Z., Hossain, M.J., Cheng, J., Leung, V.C.M.: Hybrid RF/FSO Backhaul networks with statistical-QoS-aware buffer-aided relaying. IEEE Trans. Wireless Commun. **19**(3), 1464–1483 (2020)
16. Zhang, Z., Wang, R., Yu, F.R., Fu, F., Yan, Q.: QoS aware transcoding for live streaming in edge-clouds aided HetNets: an enhanced Actor-critic approach. IEEE Trans. Veh. Technol. **68**(11), 11295–11308 (2019)
17. Chen, L., Zhang, L.: Spectral efficiency analysis for massive MIMO system under QoS constraint: an effective capacity perspective. Mob. Netw. Appl. (2020). https://doi.org/10.1007/s11036-019-01414-4
18. Wang, F., Jiang, D., Qi, S., Qiao, C., Shi, L.: A dynamic resource scheduling scheme in edge computing satellite networks. Mob. Netw. Appl. 1–12 (2020). https://doi.org/10.1007/s11036-019-01421-5
19. Jiang, D., Huo, L., Lv, Z., et al.: A joint multi-criteria utility-based network selection approach for vehicle-to-infrastructure networking. IEEE Trans. Intell. Transp. Syst. **19**(10), 3305–3319 (2018)
20. Lee, Y., Kim, Y., Park, S.: A machine learning approach that meets axiomatic properties in probabilistic analysis of LTE spectral efficiency. In: 2019 International Conference on Information and Communication Technology Convergence (ICTC), Jeju Island, Korea (South), pp. 1451–1453 (2019)
21. Ji, H., Sun, C., Shieh, W.: Spectral efficiency comparison between analog and digital RoF for mobile Fronthaul transmission link. J. Lightwave Technol. **38**, 5617–5623 (2020)
22. Hayati, M., Kalbkhani, H., Shayesteh, M.G.: Relay selection for spectral-efficient network-coded multi-source D2D communications. In: 2019 27th Iranian Conference on Electrical Engineering (ICEE), Yazd, Iran, pp. 1377–1381 (2019)
23. You, L., Xiong, J., Zappone, A., Wang, W., Gao, X.: Spectral efficiency and energy efficiency tradeoff in massive MIMO downlink transmission with statistical CSIT. IEEE Trans. Signal Process. **68**, 2645–2659 (2020)
24. Jiang, D., Zhang, P., Lv, Z., et al.: Energy-efficient multi-constraint routing algorithm with load balancing for smart city applications. IEEE Internet Things J. **3**(6), 1437–1447 (2016)
25. Jiang, D., Li, W., Lv, H.: An energy-efficient cooperative multicast routing in multi-hop wireless networks for smart medical applications. Neurocomputing **220**, 160–169 (2017)
26. Jiang, D., Wang, Y., Lv, Z., et al.: Intelligent optimization-based reliable energy-efficient networking in cloud services for IIoT networks. IEEE J. Sel. Areas Commun. (2019)
27. Jiang, D., Huo, L., Li, Y.: Fine-granularity inference and estimations to network traffic for SDN. Plos One **13**(5), 1–23 (2018)
28. Wang, Y., Jiang, D., Huo, L., et al.: A new traffic prediction algorithm to software defined networking. Mob. Netw. Appl. (2019). https://doi.org/10.1007/s11036-019-01423-3
29. Qi, S., Jiang, D., Huo, L.: A prediction approach to end-to-end traffic in space information networks. Mob. Netw. Appl. 1–10 (2019). https://doi.org/10.1007/s11036-019-01424-2
30. Guo, C., Liang, L., Li, G.Y.: Resource allocation for low-latency vehicular communications: an effective capacity perspective. IEEE J. Sel. Areas Commun. **37**(4), 905–917 (2019)
31. Shehab, M., Alves, H., Latva-aho, M.: Effective capacity and power allocation for machine-type communication. IEEE Trans. Veh. Technol. **68**(4), 4098–4102 (2019)

32. Cui, Q., Gu, Y., Ni, W., Liu, R.P.: Effective capacity of licensed-assisted access in unlicensed spectrum for 5G: from theory to application. IEEE J. Sel. Top. Signal Process. **35**(8), 1754–1767 (2017)
33. Xiao, C., Zeng, J., Ni, W., Liu, R.P., Su, X., Wang, J.: Delay guarantee and effective capacity of downlink NOMA fading channels. IEEE J. Sel. Top. Signal Process. **13**(3), 508–523 (2019)
34. Björnson, E., Larsson, E.G., Debbah, M.: Massive MIMO for maximal spectral efficiency: how many users and pilots should be allocated? IEEE Trans. Wireless Commun. **15**(2), 1293–1308 (2016)
35. Jiang, D., Wang, W., Shi, L., et al.: A compressive sensing-based approach to end-to-end network traffic reconstruction. IEEE Trans. Netw. Sci. Eng. **5**(3), 1–12 (2018)
36. Huo, L., Jiang, D., Qi, S., Song, H., Miao, L.: An AI-based adaptive cognitive modeling and measurement method of network traffic for EIS. Mob. Netw. Appl. 1–11 (2019). https://doi.org/10.1007/s11036-019-01419-z
37. Huo, L., Jiang, D., Zhu, X., et al.: An SDN-based fine-grained measurement and modeling approach to vehicular communication network traffic. Int. J. Commun. Syst. 1–12 (2019). https://doi.org/10.1002/dac.4092
38. Chen, L., Zhang, L.: Spectral efficiency analysis for massive MIMO system under QoS constraint: an effective capacity perspective. Mob. Netw. Appl. 1–9 (2020). https://doi.org/10.1007/s11036-019-01414-4
39. Danzi, P., Kalor, A.E., Stefanovic, C., Popovski, P.: Analysis of the communication traffic for blockchain synchronization of IoT devices. In: 2018 IEEE International Conference on Communications (ICC), Kansas City, MO, pp. 1–7 (2018)

A Quantum Classifier Based Active Machine Learning for Intelligent Interactive Service

Jiamin Cheng, Lei Chen(✉), and Ping Cui

Jiangsu Province Key Laboratory of Intelligent Industry Control Technology, Xuzhou University of Technology, Xuzhou 221000, China
chenlei@xzit.edu.cn

Abstract. The response time of interactive services depends not only on network latency, but also on computer time. Active learning algorithms are the most important methods. One problem is that these algorithms with uncertain sampling strategies propose an active learning sampling strategy on the basis of sample error correction to ensure that the efficiency and accuracy of interactive information calling are improved, and they have high computational complexity. However, due to computational complexity, this method is only suitable for smaller data sets. This article discusses the use of quantum clusters to accelerate calculations

Keywords: Response time · Quantum classifier · Active learning · Machine learning · Human–computer interaction

1 Introduction

With the acceleration of mobile phone access, the number of Internet users accessing the Internet through mobile devices is increasing rapidly, and the amount of computing is shifting to the cloud [1]. Services such as online translation, voice cloud, and information push are information calls. Based on large fuselage. The limitations of natural language recognition technology make it difficult to accurately retrieve large text databases. It is necessary to strengthen the search target through interactive query. Realizing the above interactive information search is an active learning strategy. Active learning will ask questions instead of passively accepting knowledge, which will increase the applicability of the next interaction according to the reaction tank. Active learning helps to find high-quality samples in the sample room and accurately describe user needs.

Due to the development of communication technology between the core network side and mobile terminal users [2–4], the user experience of traditional services has been improved on the basis of the new scheduling scheme [5–8]. Traffic analysis [9–12], a new planning strategy aimed at improving network performance indicators (such as resource use [13–16], frequency efficiency [17–20], energy efficiency [21, 22]). Measuring the performance of these new planning strategies can determine traffic flow. Therefore, the reconstruction of business flow has become an important topic [23–25]. In dense user communication, the effective capacity of the channel must be improved [26–30]. Within

H. Song and D. Jiang (Eds.): SIMUtools 2020, LNICST 370, pp. 323–332, 2021.
https://doi.org/10.1007/978-3-030-72795-6_26

the framework of the intelligent scheduling scheme, the accuracy of prediction starts from [3, 31, 32], and the traffic volume will seriously affect the effective capacity. Increasing the network performance index can improve the quality of these traditional services. In some emerging interactive services, the quality of human-computer interaction not only depends on communication performance. For these services, users have higher requirements for the delay and accuracy of such information country query services. In order to improve the quality of user experience, certain information request services put forward higher requirements for the accuracy and speed of calls. The limitation of computational cost and interaction time is the two main challenges of active learning in interactive information search.

Active learning is a continuously evolving interactive learning method. Lead to the uncertain sampling strategy in [33]. In their classification experiment, it will be a selection strategy that cannot be determined for each query problem, combined with a probabilistic classifier and satisfactory results. The above-mentioned active learning strategy has become a common learning method. Machine learning strategies are gradually being developed using different learning algorithms in the fields of natural language processing and information retrieval [34]. Common active learning strategies include committee questions, marginal sampling, and verifiability [35, 36]. These methods have produced good results in the field of information acquisition.

However, to obtain a good user experience in the interactive information search, building a high-quality training toolkit and reducing the computational complexity of the text classifier are two key issues. Active learning algorithms are the key method for learning interactive information. In recent years, deep drilling in the field of mechanical engineering has had huge practical advantages. However, Carrier Vector Machine (SVM) provides a solid theoretical foundation. Combined with the support vector machine, based on the analysis of the uncertainty and influence of the sample on the version space, several active learning strategies have been developed, including simple margins, maximum and minimum ranges and quotation ranges [37]. This is a hot topic. Therefore, the influence of labeled samples on the classification model is very important. It is very important to choose the most informative example [38]. Due to the large amount of storage space and time, SVM cannot support online services. Although multiplexing can shorten the training time [39], it will face high computational load under a larger gap.

The method of reinforcement learning was proposed in [40]. Since the latent semantic model is becoming more and more popular in text classification [42, 43], it has been widely used in the field of text classification [41]. Their version 2 of AdaBoost.MHAufgrund is more suitable for online information consulting environment. However, it is necessary to build a huge training set to increase the number of interactions and deteriorate the user's online service experience.

Interactive retrieval services require active learning algorithms with low complexity and high configuration Zision. The key to improving the accuracy of active learning algorithms is to select samples with a large amount of information [44]. The traditional active learning algorithm first selects a representative initial training set from cluster analysis [45], and then marks it. The most uncertain sample. In this case, the learning process will run through the initial set. In the case of limited initial information: In the early learning stage, the classifier often "misunderstands" the retreat target, which requires us to spend a lot of time. A method is proposed on AdaBoost.MHVersion [46]. In this research, we discussed the application of quantum mechanisms to the active learning model proposed in [47]. accelerate. This article is divided into five parts. The second part reviews the interactive retrieval method proposed in Reference [47]. The third part discusses: How to combine the retrieval method with the quantum classifier proposed in Reference [48]. The fourth part is the conclusion.

2 Survey on Interactive Methods for Querying Information

Interactive information retrieval has the following three characteristics: (1) There is no sample selected first. (2) Through interactive selection of high-quality samples, a high-quality training set is established to form a high-precision classifier. (3) Services are sensitive to delays; high-quality training courses must be established within a limited time.

In order to meet these challenges, the interactive retrieval algorithm can be designed as follows:

Step 1) The user sends query information.
Step 2) Calculate the degree of association between the query information and different documents, and create an association model.
Step 3) Use active learning strategies to improve the association model through multiple interactions with users.
Step 4) Sort the documents according to the association model, and output the query results in descending order of association level.

The interactive retrieval system model can be expressed as $A = (C, L, S, Q, U)$, where C is classifier, $L = R_t \cup N_t$, R_t and N_t are relevant and irrelevant document set which are identified through interaction, S is classifier, Q is evaluation function that acquires simplified training set by identifying high-value samples, and U is the unlabeled sample set and evaluation object of Q.

Let C_L be the classifier formed by using L as the training set, and $C_L(x)$ be the classifier calculated value of sample x. In binary classification, classification results are generally judged by sign of values. If $x \in U$, $|C_L(x)|$ is proportional to the certainty degree of categorization of x and samples with low certainty degree are usually chosen into the training set in active learning. Additionally, if $y(x)$ is the real category of sample x, the $\left|\left\{C_L^{'}, \forall C_L^{'}(x) = y(x)\right\}\right|$ is size of the version size (where $C_L^{'}$ is set of all candidate classifiers), the samples that can reduce the version space to the greatest extent can be used as high-value samples.

If these samples are recognized as different categorization by human, namely $y \cdot C_L(x) < 0$, then the high value of $|C_L(x)|$ means the high error-correcting capacity to the classifier C_L. As x is added, the new training set $C_{L\cup\{x\}}$ will be closer to real user demands. Samples that can correct classifiers have higher values when interactions are few and the early cognition of classifier significantly deviates from the retrieval target.

To evaluate the unlabeled samples' capacity to correct classifier, we designed an expression to select the high quality samples:

$$\alpha \cdot po \cdot \left(\frac{(S_{Max} - S_d)}{(S_{Max} - S_{Min})} \right) + \beta \cdot ne \cdot \left(\frac{(S_d - S_{Min})}{(S_{Max} - S_{Min})} \right) \quad (1)$$

where α and β are coefficients, *po* is expected contribution (correcting capacity) of documents judged as positive sample to the classifier, *ne* is expected contribution (correcting capacity) of documents judged as negative sample to the categorization, S_d is the score given by classifier to the current document d (a higher score means a higher expectation for the document belonging to positive sample and a lower score translates to lower expectation), S_{Max} and S_{Min} are the highest and lowest scores of the classifier for unlabeled documents.

In Eq. (1), $\frac{S_{Max} - S_d}{S_{Max} - S_{Min}}$ reflects the probability for document to be judged by the current classifier C_L as positive samples and $\frac{S_d - S_{Min}}{S_{Max} - S_{Min}}$ is the probability for document to be judged by the current classifier C_L as negative sample.

To each unlabeled document sample, the calculation formula determining its contribution coefficients (correcting capacity) for current classifier:

$$po = \sum_{\forall w \in W} c(w) \cdot idf(w) \quad (2)$$

and

$$ne = \sum_{\forall w \notin W} tf - idf(w, d), \quad (3)$$

where $c(w)$ is the relevancy between term w given by the classifier and the target query document. This score can measure the consistence between sample and the retrieval target. And W is a set of key terms in the current document d. Let D be the document set, $d \in D$ is the current document and $Tr \subset D$ is the labeled document set. Let $|Tr|$ stand for total number of labeled document, $\#Tr(w)$ stand for number of labeled documents containing the word w and $\#(w, d)$ is the frequency of w in document d. The calculation formula of the *idf* function becomes $idf(w) = \log\left(|Tr| / \#Tr(w)\right)$, so the $tf - idf$ function formula is $tf - idf(w, d) = (\#(w, d)) \cdot idf(w)$.

3 Redesign Method Based on a Quantum Enclosure

This part is explained by the mechanical authentic. By Please write that model of the quantum capsule FOI proposed in [48]. Let A_0 is the hypothesis that the sample does not belong to the wanted documents, and A_1 is the hypothesis that the document contains some wanted content. The choice is represented as a value in the interval [0, 1]. The decision is A_0 for $\Delta = 0$ and A_1 for $\Delta = 1$, respectively.

The detection Δ operator yields 1 under A_0 with the following probability:

$$Q_0 \hat{=} P(\Delta = 1|A_0) = tr(\rho_0 \Delta). \quad (4)$$

$$Q_1 = P(\Delta = 1|A_1) = tr(\rho_1 \Delta). \tag{5}$$

The average cost can be written as

$$\bar{K} = \xi K_{00} + (1-\xi)K_{01} - (1-\xi)(K_{01} - K_{11})tr(\rho_1 - \lambda\rho_0)\Delta \tag{6}$$

where

$$\lambda = \frac{\xi(K_{10} - K_{00})}{(1-\xi)(K_{01} - K_{11})}. \tag{7}$$

If $K_{01} > K_{11}$, $\bar{K}$ will be minimum if $tr((\rho_1 - \lambda\rho_0)\Delta)$ can be maximized.

The best detection operator is provided by the eigenstates $|e_1\rangle$ if the operator $\rho_1 - \lambda\rho_0$ corresponding to the positive eigenvaues, where the eigensystem is provided by

$$(\rho_1 - \lambda\rho_0)|e_l\rangle = e_l|e_l\rangle \quad l = 1, ..., (\text{rank of } \rho_1 - \lambda\rho_0) \tag{8}$$

So it is essential to maximize

$$tr(\rho_1 - \lambda\rho_0)\Delta = \sum_l e_l\langle e_l|\Delta|e_l\rangle \tag{9}$$

and this can be obtained if

$$e_k\langle e_l|\Delta|e_l\rangle = 1, e_l \geq 0 \quad \text{ and } \quad e_l\langle e_l|\Delta|e_l\rangle = 0, e_l < 0 \tag{10}$$

The estimation of the optimal projection operator between A_0 and A_1 can be written as

$$\Delta = \sum_{l:e_l \geq 0} |e_l\rangle\langle e_l| \tag{11}$$

So the error probabilities is given in [48]:

$$Q_0 = \sum_{l:e_l \geq 0} \langle e_l|\rho_0|e_l\rangle \qquad \text{and} \qquad Q_1 = 1 - \sum_{l:e_l \geq 0} \langle e_l|\rho_1|e_l\rangle \tag{12}$$

In [48], the minimum average cost is formulated as:

$$K^{-}_{min} = \xi K_{00} + (1-\xi)K_{01} - (1-\xi)(K_{01} - K_{11})\sum_{l:e_l > 0} e_l \tag{13}$$

Consider the vector $|y\rangle$, which is the input query of an IR system.

In this model, each document can being represented as a word vector $|x\rangle$. The density operator ρ_0 can assess the relevance between unlabelled sample and negative sample. The density operator ρ_1 can assess the relevance between unlabelled sample and positive sample. The quantum SDT projects both the query vector $|y\rangle$ and the document vectors $|x\rangle$ by means of the optimal detection. The ranking is computed by:

$$\langle x|\Delta|y\rangle \tag{14}$$

The operators ρ_0 and ρ_1 are given as [48]:

$$\rho_0 = \frac{|v_0\rangle\langle v_0}{tr(|v_0\rangle\langle v_0)} \qquad \rho_1 = \frac{|v_1\rangle\langle v_1}{tr(|v_1\rangle\langle v_1)} \tag{15}$$

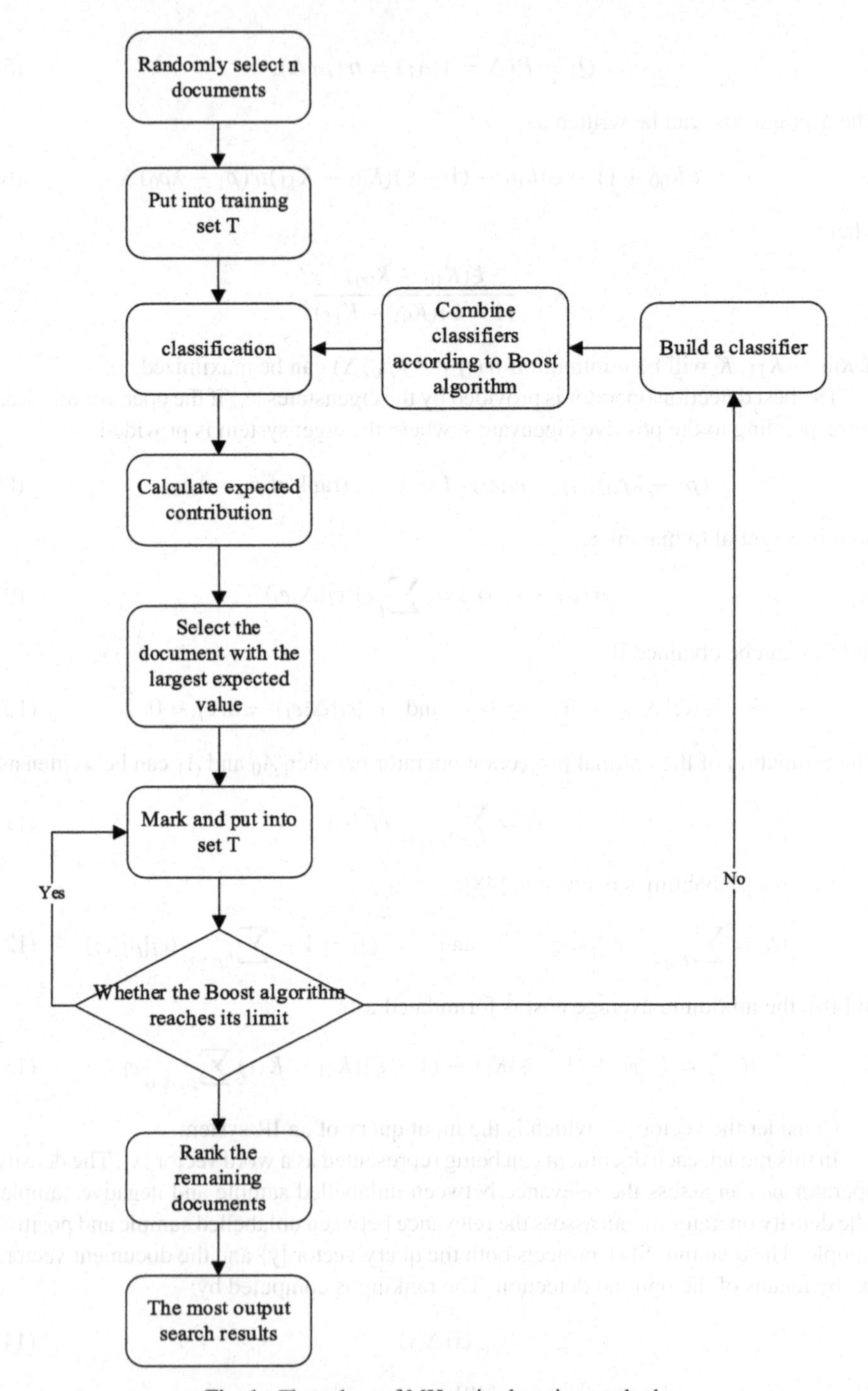

Fig. 1. Flow chart of MH active learning method

In combination with Eq. (1) and Eq. (15), the potential proportion of uncertain samples is defined. Most uncertain sample and the sample determined, if marked in the unexpected class, will provide important information.

An interactive retrieval algorithm is developed by combining the Boost algorithm with the classifier with low computational complexity, using the active learning strategy based on the error correction capability, incrementally increase the selected samples of the users and the performance of the classifier to improve. Algorithms and document sets can be used in the cloud are. To reduce the delay in interaction, adopting AdaBoost. As shown in the Fig. 1, MH Active learning methods can be redesigned as follows:

Step 1) randomly select N documents for the training set t.
Step 2) a classifier by Eq. (15).
Step 3) Combine classifiers by Boost algorithm.
Step 4) Use combination classifier to classify unlabeled documents.
Step 5) calculates the expected contribution of each unmarked document in this loop by formula (1).
Step 6) Select the document with the highest expected contribution
Step 7) provide the selected patterns for identification to the user and insert them into set t.
Step 8) If the Boost algorithm does not reach the iteration limit, return to step 2, otherwise go to step 7.
Step 9) Sort the remaining documents by Eq. (14).
Step 10) publish most search results.

4 Conclusions

Active learning is an effective way to achieve the performance of interactive information search improve. The number of iterations and response delay affect the end user's experience. We are using concept of expected contribution to evaluate the quality of the samples, to the number of iterations to reduce. Quantum-transformation is used to react the classifier to accelerate. Combined these two methods suggest an active learning algorithm.

Acknowledgements. This work is partly supported by Jiangsu technology project of Housing and Urban-Rural Development (No. 2018ZD265) and Jiangsu major natural science research project of College and University (No. 19KJA470002).

References

1. Toumi, H., Brahmi, Z., Benarfa, Z., Gammoudi, M.M.: Server load prediction using stream mining. In: 31st International Conference on Information Networking, Da Nang, Vietnam, pp. 653–661. IEEE (2017)
2. Wang, F., Jiang, D., Qi, S.: An adaptive routing algorithm for integrated information networks. China Commun. **7**(1), 196–207 (2019)

3. Huo, L., Jiang, D., Lv, Z., et al.: An intelligent optimization-based traffic information acquirement approach to software-defined networking. Comput. Intell. **36**, 151–171 (2020)
4. Zhang, K., Chen, L., An, Y., et al.: A QoE test system for vehicular voice cloud services. Mob. Netw. Appl. (2019). https://doi.org/10.1007/s11036-019-01415-3
5. Barakabitze, A.A., et al.: QoE management of multimedia streaming services in future networks: a tutorial and survey. IEEE Commun. Surv. Tutor. **22**(1), 526–565 (2020)
6. Orsolic, I., Skorin-Kapov, L.: A framework for in-network QoE monitoring of encrypted video streaming. IEEE Access **8**, 74691–74706 (2020)
7. Song, E., et al.: Threshold-oblivious on-line web QoE assessment using neural network-based regression model. IET Commun. **14**(12), 2018–2026 (2020)
8. Seufert, M., Wassermann, S., Casas, P.: Considering user behavior in the quality of experience cycle: towards proactive QoE-aware traffic management. IEEE Commun. Lett. **23**(7), 1145–1148 (2019)
9. Chen, L., Jiang, D., Bao, R., Xiong, J., Liu, F., Bei, L.: MIMO scheduling effectiveness analysis for bursty data service from view of QoE. Chin. J. Electron. **26**(5), 1079–1085 (2017)
10. Jiang, D., Wang, Y., Lv, Z., et al.: Big data analysis-based network behavior insight of cellular networks for industry 4.0 applications. IEEE Trans. Ind. Inform. **16**(2), 1310–1320 (2020)
11. Jiang, D., Huo, L., Song, H.: Rethinking behaviors and activities of base stations in mobile cellular networks based on big data analysis. IEEE Trans. Netw. Sci. Eng. **1**(1), 1–12 (2018)
12. Chen, L., Jiang, D., Song, H., Wang, P., Bao, R., Zhang, K., Li, Y.: A lightweight end-side user experience data collection system for quality evaluation of multimedia communications. IEEE Access **6**(1), 15408–15419 (2018)
13. Chen, L., Zhang, L.: Spectral efficiency analysis for massive MIMO system under QoS constraint: an effective capacity perspective. Mob. Netw. Appl. (2020). https://doi.org/10.1007/s11036-019-01414-4
14. Wang, F., Jiang, D., Qi, S., et al.: A dynamic resource scheduling scheme in edge computing satellite networks. Mob. Netw. Appl. (2019). https://doi.org/10.1007/s11036-019-01421-5
15. Jiang, D., Huo, L., Lv, Z., et al.: A joint multi-criteria utility-based network selection approach for vehicle-to-infrastructure networking. IEEE Trans. Intell. Transp. Syst. **19**(10), 3305–3319 (2018)
16. Jiang, D., Zhang, P., Lv, Z., et al.: Energy-efficient multi-constraint routing algorithm with load balancing for smart city applications. IEEE Internet Things J. **3**(6), 1437–1447 (2016)
17. Lee, Y., Kim, Y., Park, S.: A machine learning approach that meets axiomatic properties in probabilistic analysis of LTE spectral efficiency. In: 2019 International Conference on Information and Communication Technology Convergence (ICTC), Jeju Island, South Korea, pp. 1451–1453 (2019)
18. Ji, H., Sun, C., Shieh, W.: Spectral efficiency comparison between analog and digital RoF for mobile fronthaul transmission link. J. Lightwave Technol. **38**, 5617–5623 (2020)
19. Hayati, M., Kalbkhani, H., Shayesteh, M.G.: Relay selection for spectral-efficient network-coded multi-source D2D communications. In: 2019 27th Iranian Conference on Electrical Engineering (ICEE), Yazd, Iran, pp. 1377–1381 (2019)
20. You, L., Xiong, J., Zappone, A., Wang, W., Gao, X.: Spectral efficiency and energy efficiency tradeoff in massive MIMO downlink transmission with statistical CSIT. IEEE Trans. Signal Process. **68**, 2645–2659 (2020)
21. Jiang, D., Li, W., Lv, H.: An energy-efficient cooperative multicast routing in multi-hop wireless networks for smart medical applications. Neurocomputing **220**, 160–169 (2017)
22. Jiang, D., Wang, Y., Lv, Z., et al.: Intelligent optimization-based reliable energy-efficient networking in cloud services for IIoT networks. IEEE J. Sel. Areas Commun. (2019)
23. Jiang, D., Wang, W., Shi, L., et al.: A compressive sensing-based approach to end-to-end network traffic reconstruction. IEEE Trans. Netw. Sci. Eng. **5**(3), 1–12 (2018)

24. Jiang, D., Huo, L., Li, Y.: Fine-granularity inference and estimations to network traffic for SDN. PLoS ONE **13**(5), 1–23 (2018)
25. Huo, L., Jiang, D., Qi, S., et al.: An AI-based adaptive cognitive modeling and measurement method of network traffic for EIS. Mob. Netw. Appl. (2019). https://doi.org/10.1007/s11036-019-01419-z
26. Guo, C., Liang, L., Li, G.Y.: Resource allocation for low-latency vehicular communications: an effective capacity perspective. IEEE J. Sel. Areas Commun. **37**(4), 905–917 (2019)
27. Shehab, M., Alves, H., Latva-aho, M.: Effective capacity and power allocation for machine-type communication. IEEE Trans. Veh. Technol. **68**(4), 4098–4102 (2019)
28. Cui, Q., Gu, Y., Ni, W., Liu, R.P.: Effective capacity of licensed-assisted access in unlicensed spectrum for 5G: from theory to application. IEEE J. Sel. Areas Commun. **35**(8), 1754–1767 (2017)
29. Xiao, C., Zeng, J., Ni, W., Liu, R.P., Su, X., Wang, J.: Delay guarantee and effective capacity of downlink NOMA fading channels. IEEE J. Sel. Top. Signal Process. **13**(3), 508–523 (2019)
30. Björnson, E., Larsson, E.G., Debbah, M.: Massive MIMO for maximal spectral efficiency: how many users and pilots should be allocated? IEEE Trans. Wirel. Commun. **15**(2), 1293–1308 (2016)
31. Qi, S., Jiang, D., Huo, L.: A prediction approach to end-to-end traffic in space information networks. Mob. Netw. Appl. (2019). https://doi.org/10.1007/s11036-019-01424-2
32. Wang, Y., Jiang, D., Huo, L., et al.: A new traffic prediction algorithm to software defined networking. Mob. Netw. Appl. (2019). https://doi.org/10.1007/s11036-019-01423-3
33. Lewis, D.D., Gale, W.A.: A sequential algorithm for training text classifiers. In: Proceedings of the Seventeenth Annual International ACM-SIGIR Conference on Research and Development in Information Retrieval, Dublin, Ireland, vol. 29, no. 2, pp. 13–19. ACM (1994)
34. Du, B., Wang, Z., Zhang, L., Zhang, L., Liu, W., Shen, J., Tao, D.: Exploring representativeness and informativeness for active learning. IEEE Trans. Cybern. **47**(1), 14–26 (2017)
35. Tuia, D., Ratle, F., Pacifici, F., Kanevski, M.F., Emery, W.J.: Active learning methods for remote sensing image classification. IEEE Trans. Geosci. Remote Sens. **47**(7), 2218–2232 (2009)
36. Liu, K., Qiang, X., Wang, Z.: Survey on active learning algorithm. Comput. Eng. Appl. **48**(34), 1–4 (2012)
37. Tong, S., Koller, D.: Support vector machine active learning with applications to text classification. J. Mach. Learn. Res. **2**(1), 45–66 (2002)
38. Kremer, J., Pedersen, K.S., Igel, C.: Active learning with support vector machines. Wiley Interdiscip. Rev. Data Min. Knowl. Discov. **4**(4), 313–326 (2014)
39. Hu, R., Mac, N.B., Delany, S.J.: Active learning for text classification with reusability. Expert Syst. Appl. **45**(C), 438–449 (2015)
40. Freund, Y., Schapire, Robert E.: A decision-theoretic generalization of on-line learning and an application to boosting. In: Vitányi, P. (ed.) EuroCOLT 1995. LNCS, vol. 904, pp. 23–37. Springer, Heidelberg (1995). https://doi.org/10.1007/3-540-59119-2_166
41. Jiang, Y., Zhou, Z.H.: A text classification method based on term frequency classifier ensemble. J. Comput. Res. Dev. **43**(10), 1681–1687 (2006)
42. Al-Salemi, B., Ab-Aziz, M.J., Noah, S.A.: LDA-AdaBoost.MH: accelerated AdaBoost.MH based on latent Dirichlet allocation for text categorization. J. Inf. Sci. **41**(1), 27–40 (2015)
43. Omar, M., On, B.W., Lee, I., Choi, G.S.: LDA topics: representation and evaluation. J. Inf. Sci. **41**(5), 1–4 (2015)
44. Forestier, G., Wemmert, C.: Semi-supervised learning using multiple clusterings with limited labeled data. Inf. Sci. **361**(C), 48–65 (2016)
45. Zhu, J., Wang, H., Tsou, B.K., Ma, M.: Active learning with sampling by uncertainty and density for data annotations. IEEE Trans. Audio Speech Lang. Process. **18**(6), 1323–1331 (2010)

46. Zhu, J., Wang, H., Yao, T., Tsou, B.K.: Active learning with sampling by uncertainty and density for word sense disambiguation and text classification. In: Proceedings of the 22nd International Conference on Computational Linguistics, Manchester, UK, pp. 1137–1144. ACL (2008)
47. Chen, L., Bao, R., Li, Y., Zhang, K., An, Y., Van, N.N.: An interactive information-retrieval method based on active learning. J. Eng. Sci. Technol. Rev. **10**(3), 1–6 (2017)
48. Tiwari, P., Melucci, M.: Towards a quantum-inspired binary classifier. IEEE Access **7**, 42354–42372 (2019). https://doi.org/10.1109/ACCESS.2019.2904624

Analysis on Relationship Between Fractional Calculus Fluid Model and Effective Capacity of Bursty Data Service in Multi-hop Wireless Networks

Lei Chen(✉), Ping Cui, Kailiang Zhang, and Yuan An

Jiangsu Province Key Laboratory of Intelligent Industry Control Technology, Xuzhou University of Technology, Xuzhou 221018, China
chenlei@xzit.edu.cn

Abstract. A fractional calculus fluid model can be used to better explain the traffic of bursty data service. It is long-range dependence and has a fractal-like feature of network data flow. This paper builds a fluid model to describe the traffic of multi-hop wireless networks with QoS constraint. We use effective capacity model to depict the performance of bursty data service in wireless networks with QoS constraint. Finally, experiment results show that the heavy-tailed delay distributions, the hyperbolically decay of the packet delay auto-covariance function and fractional differential equations are formally related. Our method is effective and feasible.

Keywords: Fluid model · QoS · Effective capacity · Multi-hop wireless networks

1 Introduction

Because user behavior become more and more complicated and many kinds of traffic flow exist in the network, researchers proposed much more routing approaches and measurement methods for new network architecture [1–3]. Many researches concentrate on how to measure the traffic, build the traffic model and improve the scheduling by using the traffic model [4–7]. For guaranteeing the QoS of emerging services, new scheduling strategies are proposed [8–11]. Due to the situation of the crowed mobile communication, resources utilization [12–15], spectral efficiency [16–19] and energy-efficiency [20–22] become hotspots. A large number of researches also focus on flow level of traffic reconstruction [23–25]. New scheduling schemes, which based on the traffic characters, effectively improve the experience of users [26–29].

Researchers have explored the relationship between the number of scheduled user and the QoS requirement by the effective capacity model of wireless communication theory [12]. To some extent, the QoS is influenced by the traffic flow. Unlike the other services, the bursty data service has more sophisticated flow which is very difficult

H. Song and D. Jiang (Eds.): SIMUtools 2020, LNICST 370, pp. 333–341, 2021.
https://doi.org/10.1007/978-3-030-72795-6_27

to predict [30–32]. The measurements of network traffic of bursty data service have shown that traffic characteristics include features which are more efficiently described in terms of fractal rather than conventional stochastic processes [33]. We found fractal dimension and long-range dependence in statistical moments that exist in corporate, local, and wide-area networks.

The rest of this paper is organized as follows. In the second section, we introduce the system model. In the third section, we present the traffic model under QoS constraint for multi-hop wireless networks. Simulation parameters and results are described in the fourth section. Finally, the conclusions are presented in the fifth section.

2 System Model

In a communication system with stable links, the switching and routing devices usually employ a large buffer to prevent the loss of packets when the arriving rate from the service source is higher than transmission rate over a short time. The key problem of QoS guarantee thus lies in analyzing the arriving queue. However, during the transmissions of these emerging bursty data services, such as speech cloud and remote control based on real-time video, we must consider the low reliability, time-varying channel and moving users of wireless environment. The effective capacity concept is a function of the probability of nonempty buffer and the QoS exponent of connection. In many researches, effective capacity is more suitable to measure the transmission capacity of time-varying channels [34–38]. Effective capacity model is shown in Fig. 1.

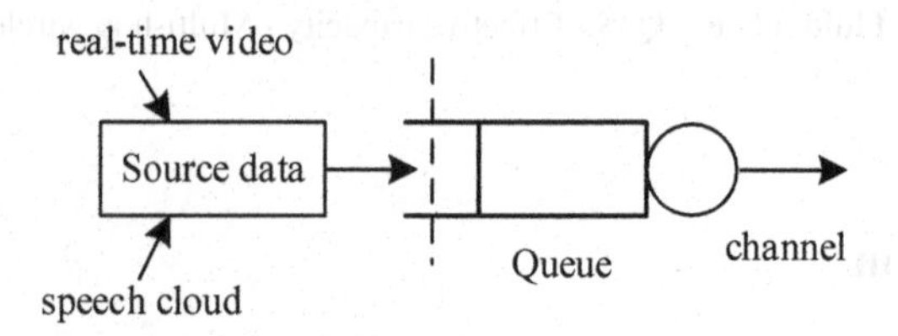

Fig. 1. Effective capacity model

The QoS requirement can be formulated as

$$P_r\left(\max_{1\leq i\leq N} Q_i(0) > B\right) \leq \in \cdot \tag{1}$$

With a large buffer size B, given a QoS constraint $\in$ and by choosing $\theta = -\log(\in)/B$, the QoS requirement can be expressed as an effective capacity problem:

$$\lambda \leq \min_{1\leq j\leq N} C_k(\theta) \tag{2}$$

where

$$C_k(\theta) = \frac{1}{\theta}\lim_{n\to\infty}\frac{-1}{n}\ln \mathrm{E}\left(e^{-\theta\sum_{t=1}^{n} r_k(t)}\right) \tag{3}$$

and $r_k(t)$ is the rate allocated to user k in cell j at time t. We assume that the scheduling scheme at the base station picks the K users out of a set of N active users for stochastic transmission with the same probability. Thus, the r_k can be written as

$$\mathrm{r}_k(t) = \begin{cases} \frac{N_f}{K}\left(1 - \frac{B}{S}\right)\log_2\left(1 + \frac{1}{I_j^{scheme}}\right), w.p.\frac{K}{N}, \\ 0, w.p.1 - \frac{K}{N}. \end{cases} \tag{4}$$

In a multi-hop wireless networks, the average packet latency at a site meets the condition [33]:

$$< t >= \int_0^{\infty} \mathrm{t} \cdot \mathrm{f(t)dt} = \infty \tag{5}$$

where, $\mathrm{f(t)} > 0$; $\int_0^{\infty} \mathrm{f(t)dt} = 1$. The corresponding expression for f(t) is a PDF:

$$\mathrm{f(t)} = \frac{\gamma}{(1+\mathrm{t})^{\gamma+1}}, 0 < \gamma < 1 \tag{6}$$

This probability density function characterizes the long-range statistical dependence in bursty data service traffic model. A CDF function can be introduced as $\mathrm{F}(\tau)$ [33]:

$$\mathrm{F}(\tau) = 1 - \int_0^{\tau} \mathrm{f(t)dt} = \frac{1}{(1+\tau)^{\gamma}} \tag{7}$$

where τ is the time a packet stays at an intermediate site x of the virtual connection. We suppose that the site is a connection device with an infinite buffer, the most probable number of packets in site x at the moment t can be denoted as:

$$\mathrm{n(x; t)} = \int_0^{\mathrm{t}} \mathrm{n(x-1; t-\tau)f(\tau)d\tau} + \mathrm{n_0(x)F(t)} \tag{8}$$

where $n_0(x)$ is the initial number of packets in the buffer of site x before the packet's arrival from site *x-1*. In this notation, the equation of packet migration can be presented as

$$\Gamma(1-\gamma)\mathrm{D}_{\mathrm{t}}^{\gamma}[\mathrm{n(x; t)}] = -\frac{\partial \mathrm{n(x; t)}}{\partial \mathrm{x}} + \frac{\mathrm{n_0(x)}}{\mathrm{t}^{\gamma}} \tag{9}$$

Where the left part of Eq. (9) is the fractional derivative of function n(x; t) with an exponent parameter γ, and

$$\mathrm{D}_{\mathrm{t}}^{\gamma}[\mathrm{n(x; t)}] = \frac{1}{\Gamma(1-\gamma)} \int_0^{\mathrm{t}} \frac{\mathrm{n(x; t)}}{(\mathrm{t}-\tau)^{\gamma}} \mathrm{d}\tau\cdot \tag{10}$$

Taking into account the discrete character of change of the variable x, [33] solves Eq. (17) subject to the following initial conditions: $\mathrm{n_0(0)} = \mathrm{n_0}$ and $\mathrm{n_0(k)} = 0$, $\mathrm{k} = 1, 2, \ldots$ In [33], the Eq. (8) can be rewritten as:

$$\mathrm{n(k;t)} = \mathrm{n_0}\left\{\frac{1}{\mathrm{t}^{\gamma}} - \mathrm{k}\frac{\Gamma^2(1-\gamma)}{\Gamma(1-2\gamma)t^{2\gamma}} - \mathrm{k}\Gamma(1-\gamma)\cdot\frac{1}{\Gamma(-\gamma)\mathrm{t}^{\gamma+1}}\right\} \tag{11}$$

$$n(k;t) = n_0\left\{\frac{1}{t^\gamma} - k\left[\frac{\Gamma^2(1-\gamma)}{\Gamma(1-2\gamma)}\cdot\frac{1}{t^{2\gamma}} + \frac{\Gamma(1-\gamma)}{\Gamma(-\gamma)}\cdot\frac{1}{t^{\gamma+1}}\right]\right\} \tag{12}$$

Taking into account the asymptotic property of the obtained solution, we can get

$$n(0;t) = n_0\left\{\frac{1}{t^\gamma}\right\} \tag{13}$$

For k = 1, the following expression is obtained:

$$n(1;t) = n_0\left\{\frac{1}{t^\gamma} - \left[\frac{\Gamma^2(1-\gamma)}{\Gamma(1-2\gamma)}\cdot\frac{1}{t^{2\gamma}} + \frac{\Gamma(1-\gamma)}{\Gamma(-\gamma)}\cdot\frac{1}{t^{\gamma+1}}\right]\right\} \tag{14}$$

Finally in [33], with the initial conditions $n(0;t) = n_0 \cdot \delta(t)$, the cumulative number of blogged packets is presented as:

$$\begin{aligned} c(m;t) &= n_0^2\Gamma(1-\gamma)\left[\frac{1}{\Gamma(1-2\gamma+1)}\cdot\frac{1}{t^{2\gamma-1}} - m\frac{\Gamma(1-\gamma)}{\Gamma(1-3\gamma+1)}\cdot\frac{1}{t^{3\gamma-1}}\right] \\ &= n_0^2\Gamma(1-\gamma)t^{1-2\gamma}\left[\frac{1}{\Gamma(2-2\gamma)} - m\frac{\Gamma(1-\gamma)}{\Gamma(2-3\gamma)}\cdot\frac{1}{t^\gamma}\right] \end{aligned} \tag{15}$$

This correlation function decays hyperbolically with increasing *t*. Therefore for $\gamma < 1$, such random processes have a fractal-like scaling behavior. Set m = 0, we get:

$$D(t) = c(0;t) = \frac{n_0\Gamma(1-\gamma)}{\Gamma(2-2\gamma)}t^{1-2\gamma} \tag{16}$$

3 Traffic Model Under QoS Constraint

Effective capacity theory provides a powerful framework to describe the relationship between transmission rate fluctuations and QoS constraints. The proposed model adopts only one QoS exponent parameter that represents two parameters, namely a delay constraint and buffer size. When the QoS exponent θ is equal to 0, rate fluctuations do not affect the effective capacity, which is equal to the average transmission rate. As the QoS requirement becomes stricter, the transmission rate fluctuations lead to a decline in effective capacity, which corresponds to the probability of delay violation. More specifically, higher fluctuation levels lead to lower effective capacity, which means higher probability of delay violation. The transmission rate fluctuations and QoS constraints is shown in Fig. 2.

With a fixed buffer size B and QoS constraint:

$$\theta = \frac{-\ln(\epsilon)}{B} \tag{17}$$

The ϵ represents the violation probability:

$$\Pr\left(t > \frac{B}{C(\theta)}\right) < \epsilon \tag{18}$$

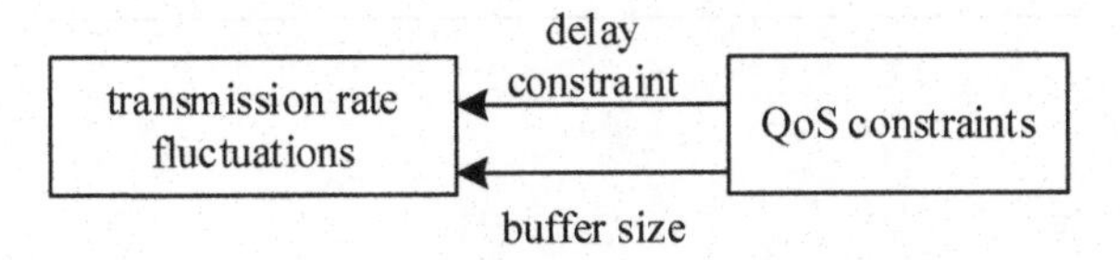

Fig. 2. The transmission rate fluctuations and QoS constraints

Combining the Eq. (7), we obtain:

$$\epsilon = F(\tau) = \frac{1}{(1+\tau)^{\gamma}} \tag{19}$$

Finally, we get a rough relationship between QoS parameter and the fluid model parameter γ:

$$\gamma = \log_{\left(1+\frac{B}{C(\theta)}\right)}\left(\frac{1}{\epsilon}\right) \tag{20}$$

4 Simulation and Results

We make simulation with fixed buffer size and QoS constraint to investigate the relationship between the QoS requirement and traffic model. Considering 10 intermediate sites, the violation probability is set as 0.001. The buffer size and bandwidth are set as 10Mbits and 10 MHz, respectively.

The simulation results are shown in Fig. 3. It shows the relationship between latency constraint and the distribution character of the fluid model. The expected blogged packet in one site is presented in Fig. 4. It based on the dynamic channel state that can achieve a certain QoS.

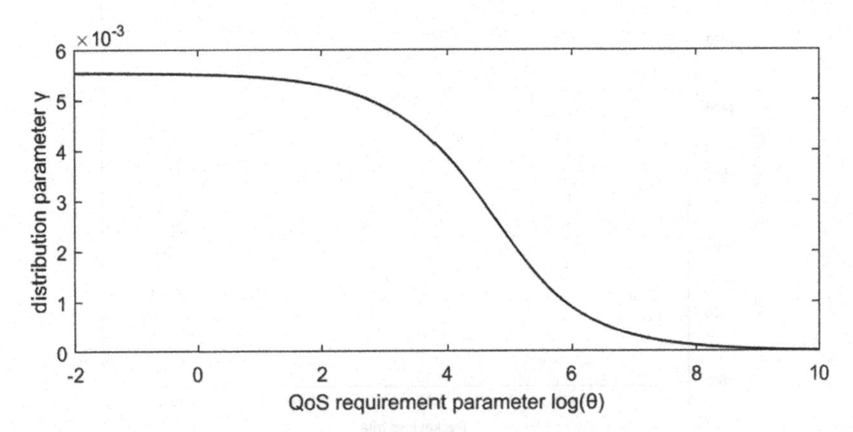

Fig. 3. QoS requirement and distribution of fluid model.

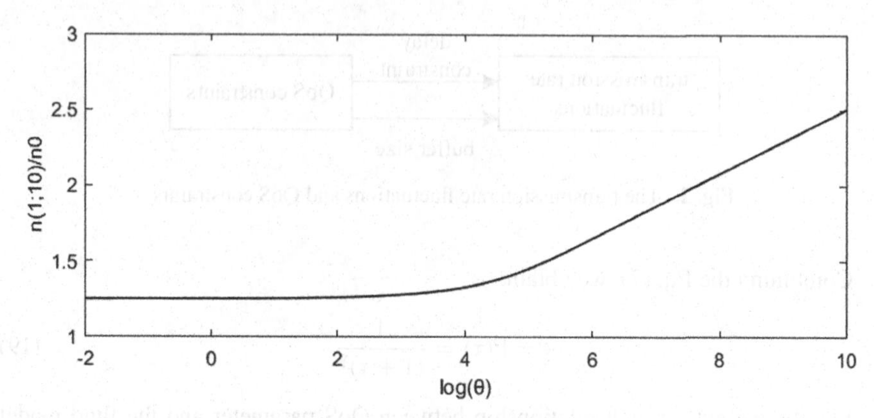

Fig. 4. QoS requirement and blogged length.

As shown in Fig. 3, When the latency constraint is between [−2, 2], the allowable arrival rate remains constant; when the latency constraint is between [2, 8], the allowable arrival rate gradually decreases; when the latency constraint is between [8, 10], the allowable arrival rate almost approaches 0. As shown in Fig. 4, When the latency constraint is between [−2, 4], the expected blogged packet number remains constant, When the latency constraint is between [4, 10], the expected blogged packet number gradually increase.

As shown in Fig. 5, in the scope of our simulation, we can find that the probability of data loss increases gradually with the decrease of buffer capacity. So in order to reduce packet loss, we should increase the buffer size as much as possible.

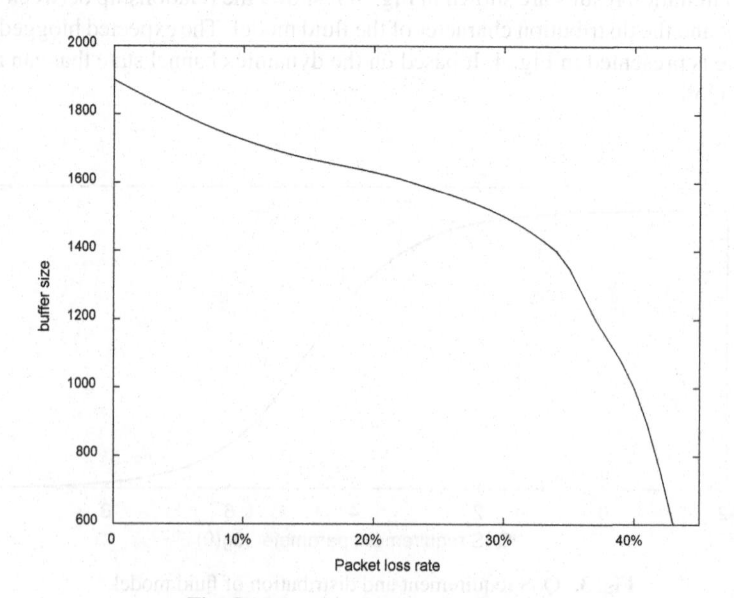

Fig. 5. Packet loss rate and buffer size

5 Conclusion

We explore the possibility of an indefinitely long packet delay at an intermediate node in Multi-hop wireless networks with QoS constraint. The QoS constraint is described as effective capacity, which is related to violation probability. To describe the traffic model of busrty data service, we adopt fractional calculus fluid model. The simulation results show that the effective capacity affects the traffic model. As the latency constraint increases, the allowable arrival rate remains constant. It will gradually decreases after specific number of the latency constraint. On the other hand, the packet loss rate increases as the buffer gets smaller, so we should try to maximize the buffer size.

Acknowledgements. This work is partly supported by Jiangsu technology project of Housing and Urban-Rural Development (No.2018ZD265) and Jiangsu major natural science research project of College and University (No. 19KJA470002).

References

1. Huo, L., Jiang, D., Lv, Z., et al.: An intelligent optimization-based traffic information acquirement approach to software-defined networking. Computational Intelligence, pp. 1–21 (2019)
2. Wang, F., Jiang, D., Qi, S.: An adaptive routing algorithm for integrated information networks. Chin. Commun. **7**(1), 196–207 (2019)
3. Zhang, K., Chen, L., An, Y., et al.: A QoE test system for vehicular voice cloud services. Mobile Network Application (2019). 10.1007/s11036-019-01415-3
4. Chen, L., Jiang, D., Bao, R., Xiong, J., Liu, F., Bei, L.: MIMO scheduling effectiveness analysis for bursty data service from view of QoE. Chin. J. Electron. **26**(5), 1079–1085 (2017)
5. D. Jiang, Y. Wang, Z. Lv, et al.: Big data analysis-based network behavior insight of cellular networks for industry 4.0 applications. IEEE Transactions on Industrial Informatics, 16(2):1310–1320, (2020)
6. Jiang, D., Huo, L., Song, H.: Rethinking behaviors and activities of base stations in mobile cellular networks based on big data analysis. IEEE Trans. Netw. Sci. Eng. **1**(1), 1–12 (2018)
7. Chen, L., et al.: A lightweight end-side user experience data collection system for quality evaluation of multimedia communications. IEEE Access **6**(1), 15408–15419 (2018)
8. Tan, J., Xiao, S., Han, S., Liang, Y., Leung, V.C.M.: QoS-aware user association and resource allocation in LAA-LTE/WiFi Coexistence systems. IEEE Trans. Wireless Commun. **18**(4), 2415–2430 (2019)
9. Wang, Y., Tang, X., Wang, T.: A unified QoS and security provisioning framework for wiretap cognitive radio networks: a statistical queueing analysis approach. IEEE Trans. Wireless Commun. **18**(3), 1548–1565 (2019)
10. Hassan, M.Z., Hossain, M.J., Cheng, J., Leung, V.C.M.: Hybrid RF/FSO backhaul networks with statistical-QoS-aware buffer-aided relaying. IEEE Trans. Wireless Commun. **19**(3), 1464–1483 (2020)
11. Zhang, Z., Wang, R., Yu, F.R., Fu, F., Yan, Q.: QoS aware transcoding for live streaming in edge-clouds aided hetnets: an enhanced actor-critic approach. IEEE Trans. Veh. Technol. **68**(11), 11295–11308 (2019)
12. Chen, L., Zhang, L.: Spectral efficiency analysis for wireless network system under QoS constraint: an effective capacity perspective. Mobile Network Application (2020). 10.1007/s11036-019-01414-4

13. Wang, F., Jiang, D., Qi, S., et al.: A dynamic resource scheduling scheme in edge computing satellite networks. Mobile Networks and Applications Online Available (2019)
14. Jiang, D., Huo, L., Lv, Z., et al.: A joint multi-criteria utility-based network selection approach for vehicle-to-infrastructure networking. IEEE Trans. Intell. Transp. Syst. **19**(10), 3305–3319 (2018)
15. Jiang, D., Zhang, P., Lv, Z., et al.: Energy-efficient multi-constraint routing algorithm with load balancing for smart city applications. IEEE Internet Things J. **3**(6), 1437–1447 (2016)
16. Lee, Y., Kim, Y., Park, S.: A machine learning approach that meets axiomatic properties in probabilistic analysis of LTE spectral efficiency. In: 2019 International Conference on Information and Communication Technology Convergence (ICTC), Jeju Island, Korea (South), pp. 1451–1453 (2019)
17. Ji, H., Sun, C., Shieh, W.: Spectral efficiency comparison between analog and digital RoF for mobile fronthaul transmission link. J. Lightwave Technol. **38**(20), 5617–5623 (2020)
18. Hayati, M., Kalbkhani, H., Shayesteh, M.G.: Relay selection for spectral-efficient network-coded multi-source D2D communications. In: 2019 27th Iranian Conference on Electrical Engineering (ICEE), Yazd, Iran, pp. 1377–1381 (2019)
19. You, L., Xiong, J., Zappone, A., Wang, W., Gao, X.: Spectral efficiency and energy efficiency tradeoff in massive MIMO downlink transmission with statistical CSIT. IEEE Trans. Sign. Process. **68**, 2645–2659 (2020)
20. Jiang, D., Li, W., Lv, H.: An energy-efficient cooperative multicast routing in multi-hop wireless networks for smart medical applications. Neurocomputing **220**, 160–169 (2017)
21. Wiatr, P., Chen, J., Monti, P., Wosinska, L.: Energy efficiency versus reliability performance in optical backbone networks. IEEE/OSA J. Opt. Commun. Netw. **7**(3), A482–A491 (2015)
22. Jiang, D., Wang, Y., Lv, Z., et al.: Intelligent optimization-based reliable energy-efficient networking in cloud services for IIoT networks. In: IEEE Journal on Selected Areas in Communications, Online Available (2019)
23. Jiang, D., Wang, W., Shi, L., et al.: A compressive sensing-based approach to end-to-end network traffic reconstruction. IEEE Trans. Netw. Sci. Eng. **5**(3), 1–12 (2018)
24. Jiang, D., Huo, L., Li, Y.: Fine-granularity inference and estimations to network traffic for SDN. Plos One **13**(5), 1–23 (2018)
25. Wang, Y., Jiang, D., Huo, L., et al.: A new traffic prediction algorithm to software defined networking. Mobile Networks and Applications, Online Available (2019)
26. Barakabitze, A.A., et al.: QoE management of multimedia streaming services in future networks: a tutorial and survey. IEEE Commun. Surv. Tutorials **22**(1), 526–565 (2020)
27. Orsolic, I., Skorin-Kapov, L.: A framework for in-network QoE monitoring of encrypted video streaming. IEEE Access **8**, 74691–74706 (2020)
28. Song, E., et al.: Threshold-oblivious on-line web QoE assessment using neural network-based regression model. IET Commun. **14**(12), 2018–2026 (2020)
29. Seufert, M., Wassermann, S., Casas, P.: Considering user behavior in the quality of experience cycle: towards proactive QoE-aware traffic management. IEEE Commun. Lett. **23**(7), 1145–1148 (2019)
30. Qi, S., Jiang, D., Huo, L.: A prediction approach to end-to-end traffic in space information networks. Mobile Networks and Applications, Online Available (2019)
31. Huo, L., Jiang, D., Qi, S., et al.: An AI-based adaptive cognitive modeling and measurement method of network traffic for EIS. Mobile Networks and Applications, Online Available (2019)
32. Huo, L., Jiang, D., Zhu, X., et al.: An SDN-based fine-grained measurement and modeling approach to vehicular communication network traffic. International Journal of Communication Systems, Online Available, pp. 1–12 (2019)
33. Zaborovsky, V., Meylanov, R.: Informational network traffic model based on fractional calculus. In: International Conferences on Info-tech & Info-net (2001)

34. Guo, C., Liang, L., Li, G.Y.: Resource allocation for low-latency vehicular communications: an effective capacity perspective. IEEE J. Sel. Areas Commun. **37**(4), 905–917 (2019)
35. Shehab, M., Alves, H., Latva-aho, M.: Effective capacity and power allocation for machine-type communication. IEEE Trans. Veh. Technol. **68**(4), 4098–4102 (2019)
36. Cui, Q., Gu, Y., Ni, W., Liu, R.P.: Effective capacity of licensed-assisted access in unlicensed spectrum for 5G: from theory to application. IEEE J. Sel. Areas Commun. **35**(8), 1754–1767 (2017)
37. Xiao, C., Zeng, J., Ni, W., Liu, R.P., Su, X., Wang, J.: Delay guarantee and effective capacity of downlink NOMA fading channels. IEEE J. Sel. Top. Sign. Process. **13**(3), 508–523 (2019)
38. Björnson, E., Larsson, E.G., Debbah, M.: Massive MIMO for maximal spectral efficiency: how many users and pilots should be allocated? IEEE Trans. Wireless Commun. **15**(2), 1293–1308 (2016)

Research of Image Dehazing Based on Image Gradient Distortion Prior

Chuangeng Tian(✉), Jiamin Cheng, and Lei Chen

School of Information and Electrical Engineering, Xuzhou University of Technology, Xuzhou 221004, China

Abstract. In view of the problem of image quality degradation caused by the scattering of outdoor atmospheric particles in dusty weather, this paper proposed a research of dehazing image based on gradient distribution prior. Firstly, the method is to obtain a prior model of the gradient distribution of images from a large number of high quality natural image data sets. Secondly, the gradient distribution of hazy image is changed to make the infinite approximation the learned prior model. Finally, the Poisson equation is used to solve the reconstructed image to obtain dehazing image. According to a large number of experimental data, it shows that this research can remove the haze of image effectively. Compared with the most advanced methods, the image processed by this algorithm not only retains more details of the original image, but also improves the image definition largely.

Keywords: Dehazing image · Prior model · Gradient distortion model · Laplace distortion model · Poisson equation

1 Introduction

In haze weather, the image degraded because of the scattering of dust, fog and other particles. The contrast and definition of the acquired image are often poor, which affects the visual effect of the image. Therefore, it is of great significance to study a fast and effective method to make the haze image clear [1, 2]. Performance of the algorithm is very important for certain application including safety monitoring systems and power distribution management.

There are many existing methods to remove the haze image, but there are some shortcomings. Grewe [3] et al. put forward the method of defogging based on wavelet transform. The visual effect of the processed image is better. The disadvantage of this method is that the efficiency of image fusion is low. Tan [4] et al. proposed to realize dehazing by enhancing local color contrast. This method not only enhances the image contrast, but also enhances the image noise. Tarel [5] et al. Proposed to achieve a certain effect of defogging through the pre-processing of foggy image and median filtering. With the requirement of real-time, the disadvantage of this method is that it can blur the boundary with large gray change and lose a lot of detail information. He [6] et al. Put forward the method of fog removal based on the prior of dark channel, using the

H. Song and D. Jiang (Eds.): SIMUtools 2020, LNICST 370, pp. 342–350, 2021.
https://doi.org/10.1007/978-3-030-72795-6_28

prior knowledge to estimate the air transmittance, and then through the principle of soft matting to optimize the estimation of the air transmittance, to achieve the purpose of image fog removal. However, due to the large amount of computation required in the process of soft matting, it is difficult for this method to have the real-time demand. In recent years, many similar defogging algorithm models have been proposed, such as [7–10]. But it is difficult to achieve real-time and effective defogging. Priori model have been verified efficiently in image dehazing, with the above observation, this paper proposes a method of defogging based on the prior of gradient distribution. Firstly, a priori model of gradient distribution is obtained from a large number of high-quality natural image data. Secondly, the gradient distribution of the fog image is changed so that the prior model can be obtained by infinite approximation learning. Finally, poisson equation is used to solve the reconstructed image to get the defog image.

2 Related Work

2.1 Prior Model

2.1.1 First Order Prior

On the logarithmic scale, the gradient distribution of natural images has a heavy tail [11–13]. Traditionally, the gradient distribution is modeled as a generalized Laplace distribution:

$$\log(p(x)) = -k\|x\|^{\alpha} + \beta \tag{1}$$

Where, k, α, are parameters, x represent gradients, p are gradient probability distributions. This model includes some commonly used priors, such as total variation (TV) ($\alpha = 1$) [14], super Laplace ($\alpha = 0.6$) [12]. But the model has many shortcomings. First of all, the model and the actual distribution data can not be well consistent. Secondly, the previous work, k, α, β used as independent variables, which violated the standardized distribution ($\int_{-\infty}^{+\infty} pdx \neq 1$). In addition, the generalized Laplace model is computationally expensive. In this paper, we design a new model to overcome the above shortcomings.

2.1.2 Two Order Prior

In addition to first-order statistics (such as gradients), second-order statistics such as mean curvature (MC) [12] and Gaussian curvature [15, 16], which can also be applied to priors. Different orders of a priori can also be combined, for example, reference [14]. But in most cases, high-order statistics is not necessary. As shown in Fig. 1, $I(x)$ and $I(x+r)$ the relationship when the derivative is of different order:

$$Corr(d, r) := correlation\left(\nabla^d I(x), \nabla^d I(x+r)\right) \tag{2}$$

where $r = (r, 0)$, The larger the order, the more obvious the correlation decreases. This shows that the first-order and second-order priors are very effective for image processing, but higher order derivatives are not necessary to improve this result. One reason is that discrete images may not be high-order differentiable. For the task of image processing, the second order can already meet [16]. Therefore, we only consider the second derivative. All kinds of second-order differential operations, Laplace operation is the most effective, because it can generate Poisson equation like gradient distribution. Therefore, these two priors can realize image reconstruction by solving a Poisson equation at the same time [17–19].

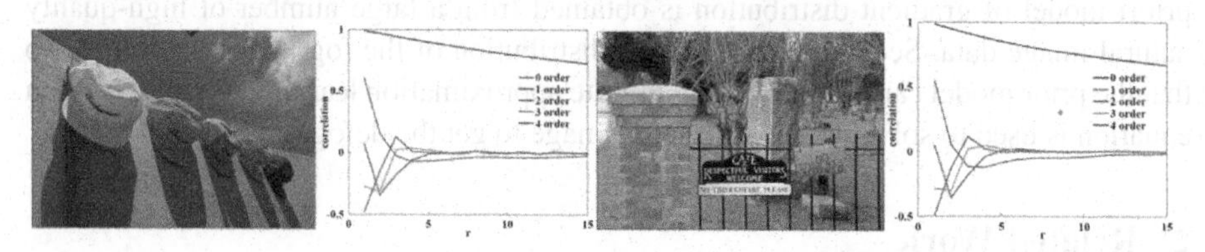

Fig. 1. Rapid reduction of higher derivative

2.2 Gradient Distribution Model

Considering that both G^x and G^y have heavy tails on the logarithmic scale, this feature can be modeled as a super Laplace distribution [20]. The traditional one-dimensional model is consistent with $p(G^x)$ on the logarithmic scale, but it can not meet this requirement $\int_{-\infty}^{+\infty} p(G^x)dG^x = 1$. Considering that cumulative distribution function (CDF) has been verified efficiently in image processing, we use CDF instead of probability density function (PDF). CDF of gradient is defined as:

$$C(G) = \int_{-255}^{G^x} \int_{-255}^{G^y} p(u, v)dudv \tag{3}$$

By observing the definition characteristics of $C(G)$, a parameter model for approaching CDF is proposed:

$$\tilde{C}(G) = \left(\frac{a\tan(T_1 G^x)}{\pi} + \frac{1}{2}\right)\left(\frac{a\tan(T_1 G^y)}{\pi} + \frac{1}{2}\right) \tag{4}$$

Among them, the choice of atan function is based on T distribution or Cauchy distribution. The model of formula (4) has only one parameter and T_1. The selection of atan function is based on t distribution or Cauchy distribution. The edge model with the same distribution of G^x is as follows:

$$\log(P(G^x)) = \log(\frac{T_1}{\pi}) - \log(1 + \left(T_1 G^x)^2\right) \tag{5}$$

2.3 Laplace Distribution Model

In order to get the distribution of Laplacian operation response, this paper uses CDF to model, as follows:

$$L(t) = \int_{-\infty}^{t} p(\Delta I(x))\, d\Delta I(x) \tag{6}$$

For Laplace CDF, we propose a parameter model:

$$\tilde{L}(t) = \frac{a\tan(T_2 t)}{\pi} + \frac{1}{2} \tag{7}$$

Where T_2 is the only free parameter.

3 Image Defogging Algorithm

3.1 Priori Model Proposed

The prior model proposed in this paper is a linear combination of the prior of gradient distribution and the prior of Laplace distribution. First of all, we learn the prior information from the high quality natural image data set. Between these two priors, a new parameter model is provided, which is effective for improving the quality of haze image. The specific operation steps are as follows:

As shown in Table 1, there are 7 high-quality natural image data sets, and image x is grayed. Gradient is defined as:

$$G(x, y) = \left(\nabla_x I(x, y), \nabla_y(x, y)\right) \tag{8}$$

Among them, the first-order finite difference approximation: $\nabla_x I = I(x+1, y) - I(x, y)$ and $\nabla_y I = I(x, y+1) - I(x, y)$. On the boundary of the image, we use the homogeneous Dirichlet boundary condition. Because the gray image is processed, the gradient value range is [−255, 255] * [−255, 255]. Where, G^x and G^y are used to represent gradients G on the X and Y axes, respectively.

Laplace L is defined as:

$$L(x, y) = \Delta I(x, y) \tag{9}$$

Δ is Laplace operation, which is discretized by using the second-order 5-point finite-difference template. The value range is [−1020, 1020].

In order to transform histogram into probability distribution, all pixels m * n of the image are divided into feet, where m and N are the number of pixels along the X and Y axis of the image, respectively. For the centralized image processing in the data set, histogram is defined for each image, and the average distribution p_1^{pr} and p_2^{pr} of gradient and Laplace operation is calculated. For color image, the learned priori is applied to each color channel.

Table 1. Comparison of indexes of each algorithm in different background

	Algorithm	Contrast	AG	MSSIM	PSNR
Haze image 1	Original image	0.0049	9.9505	–	–
	Tarel [3]	0.0047	10.999	0.3118	4.5395
	He [4]	0.0065	11.2821	0.5185	15.1796
	Proposed	0.0095	14.3483	0.5702	15.8384
Haze image 2	Original image	0.0013	5.7416	–	–
	Tarel [3]	0.0033	8.5067	0.4245	5.3348
	He [4]	0.0028	6.429	0.5659	15.0917
	Proposed	0.0049	11.4446	0.8146	16.8252
Dense fog image 1	Original image	0.0001	1.073	–	–
	Tarel [3]	0.0019	4.389	0.3338	2.9309
	He [4]	0.0016	3.6113	0.4724	10.3185
	Proposed	0.0042	6.2846	0.7118	14.2116
Dense fog image 2	Original image	0.0011	10.561	–	–
	Tarel [3]	0.0016	12.0301	0.1965	4.5153
	He [4]	0.0019	11.8551	0.4041	14.9822
	Proposed	0.0031	16.248	0.6457	17. 9894

3.2 Proposed Model

Figure 2 is the flow chart of the defogging method in this paper. For the gradient distribution of image, a new parameter model is proposed. In logarithmic scale, we use cumulative distribution function (CDF) instead of probability density function (PDF) to model. By adjusting the gradient distribution of smog image, it infinitely approximates the gradient distribution prior model learned from high-quality image. Finally, we can reconstruct the image by solving the Poisson equation, so as to achieve the goal of image defogging.

4 Experimental Results and Analysis

4.1 Subjective Evaluation Analysis

Figure 3 is the defogging effect diagram of the algorithm and the schematic diagram of detail processing. In Fig. 4, the first two different fog images are taken under the condition of haze weather. From the image comparison of the first line, the algorithm in this paper can better restore the image degradation caused by haze, and make the color of the image more realistic. From the second line is image comparison, it can be verified that the algorithm can be applied to many types of scenes. The last two lines are images taken in dense fog. It can be found that the algorithm based on the prior

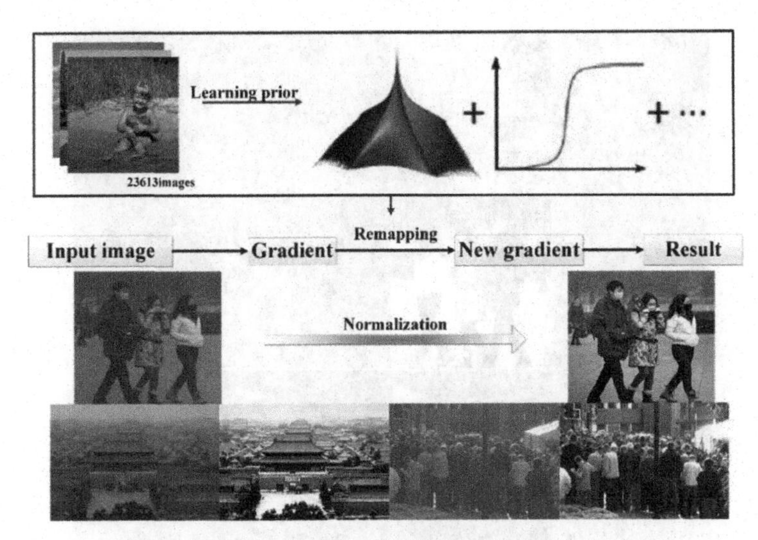

Fig. 2. Flow chart of the the proposed

gradient distribution proposed in this paper has significantly improved the visibility and contrast of haze images, especially in the processing of image details, and has achieved significant results.

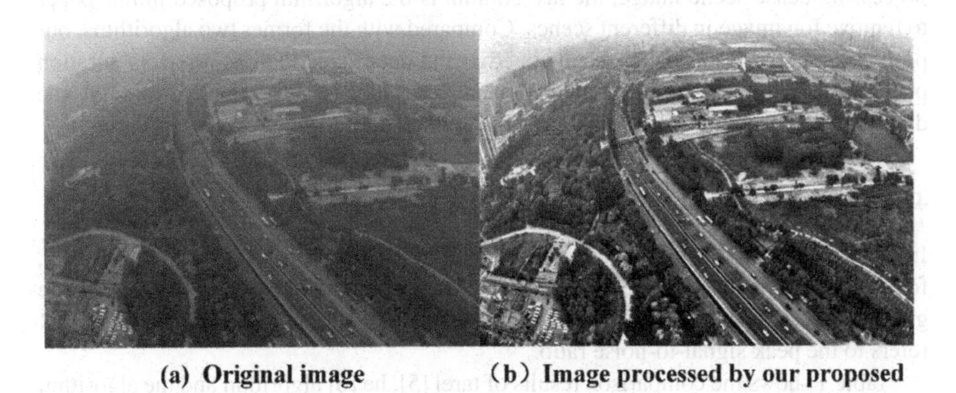

(a) Original image **(b) Image processed by our proposed**

Fig. 3. Image dehazing with proposed method

In Fig. 4, from left to right, there are the original image, the tarel [5] algorithm processing image, the he [6] algorithm processing image, and the defog image based on the prior algorithm of image gradient distribution proposed in this paper. It can be seen from the second column that the tarel [5] algorithm is oversaturated when processing the foggy image, and the color of many areas in the image is biased; the third column is the image processed by the he [6] algorithm in different scenes. It can be seen that these two algorithms are not good for the overall effect of haze image processing. At the same time, we find that he [6] algorithm is easy to appear local blur phenomenon when

(a) Original image. (b) Fattal Result. (c) Algorithm result of He. (d) Proposed.

Fig. 4. Experimental results with different methods

processing dense scene image; the last column is the algorithm proposed in this paper to remove fog image in different scenes. Compared with the former two algorithms, our proposed algorithm has better results in edge and detail, and there is no oversaturation phenomenon. It shows that the algorithm in this paper is very effective to solve the distortion problems of dense fog area, scene light and atmospheric light similar area.

4.2 Objective Evaluation and Analysis

In order to evaluate the effectiveness and superiority of the algorithm more objectively, four general objective indexes are selected for evaluation. (1) Contrast; (2) Average gradient (AG) [21]; (3) MSSIM is the average value of structural similarity; (4) PSNR refers to the peak signal-to-noise ratio.

Table 1 shows the comparison results of tarel [5], he [6] algorithm and the algorithm in this paper for different images in four indicators, namely, contrast, average gradient, MSSIM, PSNR. The test results show that compared with the other two algorithms, the contrast and average gradient of the defog image processed by this algorithm are improved to a certain extent, which shows that the defog result graph of this algorithm not only highlights the edge details and other information in the original image, but also has a good visual effect. The value of MSSIM and PSNR has also been greatly improved, which shows that the image processed by the algorithm in this paper keeps a good structural similarity, at the same time, it can remove the haze more effectively and make the image clearer.

5 Conclusion

There is serious distortion in the haze image obtained in different scenes. In this paper, a method based on the priori of image gradient distribution is proposed. Firstly, the priori model of image gradient distribution is learned from a large number of high-quality natural image data. Secondly, the gradient distribution of the fog image is changed so that the prior model can be obtained by infinite approximation learning. Finally, the reconstructed image is solved by Poisson's equation to get the defog image, and the defog effect is better. Experimental data show that this method can effectively remove fog from haze image. Compared with the latest defog algorithm, the image processed by this method retains more detail information and greatly improves the clarity of the image.

Acknowledgment. This research was financially supported by the Jiangsu technology project of Housing and Urban-Rural Development (No. 2019ZD040).

References

1. Borkar, K., Mukherjee, S.: Single image dehazing by approximating and eliminating the additional airlight component. Neurocomputing **400**, 294–308 (2020)
2. Zheng, M.Y., Qi, G.Q., Zhu, Z.Q., Li, Y.Y., Wei, H.Y., Liu, Y.: Image dehazing by an artificial image fusion method based on adaptive structure decomposition. IEEE Sens. J. **20**, 8062–8072 (2020)
3. Grewe, L.L., Brooks, R.R.: Atmospheric attenuation reduction through multi sen sor fusion. Sens. Fusion Architectures Algorithms Appl. **3376**(10), 102–109 (1998)
4. Tan, R.: Visibility in bad weatehr from a single image. IEEE Conference on Computer Vision and Pattern Recognition. Piscataway, pp. 1–8. IEEE Press (2008)
5. Tarel, J.P., Hautiere, N.: Fast visibility restoration from a single color or gray level image. In: IEEE 12th International and Conference on Computer Vision. Piscataway, pp. 2201–2208. IEEE Press (2009)
6. He, K.M.: Single image haze removal using dark channel prior. IEEE Trans. Actions Pattern Anal. Mach. Intell. **33**(12), 2341–2353 (2011)
7. Wang, J.B., He, N., Zhang, L.L., et al.: Single image dehazing with a physical model and dark channel prior. Neurocomputing, **149**, 718–728 (2015)
8. Choi, L.K., You, J., Bovik, A.C.: Referenceless prediction of perceptual fog density and perceptual image defogging. Image Process. IEEE Trans. **24**(11), 3888–3901 (2015)
9. Lai, Y.H., Chen, Y.L., Chiou, C.J., et al.: Single-image dehazing via optimal trans mission map under scene priors. Circuits Syst. Video Technol. IEEE Trans. **25**(1), 1–14 (2015)
10. Gao, Y., Hu, H.M., Wang, S., et al.: A fast image dehazing algorithm based on negative correction. Sign. Process. **103**, 380–398 (2014)
11. Shan, Q., Jia, J., Agarwala, A.: High-quality motion deblurring from a single im age. ACM Trans. Graph. **27**, 73 (2008)
12. Krishnan, D., Fergus, R.: Fast image deconvolution using hyper-Laplacian priors. Adv. Neural Inform. Proc. Sys. **22**, 1–9 (2009)
13. Cho, T.S., Zitnick, C.L., Joshi, N., Kang, S.B., Szeliski, R., Freeman, W.T.: Image restoration by matching gradient distributions. IEEE Trans. Pattern Anal. Mach. Intell. (PAMI) **34**, 683–694 (2012)

14. Rudin, L.I., Osher, S., Fatemi, E.: Nonlinear total variation based noise removal algorithms. Phys. **60**, 259–268 (1992)
15. Lee, S.H., Seo, J.K.: Noise removal with Gauss curvature-driven diffusion. IEEE Trans. Image Proc. **14**, 904–909 (2005)
16. Gong, Y., Sbalzarini, I.F.: Local weighted Gaussian curvature for image processing. In: 2013 IEEE International Conference on Image Processing (ICIP), pp. 534–538 (2013)
17. Fattal, R., Lischinski, D., Werman, M.: Gradient domain high dynamic range compression. ACM Trans. Graph. **21**, 249–256 (2002)
18. Agrawal, A., Raskar, R.: Gradient domain manipulation techniques in vision and graphics. ICCV short course (2007)
19. Gong, Y., Paul, G., Sbalzarini, I.F.: Coupled signed-distance functions for im plicit surface reconstruction. In: 2012 9th IEEE International Symposium on Biomedical Imaging (ISBI), pp. 1000–1003 (2012)
20. http://lear.inrialpes.fr/%20jegou/data.php
21. Wang, R., Du, L., Yu Z., Wan, W.: Infrared and visible images fusion using compressed sensing based on average gradient. In: 2013 IEEE International Conference on Multimedia and Expo Workshops (ICMEW), San Jose, CA, pp. 1–4 (2013)

Preconditioned Iteration Method for the Nonlinear Space Fractional Complex Ginzburg-Landau Equation

Lu Zhang[1](✉), Lei Chen[2], and Xiao Song[1]

[1] School of Mathematics and Statistics, Xuzhou University of Technology, Xuzhou 221018, Jiangsu, China

[2] School of Information Engineering, Xuzhou University of Technology, Xuzhou 221018, Jiangsu, China

chenlei@xzit.edu.cn

Abstract. In this work, we give a fast preconditioned numerical method to solve the discreted linear system, which is obtained from the nonlinear space fractional complex Ginzburg-Landau equation. The coefficient matrix of the discreted linear system is the sum of a complex diagonal matrix and a real Toeplitz matrix. The new method has a superiority in computation because we can use the circulant preconditioner and the fast Fourier transform (FFT) to solve the discreted linear system. Numerical examples are tested to illustrate the advantage of the preconditioned numerical method.

Keywords: Nonlinear fractional ginzburg-landau equation · Toeplitz matrix · Circulant preconditioner · Fast fourier transform

1 Introduction

In this work, we solve the nonlinear space fractional complex Ginzburg-Landau equation as follows [13]

$$\frac{\partial v}{\partial t} + (\nu_1 + \mathbf{i}\eta_1)(-\Delta)^{\frac{\beta}{2}} v + (\kappa_1 + \mathbf{i}\zeta_1)|v|^2 v - \gamma_1 v = 0, \tag{1}$$

$$v(x, 0) = v_0(x), \tag{2}$$

where $x \in \mathbb{R}$, $0 < t \leqslant T_1$, $\mathbf{i}$ is the imaginary unit, $1 < \beta \leqslant 2$, $v(x,t)$ is a complex-value function, $\nu_1 > 0$, $\kappa_1 > 0$, η_1, ζ_1, and γ_1 are real constants, and $v_0(x)$ is an initial function. Furthermore, the operator $(-\Delta)^{\frac{\beta}{2}} v(x,t)$ in (1) denotes the derivative operator ($1 < \beta \leqslant 2$) [10] as follows

$$-(-\Delta)^{\frac{\beta}{2}} v(x,t) = -\frac{\frac{\partial^2}{\partial x^2}\int_{-\infty}^{\infty} |x-\xi|^{1-\beta} v(\xi,t)\mathrm{d}\xi}{2\cos(\frac{\beta\pi}{2})\Gamma(2-\beta)}. \tag{3}$$

Supported by the "Peiyu" Project from Xuzhou University of Technology (Grant Number XKY2019104).

H. Song and D. Jiang (Eds.): SIMUtools 2020, LNICST 370, pp. 351–362, 2021.
https://doi.org/10.1007/978-3-030-72795-6_29

The fractional Ginzburg-Landau equations have been used to describe a lot of physical phenomena; see [27,28,33,34] for example. In theory, some scholars have studied the fractional Ginzburg-Landau equations [11,30,35]. In the numerical computation, there exists few studies for the nonlinear fractional equations (1)–(2); see [13,28,36,40] and reference therein. Therefore, based on the extensive applied fields of these equations, it is interesting to study the numerical methods for solving the nonlinear fractional equations (1)–(2).

Recently, some new techniques are proposed to improve network routing and performance measurement [14,37,41]. Based on effective user behavior and traffic analysis approaches [5,6,23,24], we can design more effective scheduling strategies to raise resources utilization [7,17,18,38] and energy-efficiency [19,20]. To test new scheduling strategies, traffic must be reconstructed in test bed [15,16,21,22,31,39]. Fluid model is effective model to reconstruct the bursty data traffic. Furthermore, fractional differential equation can be used to build the fluid model. The main aim of our paper is to give a fast preconditioned numeical method to solve the discreted linear system, which is obtained from the nonlinear fractional equations (1)–(2). The superiority of our method is that in the process of solving the nonlinear fractional equations (1)–(2), we can fast solve the complex linear systems by using the circulant preconditioner and the FFT at each step due to the special structure of coefficient matrices.

2 A Finite Difference Scheme for the Fractional Equations

In this part, we use the linearized difference scheme [13] to discretize the the nonlinear fractional equations (1)–(2). And the fractional operator $(-\Delta)^{\frac{\beta}{2}}$ is discretized in the nonlinear fractional equations (1)–(2) by the centered difference method [2] which is given as follows

$$\Delta_{\hat{h}}^{\beta} v(x) = \frac{\sum\limits_{i=-\infty}^{+\infty} d_i^{(\beta)} v(x - i\hat{h})}{\hat{h}^{\beta}},$$

where $d_i^{(\beta)}$ can be obtained by the following formulation

$$d_i^{(\beta)} = \frac{(-1)^i \Gamma(\beta+1)}{\Gamma(\frac{\beta}{2} - i + 1)\Gamma(\frac{\beta}{2} + i + 1)}, \quad i \in \mathbb{Z}. \tag{4}$$

In the following, we give the numerical finite difference scheme of the nonlinear fractional equations (1)–(2) in $\Pi = [a_1, a_2]$. We separate the domain $\{(x,t)|a_1 \leqslant x \leqslant a_2, 0 \leqslant t \leqslant T_1\}$ by a uniform grid $\{(x_j, t_i)|x_j = a_1 + j\hat{h}, t_i = i\hat{\tau}, j = 0, \ldots, M_1, i = 0, \ldots, N_1\}$, with $\hat{h} = \frac{a_2 - a_1}{M_1}, \hat{\tau} = \frac{T_1}{N_1}$, where M_1 and N_1 are two positive integers.

Denote v_j^i the numerical solution to the exact solution $v(x_j, t_i)$. For any grid function $v^i = (v_0^i, v_1^i, \ldots, v_{M_1}^i)$, According to [13], we define

$$v_j^{\bar{i}} = \frac{v_j^{i+1} + v_j^{i-1}}{2}, \quad \delta_t v_j^i = \frac{v_j^{i+1} - v_j^{i-1}}{2\hat{\tau}}.$$

Denote

$$\tilde{Z}_{\hat{h}}^0 = \{V|V = \{V_j\}, j = 0, 1, \ldots, M_1, V_0 = V_{M_1} = 0\}.$$

For any $v \in \tilde{Z}_{\hat{h}}^0$, we have

$$\Delta_{\hat{h}}^{\beta} v_j = \frac{1}{\hat{h}^{\beta}} \sum_{i=j-1}^{j-M_1+1} d_i^{(\beta)} v_{j-i}.$$

The linearized implicit method [13] for the nonlinear fractional equations (1)–(2) in $\Pi = [a_1, a_2]$ is as follows

$$\begin{aligned}
&\delta_t v_j^{i+1} + (\nu_1 + \mathbf{i}\eta_1)\Delta_{\hat{h}}^{\beta} v_j^{\bar{i}} + (\kappa_1 + \mathbf{i}\zeta_1)|v_j^i|^2 v_j^{\bar{i}} - \gamma_1 v_j^{\bar{i}} = 0,\\
&0 < j < M_1,\ 1 < i < N_1, && (5)\\
&v_j^0 = v_0(x_j), 0 < j < M_1, && (6)\\
&v_0^i = v_{M_1}^i = 0, 0 \leqslant i \leqslant N_1. && (7)
\end{aligned}$$

In the practical computation, we can obtain v^1 from the following scheme [13]

$$v_j^1 = v_j^0 - \hat{\tau}\left((\nu_1 + \mathbf{i}\eta_1)\Delta_{\hat{h}}^{\beta} v_j^0 + (\kappa_1 + \mathbf{i}\zeta_1)|v_j^0|^2 v_j^0 - \gamma_1 v_j^0\right), 0 < j < M_1. \quad (8)$$

Denote

$$\tilde{v}^i = [v_1^i, \ldots, v_{M_1-1}^i]^T,$$

$$Q_1^1 = (1 + \gamma_1\hat{\tau})I - \hat{\tau}(\kappa_1 + \mathbf{i}\zeta_1)\begin{bmatrix} |v_1^0|^2 & 0 & \ldots & 0\\ 0 & |v_2^0|^2 & \ldots & 0\\ \vdots & \vdots & \ddots & \vdots\\ 0 & 0 & \ldots & |v_{M_1-1}^0|^2 \end{bmatrix},$$

$$Q_1^i = (1 - \gamma_1\hat{\tau})I + \hat{\tau}(\kappa_1 + \mathbf{i}\zeta_1)\begin{bmatrix} |v_1^i|^2 & 0 & \ldots & 0\\ 0 & |v_2^i|^2 & \ldots & 0\\ \vdots & \vdots & \ddots & \vdots\\ 0 & 0 & \ldots & |v_{M_1-1}^i|^2 \end{bmatrix},$$

$$Q_2^i = (1 + \gamma_1\hat{\tau})I - \hat{\tau}(\kappa_1 + \mathbf{i}\zeta_1)\begin{bmatrix} |v_1^i|^2 & 0 & \ldots & 0\\ 0 & |v_2^i|^2 & \ldots & 0\\ \vdots & \vdots & \ddots & \vdots\\ 0 & 0 & \ldots & |v_{M_1-1}^i|^2 \end{bmatrix},$$

and

$$D_\beta = \begin{bmatrix} d_0^{(\beta)} & d_1^{(\beta)} & d_2^{(\beta)} & \cdots & d_{M_1-2}^{(\beta)} & d_{M_1-1}^{(\beta)} \\ d_1^{(\beta)} & d_0^{(\beta)} & d_1^{(\beta)} & \cdots & \cdots & d_{M_1-2}^{(\beta)} \\ d_2^{(\beta)} & d_1^{(\beta)} & d_0^{(\beta)} & \ddots & \ddots & \vdots \\ \vdots & \ddots & \ddots & \ddots & \ddots & d_2^{(\beta)} \\ d_{M_1-2}^{(\beta)} & \ddots & \ddots & \ddots & d_0^{(\beta)} & d_1^{(\beta)} \\ d_{M_1-1}^{(\beta)} & d_{M_1-2}^{(\beta)} & \cdots & d_2^{(\beta)} & d_1^{(\beta)} & d_0^{(\beta)} \end{bmatrix}, \tag{9}$$

then the linearized implicit method (5)–(8) has the following formulation

$$\tilde{v}^1 = (Q_1^1 - (\nu_1 + \mathbf{i}\eta_1)A_\beta)\tilde{v}^0, \tag{10}$$

$$\left(\frac{Q_1^i}{\nu_1 + \mathbf{i}\eta_1} + A_\beta\right)\tilde{v}^{i+1} = \left(\frac{Q_2^i}{\nu_1 + \mathbf{i}\eta_1} - A_\beta\right)\tilde{v}^{i-1}, 1 < i < N_1, \tag{11}$$

where $A_\beta = \frac{\hat{\tau}}{h^\beta} D_\beta$.

3 A Preconditioned Iterative Method for the Linearized Implicit Method

In this section, we propose a fast preconditioned GMRES (PGMRES) method [32] where the Strang preconditioner [3] is used to solve the linear system in the matrix-vector form (10)–(11).

3.1 Toeplitz Matrix and GMRES Method

It is known that an $n_1 \times n_1$ Toeplitz matrix B_{n_1} satisfies $(B_{n_1})_{ij} = b_{i-j}$ for $i, j = 1, 2, \ldots, n_1$. And the Toeplitz system is as follows

$$B_{n_1} u = \tilde{b},$$

where $\tilde{b}$ is a given vector. Toeplitz systems are widely used in various fields; see [1,4,8,12,25,26,29] for example.

An $n_1 \times n_1$ circulant matrix C_{n_1} is a Toeplitz matrix and its elements satisfy $c_{-j} = c_{n_1-j}$ for $1 \leqslant j \leqslant n_1 - 1$. According to the result of the reference [9], it is worth noting that if we want to compute the matrix-vector products $C_{n_1}u$ and $C_{n_1}^{-1}u$ efficiently by the fast Fourier transform, the computation complexity will be $\mathcal{O}(n_1 \log n_1)$ operations. We also note that [3] one can compute the matrix-vector product $B_{n_1}u$ by the FFT in $\mathcal{O}(2n_1 \log(2n_1))$. These important properties could be exploited to fast solve the linear system in the matrix-vector form (10)–(11).

The GMRES method [32] is a very popular numerical iterative method for solving the following non-Hermitian linear systems

$$Au = \tilde{b},$$

where A is a non-Hermitian matrix. However, the convergent rate of the GMRES method is very slow in general because of the large condition number of the coefficient matrix of linear systems. To deal with this drawback, we can use the preconditioned matrix to accelerate the convergent rate of the GMRES method. Please refer to [32] for the PGMRES method.

3.2 A Preconditioner for the Linearized Implicit Difference Scheme

From Sect. 2, It is noted that the matrix A_β is a Toeplitz matrix in the matrix-vector form (10)–(11) of the linearized implicit difference scheme. According to Sect. 3.1, it is seen that for an $M_1 \times M_1$ Toeplitz matrix B_{M_1}, we could store this matrix in $\mathcal{O}(M_1)$ of memory and compute the matrix-vector product $B_{M_1}u$ in $\mathcal{O}(M_1 \log M_1)$ by using the FFT. On the other hand, it is worth noting that the coefficient matrix $\frac{Q_1^i}{\nu_1+\mathbf{i}\eta_1} + A_\beta$ of linear systems (11) is non-Hermitian. Therefore, we exploit the PGMRES method to solve the linear systems of the linearized implicit difference scheme (10)–(11). Strang's circulant matrix [3,4] is a circulant matrix and can be applied to accelerate the convergent rate of the GMRES method.

According to the linear system (11), we note that the matrix $\frac{Q_1^i}{\nu_1+\mathbf{i}\eta_1}+A_\beta$ is not a Toeplitz matrix. In this situation, standard Strang's circulant preconditioner can not be used to solve the linear systems (11) directly. To deal with this problem, we will give a new effective preconditioner for the coefficient matrix $\frac{Q_1^i}{\nu_1+\mathbf{i}\eta_1} + A_\beta$.

For the coefficient matrix

$$\frac{Q_1^i}{\nu_1 + \mathbf{i}\eta_1} + A_\beta$$

in (11), let

$$\tilde{a} = \frac{1}{M_1 - 1} \sum_{j=1}^{M_1-1} \frac{1 - \gamma_1\hat{\tau} + \hat{\tau}(\kappa_1 + \mathbf{i}\zeta_1)|v_j^i|^2}{\nu_1 + \mathbf{i}\eta_1},$$

then the preconditioned matrix is $P = \tilde{a}I + s(A_\beta)$ for coefficient matrix $\frac{Q_1^i}{\nu_1+\mathbf{i}\eta_1} + A_\beta$, where $s(A_\beta)$ denotes the Strang circulant preconditioner for the Toeplitz matrix A_β. Therefore, we obtain a circulant matrix P. In the next section, we will see that the proposed preconditioner is very effective for the PGMRES method.

4 Numerical Examples

In this part, we carry out two numerical examples for the nonlinear fractional equations (1)–(2). We will show the computational advantage of the proposed proconditioner for GMRES method. We denote "GE" by the direct method, which is implemented by left divided in MATLAB. For the PGMRES method with Strang's circulant preconditioner, we denote by "CPGMRES". We stop the PGMRES method if the following condition satisfies

$$\frac{\|\mathrm{res}^k\|_2}{\|\mathrm{res}^0\|_2} < 10^{-7},$$

where res^k denotes the k-th residual vector for the PGMRES method. In the tables of numerical examples, "Icpu" is the computational time with seconds for GE and CPGMRES, and "It" is the numbers of iterations for CPGMRES.

Example 1. In the first example, the parameters in the nonlinear fractional Eq. (1) and (2) are same as these of Example 1 in [13].

Furthermore, similar to [13], for $1 < \beta < 2$, we calculate the exact solution v with $\hat{\tau} = \frac{1}{256}$ and $\hat{h} = \frac{1}{256}$. We use the error Err$= v - V(\hat{h}, \hat{\tau})$ as the numerical accuracy at $T_1 = 1$ with l^∞-norm.

Table 1. Numerical results for Example 1

β	$(\hat{\tau}, \hat{h})$	CPGMRES			GE	
		Err_1	It	Icpu	Err_2	Icpu
1.2	$(\frac{1}{8}, \frac{1}{8})$	9.4621e−3	6.0	0.0150	9.4621e−3	0.0160
	$(\frac{1}{16}, \frac{1}{16})$	2.3726e−3	5.0	0.0160	2.3725e−3	0.1090
	$(\frac{1}{32}, \frac{1}{32})$	5.8698e−4	5.0	0.0940	5.8697e−4	1.3910
	$(\frac{1}{64}, \frac{1}{64})$	1.4043e−4	4.0	0.2340	1.3980e−4	16.4700
	$(\frac{1}{128}, \frac{1}{128})$	2.8145e−5	4.0	1.0470	2.7961e−5	187.0640
1.5	$(\frac{1}{8}, \frac{1}{8})$	9.4826e−3	6.0	0.0160	9.4825e−3	0.0160
	$(\frac{1}{16}, \frac{1}{16})$	2.3620e−3	5.0	0.0160	2.3620e−3	0.1090
	$(\frac{1}{32}, \frac{1}{32})$	5.8343e−4	5.0	0.0930	5.8342e−4	1.3750
	$(\frac{1}{64}, \frac{1}{64})$	1.3866e−4	4.3	0.2660	1.3889e−4	16.2380
	$(\frac{1}{128}, \frac{1}{128})$	2.7810e−5	4.0	1.0320	2.7760e−5	184.3910
1.8	$(\frac{1}{8}, \frac{1}{8})$	8.2560e−3	6.0	0.0160	8.2560e−3	0.0160
	$(\frac{1}{16}, \frac{1}{16})$	2.0541e−3	5.0	0.0150	2.0540e−3	0.1400
	$(\frac{1}{32}, \frac{1}{32})$	5.0720e−4	5.0	0.0940	5.0720e−4	1.4070
	$(\frac{1}{64}, \frac{1}{64})$	1.2084e−4	4.8	0.2810	1.2074e−4	16.6880
	$(\frac{1}{128}, \frac{1}{128})$	2.4174e−5	4.2	1.1100	2.4136e−5	187.5820

We list the numerical results in Table 1. In this table, Err_1 and Err_2 are calculated by the CPGMRES method and the GE method, respectively. From the computational results of Table 1, it can be seen that Err_1 and Err_2 of the CPGMRES method and the GE method are almost the same. However, if the order of the coefficient matrix is very large in the complex linear systems (10)–(11), the computational times of the GE method are much more than these of the CPGMRES method. Furthermore, according to Table 1, we can see that the numbers of iterations for the CPGMRES method do not change when the order of the coefficient matrix increases. On the other hand, Fig. 1, 2, 3 show the distribution of the eigenvalues for the matrices $\frac{Q_1^i}{\nu_1+\mathrm{i}\eta_1}+A_\beta$ and $P^{-1}(\frac{Q_1^i}{\nu_1+\mathrm{i}\eta_1}+A_\beta)$ at $T_1=1$, respectively, when the order of the matrix is 640, and $\beta = 1.2, 1.5, 1.8$. In these figures, the blue points show that most of the eigenvalues of the matrix $P^{-1}(\frac{Q_1^i}{\nu_1+\mathrm{i}\eta_1}+A_\beta)$ approach to 1, while the eigenvalues of the matrix $\frac{Q_1^i}{\nu_1+\mathrm{i}\eta_1}+A_\beta$ do not approach to 1. These figures show that our new circulant preconditioner is very effective for solving the complex linear systems (10)–(11).

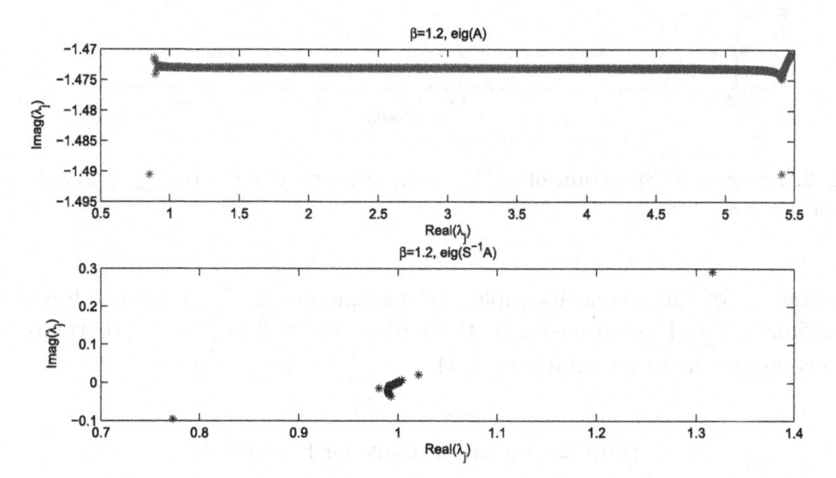

Fig. 1. Example 1: Spectrum of $\frac{Q_1^i}{\nu_1+\mathrm{i}\eta_1}+A_\beta$ (upper) and $P^{-1}(\frac{Q_1^i}{\nu_1+\mathrm{i}\eta_1}+A_\beta)$(lower), when $\beta = 1.2$.

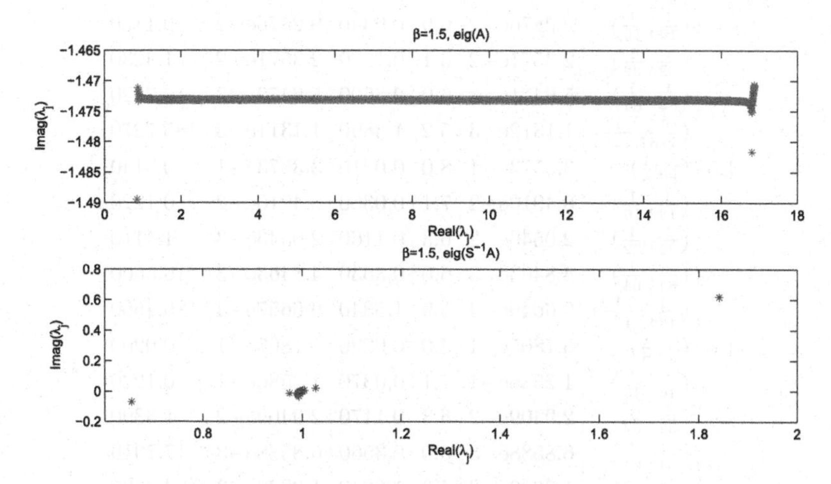

Fig. 2. Example 1: Spectrum of $\frac{Q_1^i}{\nu_1+\mathrm{i}\eta_1}+A_\beta$ (upper) and $P^{-1}(\frac{Q_1^i}{\nu_1+\mathrm{i}\eta_1}+A_\beta)$(lower), when $\beta = 1.5$.

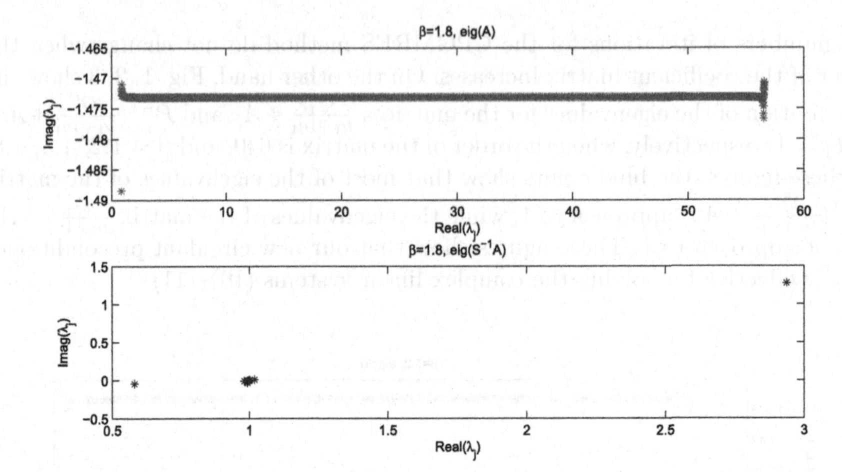

Fig. 3. Example 1: Spectrum of $\frac{Q_1^i}{\nu_1+i\eta_1}+A_\beta$ (upper) and $P^{-1}(\frac{Q_1^i}{\nu_1+i\eta_1}+A_\beta)$(lower), when $\beta=1.8$.

Example 2. In the second example, the parameters in the nonlinear fractional equations (1) and (2) are same as these of Example 2 in [13]. Furthermore, we also compute the exact solution v with $\hat{\tau}=\frac{1}{256}$ and $\hat{h}=\frac{1}{256}$.

Table 2. Numerical results for Example 2

β	$(\hat{\tau},\hat{h})$	CPGMRES			GE	
		Err_1	It	Icpu	Err_2	Icpu
1.2	$(\frac{1}{8},\frac{1}{8})$	3.3383e−1	8.1	0.0210	3.3383e−1	0.0190
	$(\frac{1}{16},\frac{1}{16})$	9.2876e−2	6.9	0.0340	9.2876e−2	0.1150
	$(\frac{1}{32},\frac{1}{32})$	2.3571e−2	6.1	0.1230	2.3571e−2	1.4230
	$(\frac{1}{64},\frac{1}{64})$	5.6471e−3	6.0	0.3600	5.6472e−3	16.7920
	$(\frac{1}{128},\frac{1}{128})$	1.1312e−3	5.2	1.3890	1.1311e−3	185.7270
1.5	$(\frac{1}{8},\frac{1}{8})$	3.3573e−1	8.0	0.0210	3.3573e−1	0.0140
	$(\frac{1}{16},\frac{1}{16})$	8.4910e−2	7.1	0.0360	8.4910e−2	0.1220
	$(\frac{1}{32},\frac{1}{32})$	2.0546e−2	6.3	0.1160	2.0546e−3	1.4110
	$(\frac{1}{64},\frac{1}{64})$	4.8463e−3	6.0	0.3530	4.8463e−3	16.5600
	$(\frac{1}{128},\frac{1}{128})$	9.6640e−4	5.9	1.5340	9.6657e−4	186.4660
1.8	$(\frac{1}{8},\frac{1}{8})$	5.1805e−1	8.0	0.0220	5.1805e−1	0.0200
	$(\frac{1}{16},\frac{1}{16})$	1.2586e−1	7.1	0.0370	1.2586e−1	0.1220
	$(\frac{1}{32},\frac{1}{32})$	2.9409e−2	6.3	0.1170	2.9409e−2	1.3890
	$(\frac{1}{64},\frac{1}{64})$	6.8588e−3	6.0	0.3560	6.8588e−3	17.7910
	$(\frac{1}{128},\frac{1}{128})$	1.3639e−3	5.9	1.6640	1.3639e−3	194.5380

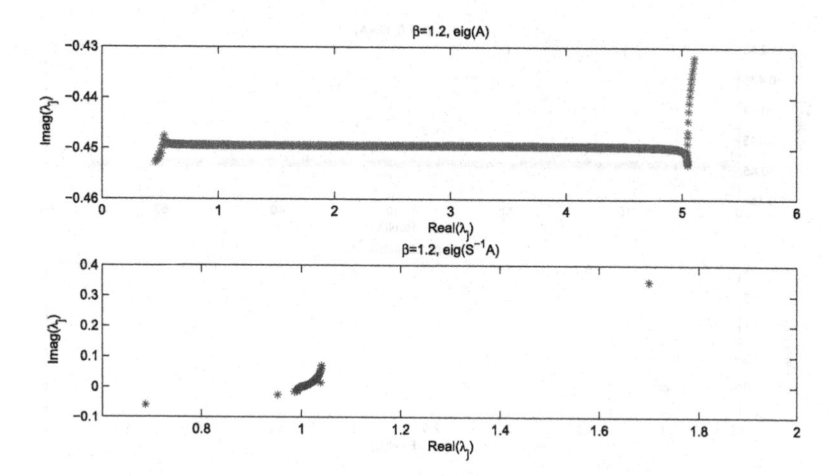

Fig. 4. Example 2: Spectrum of $\frac{Q_1^i}{\nu_1+i\eta_1} + A_\beta$ (upper) and $P^{-1}(\frac{Q_1^i}{\nu_1+i\eta_1} + A_\beta)$(lower), when $\beta = 1.2$.

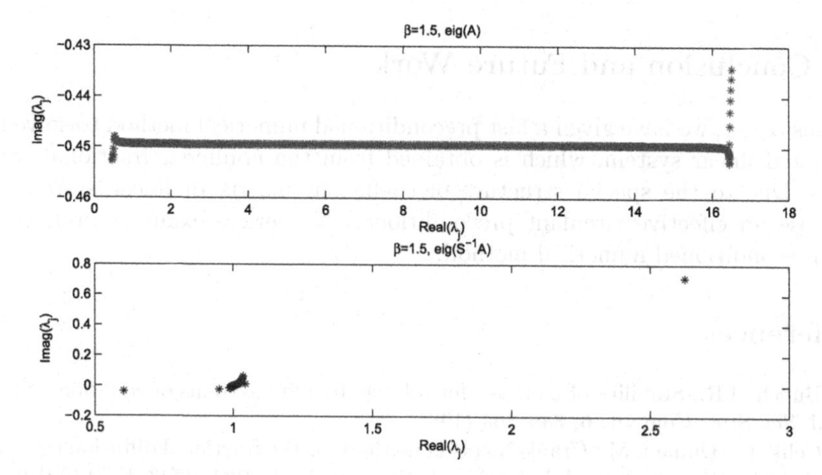

Fig. 5. Example 2: Spectrum of $\frac{Q_1^i}{\nu_1+i\eta_1} + A_\beta$ (upper) and $P^{-1}(\frac{Q_1^i}{\nu_1+i\eta_1} + A_\beta)$(lower), when $\beta = 1.5$.

Table 2 gives the numerical results and Fig. 4, 5, 6 show the distribution of the eigenvalues for the matrices $\frac{Q_1^i}{\nu_1+i\eta_1} + A_\beta$ and $P^{-1}(\frac{Q_1^i}{\nu_1+i\eta_1} + A_\beta)$ at $T_1 = 1$, respectively, when the order of the matrix is 640, and $\beta = 1.2,\ 1.5,\ 1.8$. Similar to Example 1, the numerical results and figures indicate the advantage of the proposed circulant preconditioner.

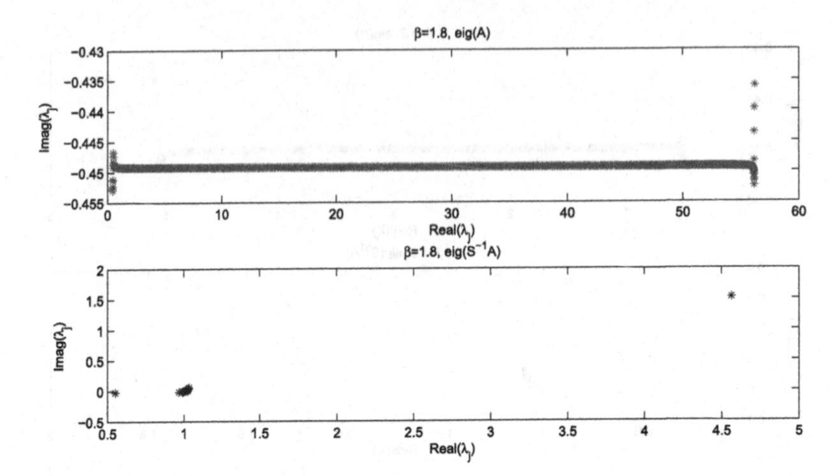

Fig. 6. Example 2: Spectrum of $\frac{Q_1^i}{\nu_1+\mathrm{i}\eta_1} + A_\beta$ (upper) and $P^{-1}(\frac{Q_1^i}{\nu_1+\mathrm{i}\eta_1} + A_\beta)$(lower), when $\beta = 1.8$.

5 Conclusion and Future Work

In this paper, we have given a fast preconditioned numerical method to solve the discreted linear system, which is obtained from the nonlinear fractional equations. Due to the special structure of coefficient matrix of linear system, we propose an effective circulant preconditioner. Numerical examples well verify the preconditioned numerical method.

References

1. Bunch, J.R.: Stability of methods for solving Toeplitz systems of equations. SIAM J. Sci. Stat. Comput. **6**, 349–364 (1985)
2. Çelik, C., Duman, M.: Crank-Nicolson method for the fractional diffusion equation with the Riesz fractional derivative. J. Comput. Phys. **231**, 1743–1750 (2012)
3. Chan, R., Jin, X.: An Introduction to Iterative Toeplitz Solvers. SIAM, Philadelphia (2007)
4. Chan, R., Ng, M.: Conjugate gradient methods for Toeplitz systems. SIAM Rev. **38**, 427–482 (1996)
5. Chen, L., Jiang, D., Bao, R., Xiong, J., Liu, F., Bei, L.: MIMO Scheduling effectiveness analysis for bursty data service from view of QoE. Chin. J. Electron **26**(5), 1079–1085 (2017)
6. Chen, L., Jiang, D., Song, H., Wang, P., Bao, R., Zhang, K., Li, Y.: A lightweight end-side user experience data collection system for quality evaluation of multimedia communications. IEEE Access **6**(1), 15408–15419 (2018)
7. Chen, L., Zhang, L.: Spectral efficiency analysis for massive MIMO system Under QoS constraint: an effective capacity perspective. Mobile Netw. Appl. (2020). https://doi.org/10.1007/s11036-019-01414-4
8. Ching, W.-K.: Iterative Methods for Queuing and Manufacturing Systems. Springer-Verlag, London (2001)

9. Davis, P.: Circulant Matrices, 2nd edn. AMS Chelsea, Providence, RI (1994)
10. Gorenflo, R., Mainardi, F.: Random walk models for space-fractional diffusion processes. Fractional Calc. Appl. Anal. **1**(2), 167–191 (1998)
11. Guo, B.-L., Huo, Z.-H.: Well-posedness for the nonlinear fractional Schrödinger equation and inviscid limit behavior of solution for the fractional Ginzburg-Landau equation. Fract. Calc. Appl. Anal. **16**(1), 226–242 (2012)
12. Hansen, P.C., Nagy, J.G., O'Leary, D.P.: Deblurring Images: Matrices, Spectra, and Filtering. SIAM, Philadelphia (2006)
13. He, D., Pan, K.: An unconditionally stable linearized difference scheme for the fractional Ginzburg-Landau equation. Numer. Algor. **79**, 899–925 (2018)
14. Huo, L., Jiang, D., Lv, Z., et al.: An intelligent optimization-based traffic information acquirement approach to software-defined networking. Comput. Intell. **1–21** (2019)
15. Huo, L., Jiang, D., Qi, S., et al.: An AI-based adaptive cognitive modeling and measurement method of network traffic for EIS. Mob. Netw. Appl. 1–11 (2019)
16. Huo, L., Jiang, D., Zhu, X., et al.: An SDN-based fine-grained measurement and modeling approach to vehicular communication network traffic. Int. J. Commun. Syst. e4092 (2019)
17. Jiang, D., Huo, L., Lv, Z., et al.: A joint multi-criteria utility-based network selection approach for vehicle-to-infrastructure networking. IEEE Trans. Intell. Transp. Syst. **19**(10), 3305–3319 (2018)
18. Jiang, D., Zhang, P., Lv, Z., et al.: Energy-efficient multi-constraint routing algorithm with load balancing for smart city applications. IEEE Internet Things J. **3**(6), 1437–1447 (2016)
19. Jiang, D., Li, W., Lv, H.: An energy-efficient cooperative multicast routing in multi-hop wireless networks for smart medical applications. Neurocomputing **220**, 160–169 (2017)
20. Jiang, D., Wang, Y., Lv, Z., et al.: Intelligent Optimization-based reliable energy-efficient networking in cloud services for IIoT networks. IEEE J. Select. Areas Commun. (2019)
21. Jiang, D., Wang, W., Shi, L., et al.: A compressive sensing-based approach to end-to-end network traffic reconstruction. IEEE Trans. Netw. Sci. Eng. **5**(3), 1–12 (2018)
22. Jiang, D., Huo, L., Li, Y.: Fine-granularity inference and estimations to network traffic for SDN. Plos One **13**(5), 1–23 (2018)
23. Jiang, D., Wang, Y., Lv, Z., et al.: Big data analysis-based network behavior insight of cellular networks for industry 4.0 applications. IEEE Trans. Ind. Inform. **16**(2), 1310–1320 (2020)
24. Jiang, D., Huo, L., Song, H.: Rethinking behaviors and activities of base stations in mobile cellular networks based on big data analysis. IEEE Trans. Netw. Sci. Eng. **1**(1), 1–12 (2018)
25. Jin, X.-Q.: Developments and Applications of Block Toeplitz Iterative Solvers. The Netherlands, and Science Press, Beijing, China, Kluwer Academic Publishers, Dordrecht (2002)
26. Kailath, T., Sayed, A.H. (eds.): Fast Reliable Algorithms for Matrices with Structure. SIAM, Philadelphia (1999)
27. Milovanov, A., Rasmussen, J.: Fractional generalization of the Ginzburg-Landau equation: an unconventional approach to critical phenomena in complex media. Phys. Lett. **337**, 75–80 (2005)

28. Mvogo, A., Tambue, A., Ben-Bolie, G., Kofane, T.: Localized numerical impulse solutions in diffuse neural networks modeled by the complex fractional Ginzburg-Landau equation. Commun. Nonlinear Sci. **39**, 396–410 (2016)
29. Ng, M.K.: Iterative Methods for Toeplitz Systems. Oxford University Press, Oxford, UK (2004)
30. Pu, X., Guo, B.: Well-posedness and dynamics for the fractional Ginzburg-Landau equation. Appl. Anal. **92**, 318–334 (2013)
31. Qi, S., Jiang, D., Huo, L.: A prediction approach to end-to-end traffic in space information networks. Mob. Netw. Appl. 1–10 (2019)
32. Saad, Y.: Iterative Methods for Sparse Linear Systems. SIAM, Philadelphia (2003)
33. Tarasov, V., Zaslavsky, G.: Fractional Ginzburg-Landau equation for fractal media. Phys. **354**, 249–261 (2005)
34. Tarasov, V., Zaslavsky, G.: Fractional dynamics of coupled oscillators with long-range interaction. Chaos **16**(2), 023110 (2006)
35. Tarasov, V.: Psi-series solution of fractional Ginzburg-Landau equation. J. Phys. A-Math. Gen. **39**, 8395–8407 (2006)
36. Wang, P., Huang, C.: An implicit midpoint difference scheme for the fractional Ginzburg-Landau equation. J. Comput. Phys. **312**, 31–49 (2016)
37. Wang, F., Jiang, D., Qi, S.: An adaptive routing algorithm for integrated information networks. China Commun. **7**(1), 196–207 (2019)
38. Wang, F., Jiang, D., Qi, S., et al.: A dynamic resource scheduling scheme in edge computing satellite networks. Mob. Netw. Appl. 1–12 (2019)
39. Wang, Y., Jiang, D., Huo, L., et al.: A new traffic prediction algorithm to software defined networking. Mob. Netw. Appl. 1–10 (2019)
40. Wang, P., Huang, C.: An efficient fourth-order in space difference scheme for the nonlinear fractional Ginzburg-Landau equation. BIT **58**, 783–805 (2018)
41. Zhang, K., Chen, L., An, Y., et al.: A QoE test system for vehicular voice cloud services. Mobile Netw. Appl. (2019). https://doi.org/10.1007/s11036-019-01415-3

Design of Circular Polarization Array Antenna Based on Uniform Rotating Feed Network

Qijia Zhou[1], Lulu Bei[1(✉)], Lei Chen[1], and Kai Huang[2]

[1] School of Information and Electrical Engineering, Xuzhou Institute of Technology, Xuzhou, China

[2] JiangSu XCMG Information Technology Co., Ltd., Xuzhou 221008, China

Abstract. In order to improve the quality of the receiving and transmitting signals in the underground wireless communication system, the circular polarization antenna located at the front of the communication system is studied. A 16 array circularized reading and writing antenna with a center frequency of 2.4 GHz is proposed for the formation of a center-shaped square-shaped circular-polarized patch radiation structure based on two-stage rotation. The structural composition and working principle of the microband antenna are introduced, and the function differential theory is applied to the feed antenna network. Finally, the simulation and experiments proved that the antenna has the advantage of having a high front-to-back ratio, and improved the performance of the antenna.

Keywords: Communication systems · Circular polarization · Microband antennas · Feed networks

1 Introduction

The circularly polarized microstrip antenna is thin in profile, small in volume, light in weight, conformable and easy to integrate. It is widely used in mobile communication, satellite communication, radar, WLAN, RFID and other communication occasions [1–6]. In coal mine wireless communication, it is necessary for the antenna to have good penetration and anti-multipath ability to coal dust and water mist. The circularly polarized antenna has two orthogonal fields, horizontal and vertical. The circularly polarized antenna has better penetrability than the linearly polarized antenna in coal seam water mist under the same power [7–12]. Circular polarization can be divided into left-handed polarization and right-handed polarization. The antenna can only receive signals in the same polarization direction. In the process of underground transmission, the signal is easy to be reflected by obstacles, changes the polarization direction, and is not received by the original polarization direction antenna. In the underground environment of coal mine, it can be used as an effective means of anti-multipath [13–17]. To sum up, the design of microstrip circularly polarized antenna with small size, easy integration, wide frequency band and high performance is the research direction of coal mine underground wireless communication [18–25].

H. Song and D. Jiang (Eds.): SIMUtools 2020, LNICST 370, pp. 363–370, 2021.
https://doi.org/10.1007/978-3-030-72795-6_30

In order to improve the quality of receiving and transmitting signals of underground wireless communication system, the circular polarization antenna at the front end of the communication system is studied. A 16-array circular polarization reading and writing antenna with a central frequency of 2.4 GHz is proposed based on the two-stage rotation of the corner cut square circular polarization patch radiation structure [26–30]. The power divider theory is applied to the antenna feed network to improve the performance of the antenna. In this paper, the design method of circularly polarized antenna is proposed, and its simulation and test are carried out [31–35].

2 Structure of Array Circular Polarization Antenna

The circularly polarized array antenna designed in this paper is shown in Fig. 1. It can be seen that the antenna is composed of two-stage rotary feed network inside and between subarrays and 16 square tangent circularly polarized radiation patches. The RF signal is fed into the energy signal through the central feeding point, and then fed through the internal one quarter T-type network. The signal is divided into four channels with the same amplitude and the phase lag of 90° in turn. Then each signal is fed into four tangent circular polarized radiation patches through a four-part T- type feed network. It can be seen that the antenna consists of an internal T-type feed network and four external T-type feed networks to form the whole two-stage rotating feed network, and then 16 channels of signals are fed into 16 square tangent circular polarized radiation patches [36–39].

The array antenna is designed with microwave sheet F4B, the thickness of the media substrate is $H_0 s = 1.5$ mm. The dielectric constant is $\varepsilon_r = 2.65$, and the media loss angle tangent value is 0.001. The array unit uses a cut-angle square patch for circular polarization radiation, with the right-angle driven by the cut triangle being $T_0 = 2.9$ mm, and the patch cell edge length being $W_0 = 36$ mm. Inside the subarray, the distance between the cell patch is $D_1 = 48$ mm, between the two subarrays, the distance between the adjacent cell patch is $D_1 = 48$ mm, and the edge length of the entire array antenna is $K_0 = 340$ mm.

3 Antenna Production and Experiments

Model and optimize the designed circularly polarized microband antenna based on the T-junction 16 array using hf-frequency electromagnetic simulation software HFSS, and then use a printed circuit board to make the physical image of the antenna as shown in Fig. 2.

The echo loss and far-field radiation performance of the antenna were tested using Agilent's vector network analyzers E8363B and SATIMO's darkroom test system SG24, respectively, with results shown in Fig. 3, 4 and 5. Figure 3 is the echo loss of the antenna, Fig. 4 is the radiation gain in the positive direction of the antenna z axis, and Fig. 5 is the axis ratio of the antenna z axis in the positive direction.

As shown in Fig. 3, the simulation frequency range of antenna return loss greater than 10 dB is 2.41–2.53 GHz, and the relative bandwidth is 4.9%; the corresponding test frequency range is 2.35–2.55 GHz, and the relative bandwidth is 8.2%. In terms of return loss, the test result is slightly better than the simulation result. It can be seen

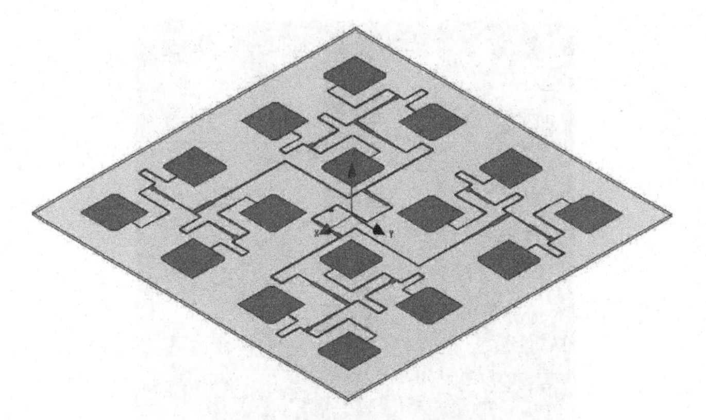

(a) Antenna 3D structural diagram

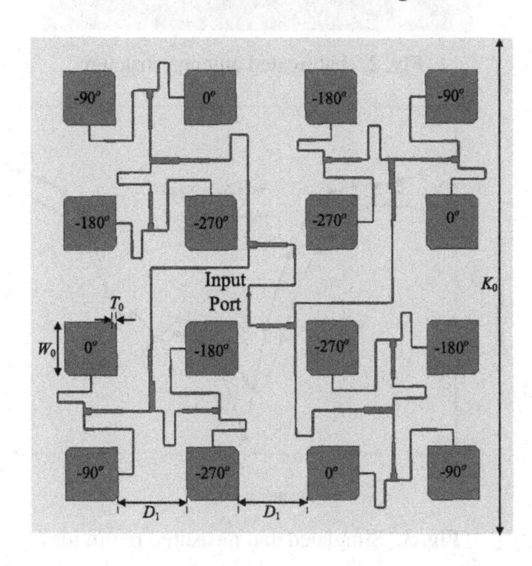

(b) Antenna Top View

Fig. 1. Antenna structure

from Fig. 4 that the simulated peak gain of the antenna is 18.75 dBi, the corresponding frequency point is 2.465 GHz, the tested peak gain is 17.60 dBi, and the corresponding frequency point is 2.472 GHz. In the frequency band 2.35–2.55 GHz, the simulated and tested gains are greater than 9.5 dbi. It can be seen from Fig. 5 that the simulation-3 dB axial ratio frequency range of the antenna is 2.42–2.51 GHz, with a relative bandwidth of 3.7%; the corresponding test working frequency range is 2.412–2.516 MHz, with a relative bandwidth of 4.2%; the above simulation and test results are basically consistent. The antenna is simulated and tested at 2.45 GHz, and the radiation patterns of XOZ and YOZ are obtained, as shown in Fig. 6. The test results show that in the XOZ plane and YOZ plane, the half power beam of the antenna is the same, both are 18° (−10°–8°). It can be seen that the simulation and test results of the antenna are very similar.

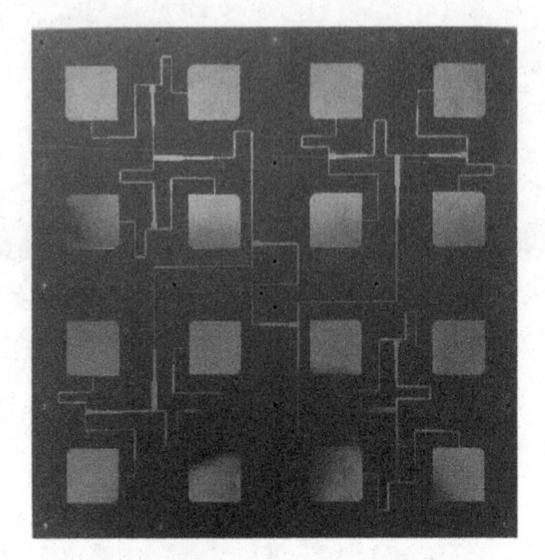

Fig. 2. Fabricated antenna structure

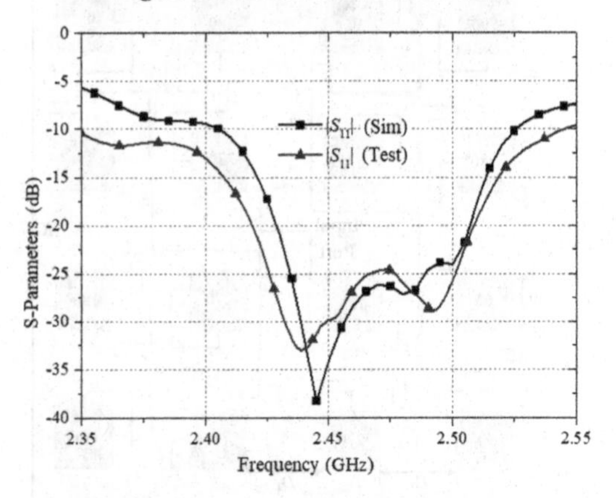

Fig. 3. Simulated and measured return loss

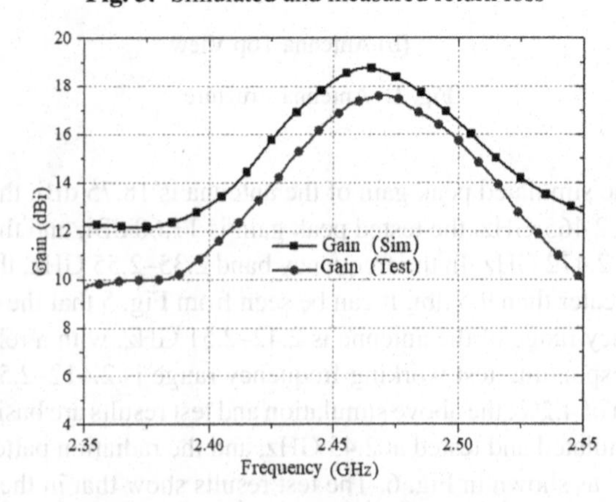

Fig. 4. Simulated and measured antenna gain

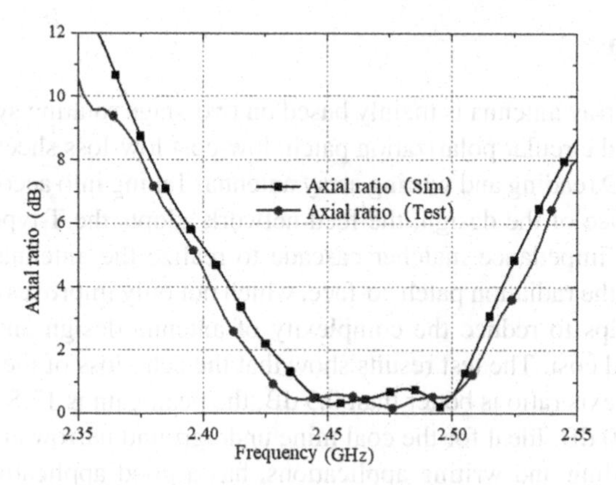

Fig. 5. Simulated and measured axial ratio

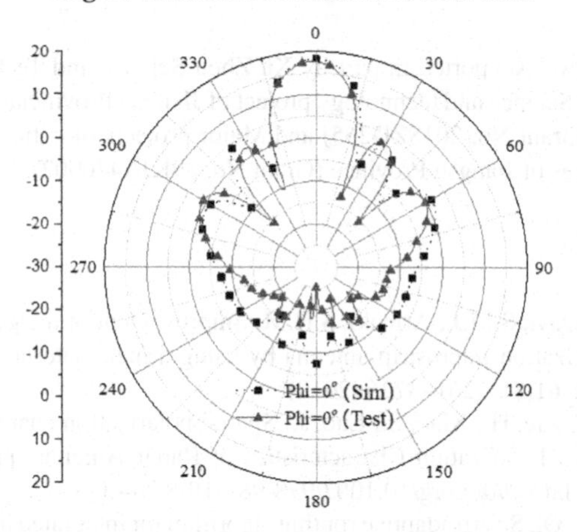

(a) Radiation pattern of XOZ plane at 2.45GHz

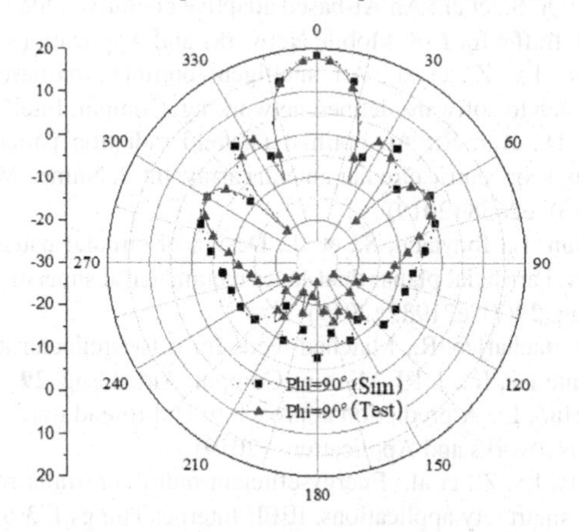

(b) Radiation pattern of YOZ plane at 2.45GHz

Fig. 6. Simulated and measured results of radiation pattern

4 Conclusion

In summary, the array antenna is mainly based on two-stage rotating symmetrical structure, square angled circular polarization patch, low-cost low-loss sheet to achieve high-performance RFID reading and writing array antenna. Taking into account the complexity and performance of the design, the feed network adopts the T-type Power Divider, phase shifter and impedance matcher cascade to realize the antenna structure of the feed network and the radiation patch co-face, which not only improves the front-to-front ratio, but also helps to reduce the complexity of antenna design and processing and assembly time and cost. The test results show that the echo loss of the antenna is better than 24.5 dB, the axis ratio is better than 0.5 dB, the peak gain is 17.8 dB, and the front and rear ratio is 30 dB. Ideal for the coal mine underground narrow environment RFID long-distance reading and writing applications, has a good application prospects and economic benefits.

Funding. This work was supported in part by Xu Zhou Science and Technology Plan Project (Grant No. KC19003), Science and technology project of Jiangsu Provincial Department of housing and construction (Grant No. 2018ZD265) and Major projects of natural science research in Colleges and universities of Jiangsu Province (Grant No. 19KJA470002).

References

1. Dehnavi, M.S., Razavi, S.M.J., Armaki, S.H.M.: Improvement of the gain and the axial ratio of a circular polarization microstrip antenna by using a metamaterial superstrate. Microw. Opt. Technol. Lett. **61**(8), 2261–3226 (2019)
2. Dai, H., Wang, X., Xie, H., Xiao, S., Luo, J.: Spatial polarization characteristics of aperture antenna. In: Spatial Polarization Characteristics of Radar Antenna, pp. 57–131. Springer, Singapore (2019). https://doi.org/10.1007/978-981-10-8794-3_3
3. Wang, F., Jiang, D., Qi, S.: An adaptive routing algorithm for integrated information networks. China Commun. **7**(1), 196–207 (2019)
4. Huo, L., Jiang, D., Qi, S., et al.: An AI-based adaptive cognitive modeling and measurement method of network traffic for EIS. Mobile Networks and Applications (2019)
5. Huo, L., Jiang, D., Lv, Z., et al.: An intelligent optimization-based traffic information acquirement approach to software-defined networking. Comput. Intell. **36**, 1–21 (2019)
6. Das, A., Mandal, D., Kar, R.: An optimal far-field radiation pattern synthesis of time-modulated linear and concentric circular antenna array. Int. J. Numer. Model. Electron. Netw. Devices Fields **32**(3), e2658 (2019)
7. Yassini, A.E., Aguni, L., Ibnyaich, S., et al.: Design of circular patch antenna with a high gain by using a novel artificial planar dual-layer metamaterial superstrate. Int. J. RF Microw. Comput. Aided Eng. **29**(7), e21939 (2019)
8. Debbarma, K., Bhattacharjee, R.: Matched feeds for offset reflector antenna using circular microstrip patch antenna. Int. J. RF Microw. Comput. Aided Eng. **29**(3), e21909 (2019)
9. Qi, S., Jiang, D., Huo, L.: A prediction approach to end-to-end traffic in space information networks. Mobile Networks and Applications (2019)
10. Jiang, D., Zhang, P., Lv, Z., et al.: Energy-efficient multi-constraint routing algorithm with load balancing for smart city applications. IEEE Internet Things J. **3**(6), 1437–1447 (2016)
11. Jiang, D., Li, W., Lv, H.: An energy-efficient cooperative multicast routing in multi-hop wireless networks for smart medical applications. Neurocomputing **220**(2017), 160–169 (2017)

12. Hatami, A., Naser-Moghadasi, M.: Split ring resonator metamaterial loads for compact Hepta band printed inverted F antenna with circular polarization. Microw. Opt. Technol. Lett. **60**(10), 2552–2559 (2018)
13. Bagheroghli, H., Zaker, R.: Double triangular monopole-like antenna with reconfigurable single/dual-wideband circular polarization. Int. J. RF Microw. Comput. Aided Eng. **28**(6), e21267 (2018)
14. Jiang, D., Huo, L., Lv, Z., Song, H., Qin, W.: A joint multi-criteria utility-based network selection approach for vehicle-to-infrastructure networking. IEEE Trans. Intell. Transp. Syst. **19**(10), 3305–3319 (2018)
15. Jiang, D., Huo, L., Li, Y.: Fine-granularity inference and estimations to network traffic for SDN. PLoS ONE **13**(5), 1–23 (2018)
16. Wang, Y., Jiang, D., Huo, L., Zhao, Y.: A new traffic prediction algorithm to software defined networking. Mobile Networks and Applications (2019)
17. Yuwono, R., Hidayatullah, B.R., Dahlan, E.A., et al.: Design of electromagnetic wave pollutant reducing device using rectenna and circular polarization microstrip antenna. Adv. Sci. Lett. **24**(1), 576–580 (2018)
18. Li, W., Leung, K.W., Yang, N.: Omnidirectional dielectric resonator antenna with a planar feed for circular polarization diversity design. IEEE Trans. Antennas Propag. **66**(3), 1189–1197 (2018)
19. Zhang, L., Gao, S., Luo, Q., et al.: Inverted-S antenna with wideband circular polarization and wide axial ratio beamwidth. IEEE Trans. Antennas Propag. **65**(4), 1740–1748 (2017)
20. Jiang, D., Wang, Y., Lv, Z., Wang, W., Wang, H.: An energy-efficient networking approach in cloud services for IIoT networks. IEEE J. Sel. Areas Commun. **38**(5), 928–941 (2020)
21. Jiang, D., Wang, W., Shi, L., Song, H.: A compressive sensing-based approach to end-to-end network traffic reconstruction. IEEE Trans. Netw. Sci. Eng. **7**(1), 507–519 (2020)
22. Jiang, D., Huo, L., Song, H.: Rethinking behaviors and activities of base stations in mobile cellular networks based on big data analysis. IEEE Trans. Netw. Sci. Eng. **7**(1), 80–90 (2020)
23. Shahid, H., Khan, M.T.A., Tayyab, U., et al.: RFID antenna design for circular polarization in UHF band. SPIE Defense + Security (2017)
24. Fukusako, T., Yamauchi, R.: Broadband waveguide antenna using L-shaped probe for wide-angle circular polarization radiation. IEICE Commun. Express **6**, 53–54 (2017)
25. Jiang, D., Wang, Y., Lv, Z., Qi, S., Singh, S.: Big data analysis based network behavior insight of cellular networks for industry 4.0 applications. IEEE Trans. Ind. Inform. **16**(2), 1310–1320 (2020)
26. Chen, L.N., Xie, H.H., Jiao, Y.C., Zhang, F.S.: A novel 4:1 unequal dual-frequency Wilkinson power divider. In: 2010 International Conference on Microwave and Millimeter Wave Technology (ICMMT), pp. 1290–1293. IEEE (2010)
27. Mohra, A., Alkanhal, M.: Dual band Wilkinson power dividers using T-sections. J. Microw. Optoelectron. Electromagn. Appl. **7**(2), 83–90 (2008)
28. Li, X., et al.: A novel design of dual-band unequal Wilkinson power divider. Prog. Electromagn. Res. C **12**, 93–100 (2010)
29. Wu, L., Yilmaz, H., Bitzer, T., et al.: A dual-frequency Wilkinson power divider: for a frequency and its first harmonic. IEEE Microw. Wirel. Compon. Lett. **15**(2), 107–109 (2005)
30. Sakagami, I., Wang, X., Takahashi, K., et al.: Generalized two-way two-section dual-band Wilkinson power divider with two absorption resistors and its miniaturization. IEEE Trans. Microw. Theory Tech. **59**(11), 2833–2847 (2011)
31. Ahn, H., Kim, B.: Toward integrated circuit size reduction. Microw. Mag. **9**(1), 65–75 (2008)
32. Monzon, C.: A small dual-frequency transformer in two sections. IEEE Trans. Microw. Theory Tech. **51**(4), 1157–1161 (2003)

33. Oraizi, H., Sharifi, A.R.: Optimum design of a wideband two-way Gysel power divider with source to load impedance matching. IEEE Trans. Microw. Theory Tech. **57**(9), 2238–2248 (2009)
34. Zysman, G., Johnson, A.K.: Coupled transmission line networks in an inhomogeneous dielectric medium. IEEE Trans. Microw. Theory Tech. **17**(10), 753–759 (1969)
35. Chongcheawchamnan, M., Patisang, S., Srisathit, S., et al.: Analysis and design of a three-section transmission-line transformer. IEEE Trans. Microw. Theory Tech. **53**(7), 2458–2462 (2005)
36. Huang, J.: A technique for an array to generate circular polarization with linearly polarized elements. IEEE Trans. Antennas Propag. **34**(9), 1113–1124 (1986)
37. Zhi, N.C., Xianming, Q., Hang, L.C.: A universal UHF RFID reader antenna. IEEE Trans. Microw. Theory Tech. **57**(5), 1275–1282 (2009)
38. Pan, Y., Zheng, L., Liu, H.J., et al.: Directly-fed single-layer wideband RFID reader antenna. Electron. Lett. **48**(11), 607–608 (2012)
39. Sim, C., Chi, C.: A slot loaded circularly polarized patch antenna for UHF RFID reader. IEEE Trans. Antennas Propag. **60**(10), 4516–4521 (2012)

Simulation Analysis of the Impact of the Rural Financial Efficiency on the Rural Economic Fluctuations

Chenggang Li[1], Hongye Jia[1], and Bing Yang[2](✉)

[1] School of Big Data Application and Economics, Guizhou University of Finance and Economics, Guiyang 550000, People's Republic of China

[2] School of Economics, Guizhou University of Finance and Economics, Guiyang 550000, Guizhou, People's Republic of China

Abstract. This paper constructs a Dynamic Stochastic General Equilibrium model (DSGE) which includes external shocks such as technology shock, cost push shock and monetary policy shock. The static and dynamic parameters of DSGE model are estimated by using calibration method and Bayesian estimation respectively, and the financial efficiency of China's rural areas is measured. Using impulse response analysis and social welfare loss of rural residents, we studied the impact of external shocks on rural economy under the level of rural financial efficiency. The results showed that: the current rural financial efficiency is 0.64; in response to external shocks, when the rural financial efficiency is 0.85, it can most effectively iron the fluctuations of the external shocks on the rural economy.

Keywords: Simulation analysis · Rural financial efficiency · External shocks · Rural economic fluctuations · Dynamic Stochastic General Equilibrium model

1 Introduction

Agriculture is the foundation of our national economic development, rural financial system is not only an important part of our financial system, but also an important guarantee for the sustainable development of rural finance. For a long time, the level of rural financial efficiency has been an important factor affecting the development of rural economy, but on the whole, the current development of rural finance in China is relatively slow. The low efficiency of rural finance is the main obstacle to the development of rural economy, the transformation and upgrading of agricultural industry, and the growth of farmers' income. Chinese rural system has not fundamentally solved the problem of low efficiency of rural finance, the development of agricultural modernization is still not up to the requirements, and the rural economy is in urgent need of greater development. Therefore, the study of agricultural economic growth has an important practical significance.

Financial efficiency refers to the efficiency of banks and other financial institutions in converting deposits into loans (Yan, 2012) [1], while rural financial efficiency

H. Song and D. Jiang (Eds.): SIMUtools 2020, LNICST 370, pp. 371–384, 2021.
https://doi.org/10.1007/978-3-030-72795-6_31

mainly refers to the efficiency of rural financial institutions in converting rural deposits into rural loans (Zhao and Zhu, 2015) [2]. Based on the window reference Malmquist index method, Wu and Zhang (2019) [3] used panel sample data from 31 provinces and cities in China from 2009 to 2016 to empirically test the development of rural financial efficiency on agricultural economy. The study found that rural financial efficiency has varying degrees of impact on current rural economic growth. The impact of rural financial efficiency on rural economically developed areas and rural economically backward areas is relatively weak. Sui et al. (2019) [4] analyzed the efficiency of rural financial services in China's major grain-producing regions, and research shows that the efficiency of rural financial services in China's major grain-producing regions is generally stable and effective. The growth of total factor productivity is mainly attributed to the improvement of technical efficiency. There is a positive correlation between the development of the rural economy and the efficiency of financial services. Jugnaru (2019) [5] studied the relationship between rural human resources development and financial efficiency in ten rural areas of Constanta County in 2016. The research results show that the effective combination of rural financial capital and labor development, that is, the improvement of rural financial efficiency will promote Rural economic development.

Most scholars mainly focus on the following two aspects: First, the research on the relationship between rural finance and rural economy, while the research on the relationship between rural financial efficiency and rural economy is less; second, most of the research methods adopt the traditional quantitative analysis methods such as co-integration and VAR model, but lacking of micro theory (Yu, Wang 2011) [6], and less consideration of the impact of external shocks. In view of this, this paper constructs a Dynamic Stochastic General Equilibrium model (DSGE) which includes external shocks such as technology shock, cost push shock and monetary policy shock, solves the shortcomings of quantitative analysis and micro foundation of traditional econometric model, and then introduces the Dynamic Stochastic General Equilibrium model into the field of rural economic research to simulate and analyze the external factors under different rural financial efficiency, the impact of shock on rural economy, using impulse response analysis and rural residents' social welfare loss analysis to find the best rural financial efficiency.

2 Model Construction

2.1 Rural Families

A rural economy is made up of countless families. The utility of each rural household is affected by the consumption of rural residents, the cash held by rural residents and the labor provided by rural residents. The utility function of rural family is assumed to be:

$$u(C_t, m_t, L_t) = \frac{C_t^{1-\sigma}}{1-\delta} + \frac{(M_t/P_t)^{1-\nu}}{1-\nu} - \frac{L_t^{1-\varphi}}{1-\varphi} \tag{1}$$

Among them, C_t refers to the actual consumption demand of rural residents, M_t is the amount of the nominal monetary balance of rural households, and P_t is the price level $m_t = M_t/P_t$, m_t refers to the actual monetary balance of rural households; L_t refers to the working time of rural residents; σ indicates the reciprocal of the intertemporal substitution elasticity of rural residents' consumption; v indicates the reciprocal of rural monetary demand to interest rate elasticity, and φ indicates the reciprocal of rural labor supply to real wage elasticity. In the t period, the budget constraints faced by rural families are:

$$W_tL_t + r_t^k(i_{t-1} + (1-\sigma)k_{t-1}) + R_tB_{t-1}/\pi_t + (M_{t-1}/P_{t-1})/\pi_t = C_t + I_t + m_t + B_t \tag{2}$$

Among them, W_t means that the rural residents get real wages in the period; R_t means the nominal deposit interest rate of the rural credit cooperatives in the period; r_t^k means the real return on capital of the rural residents. i_t represents the actual investment of rural residents, π_t represents the inflation rate in rural areas, and B_t represents the actual deposit balance of rural residents. The following first-order conditions can be obtained by deriving the consumption of rural residents, the amount of real money balance of rural residents, the labor time of rural residents, the real capital stock of rural residents and the real deposit balance of rural residents:

$$\frac{\partial LN}{\partial C_t} = o \Rightarrow \frac{1}{C_t^{\sigma}} = \lambda_t \tag{3}$$

$$\frac{\partial LN}{\partial m_t} = o \Rightarrow \frac{V_t}{m_t^{v}} + \beta E_t \frac{\lambda_{t+1}}{\pi_{t+1}} = \lambda_t \tag{4}$$

$$\frac{\partial LN}{\partial L_t} = o \Rightarrow L_t^{\phi} = \lambda_t W_t \tag{5}$$

$$\frac{\partial LN}{\partial B_t} = o \Rightarrow \beta E_t \lambda_{t+1} \frac{R_t}{\pi_{t+1}} = \lambda_t \tag{6}$$

$$\frac{\partial LN}{\partial k_t} = o \Rightarrow \beta E_t \lambda_{t+1}((1-\delta) + r_t^k) = \lambda_t \tag{7}$$

2.2 Final Township Enterprises

The final products consumed by rural residents and government departments are provided by the final township enterprises. Under the condition that the scale of production technology remains unchanged, when the nominal price of the intermediate commodity of i is P_{it}, the production function of the final township enterprise is $y_t = \left[\int_0^1 y_{it}^{\frac{\theta-1}{\theta}} di\right]^{\frac{\theta}{\theta-1}}$. Among them, θ represents the substitution elasticity of intermediate products, and the fixed price of the final product township enterprises facing the market is P_t, then the first-order condition for the profit maximization of the final product township enterprises is $y_t(i) = (\frac{P_t(i)}{P_t})^{-\theta} y_t$, the first-order condition function represents the demand function of the intermediate product of i, and under the first-order condition for the profit maximization of the final township enterprises, θ is the demand price elasticity of the intermediate product. The demand price elasticity of intermediate goods reflects the competition intensity of intermediate goods. The larger the price elasticity θ is, the more incentive the market competition of intermediate goods is. According to the definition of perfect competitive market, the profit of township enterprises in perfect competitive market is 0. Therefore, the price P_t in the period t must be met: $P_t = (\int_0^1 P_t(i)^{1-\theta} di)^{\frac{1}{1-\theta}}$.

2.3 Intermediate Township Enterprises

Referring to the research of Ball (1999) [7] and Walsh (2010) [8], it is assumed that the production of products is monopolistic, and the production technology will have an impact on the output of intermediate products, and its production function is $y_{it} = Z_t K_{it-1}^{\alpha} N_{it}^{1-\alpha}$. Among them, K_t means the capital required for the production of intermediate products, N_t means the quantity of labor required for the production of intermediate products, and Z_t means the level of production technology, and it has: $\ln(Z_t) = (1-\rho_Z)\ln(Z) + \rho_Z \ln(Z_{t-1}) + \varepsilon_Z, \rho_Z \in (-1, 1), \varepsilon_Z \sim N(0, \delta_z^2)$.

For township enterprises, the choice of the optimal pricing model and the optimal factor demand is mainly based on the cost minimization. Assuming that the wage rate faced by township enterprises is the same as the marginal cost of labor, the total cost of township enterprises is minimized to $\min_{N_t, K_{t-1}} W_t * N_{it} + r_t^k * K_{it-1}$. Therefore, according to the cost minimization of the enterprise of i under the constraint of production function, the following equation can be obtained:

$$\frac{N_{it}}{K_{it-1}} = \frac{(1-\alpha) * r_t^k}{\alpha * W_t} \tag{8}$$

In addition, the actual marginal cost of the township enterprise of i is:

$$mc_t(i) = \frac{W_t * N_t}{(1-\alpha) * y_t(i)} = \frac{W_t^{1-\alpha} * (r_t^k)^{\alpha} * \alpha^{-\alpha} * (1-\alpha)^{\alpha-1}}{Z_t} \tag{9}$$

It can be seen from Eq. (9) that under the same conditions of wages, return on capital and production technology in the same cycle, the marginal cost of all township enterprises is the same.

In addition, according to the marginal cost equation, the capital equation and labor equation needed by township enterprises can also be obtained, as shown in Eqs. (10) and (11):

$$r_t^k * K_{it-1} = \alpha * mc_t * y_{it} \tag{10}$$

$$W_t * N_{it} = (1-\alpha) * mc_t * y_{it} \tag{11}$$

It is assumed that the intermediate township enterprises have the nature of monopoly, the monopoly competition township enterprises have the ability to price the goods, and the price of the goods will produce stickiness. According to the enterprise pricing rule of Calvo (1983) [9], in each period, only $1-\rho$ township enterprises can optimize the price, and the rest of the township enterprises can only maintain the price of the previous period, and $P_{it} = P_{i.t-1}$. When the intermediate township enterprises can specify the commodity price separately, assuming that all township enterprises choose the same price P_{it}^*, the problem of maximizing the number of township enterprises can be expressed as follows:

$$Max\, E_t \sum_{j=0}^{\infty} \beta^j \rho^j Q_{t+j}(P_{it}^* y_{i,t-1} - W_t N_{it} - r_t^k K_{i,t-1}) \tag{12}$$

Among them, the marginal utility of wealth is $Q_{t+j} = C_{t+j}^{-\delta}/C_t^{-\delta} * P_t/P_{t+k}$. The demand equation and cost equation are brought into the above equation, and the optimal price of intermediate township enterprises is obtained by solving the first-order reciprocal:

$$P_{it}^* = \frac{\theta}{\theta - 1} * \frac{E_t \sum_{j=0}^{\infty} (\beta\rho)^j C_{t+k}^{-\delta} y_{t+j} P_{t+j}^{\beta} mc_{t+j}}{E_t \sum_{j=0}^{\infty} (\beta\rho)^j C_{t+k}^{-\delta} y_{t+j} P_{t+j}^{\theta-1}} \tag{13}$$

At the same time, the equation for price level movement is:

$$P_{it}^{1-\theta} = \rho(\pi P_{t-1})^{1-\theta} + (1-\rho)P_t^{*(1-\theta)} \tag{14}$$

Because all township enterprises have the same marginal cost, so in each period, P_{it}^* are same for all township and village enterprises that can adjust prices. Gali and Gertler (1999) [10] on the basis of Calvo (1983) [9] cross price adjustment model, assuming that there are $1 - \zeta$ proportion of the township enterprises that adjust the price follow the thumb rule of forward behavior, and the remaining township enterprises follow the backward expectation rule, we can get the hybrid New Keynesian Phillips Curve (NKPC) in the following form:

$$\hat{\pi} = \frac{\beta\rho}{\zeta + \rho + \zeta\rho(\beta - 1)} E_t \hat{\pi}_{t+1} + \frac{\zeta}{\zeta + \rho + \zeta\rho(\beta - 1)} \hat{\pi}_{t-1} + \frac{(1-\zeta)(1-\rho)(1-\beta\rho)}{\zeta + \rho + \zeta\rho(\beta - 1)} \hat{mc}_t + \varepsilon_t^c \tag{15}$$

Among them, $\hat{\pi}_t$ represents the deviation of rural inflation rate from the steady state, $E_t\hat{\pi}_{t+1}$ represents the expected rural inflation, $\hat{\pi}_{t-1}$ represents the inertia of rural inflation, $\hat{mc}_t$ represents the deviation percentage of actual marginal cost from the steady state, respectively, ζ, ρ, β represent the proportion of backward pricing, the proportion of maintaining price and the discount factor. ε_t^c represents cost driven impact, $\varepsilon_t^c \sim N(0, \delta_c^2)$, and follows the AR (1) process: $\hat{c}_t = \rho_c \hat{c}_{t-1} + \varepsilon_t^c$.

2.4 Rural Financial Institutions

The management system of township banks is consistent with that of other banks. Assuming that the financial market is a completely competitive market and the financial intermediary in the economy is the banking system, the following methods are adopted to convert the deposits of rural residents into loans needed for the production of township enterprises:

$$I_t = \Gamma * (Y_t/Y)^{\gamma} * B_t \tag{16}$$

Among them, the parameter Γ represents the ratio of rural deposits to rural loans under the condition of steady-state equilibrium, which reflects the efficiency of township banks in transforming rural deposits into rural loans under the condition of steady-state equilibrium. The larger Γ is, the financial efficiency is, the smaller the friction in the financial market is; on the contrary, the smaller Γ is, the financial efficiency is, the greater the friction in the financial market is. γ indicates the degree of sensitivity parameter of financial credit volume to economic state. $(Y_t/Y)^\gamma$ indicates that the actual loan volume is affected by the economic status. When the rural economic form is better than the steady-state level, the loan amount of financial institutions such as banks is higher than the steady-state level; otherwise, it is lower than the steady-state level. Assuming that the financial market is a completely competitive market, under the equilibrium condition, the profit of banks and other financial institutions is 0, then there is $(r_t^I - 1)^* I_t = (r_t^B - 1)^* B_t$.

2.5 Monetary Authority

Drawing on the research results of Wang et al. (2016) [11] and improving, the expression of monetary policy rules is as follows:

$$\hat{r}_t^B = \rho_r \hat{r}_{t-1}^B + (1 - \rho_r)(\varphi_\pi E\hat{\pi}_{t+1} + \varphi_y \hat{y}_t) + \varepsilon_t^r \quad (17)$$

Among them, ε_t^r represents exogenous interest rate shock, $\varepsilon_t^r \sim N(0, \delta_r^2)$, and follows the AR (1) process: $\hat{r}_t = \rho_u \hat{r}_{t-1} + \varepsilon_t^r$.

2.6 Market Equilibrium

When the market is balanced, the market in the model can be cleared at the same time. Among them, the conditions for clearing the labor market is $N_{it} = N_t$, the condition for the capital market is $K_{it} = K_t$, the product market price relationship is $P_{it} = P_t$, the condition for the money market is $M_t^s = M_t$, and the condition for the overall equilibrium is $y_t = C_t + I_t$.

3 Data Processing, Processing and Parameter Estimation

3.1 Data Source and Processing

The data of this paper mainly come from China Financial Yearbook, China Financial Yearbook, China Rural Statistical Yearbook, website of National Bureau of statistics, website of people's Bank of China and wind database. The time interval is from the first quarter of 2002 to the third quarter of 2017. This paper constructs a DSGE model including technology shock, cost driven shock and monetary policy shock. According to the principle of Bayesian estimation, the number of observation variables should be less than or equal to the number of exogenous shocks. In order to ensure that exogenous shocks can be well identified and estimated, and to make the results of subsequent analysis more reliable, the rural output gap (Y) and rural residents' price consumption index (CPI) are used as observation variables in Bayesian estimation. Rural GDP, referring to the research of sun Yu et al. (2014) [12], is characterized by the sum of total agricultural output value and added value of output value of township enterprises. According to Hu (2015) [13], the actual rural GDP is equal to the ratio of the nominal rural GDP to the rural GDP deflator.

The data processing process of this paper is as follows: firstly, take the logarithm of the actual rural GDP and CPI; secondly, take the logarithm of the actual rural GDP and CPI to remove the seasonality by using X-12; finally, use HP filter to remove the actual rural GDP and CPI to remove the seasonality In the long run, we can get the gap value of rural output and the gap value of rural consumer price index.

3.2 Parameter Estimation

The parameters in DSGE model are generally divided into two categories: one is the parameters reflecting the steady-state characteristics of the model, the other is the parameters describing the dynamic characteristics of the model.

Using the existing literature for reference to calibrate the steady-state parameters of the model, referring to the research of He and Li (2017) [14], the discount factor of rural residents β is set as 0.99, and the corresponding actual annual interest rate of rural financial institutions is 4%. Referring to the research of the army and Zhong (2003) [15], the reciprocal of the elasticity of labor supply to real wages of rural residents ϕ is taken as 1, the reciprocal of the intertemporal substitution elasticity of rural residents' consumption σ is 1.5, and the reciprocal of the elasticity of money demand to interest rate of rural residents v is taken as 2. According to the research results of Chen (2017) [16], Yang Yuanyuan and Yu (2017) [17], the capital output elasticity is 0.5, and the stable value of the smoothing index of technological shock Z is 1. Based on Xie P and Luo X (2002) [18], the steady-state rural inflation δ is set as 1. For the capital depreciation rate, referring to the research of Gong and Xie (2004) [19], the annual depreciation rate value of capital is set as 10%, and the corresponding quarterly value is 2.5%. Based on the research of Liu Bin (2008) [20], the marginal cost parameter of township enterprises mc is set as 0.91. Referring to the research of Hu and Zhang (2015) [13], we set the substitution relationship between leisure and consumption φ as 1, the price forward probability ζ as 0.25, and the price proportion ρ as 0.85. According to Yan (2012) [21], the sensitivity parameter of financial intermediary credit volume to economic state γ is set as 1.12.

This paper focuses on the analysis of the impact of different rural financial efficiency on rural economy. Parameters Γ are the key indicators to measure rural financial efficiency. The larger Γ is, the rural financial efficiency is, the higher the rural financial efficiency is; the smaller Γ is, the rural financial efficiency is, the lower the rural financial efficiency is. From the function formula (16) of rural financial institutions, it can be seen that Γ is equal to the ratio of deposits of rural financial institutions to loans of rural financial institutions, because the efficiency of rural financial institutions does not have a fixed value. Therefore, this paper chooses the ratio of rural financial institution deposits to rural financial institution loans to calibrate, since 2002, the ratio of deposits of rural financial institutions to loans of rural financial institutions (i.e., rural financial efficiency) has changed greatly, so it is difficult to find a value to represent the steady state deposit and loan ratio. Therefore, the base value of this paper is 64%, and the other values are

50%, 85% and 95% to compare the effects of different rural financial efficiency on rural economy, so as to determine the optimal value of rural financial efficiency.

For the dynamic parameter estimation of DSGE model, the Bayesian estimation method is mainly used. Firstly, the prior step-by-step determination of the estimated parameters is needed. Referring to the research of Liu and Yao (2016) [22], etc., the prior probability distribution of parameters with values between 0 and 1 is set as beta distribution, the prior probability distribution of parameters with values between 0 and infinity is set as gamma distribution, and the prior probability distribution of standard deviation of exogenous impact is set as Inv-Gamma distribution. According to Coenen and Straub (2005) [23], the smoothing index of exogenous shock ρ is set to obey the beta distribution with the mean value of 0.5 and the standard deviation of 0.1; the standard deviation parameter of exogenous shock σ is set to obey the Inv-gamma distribution with the mean value of 0.01.

When using Bayesian estimation method to estimate the parameters of DSGE model, considering the impact of technology and money supply in the model, we should select the corresponding rural output gap value and the fluctuation of real money balance as samples to estimate. However, in the estimation process, it is found that there is a problem of unrecognized parameters φ_π. Therefore, this paper selects the gap value of rural output and the gap value of rural consumer price index to estimate, and finds that all parameters can be identified. The Bayesian estimation results are shown in Table 1.

Table 1. Bayesian estimation results.

Parameter	Index meaning	Distribution type	Bayesian estimation	
			Average	90% confidence interval
ρ_Z	Technical impact smoothing coefficient	Beta[0.5, 0.2]	0.8408	[0.7606, 0.9286]
ρ_C	Cost driven impact smoothing factor	Beta[0.5, 0.2]	0.8475	[0.7058, 0.9893]
ρ_R	Smoothing coefficient of monetary policy impact	Beta[0.5, 0.2]	0.6985	[0.6137, 0.7743]
σ_Z	Standard deviation of technical shock	Inv-gamma[0.01, ∞]	0.0379	[0.0298, 0.0456]
σ_C	Standard deviation of cost driven impact	Inv-gamma[0.01, ∞]	0.0110	[0.0022, 0.0225]

(*continued*)

Table 1. (*continued*)

Parameter	Index meaning	Distribution type	Bayesian estimation	
			Average	90% confidence interval
σ_R	Standard deviation of monetary policy shock	Inv-gamma[0.01, ∞]	0.0082	[0.0065, 0.0098]
φ_π	Inflation response coefficient	Gamma[1.5, 0.2]	1.2469	[1.1179, 1.3648]
φ_y	Output response coefficient	Gamma[0.5, 0.2]	0.6612	[0.5026, 0.8548]

4 Simulation Result Analysis

4.1 Impulse Response of Technology Shock to Rural Output, Inflation, Consumption and Investment

Figure 1 shows the response of rural output, inflation, investment and consumption to external shocks under the four rural financial efficiency with 1% increase in technological progress. In the sense of economics, the progress of industrial and agricultural technology leads to the improvement of the production efficiency of township enterprises, which will reduce the marginal cost of agricultural products. In the short term, the decrease of the marginal cost of agricultural products will produce downward pressure on the price of agricultural products. The price level will fall, and the inflation in rural areas will be restrained, which is shown in Fig. 1 (b) the negative fluctuation of rural inflation. With the development of industrial and agricultural technology, the scale of township enterprises will be adjusted to the optimal production scale after a period of development. Therefore, township enterprises have the ability to increase agricultural output at the original level of employment. At this time, the best choice of manufacturers is to increase investment and expand production scale by using the advantage of actual marginal cost reduction, which is shown in the negative and positive fluctuations of rural output in Fig. 1 (a) and rural investment in Fig. 1 (c). Technological progress reduces the marginal cost of products, but also reduces the demand for rural labor. The decline of rural labor demand leads to the decline of rural consumption, which is shown as the negative fluctuation of rural consumption in Fig. 1 (d). From Fig. 1 (a), Fig. 1 (b), Fig. 1 (c) and Fig. 1 (d), it can be seen that in the case of rural financial efficiency, the fluctuation of technological shock to rural output, inflation, investment and consumption is the smallest and the speed of returning to the steady-state equilibrium is the fastest, which shows that when the rural financial efficiency is $\Gamma = 0.85$, there are more effective to cope with exogenous shocks and smooth rural economic fluctuations. Therefore, in response to the impact of technology, the optimal rural financial efficiency value is 0.85.

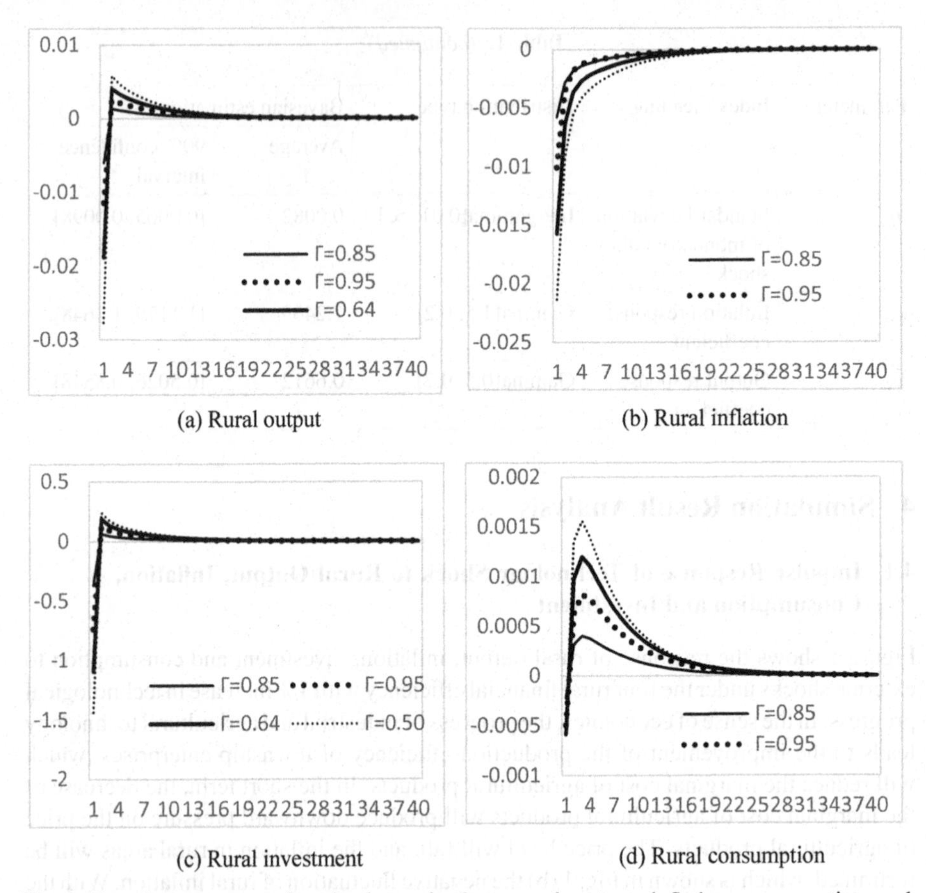

Fig. 1. Impulse response of technology shock to rural output, inflation, consumption and investment under four financial efficiency

4.2 Impulse Response of Cost Driven Shocks to Rural Output, Inflation, Consumption and Investment

Figure 2 shows the response of rural output, inflation, investment and consumption under four different rural financial efficiency in the face of the impact of cost promotion. In the sense of economics, the rise of agricultural product cost will first promote the rise of rural inflation, which is shown in Fig. 2 (b) as the positive fluctuation of rural inflation. When the cost of products increases, the output of the whole market is restrained, resulting in the decline of output level, which is shown as the negative fluctuation of rural output in Fig. 2 (a). With the increase of cost, enterprises spend more money on raw materials of products and occupy the investment in other industries, which leads to the decrease of rural investment level, which is shown as the negative fluctuation of rural investment in Fig. 2 (c). The decrease of rural investment makes the liquidity in the hands of rural residents increase, and the consumption of rural residents increases rapidly in the short term. However, with the passage of time, the impact of rural inflation on the agricultural product market is more and more obvious. The price of agricultural products is rising, which makes rational consumers make cross period choices to reduce current consumption and increase future consumption driven by the maximization of lifetime utility. Therefore, it is shown as the first positive and then negative fluctuation of rural consumption in Fig. 2 (d). From Fig. 2 (a), Fig. 2 (b), Fig. 2 (c) and Fig. 2 (d), it can be seen that when the rural financial efficiency is $\Gamma = 0.85$, the impact of cost promotion should be dealt with, and the fluctuation of rural output, inflation, investment and consumption should be the smallest and the speed of returning to equilibrium state should be the fastest. It shows that when the optimal rural financial efficiency is 0.85, the impact of cost promotion can be dealt with more effectively and the fluctuation of rural economy can be ironed out.

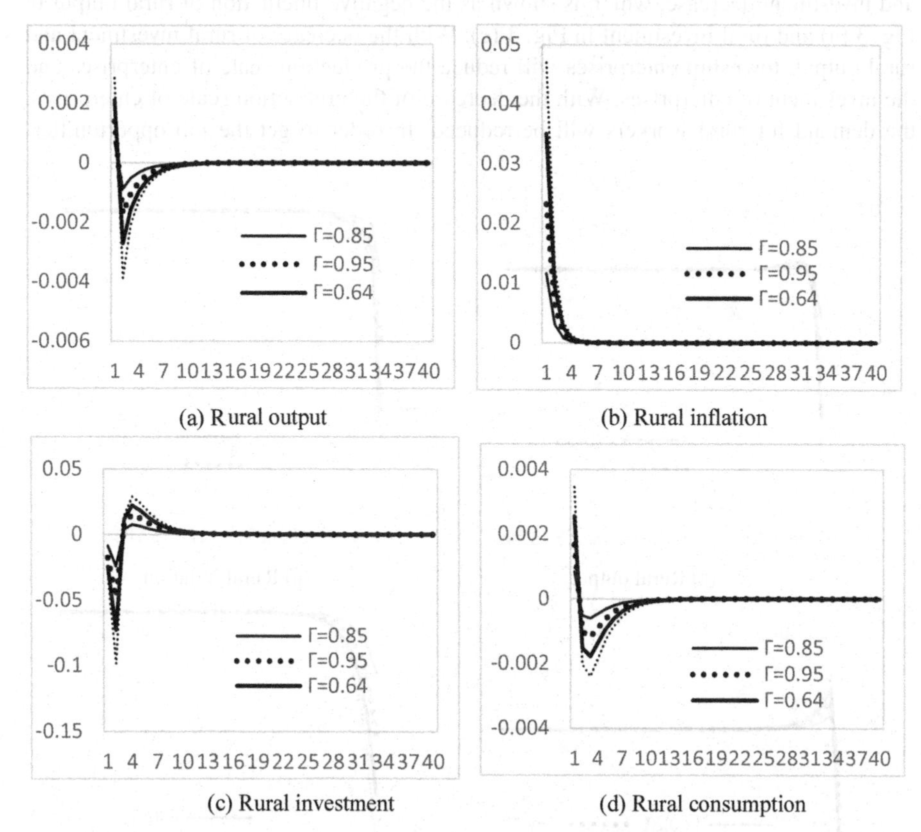

Fig. 2. Impulse response of cost driven shocks to rural output, inflation, consumption and investment under four financial efficiency

4.3 Impulse Response of Monetary Policy Shock to Rural Output, Inflation, Consumption

Figure 3 shows the response of rural output, inflation, investment and consumption to the impact of monetary policy. The monetary policy equation in DSGE model is Taylor equation, and the impact of monetary policy is mainly reflected in the impact of interest rate impact on rural economy. Therefore, in the economic sense, the increase of interest rate of rural financial institutions first impacts on the township enterprises that have loans in the bank, and the increase of interest rate makes the financing cost of township enterprises rise. At the same time, the increase of interest rate makes the investment profit space of township enterprises be compressed. In the case of no profit, township enterprises will reduce the production scale and investment, which makes the rural output and investment decrease, which is shown as the negative fluctuation of rural output in Fig. 3 (a) and rural investment in Fig. 3 (c). With the decrease of rural investment and rural output, township enterprises will reduce the production scale of enterprises and the investment of enterprises. With the decrease of the production scale of enterprises, the demand for rural workers will be reduced. In order to get the job opportunities,

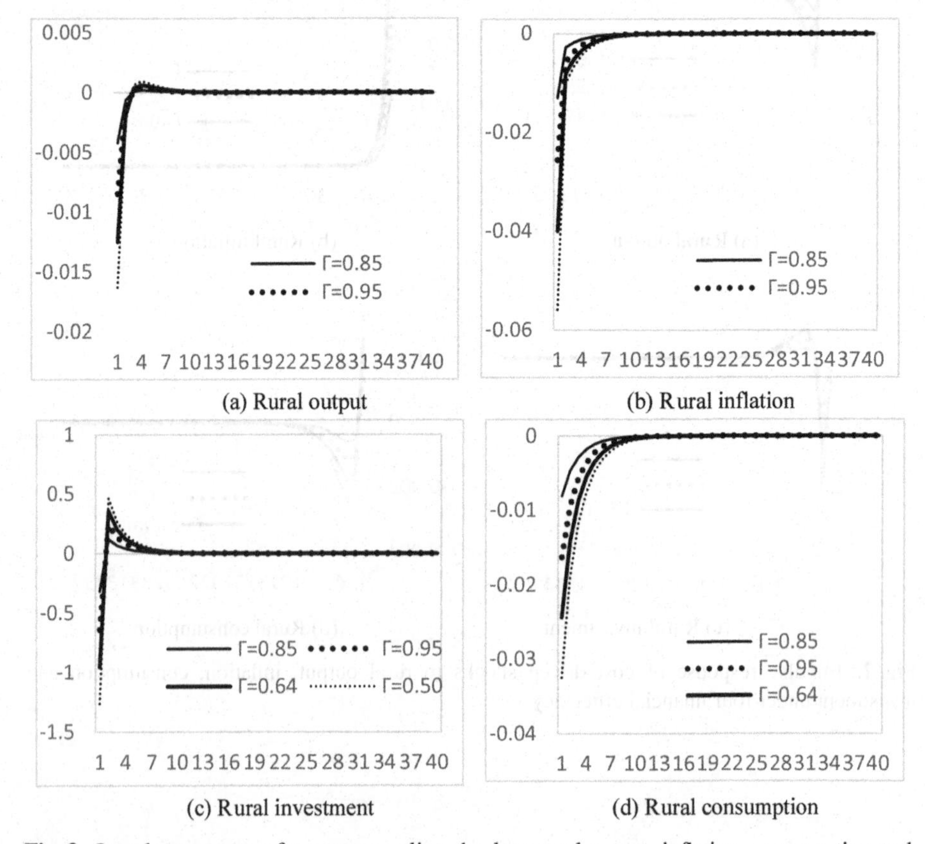

Fig. 3. Impulse response of monetary policy shock to rural output, inflation, consumption and investment under four financial efficiency

workers have to reduce the demand for wage remuneration, so that the wages will be reduced. Residents do not have more funds for consumption, which is shown in Fig. 3 (d) The negative fluctuation of consumption. With the decrease of consumption demand, there will be a phenomenon that the supply of goods is greater than the demand in the agricultural product market. In order to reduce inventory, township enterprises have to reduce the price of agricultural products to sell more goods, which makes the inflation driven by consumption demand lower, which is shown as the negative fluctuation of rural inflation in Fig. 3 (b). From Fig. 3 (a), Fig. 3 (b), Fig. 3 (c) and Fig. 3 (d), it can be seen that when the rural financial efficiency is 0.85, the impact of monetary policy (i.e. the impact of interest rate) should be dealt with, and the fluctuation of rural output, inflation, investment and consumption is the smallest and the fastest to return to the equilibrium state, indicating that when the rural financial efficiency is $\Gamma = 0.85$ can more effectively deal with the exogenous impact and iron out the rural economic fluctuation.

5 Conclusions

This paper first constructs a Dynamic Stochastic General Equilibrium model including technology shock, cost push shock and monetary policy shock, estimates the static parameters and dynamic parameters respectively through calibration method and Bayesian method, and calculates the current rural financial efficiency value; secondly, uses impulse response analysis method and social welfare loss analysis to investigate different financial effects respectively The impact of technology shock, cost push shock and monetary policy shock on rural output, inflation, investment and consumption, as well as the loss of social welfare of rural residents, in order to seek the optimal rural financial efficiency. The results show that: first, the current rural financial efficiency is 0.64; second, from the four rural financial efficiency of rural output, inflation, investment and consumption response to external shocks, when the rural financial efficiency is 0.85, it can more effectively respond to external shocks and quickly iron the fluctuations of external shocks on the rural economy; third, from the four rural financial efficiency, when the rural financial efficiency is 0.85, it can more effectively stabilize the rural economy.

References

1. Yan, L.L.: The impact of financial intermediary efficiency on the effect of monetary policy - a study based on dynamic stochastic general equilibrium model. Stud. Int. Financ. (6), 4–11 (2012)
2. Zhao, H.D., Zhu, X.P.: Rural financial scale, rural financial efficiency and rural economic growth: evidence from Jilin province. Econ. Surv. **32**(3), 28–34 (2015)
3. Wu, L.J., Zhang, J.Q.: Rural finance supporting agricultural efficiency analysis based on window reference Malmquist index. Jiangxi Soc. Sci. **39**(08), 64–74 (2019)
4. Xin, S., Hong, M.W., Feng, G.Y.: Analysis of rural financial service efficiency in major grain producing areas. J. Phys. Conf. **1**(1), 57–68 (2019)
5. Jugnaru, M.: A dynamic analysis of economic and financial efficiency, correlated with the dimension of the Human resources used by companies from the Rural Area of Constanta country, Romania. Ovidius Univ. Ann. Econ. Sci. Ser. **19**, 233–237 (2019)

6. Yu, Y.J., Wang, J.H.: Research on the relationship between rural financial development and rural economic growth in China based on VAR model. Econ. Probl. (12), 106–110 (2011)
7. Ball, L.: Efficient rules for monetary policy. Int. Financ. **2**(1), 63–83 (1999)
8. Walsh, C.E.: Monetary Theory and policy, pp. 200–214. Massachusetts Institute of Technology Press, Cambridge (2010)
9. Calvo, C.A.: Staggered prices in a utility maximizing framework. J. Monet. Econ. **12**(3), 383–398 (1983)
10. Ga, L.J., Gertler, M.: Inflation dynamics: a structural economic analysis. J. Monet. Econ. **44**(2), 195–222 (1999)
11. Wang, X., Wang, Q., Cheng, Z.F.: Monetary policy expectations and inflation management: a DSGE analysis based on news shock. Econ. Res. J. **51**(2), 16–29 (2016)
12. Sun, Y.K., Zhou, N.Y., Li, P.D.: Research on the impact of rural financial development on rural residents income. Stat. Res. **31**(11), 90–95 (2014)
13. Hu, X.W., Zhang, S.F.: The impact of interest rate liberalization on the effects of quantitative and price-based monetary policies. Financ. Forum **20**(4), 26–35 (2015)
14. He, G, H., Li, J.: Cross-border capital flows, financial volatility and monetary policy choices. Studies of International Finance (9), 3–13(2017).
15. Lu, J., Zhong, D.: Cointegration test of Taylor rule in China. Econ. Res. J. **8**, 76–93 (2003)
16. Chen, L.F.: News, real estate prices and monetary policy. Contemp. Financ. Econ. (6), 3–17 (2017)
17. Yang, Y.Y., Yu, J.P.: The choice of China's optimal monetary regulation paradigm in the new normal—based on the perspective of fiscal and monetary policy interaction. World Econ. Pap. (2), 72–86 (2017)
18. Xie, P., Luo, X.: Taylor Rule and its test in Chinese monetary policy. Econ. Res. J. **7**(3), 3–12 (2002)
19. Gong, L.T., Xie, D.Y.: Analysis of differences in factor flows and marginal productivity among provinces in China. Econ. Res. J. **1**, 45–53 (2004)
20. Liu, B.: Development of China DSGE model and its application in monetary policy analysis. J. Financ. Res. **10**, 1–21 (2008)
21. Yan, L.L.: The impact of financial intermediary efficiency on the effect of monetary policy—a study based on dynamic stochastic general equilibrium model. Stud. Int. Financ. (6), 4–11 (2012)
22. Liu, X.X., Yao, D.B.: Financial disintermediation, asset price and economic fluctuation: analysis based on DNK-DSGE Model. World Econ. **39**(6), 29–53 (2016)
23. Coenen, G., Straub, R.: Does government spending crowd in private consumption? Theory and empirical evidence for the euro area. Int. Financ. **8**(3), 435–470 (2005)

Finding Good Mobile Sink Information Collection Paths with Quicker Search Time: A Single-Particle Multi-dimensional Search Algorithm-Based Approach

Pengyu Huang[1], Fuping Wu[2], Wei Wang[3(✉)], Haiyan Liu[4], and Qin Liu[1]

[1] School of Telecommunications Engineering, State Key Labs of ISN, Xidian University, Xi'an 710071, China
{pyhuang,qinliu}@mail.xidian.edu.cn

[2] School of Physics and Optoelectronic Engineering, Xidian University, Xi'an 710071, China
fpwu@xidian.edu.cn

[3] School of Information Science and Technology, Northwest University, Xi'an 710127, China
wwang@nwu.edu.cn

[4] University of Leeds, Leeds LS2 9JT, UK
h.liu1@leeds.ac.uk

Abstract. Energy consumption is the first-class constraint for the battery-powered Internet of Things (IoT) devices and sensors. By moving around a sensor network to gather data, a mobile sink (MS) can greatly save sensor energy for multi-hop communication. To unlock the potential of mobile sinks, we need to carefully plan the path a mobile sink moves within the network for collecting information without compromising its coverage and its battery life. This paper presents a new way to find the optimal information collection path for mobile sinks. We achieve this by formulating the optimization problem as a classical Traveling Salesman Problem with Neighborhoods (TSPN). We then design a novel solver based on the particle multi-dimensional search algorithm to quickly locate a good path schedule in the TSPN optimization space. As a significant departure from prior work which uses multiple particles to explore multiple potential solutions, our method uses only one particle for problem-solving. Doing so significantly reduces the complexity of the algorithm, allowing it to scale to a larger sensor network. To ensure the quality of the chosen solution, we have carefully designed the evolutionary process for problem-solving. We show that our approach finds a solution with similar quality as those given by a multi-particle-based search, but with significantly less time. Simulation results show that our

Supported in part by the Key Research and Development Project of Shaanxi Province under Grant 2018GY-017, 2017GY-191 and 2019GY-146, in part by the Foundation of Education Department of Shaanxi Province Natural Science (15JK1742), and in part by the Foundation of Science Project of Xi'an (201805029YD7CG13(4)).

H. Song and D. Jiang (Eds.): SIMUtools 2020, LNICST 370, pp. 385–398, 2021.
https://doi.org/10.1007/978-3-030-72795-6_32

approach can find a high-quality path schedule compared to the state-of-the-art algorithm in a large sensor network.

Keywords: Internet of Things · Mobile sink · TSPN · Single particle · Shortest path

1 Introduction

With the explosive growth of the Internet of Things (IoT), it has been widely used in environmental monitoring, fire monitoring, battlefield exploration, and other application scenarios [1]. In these applications, a large number of sensor nodes are typically deployed over a wide area. Whenever a target event happens, the sensor nodes of this field could perceive it, and send the relevant data to the static sink node by multi-hop routing. We can see, these data will gradually converge to the static sink node. And as time goes on, the sensor nodes that close to the static sink node will inevitably consume more energy due to the heavy burden of data forwarding. Whenever these sensor nodes run out of energy, the static sink node also can't receive data anymore. This phenomenon is called as energy hole [2], and it could greatly decrease the lifetime of the whole network.

To address the 'energy hole' problem, researchers introduce mobile sinks (MS) into sensor networks. The mobile sink is a sink node been installed on mobile equipment, such as drones. The mobile sink collects data from sensor nodes along the tour that composed of multiple collection sites (CSs). While the mobile sink passing through a CS could collect the data from the sensor nodes close to it directly. As gathering data in this way, the sensor node no longer needs to forward data for each other. Therefore, by using the mobile sink could greatly reduce the energy-consuming of data forwarding and simplify the process of data collection. So the mobile sink is a good solution to balance the energy consumption among the sensor nodes and expand the lifetime of the whole networks [3–5], as compared to the static sink node. Moreover, the mobile sink is closer to the sensor node during communication than the static one, so it could achieve lower frame error rate and better data collection quality. Besides, the mobile sink can be used for node positioning and gain a significant improvement in accuracy [6,7].

But in practice, IoT is often been used in a wide-open area. And for the mobile sink, its ability for data collection is limited by the speed range and energy efficiency of the carrier platform. Compared with the static sink node, the mobile sink may need more time to complete the data collection of the whole network. But for now, it's still difficult to improve the capability (including speed or power supply) of the carrier platform. So, the most efficient way is to shorten the length of the data collection tour. The shorter the data collection tour, the lower the data collection latency for the whole network. So, it's an urgent issue to discover the optimal tour planning method for the mobile sink to minimize the data collection latency. In this paper, we focus on how to obtain the shortest tour of the mobile sink, so as to implement the minimum data collection latency.

During the data collection, the mobile sink travels along a pre-planned tour. Usually, the transceiver of the mobile sink is off for most of the time to save energy, and it will be available for data collection while reaching the collection sites. Base on this, we can see that how to determine the positions of the CS is the key problem. So we can divide the tour planning algorithms of mobile sink into two types according to whether use the sensor node's location as a CS:

(1) Only using the sensor node's location as the CS, the tour planning method of mobile sink could be reduced to a traveling salesman problem (TSP) problem [8–13,17–19].
(2) Without limiting the position of the CS, this tour planning problem could be reduced to a traveling salesman problem with neighborhoods (TSPN) [14,20–23].

For the TSP-based data collection methods, the mobile sink directly traverses the position of each sensor node, without considering the communication range of the sensor node. This method could simplify data collection. However, TSP is an NP-Hard problem. Especially for large-scale IoT, TSP-based method would consume a lot of computing resources, and this makes it difficult to be applied in real applications.

Moreover, for TSP-based methods, the mobile sink must visit every sensor node one by one. This means even with the fixed deployment field, the length of the data collection tour will increase with a growing number of the sensor nodes, and leading to greater data collection latency and energy consumption. As a result, the TSP-based method has many problems in scalability and practicability.

In contrast to the TSP-based method, TSPN-based methods will be better. In this kind of method, the mobile sink doesn't directly traverse the positions of sensor nodes, replaced by firstly finds out some collection sites (CSs) and only traverses them. And the CS could be any location of the deployment area. We can see, TSPN-based methods could be more flexible.

At the CS point, the MS can collect data from every sensor node within its communication range. So TSPN-based method only needs to traverse fewer CSs to complete the data collection, which means shorter data collection tour length and shorter data collection latency. So the TSPN-based method could make full use of the wireless communication capability and achieve high data collection efficiency. Furthermore, in TSPN-base methods, the mobile sink directly gathers data from every sensor node, so there are no more needs for support of network structure or protocol. So it could extremely simplify the network structure and reduce energy consumption.

To realize the TSPN-based method, the key challenge is how to choose the CSs. To address this challenge, we propose a novel multi-dimensional searching algorithm base on a single particle.

This algorithm inherits the concept of the particle from the particle swarm optimization (PSO) algorithm. The particle is composed of a series of dimensions, and each dimension represents the position of a CS. In this way, a particle could represent a complete data collection tour.

The traditional PSO algorithms usually use hundreds of particles, and each particle needs to keep tracking the historical optimal solution of its own and the population. Which requires extensive computation, then will seriously affect the effectiveness of it in practice. And it will be worse to apply PSO for resource-constrained IoT nodes.

To perform the advantages of the PSO and bypass the above problems, this paper innovatively proposes a method with only 'one' particle to solve the TSPN problem. Based on one particle, we transform the searching problem of optimal data collection tour for the mobile sink into the searching problem of a single particle with different dimensions. Besides, to further reduce the complexity of the algorithm, we employ a predefined non-uniform searching grid for each dimension. In this way, the continuous searching for each dimension is converted into a discrete searching, which could greatly reduce the complexity and computation of the algorithm.

The main contributions of this paper can be summarized as follows.

(1) In the process of using PSO to solve the best path of mobile sink, 'one' particle is used to reduce the computation.
(2) The global search for PSO is transformed into a local search for each dimension in 'one' particle.
(3) Preset grid is used in local search to further reduce the complexity of the search.

With these approaches, our algorithm could greatly reduce computation and energy consumption and gain availability in real scenarios. Simulation results show that our algorithm could dramatically reduce the computation, at the same time, it could guarantee the quality of data collection tour. So it could reduce the data collection latency significantly.

2 Related Work

As a TSP-based method, [8] studies the relationship between the moving speed of the mobile sink and the data collection efficiency in a random deployment. MASP (Maximum Amount Shortest Path) is proposed to transform the finite-time information collection problem into an integer linear programming problem. The optimal path solution will be obtained by using the genetic algorithm and a distributed approximation algorithm. [9] proposes a data acquisition protocol DCPD for the mobile sink. This protocol selects collection points according to the distribution of the sensor nodes and obtains the shortest data collection route through these collection points based on the quantum genetic algorithm. [10] summarizes the problem of data collection path length minimization as a single hop data collection problem (SHDGP), then solve it by transforming SHDGP into a mixed integer programming problem. [11] proposes an optimal selection method of the mobile sink path based on the priority of virtual points. This method could achieve the optimization objective that minimizes the energy consumption of the whole network with the restriction of time delay. It could reduce the time complexity of the algorithm in case of a small increase in energy consumption. [12,13] propose mobile sink data collection mechanisms based on the information RP (rendezvous point). In these mechanisms, the sensor node sends the collected information to the nearest RP node in a multi-hop manner, and the mobile sink traverses each PR successively to collect data. [17] proposes a routing algorithm called cluster-chain mobile agent routing (CCMAR). CCMAR introduces data compression in the process of constructing the mobile sink path. [18] proposes a path planning which is called EARTH. The main idea of this algorithm is to consider the relationship between path planning and data produce speed and buffer size of sensors. In [19], considering the relationship between the packet loss rate and mobile sink's speed, a path planning method with effective energy and the minimum packet loss rate is proposed.

In the methods based on TSPN, in [14], the shortest information collection path problem is regarded as a traveling salesman problem with the adjacent area (TSPN). This paper proposes a heuristic algorithm to construct an information collection path. [20] introduces MC-WSN architecture. Through active network deployment and topology design, the number of hops from the sensor to the moving sink is limited, thus improving the effect of path planning. [21] proposes an algorithm is called BR-CTR. In this algorithm, the multi-hop transmission is considered to reduce the delay time of data gathering. [22] considers an energy balanced tree-based data collection strategy to improve the sensing efficiency of the sensor network. To solve the problem of the coverage of the sensing area and the coverage holes problem in WSN, [23] proposed an energy-efficient solution based on the mobile sink.

3 Single Particle Multi-dimensional Particle Swarm Optimization Algorithm

3.1 Algorithm Overview

In this section, to simplify the problem, we assume the communication radius of the sensor nodes and mobile sinks are the same and the distribution of sensor nodes is already known. Furthermore, we assume the transmission time between the MS and sensor nodes is negligible as compared with the traveling time of the MS [15]. With these assumptions, the data collection mission can be accomplished as the traveling tour been finished.

For now, the key to achieving the MS's optimal data collection tour depends on the answer of two challenges: how to choose the optimal CSs and how to connect these CSs to form a shortest data collection tour. So the focus of this paper can be summarized as follows: under the condition of given sensor node distribution, select the appropriate CSs, and then connect these CSs to form a shortest data collection tour. Mobile sink visits all CSs along this tour to collect data and finally achieves the minimal data collection latency. The tour construction problem, in this case, can be formulated as:

$$min\,|T|\; s.t. \forall s_i \in S, \exists st_j \in ST, |s_i, st_j| \leq d \quad T \in \Gamma \tag{1}$$

- $|T|$ is the length of the tour;
- S is the set of sensor nodes;
- ST is the set of collection sites;
- Γ is the set of possible tours;
- $|s_i, st_j|$ is the Euclidean distance from s_i to st_i;
- d is the effective communication distance of the sensors;

According to the above discussion, the first step is to select some CSs that could cover all the sensor nodes within their communication range. Then the second step is to figure out the optimal TSP tour involves all the CSs. Since there are many mature methods to calculate the optimal TSP tour, so how to realize the first step is the key challenge for us.

According to relevant studies, we know different sets of CSs will directly cause the different length of MS's tour and the size of data collection latency. Therefore, the key to obtaining the optimal data collection tour is to find the best number and the positions of CSs. However, the distribution of sensor nodes is random in the real application. Moreover, even if the distribution of sensor nodes is fixed, there are infinite possibilities for the sets of CSs.

For the data collection tour is a multi-dimensional structure composed of CSs, so finding the best set of CSs is a typical multi-dimensional searching problem. The most difficult part of this problem is how to find the best one from the infinite combinations of CSs. We find that the PSO algorithm is one of the most effective methods to solve this kind of problem. In the traditional PSO algorithm, a particle is used to represent a feasible tour of the mobile sink. Each particle involves several dimensions, and each dimension represents the coordinates of a CS.

As shown in formula 2 and 3, in the traditional PSO-based information collection tour solution, x is an ordered vector consisting of the positions of CHs to represent a feasible solution. According to formula 2 and 3, x is iterated continuously to obtain the optimal solution.

$$\nu_{id}^{k+1} = \omega \times \nu_{id}^{k} + c_1 \times \gamma_1 \times (pbest_{id}^{k} - x_{id}^{k}) + c_2 \times \gamma_2 \times (gbest_{id}^{k} - x_{id}^{k}) \quad (2)$$

$$x_{id}^{k+1} = x_{id}^{k} + \nu_{id}^{k} \quad (3)$$

- x_{id}^{k}: The d-th dimension (the location of the d-th CH) of particle i at time k;
- ν_{id}^{k}: The speed of d-th dimension of particle i at time k;
- ω: Inertia weight;
- c_1: The step-length for the particle which follows to the best solution of itself;
- c_2: The step-length that the particle which follows to the best solution of the group;
- γ_1 and γ_2: Random values between [0,1];

Therefore, a particle can be considered as a multi-dimensional vector. The data collection can be achieved by gradually moving the mobile sink along the dimensions (CSs) of the particle.

However, the traditional PSO algorithm usually involves hundreds of particles, and each particle needs to track the historical optimal solution of its own and the entire swarm. These methods will consume more memory space and more energy. But in the realistic application, MS's computing capacity and power supply are very limited, so it is difficult for them to implement the tour searching by PSO. To address this problem, we improve the traditional PSO algorithm from 4 aspects.

(1) We don't use a large number of particles but only one particle, and it becomes unavailable to use the tracking method of the traditional PSO algorithm. This means the particle does not need to record the historical optimal solution of the other particle anymore. In this way, it can greatly reduce the complexity of the algorithm, and the requirement of computing resource.
(2) Using a single particle, the main challenge is how to realize multi-dimensional searching without traditional evolution proceeding. Here we propose a novel multi-dimensional searching method that evolves from random searching. In this method, we repeatedly select a dimension randomly, then search a better position for it, iterating until finding the optimal data collection tour.
(3) While searching in each dimension, to further reduce the complexity, we introduce a pre-built searching grid template. After selecting a dimension randomly from the particle, we use the searching grid template to determine the alternative dimension set. Since a dimension represents the position of a CS, it is necessary to ensure that the mobile sink can still cover the same or more sensor nodes when moving from the current position to the alternative position. So, we must decide the alternative dimension positions one by one and select these valid positions to form the final alternative dimension set. After that, calculate the length of all TSP tour when the current dimension is replaced by the dimension in the final alternative dimension set. Then select the alternative dimension with the shortest tour length to update the current dimension. Iterate the above process to get the best tour.
(4) As the algorithm runs, some of the dimensions in the particle may get closer. Because this phenomenon could reduce the quality of the data collection tour, we design a dimension merge method to merge the dimensions that too closed. Through this method, we can optimize the data collection tour.

3.2 Algorithm Flow

Without loss of generality, the communication range of mobile sink and sensor nodes is set as R. That is, while in the circle with the mobile sink as the center and the R as the radius, the sensor nodes can transfer data to the mobile sink. We assume the position of the sensor nodes is known.

At first, by using the traditional TSP algorithm, we get a TSP tour and use it as the initial data collection tour. The next step is to randomly select a dimension (CS) from the initial data collection tour. Then performing the third step that the local searching for the selected CS. And this is right the key of the whole algorithm. In order to simplify the searching process and reduce complexity, we design a pre-build search grid template. Its principle is shown in Fig. 1.

First of all, we assume the randomly selected CS is RP_i and $\widehat{RP}_i$ is the new position that RP_i will move to. The black dots in Fig. 1 represent the sensors in the WSN. From Fig. 1, we can see that the greater the distance between the RP_i and $\widehat{RP}_i$, the less overlap their coverage. Furthermore, the smaller the overlap area, the fewer sensor nodes in the overlap area.

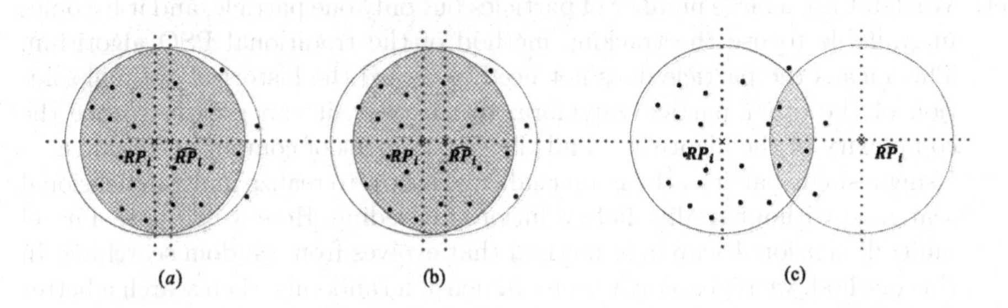

Fig. 1. The relationship between the distance of CSs and overlapping areas.

MS only collects data from sensor nodes at CSs. So we can see, if the MS moves from RP_i to $\widehat{RP}_i$, as shown in Fig. 1(a), the probability of it to cover the sensor nodes becomes greater. In the case of Fig. 1(c), the probability is small. Therefore, we can move the position of CS bit by bit while maintaining the same coverage, and at the same time optimize the data collection tour gradually.

The next challenge is how to find the best $\widehat{RP}_i$. Here we propose a searching grid template to solve this problem. Since we take the CS's coverage as a circle, we set the area covered by RP_i as S1 and the overlap area of RP_i and $\widehat{RP}_i$ as S2. According to the percentage S2 over S1, a searching grid template with 4 layers is constructed, and the searching circles from inside to outside represent the percentage 95%, 90%, 80%, and 70%. The radius of the searching circle can be calculated from the percentage and R. The same number of grid points is set on every searching circle. Finally, we construct a searching grid template with 4 layers and 32 grid points, as shown in Fig. 2.

As shown in Fig. 2, we select the points in 8 directions of each layer as grid points. For convenience, we mark the grid points as $gp_i^k, i \in \{0.95, 0.9, 0.8, 0, 7\}, k \in \{1, 2, ..., 8\}$. Where subscript i represents the percentage of overlap, superscript k represents the index of the grid point and the value from 1 to 8.

So the algorithm flow can be described as follows:

Step1: Generate a TSP tour based on all sensor nodes as the initial tour.

At this time, the position of every sensor node serves as the CS. The tour is mapped to a single particle. Generate the searching grid template as shown in Fig. 2.

Step2: Multi-dimensional searching for single particle.

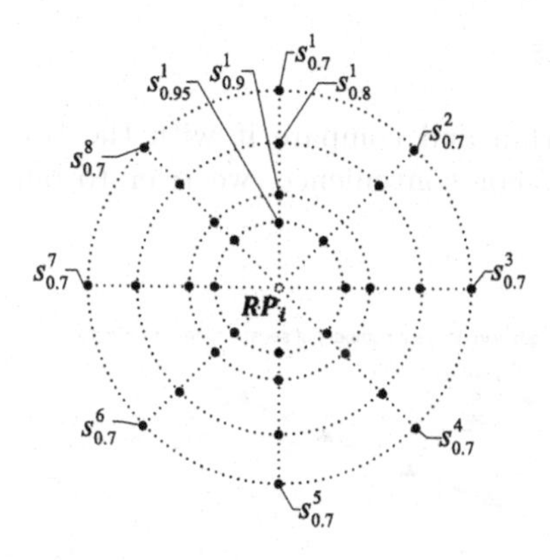

Fig. 2. The search grid template of RP_i.

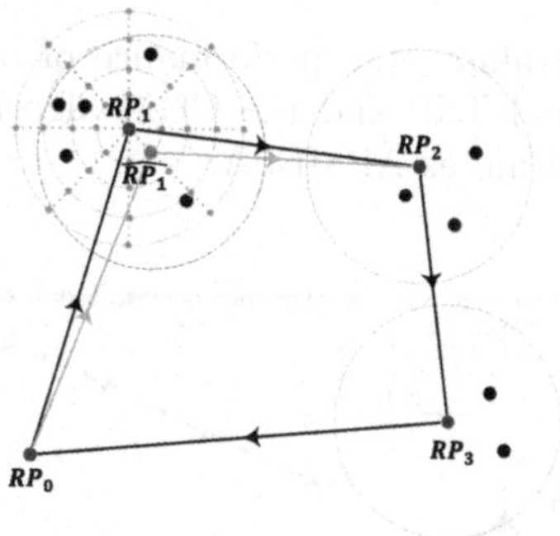

Fig. 3. The flow of the single particle multi-dimensional searching algorithm.

Select a random dimension (CS) in the particle;

The searching grid template is superimposed on the position of the CS selected in the previous step to obtain the alternative position set according to the searching grid template.

Search for each position in the alternative position set. Maintaining coverage, determine which position will reduce the tour length most. Select this position to update the selected CS.

Step3: Merge CSs.

Calculate the Euclidean distance between adjacent CSs. If the distance is less than 0.1R, arrange the sensor node sets covered by these two CSs respectively. If there is one CS that could cover all the sensor nodes set by the others, then delete the other CSs from the tour.

Step4: Calculate the fitness value.

Fitness value is used to define the effectiveness of the tour. In this algorithm, we select the tour length as the fitness value of the particle. The smaller the fitness value, the better the tour.

Step5: Exit condition judgment.

In order to improve the computing speed of the algorithm, we set the algorithm exit condition as followed: if the rate of change of fitness value is less than 0.1% in 100 iterations, the algorithm exits. If the algorithm doesn't terminate, return to Step2.

Step6: Output the optimal data collection tour and the fitness value.

Figure 3 is an example of our algorithm.

4 Analysis of Simulation Results

We evaluate the performance of our algorithm and compare it with the traditional TSP and the COM algorithm [16]. For convenience, we refer to our algorithm as APMDSA.

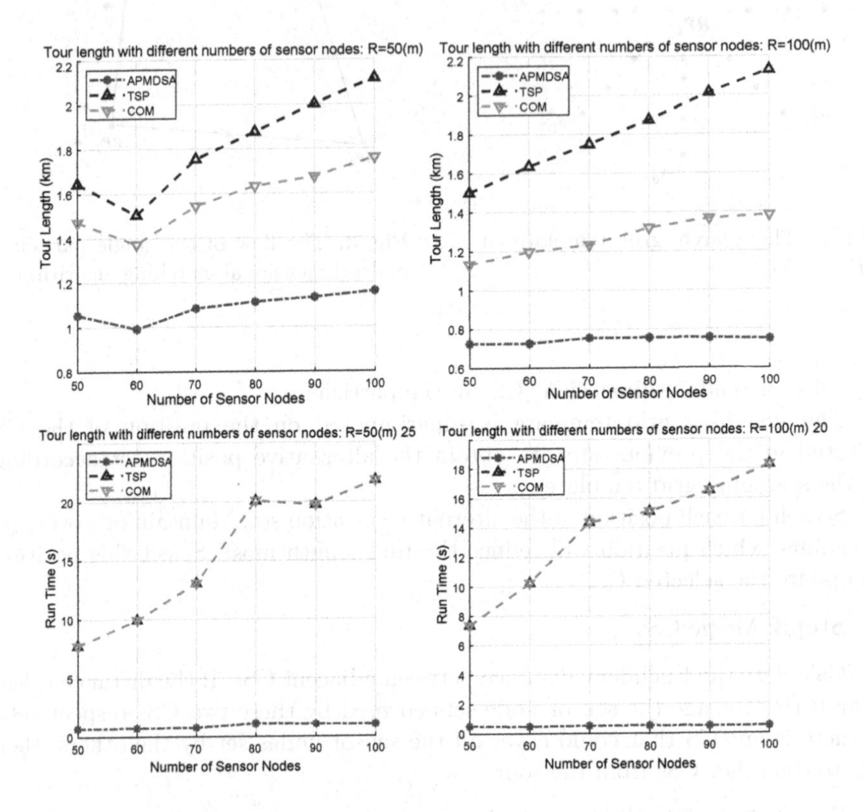

Fig. 4. Tour length and run time with different communication radiuses.

In the simulation, we use the same settings as in [16]. We consider a sparse square sensing field with size 500 m 500 m, where 50 to 100 nodes are uniformly deployed at random. The sensor node and the mobile sink have the same communication range, which is set from 20 m to 100 m. We generate 100 topologies for each case, and each topology is simulated 50 times. Our simulation is run on a 3.00 GHz CPU and 8G memory.

At first, the communication radius is 50 m and 100 m, and the number of sensor nodes increases from 50 to 100 step by step. Simulation is performed for these three algorithms, and the results are shown in Fig. 4. Then we fix the number of sensor nodes as 50 and 100, and the communication radius of sensor nodes gradually increases from 20 m to 100 m. The three algorithms were simulated again, and the results are shown in Fig. 5.

In Fig. 4, fixing the communication radius, the optimal tour length obtained by these three algorithms gradually increases no matter the number of sensor nodes is 50 or 100. Because as the number of sensor nodes increases, the number of CSs that the mobile sink needs to traverse also increases simultaneously, and this will increase the tour length. With a different number of sensor nodes, we can see that the performance of TSP is the worst. Because in the TSP, the position of each sensor node is the CS, which means the mobile sink needs to traverse every sensor node's position to perform the data collection. So the tour length obtained by TSP is the longest.

Compared with the TSP, the performance of the COM has been greatly improved. Since the COM merges adjacent CSs according to the overlap degree of adjacent CSs' coverage. The tour generated by the COM comprises fewer CSs, so the tour length is relatively better. Our APMDSA algorithm has the best performance. In the case that the communication radius is 50 m and the node's number is 50, the tour length generated by APMDSA is 30% shorter than the COM algorithm. If the number of sensor nodes increases to 100, the advantage of APMDSA expands to 33%. As evaluate time consumption, while the number of nodes is 50 and the communication radius is 50m, the running time of APMDSA is 12.5% of COM. While the number of sensor nodes increases to 100, the running time of APMDSA is only 4.6% of COM.

The number of sensor nodes is fixed as 50 and 100, while the communication radius increased from 20 m to 100 m, the simulation results are also similar and the TSP is still the worst. With the communication radius increase, since both COM and APMDSA take into account the communication range, their performance will be improved accordingly. Moreover, with a larger communication radius, the mobile sink can collect data from the sensor nodes within a larger range. Therefore, with the increase of communication radius, the performance improvement of COM and APMDSA expands, and APMDSA improves more. As can be seen from Fig. 5, when the number of sensor nodes is 50 and the communication radius is 100 m, the tour length of APMDSA is reduced by 35% as compare to COM. If the number of sensor nodes increases to 100, the advantage of APMDSA increases to 45%. In terms of time consumption, the advantages of our algorithm are more obvious. While the number of sensor nodes is 50 and the communication radius is 100 m, the operation time of APMDSA is 7% of COM. While the number of sensor nodes increases to 100, the running time of APMDSA only is 1% of COM.

These results show that APMDSA has achieved excellent performance under different simulation settings. APMDSA can effectively reduce the length of the data collection tour of the mobile sink, thus reducing the data collection latency of the whole network.

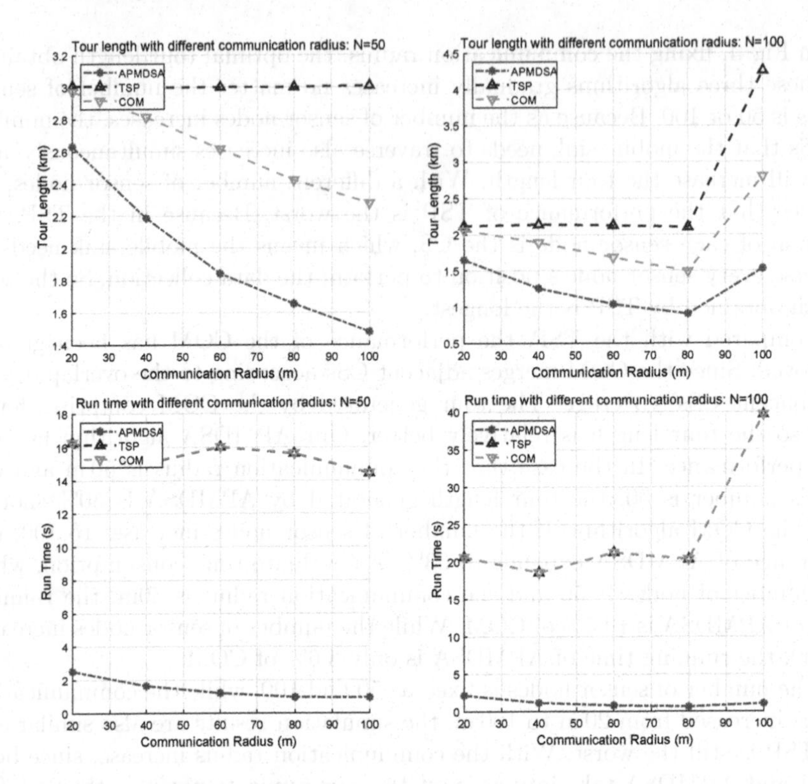

Fig. 5. Tour length and run time with different numbers of sensor nodes.

5 Conclusions

Data collection is very important for IoT applications. The use of the mobile element makes it possible to collect data within the communication range during the movement of the mobile sink. At the same time, this method can reduce the complexity of the network protocol. In this paper, we propose a novel algorithm to finding the optimal data collection tour for the mobile sink in IoT applications. We fully consider the wireless communication ability of the sensor node. Our algorithm selects several CSs and obtains the shortest tour by the multi-dimensional searching method within single-particle. This algorithm greatly reduces the complexity and computation of the searching method. The simulation results show that as compared with the TSP and COM, it can obtain a better data collection tour and further extend network lifetime. As compared to the other TSPN-based algorithms, the proposed algorithm has a simpler structure and better expansibility. Therefore, our algorithm is very suitable for large-scale networks.

References

1. Ou, C.H.: A localization scheme for wireless sensor networks using mobile anchors with directional antennas. IEEE Sens. J. **11**(7), 1607–1616 (2011)
2. Olariu, S., Stojmenovi, I.: Design guidelines for maximizing life time and avoiding energy holes in sensor networks with uniform distribution and uniform reporting. In: IEEE INFOCOM, pp. 1–12. Society Press (2006)
3. Zhao, M., Yang, Y.: Bounded relay hop mobile data gathering in wireless sensor networks. IEEE Trans. Comput. **61**(2), 265–277 (2012)
4. Lin, K., Chen, M., Zeadally, S., et al.: Balancing energy consumption with mobile agents in wireless sensor networks. Future Gener. Comput. Syst. **28**(2), 446–456 (2012)
5. Guo, S., Wang, C., Yang, Y.: Joint mobile data gathering and energy provisioning in wireless rrechargeable sensor networks. IEEE Trans. Mob. Comput. **13**(12), 2836–2852 (2014)
6. Liao, W.H., Lee, Y.C., Kedia, S.P.: Mobile anchor positioning for wireless sensor networks. IET Commun. **5**(7), 914–921 (2011)
7. Chen, H., Shi, Q., Tan, R., et al.: Mobile element assisted cooperative localization for wireless sensor networks with obstacles. IEEE Trans. Wirel. Commun. **9**(3), 956–963 (2010)
8. Gao, S., Zhang, H., Das, S.K.: Efficient data collection in wireless sensor networks with path-constrained mobile sinks. IEEE Trans. Mob. Comput. **10**(4), 592–608 (2010)
9. Jian, G., Lijuan, S., Ruchuan, W., Xiao, F.: Path planning of mobile sink for wireless multimedia sensor networks. J. Comput. Res. Dev. **47**(z2), 184–188 (2010)
10. Ma, M., Yang, Y., Zhao, M.: Tour planning for mobile data-gathering mechanisms in wireless sensor networks. IEEE Trans. Veh. Technol. **62**(4), 1472–1483 (2013)
11. Shuai, G., Hongke, Z.: Optimal path selection for mobile sink in delay-guaranteed. ACTA Electronica Sinica **39**(4), 742–747 (2011)
12. Xing, G., Wang, T., Jia, W., et al.: Rendezvous design algorithms for wireless sensor networks with a mobile base station. In: ACM International Symposium on Mobile Ad Hoc Networking and Computing 2008, pp. 231–240. ACM (2008)
13. Xing, G., Wang, T., Xie, Z., et al.: Rendezvous planning in wireless sensor networks with mobile elements. IEEE Trans. Mob. Comput. **7**(12), 1430–1443 (2008)
14. Yuan, Y., Yuxing, P., Shanshan, L.: Efficient heuristic algorithm for the mobile sink routing problem. J. Commun. **32**(10), 107–117 (2011)
15. Chipara, O., et al.: Real-time power-aware routing in sensor networks. In: 14th IEEE International Workshop on Quality of Service 2006, pp. 83–92. IEEE (2006)
16. He, L., Pan, J., Xu, J.: A progressive approach to reducing data collection latency in wireless sensor networks with mobile elements. IEEE Trans. Mob. Comput. **12**(7), 1308–1320 (2013)
17. Sasirekha, S., Swamynathan, S.: Cluster-chain mobile agent routing algorithm for efficient data aggregation in wireless sensor network. J. Commun. Netw. **19**(4), 392–401 (2017)
18. Wang, Y.C., Chen, K.C.: Efficient path planning for a mobile sink to reliably gather data from sensors with diverse sensing rates and limited buffers. IEEE Trans. Mob. Comput. **18**(7), 1527–1540 (2018)
19. Alami, H.E., Najid, A.: MS-routing-G i: routing technique to minimize energy consumption and packet loss in WSNs with mobile sink. IET Netw. **7**(6), 422–428 (2018)

20. Abdelhakim, M., Liang, Y., Li, T.: Mobile access coordinated wireless sensor networks - design and analysis. IEEE Tran. Signal Inf. Process. Over Netw. **3**(1), 172–186 (2016)
21. Chien-Fu, C., Chao-Fu, Y.: Mobile data gathering with bounded relay in wireless sensor networks. IEEE Internet Things J. **5**(5), 3891–3907 (2018)
22. Sha, C., Song, D., Yang, R., et al.: A type of energy-balanced tree based data collection strategy for sensor network with mobile sink. IEEE Access **7**, 85226–85240 (2019)
23. Gharaei, N., Malebary, S.J., Bakar, K.A., et al.: Energy-efficient mobile-sink sojourn location optimization scheme for consumer home networks. IEEE Access **7**, 112079–112086 (2019)

Information Intelligent Acquisition Generated by Matrix Reasoning of Inverse P-Set

Shouwei Li(✉) and Zongjun You

Guiyang Institute for Big Data and Finance, Guizhou University of Finance and Economics, Guiyang 550025, China
{lishouwei,youzongjuun}@mail.gufe.edu.cn

Abstract. More and more attention has been paid to the intelligent acquisition of information in the field of data mining in big data, Internet of things and so on. Unlike popular methods such as machine learning, this paper attempts to propose a matrix reasoning method for intelligent information acquisition from the perspective of set theory. The inverse packet set (P-set) is a new set model with dynamic features. In the inverse P-set, the attribute α_i of the element x_i satisfies the expansion or contraction paradigm for attribute extraction. Based on the concept of inverse P-set, this paper presents some new concepts as α^F-information equivalence class, $\alpha^{\overline{F}}$-information equivalence class, and $(\alpha^F, \alpha^{\overline{F}})$-information equivalence class. Then, this paper gives some theorems as internal inverse P-augmented matrix inference, outer inverse P-augmented matrix inference, and inverse P-augmented matrix inference, which are generated by the above information equivalence class. At last, this paper gives the application of intelligent acquisition of information.

Keywords: Inverse P-set · Matrix inference · Information equivalence class · Intelligent acquisition algorithm

1 Introduction

Information widely exists in real life, and becomes more and more [17]. Sometimes there is "redundancy" in the data. Although machine learning method has been widely used and achieved in information intelligent acquisition [4,13], the research on the mathematical logic of information intelligent acquisition is still relatively small. More "intelligent" information acquisition methods are urgently needed to find target information or "useful" information from massive data [1,5].

Generally, data are stored in various documents in the form of "records" with multiple attributes, such as databases, data warehouses, etc. [2]. Attributes are very important in distinguishing each record and determining the number of

Supported by National Natural Science Foundation of China (Grant No. 71663010), and Guizhou Science and Technology platform talents ([2017] No. 5736-018).

H. Song and D. Jiang (Eds.): SIMUtools 2020, LNICST 370, pp. 399–408, 2021.
https://doi.org/10.1007/978-3-030-72795-6_33

elements in the dataset, the set which has dynamic characteristics in the process of adding or reducing attributes to them. The inverse P-Set [10] is one of the theories and methods to describe this dynamic feature.

The inverse packet set (P-set) [16,20] is a new set model obtained by introducing dynamic features. The function in the function space represents a kind of law, and the function space F represents a set of laws to be selected by decision makers. Function space F is very important for the construction of inverse P-sets, which is related to specific goals, such as data mining, information hiding, knowledge reasoning, etc. Applying function space to inverse P-Set can enlarge the application scope of inverse P-Set and provide better choice for decision makers.

A propositional formula is called conjunctive normal form if and only if it has the form: $A_1 \wedge A_2 \wedge \cdots \wedge A_n$, $A_1, A_2, \cdots, A_n$ are called sub normal form, which are all disjunctive forms of propositional arguments or their negations. A propositional formula is called disjunctive normal form, if and only if it has the form: $A_1 \vee A_2 \vee \cdots \vee A_n$, $A_1, A_2, \cdots, A_n$ are called a sub normal form, which are all conjunctions of propositional arguments or their negation. Any conjunctive normal form can be transformed into a disjunctive normal form through calculus.

Taking knowledge mining as an example, Knowledge (x) and its attribute set α appear in pairs or knowledge (x) has attribute set α; conversely, attribute set α corresponds to knowledge (x). This conclusion comes from a popular example: $X = \{x_i\}$ $(i = 1, 2, 3)$ is a set of toy, $\alpha = \{\alpha_j\}$ $(j = 1, 2, 3)$ is its attribute set, such as, $\alpha_1 = \mathit{Red}$, $\alpha_2 = \mathit{Blue}$, $\alpha_3 = \mathit{Green}$. conversely, the attribute α_j satisfies: $\alpha_i = \alpha_1 \vee \alpha_2 \vee \alpha_3 = \bigvee_{t=1}^{3} \alpha_t$. If toy x_4 with purple color is added into X, then $\alpha_i = \bigvee_{t=1}^{4} \alpha_t$. From this popular example, we get two facts: (I) Supplementing element x_i in (x) is equivalent to add attribute in attribute set α of (x) or supplementing element x_i in (x) is equivalent to expansion of disjunctive normal form of attribute α_i of element x_i. (II) Deleting element x_i from (x) is equivalent to deleting attributes, which is equivalent to contraction of disjunctive normal form of attribute α_i of element x_i. This popular example is the factual basis of this paper.

Some new analyses are given in this paper: (a) the logic features of the inverse P-augmented matrix which are the base of this paper are given; (b) the equivalence class is generated by inverse P-aggregate; (c) the theorems of matrix inference are presented as the methods for obtaining the information equivalence class; (d) the simple application of intelligence is shown in this paper. All the results given in this paper are very new.

2 Literature Review

Inverse P-sets are firstly put forward by the authors of [10] and has been studied in the field of information [6] and image processing [15].

In terms of information fusion, the authors of [7] gave the structures, the separations, and the equivalence class characteristics of P-sets by through the reasoning discovery is presented by the author of [12]; moreover, the application of intelligent fusion is presented by through P-information fusion[8].

In terms of information hiding, the information hiding generated by inverse P-intelligent fusion are put forward in [11]. On the other hand, the restoration of information hiding is analyzed in [18].

In terms of information mining, the attributes of the elements in inverse P-sets satisfy disjunctive characteristics [3]. Using the structure, the intelligent mining under the condition of disjunctive attribute expanded were presented in [9]. In addition, the intelligent mining-discovery of complete information with satisfying internal inverse P-reasoning and/or complete information condition was also put forward by [19]. Moreover, some applications of the information with expanded characteristic of the disjunctive attribute was analyzed by [20].

In terms of information reasoning, on the basis of inverse P-reasoning, the theorem of inverse P-reasoning was given by [14]. Later, The geometric characteristics of inverse P-reasoning have been given [15]. Moreover, unknown information search by using inverse P-reasoning have been given [16].

Compared with the previous literature, this paper proposes the concept inverse P-Set, and proves several theorems of the logical characteristics of information equivalence class. Based on the information equivalence class, this paper presents several theorems of matrix reasoning and information intelligent acquisition algorithm. These research contents are all new.

3 The Information Equivalence Class

For set $X = \{x_1, x_2, \cdots, x_q\} \subset U$ and its attribute $\alpha = \{\alpha_1, \alpha_2, \cdots, \alpha_q\} \subset V$, the following set pair $(\overline{X}^F, \overline{X}^{\overline{F}})$ is called as inverse P-set [10], where, $\overline{X}^F = X \cup X^+$, $\overline{X}^{\overline{F}} = X - X^-$. There are two special set $X^+ = \{u_i | u_i \in U, u_i \overline{\in} X, f(u_i) = x'_i \in X, f \in F\}$ and $X^- = \{x_i | x_i \in X, \overline{f}(x_i) = u_i \overline{\in} X, \overline{f} \in \overline{F}\}$. For the attribute set, they satisfy $\alpha^F = \alpha \bigcup \{f(\beta_i) = \alpha'_i \in \alpha, f \in F\}$ and $\alpha^{\overline{F}} = \alpha - \{\overline{f}(\alpha_i) = \beta_i \overline{\in} \alpha, \overline{f} \in \overline{F}\}$. F and $\overline{F}$ mean math function. Obviously, $\alpha_1^F \subseteq \alpha_2^F \subseteq \cdots \subseteq \alpha_{n-1}^F \subseteq \alpha_n^F$, $\alpha_n^{\overline{F}} \subseteq \alpha_{n-1}^{\overline{F}} \subseteq \cdots \subseteq \alpha_2^{\overline{F}} \subseteq \alpha_1^{\overline{F}}$, so $\overline{X}_1^F \subseteq \overline{X}_2^F \subseteq \cdots \subseteq \overline{X}_{n-1}^F \subseteq \overline{X}_n^F$ and $\overline{X}_n^{\overline{F}} \subseteq \overline{X}_{n-1}^{\overline{F}} \subseteq \cdots \subseteq \overline{X}_2^{\overline{F}} \subseteq \overline{X}_1^{\overline{F}}$. A matrix pair $(\overline{A}^F, \overline{A}^{\overline{F}})$ is defined as inverse P-augmented matrix [20] generated by $(\overline{X}^F, \overline{X}^{\overline{F}})$.

Definition 1. *For the inverse P-information* $((\overline{x})^F, (\overline{x})^{\overline{F}})$, $[[\overline{x}]^F, [\overline{x}]^{\overline{F}}]$ *is called as the* $(\alpha^F, \alpha^{\overline{F}})$*-information equivalence class, in which* $(\alpha^F, \alpha^{\overline{F}})$ *is its attribute set. For the inverse P-information family* $\{((\overline{x})_i^F, (\overline{x})_j^{\overline{F}}) | i \in I, j \in J\}$, $\{[[\overline{x}]_i^F, [\overline{x}]_j^{\overline{F}}] | i \in I, j \in J\}$ *is referred to as the* $(\alpha_i^F, \alpha_j^{\overline{F}})$*-information equivalent class family, in which* $\{(\alpha_i^F, \alpha_j^{\overline{F}}) | i \in I, j \in J\}$ *is their attribute set family.*

From the above definition, we get the following propositions.

Proposition 1. $((\overline{x})^F, (\overline{x})^{\overline{F}})$ *and* $(\alpha^F, \alpha^{\overline{F}})$ *are equivalence. The former is the inverse P-information, and the later is the information equivalence class.*

Proof. Given information (x) and its attribute α, let α be the relationship R between $(x) \times (x)$, $R = \alpha$. It is easy to get the followings:

1. $\forall x_i \in (x)$, x_i has a relationship R with x_i, or $x_i R x_i$, which satisfies reflexivity.
2. $\forall x_i, x_j \in (x)$, x_i has a relationship R with x_j, x_j has a relationship R with x_i or $x_i R x_j \Rightarrow x_j R x_i$, which satisfies symmetry.
3. $\forall x_i, x_j, x_k \in (x)$, $x_i R x_j, x_j R x_k \Rightarrow x_i R x_k$, which satisfies transitivity.

From Items 1–3, we get that the information (x) is an equivalence class; similarly, $[[\overline{x}]^F, [\overline{x}]^{\overline{F}}]$ is also an equivalence class. thus, Proposition 1 is proved.

Proposition 2. *$\{((\overline{x})_i^F, (\overline{x})_j^{\overline{F}}) | i \in I, j \in J\}$ and (α_i^F, α_j^F) are two equivalent concepts. The former are the The inverse P-information family. The later are the information equivalence class family.*

Similarly, The Propositions 2 could be proofed, so we omit it.

Theorem 1. *(The theorem of attribute disjunctive normal form for α-information equivalence class $[x]$) For the α-information equivalence class $[x]$, the attribute α_i satisfies*

$$\alpha_i = \bigvee_{t=1}^{q} \alpha_t \tag{1}$$

Proof. Given information (x) and its attribute set α, from the concepts of inverse P-set, we get that, for $\forall x_i \in (x)$, the attribute $\alpha_i = \alpha_1 \vee \alpha_2 \vee \cdots \vee \alpha_q = \vee_{t=1}^{q} \alpha_t$, thus we get Eq. (1).

Theorem 2. *(The expansion theorem of attribute disjunctive normal form) If the attribute α_i of information element $x_i \in [\overline{x}]^F$ satisfies the expansion of attribute disjunctive normal form, then the attribute α_i of x_i satisfies*

$$\alpha_i = (\bigvee_{t=1}^{q} \alpha_t) \bigvee_{t=q+1}^{\lambda} \alpha_t \tag{2}$$

Proof. Given information (x) and its attribute set α, the attribute α_i of $\forall x_i \in [x]$ satisfies $\alpha_i = \alpha_1 \vee \alpha_2 \vee \cdots \vee \alpha_q = \vee_{t=1}^{q} \alpha_t$. By using Eqs. (1) in Sect. 3, $(\overline{x})^F = \{x_1, x_2, \cdots, x_q, x_{q+1}, \cdots, x_\lambda\}$, $\alpha^F = \{\alpha_1, \alpha_2, \cdots, \alpha_q, \alpha_{q+1}, \cdots, \alpha_\lambda\}$ is the attribute set of $(\overline{x})^F$, the attribute α_i of $\forall x_i \in (\overline{x})^F$ satisfies that $\alpha_i = (\alpha_1 \vee \alpha_2 \vee \cdots \vee \alpha_q) \vee \alpha_{q+1} \vee \cdots \vee \alpha_\lambda = (\vee_{t=1}^{q} \alpha_t) \vee_{t=Q+1}^{\lambda} \alpha_t$. From Definition, we get that $[\overline{x}]^F = (\overline{x})^F$. Obviously, the attributes α_i of $\forall x_i \in [\overline{x}]^F$ satisfy $\alpha_i = (\alpha_1 \vee \alpha_2 \vee \cdots \vee \alpha_q) \vee \alpha_{q+1} Vee \cdots \vee \alpha_\lambda = (\vee_{t=1}^{q} \alpha_t) \vee_{t=q+1}^{\lambda} \alpha_t$, thus we get Eq. (2).

Theorem 3. *(The contraction theorem of attribute disjunctive normal form) If $\alpha^{\overline{F}}$ is the attribute set of $\alpha^{\overline{F}}$-information equivalence class, then the attribute α_i of x_i satisfies*

$$\alpha_i = (\bigvee_{t=1}^{q} \alpha_t) - \bigvee_{t=p+1}^{q} \alpha_t \tag{3}$$

Similarly, the proof of Theorem 3 is easy, so it is omitted. By using Theorems 1–3, we directly get Theorem 4,

Theorem 4. *(The expansion and contraction theorem of disjunctive normal form) IF the $(\alpha^F, \alpha^{\overline{F}})$-information equivalence class satisfy expansion and contraction, then the attributes satisfy*

$$(\alpha_i, \alpha_j) = ((\vee_{t=1}^{q}\alpha_t) \vee_{t=q+1}^{\lambda} \alpha_t, (\vee_{t=1}^{q}\alpha_t) - \vee_{t=p+1}^{q}\alpha_t) \tag{4}$$

In Eq. (4), $\alpha_i = (\vee_{t=1}^{q}\alpha_t) \vee_{t=q+1}^{\lambda} \alpha_t$, $\alpha_j = (\vee_{t=1}^{q}\alpha_t) - \vee_{t=p+1}^{q}\alpha_t$.

By using the preparatory concepts and the information equivalence class concepts in Sect. 3, we give the intelligent acquisition algorithm for information equivalence classes in Sect. 4.

4 Intelligent Acquisition Algorithm for Information Equivalence Classes

Definition 2. *For $\overline{A}_k^F$ and $\overline{A}_{k+1}^F$, if the attribute set α_k^F and α_{k+1}^F satisfy the following*

$$if \quad \overline{A}_k^F \Rightarrow \overline{A}_{k+1}^F, \quad then \quad \alpha_k^F \Rightarrow \alpha_{k+1}^F \tag{5}$$

Then, Eq. (5) is called as internal inverse P-matrix inference.

In Eq. (5), $\overline{A}_k^F \Rightarrow \overline{A}_{k+1}^F$ represents $\overline{A}_k^F \subseteq \overline{A}_{k+1}^F$ and $\alpha_k^F \Rightarrow \alpha_{k+1}^F$ represents $\alpha_k^F \subseteq \alpha_{k+1}^F$.

Definition 3. *For the outer inverse P-matrix $\overline{A}_k^{\overline{F}}$ and $\overline{A}_{k+1}^{\overline{F}}$, if the attribute set $\alpha_k^{\overline{F}}$ and $\alpha_{k+1}^{\overline{F}}$ satisfy the following*

$$if \quad \overline{A}_{k+1}^{\overline{F}} \Rightarrow \overline{A}_k^{\overline{F}}, \quad then \quad \alpha_{k+1}^{\overline{F}} \Rightarrow \alpha_k^{\overline{F}} \tag{6}$$

then, Eq. (6) is called as outer inverse P-matrix inference.

Definition 4. *For the inverse P-matrix $(\overline{A}_k^F, \overline{A}_{k+1}^{\overline{F}})$ and $(\overline{A}_{k+1}^F, \overline{A}_k^{\overline{F}})$, if the attribute set $(\alpha_k^F, \alpha_{k+1}^{\overline{F}})$ and $(\alpha_{k+1}^F, \alpha_k^{\overline{F}})$ satisfy the following*

$$if \quad (\overline{A}_k^F, \overline{A}_{k+1}^{\overline{F}}) \Rightarrow (\overline{A}_{k+1}^F, \overline{A}_k^{\overline{F}}), \quad then \quad (\alpha_k^F, \alpha_{k+1}^{\overline{F}}) \Rightarrow (\alpha_{k+1}^F, \alpha_k^{\overline{F}}) \tag{7}$$

then, Eq. (7) is called as inverse P-matrix inference.

In Eq. (7), $(\alpha_k^F, \alpha_{k+1}^{\overline{F}}) \Rightarrow (\alpha_{k+1}^F, \alpha_k^{\overline{F}})$ represents $\alpha_k^F \Rightarrow \alpha_{k+1}^F$ and $\alpha_{k+1}^{\overline{F}} \Rightarrow \alpha_k^{\overline{F}}$.

By using Definitions 2–4, we get Theorem 5.

Theorem 5. *(The intelligent acquisition theorem for α^F-information equivalence class) If the internal inverse P-matrix $\overline{A}_k^F$ and $\overline{A}_{k+1}^F$, α_k^F-information equivalence class $[\overline{x}]_k^F$ and $[\overline{x}]_{k+1}^F$ satisfy*

$$if \quad \overline{A}_k^F \Rightarrow \overline{A}_{k+1}^F, \quad then \quad [\overline{x}]_k^F \Rightarrow [\overline{x}]_{k+1}^F, \tag{8}$$

then $[\overline{x}]_{k+1}^F$ is intelligently acquired from the outer of $[\overline{x}]_k^F$.

Proof. From the preliminary concepts in Sect. 3, we get that $\overline{A}_k^F$ and $\overline{A}_{k+1}^F$ are generated by the inverse P-sets $\overline{X}_k^F$ and $\overline{X}_{k+1}^F$, respectively. $\overline{A}_k^F$ and $\overline{A}_{k+1}^F$ satisfy $\overline{A}_k^F \Rightarrow \overline{A}_{k+1}^F$; $[\overline{x}]_k^F$ and $[\overline{x}]_{k+1}^F$ are the α^F- information equivalence class generated by $\overline{X}_k^F$ and $\overline{X}_{k+1}^F$, respectively; and $[\overline{x}]_k^F = \overline{X}_k^F$, $[\overline{x}]_{k+1}^F = \overline{X}_{k+1}^F$ or $\overline{X}_k^F \subseteq \overline{X}_{k+1}^F$. Obviously, under the condition of the internal inverse P-matrix reasoning $\overline{A}_k^F \Rightarrow \overline{A}_{k+1}^F$, we get that $[\overline{x}]_k^F \subseteq [\overline{x}]_{k+1}^F$, that is, $[\overline{x}]_{k+1}^F$ is intelligently acquired from the outer of $[\overline{x}]_k^F$, so we get Theorem 5.

Theorem 6. *($\alpha^{\overline{F}}$-information equivalence class intelligent acquisition theorem) If the outer inverse P-matrix $\overline{A}_{k+1}^{\overline{F}}$, $\overline{A}_k^{\overline{F}}$, and $\alpha^{\overline{F}}$-information equivalence class satisfy*

$$if \quad \overline{A}_{k+1}^{\overline{F}} \Rightarrow \overline{A}_k^{\overline{F}}, \quad then \quad [x]_{k+1}^{\overline{F}} \Rightarrow [x]_k^{\overline{F}}, \tag{9}$$

then $[x]_{k+1}^{\overline{F}}$ is intelligently acquired from the internal of $[x]_k^{\overline{F}}$.

The proof of Theorem 6 is similar to that of Theorem 5, thus it is omitted. Theorem 7 in the following are directly obtained from Theorems 5 and 6.

Theorem 7. *($(\alpha^F, \alpha^{\overline{F}})$-information equivalence class intelligent acquisition theorem) If the inverse P-matrix $(\overline{A}_k^F, \overline{A}_{k+1}^{\overline{F}})$ and $(\overline{A}_{k+1}^F, \overline{A}_k^{\overline{F}})$, and $(\alpha^F, \alpha^{\overline{F}})$-information equivalence class satisfy*

$$if \quad (\overline{A}_k^F, \overline{A}_{k+1}^{\overline{F}}) \Rightarrow (\overline{A}_{k+1}^F, \overline{A}_k^{\overline{F}}), \quad then \quad [[\overline{x}]_k^F, [\overline{x}]_{k+1}^{\overline{F}}] \Rightarrow [[\overline{x}]_{k+1}^F, [\overline{x}]_k^{\overline{F}}], \tag{10}$$

then $[\overline{x}]_{k+1}^F$ is intelligently acquired from the outer of $[\overline{x}]_k^F$, and, at the same time, $[\overline{x}]_{k+1}^{\overline{F}}$ is intelligently acquired from the inner of $[\overline{x}]_k^{\overline{F}}$. $[\overline{x}]_{k+1}^F$ and $[\overline{x}]_{k+1}^{\overline{F}}$ compose $(\alpha^F, \alpha^{\overline{F}})$-information equivalence class $[[\overline{x}]_{k+1}^F, [\overline{x}]_{k+1}^{\overline{F}}]$.

By using Definitions 2–4 and Theorem 5–7, we get the following algorithm. The algorithm is shown in Fig. 1.

Special Note: The intelligent acquisition algorithm of $\alpha_k^{\overline{F}}$-information equivalence class $[\overline{x}]_k^{\overline{F}}$ in Fig. 1 is a part of the intelligent acquisition algorithm of information equivalence class. The intelligent acquisition algorithm of α_k^F-information equivalence class $[\overline{x}]_k^F$ is omitted in Fig. 1. The intelligent acquisition algorithm of α_k^F-information equivalence class $[\overline{x}]_k^F$ is similar to Fig. 1.

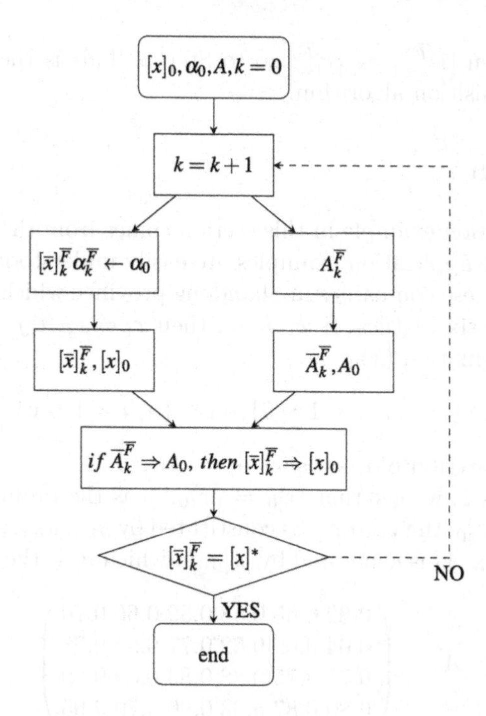

Fig. 1. Flow chart of intelligent acquisition algorithm of information equivalence class.

There are four steps in the process of intelligent acquisition algorithm of information equivalence class.

Step 1. Given the original information $(x)_0 = \{x_1, x_2, \cdots, x_q\}$, $[x]_0$ is the α-information equivalence class generated by $(x)_0$, $[x]_0 = (x)_0$; $\alpha = \{\alpha_1, \alpha_2, \cdots, \alpha_q\}$ is an attribute set of $[x]_0$, and A_0 is a matrix of information values generated by $[x]_0$; A_0 is A in Sect. 3.

Step 2. Delete several attributes α_i from α_0, α_0 generates $\alpha_k^{\overline{F}}$, $\alpha_k^{\overline{F}} \subseteq \alpha_0$; $[x]_0$ generates outer inverse P-information $(\overline{x})_k^{\overline{F}}$, $(\overline{x})_k^{\overline{F}} \subseteq (x)_0$; $[\overline{x}]_k^{\overline{F}}$ is the $\alpha_k^{\overline{F}}$-information equivalent class generated by $(\overline{x})_k^{\overline{F}}$, $[\overline{x}]_k^{\overline{F}} = (\overline{x})_k^{\overline{F}}$; from $[\overline{x}]_k^{\overline{F}}$, $[x]_0$ and $\overline{A}_k^{\overline{F}}$, A_0 in Step 1 constitutes an outer inverse P-matrix inference: $if\ \overline{A}_k^{\overline{F}} \Rightarrow A_0, then[\overline{x}]_k^{\overline{F}} \Rightarrow [x]_0$.

Step 3. Compare the $[\overline{x}]_k^{\overline{F}}$ obtained by intelligent acquisition in Step 2 with the standard $[x]^*$, if $[\overline{x}]_k^{\overline{F}} = [x]^*$, then the intelligent acquisition process of $[\overline{x}]_k^{\overline{F}}$ ends.

Step 4. If $[\overline{x}]_k^{\overline{F}} \neq [x]^*$, the process of intelligent acquisition continues. Delete several attributes α_j from $\alpha_k^{\overline{F}}$, $\alpha_k^{\overline{F}}$ generates $\alpha_{k+1}^{\overline{F}}$, $\alpha_{k+1}^{\overline{F}} \subseteq \alpha_k^{\overline{F}}$; $[\overline{x}]_k^{\overline{F}}$ generates outer inverse P-information $(\overline{x})_{k+1}^{\overline{F}}$; $[\overline{x}]_{k+1}^{\overline{F}}$ generate outer inverse P-matrix $\overline{A}_{k+1}^{\overline{F}}$. From $\overline{A}_{k+1}^{\overline{F}}$, the $\overline{A}_k^{\overline{F}}$ in Step 2 constitutes the outer inverse P-matrix inference:

if $\overline{A}_{k+1}^{F} \Rightarrow \overline{A}_{k}^{F}$, then $[\overline{x}]_{k+1}^{F} \Rightarrow [\overline{x}]_{k}^{F}$; go to Step 3. This is the cyclic process in the intelligent acquisition algorithm.

5 Application

The simple application example in this section comes from the commodity trading system, and the application examples are easily understood. w_1, w_2, w_3, w_4, w_5, and w_6 are Chinese companies in Shandong province which produce different textiles. $x_1 \sim x_6$ are six textiles. If $w_i \neq w_j$, then $x_i \neq x_j$, $i, j = 1, 2, \cdots 6$; x_1–x_6 constitute the information $(x)_0$.

$$(x)_0 = \{x_i | i = 1 \sim 6\}, \alpha_0 = \{\alpha_i | i = 1 \sim 6\} \tag{11}$$

Obviously, the attribute α_i satisfies $\alpha_i = \vee \alpha_i$.

From Definition 1, we get that $[x]_0 = (x)_0$. y_j is the profit during January–April 2018 of $x_j \in [x]_0$: the vector y_j is constituted by $y_{1,j}, y_{2,j}, y_{3,j}, y_{4,j}$, the information value matrix A_0 is generated by $[x]_o$, in which y_j is the column.

$$A_0 = \begin{pmatrix} 0.82\ 0.65\ 0.58\ 0.59\ 0.66\ 0.70 \\ 0.64\ 0.69\ 0.62\ 0.77\ 0.94\ 0.78 \\ 0.71\ 0.75\ 0.88\ 0.83\ 0.59\ 0.89 \\ 0.80\ 0.87\ 0.63\ 0.90\ 0.70\ 0.95 \end{pmatrix} \tag{12}$$

Here, for the sake of trade secrets, the value of each column of A_0 in Eq. (12) obtained by transforming the real profit value using the technical method. The value obtained by the technical method transformation does not affect the analysis of the application example.

In August 2018, the textiles x_3 and x_5 produced by the companies W_3 and W_5, respectively, were saturated in the market, thus companies W_3 and W_5 were closed. x_3 and x_5 in $(x)_0$ are deleted to generate $(\overline{x})^{\overline{F}} \subset (x)_0$.

$$(\overline{x})^{\overline{F}} = \{x_1, x_2, x_4, x_6\} \tag{13}$$

α_3 and α_5 are deleted from α in Eq. (11), thus α generates $\alpha^{\overline{F}}$

$$\alpha^{\overline{F}} = \alpha - \{\alpha_3, \alpha_5\} = \{\alpha_1, \alpha_2, \alpha_4, \alpha_6\} \tag{14}$$

Delete the third column and the fifth column from A_0 in Eq. (12), A_0 generates $\overline{A}^{\overline{F}}, \overline{A}^{\overline{F}} \subset A_0$

$$\overline{A}^{\overline{F}} = \begin{pmatrix} 0.82\ 0.65\ 0.59\ 0.70 \\ 0.64\ 0.69\ 0.77\ 0.78 \\ 0.71\ 0.75\ 0.83\ 0.89 \\ 0.80\ 0.87\ 0.90\ 0.95 \end{pmatrix} \tag{15}$$

The attribute α_i satisfies $\alpha_i = (\vee \alpha_i) - \alpha_3 \vee \alpha_5 = \alpha_1 \vee \alpha_2 \vee \alpha_4 \vee \alpha_6$. Equations (12), (15), (13), and (11) constitute the outer inverse P-matrix inference (shown in Eq. (6)), or, if $\overline{A}^{\overline{F}} \Rightarrow A_0$, then $[\overline{x}]^{\overline{F}} \Rightarrow [x]_0$.

We conducted a four-month survey and inquired about the user on the sales market of textiles x_3 and x_5. The results were the same as those given in the example.

Due to the limitation of the length of the paper, we use technical means to simplify the application examples without affecting the interpretation of the research content. We will show more complex application examples in further research.

6 Discussions and Conclusions

This article is inspired by the literature [1,5]. Given set X and its attribute set α, α generates α^F by adding some attributes into α, $\alpha \subseteq \alpha^F$; X generates $\overline{X}^F$, then $X \subseteq \overline{X}^F$. if deleting the attribute in α, α generates $\alpha^{\overline{F}}$, $\alpha^{\overline{F}} \subseteq \alpha$; X generates $\overline{X}^{\overline{F}} \subseteq X$. Then inverse P-set $(\overline{X}^F, \overline{X}^{\overline{F}})$ is obtained. By using the inference condition $A \Longrightarrow \overline{A}^F$, the inner inverse P-information $(\overline{x})^F$ is found intelligently from outer of (x), $(x) \subseteq (\overline{x})^F$. By using outer inverse P-matrix inference conditions: $\overline{A}^{\overline{F}} \Rightarrow A$, the outer inverse P-information $(\overline{x})^{\overline{F}}$ is found intelligently from the inner of (x), $(\overline{x})^{\overline{F}} \subseteq (x)$.

The new methods and new theories about the dynamic intelligent retrieval of information are presented in this paper. There is a fact: on a train, passenger $x_1, x_2, \cdots, x_n$ constitutes passenger information $(x) = \{x_1, x_2, \cdots, x_n\}$, each traveler $x_i \in (x)$ has attributes α_i, $\alpha_i=$ tickets; We get that if $\forall x_i, x_j \in (x)$, $x_i \neq x_j$, then $\alpha_i \neq \alpha_j$, $\alpha_i, \alpha_j \in \alpha$, or if $\forall x_i, x_j \in (x)$, $x_i = x_j$, then $\alpha_i = \alpha_j$, $\alpha_i, \alpha_j \in \alpha$.

Inverse P-sets include not only all kinds of elements, but also the attributes of elements. They are in pairs. In prediction or reasoning, elements and their attributes cooperate, interact, and influence each other to complete tasks together.

Inverse P-Set theory can be used to study the characteristics of dynamic information law. It is very different from other traditional methods, such as clustering, regression analysis, and so on. They are two different theoretical methods.

The theorems, propositions, and methods proposed in this paper are mainly applied to the discrete system composed of many elements. The research results of this paper are more applied to the analysis of the relationship between elements and systems, but less to the analysis of the relationship between elements.

References

1. Ajao, O., Hong, J., Liu, W.: A survey of location inference techniques on twitter. J. Inf. Sci. **41**(6), 855–864 (2015)
2. Budd, J.M.: Revisiting the importance of cognition in information science. J. Inf. Sci. **37**(4), 360–368 (2011)
3. Chengxian, F., Shunliang, H.: Inverse P-reasoning discovery identification of inverse P-information. Int. J. Digit. Content Technol. Appl. **6**(20), 735–744 (2012)

4. Din, I.U., Guizani, M., Rodrigues, J.J., Hassan, S., Korotaev, V.V.: Machine learning in the Internet of Things: designed techniques for smart cities. Future Gener. Comput. Syst. **100**, 826–843 (2019)
5. Hao, T., Zhu, C., Mu, Y., Liu, G.: A user-oriented semantic annotation approach to knowledge acquisition and conversion. J. Inf. Sci. **43**(3), 393–411 (2017)
6. Hongkang, L., Chengxian, F.: Embedding-camouflage of inverse P-information and application.Int. J. Converg. Inf. Technol **7**(20), 471–480 (2012)
7. Hualong, G., Baohui, C., Jihua, T.: Inverse P-sets and intelligent fusion mining-discovery of information. J. Shandong Univ. (Nat. Sci.) **48**(8), 97–103, 110 (2013)
8. Jihua, T., Bao-hui, C.: Intelligent fusion of internal inverse packet information and expansion relationship of attribute disjunction. J. Shandong Univ. (Nat. Sci.) **49**(2), 89–92, 97 (2014)
9. Jihua, T., Ling, Z., Baohui, C., Kaiquan, S., HsienWei, T., Yilun, C.: Outer P-information law reasoning and its application in intelligent fusion and separating of information law. Microsyst. Technol. **24**(10), 4389–4398 (2018)
10. Kaiquan, S.: Inverse P-sets. J. Shandong Univ. (Nat. Sci.) **47**(1), 98–109 (2012)
11. Shi, K.-Q.: Function inverse P-sets and the hiding information generated by function inverse P-information law fusion. In: Li, H., Mäntymäki, M., Zhang, X. (eds.) I3E 2014. IAICT, vol. 445, pp. 224–237. Springer, Heidelberg (2014). https://doi.org/10.1007/978-3-662-45526-5_22
12. Kaiquan, S., Jihua, T., Ling, Z.: Intelligent fusion of inverse packet information and recessive transmission of information's intelligent hiding. Syst. Eng. Electron. **37**(3), 599–605 (2015)
13. Li, D., Deng, L., Lee, M., Wang, H.: IoT data feature extraction and intrusion detection system for smart cities based on deep migration learning. Int. J. Inf. Manag. **49**, 533–545 (2019)
14. Ling, Z., Xuefang, R., Kaiquan, S.: Inverse P-data models and data intelligent separation. In: 2016 International Conference on Electronic Information Technology and Intellectualization (ICEITI), pp. 1–9 (2016)
15. Ling, Z., Xuefang, R., Kaiquan, S.: Inverse P-information law models and the reality-camouflage intelligent transformations of information image. In: 2016 International Conference on Network and Information Systems for Computers (ICNISC), pp. 337–341. IEEE (2016)
16. Zhang, L., Ren, X., Shi, K., Tseng, H.-W., Chen, Y.-L.: Inverse packet matrix reasoning model-based the intelligent dynamic separation and acquisition of educational information. Microsyst. Technol. **24**(10), 4415–4421 (2018). https://doi.org/10.1007/s00542-018-3894-2
17. Xu, T., Peng, Q.: Extended information inference model for unsupervised categorization of web short texts. J. Inf. Sci. **38**(6), 512–531 (2012)
18. Xuefang, R., Ling, Z.: Boundary characteristics of inverse P-sets and system condition monitoring. Comput. Sci. **43**(10), 211–213, 255 (2016)
19. Xuefang, R., Ling, Z.: Perturbation theorems of inverse P-sets and perturbation-based data mining. J. Shandong Univ. (Nat. Sci.) **51**(12), 54–60 (2016)
20. Xuefang, R., Ling, Z., Kaiquan, S., Tseng, H., YiLun, C.: Inverse P-augmented matrix method-based the dynamic findings of unknown information. Microsyst. Technol. **24**(10), 4187–4192 (2018)

A New Modeling and Analysis Approach of Overseas Tax Reform Requirements

Chao Wang[1], GuangLi Yang[2](✉), and Wenmin Kuang[3]

[1] Zibo Local Financial Supervision Bureau of Shandong Province, Beijing, China
[2] School of Accountancy, Xinhua College of Sun Yat-Sen University, Dongguan, China
491082948@qq.com
[3] Beijing Global Industrial Safety Technology Co., Ltd., Beijing, China

Abstract. Along with the explosive development of the Internet, the domestic consumer's consumption level unceasing enhancement, more and more to pursue high quality life, in order to meet the needs of their own and vanity and living out of a way of shopping - overseas procurement agent. Because of the explosive growth of overseas procurement agent, there are also many problems, so it is very important to standardize law and institutionalize the overseas procurement agent behaviors. Based on overseas procurement agent tax reform the necessity of the implementation of the new deal as the research subject, firstly describes the present situation of overseas procurement agent at home and abroad, mainly through the analysis of existing problems for overseas procurement agent behavior and overseas procurement agent tax reform the superiority of the new deal implementation and influence, clearly put forward the necessity of overseas procurement agent tax reform new deal implementation. Second, to overseas procurement agent problem which caused by the tax reform after implementation of the new deal to make further analysis, and cross-border electricity enterprise itself from the government management and so on different levels and a new modeling and analysis method for foreign tax reform demand is established. Finally, the simulation results show the effectiveness of the method.

Keywords: Overseas procurement agent · Tax reform new deal · Cross-border

1 Introduction

With the booming development of the international market, the living standards of domestic consumers are constantly improving. To meet their own needs and vanity, "overseas purchasing" has emerged as a shopping method. In April 2016, China implemented a new tax policy on cross-border e-commerce retail imports [1]. Overseas purchasing tax is a kind of tax levied on overseas purchasing behavior [3]. Since the introduction of the new policy, many goods that are not supported by direct mail to China can now be directly mailed without transshipment, so that consumers can purchase them more directly and enjoy all after-sales services.

H. Song and D. Jiang (Eds.): SIMUtools 2020, LNICST 370, pp. 409–419, 2021.
https://doi.org/10.1007/978-3-030-72795-6_34

R. Doernberg proposed a tax scheme on cross-border e-commerce from the perspective of tax jurisdiction. He advocated that the business income of cross-border e-commerce should be taxed with reference to the investment income. Blair Hergman proposed to separate e-commerce sellers into smaller sellers, and exempted enterprises with annual income less than $500,000 from Sales Tax [2]. Ning Lizhen (2017) believes that there are many problems in existing cross-border e-commerce, and relevant regulatory policies are not formulated and implemented in a timely manner, resulting in poor product quality and infringement of consumers' rights and interests. F. Mengyan (2016) proposed the impact of the tax reform on cross-border e-commerce enterprises after the implementation of the new policy, as well as its future development direction. Yang Yuanjian (2014) proposed the necessity of taxation on e-commerce from three aspects. Li Mengzhe (2016) analyzed the difficulties faced by individuals, enterprises and government agencies, and proposed to increase product categories and improve service quality for cross-border e-commerce, while regulators should strictly implement regulations and build a comprehensive information system, so as to better cope with the new environment brought by the new policy.

Overseas purchasing refers to that individuals post information through WeChat, Weibo or e-commerce platforms like Taobao to buy foreign products for domestic consumers [3]. With the development of the Internet, domestic consumers can learn about and buy foreign goods through online platforms. And with the increase of people's disposable income, many people pay more and more attention to and pursue high-quality goods to meet their own life and the sense of enjoyment it brings. Therefore, in order to meet their own needs and vanity of a form of shopping - overseas purchasing. This model relies on the rapid development of various electronic network platforms based on the Internet, while overseas purchasing has gradually become a full-time job for some people. Therefore, we need to establish a model to describe the overseas tax strategy.

2 Modeling Method for New Tax Reform

Comparison of the Old and New Policies. Before the tax reform and new policy, in order to vigorously develop the cross-border e-commerce industry, the country mainly implemented the postal tax policy, and levied the postal tax for individual purchasing on behalf of individuals. At that time, the cross-border e-commerce tax was also levied according to the postal tax.

After the tax reform policy, China will implement the cross-border e-commerce retail import tax policy. The specific contents of the policy are as follows:(1) the change of the postal tax policy is to raise the limit of a single transaction from 1000 yuan to 2000 yuan (800 yuan in Hong Kong, Macao and Taiwan) [3]; (2) set the annual limit of individual trading volume as 20,000 yuan. Within the limit of the trading volume, the tariff rate shall be zero. A single transaction exceeding a single limit value, a single transaction exceeding an individual's annual limit value after accumulation, and a single indivisible commodity whose customs value exceeds the limit value of 2000 yuan will be taxed in full according to the general trade method [1].

The classification of postal and postal taxes has also changed from the original four categories to three categories. The categories before the adjustment include: 10%, 20%, 30% and 50%, while the categories after the adjustment include: 15%, 30% and 60%. Only the categories and tax rates are adjusted, and the 50-yuan tax exemption remains unchanged [4]. As can be seen from Table 1, although the tax reform and new policies are mainly introduced for cross-border e-commerce, increasing the pressure on small and medium-sized enterprises of cross-border e-commerce, from the perspective of long-term development, it is conducive to the future development of cross-border e-commerce industry. And for the individual purchase, this tax reform New Deal may be a small benefit. As the following table shows:

Table 1. Comparison of tariff items before and after the adjustment of postal tariff and cross-border e-commerce retail

Commodity category	Price	Tax rate changes		
		Line post tax		Cross-border e-commerce tax
		Before tax reform	After tax reform	After tax reform
Daily necessities, food, etc.	Less than 500 yuan	Tax rate of 10% Less than 50 yuan is exempted	15%	VAT: 70% of 17% levy
	More than 500 yuan	Tax rate of 10%	15%	VAT: 70% of 17% levy
Cosmetics	Less than 100 yuan	Tax rate of 50% Less than 50 yuan is exempted	60%	VAT: 70% of 17% levy Consumption tax: 70% of 30% Value added tax: the sum of customs duty and consumption tax multiplied by 17% Consolidated tax: about 52.51%
	More than 100 yuan	Tax rate of 50%	60%	VAT: 70% of 17% levy Consumption tax: 70% of 30% Value added tax: the sum of customs duty and consumption tax multiplied by 17% Consolidated tax: about 52.51%
Skin care cosmetics, personal care	Less than 100 yuan	Tax rate of 50% Less than 50 yuan is exempted	30%	VAT: 70% of 17% levy
	More than 100 yuan	Tax rate of 50%	30%	VAT: 70% of 17% levy
Light luxury clothing, bedding fabrics, electrical appliances	Less than 250 yuan	Tax rate of 20% Less than 50 yuan is exempted	30%	VAT: 70% of 17% levy
	More than 250 yuan	Tax rate of 20%	30%	VAT: 70% of 17% levy

Advantages of Overseas Purchasing Tax Reform. Purchased overseas tax reform after the implementation of the New Deal, small cross-border e-commerce platform and individual purchased overseas gradually lose their respective advantages, making large cross-border e-commerce situation, to form the benign competition between businesses and greatly struck in order to obtain profits dividend not only attaches importance to product quality and after-sales service of act as purchasing agency. The implementation of the new policy is conducive to the stable development of large cross-border e-commerce enterprises, and standardizes the consumer market in the overseas purchasing industry.

We use linear regression analysis to model the problem as follows:

$$\hat{Y} = a + bx \tag{1}$$

where x and $\hat{Y}$ denotes the independent and dependent variables, respectively; a and b are parameters of linear regression equation with one variable. We can obtain the following formula:

$$\begin{cases} a = (\sum_{i=1}^{n} Y_i/n) - b(\sum_{i=1}^{n} X_i/n) \\ b = \dfrac{n\sum_{i=1}^{n} X_i Y_{ii} - \sum_{i=1}^{n} X_i \sum_{i=1}^{n} Y_i}{n\sum_{i=1}^{n} X_i^2 - \left(\sum_{i=1}^{n} X_i\right)^2} \end{cases} \tag{2}$$

For the convenience of calculation, we make the following definitions:

$$\begin{cases} S_{xx} = \sum_{i=1}^{n} (X_i - \overline{X})^2 = \sum_{i=1}^{n} X_i^2 - (\left(\sum_{i=1}^{n} X_i\right)^2/n) \\ S_{yy} = \sum_{i=1}^{n} (Y_i - \overline{Y})^2 = \sum_{i=1}^{n} Y_i^2 - (\left(\sum_{i=1}^{n} Y_i\right)^2/n) \\ S_{xy} = \sum_{i=1}^{n} (X_i - \overline{X})^2 (Y_i - \overline{Y}) = \sum_{i=1}^{n} X_i Y_i - (\sum_{i=1}^{n} X_i \sum_{i=1}^{n} Y_i/n) \end{cases} \tag{3}$$

where $\overline{X} = \frac{\sum_{i=1}^{n} X_i}{n}$, $\overline{Y} = \frac{\sum_{i=1}^{n} Y_i}{n}$, based on the definition of a, b we can get the solution as follows:

$$\begin{cases} a = \overline{Y} - b\overline{X} \\ b = X_{xy}/S_{xx} \end{cases} \tag{4}$$

AB is introduced into the equation $\hat{Y} = a + bx$. As long as x is given, the $\hat{Y}$ value can be predicted. The above cases are suitable for linear model. But for nonlinear models, curves are needed for modeling. The least square method is a kind of mathematical optimization technology. It finds the best function matching of data by minimizing the sum of squares of errors. Using the least square method, unknown data can be easily obtained, and the square sum of the errors between the obtained data and the actual data is minimized.

Given the function $y = f(x)$ and the function values $y_1, y_2, \ldots, y_n$ at points $x_1, x_2, \ldots, x_n$ are given.
The polynomial $p(x) = a_0 + a_1x + a_2x^2 + \ldots + a_nx^k$ makes

$$\min = \sum_{i=1}^{k} (p(x_i) - y_i)^2 \tag{5}$$

In order to obtain a value of load condition, the right side of equation $a_i(i = 0, 1, 2, \ldots, k)$ to find the partial derivative, the result is $k + 1$ equations

$$\begin{aligned} &-2\sum_{i=1}^{n}\left[y - \left(a_0 + a_1x + \ldots + a_kx^k\right)\right] = 0 \\ &-2\sum_{i=1}^{n}\left[y - \left(a_0 + a_1x + \ldots + a_kx^k\right)\right]x = 0 \\ &\ldots \\ &-2\sum_{i=1}^{n}\left[y - \left(a_0 + a_1x + \ldots + a_kx^k\right)\right]x^k = 0 \end{aligned} \tag{6}$$

By sorting out the equations, the following results are obtained

$$\begin{cases} na_0 + \left(\sum_{i=1}^{k} x_i\right)a_1 + \ldots + \left(\sum_{i=1}^{k} x_i^k\right)a_k = \sum_{i=1}^{k} y_i \\ \left(\sum_{i=1}^{k} x_i\right)a_1 + \left(\sum_{i=1}^{k} x_i^2\right)a_2 \ldots + \left(\sum_{i=1}^{k} x^{k+1}\right)a_k = \sum_{i=1}^{k} x_iy_i \\ \left(\sum_{i=1}^{k} x_i^k\right)a_1 + \left(\sum_{i=1}^{k} x_i^{k+1}\right)a_2 \ldots + \left(\sum_{i=1}^{k} x^{2k}\right)a_k = \sum_{i=1}^{k} x_i^k y_i \end{cases} \tag{7}$$

The following equation can be obtained:

$$\begin{bmatrix} n & \sum_{i=1}^{n} x_i & \cdots & \sum_{i=1}^{n} x_i^k \\ \sum_{i=1}^{n} x_i & \sum_{i=1}^{n} x_i^2 & \cdots & \sum_{i=1}^{n} x_i^{k+1} \\ \vdots & \vdots & \ddots & \vdots \\ \sum_{i=1}^{n} x_i^k & \sum_{i=1}^{n} x_i^{k+1} & \cdots & \sum_{i=1}^{n} x_i^{2k} \end{bmatrix} \begin{bmatrix} a_0 \\ a_1 \\ \vdots \\ a_k \end{bmatrix} = \begin{bmatrix} \sum_{i=1}^{n} y_i \\ \sum_{i=1}^{n} x_iy_i \\ \vdots \\ \sum_{i=1}^{n} x_i^k y_i \end{bmatrix} \tag{8}$$

A nonlinear fitting equation can be obtained by the above method.

We can see from the above relationship. The higher the price of goods, the lower the tax rate charged. On the contrary, the lower the price of imported goods, the higher the tax rate. Through the simulation of the model, we can determine the tax rate of import goods at different prices.

3 Problem Statements and Analysis Method

There are many imperfections in the tax policy of overseas procurement. Especially after the implementation of the new policy, the workload of customs officers will increase and the pressure will increase. Moreover, relevant tax laws and regulations have not been formulated and improved in time, and the implementation efficiency is not high. With the implementation of the new tax reform, the cost of goods purchased overseas will also increase, and the price of goods will also rise, which may lead to the imbalance of consumers' psychology and reduce the amount of goods purchased. Therefore, to regain the confidence of consumers, cross-border enterprises need to improve the quality of their products and their services.

One of the main development drivers of overseas purchasing is the relatively low price of goods abroad compared with that at home. The attention to commodity price indicates that the consumption of domestic consumers is still in a stage of emphasizing commodity price. So once the implementation of overseas purchasing tax reform New Deal, foreign commodity prices imported to the domestic will increase accordingly. Personal overseas purchasing and overseas purchasing platforms import goods of the same quality from abroad compared with the goods of domestic import brand agency physical stores, in terms of price, there is no difference between the two. At the same time, domestic consumers also pay attention to the type and quality of goods, they go abroad to improve their living standards and satisfy their inner vanity. Therefore, once the tax reform of overseas purchasing is implemented, the increased tax costs of cross-border e-commerce enterprises will eventually be passed on to consumers through product prices [7]. In order not to lose money, the enterprises have to increase the price of their products, which will eventually lead to the reduction of consumers' purchase, but will not have a great impact on the domestic retail of consumer goods. Therefore, in the short term, it will only restrain consumers' consumption of cross-border online goods and limit their behavior of overseas purchasing.

$$y = f(x) + \varepsilon \tag{9}$$

where ε is the error term satisfying Gaussian distribution. After that, the least square method is still used to obtain the best model curve.

Define and standardize the rules after the new tax reform. The implementation of the new policy will inevitably contradict the old policy, that is, the implementation of the new policy needs to constantly formulate other rules to supplement. The new tax reform policy for overseas purchasing agents is quite different from the original postal tax policy in terms of taxation, supervision and industry norms, which requires a relatively smooth transition period and timely formulation of detailed regulations to supplement, so as to facilitate the understanding of customs staff and consumers and ensure the correct implementation of the new tax reform policy. However, the new policy was announced in the form of a "notice" [10], and no official document was issued, nor specific operation and implementation rules of the customs were formulated. In addition, the limit value of imported goods and the starting time of the limit are not clearly specified, and the treatment of goods beyond the limit is not specified. If the detailed rules of relevant regulations are not issued, it may lead to different understandings in the collection of customs in different places, resulting in inconsistent basis for collection, and ultimately leading to inconsistent implementation methods.

Optimize the import supervision mode and improve the efficiency of customs enforcement. Customs import regulations at present stage, mainly through artificial means of regulation in the form of mail, parcels of entry, such as at a reasonable number (mainly for personal use) standard, and the number of reasonable strong subjectivity exists in the judgment, therefore, will appear different customs officers on the same piece of imported items may appear different results. Therefore, to avoid the above situation, it is necessary to optimize the import supervision mode, so as to reduce the working pressure of customs officers and improve the efficiency of customs enforcement. First, we need to change the traditional way of supervision from manual supervision to equipment supervision. We can introduce advanced technology or equipment from abroad, or conduct independent research and development. Connected with the equipment through the internal network of the customs, parcel post can be tracked from entry, port of arrival, unloading, inspection and a series of other links, providing information records, so as to reduce the pressure on the customs staff and improve the working efficiency of the customs [13]. The second is to the New Deal reform, the first to vigorously publicize the latest policy knowledge, the second to the relevant staff for professional training, improve work efficiency, to avoid staff on the relevant business of ambiguity, error.

A new regression model was established by adding Poisson distribution to the right side of model 3. In Eq. (9) ε is the error term satisfying Poisson distribution.

4 Simulation Experiments and Analysis

Now we validate our method via a simulation experiment. Fig. 1 plots the impact of overseas purchasing tax reform new policy compared with existing policies. We can see that after adding the error with Gaussian distribution, the curve of the model has changed slightly. However, it can be seen that the overall curve trend has not changed significantly. However, there is a higher tax rate for high price goods and higher tax rate for low-cost goods. This also shows that in order to protect state-owned enterprises, China adopts a higher tax rate policy for low-cost imported goods. Therefore, the tax rate can be adjusted according to the simulation diagram.

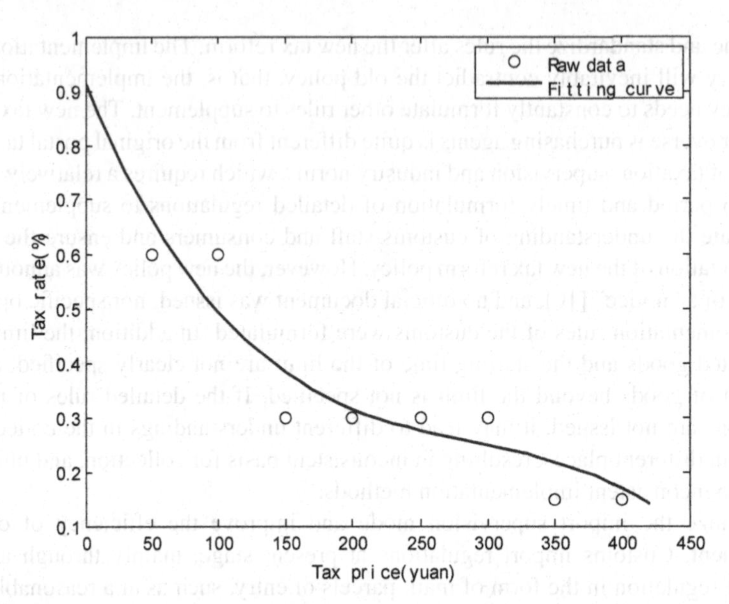

Fig. 1. Impact of overseas purchasing tax reform new policy compared with existing policies.

In Fig. 2, we simulate and analyze the error data with exponential distribution. Figure 2 shows the impact of overseas purchasing tax reform and new policy on national tax revenue. In Fig. 2 (a), it can be seen that there are obvious changes in the data, the overall trend of the model curve obtained does not change significantly. On the whole, the trade volume of import and export shows a trend of increasing year by year, and the growth rate is higher and higher, which corresponds to the current situation of China's good economic development. It shows that the model is suitable for the actual development. And Fig. 2 (b) shows that the current proportion of e-commerce is further expanding. With the rapid development of Internet companies and the rapid development of Internet companies, e-commerce has been developing rapidly. This also proves that the model can describe the current development trend. Cross platform e-commerce has a large scale, and can provide better support for international trade, can provide more tax. The above results can be used to provide the basis for future tax policy-making.

Figure 3 plots the impact of overseas purchasing tax reform and new policies on household consumption. In Fig. 3, we simulate and analyze the error data with exponential distribution. It can be seen that although there are obvious changes in the data, the overall trend of the model curve obtained does not change significantly. On the whole, the trade volume of import and export shows a trend of increasing year by year, and the growth rate is higher and higher, which corresponds to the current situation of China's good economic development. It shows that the model is suitable for the actual development. And figure B shows that the current proportion of e-commerce is further expanding. With the rapid development of Internet companies and the rapid development of Internet companies, e-commerce has been developing rapidly. This also proves that the model can describe the current development trend. Cross platform e-commerce has a large scale, and can provide better support for international trade, can provide more tax. The above results can be used to provide the basis for future tax policy-making.

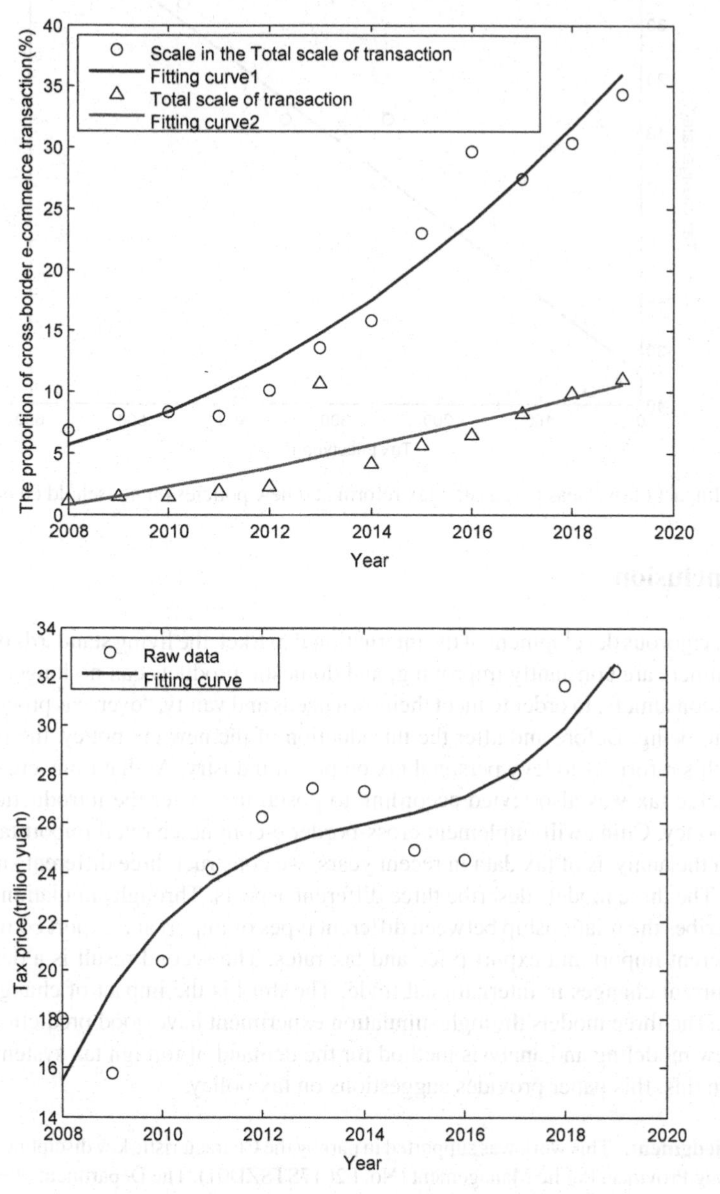

Fig. 2. Impact of overseas purchasing tax reform and new policy on national tax revenue.

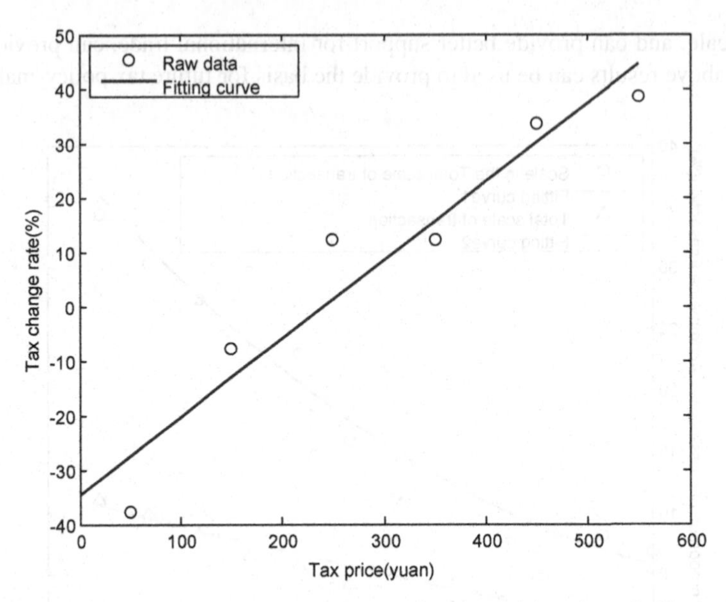

Fig. 3. Impact of overseas purchasing tax reform and new policies on household consumption.

5 Conclusion

With the vigorous development of the international market, the living standards of domestic consumers are constantly improving, and domestic products can no longer meet the needs of consumers. In order to meet their own needs and vanity, "overseas procurement" came into being. Before and after the introduction of the new tax policy, the main purpose of this reform is to levy personal tax on postal industry. At that time, cross-border e-commerce tax was also levied according to postal tax. After the introduction of tax reform policy, China will implement cross-border e-commerce retail import tax policy. Through the analysis of tax data in recent years, we construct three different simulation models. The three models describe three different aspects. Through simulation, the first one describes the relationship between different types of import and export commodities and different import and export price and tax rates. The second result is a description of the current changes in international trade. The third is the impact of changes in tax policies. The three models through simulation experiment have good prediction ability. It is a new modeling and analysis method for the demand of foreign tax system reform. Based on this, this paper provides suggestions on tax policy.

Acknowledgment. This work was supported in part by the Characteristic key discipline project of Guangdong Province, Public Management (No. F2017STSZD01), The Department of Finance of Guangdong Province, Research on social Welfare Effect of Tax Reduction and Fee Reduction (No. Z2020121), Research on the Long-term Mechanism and implementation of Vocational education to Consolidate the achievements of poverty alleviation (No. 412/31512000106). The authors wish to thank the reviewers for their helpful comments.

References

1. Qi, X., Ke, Y., Wang, S.: Analysis of main risk factors of China's PPP project based on case. China Soft Sci. (05), 107–113 (2009)
2. Yang, W.: Characteristics of contractual relationship between government and social capital under PPP mode. China Government Procurement News (003), May 10, 2019
3. Ou, C., Jia, K.: Solving the financing constraints of basic public services with PPP innovation. Econ. Manag. Res. (4), 85–94 (2017)
4. Zhang, P.: A study on the repayment pressure of various debts in China during the 13th five year plan period. Econ. Syst. Reform (1), 5–10 (2017)
5. Han, J., Lv, Y., Xu, Y.: Study on the cooperation mode of government and social capital. Shanghai Econ. Res. (2), 106–110 (2017)
6. Darrin, G., Mervyn, K.L.: The Worldwide Revolution in Infrastructure Provision and Project Finance. China Renmin University Press, Beijing (2016)
7. Bai, D.: Regulating PPP development, preventing and resolving local government debt risk. Theoret. Discuss. (03), 88–94 (2018)
8. Zhang, T.: Analysis of financial institutions to prevent the hidden debt risk of PPP government – based on the interpretation of Caijin 2019 No. 10. J. Financ. Account. (17), 172–176 (2019)
9. Dong, Z.: Analysis of types and characteristics of government debt in PPP mode in China. Local Financ. Res. (09), 61–66 (2016)
10. Luo, Y.: On the normalization of PPP project mode. Archit. Budg. (12), 16–21 (2018)
11. Jiang, D., Huo, L., Lv, Z., Song, H., Qin, W.: A joint multi-criteria utility-based network selection approach for vehicle-to-infrastructure networking. IEEE Trans. Intell. Transp. Syst. **19**(10), 3305–3319 (2018)
12. Mu, Z., Zheng, L., Cheng, Y.: Study on performance appraisal system of PPP project. J. Eng. Manag.

Exploring Users' Temporal Characteristics of "Purchase and Comment" Behaviors Based on Human Dynamics

Liangqiang Li[1], Bo Lv[2], Liang Wu[2](✉), and Hongwei Lin[3]

[1] Business School, Sichuan Agricultural University, Dujiangyan, Sichuan, China
lilq@sicau.edu.cn
[2] School of Economics and Management, Guizhou Normal University, Guiyang, Guizhou, China
wuliangmail@yeah.net
[3] School of Public Health and Management, Hubei University of Medicine, Shiyan, China

Abstract. This paper carried out an empirical study on the formation rules underlying online user comments. Using the human dynamics method, which is used for describing human behavior, this paper explored two typical consumer human behaviors related to "purchase-comment" behaviors within an electronic commerce environment. Using massive real data crawled by Python web crawler, this paper analyzes three types of temporal characteristics (time interval, burstiness index, and memory index) of post-purchase comments by users, explores when consumers choose to disclose their shopping experiences on the Internet after purchase, and studies the formation rules of online user comments. The study also verified that the temporal characteristics of consumers' "purchase-comment" behaviors in the e-commerce environment also conformed to power-law distribution principles, thus providing new empirical insights for the study of online human behavior dynamics.

Keywords: Online comments · Human dynamics · Temporal characteristics · Purchase-comment time interval

1 Introduction

Online user comments often post online consumer experiences, opinions, and emotional tendencies regarding products, services, brands, which consumers share with other consumers through Web 2.0 technology in the form of texts, videos, symbols, and so on [1–3]. It has also become a habit for many consumers to view online comments before making purchase decisions and to share post-purchase experiences. Massive amounts of online shopping-related comments and user generated content (UGC) with regard to

This research is support by Major Research Project of Innovative Groups in Guizhou Provincial Education Department (No. Qian Jiao He KY Zi [2017]034).

H. Song and D. Jiang (Eds.): SIMUtools 2020, LNICST 370, pp. 420–438, 2021.
https://doi.org/10.1007/978-3-030-72795-6_35

sharing experiences through social media constitute the main source of big data, and these have gradually begun to dominate the generation of website content.

Online user comments benefit increase interaction between enterprise platforms and customers; furthermore, they can create a good reputation and gain profits for enterprises. Online user comments have also played an important role in promoting the development of e-commerce. On one hand, online user comments influence consumers by eliminating the uncertainty of product information, thus affecting their online shopping decisions. In the case of new products or experiences, online user comments are an important information disclosure channel that help to eliminate potential consumers' uncertainty regarding the product. It is also used to educate potential customers. Being introduced to the experiences of other consumers can help a potential consumer use the product well. On the other hand, online user comments can influence the adjustments made to e-commerce enterprise platforms' operation strategies and the product perfection and service improvements undertaken by manufacturers.

1.1 Importance of Research on Temporal Characteristics of Users' "purchase and Comment" Behaviors

Scholars have conducted extensive research on online user comments from multiple perspectives. Previous studies have mainly focused on consumers' sentiment, usefulness [1, 3], influencing factors [4, 5], and the application of online comment corpus mining with regard to usefulness, emotional analysis, recommendation systems, and other aspects [7, 8]. However, little research has been conducted regarding the formation rules underlying the timing of online user comments (i.e. at what point users post comments after purchasing products or services). Such formation rules have great significance for theory as well as practice.

Online comments are an important source of product information, and they can affect consumers' product attitudes and purchasing behaviors [9]. Making online user comments is a voluntary contribution by consumers; undertaking this activity requires a certain amount of time and energy, and it has both cost and benefit effects. The point at which users choose to post comments on the Internet is restricted by various motivations and certain constraints. Through experimental research, McGuire and Kable found that human made decisions respond according to certain statistical time clues and that people adjust their persistence level according to the distribution of delay time intervals they encounter [10]. Therefore, consumers also tend to adopt certain strategies in order to obtain the corresponding benefits when deciding when to release a comment [11].

For potential consumers, online user comments disclose user experience-related information that is not offered by e-commerce platforms' product introductions or advertisements. The disclosure of such information plays an important role in easing any uncertainty regarding potential consumers' purchase decisions. The points of time (time points) at which any existing consumer comments are published will affect the decision-making time points of potential consumers. Some studies show that the time between the "reading time" of potential users and the "release time" of comments has a positive impact on the consumer awareness of usefulness regarding the comments [12]; the release time points of consumers' comments also have different influences on potential consumers' purchasing decisions. Recently published comments have a greater influence on purchasing decisions to be made in the near future. Older comments are considered more important for making long-term purchase decisions. Purchase-related comments that are released after a longer consumer experience period can also make people feel that the comments are more credible [13]. If an e-commerce platform, which has mastered the general temporal rules underlying users' comments, chooses appropriate time points to motivate consumers and promotes the formation of online user comments, it can persuade consumers to generate effective comments within a specified time range. Therefore, studying the temporal characteristics of the users' "purchase-comment" behaviors and the rules underlying comment formation is important.

1.2 Using Human Behavior Dynamics to Characterize the Temporal Characteristics of User Comment Behaviors

Users purchase goods and release comments online are two typical online behavior modes. In order to understand the activity rules underlying these two behaviors, it may prove useful to discuss them using the dynamic mechanisms underlying human behaviors. The Poisson process framework is widely used to quantitatively describe the behavior rules and influences underlying human behavior activities [14]; that is, within a given time interval $[t, t+\tau]$, the time interval τ between two successive behaviors follows exponential distribution: $p(\tau) = \lambda e^{-\lambda\tau}$. The probability of occurrence of these two behaviors is not related to time t but only to their interval τ.

With the development of computer and Internet technologies, extensive data related to various human activities can be collected and stored. Thus, it is now possible to use real data to illustrate human activities. In a 2005 Nature article, Barabási dscribed an empirical counting of the time intervals between users' sending and replying behaviors regarding e-mails and ordinary mails and found that this human behavior follows a power-law distribution: $p(\tau) = \tau^{-\nu}$; this finding has opened up a new direction for modern "human dynamics" research [15]. This study found that the distribution of human behavior rules was uneven and deviated from the Poisson process in terms of classical assumptions. This uneven distribution of human behavior rules is characterized by both "short-term outbreak" and "long-term silence."

With the emergence of Web 2.0, the concept of user participation has become increasingly popular. Online life has become an important part of modern life. People's behavior motivation patterns on the Internet have also attracted more and more scholars' strong attention and research interest. Scholars have studied the various dynamic modes underlying people's online behaviors such as web browsing [16], online movies watching [17], online game and music [18], online instant messaging (QQ) [19], behaviors in online games [20], editing entries on Wikipedia [21, 22], login behavior of information system users [23], topic-related reply behaviors on social networks [24], and customer behaviors in e-commerce transactions [25].

Empirical data analysis of these studies shows that the time intervals involved in online human behaviors follow a power-law distribution with an index of 1–3. Vázquez et al. [26] divided the distribution of human behavior rules into two universal classes with indices of -1 and -1.5, respectively, but several empirical research results have gradually expanded beyond the scope of these two universal classes [17].

Therefore, as an important part of online human behaviors, the two typical online human behavior laws underlying consumer purchase and comment behaviors within an e-commerce environment can also be characterized by using the human behavior dynamic mechanism method as a reference. A large amount of data on consumers' purchase and review behavior accumulated on e-commerce platforms; these data are useful for studying the behavioral dynamics underlying people's buying-related comments. This article also used the time interval characteristics underlying users' purchase behaviors and comment behaviors within e-commerce environments to explore whether they also conformed to the general rules of online human behaviors.

The rest of this paper is organized as follows: Sect. 2 introduces related works; Sect. 3 proposes three kinds of indicators for analyzing the temporal characteristics of "purchase-comment" behaviors based on human behavior dynamics and provides the whole experiment's framework; Sect. 4 reports the experiment results with real data; and finally, Sect. 5 concludes this paper.

2 Relevant Works

This paper studies the temporal characteristics of users' "purchase-comments" within e-commerce environments is mainly based on the theories and methods that describe the mechanisms underlying online human behavior dynamics. One of the most commonly used distribution theories with regard to online human behavior dynamics is the power-law distribution theory (also called long tail theory).

Pareto [27] and Zipf [28] have conducted pioneering work on power-law distribution. If the probability distribution or density function of a random variable is in the form of $p(k) = ck^r$, where c and r are constants, the random variable can be said to conform to the power-law distribution. Most real-world distributions conform to the negative exponent of the power law distribution—that is, $r < 0$. Power-law distribution has an important characteristic—scale-invariance; that is, $f(cx) = \alpha(cx)^k = c^k f(x) \propto f(x)$. By considering logarithms on both sides of the power-law distribution function, $ln\, p(k) = r\, ln\, k + ln\, c$ is obtained. This function is represented as a straight line in the double logarithmic coordinates, and this graphical feature is utilized for judging whether the random variable conforms to the power-law [29]. Under power-law distribution, the frequency of most events is small, while the frequency of a few events can be large. When k increases, $P(k)$ does not rapidly approach 0 in exponential form but gradually approaches 0 in a relatively gentle power-law form, thus showing "long tail" and "wide tail" properties [30].

Power-law distribution takes various forms in various professional fields. In different subject areas, scholars have also identified corresponding characteristic rules, thus gradually setting off an upsurge of empirical research on the power-law phenomenon. For example, scholars have discussed the occurrence laws underlying extreme events such as the distribution of lunar crater diameters [31], the distribution of earthquake levels and frequencies [32], and the distribution of war scales and frequencies [33]. Simultaneously, other distributions, such as surname distribution [34], the distribution of numbers of papers published [35], distribution of numbers of web pages in websites [36] and the time interval distribution of online human behaviors can also conform to power-law distribution and thus show a typical "heavy tail" phenomenon.

3 Analysis of the Temporal Characteristics of "Purchase and Comment" Behaviors Based on Human Behavior Dynamics

This section uses a method based on the dynamic mechanisms underlying human behaviors to analyze the temporal characteristics of the two typical behaviors underlying users' "purchase-comment" behaviors within e-commerce environments. First, the paper discusses the correlation indexes (time interval, burstiness index, and memory index) of time characteristic analysis. Second, the main method used for describing time characteristic rules is introduced, and the stage division method of time interval series is provided. Finally, the paper provides an experimental scheme for analyzing "purchase-comment" behaviors from a time characteristic perspective based on human behavior dynamics.

3.1 Relevant Indicators for Analyzing the Temporal Characteristics of "Purchase and Comment" Behaviors

Utilizing the human dynamics method for depicting human behavior rules, this paper mainly uses three related indicators in its analysis of the temporal characteristics underlying users' "purchase-comment" behaviors within e-commerce environments: (1) Time interval; (2) burstiness index; and (3) memory index.

3.1.1 Time Interval

"Time interval" represents the time difference between the occurrences of two consecutive events within a certain time range. For example, if a consumer buys a product on "January 2, 2016" and makes a comment on "January 8, 2016" after receiving the product or experiencing it, the time interval between the consumer's two consecutive behaviors (i.e., buying and commenting) is 6 days. If this product receives n comments on the Internet, this indicates that there are n "buy-evaluate" time intervals. In general, the e-commerce environment includes three kinds of time intervals related to consumers' purchase and comment behaviors.

Time Interval Between Adjacent Purchases
The time interval between adjacent purchases is the interval between two purchases made by a consumer. The consumer's purchase cycle can be perceived from the perspective of the individual. At the group level, it is used to depict the time difference between the purchase times for a product among different consumers; this reflects the sales situation of the product.

In order to calculate the time intervals between the adjacent purchasing behaviors of consumers, all purchasing behaviors are first sorted based on time to form a purchasing behavior occurrence time series; second, the purchase time interval series is obtained by subtracting between two purchase time series. Finally, the interval sequence is calculated. Similarly, for the time series analysis regarding purchase behaviors, the distribution in different time cycles can be considered.

Time Interval Between Adjacent Comments
The interval between the release times for adjacent comments can also be viewed from two perspectives; on the one hand, it is aimed at all comments made by a consumer,

thus describing the distribution law for individual consumers at the time point of the comments; on the other hand, it reflects the public praise of all consumers for a given product.

For calculating the release time intervals for adjacent comments, first, all users' comments on a certain product $C = \{r_1, r_2, ..., r_n\}$ are sorted according to their release time in order to form a comment release time series (from the earliest release time to the latest release time) $\{t_1, t_2, ...t_n\}$; second, two adjacent release times are subtracted to form a time interval series for releasing comments $\{\tau_1, \tau_2, ...\tau_n\}$; and finally, the time interval series is used for calculating probability distribution and ensuring the power-law fitting of the main part of time interval sequence.

Time Intervals Between Consumers' "Purchase-Comment" Behaviors

The comment page not only provides the consumers' comment time for a given product but also provides the consumers' purchase time for that product. To some extent, the calculation of the time interval between the two behaviors ("purchase-comment") reflects the motivation underlying the consumer's comment release and the factors that influence online comment generation. Therefore, by capturing all the consumer comments on the comment page of the target product C, the "purchase-comment" time interval for consumers can be calculated for each comment c_i; this method is used to measure how long consumers will post comments after making a purchase. The calculation formula is: $x_i = reviewing\_time - purchasing\_time$, where *reviewing_time* is the time point at which the consumer releases comments, and *purchasing_time* is the time point at which the consumer purchases the product.

After calculating the time interval for consumer behaviors, it is necessary to describe the rules governing these intervals. The main description method is as follows. By calculating the frequency $y_i = frequency(x_i)$ of the time interval x_i, the data series for the "purchase-comment" time interval of consumers is obtained: $T = \{(x_i, y_i)\}_{i=1,...,|C|}$. This data series is used to depict the time distribution for consumers' purchases and comments ("purchase-comment" behavior).

3.1.2 Burstiness Index

Intermittence is a physical statistic that characterizes the frequency degree of human activities within a short time period and the degree of silence within a long time perod [15, 26]. Goh and Barabasi define intermittence as follows [37]:

$$B = \frac{(\sigma_\tau / m_\tau - 1)}{(\sigma_\tau / m_\tau + 1)} = \frac{\sigma_\tau - m_\tau}{\sigma_\tau + m_\tau} \tag{1}$$

Where σ_τ represents the standard deviation of the event interval sequence τ, and m_τ represents the average value of the time interval series τ. If the standard deviation σ_τ and the mean value m_τ are equal, the statistical value of intermittence is $B = 0$. For example, if the mean and standard deviation in the exponential distribution are equal, the intermittence value is $B = 0$. For the "heavy tail" distribution, the standard deviation is much larger than the mean, and the final intermittence value approaches [23].

The definition of burstiness index can also be used to describe the frequency of human behavior—that is, whether there will be many purchases and comments within a short period of time. For example, it can be used for predicting whether people will generate a large number of purchasing activities in the Double Eleven Shopping Canival of China; it can also be used for focusing on comments made for purchased goods after receiving the goods, while there is no purchasing behavior and comment behavior for a long period during which there is no promotion activity.

3.1.3 Memory Index

The memory index is used to describe the correlation of adjacent subsequences in a given time interval series [14, 25]. The specific process for calculating memory index is as follows.

First, the time series is formed according to the time series of the occurrence of events $\{t_1, t_2, \ldots, t_{n+1}\}$, and every two adjacent time points are subtracted to obtain a time interval series $\{\tau_1, \tau_2, \ldots \tau_n\}$, with n_τ representing the total number of time interval series; a time interval series is divided into two time interval subsequences: $\{\tau_1, \tau_2, \ldots \tau_{n-1}\}$ and $\{\tau_2, \tau_3, \ldots \tau_n\}$. σ_1, σ_2 are the standard deviations of the two subsequences, and m_1, m_2 are the mean values of these two subsequences. The calculation formula is as follows [27]:

$$M = \frac{1}{n_\tau - 1} \sum_{i=1}^{n_\tau - 1} \frac{(\tau_i - m_1)(\tau_{i+1} - m_2)}{\sigma_1 \sigma_2} \tag{2}$$

The memory index is used to describe the laws underlying consumers' purchase and comment behavior in e-commerce environments. It mainly reflects whether consumers will experience a longer or shorter time interval after purchasing or commenting for a longer or shorter time interval. This indicator can help e-commerce platforms master the principles of consumer purchasing behavior and online comments generation; this will aid them in selecting appropriate time points to stimulate purchases and generate comments.

3.2 Characterization Methods for Time Characteristic Rules

In the world of online human behavior dynamics, the distribution rules and characteristics of the time interval between two consecutive behaviors are generally used to describe the activity rules governing human behavior. The most common characterization method for time interval distribution rules is to fit their distribution with power-law distribution and use statistical tests to verify whether they conform to power-law distribution.

The fitting of power-law distribution mainly involves the fitting of the power-law index. Clauset et al. gave a detailed derivation and explanation for the power-law distribution fitting [38]. This paper takes Clauset's method as a reference, mainly using linear regression and least square method to complete the curve fitting of the power-law function, thus obtaining the fitting straight line slope for the main data, namely the power index [37, 39].

The specific method for determining the power-law distribution fitting is as follows:

The distribution curve of the general power-law function is $y = ax^{-b}$, where $a > 0$; in this paper, y represents the probability of occurrence for the time interval, and x represents the time interval between consumer purchase and comment.

- Taking the logarithm of both sides of $y = ax^{-b}$, $\ln y = (-b) \ln x + \ln a$ can be obtained;
- Certain substitutions are performed ($u = \ln x$, $v = \ln y$, $A = \ln a$), and the curve equation of the power-law distribution function becomes a straight line equation: $v = A + (-b)u$.
- Using the empirical data value (x_i, y_i), we obtain (u_i, v_i) and $i = 1, 2, \ldots, n$.
- Using the least square regression straight line to calculate u and v, the following equation can be obtained:

$$\hat{b} = -\frac{\sum_{i=1}^{n} u_i v_i - \left(\sum_{i=1}^{n} u_i\right)\left(\sum_{i=1}^{n} v_i\right)}{n \sum_{i=1}^{n} u_i^2 - \left(\sum_{i=1}^{n} u_i\right)^2}$$

- $A = \overline{v} + \hat{b}\overline{u}$.
- Because $a = e^{\hat{A}}$, the value of a is obtained. Therefore, $y = ax^{-b}$.

3.3 Division Method of Time Interval Series

In order to explore when and for what reasons users choose to release online comments after making a purchase, this paper divides the "purchase-comment" time interval series for consumers into stages and explores the behavior characteristics of users at different stages [40, 41].

The division of the time interval stages is implemented on the interval sequence $T = \{(x_i, y_i)\}_{i=1,\ldots,|C|}$, and the specific division method implementation rules are as follows.

First, the starting point $p_1 = (x_1, y_1)$ and the ending point $p_{|C|} = \left(x_{|C|}, y_{|C|}\right)$ are selected as the initial points of division.

Next, the distance between them and the starting point and the ending point $d_i = \frac{|(p_{|C|} - p_1) \times (p_1 - p_i)|}{|p_{|C|} - p_1|}$ is calculated for other points $p_i = (x_i, y_i)$ and lines $\overline{p_1 p_{|C|}}$ in the interval sequence (excluding the starting point and the ending point) [42, 43].

The dividing point $i^* = \arg_{i \in \{2,\ldots,|C|-1\}} \max\{d_i\}$ is determined according to a predefined threshold value, and the number of stages are determined based on stage division, where the value of d_{i^*} is greater than the predefined distance value d_0; this process is repeated until all points with distances greater than d_0 are found.

An example of the time series stage divisions is shown in Fig. 1.

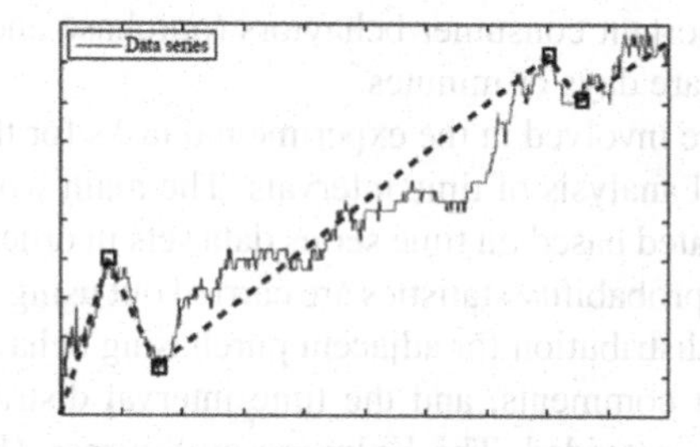

Fig. 1. Example of stage division of data series

3.4 Experimental Plan for Analyzing the Temporal Characteristics of "Purchase and Comment" Behaviors

The overall experimental framework of this paper is shown in Fig. 2. Which is mainly composed of three task stages. The first stage is the online user comment data acquisition stage. The second stage is the comment data preprocessing stage. The third stage is the time series data analysis stage.

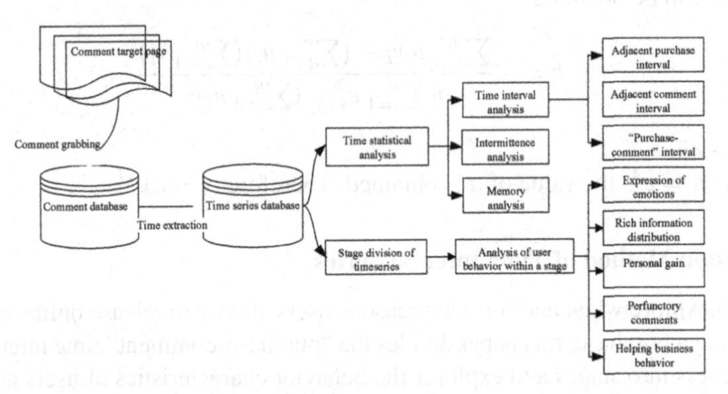

Fig. 2. The analysis framework for users' "purchase-comment" behavior dynamics

The main tasks for data acquisition follow 5 stages: (1) select target website and target comment webpage containing the e-commerce online user comments; (2) analysis the source code of webpage; (3) the determination of the presentation rules for online user comments on the selected webpage; (4) a combination of Python's web crawler toolkit and offline web crawler software (Locoy Spider) for customizing the regular expressions for web crawling rules; and (5) utilization of the web crawler for completing the grabbing of comment page data and forming the initial comment database [44].

The specific preprocessing work for comment data is as follows. First, the initial database is cleaned up to form an effective comment data set. The main operations include clearing duplicate comment records in the initial database and deleting blank data records that may not have been successfully captured due to inconsistent rules. Second, time data related to "purchase and comment" behavior are extracted from the effective data set to form a time data set. Table 1 shows an example of the time interval data set for the two important consumer behaviors (purchase and comment); the units used for recording time are days or minutes.

Two major blocks are involved in the experimental tasks for the data analysis stage. The first is the statistical analysis of time intervals. The main work is as follows. First, time intervals are calculated based on time series data sets in order to form time interval series data sets; second, probability statistics are carried out using the time intervals; and third, the event interval distribution for adjacent purchasing behaviors, the time interval distribution for adjacent comments, and the time interval distribution for "purchase-comment" behaviors are provided. The Kolmogorov-Smirnov (KS) test is carried out on each distribution to verify whether it conforms to the power-law distribution. The intermittence and memory of interval sequences are calculated again. The second major

Table 1. Example of time data set

ID	Purchasing time	Commenting time	Time interval (days)
ID1	6-30-2011	7-1-2011	1
ID2	08-01-2012	08-08-2012…	7
…	…	…	…

task is to divide the stages of the "purchase-comment" time interval series and analyze the user comment behavior at each stage.

4 Experiment Results

4.1 Data Description

According to the product classification used at the target website, comments from popular categories that contained numerous comments were selected as crawling objects. Considering representativeness, snowball sampling method was used to extracting the comment data of representative individual products in different categories. At the same time, product type like search-type and experience-type was also considered. Due to privacy protection proxy, after conducting this experiment, the comment page displayed in the online mall had been revised and that it no longer provided purchase time information. Data selected for this experiment were the comment data captured before the revision, and the time span of the selected products was relatively long.

The selected data were divided into three types. The first type was for personal historical purchase records and comment records; the second type involved selecting data sets for six major product types; and the third type included search-type and experience-type products, though it was the comment data set for each individual product.

The data set for personal historical purchases and comments used a random sampling method to capture 11,081 purchase records from 152 users of the selected online mall (name: JD) from December 1, 2008 to April 14, 2014.

For popular categories selection, this study selected six categories (office electronic products, notebook computers, routers and network cards, mobile phones, decorations, and cosmetics) for the comments. Product type, product category, number of products included in the category, and the total number of comments for each category are shown in Table 2. For this category, "purchase-comment" interval was mainly considered.

For single product selection, only the product names that were used frequently in the four experiments are listed here. The product name, product type, time span, and number of comments for these four individual products are shown in Table 3. In addition, comment data for another 55 individual products were randomly captured, but their detailed description information is not repeated here due to space reasons. For individual products, not only was the time interval for their "purchase-comment" behaviors considered in the data, but their comments were also sorted; furthermore, the time interval distribution between adjacent comments was explored.

Table 2. Data description of six product categories

Product type	Product category	Number of products	Number of comments
Search type	Office electronic products	10	15 928
	Notebook computers	12	34 504
	Routers and network cards	9	285 535
	Mobile phones	71	487 817
Experience type	Decorations	10	3450
	Cosmetics	11	42 545

Table 3. Data set description for four individual products

Product type	Product name	Time span	Number of comments
Search-type	D90 Single-lens reflex digital camera	November 6, 2008 to December 31, 2011	5460
	Double Happiness badminton racket	May 22, 2010 to December 8, 2013	13 733
Experience-type	Books: One Hundred Years of Solitude	June 1, 2011 to December 8, 2013	49 422
	Red wine	February 14, 2011 to July 25, 2012	4246

4.2 Analysis Results for the Temporal Characteristics of Purchase Behavior

4.2.1 Data Description

For dynamic analysis of users' purchase behavior, this study used a data set of 152 users' purchase history, with a total of 11 081 purchase records and 73.87 purchase records per person on average; these included at least 2 purchase records and at most 1063 purchase records per person.

4.2.2 Time Interval Characteristics of Adjacent Purchasing Behaviors

In order to verify the temporal characteristics shared by users' adjacent purchasing behaviors, the purchase records of a consumer (ID: Kom_isun) from a purchase database of 152 consumers (with a total of 857 records) were selected, and a time span from May 19, 2009 to October 25, 2013 was chosen. Figure 3 shows the online purchase sequence for the consumer with the ID Kom_isun. The horizontal axis of the series indicates the interval (in days) between the first purchase and the current purchase, and the vertical axis indicates that the purchase occurred on the same day.

Figure 4 gives the scatter plot for the time interval of Kom_isun's adjacent purchase behavior and the power-law distribution fitting for the main part data. The exponent of power-law fitting is -1.372, and the goodness of fit is 0.853.

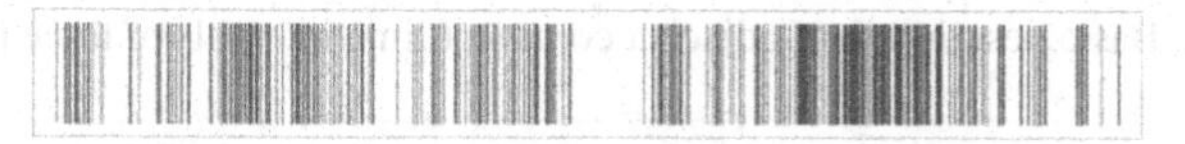

Fig. 3. Online shopping series (Kom_isun)

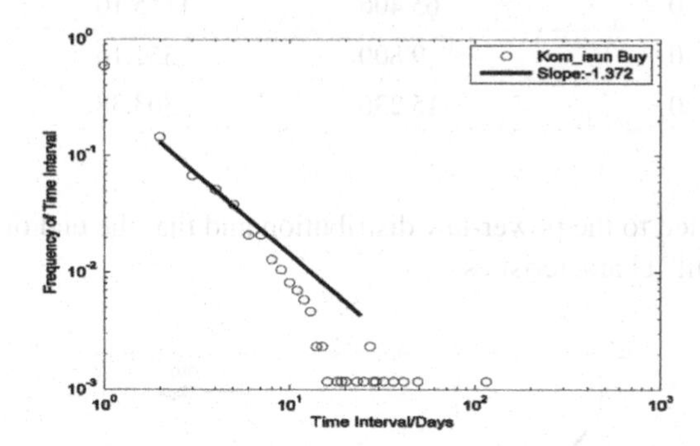

Fig. 4. Statistical double logarithmic chart for adjacent purchase time interval (Kom_isun).

Using Sect. 3.1 to introduce the calculation formula for intermittence and memory of purchase behavior, it was found that the burstiness index value for the adjacent purchasing behaviors of Kom_isun was 0.50128, thus showing intensive purchase behavior for a short time and non-purchase behavior for a long time. Statistics regarding its purchasing behaviors showed that 58.52% of all the purchasing records for the user had no interval (i.e., 0 day) between adjacent purchasing behaviors. The proportion within 3 days was 84.8%; the longest interval between two adjacent purchases was 114 days. The M value calculated by depicting the memory of purchasing behavior was 0.1264, which indicates that the purchasing behaviors of this user were characterized based on memory. That is, after adjacent purchasing behaviors were conducted within a short time period, some adjacent purchasing behaviors continued within a short time period.

4.3 Analysis Results for Temporal Characteristics of Comment Behaviors

According to the temporal characteristics' statistics for comment release through online user comment behaviors, the time intervals for pairs of adjacent comments released by consumers in the product comments category were depicted based on the data of three separate products (D90 single-lens reflex digital camera, red wine, and badminton racket). The shortest time interval, longest time interval, standard deviation of time interval, and average value for the three groups of data are shown in Table 4.

Figures 5, 6, and 7 show the probability distributions for the time intervals of the three products under the double logarithmic coordinate and the power-law fitting of the main part's curve, respectively. The circle in the figure represents the original curve of the release time interval between two adjacent comments, and the black straight line represents the fitting curve. The power-law indexes of the three data groups were −1.092, −0.935, and −0.878, respectively, and the goodness of fit was above 85%. The Kolmogorov-Smirnov (KS) test results showed that the release time intervals for adjacent

Table 4. Basic description of adjacent comment time intervals of three products

Product type	Minimum interval (minutes)	Maximum interval (minutes)	Standard deviation	Mean value
D90 digital camera	0	65 406	1115.10	271.27
Red wine	0	9 800	351.16	178.36
Badminton racket	0	15 236	303.38	128.68

comments conformed to the power-law distribution and that the end of the distribution had obvious "fat tail" characteristics.

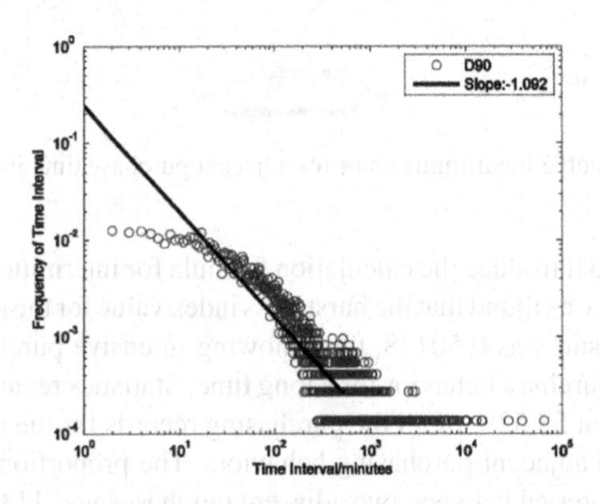

Fig. 5. Time interval distribution for adjacent comments regarding D90 digital camera

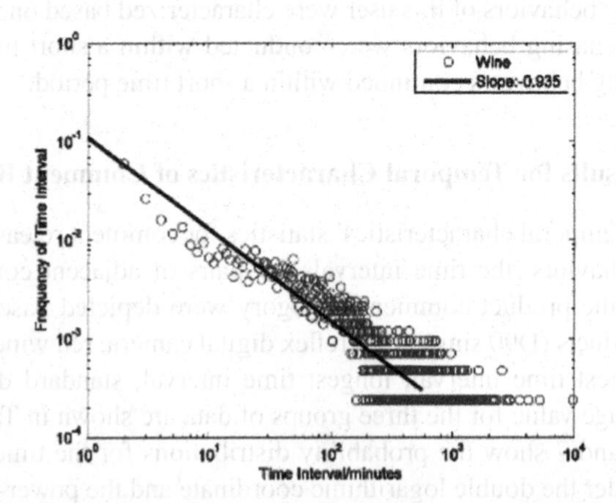

Fig. 6. Time interval distribution for adjacent comments regarding red wine.

This study calculated the intermittence and memory indexes for the time interval between two adjacent consumers' comments for three online products by using the previous formulas (3–1) and (3–2). The calculation results are shown in Table 5.

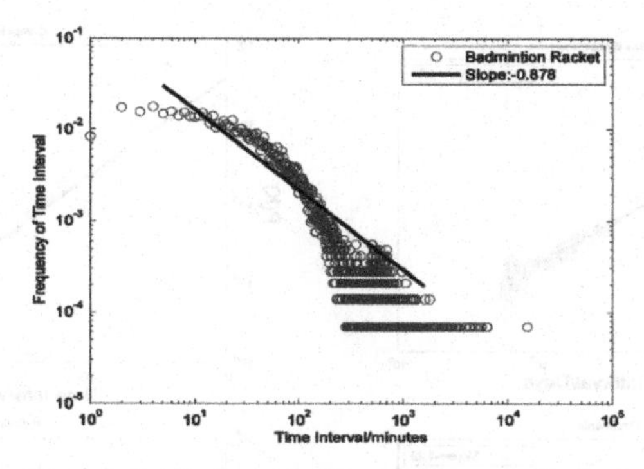

Fig. 7. Time interval distribution of adjacent comments for badminton racket.

Table 5. Burstness and memory for time intervals of consumer comments for three products

Products	Number of comments	Power exponent	Burstness	Memory
D90 digital camera	5460	−1.092	0.6087	2.206e−22
Badminton racket	13 733	−0.878	0.4044	2.755e−22
Red wine	4246	−0.935	0.3263	1.432e−22

Burstness index to comments released for the three products sold online were 0.6087, 0.4044, and 0.3263, respectively; this indicated that the time interval for the adjacent consumer comment released for each product had a strong intermittence. However, the memory values for the three products were almost zero. In other words, the adjacent comment behaviors for online products showed a strong intermittence and weak memory.

Based on the actual situation, the possible reason for this phenomenon is that, during the specific sales activities for the products or when the e-commerce platform was urging consumers to evaluate products that had not been evaluated, consumers tended to release comments within a short time period, and this resulted in a short interval between adjacent comments made within a certain time period, thus resulting in intense intermittence. On the e-commerce platform, each consumers' "buying and commenting"-related behavior was independent. Consumers' tend to comment on a product according to their own behavioral preferences or rules. The time point at which a consumer releases a comment has nothing to do with the time point at which the persons before him/her released their comments; therefore, the weak memory can be explained well.

4.4 Analysis Results for Temporal Characteristics of "Purchase and Comment" Behaviors

In order to verify the time interval distribution of consumers' "purchase-comments" macroscopically, the experimental data sets of six major categories were used for verification. The basic description of the "purchase-comment" time intervals for the consumers in the six product categories is shown in Table 6. Among them, the shortest time interval for "purchase-comment" was 0 (day)—that is, purchase and comment on the same

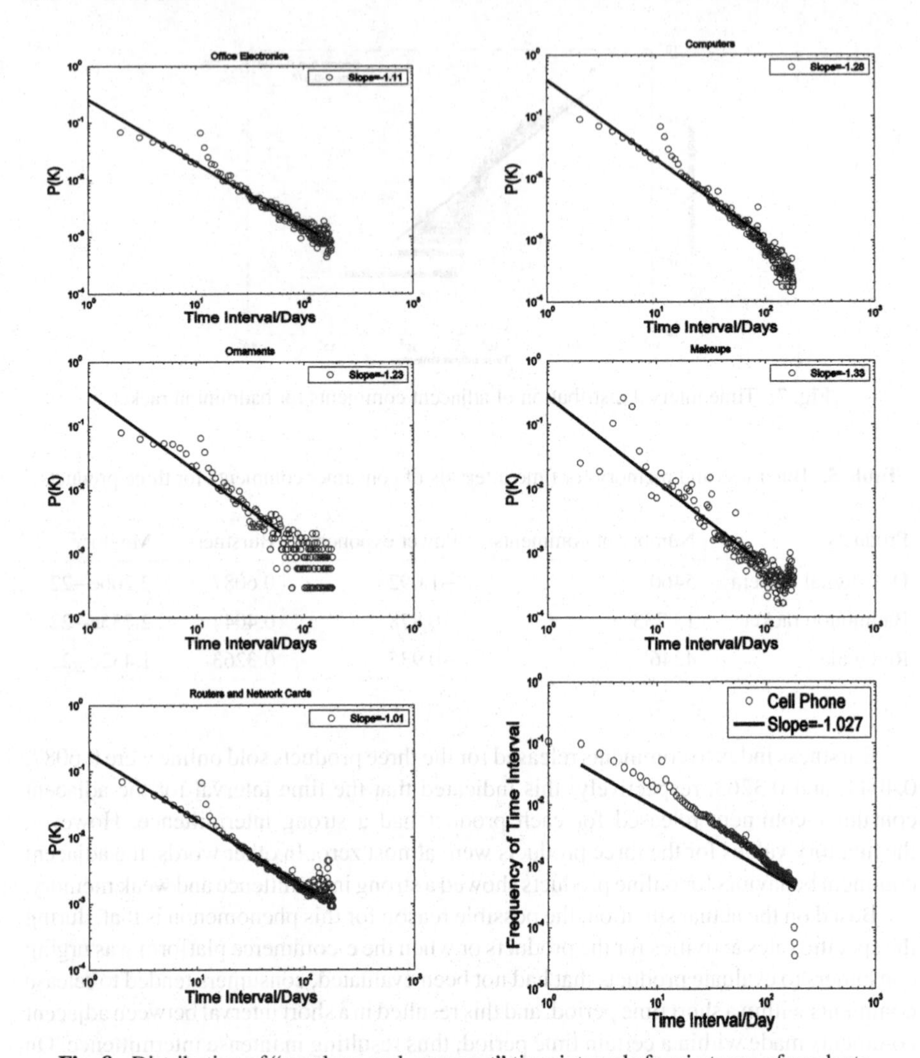

Fig. 8. Distribution of "purchase and comment" time intervals for six types of products.

day. The longest time interval was 183 days, and it was related to the platform's comment policy. The JD online mall only allowed consumers to post comments on products purchased within half a year.

The double logarithmic scatter plot for the "purchase-comment" time interval distribution of the six product categories and the curve fitting of the main part are shown in Fig. 8. The power-law indexes of these six product categories were as follows: Office electronic products: −1.11; notebook computers: −1.28; decorations: −1.23; cosmetics: −1.33; routers and network cards: −1.01; and mobile phones: −1.027.

In order to verify the distribution rules followed by consumers within the "purchase-comment" time interval for a single product, in this experiment, the time intervals of the selected products for two types (experience-type and search-type) were selected. Two

Table 6. Basic description of "purchase and comment" time intervals for six types of products

	Shortest interval	Longest interval	Standard deviation	Mean value
Office electronic products	0	180	28.94	19.83
Notebook computers	0	180	29.84	39.97
Routers and network cards	0	180	46.49	36.15
Mobile phones	0	183	35.87	24.58
Decorations	0	180	33.10	21.49
Cosmetics	0	182	28.32	16.18

of the selected products were search-type products (D90 digital camera and badminton racket), and the other two were experience-type products (One Hundred Years of Solitude [book] and red wine). Their interval distribution is shown in Fig. 9. Their power-law indices were −1.17, −1.03, −0.95, and −1.14, respectively.

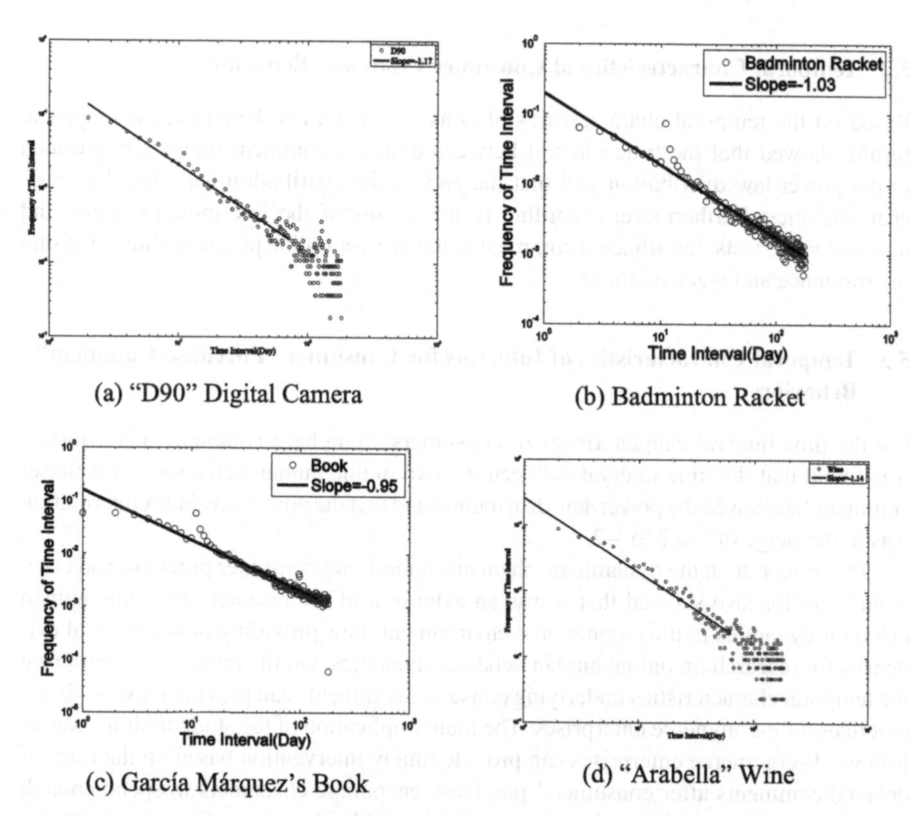

(a) "D90" Digital Camera

(b) Badminton Racket

(c) García Márquez's Book

(d) "Arabella" Wine

Fig. 9. Double logarithmic diagram of time interval distribution for four products.

5 Conclusion

This paper used the theories and methods of human behavior dynamics to describe the temporal characteristics of consumers' online purchasing and commenting behaviors. The main contents included the following. (1) By introducing the rules of online human behaviors, the form, cause, application scope, and fitting method of the power-law distribution theory utilized in this study were introduced; (2) three indexes describing the temporal rules for human behaviors were introduced (time interval, intermittence, and memory); (3) a method for dividing time interval series into stages was proposed; (4) based on online user comment behavior data, this empirical study examined three indicators used for describing the temporal rules underlying purchase behaviors, comment behaviors, and "purchase-comment" behaviors. The following conclusions were drawn.

5.1 Temporal Characteristics of Consumer Purchase Behavior

Based on the empirical results of the burstiness index and memory index tests, the temporal characteristics of consumers' purchasing behavior showed intensive purchasing behavior within a short time period, non-purchasing behavior within a long time period, and memory of purchasing behavior.

5.2 Temporal Characteristics of Consumer Comment Behavior

Based on the temporal characteristics of consumer comment behavior, the empirical results showed that the time interval between adjacent comment releases conformed to the power-law distribution and that the end of the distribution had clear "fat tail" characteristics. Furthermore, according to the results of the intermittence value and memory value tests, the adjacent comment behaviors for online products showed strong intermittence and weak memory.

5.3 Temporal Characteristics of Intervals for Consumer "Purchase-Comment" Behaviors

For the time interval characteristics of consumers' "purchase-comment" behaviors, it was found that the time interval between the two online human behaviors ("purchase-comment") followed the power-law distribution and that the power-law index distribution was in the range of -0.7 to -2.

This research on the dynamic mechanisms underlying consumer purchase and comment behavior also showed that it was an extension of the research on online human behavior dynamics in the e-commerce environment, thus providing new empirical evidences for research on online human behavior dynamics. On the other hand, mastering the temporal characteristics underlying consumer comments can provide good guidance practices for e-commerce enterprises. The main implication of the study findings was as follows. E-commerce enterprises can provide timely intervention based on the rules of delayed comments after consumers' purchase, encourage consumers to spend enough time and, at the same time, make real comments quickly, thus generating good publicity for the selected product or service. In addition, such enterprises can allocate reasonable resources to customer service in order to better serve customers and improve customer satisfaction and loyalty.

References

1. Wang, T., Wang, K., Chen, H.: The impact of temporal distance on the increase of the perceived usefulness of online reviews: from the perspective of the attribution theory. J. Bus. Econ. **280**(2), 46–55 (2015)
2. Jiang, D., Huo, L., Song, H.: Rethinking behaviors and activities of base stations in mobile cellular networks based on big data analysis. IEEE Trans. Netw. Sci. Eng. **7**(1), 80–90 (2020)
3. Jiang, D., Wang, Y., Lv, Z., et al.: Big data analysis-based network behavior insight of cellular networks for industry 4.0 applications. IEEE Trans. Ind. Inform. **16**(2), 1310–1320 (2019)
4. Shen, C.C., Chiou, J.S.: The impact of perceived ease of use on Internet service adoption: the moderating effects of temporal distance and perceived risk. Comput. Hum. Behav. **26**(1), 42–50 (2010)
5. Shi, W., Gong, X., Zhang, Q., et al.: A comparative study on the first-time online reviews and appended online review. J. Manag. Sci. **29**(4), 45–58 (2016)
6. Min, Q., Qin, L., Zhang, K., Qing, F.M., Liang, Q., Keliang, Z.: Factors affecting the perceived usefulness of online reviews. Manag. Rev. **29**(10), 95–107 (2017)
7. Bauman, K., Liu, B., Tuzhilin, A.: Aspect based recommendations: recommending items with the most valuable aspects based on user reviews. In: Proceedings of the 23rd ACM SIGKDD International Conference on Knowledge Discovery and Data Mining, pp. 717–725 (2017)
8. Ren, X., Lv, Y., Wang, K., et al.: Comparative document analysis for large text corpora. In: Proceedings of the Tenth ACM International Conference on Web Search and Data Mining, pp. 325–334 (2017)
9. Pang, J., Qiu, L.: Effect of online review chunking on product attitude: the moderating role of motivation to think. Int. J. Electron. Commer. **20**(3), 355–383 (2016)
10. McGuire, J.T., Kable, J.W.: Decision makers calibrate behavioral persistence on the basis of time-interval experience. Cognition **124**(2), 216–226 (2012)
11. Shen, W., Hu, Y.J., Rees, J.: Competing for attention: aAn empirical study of online reviewers' strategic behaviors. MIS Q. **39**(3), 683–696 (2015)
12. Gong, Y.P., Huang, K., Zhang, Q., Gu, H.P.: Study on the relationship between time and distance of new product forecast, consumer online comments and their purchase targets. R&D Manag. **04**, 36–44 (2015)
13. Jin, L., Hu, B., He, Y.: The recent versus the outdated: an experimental examination of the time-variant effects of online consumer reviews. J. Retail. **90**(4), 552–566 (2014)
14. Santhanam, M.S., Kantz, H.: Return interval distribution of extreme events and long-term memory. Phys. Rev. E **78**(5), 051113 (2008)
15. Barabasi, A.L.: The origin of bursts and heavy tails in human dynamics. Nature **435**(7039), 207–211 (2005)
16. Gonçalves, B., Ramasco, J.J.: Human dynamics revealed through web analytics. Phys. Rev. E **78**(2), 026123 (2008)
17. Zhou, T., Kiet, H.A.T., Kim, B.J., et al.: Role of activity in human dynamics. EPL (Europhys. Lett.) **82**(2), 28002 (2008)
18. Hu, H.B., Han, D.Y.: Empirical analysis of individual popularity and activity on an online music service system. Phys. A **387**(23), 5916–5921 (2008)
19. Chen, G., Han, X., Wang, B.: Multi-level scaling properties of instant-message communications. Phys. Procedia **3**(5), 1897–1905 (2010)

20. Henderson, T., Bhatti, S.: Modelling user behavior in networked games. In: Proceedings of the Ninth ACM International Conference on Multimedia, pp. 212–220. ACM (2001)
21. Zhao, F., Liu, J.H., Zha, Y.L., et al.: Human dynamics analysis of online collaborative writing. Acta Physica Sinica **60**(11), 118902–118902 (2011)
22. Sumi, R., Yasseri, T., Rung, A., et al.: Edit wars in Wikipedia. arXiv preprint arXiv:1107.3689 (2011)
23. Ren, J.J., Wang, N.X., Ge, S.L.: Information system user access law based on human behavior dynamics. Comput. Mod. **10**, 10–15 (2015)
24. Wu, Y., Zhou, C., Chen, M., et al.: Human comment dynamics in on-line social systems. Phys. A **389**(24), 5832–5837 (2010)
25. Wang, J., Gao, K., Li, G.: Empirical analysis of customer behaviors in Chinese e-commerce. J. Netw. **5**(10), 1177–1184 (2010)
26. Vázquez, A., Oliveira, J.G., Dezsö, Z., et al.: Modeling bursts and heavy tails in human dynamics. Phys. Rev. E **73**(3), 036127 (2006)
27. Chris, A.: The Long Tail, pp. 1–100. CITIC Press, China (2006)
28. Zipf, G.K.: Human Behavior and the Principle of Least Effort, pp. 1–573. Addisom-Wesley Press, Cambridge (1949)
29. Hu, H.B., Wang, L.: Brief research history of power-law distribution. Physics **34**(12), 889–896 (2005)
30. Newman, M.E.J.: Power laws, Pareto distributions and Zipf's law. Contemp. Phys. **46**(5), 323–351 (2005)
31. Zhang, J.Z.: Fractals, pp. 300–326. Tsinghua University Press, Beijing (1997)
32. Gutenberg, B., Richter, C.F.: Frequency of earthquakes in California. Bull. Seismol. Soc. Am. **34**, 185–188 (1944)
33. Roberts, D.C., Turcotte, D.L.: Fractality and self organized criticality of wars. Fractals **6**, 351–357 (1998)
34. Zanette, D.H., Manrubia, S.C.: Vertical transmission of culture and the distribution of family names. Phys. A **259**, 1–8 (2001)
35. Lotka, A.J.: The frequency distribution of scientific productivity. J. Wash. Acad. Sci. **16**, 317–323 (1926)
36. Crovella, M., Bestavros, A.: Self-similarity in world wide web traffic: evidence and possible causes. IEEE/ACM Trans. Netw. **5**(6), 835–846 (1997)
37. Goh, K.I., Barabási, A.L.: Burstiness and memory in complex systems. EPL (Europhys. Lett.) **81**(4), 48002 (2008)
38. Clauset, A., Shalizi, C.R., Newman, M.E.J.: Power-law distributions in empirical data. SIAM Rev. **51**(4), 661–703 (2009)
39. Jiang, D., Wang, Y., Lv, Z., et al.: An energy-efficient networking approach in cloud services for IIoT networks. IEEE J. Sel. Areas Commun. **38**(5), 928–941 (2020)
40. Jiang, D., Wang, W., Shi, L., et al.: A compressive sensing-based approach to end-to-end network traffic reconstruction. IEEE Trans. Netw. Sci. Eng. **7**(1), 507–519 (2020)
41. Jiang, D., Huo, L., Lv, Z., et al.: A joint multi-criteria utility-based network selection approach for vehicle-to-infrastructure networking. IEEE Trans. Intell. Transp. Syst. **19**(10), 3305–3319 (2018)
42. Jiang, D., Huo, L., Li, Y.: Fine-granularity inference and estimations to network traffic for SDN. PLoS ONE **3**(5), 1–23 (2018)
43. Jiang, D., Li, W., Lv, H.: An energy-efficient cooperative multicast routing in multi-hop wireless networks for smart medical applications. Neurocomputing **220**, 160–169 (2017)
44. Jiang, D., Zhang, P., Lv, Z., et al.: Energy-efficient multi-constraint routing algorithm with load balancing for smart city applications. IEEE Internet Things J. **3**(6), 1437–1447 (2016)

Enhanced Shortest Path Routing Algorithm for Named Data Mobile Ad-Hoc Network

Junyu Lai(✉), Zhengyin Han, Yingbing Sun, Han Xiao, and Xiaohui Zheng

School of Aeronautics and Astronautics, University of Electronic Science and Technologies of China, Xiyuan Ave No. 2006, Chengdu, China
laijy@uestc.edu.cn

Abstract. Named Data Network (NDN) technology can efficiently mitigate the negative influence due to instable links and dynamic topology on MANET performance. Considering MANET nodes' mobility, this paper enhances NDN's shortest path routing algorithm by storing the interest packets generated by the consumers in the pending interest tables (PITs) of the shortest path's neighboring nodes, aiming at building the backup paths for transmitting back the requested data packets, so as to mitigate the shortest path disruption problem caused by MANET topology dynamicity. This paper designs and implements a simulation platform based on NS3/ndnSIM, and the experimental results indicate that, the enhanced shortest path plus backup (SPPB) algorithm can effectively decrease the average request response delay and improve successful request response ratio.

Keywords: Named data network (NDN) · Mobile ad-hoc network (MANET) · Routing algorithm · Quality of service · Network simulation

1 Introduction

MANET is a peer-to-peer wireless network composed of multiple mobile nodes with topology changing dynamically. Its performance is vulnerable to packet loss and transmission errors. It is required that at least one complete route existing between two communicating nodes, to successfully transmit control and data information. Efficient networking protocol is crucial to guarantee MANET performance. TCP/IP protocol relies on IP address to identify and locate network nodes, as well as to build P2P connections for reliable transmissions. However, considering the mobility feature of MANET nodes, it is impossible to locate network nodes based on IP addresses alone. For the connection-oriented services, IP based MANET has the problems of short connection duration with frequent interruptions. While for the best-effort services, a high packet loss ratio will significantly degrade network performance. Therefore, IP protocol is not suitable for highly dynamic MANETs.

Information Centric Networking (ICN) is one major future networking architecture, and NDN is its most representative technical implementation [1]. NDN adopts the

H. Song and D. Jiang (Eds.): SIMUtools 2020, LNICST 370, pp. 439–450, 2021.
https://doi.org/10.1007/978-3-030-72795-6_36

receiver driven pattern and can realize asynchronous communications between two network nodes without constructing and maintaining a dedicated connection. In addition, the in-networking caching mechanism employed by NDN enables the network nodes to make full use of the wireless broadcast feature, which can effectively reduce redundant information and can shorten the request response delay. Therefore, NDN are more suitable for the data transmission needs in highly dynamic MANETs. In the following of this paper, NDN based MANET is called NDMANET.

Currently, naming, routing/forwarding, caching, security, and transport are the five major research areas of NDMANET [2]. Among them, routing algorithm is vital to networking performance. NDMANET's flooding and shortest path routing (SPR) algorithms are directly adopted from the fixed NDN networks, and are not optimized for MANET. Considering MANET's frequent link interruptions and topology changes, this paper focuses on enhancing the SPR algorithm for NDMANET. The principle is that, all the neighboring nodes along the shortest path determined by the original SPR algorithm will store the received interest packet in PIT, so as to create backup paths for transmitting the requested data packet, alleviating the shortest path interruption problems caused by frequent topology changes. In this way, the transmission failure probability can be reduced. The major contribution of this work includes:

- Elaborates an innovative shortest path plus backup (SPPB) routing algorithm to enhance the legacy algorithm, without changing NDMANET's protocol architecture;
- Based on the widely used NS3/ndnSIM simulation framework, designs and implements a NDMANET simulator to effectively evaluate the performance of different routing algorithms;
- Based on the derived simulator, comparative simulation experiments for the default flooding routing algorithm, the legacy shortest path routing algorithm and the proposed SPPB routing algorithm are conducted.

The remaining part of this paper goes as follows. Section 2 investigates the state-of-the-art on NDMANET routing algorithms. Section 3 briefly introduces NDMANET, and two legacy routing algorithms. A novel shortest path plus backup routing algorithm for NDMANET is derived in Sect. 4. In Sect. 5, based on NS3/ndnSIM framework, a dedicated simulator is developed for evaluating NDMANET performance, and comprehensive simulation experiments are carried out, and the comparative results are carefully analyzed and discussed. Finally, Sect. 6 concludes the paper and provides the outlook.

2 Related Work

Typically, NDMANET's routing algorithms can be categorized into three types, namely reactive, active and opportunistic classes. **Reactive routing algorithms** do not generate the routes in advance, while only calculate them when it is needed, which can effectively reduce network bandwidth consumption and computing overheads caused by flooding interest packets. Zhang et al. [3] proposed a forwarding strategy based on neighboring nodes perception. Each node maintains the information of its adjacent nodes and their request information. Rehman et al. [4] proposed an on-demand multi-path caching and forwarding strategy based on location awareness to solve the problems of packet conflict, excessive bandwidth consumption, and packet redundancy in NDMANET. Rehman et al. [5] proposed a robust and effective multi-path interest forwarding strategy to reduce NDMANET's unnecessary interest packets broadcasting by using controlled flooding mechanism. Cui et al. [6] elaborates an algorithm to balance network overhead and success rate of content retrieval.

To establish a route to deliver content, **active routing algorithms** require each node to actively broadcast its location information or the content name prefix. In order to achieve mobility robustness in the high speed situation of ICN-MANET, HBFR [7] utilizes completely distributed BF aggregation without cluster heads being in charge of. The CSAR algorithm presented by JianKuang in [8] considers that each content has its special content-scent and can be found by tracing the scent it spreads over the ICN. In [9] presents a lightweight name-based content retrieving algorithm for a multi-hop wireless CCN based on a three-tier strategy that consists of a periodic forwarder information updating, an eligible forwarder selection, and a reliable CCN message broadcasting strategy methods.

Content discovery exists in both active and reactive routing processes. To eliminate the overheads of flooding and the routing failure caused by frequent topology changes, a passive discovery mechanism called **opportunistic routing** was proposed. Yu et al. [10] derived the LER algorithm, which uses the encounters between nodes to broadcast interest packets. Anastasiades et al. [11, 12] proposed an agent-assisted content retrieval scheme, which takes one-hop neighboring nodes as potential agents to retrieve contents. Lu et al. [13] developed STCR, an opportunistic content routing scheme based on social relations. Each node maintains a social relationship table, containing relations of nodes. SACR [14] algorithm uses a group of mobile nodes to distribute random groups periodically, and calculates the centrality of nodes.

To summarize, active routing algorithms perform better for small scale MANETs with relatively stable topology. Opportunistic routing algorithms may lead to considerable transmission delays, and are suitable for applications insensitive of latency. Reactive routing algorithms are deficient in highly dynamic MANET scenarios. This paper focuses on NDMANET's shortest path routing algorithms. As a typical example, listen first broadcast later (LFBL) [15, 16] algorithm introduces the distance table (DT), which records the distance from current node to consumer and to producer, so as to determine the appropriate forwarder. Similarly, CHANET in [17, 18] introduced the content source tables CPT and PRT in each node, which only forwards the packet when it is closer to the consumer than the preferred provider. CCVN [19, 20] has enhanced CHANET for VANET scenarios. However, the aforementioned algorithms are not improved for highly dynamic MANETs.

3 NDMANET and Its Routing Algorithms

3.1 IP Based MANET

At present, the network layer of MANET usually adopts Internet protocol (IP), which has already achieved great success in fixed networks while may not be an efficient L3 protocol for highly dynamic MANETs. The reasons lie in the following three aspects:

- IP based MANET is host-centric network (HCN). IP address is not only a node identifier, but also a node locator, used to indicate its geographic position. Once a MANET node moves out of the network boundary, it cannot be located, and its path established with other nodes on IP will also disconnect;
- To keep the binding relationship with others, networking nodes of highly dynamic MANET update their location information to each other in real time, and this will generate a large amount of control information and may increase bandwidth consumption, thus to further degrade network performance;
- IP based MANET uses sender driven mode, which may not effectively utilize all the intermediate nodes to forward information, thus cannot build the optimal routing path on the basis of global information.

3.2 NDN Based MANET

NDN based MANET (NDMANET) can much better support node mobility and dynamic topology. The major reasons are:

- NDMANET belongs to ICN, which utilizes the content name to identify each packet. The networking nodes can produce and consume content based on its name. The content position is not required, neither the node address;
- The receiver driven mode based NDMANET needs not to establish connections between communicating nodes, so it will not be influenced by the frequent interruptions of wireless links and dynamic changes of network topology;
- NDMANET separates content from location, and reuses the historical received packets stored in the intermediate nodes through network caching mechanism, thus can further accelerating transmission rate and reducing bandwidth consumption.

Compared with IP protocol, NDN technology can provide better networking performance for applications and users in highly dynamic MANET scenarios.

3.3 NDMANET Routing Algorithms

Currently, most NDMANET routing algorithms are originated from NDN's practice in fixed networks, and are not optimized for MANET. They are:

- **Flood Routing Algorithm.** A consumer node broadcasts an interest packet, and all its neighboring nodes can receive it. The receivers will forward the packet again via broadcasting until the interest packet encounters a node storing the requested data packet. Then, this node will broadcast the data packet and transmit it back to the consumer nodes along the reverse propagation path of the interest packet;
- **LFBL Algorithm.** A consumer broadcast an interest packet. Each intermediate node maintains its distance to consumer as *CD*. Producer receives the interest packet and can obtain its distance to consumer, then inserts the distance value into the requested data packet. When the intermediate nodes receive the data packet, it extracts the distance value *PD* to producer, and the shortest distance *MD* between producer and consumer. When an intermediate node receives a packet, it will evaluate the inequality:

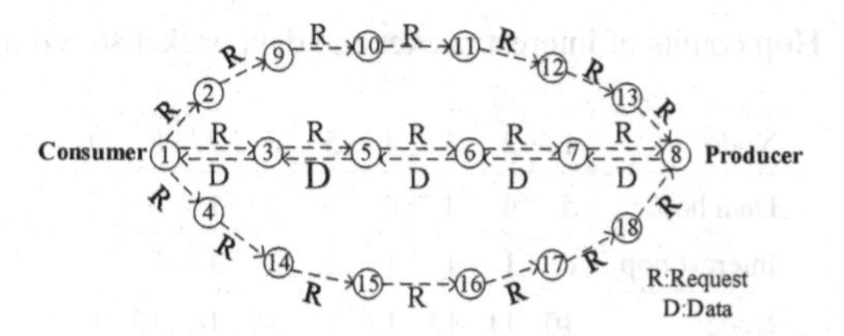

Fig. 1. Schematic of the shortest path lookup and build process

$$CD + PD \leq MD.$$

If the inequality is true, the packet will be forwarded according to the corresponding record in PIT table. Otherwise, the packet will not be forwarded.

4 Innovative Shortest Path Plus Backup Routing Algorithm

4.1 Design Principle

This paper elaborates a shortest path plus backup (SPPB) routing algorithm which enhances the legacy SPR algorithm. The basic idea is that, all the neighboring nodes along the shortest path determined by the original SPR algorithm will store the received interest packet in PIT, so as to create redundant routes for sending back the requested data packet, alleviating the shortest path interruption problems caused by frequent topology changes in highly dynamic MANET scenarios. Two major steps are included.

Identify the Shortest Transmission Path. First, a consumer in NDMANET broadcasts an interest packet for path seeking. When the producer receives the packet, the hop count of this packet is used as the distance indicator of this path. Then, the producer generates the requested data packet and sends it back to the consumer. Figure 1 demonstrates the above procedure. The consumer ① floods the first interest packet to the entire network. The producer ⑧ returns the requested data packet. When an intermediate node receives one data packet, it will record the hop count to local repository. For example, the number of intermediate nodes between ⑪ and the producer ⑧ is 2. When the data packet is sent from the producer ⑧ to the consumer ①, the consumer can finally obtain three different distance hop values, which are 5, 7, and 7, respectively. The consumer can then utilize the path with the minimum hop count as the shortest path. The hop information of the above three paths are shown in Table 1. At a certain moment, the hop counts summation of the interest packet and the data packet is a fixed value on the same data transmission path. The smaller the value, the shorter the distance between the consumer and the producer.

Constructing Backup Paths Using Shortest Path's Neighboring Nodes. As is shown in Fig. 2 (a), nodes ② and ④ as the neighboring nodes of the consumer ① are not on the shortest path. The SPPB algorithm can set the two nodes as backup intermediate nodes of node ③. Similarly, node ⑪ is selected as the backup node of node ⑦. In Fig. 2 (b), it is assumed that, node ③ moves outside the communication range of node ⑤, and node ⑦ also leaves node ⑧. While nodes ② and ⑪ can be used to build a new path

Table 1. Hop counts of interest packet and data packet stored in a node

Node	1	2	3	4	5	6	7	8	9
Data hop	5	6	4	6	3	2	1	0	5
Interest hop	0	1	1	1	2	3	4	5	2
Node	10	11	12	13	14	15	16	17	18
Data hop	4	3	2	1	5	4	3	2	1
Interest hop	3	4	5	6	2	3	4	5	6

between the consumer and the producer. Therefore, this approach can avoid data packet loss caused by routing interruption and retransmissions of interest packets.

4.2 Implementation Details

The implementation of the proposed SPPB routing algorithm includes two parts.

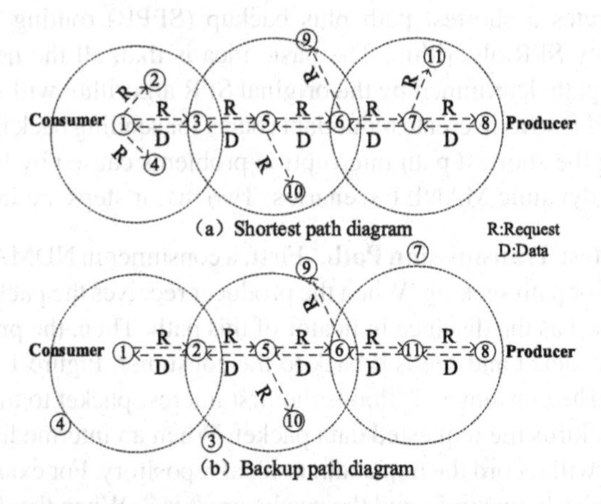

Fig. 2. Schematic of backup node selection and path construction

Interest Packet Processing Flow

- **Current node is a consumer.** When the consumer node generates the first interest packet, the distance hop information carried by the interest packet is set to 0, and the interest packet needs to be flooded to the entire network to dynamically create a shortest routing path between the consumer and the producer. After that, when the intermediate node in the network receives the subsequent interest packets sent by the consumer, it will forward it according to the generated shortest routing path. Besides, when a consumer sends an interest packet and does not receive the requested data packet within a certain period of time, the consumer can consider that the shortest routing path generated has expired, and the consumer will flood an interest packet again to rebuild another shortest route.

Table 2. Interest packet process flow

```
Interest packet processing flow
1: if current node is consumer then
2:     set interestHC 0
3:     if sending interest packet for the first time or data timeout then
4:         broadcast interest
5:     else
6:         set buildSP flag 0
7: else
8:     record interestHC
9:     interest.interestHC=interest.interestHC+1
10:    if buildSP==0 then
11:        if interestHC+dataHC≤expectedHC then
12:            forward interest
13:        else
14:            insert interest into PIT
15:    else if ds.nonce!=interest.nonce then
16:        forward interest
17:        record the nonce of interest
```

- **Current node is an intermediate node.** When an intermediate node receives an interest packet, it first records the hop count of the packet (*interestHC*). If the packet is a broadcast packet (if the *buildSP* field in this packet is 1, the packet is a broadcast packet, this field is used to identify whether it is necessary to rebuild the shortest path) and the current node has not previously forwarded the packet, then it forwards the packet. If the packet is not a broadcast packet, it will evaluate the inequality

$$interestHC + dataHC \leq expectedHC$$

where *dataHC* is the hop count of the data packet and *expectedHC* is the expected hop count of the producer to the consumer. If it is true, the interest packet is forwarded; otherwise, the packet is inserted into the PIT table of the node, and the intermediate node becomes a backup node. The pseudo code is shown in Table 2.

Data Packet Processing Flow

- **Current node is a producer.** The producer node records the number of packets sent so far as *Seq*. If the producer receives a broadcasted interest packet, its hop count information will be used as expected distance between the consumer and the producer. The producer sets the hop value of the data packet to 0, and the flag *buildSP* the same as the flag of the interest packet to ensure that subsequent processing can identify the broadcast packet. At the same time, *Seq* is stored in the current data packet to ensure that the record in the intermediate node is only updated by the newly generated data packet.

- **Current node is an intermediate node.** The intermediate node receives a data packet. If the data packet is not broadcasted or the *Seq* of the Data packet is smaller than the *Seq* in local record. The data packet is forwarded directly; otherwise, the *expectedHC* and *Seq* of the data packet need to be recorded. If the *Seq* of the data packet is not equal to the *Seq* recorded by the node or the *dataHC* recorded locally is greater than the *dataHC* of the data packet, the *dataHC* of the data packet is re-recorded in the local record. The pseudo code is given in Table 3.

Table 3. Data packet processing flow

```
Data packet processing flow
1: if current node is producer then
2:     data.dataHC=0
3:     if interest.buildSP==1 then
4:         data.expectedHC=interestHC-1
5:     record sequence number of data
6:     sequence number of data increase 1
7:     data.buildSP=interest.buildSP
8: else if data.buildSP==1 and newer then
9:     if recorded dataHC > data.dataHC then
10:        record dataHC
11:    if recorded sequence number of data ≠ data.seq then
12:        record dataHC
13:    record expectedHC
14:    record sequence number of data
15:    data.dataHC increase 1
16: if corresponding entry in PIT then
17:    forward data
```

5 Performance Evaluation

Computer simulation approach is adopted to evaluate the performance of the proposed SPPB algorithm. The performance metrics include: *network bandwidth consumption* (NBC), *average request response delay* (ARRD) and *successful request ratio* (SRR):

- **Network Bandwidth Consumption.** It refers to the amount of content sent or forwarded by all the nodes in NDMANET, during the simulation period, including all the interest and data packets;
- **Average Request Response Delay.** It is defined as the average time delay from the consumer's sending of an interest packet to the final reception of the requested data packet during the simulation period. Assuming that the number of successful packet requests is N and the delay of the i-th packet request is $delay_i$, the ARRD is:

$$ARRD = \frac{\sum_1^N delay_i}{N}$$

- **Successful Request Ratio.** It is the ratio of the number of data packets received to the number of interest packets sent by the consumer during the simulation period, which indicates that, to which extent, the network can satisfy the customer requests:

$$SRR = \frac{DataCount_{received}}{InterestCount_{sent}}$$

5.1 Developing Simulation Platform

A software simulation platform based on NS3 and ndnSIM is developed. The detailed configurations of the underlying hardware and software are shown in Table 4.

Table 4. Hardware and software configurations of the simulation platform

	Hardware configuration
CPU	AMD Ryzen 7 1700X
RAM	64 GB DDR4 2400MHz
Disk	500 GB SSD
	Software configuration
Operating system	Ubuntu 16.04
NS3 version	3.25
ndnSIM version	2.6

5.2 Simulation Assumptions and Settings

Medially to highly dynamic MANET scenarios are considered and NDMANET nodes can move randomly at a certain speed. In order to make the experimental results comparable, the same random number seed and node mobility models are used in the simulation experiments for three different routing algorithms. The parameter settings for the simulation experiments are summarized in Table 5.

5.3 Simulation Results

First, the experimental results of *network bandwidth consumption* are shown in Fig. 3. The default flood routing algorithm broadcasts interest packets and data packets, while the SPPB and LFBL algorithms can suppress the broadcasts. Compared with the LFBL algorithm, the proposed SPPB algorithm constructs a backup path besides shortest path. Nodes on this backup path may broadcast data packets, which causes the SPPB algorithm to consume slightly more bandwidth than the LFBL algorithm.

Table 5. Scenario parameter settings

Parameter	Value
Protocol	802.11a
Max transfer rate	24 Mbps
Communication radius of each node	75 m
Size of scenario	600 m * 600 m
Amount of node	60–140 (step value of 20)
Speed of node	10 m/s
The frequency of sending Interest packet	10 Interest/s
Simulation time	300 s

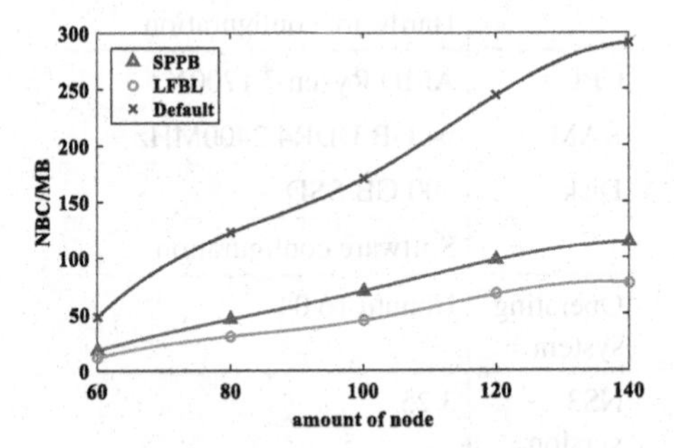

Fig. 3. The experimental results of network bandwidth consumption

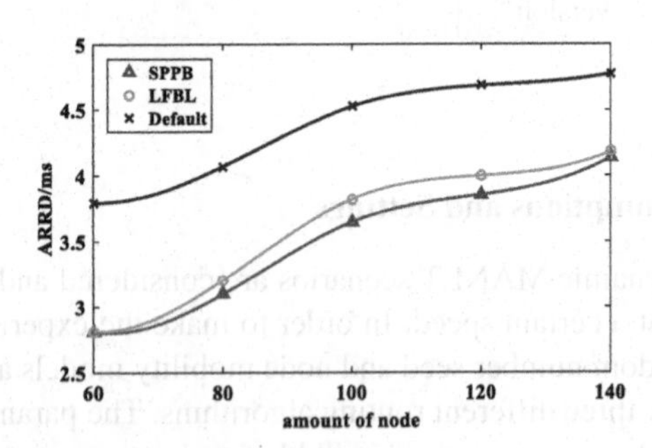

Fig. 4. The experimental results of average request response delay

Secondly, the results of the *average request response delay* are shown in Fig. 4. The default flooding algorithm employs more intermediate nodes for transmitting interest packets and data packets, so the average request response delay is the largest. In the LFBL algorithm, Data packets are transmitted back along a predetermined shortest path, which can effectively reduce the latency. The derived SPPB algorithm considers the network topology dynamicity and provides a backup path option for packet transmitted. Therefore, it can further reduce the request response delay.

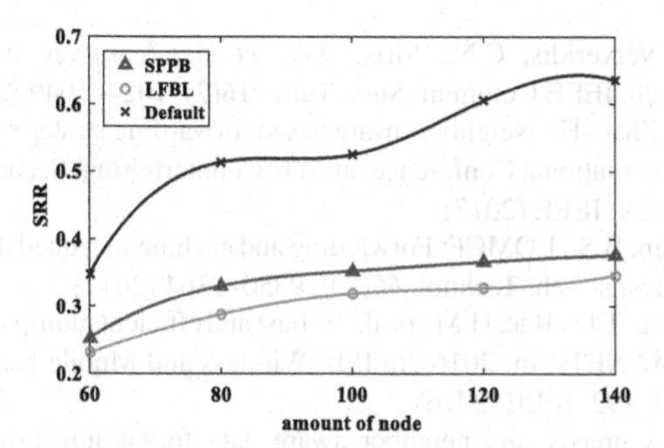

Fig. 5. The experimental results of successful request ratio

Finally, Fig. 5 illustrates the simulation results of the *successful request ratio*. The default flooding algorithm broadcast the packets, so it can lead to the highest successful request ratio. The cost is the largest network bandwidth consumption. The elaborated SPPB algorithm has a higher successful request ratio than the LFBL algorithm. The reason is that the SPPB algorithm builds the backup paths besides the shortest path, which reduces the failure probability of data packet transmitted, thus the successful request ratio gets a certain degree of promotion.

6 Conclusions

NDN technology can reduce negative influence on MANET performance due to unreliable links and dynamic changed topologies. This paper innovatively elaborates a shortest path plus backup (SPPB) routing algorithm to improve NDN's legacy shortest path routing algorithm. The proposed algorithm stores the interest packets generated by the consumer in PITs of the shortest path's neighboring nodes, aiming at building the backup path for transmitting back the requested data packets, so as to mitigate the shortest path disruption problem caused by MANET topology dynamicity, and to lower data packet drop probability. This paper also develops a simulation platform based on NS3/ndnSIM, and the experimental results indicate that the enhanced algorithm consumes limited bandwidth, with further decreasing the average request response delay and improves the request response ratio.

The topology change of MANET also affects the interest packets' transmissions by the shortest path routing algorithm. Therefore, as the next step of this work, further improving the routing performance for the interest packets is planned.

Acknowledgement. This work is supported by the National Natural Science Foundation of China (Grant No. 61402085), the Science and Technology on Communication Networks Laboratory (Grant No. XX17641X011-03), and the 54th Research Institute of CETC.

References

1. Zhang, L., Estrin, D., Burke, J., et al.: Named data networking (ndn) project. Relatório Técnico NDN-0001, Xerox Palo Alto Research Center-PARC (2010)

2. Xylomenos, G., Ververidis, C.N., Siris, V.A., et al.: A survey of information-centric networking research. IEEE Commun. Surv. Tutor. **16**(2), 1024–1049 (2013)
3. Zhang, X., Li, R., Zhao, H.: Neighbor-aware based forwarding strategy in NDN-MANET. In: 2017 11th IEEE International Conference on Anti-Counterfeiting, Security, and Identification (ASID), pp. 125–129. IEEE (2017)
4. Rehman, R.A., Kim, B.S.: LOMCF: Forwarding and caching in named data networking based MANETs. IEEE Trans. Veh. Technol. **66**(10), 9350–9364 (2017)
5. Rehman, R.A., Hieu, T.D., Bae, H.M., et al.: Robust and efficient multipath interest forwarding for NDN-based MANETs. In: 2016 9th IFIP Wireless and Mobile Networking Conference (WMNC), pp. 187–192. IEEE (2016)
6. Cui, B., Li, R.: A greedy and neighbor aware data forwarding protocol in named data MANETs. In: Ninth International Conference on Ubiquitous & Future Networks, pp. 934–939. IEEE (2017)
7. Yu, Y.-T., Gerla, M., Sanadidi, M.: Scalable VANET content routing using hierarchical bloom filters. Wirel. Commun. Mob. Comput. **15**(6), 1001–1014 (2015)
8. Kuang, J., Yu, S.Z.: CSAR: a content-scent based architecture for information-centric mobile ad hoc networks. Comput. Commun. **71**, 84–96 (2015)
9. Kim, D., Ko, Y.-B.: A novel message broadcasting strategy for reliable content retrieval in multi-hop wireless content centric networks. In: Proceedings of the 9th International Conference on Ubiquitous Information Management and Communication, p. 108. ACM (2015)
10. Yu, Y.T., Li, Y., Ma, X., et al.: Scalable opportunistic vanet content routing with encounter information. In: 2013 21st IEEE International Conference on Network Protocols (ICNP), pp. 1–6. IEEE (2013)
11. Anastasiades, C., El Alami, W.E.M., Braun, T.: Agent-based content retrieval for opportunistic content-centric networks. In: Mellouk, A., Fowler, S., Hoceini, S., Daachi, B. (eds.) WWIC 2014. LNCS, vol. 8458, pp. 175–188. Springer, Cham (2014). https://doi.org/10.1007/978-3-319-13174-0_14
12. Anastasiades, C., Braun, T., Siris, V.A.: Information-centric networking in mobile and opportunistic networks. In: Ganchev, I., Curado, M., Kassler, A. (eds.) Wireless Networking for Moving Objects. Lecture Notes in Computer Science, vol. 8611. Springer, Cham (2014). https://doi.org/10.1007/978-3-319-10834-6_2
13. Lu, Y., Li, X., Yu, Y.T., et al.: Information-centric delay-tolerant mobile ad-hoc networks. In: Workshop on Infocom, pp. 428–433. IEEE (2014)
14. Le, T., Kalantarian, H., Gerla, M.: Socially-aware content retrieval using random walks in disruption tolerant networks. In: 2015 IEEE 16th International Symposium on "A World of Wireless, Mobile and Multimedia Networks" (WoWMoM), pp. 1–6. IEEE (2015)
15. Meisel, M., Pappas, V., Zhang, L.: Ad hoc networking via named data. In: Proceedings of the Fifth ACM International Workshop on Mobility in the Evolving Internet Architecture, pp. 3–8 (2010)
16. Meisel, M., Pappas, V., Zhang, L.: Listen first, broadcast later: topology-agnostic forwarding under high dynamics. In: Annual Conference of International Technology Alliance in Network and Information Science, p. 8. (2010)
17. Molinaro, A.: CHANET: a content-centric architecture for IEEE 802.11 MANETs. In: IEEE Network of the Future (NoF) 2011, pp. 122–127. IEEE (2011)
18. Amadeo, M., Campolo, C., Molinaro, A.: CRoWN: content-centric networking in vehicular ad hoc networks. IEEE Commun. Lett. **16**(9), 1380–1383 (2012)
19. Campolo, C., Amadeo, M., Molinaro, A.: Content-centric networking: is that a solution for upcoming vehicular networks. In: ACM VANET, pp. 99–102. ACM (2012)
20. Amadeo, M., Campolo, C., Molinaro, A.: Enhancing content-centric networking for vehicular environments. Comput. Netw. **57**(16), 3222–3234 (2013)

Modeling and Simulation of Computation Offloading at LEO Satellite Constellation Network Edge

Junyu Lai(✉), Huidong Tan, Meilin He, Ying Qu, and Lei Zhong

School of Aeronautics and Astronautics, University of Electronic Science and Technologies of China, Chengdu, China
laijy@uestc.edu.cn

Abstract. Similar to terrestrial networks where edge computing facilities have already been introduced to decrease the user request response delay and to reduce the backhaul bandwidth consumption, low earth orbit (LEO) satellite constellation networks can also be benefited by adopting edge computing technologies. The computation tasks generated by ground users can be offloaded to their accessing LEO satellites to enhance network QoS and user QoE. This paper focuses on modeling and simulating computation offloading at LEO constellation network edge. A one-dimensional networking model for edge computing enabled LEO constellation networks is derived, and on that basis, a Monte Carlo simulator is developed from scratch to evaluate system performance. As a case study, three different computation offloading schemes are elaborated and implemented on the simulator. Comparative evaluation experiments have been conducted and the results indicate that, in resource restricted scenarios, allowing computation offloading to the neighbors of the access satellites can considerably reduce the request blocking probability with only slightly increasing the average request response delay.

Keywords: Satellite network · Edge computing · Computation offloading · Simulation · Performance evaluation

1 Introduction

In the past several years, numerous projects on constructing the next-generation LEO satellite constellation networks such as *Starlink* and *OneWeb* were proposed, and their satellites are currently launching to the space at an astonishingly fast speed which had never happened before. The major purpose of constructing these LEO satellite constellation networks are to complement the traditional terrestrial networks, so as to provide wireless broadband networking services directly from space for the ground users no matter where they are in the globe. On the other hand, Edge computing refers to move compute resource conventionally located in the backend cloud towards the users at the frontend. It can be realized in any nodes such as gateways, base stations, or even user terminals themselves aiming at reducing request response latency, enhancing user quality

H. Song and D. Jiang (Eds.): SIMUtools 2020, LNICST 370, pp. 451–463, 2021.
https://doi.org/10.1007/978-3-030-72795-6_37

of experience (QoE) and decreasing the backhaul bandwidth consumption. Integrating edge computing ability in a terrestrial network has already attracted tremendous research efforts. It is obvious that edge computing in LEO satellite constellation networks can gain even larger profits on network performance improvements considering the networking scales. However, to our best knowledge, few research efforts are now existing on modeling and simulating edge computing offloading in the upcoming LEO constellation networks. This paper explores this research topic, and the major contribution includes:

- Designs a simplified one-dimensional networking model for the edge computing ability enabled LEO satellite constellation network;
- Elaborates an innovative *Monte Carlo* simulator from scratch to evaluate the performance of diverse edge computing offload schemes in the LEO satellite constellation networks;
- Proposes three different edge computing offload schemes and conducts comparative simulation experiments to evaluate them as a comprehensive case study.

The remaining part of this paper goes as follows. Section 2 introduces the state-of-the-art on applying edge computing in satellite networks. Section 3 derives a system model on network architecture, user behavior, as well as network applications in LEO satellite constellation networks. An innovative Monte Carlo simulator, developed from scratch to evaluate the performance of computing offloading at the LEO satellite constellation network edge, is presented in Sect. 4, followed by Sect. 5 where three different edge computation offloading schemes are proposed and evaluated in the above simulator. Finally, Sect. 6 concludes the paper and provides the outlook.

2 Related Work

In legacy terrestrial mobile networking domain, real time applications are facing non-neglectful and large request response delays between the frontend mobile users and the backend cloud. In [2], a new paradigm named mobile edge computing (MEC) has been proposed to facilitate its influence. It refers that the edge of introducing computing and storage resources to mobile networks provide users with ultra-low latency and high-bandwidth network services. As one of the key technologies of MEC, computing offloading means that the terminal device will offload part or all of the computing tasks to edge to eliminate the shortcomings of mobile devices in terms of resource storage, computing performance, and energy efficiency. In order to reduce transmission delay, Liu et al. in [3], establish a mathematical model and design the Lyapunov optimized dynamic offload algorithm. Focus on the issue of a long response time of task offloading in multi-cloudlet, in [4], a weighted self-adaptive inertia weight particle swarm optimization algorithm based on multi-cloudlet collaboration is proposed. In [5], Long et al. aimed to minimize the delay and monetary cost on mobile devices under the limited computation resource and communication resource and design a multi-objective computation offloading and resource allocation game algorithm to achieve the optimal computational offload strategy. In [6], an energy optimization model was established and an artificial fish swarm algorithm was used to achieve the goal of reducing energy consumption.

The work in [7] aimed at the trade-off between energy consumption and time delay. It proposes a trade-off model of energy and time delay and an optimization algorithm based on the Lyapunov strategy. Considering delay and energy consumption of mobile devices jointly in a wireless access network, Hao in [8], propose a delay and energy consumption efficient offloading algorithm based on an alternating direction method of multipliers.

Edge computing has also been introduced to satellite networks to reduce propagation delays caused by long distances between satellites and ground terminals. In [9], Bradley *et al.* designed an Orbital Edge Computing System and using Nanosatellite constellations to reduce the transmission delay. Cheng *et al.* in [10], proposed a space-air-ground integrated network architecture, which utilizes the satellite as an access to connect edge servers with the terrestrial network to provide seamless and flexible network coverage and services to large areas. In [11], Cao *et al.* design the space-based cloud-fog satellite network slices, reducing the average delay of services in different slices by cloud-fog satellite which equipped edge computing servers. In [12], it represents a double edge computation offloading algorithm to optimize energy consumption and reduce latency, utilizing an access satellite to combine a satellite terrestrial network and a satellite-terrestrial transport to offer computing offloading services for the region.

The above works rarely consider the upcoming LEO satellite broadband networks, neither the impact caused by the motion of LEO satellites on edge computing. Therefore, it is urgent and meaningful to investigate edge computing in LEO satellite network by mathematical modeling and to build a dedicated simulation platform.

3 System Modeling

3.1 Modeling Network Architecture

The number of satellite orbits in the targeting LEO satellite constellation network is denoted as N_{orbit}. For simplicity, only one orbit is considered in this network model. There are N_s LEO satellites evenly distributed in the orbit. The height of these satellites is l_{gs} kilometers from the Earth surface, and the moving speed can be calculated by the Kepler's law:

$$V_s = 2\pi \frac{R'^{\frac{3}{2}}}{T' \times (R' + l_{gs})^{\frac{1}{2}}} \tag{1}$$

Where, R' is the radius of the Earth and T' is its rotation period.

A ground user can directly communicate with those LEO satellites flying above it, and there is only one satellite serving as the access satellite for each user at a random point of time. Due to different landforms of the Earth, users are unevenly distributed on the ground. This paper divides the Earth surface into M different regions, each with non-identical population distribution. For instance, the population in the mountain regions is much smaller than that in the plain areas, but is larger than that in the ocean regions. Assume that in the ith region, the number of users is denoted as P_i, which conforms to a uniform distribution with the average value U_i, its probability density function (PDF) is:

$$f(x) = \begin{cases} \frac{1}{b-a} & for\, a \le x \le b \\ 0 & elsewhere \end{cases} \quad (2)$$

Where, a and b are the boundaries of the area.

To summarize, the above network architecture model is shown in Fig. 1. The ith satellite S_i moves at speed V_s. The distance between the S_i and ground terminal is l_{gs} and l_{ss} is the length of inter-satellites link (ISL). Moreover, the earth ground is evenly divided into M areas.

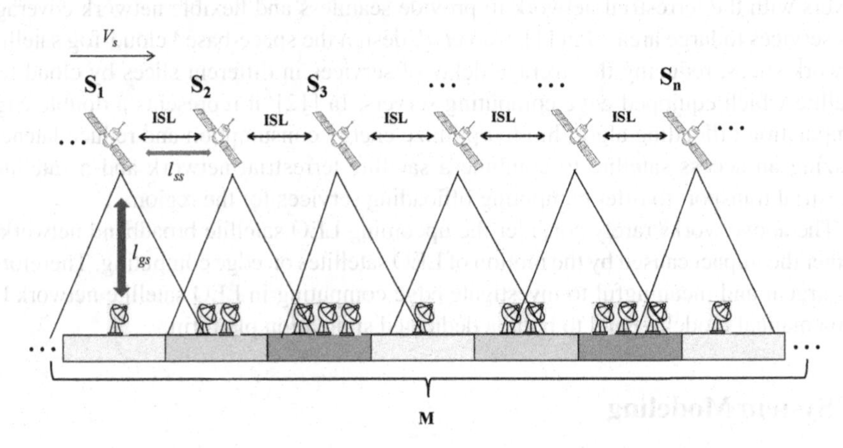

Fig. 1. Networking architecture model

3.2 Modeling User Behavior

Assume all the ground users are independent and are periodically sending computation task requests to their access LEO satellites. The computation task requesting frequencies of all the users are assumed to be the same, and the time interval between a user's two consecutive task requests follows a negative exponential distribution, with the probability density function:

$$f(x;\lambda) = \begin{cases} \lambda e^{-\lambda x}, & x \ge 0 \\ 0, & x < 0 \end{cases} \quad (3)$$

Where, λ is the average number of task request occurrences per unit time. It is worth noting that ground user's moving speed is much smaller than the speed of LEO satellites, so the users' mobility can be ignored.

3.3 Modeling Network Applications

Task Model. Due to the power and space limitations, as well as the critical working environments, computing resources installed on the LEO satellite are limited and cannot be comparable with the edge computing facilities on the ground. The computation tasks offloaded to the LEO satellite are served by micro services running on light-weighted virtual instances assigned by satellite onboard computation resource management system. Therefore, the computation tasks offloaded to the LEO satellite are modeled as atomic operations, and can be presented by $A(S, T_d)$, where S is the task computation amount and T_d is the task completion deadline [13]. Assume all the computation tasks are independent and the task computation volume conforms to normal distributed:

$$f(x) = \frac{1}{\sigma\sqrt{2\pi}} e^{-\frac{1}{2}\left(\frac{x-\mu}{\sigma}\right)^2} \quad (4)$$

Where, σ^2 is the variance of the task calculation volume and μ is the mean of the task calculation volume. A task $A(S, T_d)$ must be completed within the predefined time deadline T_d, or else the allocated edge computing resource will be released and the task will fail.

Delay Model. The computation task request response delay is denoted as *RRD*, which can be expressed by:

$$RRD = D_T + D_C + D_R \quad (5)$$

RRD includes task uploading delay D_T, task computation delay D_C, and result returning delay D_R. Task uploading delay includes the uplink transmission delay d_{up_t}, uplink propagation delay d_{up_p}, as well as possibly relaying delay $n \times d_r$.

Similarly, result returning delay consists of downlink transmission delay d_{down_t}, down link propagation delay d_{down_p}, and possibly relaying delay $n \times d_r$. The propagation delay is proportional to the distance between the LEO satellite actually running the task and the requesting user. In addition, the data transmission delay is related to the volume of the computation task and the bandwidth of the transmission link [11]. Each relaying satellite forwarding the task or computation result will introduce delay d_r, and if the number of relaying satellites for a transmission is *m*, the total number of relay times will be *n* =2 *m*. The equations to evaluate the task uploading delay D_T and the result returning delay D_R are as follows:

$$D_R = d_{down\_t} + d_{down\_p} + n \times d_r \quad (6)$$

$$D_T = d_{up\_t} + d_{up\_p} + n \times d_r \quad (7)$$

To make the above depiction clearer, the delay model is explained in Fig. 2

4 Designing and Implementing Simulation Platform

4.1 General Assumptions

In order to develop the simulation platform for the LEO satellite constellation network, some general assumptions are made. Firstly, the shape of the orbit is circular, and only

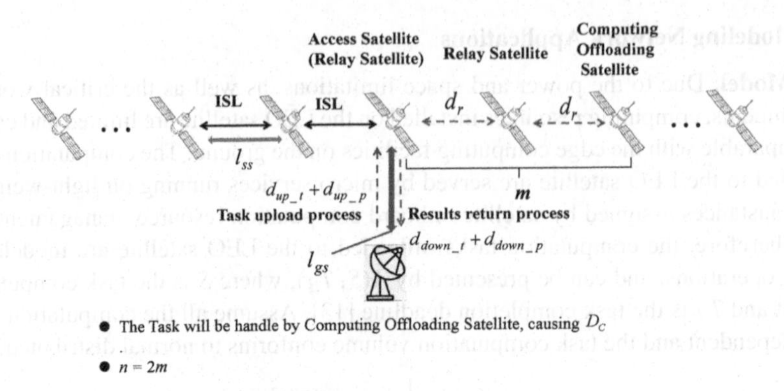

Fig. 2. Application delay model

one orbit is considered. Secondly, the LEO satellites are evenly distributed in the orbit. Thirdly, access satellite handover is seamless, and there is no communication break when during handover. Fourthly, the speed of signal transmission in the satellite network is the speed of light. Fifthly, the user moving speed is zero.

4.2 Designing Principles

The derived *Monte Carlo* simulation platform is based on an event-driven model. There are logically three steps in general:

- **STEP_I.** According to the satellite networking model and the user distribution model, establish the targeted network topology;
- **STEP_II.** Following user requesting model and task model, generate a discrete event trace file (including handover events, user requesting events, etc.) for each single user;
- **STEP_III.** Combine the trace files of all the users into a system-level trace file and dynamically insert new events during the simulation process.

4.3 Implementations

The simulation is divided into three parts: satellite section, user section, user and satellite interacting section. The satellite section mainly initializes the satellite

position and the satellite resource. The user section is to simulate the distribution of users, user parameters, and user event trace files which are the base of the system-level trace file. The satellite-user interaction part is to execute the procedure following the system-level trace file. Firstly, if there are redundant events in the list, it will check the event and policy type, and the communication access situation before occupying computing resources. Then, it will calculate every task end time according to different strategies. Lastly, the end event will be inserted into the system-level trace file followed by returning to step 1. Its flow chart is shown in Fig. 3.

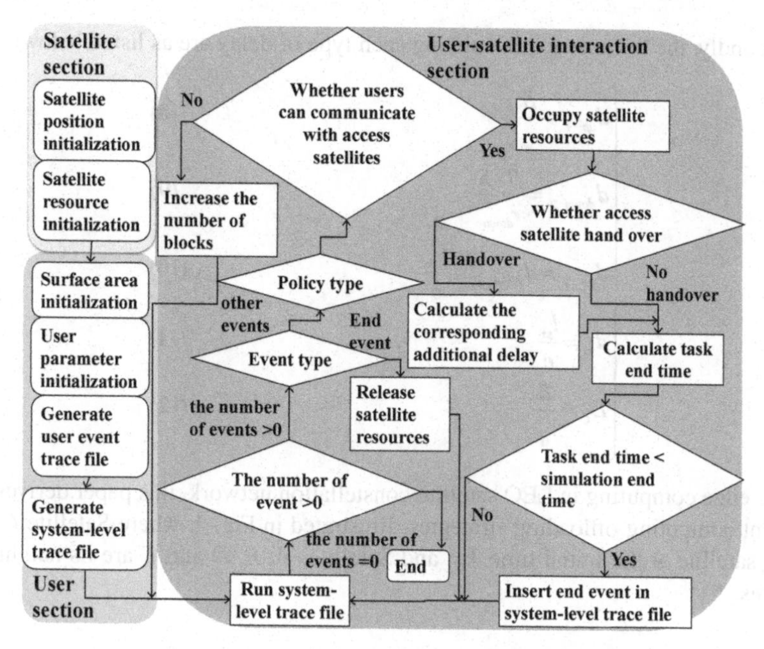

Fig. 3. Simulation platform implementation details

5 Case Study: Comparative Performance Evaluation for Three Different Computation Offloading Schemes

5.1 Three Different Computing Offloading Schemes

Firstly, notations used in this case study are defined in Table 1.

Table 1. Summary of notations.

Notation	Implication	Notation	Implication
d_r	Relay delay due to one satellite pitch	l_{ss}	Inter-satellite link length
l_{gs}	Satellite-to-ground link length	d_{ss}	Propagation delay between two adjacent satellites
S	Task calculation	r_{up}	Task upload rate
r_{down}	Result return rate	r_{cp}	Calculation rate
β	Amount of results/number of tasks	a	Completion of mission uploading when satellite hand over
b	Completeness of result returning during satellite handover	c	Light speed

Secondly, the formula for calculating each type of delay are as listed below:

$$\begin{cases} d_{up\_t} = \dfrac{S}{r_{up}} & (8) \\ d_{down\_t} = \dfrac{\beta \times S}{r_{down}} & (9) \\ d_{up\_p} = d_{down\_p} = \dfrac{l_{gs}}{c} & (10) \\ d_{ss} = \dfrac{l_{ss}}{c} & (11) \\ D_C = \dfrac{S}{r_{cp}} & (12) \end{cases}$$

For edge computing in LEO satellite constellation network, this paper derives three different computing offloading strategies. Illustrated in Fig. 4, where Satellite C is the access satellite at the initial time T_0, and Satellites A, B, D and E are its neighboring satellites.

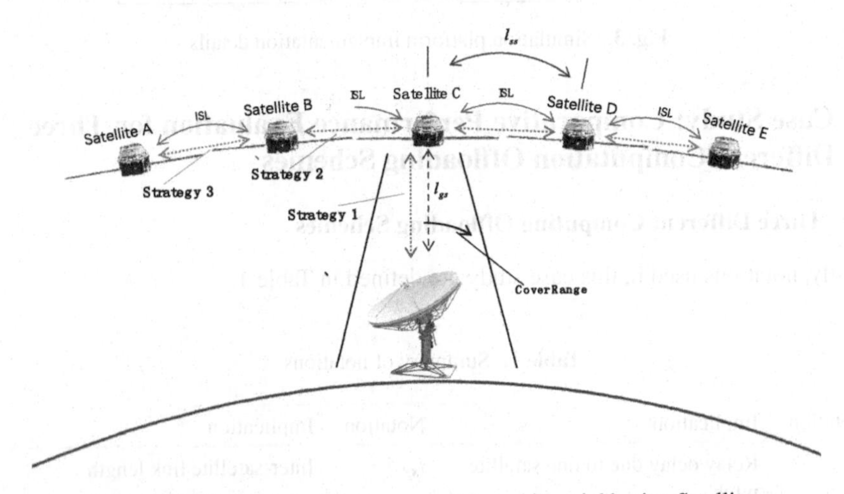

Fig. 4. The Distribution of access satellite and its neighboring Satellites

Strategy 1. Satellite C is adopted as the only satellite for conducting the offloaded computing tasks at network edge. If the computing resources of C are used up, the newly arrived user task request will be rejected. Since the LEO satellite is always in motion, the coverage time of a certain location on the earth is limited, so that the user's access satellite may have changed during the task processing duration (defined as the time interval starting from the instant when a user generates the computation task to the moment when the user receives the computation result). This process is called handover. In order to accurately evaluate the *RRD*, the time instant when the handover happens are carefully analyzed. Table 2 summarizes all the possible situations.

Table 2. Delay calculation for Strategy 1

Strategy_1		
No. handover		$D_T = d_{up\_t} + d_{up\_p},\ D_R = d_{down\_p} + d_{down\_t}$
Handover happens at	Task upload stage	$D_T = d_{up\_t} + d_{up\_p} + d_r + d_{ss}$, $D_R = d_{down\_t} + d_{down\_p} + d_r + d_{ss}$
	Task calculation stage	$D_T = d_{up\_t} + d_{up\_p},\ D_R = d_{down\_t} + d_{down\_p} + d_r + d_{ss}$
	Task result return stage	$D_T = d_{up\_t} + d_{up\_p},\ D_R = d_{down\_t} + d_{down\_p} + d_r + d_{ss}$

Strategy 2. Based on Strategy 1, if Satellite C do not have sufficient compute resource, it can further estimate whether the computing resources of satellites B and D are available. If one of the above satellites has compute resources, it will offload the computation task to that satellite, and the access satellite becomes a relay satellite to forward the content. Then, if the resource of the two nearby satellite also ran out the compute resources, the user's request is rejected, increasing the blocking counter. Handover occurs in different time periods of task processing duration will have different effects on the *RRD*. Table 3 summarizes all the possible situations.

Table 3. Delay calculation for Strategy 2

Strategy_2	No. handover		
	$D_T = d_{up\_t} + d_{up\_p} + d_r + d_{ss},\ D_R = d_{down\_t} = d_{down\_p} + d_r + d_{ss}$		
	Handover		
		Getting further	Getting closer
	Task upload stage	$D_T = d_{up\_t} + d_{up\_p} + 2(d_r + d_{ss})$ $D_R = d_{down\_t} + d_{down\_p} + 2(d_r + d_{ss})$	$D_T = \max\begin{cases} a \times d_{up\_t} + d_{up\_p} + d_r + d_{ss} \\ d_{up\_t} + d_{up\_p} \end{cases}$ $D_R = d_{down\_t} + d_{up\_p}$
	Task calculation stage	$D_T = d_{up\_t} + d_{up\_p} + d_r + d_{ss}$ $D_R = d_{down\_t} + d_{down\_p} + 2(d_r + d_{ss})$	$D_T = d_{up\_t} + d_{up\_p} + d_r + d_{ss}$ $D_R = d_{down\_t} + d_{down\_p}$
	Task results return stage	$D_T = d_{up\_t} + d_{up\_p} + d_r + d_{ss}$ $D_R = d_{down\_t} + d_{down\_p} + 2(d_r + d_{ss})$	$D_T = d_{up\_t} + d_{up\_p} + d_r + d_{ss}$ $D_R = \max\begin{cases} b \times d_{down\_t} + d_{down\_p} + d_r + d_{ss} \\ d_{down\_t} + d_{down\_p} \end{cases}$

Strategy 3. Based on strategy 2, if the computing resources of satellites B and D are running out, then the strategy will check whether there are available computing resources on satellites A and E. If one of the satellites has sufficient resources, the strategy will offload the computation task to that satellite with satellite B or D as its relay node. If satellites A and E have not enough compute resource, this task request is blocked. It is noted that, strategies 2 and 3 prefer to choose the satellites flying close to the user than the one fly away from the user. The impact of the handover event on *RRD* in Strategy 3 is shown in Table 4.

Table 4. Delay calculation for Strategy 3

Strategy 3	No. handover		
	$D_T = d_{up\_t} + d_{up\_p} + 2(d_r + d_{ss})$, $D_R = d_{down\_t} + d_{down\_p} + 2(d_r + d_{ss})$		
	Handover		
		Getting further	Getting closer
	Task upload stage	$D_T = d_{up\_t} + d_{up\_p} + 3(d_r + d_{ss})$ $D_R = d_{down\_t} + d_{down\_p} + 3(d_r + d_{ss})$	$D_T = \max\begin{cases} a \times d_{up\_t} + d_{up\_p} + 2(d_r + d_{ss}) \\ d_{up\_t} + d_{up\_p} \end{cases}$ $D_R = d_{down\_t} + d_{down\_p} + d_r + d_{ss}$
	Task calculation stage	$D_T = d_{up\_t} + d_{up\_p} + 2(d_r + d_{ss})$ $D_R = d_{down\_t} + d_{down\_p} + 3(d_r + d_{ss})$	$D_T = d_{up\_t} + d_{up\_p} + 2(d_r + d_{ss})$ $D_R = d_{down\_t} + d_{down\_p} + d_r + d_{ss}$
	Task results return stage	$D_T = d_{up\_t} + d_{up\_p} + 2(d_r + d_{down\_p})$ $D_R = d_{down\_t} + d_{down\_p} + 3(d_r + d_{down\_p})$	$D_T = d_{up\_t} + d_{up\_p} + 2(d_r + d_{ss})$ $D_R = \max\begin{cases} b \times d_{down\_t} + d_{down\_p} + 2(d_r + d_{ss}) \\ d_{down\_t} + d_{down\_p} + d_r + d_{ss} \end{cases}$

5.2 Performance Metrics

The performance metrics considered in this case study are request blocking probability (RBP) and average request response delay (ARRD). RBP is the ratio of the number of blocked task requests due to resource restriction to the total number of task requests, while ARRD is the average value of the computation task request response delay.

5.3 Simulation Assumptions

The specific assumptions for the simulation experiments are given in Table 5:

Table 5. Simulation assumptions

Parameters	Values	Parameters	Values
The satellite speed	27000 km/h	The distance between adjacent satellites	4481 km
The height of the satellite	780 km	Relay delay	1 s
The number of satellites	11	The inter-satellite transmission delay of the two adjacent satellites	14.9 ms
Rate parameter λ	1	The number of regions divided on the Earth's surface	9

5.4 Experimental Results

The results of the comparative simulation experiments for the aforementioned three computation offload strategies are as given in Fig. 5 and 6. The vertical lines represent the confidence intervals. According to the above results, it is no surprise that Strategies 2 and 3 have a better quality of RBP than Strategy 1, attributed to fuller use of satellite resources. However, as the user population gets larger, the difference between Strategy 2 and Strategy 3 is getting smaller. Because each strategy has the same amount of computing resources. Even if these two strategies are distinct, the RBP of them will become close with a sufficient utilization of computing resources. From Fig. 6, the initial ADDR of the three strategies are close, indicating that under the initial number of users, there are still surplus resources in the satellite, and users always offload to the access satellite. With the increase in the number of users, the access satellite resources gradually decreases, and users access to nearby satellites through Strategy 2 and Strategy 3. Therefore, the ADDR of Strategy 2 and Strategy 3 begin to increase. Although the trends of these strategies are various, the ARRD of Strategy 3 rise most rapidly, which implicates that the number of relay satellites should be appropriate.

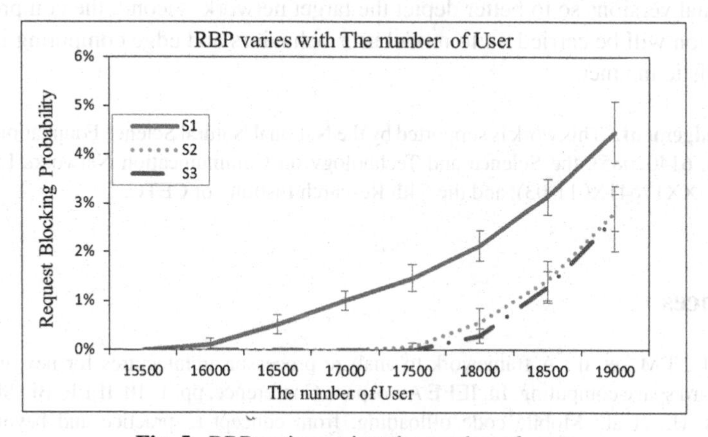

Fig. 5. RBP varies against the number of users

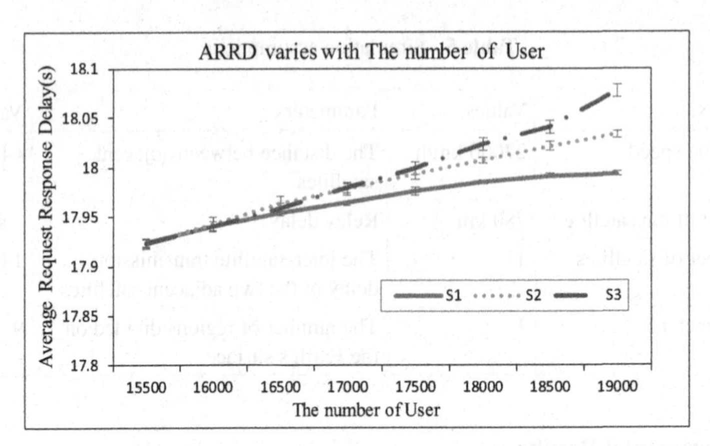

Fig. 6. ARRD varies against the number of users

6 Conclusions

Currently, the edge computing technology is widely deployed in terrestrial networks, and it also has a great potential to be applied in the upcoming broadband LEO satellite networks to effectively reduce user request response delay, as well as to decrease satellite backhaul bandwidth consumption. This paper derives a one-dimensional model for edge computing enabled LEO satellite network, and on that basis, it develops a novel Monte Carlo simulator to evaluate system performance. This paper also elaborates three different computation offloading schemes and conducts comparative evaluation experiments using the simulator. The results indicate that computation offloading to the neighboring LEO satellites can considerably reduce the request blocking probability with only slightly increasing the average request response delay.

The potential research work we have planned for the next step includes the following two points: Firstly, the current one-dimensional simulator will be extended to three dimensional versions so to better depict the target network; second, the comprehensive investigation will be carried on to model user behaviors and edge computing tasks in a more realistic manner.

Acknowledgement. This work is supported by the National Natural Science Foundation of China (Grant No. 61402085), the Science and Technology on Communication Networks Laboratory (Grant No. XX17641X011-03), and the 54th Research Institute of CETC.

References

1. Lovelly, T.M., et al.: A framework to analyze processor architectures for next-generation on-board space computing. In: IEEE Aerospace Conference, pp. 1–10. IEEE, Big Sky (2014)
2. Flores, H., et al.: Mobile code offloading: from concept to practice and beyond. IEEE Commun. Mag. **53**(3), 80–88 (2015)

3. Liu, J., et al.: Delay-optimal computation task scheduling for mobile-edge computing systems. In: 2016 IEEE International Symposium on Information Theory, pp. 1451–1455. IEEE, Barcelona (2016)
4. Wang, Q., et al.: Computation tasks offloading scheme based on multi-cloudlet collaboration for edge computing. In: 2019 Seventh International Conference on Advanced Cloud and Big Data, pp. 339–344. IEEE, Suzhou (2019)
5. Long, L., et al.: Delay optimized computation offloading and resource allocation for mobile edge computing. In: 2019 IEEE 90th Vehicular Technology Conference, pp. 1–5. IEEE, Honolulu (2019)
6. Kamoun, M., et al.: Joint resource allocation and offloading strategies in cloud enabled cellular networks. In: IEEE International Conference on Communications, pp. 5529–5534. IEEE, London (2015)
7. Wang, W., et al.: Computational offloading with delay and capacity constraints in mobile edge. In: IEEE International Conference on Communications, pp. 1–6. IEEE, Paris (2017)
8. Hao, Z., et al.: A delay and energy consumption efficient offloading algorithm in mobile edge computing system. In: 2019 IEEE 11th International Conference on Communication Software and Networks, pp. 251–257, Chongqing (2019)
9. Denby, B., et al.: Orbital edge computing: machine inference in space. IEEE Comput. Archit. Lett. **18**(1), 59–62 (2019)
10. Cheng, N., et al.: Space/aerial-assisted computing offloading for IoT applications: a learning-based approach. IEEE J. Sel. Areas Commun. **37**(5), 1117–1129 (2019)
11. Suzhi, C., et al.: Space edge cloud enabling network slicing for 5G satellite network. In: 15th International Wireless Communications & Mobile Computing Conference, pp. 787–792. IEEE, Tangier (2019)
12. Wang, Y., et al.: A computation offloading strategy in satellite terrestrial networks with double edge computing. In: IEEE International Conference on Communication Systems, pp. 450–455. IEEE, Chengdu (2018)
13. Wang, T.F., et al.: Computing as a service pattern based on edge computing. Appl. Electron. Technol. **45**(5), 80–83 (2019)

Simulation Design of Power Electronic Transformer with Dual-PWM

Jiu-yang Mu[1], En Fang[1(✉)], Guan-bao Zhang[1], and Song-hai Zhou[2]

[1] School of Information and Electrical Engineering, Xuzhou University of Technology, Xuzhou 221018, Jiangsu, China
fangen@cumt.edu.cn

[2] Scientific Research Management Department, Jiangsu Zongshen Automobile Industry Limited Company, Xuzhou 221018, Jiangsu, China

Abstract. With the continuous development of new energy power generation and smart grid, Power Electronic Transformer (PET) has a good prospect for development because of its remarkable advantages. Based on the topology of AC/DC/AC, the modulation strategy with dual-PWM (Pulse Width Modulation) is adopted to control the operation of power electronic transformers in this paper. The PET structure is consisted of three units: input portion, isolation portion and output portion. Model building and analyzing both at the input and output terminals are achieved step by step. And the simulation of the whole PET system with dual-PWM is accomplished with MATLAB/ Simulink. The simulation results show the control system stability and output voltage regulation precision are improved with feed forward voltage decoupling vector control system. The correctness and effectiveness of the control strategy are demonstrated through the simulation. PETs play an important role in enhancing power supply reliability for Power Grid and promoting the new energy power generation development.

Keywords: Power electronic transformer · Topology · Loop control · Modeling simulation

1 Introduction

Nowadays, power distribution transformers have great application prospects with the development of distributed power generation and smart grid. The transformer is not only used to provide electrical isolation, but also to connect the system at different voltage levels. Generally, the transformer size is inversely proportional to the operating frequency. The steady reduction in the cost of power electronics and the advent of advanced magnetic materials with lower loss density and high saturation flux density proclaim that power electronic transformers with high power density design are feasible and economical.

In China, research on power electronic transformers started relatively late compared with the developed countries and is still in its infancy stage. In this paper, the output of

H. Song and D. Jiang (Eds.): SIMUtools 2020, LNICST 370, pp. 464–478, 2021.
https://doi.org/10.1007/978-3-030-72795-6_38

PET based on dual-PWM (Pulse Width Modulation) is analyzed. The dual-loop decoupling control strategy is adopted both at the input and output ends, and the dynamic response speed of the control system is improved by voltage feedforward. The intermediate isolation portion is controlled with an open-loop control strategy. Through simulation verification, the output voltage waveform can be obtained, which has a high-power quality. The voltage fluctuates with the load variation.

2 Circuit Structure of Dual-PWM Power Electronic Transformer

2.1 DC Topology in PET

In this paper, AC/DC/AC conversion structure is adopted in PETs as shown in Fig. 1. The working process is as follows: the power frequency AC input is converted into DC power on the primary side, then the DC power is inverted into high frequency alternating current. Then the high frequency alternating current is transmitted to the secondary side. The conversion process is reversed [1]. There are two AC/DC/AC converters in the power electronic transformer topology [2, 3]. With a circuit consisting of two DC bus lines (low-voltage DC bus and medium-voltage DC bus), a transformer can be used to isolate two power supplies required by certain electrical safety standards.

In the structure of the PET DC-DC converter stage, single-phase dual-active-bridge (DAB) circuit with high efficiency and few passive components is studied in this paper.

2.2 Overall Structure of PET

PET works at high frequencies. The size of the transformer is inversely proportional to the frequency and saturation flux density, so the size can be reduced under high-frequency operating conditions [4–6]. The active-double-bridge topology is used in DC portion of PET, so the overall structure of PET is shown in Fig. 2 [7–9].

Among many types of PET systems, AC/DC/AC transformation is chosen in this paper. As can be seen from Fig. 2, this PET type has a three-stage structure [10–12].They are input portion (three-phase voltage source type PWM rectifier), intermediate portion (DC-DC converter) and output portion (three-phase voltage source type PWM inverter). Input portion can effectively control power factor and stabilize DC voltage, energy bidirectional flow (with fully controlled devices). Intermediate portion turns high-voltage direct current into high-frequency medium-voltage direct current while achieving electrical isolation [13–15].The output portion uses a three-phase, three-wire two-level topology [16–19].

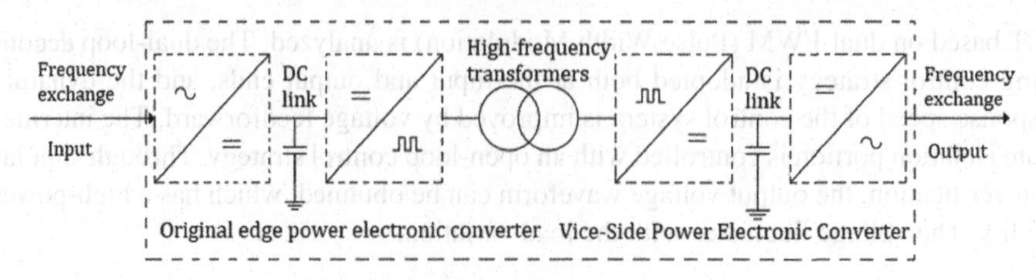

Fig. 1. PET with DC link.

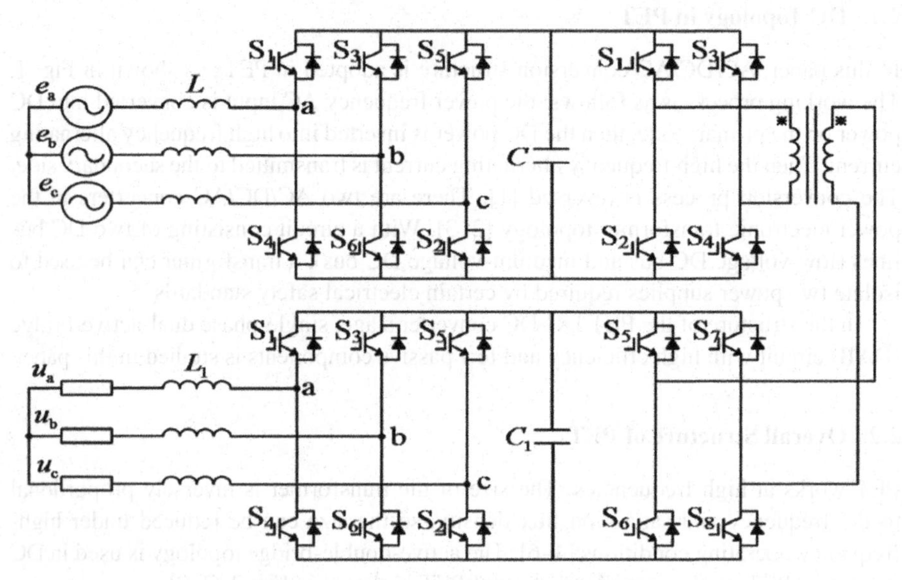

Fig. 2. PET overall circuit structure diagram.

3 Control Topology Design of Power Electronic Transformer

3.1 Control Topology of the PWM Converter on the Input Side

1) Control topology for the input side

On the input side, the rectification is accomplished with PWM. The space vector control topology is used in the converter oriented with the d-axis [20, 21]. The calculation formulas of active power P and reactive power Q in the PWM converter are as follows:

$$P = \frac{3}{2}(e_{gd}i_d + e_{gq}i_q) \tag{1}$$

$$Q = \frac{3}{2}(e_{gq}i_d - e_{gd}i_q) \tag{2}$$

Where: *P*- active power (W); *Q*- reactive power (Var); i_d- d-axis current component; i_q- q-axis current component; e_{gd}- d-axis current component; e_{gq}- q-axis current component.

As shown in Fig. 3, the active power is controlled by i_d and the reactive is by i_q. With the help of energy balance theory, it can be expressed as follows:

$$u_{dc}i_{dc} = \frac{3}{2}e_{gd}i_d \tag{3}$$

As can be seen from Eq. (3), the active power is only related to the d-axis current component. Therefore, double-loop control topology can be adopted. As shown in Fig. 3, the outer loop is a DC voltage loop, and the inner loop is a d-axis current loop.

As shown in Fig. 3, the control process can be described as follows: the main control of the voltage outer loop is to realize the stability of the DC bus voltage u_{dc}. The voltage error signal between the value of u_{dc} and the reference value u_{dc}^* is considered as the input of left PI regulator and the output of PI regulator which is the reference active current value i_d^* can be obtained for inner loop. In order to realize the unity power factor on the input side, the reference reactive current value i_q^* is set to 0. The voltage component can be obtained as the output of the bottom-right PI regulator after the difference between two values of i_q and i_q^* passing through the regulator. In Fig. 3, the currents of the two axes d and q are not completely decoupled, so the two methods of decoupling and voltage feedforward can be selected. The actual current on the d-q axes as the output of the current inner loop PI regulator can be compensated to the SPWM model by multiplying by ωL. The drive signals for the devices can be generated, which can control the power electrical devices turn on and turn off.

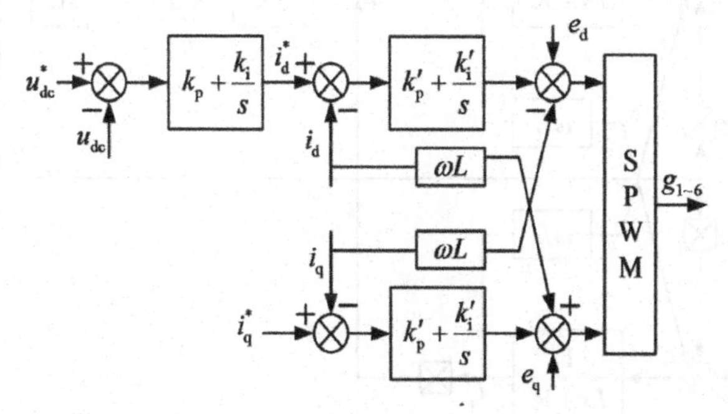

Fig. 3. Input portion control block diagram.

2) Controller design

Design the closed loop system in Fig. 3.

a) Inner loop controller design

The converter model in the d-q coordinate system is:

$$\begin{cases} C\dfrac{du_{dc}}{dt} = \dfrac{3}{2}(i_d S_d + i_q S_q) - \dfrac{u_{dc}}{R_L} \\ L\dfrac{di_d}{dt} + Ri_d - \omega L i_q = e_{gd} - u_{dc}s_d \\ L\dfrac{di_q}{dt} + Ri_q - \omega L i_d = e_{gq} - u_{dc}s_q \end{cases} \tag{4}$$

Where: Sd- the d-axis switch function; Sq- the q-axis switch function.

To effectively control the current inner loop, the current inner loop analysis should be performed first, and the Eq. (4) changes through formula variation by using the Laplace transformation:

$$Li_d(s) + Ri_d(s) - \omega L i_q(s) = e_{gd}(s) - u_{dc}(s)s_d(s) \tag{5}$$

$$Li_q(s) + Ri_q(s) - \omega L i_d(s) = e_{gq}(s) - u_{dc}(s)s_q(s) \tag{6}$$

$$Cu_{dc}(s) = \frac{3}{2}(i_d(s)S_d(s) + i_q(s)S_q(s)) - \frac{u_{dc}(s)}{R_L} \tag{7}$$

From formulas (5) and (7), the converter decoupling model can be shown in Fig. 4.

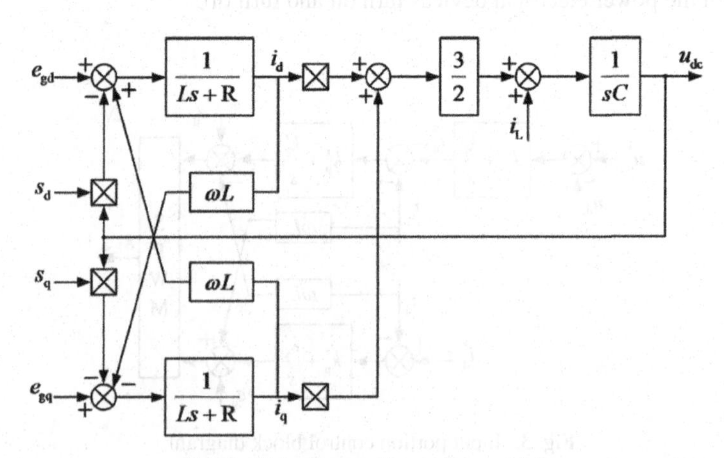

Fig. 4. Transformer decoupling model.

After decoupling the two axes d and q. The block diagram of the system closed-loop control is shown in Fig. 5 [22–24].

The system transfer function is:

$$G_{io}(s) = \left(\frac{1}{T_S S - \pi}\right)\left(K_p + \frac{k_i}{s}\right)\left(\frac{k_{PWM}}{0.5T_S S} + 1\right)\left(\frac{1}{Ls + R}\right) \tag{8}$$

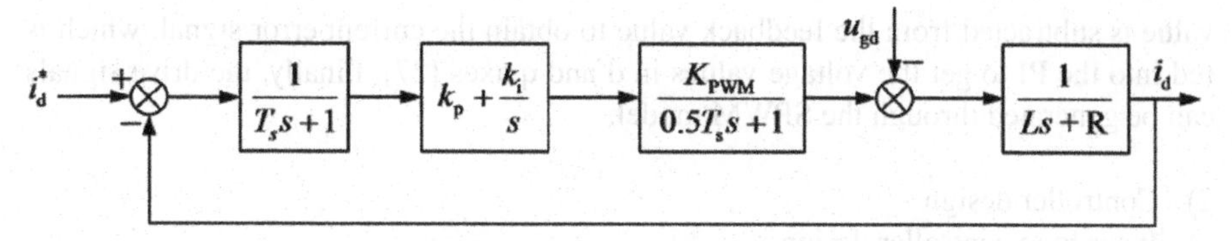

Fig. 5. Block diagram of closed-loop system control structure.

k_{PWM}- PWM gain; T_s- current inner loop current sampling period(s); k_p- current regulator scale factor; k_i- current regulator integral coefficient.

b) Outer loop controller design

The design of voltage conversion can be referenced [25, 26], and its block diagram is shown as Fig. 6.

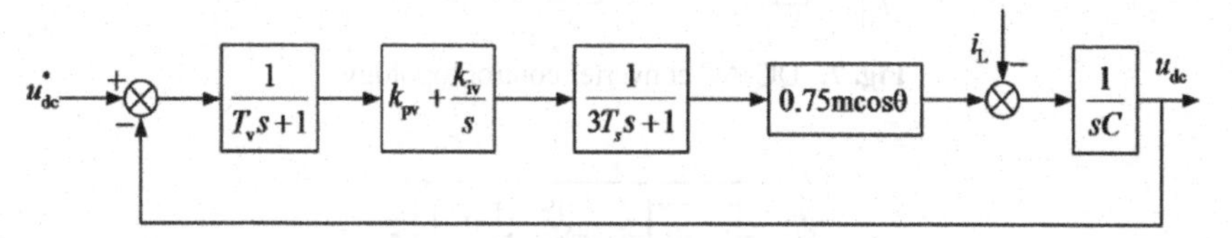

Fig. 6. Voltage closed-loop structure diagram.

The system transfer function is:

$$G_{vo}(s) = \left(\frac{1}{T_v S + 1}\right)\left(k_{pv} + \frac{k_{iv}}{s}\right)\left(\frac{1}{3T_s S + 1}\right)\left(\frac{0.75m\cos\theta}{sC}\right) \quad (9)$$

T_v- inertial time constant (s); k_{pv}- voltage regulator scale factor; k_{iv}- voltage regulator integral coefficient; $0.75m\cos\theta$- the DC side current considering the fundamental component of the switching function; m-modulation ratio.

3.2 Control Topology of the PWM Converter on the Output Side

1) Control topology for the output side

Unlike the input portion, for the output side PWM converter, it is controlled to convert high-frequency direct current into low-voltage alternating current, which requires constant output side voltage and frequency. In order to achieve the goal, in general, the constant voltage and constant frequency control topology is selected [14] as shown in Fig. 7.

The control topology on the output side is roughly the same as on the input side and consist of two modules: the voltage outer loop and the current inner loop. The difference is that the voltage outer loop on the output side is AC. The error signal of the d and q two axes are adjusted by PI to obtain the current inner loop command value. The command value is subtracted from the feedback value to obtain the current error signal, which is fed into the PI to get the voltage values in d and q axes [27]. Finally, the drive signals can be generated through the SPWM model.

2) Controller design
a) Inner loop controller design

The control block diagram of the inverter can be shown as in Fig. 8.

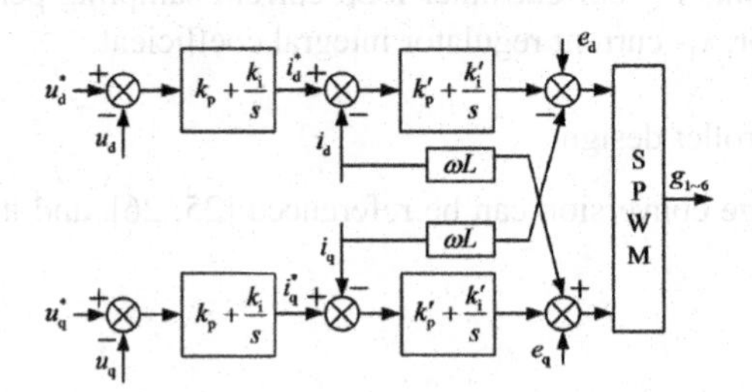

Fig. 7. DC-AC converter control topology.

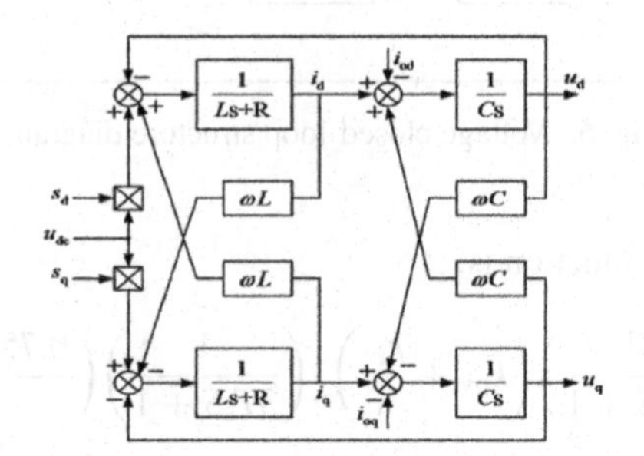

Fig. 8. Control block diagram of the inverter.

The structure block diagram that can be drawn from Fig. 8 as Fig. 9.

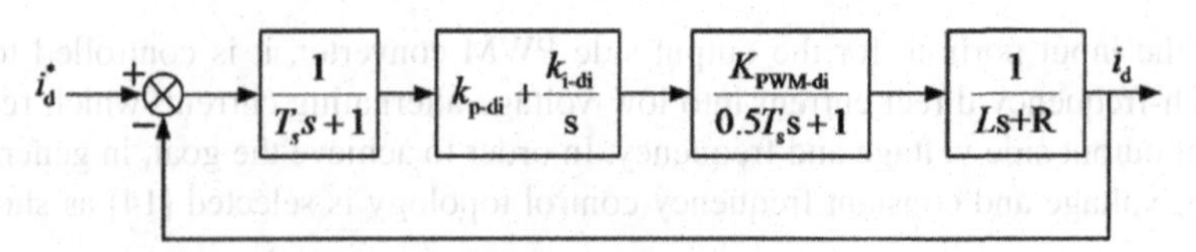

Fig. 9. Current closed-loop structure block.

b) Outer loop controller design

As shown in Fig. 8, the structures in d and q axes are the same with each other. So the analysis of voltage closed-loop parameters in d-axis is taken as an example. The block diagram is established as in Fig. 10.

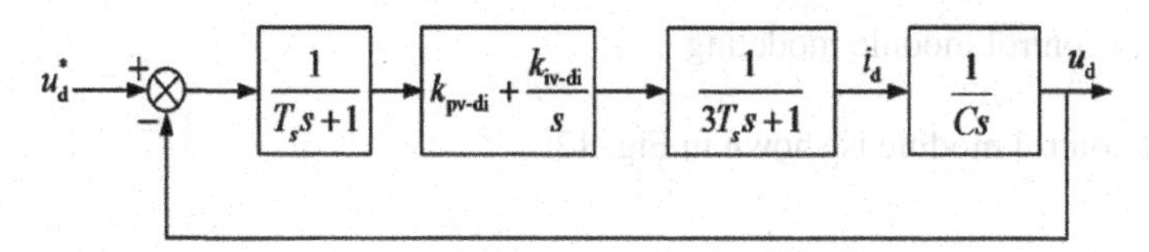

Fig. 10. Block diagram of voltage closed-loop transfer function at low voltage side.

4 Simulation Modeling and Results Analysis

4.1 Simulation Modeling

1) System modeling diagram

Based on the software MATLAB/Simulink, the PET system simulation model is built as shown in Fig. 11. The parameters include: the frequency of the grid (50 Hz), capacity (10 MVA), the switch frequency of output side inverter (1000 Hz), the output voltage of the inverter (660 V), the simulation time (0.3 s). And there is a load mutation (equivalent to half the load reduction) at 0.15 s.

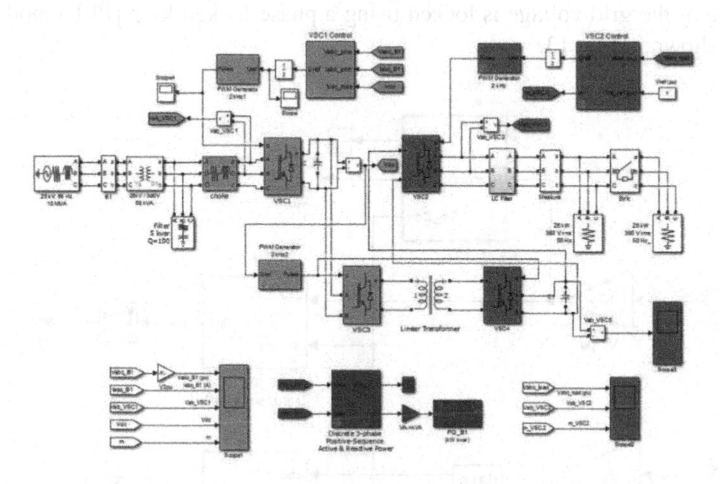

Fig. 11. Simulation model of PET system.

The simulation model is built as in Fig. 11. The common bridge circuit is selected as the converter circuit for rectifier and inverter on both primary side and secondary side.

Because of the high working frequency, the size and weight of the transformer are greatly reduced. The variation on the original side can be coupled to the secondary side through the intermediate isolation portion [19, 28]. A dual-loop control strategy is selected for the input and output portions. The intermediate isolation portion is implemented by open-loop control, and the gate signals are sent by the PWM Generator.

2) System control module modeling

PWM control module is shown in Fig. 12.

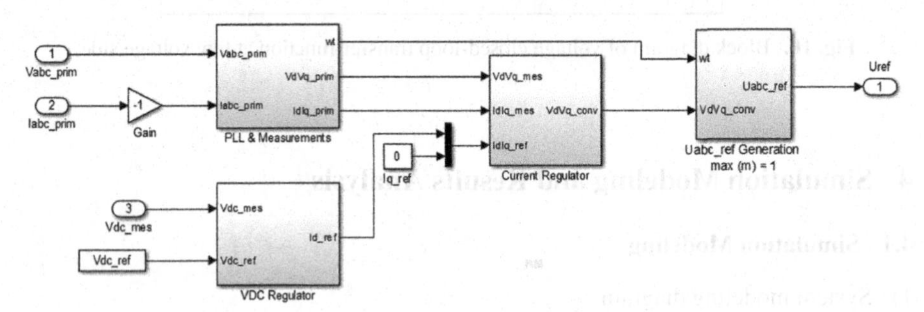

Fig. 12. PWM control module.

It consists of four various modules:

a) Phase-locked loop(PLL) module

The phase of the grid voltage is locked using a phase-locked loop (PLL) module. The circuit is shown in Fig. 13.

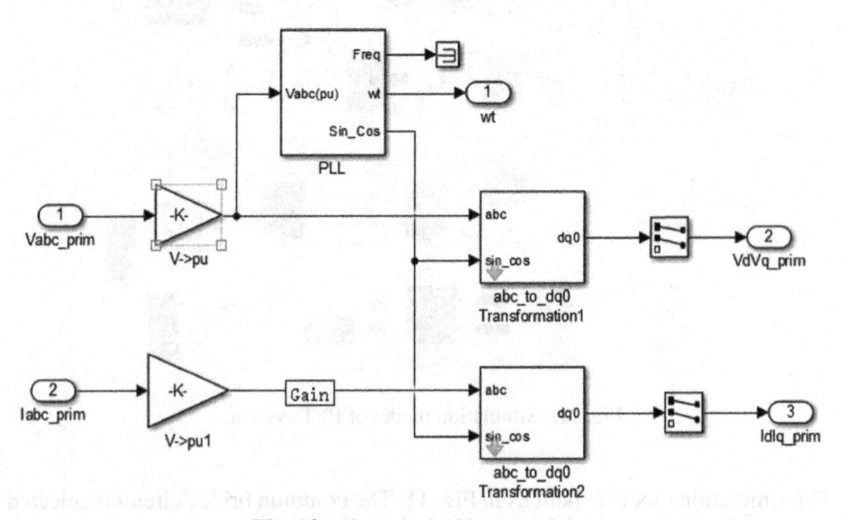

Fig. 13. Phase-locked loop module.

b) Voltage loop module

The voltage loop module is designed as described above shown in Fig. 14.

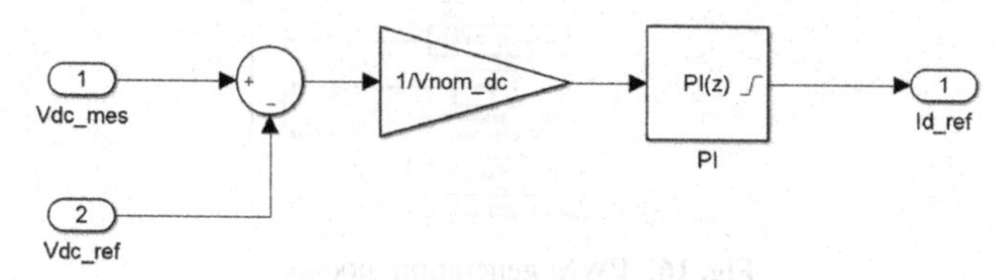

Fig. 14. Voltage loop module.

c) Decoupling module

As shown in Fig. 15, i_d is decoupled from i_q in the decoupling module constructed according to the decoupling circuit in Fig. 3.

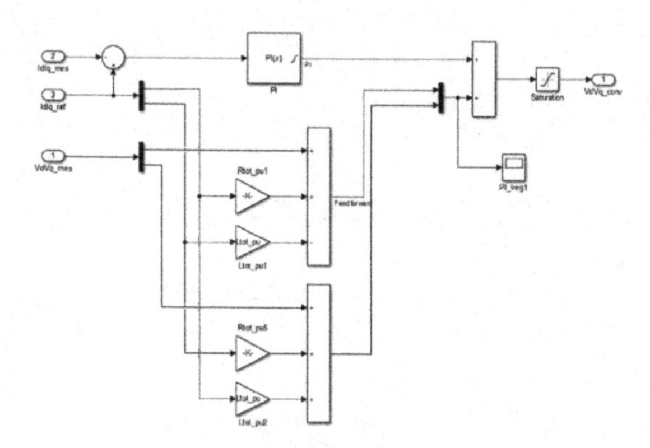

Fig. 15. Decoupling module.

d) PWM generation module

The PWM generation module is drawn as shown in Fig. 16.

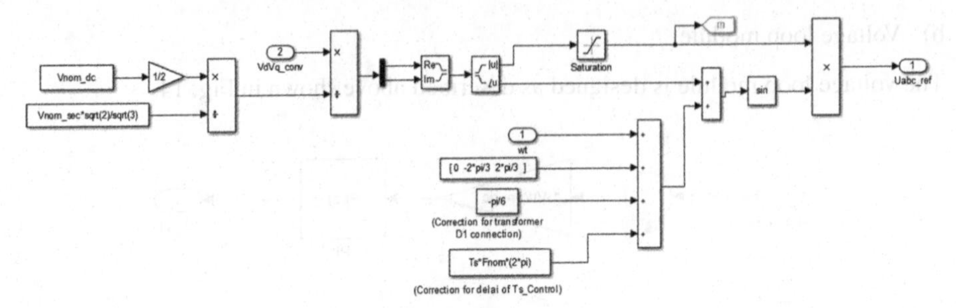

Fig. 16. PWM generation module.

4.2 Simulation Result Analysis

1) Input side simulation results

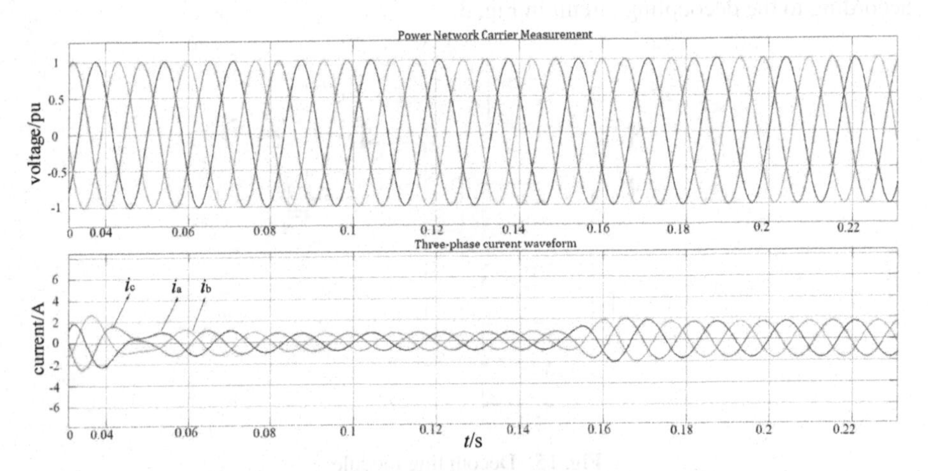

Fig. 17. Input side voltage and current simulation results.

The upper image in Fig. 17 shows the grid voltage per unit measurement waveforms, the value of which is obtained by the normalized three-phase voltage. And the following picture is a three-phase current graph. At 0.15 s, the load mutation occurred (parallel connection with another load module), so the current increased. In addition, it can be seen in Fig. 17, the phase of grid voltage and current are the same, even the load changes. The isolation transformer plays the role of improving the quality of electrical energy.

Figure 18 is the input line voltage waveform and the IGBT voltage waveform. From the figure the input variable is the alternating current. The DC voltage can be stabilized after rectification filtering as shown in Fig. 19. The voltage is rapidly reduced by 7.14% after a load mutation at 0.15 s (parallel connection with another load module). Due to the presence of the control system, the voltage gets back to the original voltage value after about 0.02 s.

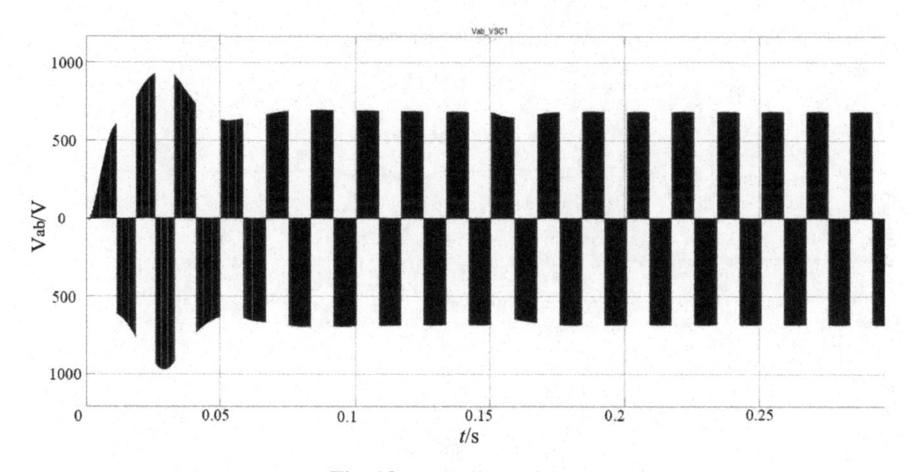

Fig. 18. Input line voltage.

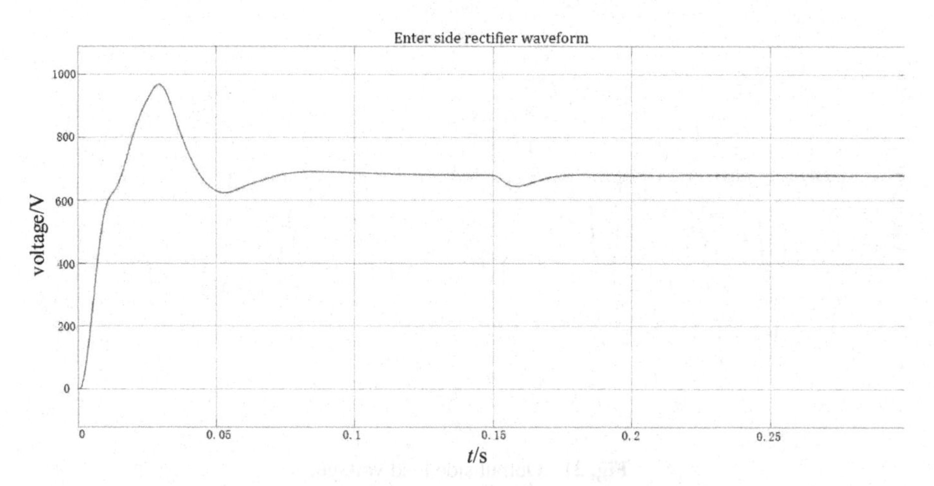

Fig. 19. The voltage after entering the side rectifier.

2) Isolated side simulation results

From Fig. 20, the voltage value of the output is the same as that of the rectifier output. The isolation transformer had not changed its output voltage.

3) Output side simulation results

Figure 21 is three-phase load voltage waveforms, which can quickly restore to the normal operating voltage after the load mutation. It is shown that the control system is good and the power factor of the system is approximately 1.

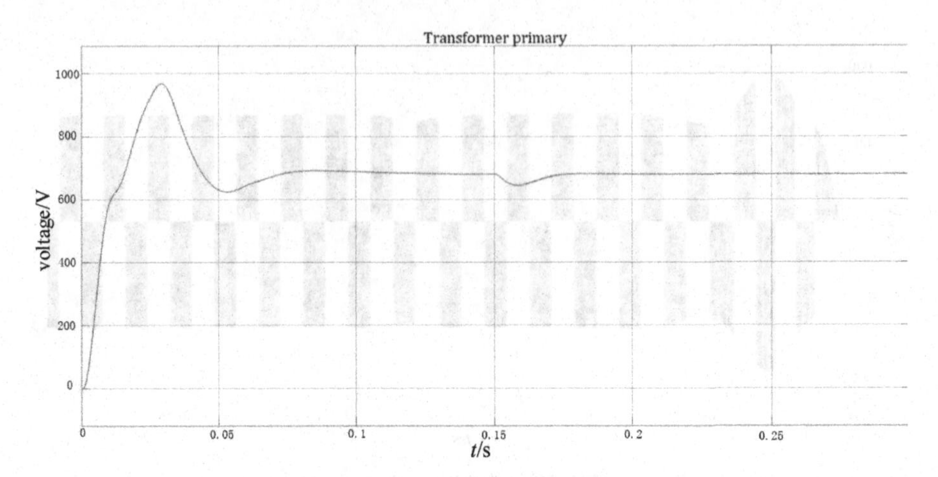

Fig. 20. Isolated side output voltage.

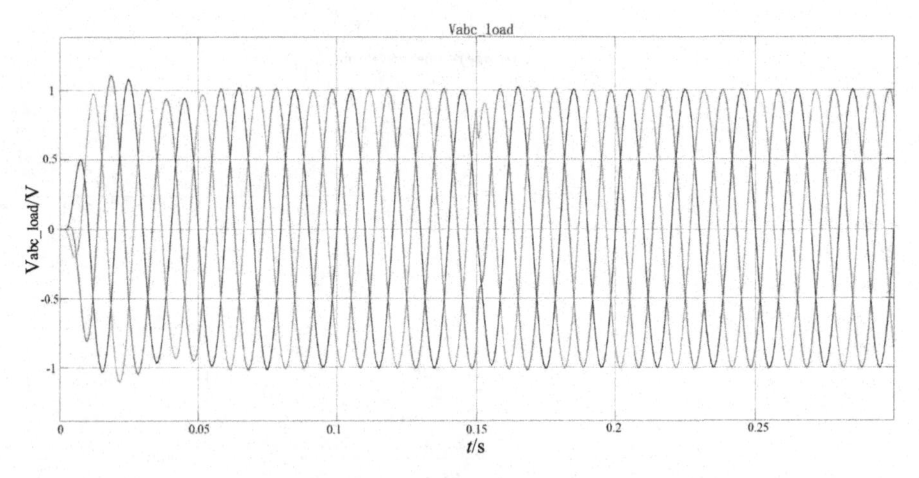

Fig. 21. Output side load voltage.

5 Conclusions

With smart substations for application, the control topologies for PETs are studied in this paper. AC/AC PET can be applied to grid-connected inverters and can also be used for power generation. DC/AC PET can be used in solar inverters while reducing the substation cost and improving the system reliability. The advantages of PET compared to traditional transformers are studied in this paper and the importance of PET for smart substations is shown.

The topological structure of input portion, isolation portion and output portion are analyzed. The converter models and control strategies are accomplished by MATLAB/Simulink.

Since the output variables of the input and output portions are different, the output variables of each portion in PET is analyzed. And the control topologies are designed for different purposes separately. The control block diagrams are given above. The correctness of the theory is proved by the simulation results of each module.

Acknowledgements. The authors acknowledge the Jiangsu University Natural Science Research Project (18KJB470024) and Provincial Construction System Science and Technology Project of Jiangsu Provincial Housing and Urban-Rural Construction Department (2018ZD088). This work is partly supported by the Natural Science Foundation of Jiangsu Province of China (No. BK20161165), the applied fundamental research Foundation of Xuzhou of China (No. KC17072). The authorized patents for invention are also the research and development of Jiangsu Province Industry-University-Research Cooperation Project (BY2019056).

References

1. Wensi, L.: Research on Modeling Method and Control Technology of Power Electronic Transformer. Shandong University, Shandong (2011)
2. Yisheng, Y., Kaixiang, M., Shiying, Y.: Large signal modeling and analysis of DC-bus voltage of power electronic transformer applied in electric locomotive. Power System Protection and Control (2019)
3. Jiang, D., Wang, Y., Lv, Z., Wang, W., Wang, H.: An energy-efficient networking approach in cloud services for IIoT networks. IEEE J. Sel. Areas Commun. **38**(5), 928–941 (2020)
4. Feng, X., Jun, J.: Simulation research on improving power quality of power electronic transformer. Electromech. Inf. **36**, 4–6 (2018)
5. Jiang, D., Wang, W., Shi, L., Song, H.: A compressive sensing-based approach to end-to-end network traffic reconstruction. IEEE Trans. Netw. Sci. Eng. **7**(1), 507–519 (2020)
6. Xiang-Long, L., et al.: Coordinating voltage regulation for AC-DC hybrid distribution network with multiple power electronic transformer. Adv. Technol. Electr. Eng. Energy (2019)
7. Jiang, D., Huo, L., Song, H.: Rethinking behaviors and activities of base stations in mobile cellular networks based on big data analysis. IEEE Trans. Netw. Sci. Eng. **7**(1), 80–90 (2020)
8. Jiang, D., Wang, Y., Lv, Z., Qi, S., Singh, S.: Big data analysis based network behavior insight of cellular networks for industry 4.0 applications. IEEE Trans. Ind. Inform. **16**(2), 1310–1320 (2020)
9. Jiang, D., Huo, L., Lv, Z., Song, H., Qin, W.: A joint multi-criteria utility-based network selection approach for vehicle-to-infrastructure networking. IEEE Trans. Intell. Transp. Syst. **19**(10), 3305–3319 (2018)
10. Tian, Z.: Design of Power Electronic Transformer. Ind. Control Comput. **30**(12), 149–150 (2017)
11. Jiang, D., Huo, L., Li, Y.: Fine-granularity inference and estimations to network traffic for SDN. PLoS ONE **13**(5), 1–23 (2018)
12. Lewicki, A., Morawiec, M.: The structure and the space vector modulation for a medium voltage power-electronic-transformer based on two seven-level cascade H-bridge inverters. IET Electric Power Appl. **13**(10), 1514-1523 (2019)
13. Wang, Y., Jiang, D., Huo, L., Zhao, Y.: A new traffic prediction algorithm to software defined networking. Mob. Netw. Appl. 110 (2019)
14. Zhang, X.: Research on PWM rectifier and its control strategy. Ph.D. Thesis, Hefei University of Technology (2003)

15. Qi, S., Jiang, D., Huo, L.: A prediction approach to end-to-end traffic in space information networks. Mob. Netw. Appl. 110 (2019)
16. Shaodi, O., et al.: DC voltage control strategy of three-terminal medium-voltage power electronic transformer-based soft normally open points. IEEE Trans. Ind. Electron. **67**(5), 3684-3695 (2019)
17. Jiang, D., Zhang, P., Lv, Z., et al.: Energy-efficient multi-constraint routing algorithm with load balancing for smart city applications. IEEE Internet Things J. **3**(6), 1437–1447 (2016)
18. Jiang, D., Li, W., Lv, H.: An energy-efficient cooperative multicast routing in multi-hop wireless networks for smart medical applications. Neurocomputing **2017**(220), 160–169 (2017)
19. Qin, H., Kimbal, J.W.: Solid-state transformer architecture using AC–AC dual-active-bridge converter. IEEE Trans. Ind. Electron. **60**(9), 3720–3730 (2013)
20. Yang, Y., Pei, W., Deng, W., et al.: Benefit analysis of using multi-port and multi-function power electronic transformer connecting hybrid AC/DC power grids. J. Eng. **2019**(16), 1076–1080 (2019)
21. Li, J.: Discussion on the control and application of electronic power transformer. Heilongjiang Sci. Technol. Inf. (23), 53 (2016)
22. Wang, F., Jiang, D., Qi, S.: An adaptive routing algorithm for integrated information networks. China Commun. **7**(1), 196–207 (2019)
23. Geng, Q., Hu, Y.: Energy storage device locating and sizing based on power electronic transformer. J. Eng. **2019**(16), 3164–3168 (2019)
24. Huo, L., et al.: An AI-based adaptive cognitive modeling and measurement method of network traffic for EIS. Mob. Netw. Appl. 110 (2019)
25. Huo, L., Jiang, D., Lv, Z., et al.: An intelligent optimization-based traffic information acquirement approach to software-defined networking. Comput. Intell. **36**(1), 151-171 (2019)
26. Zhao, J.: Simulation of power electronic transformer with constant output voltage. Power Syst. Autom. **27**(18), 30-33 (2003)
27. Guo, S., Mu, Y., Jia, H., et al.: Optimization of AC/DC hybrid distributed energy system with power electronic transformer. Energy Procedia **158**, 6687–6692 (2019)
28. Liu, Z., Lin, Y.: Design and simulation of power electronic transformer in distributed power generation system. Electr. Appl. **37**(06), 82–86 (2008)

Research and Development of Hydraulic Controlled Vibration Damper with Vibration Energy Absorption

En Fang, Ren-li Wang, Li-wen Wang(✉), Shu-yun Qiao, and Jian-zhong Xi

School of Information and Electrical Engineering,
Xuzhou University of Technology, Xuzhou 221018, Jiangsu, China
fangen@cumt.edu.cn

Abstract. Research on the core technology has obtained a patent for invention authorization of hydraulic controlled vibration damper with vibration energy absorption. The hydraulic cylinder structure is designed by the principle of liquid incompressibility. The hydraulic circuit increases and the gap distance changes due to the movement of balance plate under the pressure difference between the two chambers of the hydraulic cylinder. With the damping effect and pressure drop of liquid flow, the purpose of efficient vibration reduction is achieved by absorbing the vibration energy into heat. The invention patent of the self-tuning hydraulic vibration energy absorption is suitable for the situations with high speed, precision, light load and small and medium energy level. For example, the measurement accuracy is improved by vibration reduction with dynamic weighing in instrument. And the processing precision is achieved by vibration reduction for precision machine tools. The invention patent of high power hydraulic emergency energy absorption for multistory parking area is applicable to the conditions with high altitude fall, impact, collision and large-scale energy absorption, such as elevator emergency energy absorption safety devices, the bases of forging press machine tools, high-speed rail locomotive vehicle chassis, and so on.

Keywords: Hydraulic controlled · Vibration damper · Vibration energy absorption

1 Introduction

With the development of science and technology, modern vibration dampers have broad application prospects in some special environments and scenes, such as engineering machineries, machine tool equipment, construction machineries, high speed rail vehicles and dynamic working environments, and so on and so forth. Those are commonly used in concrete mixer truck weighing devices and high lift elevator safety devices. Beyond the working area, vibration damping and vibration elimination are usually required for non-executive components, especially for mechanical equipment working in vibration mode, such as vibratory roller, vibration drill, crusher, percussion drill, vibrating screen and so on [1–4].

H. Song and D. Jiang (Eds.): SIMUtools 2020, LNICST 370, pp. 479–488, 2021.
https://doi.org/10.1007/978-3-030-72795-6_39

Research on the core technology has obtained a patent for invention authorization of hydraulic controlled vibration damper with vibration energy absorption [5–8]. The hydraulic cylinder structure is designed by the principle of liquid incompressibility. The hydraulic circuit increases and the gap distance changes due to the movement of balance plate under the pressure difference between the two chambers of the hydraulic cylinder [9–11]. With the damping effect and pressure drop of liquid flow, the purpose of efficient vibration reduction is achieved by absorbing the vibration energy into heat. Thus, the noise pollution produced by the vibration is reduced [12–14].

2 Self-tuning Hydraulic Vibration Energy Absorption

Self-tuning method and device for hydraulic vibration energy absorption are authorized a national patent for invention of China in 2013: ZL2011 10117133.3.

2.1 Basic Structure

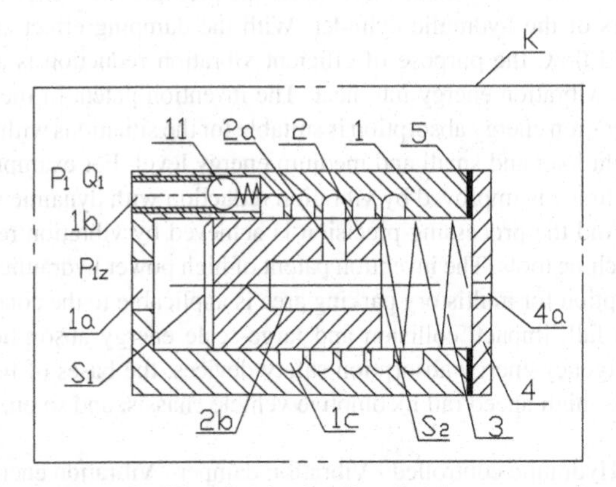

Fig. 1. Basic structure of self-tuning hydraulic vibration energy absorption device.

Where: 1-damping shell, 2-piston slider, 3-pressure differential equilibrator spring, 4-end cover, 5- prepressed adjusting shim, 1a-vibration wave receiving hole, 1b-liquid resistance channel, 11-check valve, 4a-oil return port, 1c-liquid resistance spiral groove, 2a-radial pressure drop channel, 2b-axial steady flow chamber.

As shown in Fig. 1, the self-tuning hydraulic vibration energy absorption device is composed of damping shell 1, piston slider 2, pressure difference equilibrator spring 3, end cover 4, prepressed adjusting shim 5, vibration wave receiving hole 1a, liquid resistance channel 1b, check valve 11, oil return port 4a, liquid resistance spiral groove 1c, radial pressure drop channel 2a, axial steady flow chamber 2b. The adjustable piston slide block 2 is located in the inner cavity of the damping shell 1 and seals with the inner cavity wall. The right side of piston slider 2 and the left end of the pressure differential equilibrator spring 3 fit together. The right end of pressure differential equilibrator spring 3 is connected with the end cover 4 with the oil return port 4a in the center. The prepressed adjusting shim 5 is located between the end cover 4 and the damping shell 1. And there is a check valve 11 in the liquid resistance channel 1b.

There is an inner cavity in the middle of the damping shell 1 and the liquid resistance channel 1b on one side of the damping shell. In the center of the section, the vibration wave receiving hole 1a is connected with the inner cavity. There is the liquid resistance spiral groove 1c on the side wall of the inner cavity, which links up with the liquid resistance channel 1b. The piston slider 2 is a cylinder with an axial blind hole called the axial steady flow chamber 2b in the center. On the side of the cylinder, the radial hole which is named the radial pressure drop channel 2a is connected with the axial steady flow chamber 2b. The pressure differential equilibrator spring 3 is a variable force spring or a memory shrapnel. The prepressed adjusting shim 5 can be a rigid or elastic shim.

2.2 Principles

The wave energy produced by the vibration source enters the inner cavity of the damping shell 1 through the vibration wave receiving hole 1a [15–18]. The effect of oil pressure P_1 and vibration wave pressure P_{1z} is on the left end S_1 of the piston slide block 2 simultaneously, at the same time, the energy P_1Q_1 of the vibration source can enter the liquid resistance channel 1b. The damping and depressurization of the vibration energy is accomplished through liquid resistance spiral groove 1c in the form of heat radiation. Passing through the depressurization channel into the anti-surge chamber, the vibration energy after the flow-stabilizing enters the returning line. Thus, the oil pressure P_2 is brought on the right end S_2 of the piston slider 2. The piston slider 2 moves to the right under the action of the oil pressure P_1. The flow resistance loss becomes larger due to the length of the liquid resistance spiral groove with oil increases. The pressure differential equilibrator spring 3 is compressed with the reverse thrust increasing. The instantaneous peak amplitude response of vibration waves decreases rapidly [19–21]. The piston slider 2 is adjusted to the left under the synthesis thrust of the pressure difference equilibrator spring 3 and the oil pressure P_2, which is greater than the thrust produced by the oil pressure P_1. The number of oil pressure P_2 extends and the length of the liquid resistance spiral groove with oil gradually decreases with vibration energy reduction. The equilibrium between both ends of the piston slider 2 (S_1 and S_2) is achieved. The amplitudes applying on both sides of the piston slider counteracts each other at the same time, thus the impact of the vibration source is avoided. When the vibration wave is at the negative peak, the slide block repeats the steps above automatically eliminating the adverse effects to achieve a new balance.

The self-tuning hydraulic vibration energy absorption devices can be installed in pairs with a reverse one for better damping effect [22].

The change of vibration waveforms before and after self-tuning is shown in Fig. 2 [23–25].

Working principles: The external vibration signals is divided into two parts [26–30]: one part of the signals passes through vibration wave receiving hole 1a to the left side S_1 of piston slider 2. The extent of shaking force is determined according to the effective area of the side on the left S_1. The pressure decreases from P_{1z} to P_{2z} as other part of the signals receives the right side S_2 in the damping shell 1 with the rate of flow Q_1 through the path 1b-2a-2b-S_2. According to the effective area of the right side S_2, the energy

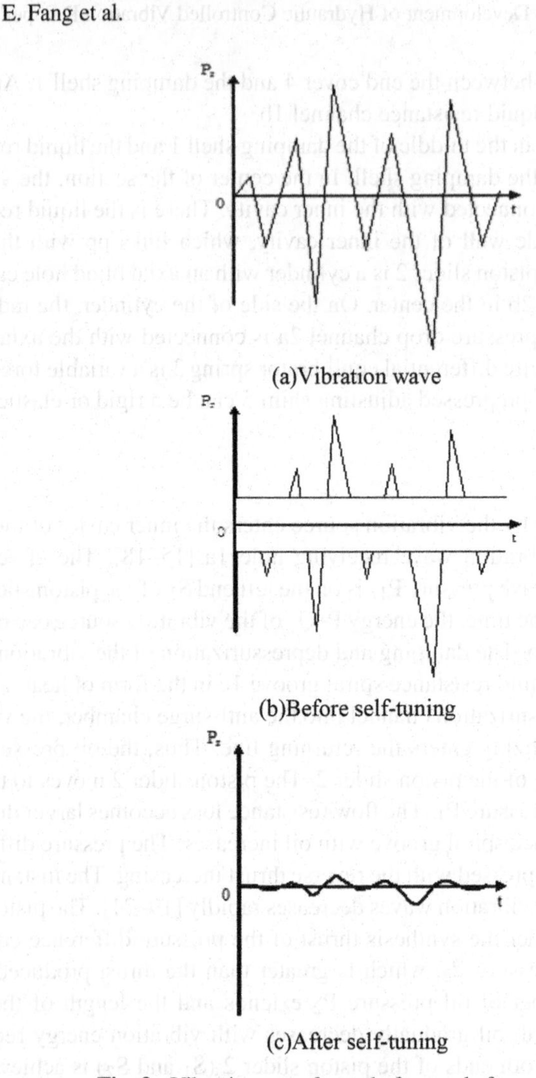

Fig. 2. Vibration waveforms before and after self-tuning.

consumption and the vibration force counteraction are calculated by the whole process of liquid flow through S_1 to S_2 [31–33]。The receiving, measuring and calculating of vibration power consumption is completed. The equilibrium position of piston slider 2 is determined under the synthesis thrust of the pressure difference between both sides (S1 and S2) and the pressure difference equilibrator spring 3. The purpose of completely eliminating vibration force by hydraulic damping is achieved and the system is in the steady state.

2.3 Characteristics and Innovation Points

The features and innovations of invention patent (Self-tuning method and device for hydraulic vibration energy absorption: ZL2011 1 0117133.3) are listed as follows [5].

(1) In the prior art, the inner diameter of the valve bore is invariable, which is not adjustable, movable or of self-setting characteristics. When the vibration energy acts on it, the piston bounces up and down in oil due to the wave shock. The reaction is played on the vibration source and the damping effect is reduced. Due to the structural design, the damping length and pressure drop of the liquid resistance spiral groove are automatically adjusted with the change of the vibration energy. The complicated and changeable vibration energy is absorbed by adjusting the positions of the piston slider 2 and pressure differential equilibrator spring 3. The impact of the vibration energy on the vibration source is ingeniously avoided as the self-adjustment and cancellation of wave transmission characteristics.
(2) The piston moving produced by the vibration squeezes the oil passing through the liquid resistance spiral groove. The huge liquid damping and pressure drop are formed to absorb the vibration energy and the shock absorption is achieved.
(3) The vibration wave is counterbalanced by the adjustable slider to avoid the interference back to the vibration source. The energy is absorbed in time by self-setting and the frequency of reciprocating vibration is reduced for the purpose of great damping effect.

3 High-Power Hydraulic Emergency Energy Absorption for Multistory Parking Area

The invention patent authorization with the method and device of high-power hydraulic emergency energy absorption for multistory parking area is obtained in 2015 (ZL2013 10217678.0) [30]. And the manual hydraulic energy absorption method and device for emergency protection of multistory parking area is authorized as a national patent for invention of China in 2015 (ZL2013 10217657.9) [34].

3.1 Structure Diagram of Hydraulic Energy Absorption Device Controlled by Analog Variables

As shown in Fig. 3, the hydraulic energy absorption device [35] is controlled with analog variables. It consists of energy absorption shell 1, elastic supporting mechanism 2, sealing strip 3, top cover 4, floating pressure equilibrium side-panel 5, front and rear end cap 6, control valve housing 7, movable support gasket 8, cone valve core 9, conical valve spring 10, regulating handle 11, return oil piping for control circuit 12, oil tank 13, pressure sensor in the upper plenum chamber 14, pressure sensor in the lower internal consumption cavity 15, servo stepper motor 16, control computer 17, potential difference comparator 18, output pulse control signal circuit 19, input control signal circuit 20, signal conversion module for pressure sensor in the upper plenum chamber 21, signal conversion module for pressure sensor in the lower internal consumption cavity 22, main oil inlet a, main oil outlet b, lower internal friction cavity C, upper plenum chamber D, connection hole between the upper and lower cavities E, lower internal friction cavity gap δ, lower inner friction cavity width L, oil intake control port K_1 and oil outlet control port K_2.

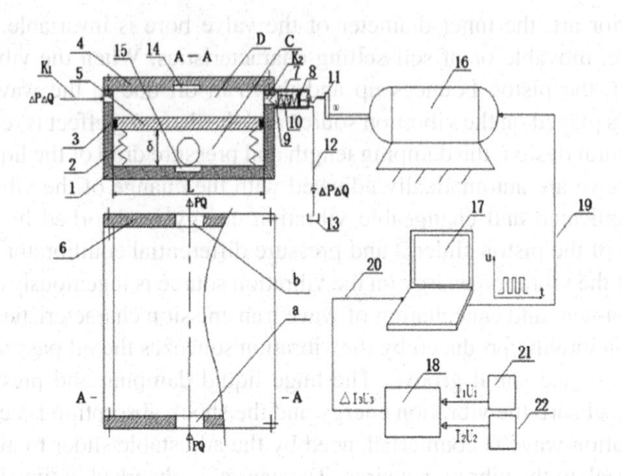

Fig. 3. Structure diagram of hydraulic energy absorption device controlled with analog variables.

Where: 1-energy absorption shell, 2-elastic supporting mechanism, 3-sealing strip, 4-top cover, 5-floating pressure equilibrium side-panel, 6-front and rear end cap, 7-control valve housing, 8-movable support gasket, 9-cone valve core, 10-conical valve spring, 11-regulating handle, 12-return oil piping for control circuit, 13-oil tank, 14-pressure sensor in the upper plenum chamber, 15-pressure sensor in the lower internal consumption cavity, 16-servo stepper motor, 17-control computer, 18-potential difference comparator, 19-output pulse control signal circuit, 20-input control signal circuit, 21-signal conversion module for pressure sensor in the upper plenum chamber, 22-signal conversion module for pressure sensor in the lower internal consumption cavity, a-main oil inlet, b-main oil outlet, C-lower internal friction cavity, D-upper plenum chamber, E-connection hole be tweeen the upper and lower cavities, δ-lower internal friction cavity gap, *L*-lower inner friction cavity width, K_1 -oil intake control port, K_2-oil outlet control port.

3.2 Working Principles for Hydraulic Energy Absorption Device Controlled with Analog Variables

(1) The main energy absorption process: the liquid PQ with energy flows within the lower internal friction cavity C to the upper plenum chamber D through the cross section formed by the lower internal friction cavity gap δ and width L after entering the main oil inlet a. And the pressure of lower internal friction cavity is reduced from P_c to ΔP_0. Regulating handle control circuit: the control oil $\Delta P \Delta Q$ enters the upper plenum chamber D through oil intake control port K_1. The cone valve core 9 moves right to open the oil outlet control port K_2 when the pressure can overcome the pre-tightening force of the conical valve spring 10. Thus, the liquid flows to the oil tank 13 through the oil outlet control port K_2 and return oil piping for control circuit 12. If the liquid remains flowing, the pressure of upper plenum chamber D is considered as a constant value. The difference between the product of the pressure and the upper area of floating pressure equilibrium side-panel 5

and that of the pressure P_0 of lower internal friction cavity C and the lower area of floating pressure equilibrium side-panel 5 decreases the lower internal friction cavity gap δ and shifts down the loading pressure equilibrium side-panel 5 with the elastic supporting mechanism 2 overcome. Thus, the energy loss with the liquid flowing into the energy absorption device y has a cubic curvilinear add along with lower internal friction cavity gap δ reducing and increases linearly with the flow of liquid into the energy absorption device Q according to formula (1). All the energy losses are fully absorbed converting into heat energy and the value of the variable y is controlled by adjusting the lower internal friction cavity gap δ with regulating handle 11.

In the main energy absorption process, an algorithm model is established by experiments to calculate the energy loss with the liquid flowing into the energy absorption device.

$$y = f(\mu \frac{Q}{\delta^3}) \tag{1}$$

where:

y–the energy loss with the liquid flowing into the energy absorption device.

Q–the flow of liquid into the energy absorption device.

δ–lower internal friction cavity gap for the liquid flowing into the energy absorption device.

μ–the experimental coefficient related to the liquid medium.

(2) the control circuit: the value of lower internal friction cavity gap δ has corresponding changes according to the adjustment of the regulating handle 11. The regulating handle 11 is controlled by the output torque angle ω of the servo stepper motor 16. The input signal $\triangle I_3 \triangle U_3$ of the control computer 17 through the input control signal circuit 20 is obtained by the comparison of $I_1 U_1$ and $I_2 U_2$ with the potential difference comparator 18, which are the output signals from the signal conversion modules for pressure sensors in the upper plenum chamber 21 and the lower internal consumption cavity 22. Then the pulsing signals generated by the control computer 17 pass through the output pulse control signal circuit 19 to control the servo stepper motor 16. Thus, the regulating handle is adjusted by the servo stepper motor 16. The control process with analog variables can be divided into two parts: several regions are divided in the arithmetical or geometric series according to the magnitude of the absorption energy and then the value of the variable y is controlled. The different energy changes can be identified by the control computer 17 and the control programs according to analog signals are achieved. According to the different working conditions of the outside world, the program is adjusted by computer.

3.3 Characteristics and Innovation Points

Based on the hydraulic energy absorption device controlled with analog variables mentioned above, the patent for invention on high-power hydraulic emergency energy absorption device and method for the multistory parking area is authorized (ZL2013 1 0217678.0) [30]. The manual regulation hydraulic energy absorption device and method for emergency protection of the multistory parking area is patented for invention (ZL2013 1 0217657.9) [34], and the hydraulic energy absorption device controlled with analog variables is patented for utility models (ZL2013 2 0586602.0) [35]. And the characteristics and innovation points are as follows:

(1) The absorbed vibration energy is of large degree and the effect of vibration damping is good.
(2) The effect of vibration damping depends on Q proportionately, and it is inversely proportional to δ^3.
(3) The magnitude of the absorbed vibration energy is adjustable and controllable and the energy is stable and counteractive at vibration source. When external vibrational energy increases, the device absorbs more energy. The absorption energy fully matches with the vibrational energy and the value of the absorption energy can be controlled.
(4) The flat structure is adopted for good heat dissipation performance due to large heat dissipation area. As the lower inner friction cavity width L of the flow section is long, the whole system is of strong anti-interference ability even with localized flow channel blocked.

4 Conclusions

The self-tuning hydraulic vibration energy absorption device and method are suitable in applications of high speed, precision, light load and small or medium energy level. For example, the measurement accuracy is improved by vibration reduction with dynamic weighing in instrument. And the processing precision is achieved by vibration reduction for precision machine tools [2, 4, 18–22].

The invention patent of high power hydraulic emergency energy absorption for multi-story parking area is applicable to the conditions with high altitude fall, impact, collision and large-scale energy absorption, such as elevator emergency energy absorption safety devices, the bases of forging press machine tools, high-speed rail locomotive vehicle chassis, and so on [27].

Acknowledgements. The authors acknowledge the Jiangsu University Natural Science Research Project (18KJB470024) and Provincial Construction System Science and Technology Project of Jiangsu Provincial Housing and Urban-Rural Construction Department (2018ZD088). This work is partly supported by the Natural Science Foundation of Jiangsu Province of China (No. BK20161165), the applied fundamental research Foundation of Xuzhou of China (No. KC17072). The authorized patents for invention are also the research and development of Jiangsu Province Industry-University-Research Cooperation Project (BY2019056).

References

1. Nakamura, H., et al.: Predicting the attenuation characteristics of a micro-vibration damper for automobile bodies using transfer-function synthesis. In: INTER-NOISE and NOISE-CON Congress and Conference Proceedings, Vol. 259, No. 5, pp. 4934–4944 (2019)
2. Xi, J., Han, C., Wang, X.: Development of a damping worktable for hydraulic excavator. Mach. Tool Hydraul. **41**(2), 85–90 (2013)

3. Xie, L., Cai, S., Huang, G., et al.: On energy harvesting from a vehicle damper. IEEE/ASME Trans. Mechatron. **25**(1), 108–117 (2020)
4. Jiang, D., Wang, Y., Lv, Z., Wang, W., Wang, H.: An energy-efficient networking approach in cloud services for IIoT networks. IEEE J. Sel. Areas Commun. **38**(5), 928–941 (2020)
5. Xi, J., Fan, Q., Han, C.: Self-tuning hydraulic vibration energy absorption method and device. China, p. 10117133.3 (2011)
6. Bachmaier, G., et al.: Lifting system, method for electrical testing, vibration damper, and machine assembly (2019)
7. Jiang, D., Wang, W., Shi, L., Song, H.: A compressive sensing-based approach to end-to-end network traffic reconstruction. IEEE Trans. Netw. Sci. Eng. **7**(1), 507–519 (2020)
8. Jehle, G., Fidlin, A.: Hydrodynamic optimized vibration damper. J. Sound Vib. **440**, 100–112 (2019)
9. Han, C., Xi, J., Zhang, N.: Development of H-type zero opening hydraulic pulse control servo valve for the vibration digging implementation of hydraulic excavator. Mach. Tool Hydraul. **40**(20), 85–95 (2012)
10. Lei-Gang, Z., Yong, Y., Qi-Chang, Y.I.: Optimal design of high power and large flow hydraulic power unit in marine engineering. Mech. Electr. Eng. Technol (2019)
11. Xi, J., Han, C.: Research and development of the self-tuning hydraulic shock wave energy absorption equipment. Mach. Des. Manuf. **10**, 125–127 (2012)
12. Jiang, D., Huo, L., Song, H.: Rethinking behaviors and activities of base stations in mobile cellular networks based on big data analysis. IEEE Trans. Netw. Sci. Eng. **7**(1), 80–90 (2020)
13. Jiang, D., Wang, Y., Lv, Z., Qi, S., Singh, S.: Big data analysis based network behavior insight of cellular networks for industry 4.0 applications. IEEE Trans. Ind. Inform. **16**(2), 1310–1320 (2020)
14. Zhang, N., Xi, J., Han, C.: Research and development of the new hydraulic shock absorber. Hydraul. Pneumatic **07,** 1–3 (2012)
15. Zhang, N., Xi, J., Han, C.: Research and development of vibration absorption device for excavator hammer. Hydraul. Pneumatic, **08**, 103–104 (2012)
16. Jiang, D., Huo, L., Lv, Z., Song, H., Qin, W.: A joint multi-criteria utility-based network selection approach for vehicle-to-infrastructure networking. IEEE Trans. Intell. Transp. Syst. **19**(10), 3305–3319 (2018)
17. Jiang, D., Huo, L., Li, Y.: Fine-granularity inference and estimations to network traffic for SDN. PLoS ONE **13**(5), 1–23 (2018)
18. Xi, J., Han, C., Xu, X.: Research and development of the damping rod of vibration excavator. Mach. Des. Manuf. **11**, 261–262 (2012)
19. Wang, Y., Jiang, D., Huo, L., Zhao, Y.: A new traffic prediction algorithm to software defined networking. Mob. Netw. Appl. (2019)
20. Qi, S., Jiang, D., Huo, L.: A prediction approach to end-to-end traffic in space information networks. Mob. Netw. Appl. (2019)
21. Wang, X., Huang, G., Xi, J.: Research and development of the vibration energy absorbing device for the weighting transducer of belt scale. Process Autom. Instrum. **08**, 34–36 (2013)
22. Xi, J., Han, C., Qiao, S.: Development of self-tuning belt scale weighing device. Instrum. Tech. Sensor **10**, 94–95 (2012)
23. Cui, Q., Han, C., Kong, L.: Development of an anti-impact device for halting vibration screens. Mining Process. Equipment **40**(6), 87–89 (2012)
24. Jiang, D., Zhang, P., Lv, Z., et al.: Energy-efficient multi-constraint routing algorithm with load balancing for smart city applications. IEEE Internet Things J. **3**(6), 1437–1447 (2016)
25. Wang, C.L., Qiu, Z.W., Zeng, Q.L., et al.: Energy dissipation mechanism and control model of a digital hydraulic damper. Shock. Vib. **2019**(9), 1–20 (2019)

26. Jiang, D., Li, W., Lv, H.: An energy-efficient cooperative multicast routing in multi-hop wireless networks for smart medical applications. Neurocomputing **2017**(220), 160–169 (2017)
27. Wang, X.: Research and development of the lift system safety device for stereoscopic parking. Mach. Des. Manuf. **07**, 228–232 (2014)
28. Wang, F., Jiang, D., Qi, S.: An adaptive routing algorithm for integrated information networks. China Commun. 7(1), 196–207 (2019)
29. Wang, L., Xi, J., Han, C.: Self-adjusting belt scale weighing device. China, p. 10232807 (2012)
30. Xi, J., Han, C.: High-power hydraulic emergency energy absorbing device and method of parking tower. China, p. 10217678.0 (2013)
31. Huo, L., et al.: An AI-based adaptive cognitive modeling and measurement method of network traffic for EIS. Mob. Netw. Appl. (2019)
32. Chao, L., Jijian, L., Jinliang, Z.: Double TMD vibration damping method for reducing the hydraulic gate accompanying vibration induced by high dam flood discharge. J. Vib. Shock (2019)
33. Huo, L., et al.: An intelligent optimization-based traffic information acquirement approach to software-defined networking. Comput. Intell. 1–21 (2019)
34. Qiao, S., Wang, X., Han, C.: Manual-adjustment hydraulic energy absorption device and method for emergency protection of parking tower. China, p. 10217657.9 (2013)
35. Hao, X., Han, C.: Device controlling hydraulic energy absorption through analog quantity. China, p. 20586602.0 (2014) Author, F.: Article title. Journal 2(5), 99–110 (2016)

Research on Influence Mechanism of e-WOM on Purchase of Followers in Virtual Community

L. U. Meili[1(✉)], G. A. O. Yujia[1], and W. A. N. Qin[2]

[1] Business Administration College, Shanxi University of Finance and Economics, Taiyuan 030006, China

[2] School of Economics and Management, Southwest Petroleum University, Chengdu 610500, China

Abstract. With more abundant forms of e-WOM, the virtual community provides consumers with an important site to share the personal comprehension on products and exchange the user experience. Electronic commerce blending into social communication has been a new direction for the development of electronic commerce. In social network perspective this paper analyzes the influence of the three related indicators upon the follower purchase, viz. the feature of e-WOM releaser, the interaction of e-WOM diffusion and the factor of commodity, and builds an influence factor model which is tested empirically by taking koubei.jumei.com as an example. The result of Logit regression demonstrates that there is a positive influence of the frequency of releaser's being followed, the total amount of released e-WOM, the number of times of e-WOM being read over follower purchase; while the influence of the frequency of releaser's following, and the quantity of replies of e-WOM and usefulness on follower purchase is not significant. These conclusions have some directive role in building the virtual community of electronic commerce and provide a new idea for the further study of the direct influence of e-WOM upon purchase behavior.

Keywords: e-WOM · Purchase behavior · Virtual community · Logit regression · Follower

1 Introduction

WOM is an important stimulus that influences purchase decision. In contrast with offline WOM, the e-WOM plays an important role within the purchase process for its even wider diffusive range and even higher diffusive speed [1]. The influence of e-WOM on consumer was first reflected in the consumer review system. In 1995, America's Amazon began to provide consumers with a function on its electronic commerce website to publish their comments about their purchased commodities, which gave a great impetus to its growth in business. And now China's electronic commerce trading platforms such as Taobao, JD, Dangdang have set up their online evaluation system for consumers to publish their comments on commodities, so it is very common for consumers to collect

H. Song and D. Jiang (Eds.): SIMUtools 2020, LNICST 370, pp. 489–504, 2021.
https://doi.org/10.1007/978-3-030-72795-6_40

the online review information before purchasing. In recent years, many third-part websites or electronic commerce websites have opened special virtual communities such as douban.com, dianping.com, koubei.jumei.com and so on. These experiential communication platforms provide consumers with a better site to share the purchase experience and exchange the use comprehension. The better interaction of virtual community has been an important platform on which consumers participate in value co-creation, which imperceptibly improves the additional value and influences consumer's purchase decision more strongly [2].

There is a great number of achievements in researching the influence of e-WOM on consumption behavior [3]. Lee et al. induced the research of e-WOM influence on two levels, viz. the market level research and individual level research [4]. From the aggregation perspective the research conclusions of the marker level research demonstrate that there is a significant correlativity between e-WOM and sale volume [5]. The individual level research probes into the influence of e-WOM on confidence, purchase intention and usefulness through the approach of questionnaire, and the resultant conclusion will be drawn that WOM facilitates decision on purchase and acceptance of information [6]. However, facing a great quantity of purchase data accumulated and provided by virtual community, can the relation of WOM and follower purchase behavior be directly discussed on the individual level in order that the degree of that WOM or the releaser of WOM influences the follower of WOM? This is a question which is lack of attention and worth paying attention to.

With the increasing and enriching of socialized function of third-part review website and electronic commerce, such situations as the network relationship between e-WOM releaser and e-WOM follower, commodity purchase, e-WOM releasing, likelihood of follower purchase and so on can be observed from the social network perspective of virtual community. Therefore, considering the consumer who follows a certain e-WOM releaser, if some of his purchase behaviors are observed after an e-WOM having been released, we can affirm that the follower purchase after a release of e-WOM is mainly owes to the e-WOM released by the corresponding releaser, although we cannot affirm it is the direct result of the very influence of e-WOM. Meanwhile, the unique data structure of e-WOM release and follower purchase from koubei.jumei.com provides a likelihood for the further research in the individual perspective. Therefore, this paper will build a research model on the influence of e-WOM of virtual community on follower purchase and empirically analyze it with the data got from koubei.jumei.com, and then verify the related hypothesis.

The paper is organized as follows. Section 2 puts forward the hypothesis and establishes the influencing factor model of e-WOM on the purchase of followers in virtual community. Section 3 describes the empirical data and research method. Section 4 gives empirical result. Section 5 concludes and gives some management implications as well as limited aspects.

2 Model Building

In general, the individual level research on e-WOM takes purchase intention or usefulness of e-WOM as dependent variable [4, 7], or takes usefulness of e-WOM and reliability of e-WOM as intermediate variable at the same time takes the acceptance of e-WOM as final dependent variable [8]. However, neither of the two approaches can get the actual purchase result. Therefore, purchase intention directly correlates with actual purchase and usefulness of e-WOM leads acceptance of e-WOM, and an actual purchase is attained in the end, but we only in an aggregation perspective observe the overall situation of purchase. Actually, the aggregated thing is the aggregation of individuals. Generally saying, the relation of sale volume and e-WOM necessarily stems from the stimulus of e-WOM to every individual, however, the influence of e-WOM on consumer behavior is only an aspect of it and the degree of influence of itself is not to be lumped together. The key of direct research on the relation between e-WOM and purchase on individual level lies in how empirically prove the relation between e-WOM and purchase behavior combining with the concrete factors that e-WOM influences purchase decision.

The analysis of consumer's WOM release behavior is as follows. After logging in the virtual community, the consumer buys commodity and forms experience about using of these commodities. Because of reading the released e-WOM, the members of the community follow these commodities because of their interests or identifications, as a result, they become followers, viz. "fans" so called on website commonly. Then these followers (fans) read e-WOM, communicate, interact, browse commodities, and buy commodities for being influenced by the releaser of this e-WOM, and maybe release new e-WOM after their buying and experiencing these commodities. The corresponding data records which shown as Fig. 1 should be generated in all these concrete procedures. If we call this purchase behavior of the corresponding fans (viz. followers of the certain e-WOM releasers) as "follower purchase" which indicates that the follower has bought the commodity described by the corresponding e-WOM. Then, through all the process of paying attention to WOM release, reading e-WOM and browsing commodities we can see the influence of the active states of e-WOM releasers, replies after browsing the current e-WOM and the concrete condition of corresponding commodities on follower's purchase.

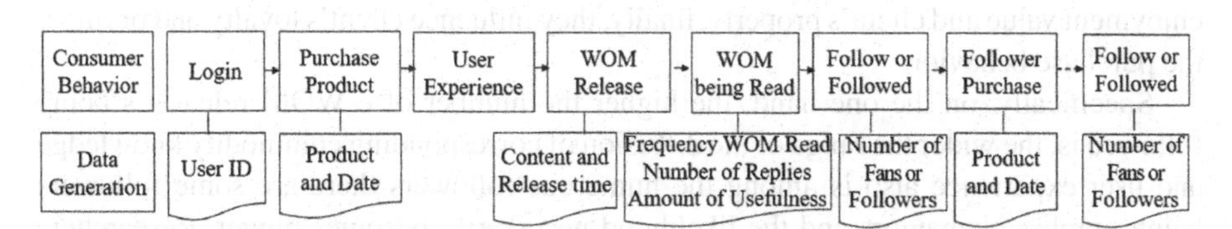

Fig. 1. e-WOM release, following process and data generation in virtual community.

2.1 The Influence of WOM Releaser

As an information disseminator the e-WOM releaser influences the convincing force of e-WOM by the identity openness, specialization and reputation of itself. In recent years, with the development of trade-type virtual community, the research on opinion leader and the feature of e-WOM releaser in the network structure perspective has attracted much attention in every possible way.

(1) Social identity theory and network formation
The social identity theory holds that every individual makes social classification according to the similarity of interest, background and value, shows more positive attitude to the group and active communication with each other, and displays stronger confidence to the members within the group. As a result, he is more apt to accept the opinions of other members within. After being put forward this theory gradually gets recognition in several fields just like marketing [9]. More and more researchers have understood that the factors such as group classification, value and reputation promote the social identity of individual [10]. Xiong et al. deems that the leader of opinion accumulates the social interactive relation through different ways such as self-identity, knowledge contribution and reciprocation and so on, meanwhile, the knowledge contribution and reciprocation have intermediary function on the status of opinion leadership and social interactive relation [11]. The formation of social network structure of trade-type virtual community produces following and being followed behavior.

(2) Social network theory and the influence of e-WOM
Liang thinks Node centrality (out-degree) analysis can be applied to identify the users with high degree of active [12]. The researchers like Cao et al. focus on member's special position in network and analyze the different contribution of different position to virtual community [13]. Yang et al. points out that the opinion leadership in community can influence widely the others' consumer attitude, belief and behavior, and bring purchase influence on majority of members [14].

By the above content it can be understood that with the social intercourse blending into electronic commerce and social identity being promoted, the influence of the features of network structure over e-WOM releaser should be paid attention to. The e-WOM releaser's network features are in-degree and out-degree, and herein the in-degree means the frequency of being followed and the out-degree means the frequency of following. "Following and being followed" are the major method of reflecting identification and interactive exchange and client's interaction helps the realization of practical value, enjoyment value and client's property, finally, they influence client's loyalty and promote the purchase behavior.

Specifically, on the one hand, the higher the number of e-WOM releaser's being followed is, the wider the range of the diffusion of corresponding commodity knowledge and user experience also is among the numerous followers there are some followers being purchase behaviors and the likelihood necessarily becomes bigger. Researchers as Susarla et al. deems that social network has important influence on the dissemination and diffusion of e-WOM [15]. The more chains are followed, the rapider the videos disseminate and the more possible the positive evaluation may be gotten. Therefore, we put forward the hypothesis below.

Hypothesis 1: there is a positive correlation between the frequency of e-WOM releaser's being followed and follower purchase.

On the other hand, if the follower who acts as the member of virtual community can effectively identify his membership and establish a good relationship of membership in community his sense of belonging to community can be enhanced and a positive attitude to community can be improved directly. Therefore, if follower can draw more attention from e-WOM releaser, he will establish further the belief of virtual role of himself [16] and reinforce the trust in the e-WOM releaser. This belief and trust will improve mutual understanding and social identity [17]. The bigger the number of e-WOM releaser's following is, the more followers this kind of trust and belief are transmitted to, consequently, there may be more followers who make purchase. So, we put forward the hypothesis as below.

Hypothesis 2: there is a positive correlation between the frequency of words-of mouth releaser's following (out-degree) and follower purchase.

In addition, the quantity of e-WOM released by releaser is a major factor of its own influence. Obviously, if there are more e-WOM, the more potential to attract consumer visits and increase the time spent online [18], the influence of dissemination will become stronger, so this condition must promote follower purchase. We put forward the study hypothesis as below.

Hypothesis 3: there is a positive correlation between the amount of e-WOM released by e-WOM releaser and follower purchase.

2.2 The Influence of Current e-WOM

The central path of the elaboration likelihood model demonstrates clearly that the information acceptor analyzes and identifies the accepted information seriously, carefully and as systematically as possible, and after meditating he attains the understanding on the accepted information and forms his own self-view, furthermore, this condition leads to an attitude change. To evaluate e-WOM being read, replies to e-WOM and possibility of usefulness is the expression of the concrete behavior in the process of accepting information, so, the number of reviews being read, the number of replies and the quantity of usefulness can directly measure the response degree of WOM readers.

Schlosser supposed 98% of online shoppers read reviews before making purchases [19]. Many studies discussed the influence of text-based WOM [20, 21], Zhang et al. proposed that consumers read multimedia expression of WOM that is a mixture of words, photos, videos, and audio before purchasing [22]. So, the frequency of WOM being read is key to purchase.

When text-based WOM or multimedia WOM have been shared by the WOM releaser, it is incessantly read by online consumers, especially the frequency of being read accumulated to a certain number, it can spur more consumers on to browse, and then it may influence browsers' purchase decisions. For that reason, we put forward the hypothesis below.

Hypothesis 4: there is a positive correlation between the frequency of e-WOM being read and follower purchase.

Moreover, if some browsers make replies on e-WOM it shows that there is core information drawing the browsers' attention and the e-WOM in itself is readable. Beside "following and being followed", the reply to the opinion on e-WOM is another major way to inform identification and interactive exchange. Interaction is an important action and content of virtual community. Among the members the interaction reduces the uncertainty of consumer's attitude to commodities and then produces a positive influence over purchase behavior [23]. Brodie and other researchers hold that the process of consumer's participation includes a series of subsystems and through interaction consumers in virtual community experience co-creation of value [24], what is more, the consumers express higher client's loyalty, satisfaction, trust and commitment. From the above it can be found that the number of replies to WOM in virtual community is a kind of approval for the information of WOM and reflects the direct data of community numbers interactive behavior as well. The discussion about commodity and interaction may make commodity to draw more attention, also, it is more possible to promote more purchase. So, we put forward the hypothesis below.

Hypothesis 5: there is a positive correlation between the number of replies and follower purchase.

In addition, having put forward a model of acceptance of information based on situation of online information communication and analyzed the concept of "usefulness of information", Sussma and Siegal found that there are two direct factors that information acceptor feels information usefulness such as the reliance of information resource and the quality of information [25], so they held that the information usefulness directly influences information acceptance. Mudambi and Schuff explicitly put forward the theoretical construct of usefulness of online review which was defined as a feeling value in the consumer's decisive process [18], they also pointed out all the online review extremity, review depth, and product type could influence the perception of the usefulness of review. The higher the usefulness of online review is, the better the convincing effect on information acceptors will be [26], in consequence, the usefulness of online review has more significant influence on purchase decision, so we bring forward the hypothesis as below.

Hypothesis 6: there is a positive correlation between the amount of helpfulness of e-WOM and follower purchase.

Besides, considering the influence of commodity itself on purchase behavior, the number of reviews, review valence and commodity price have significant influence on purchase behavior [5, 27]. This research takes the three variables which come from the overall commodity condition as control variables.

2.3 Influence Factor Model

Based on the above hypotheses and analysis, we principally research the influence of the indicators of social network features, e-WOM being read, replies and helpful reviews on the e-WOM follower purchase. For this research we put forward the influence factor model shown as Fig. 2.

The dependent variable is follower purchase, the independent variables consist three types, those are e-WOM releaser, current e-WOM, and Current product factors. Specifically, e-WOM releaser factors are e-WOM releaser In-degree, e-WOM releaser Out-degree and the amount. The influence of current e-WOM factor comes from frequency WOM read, number of replies and amount of usefulness. The current product factors have the number of reviews, review valence and commodity price.

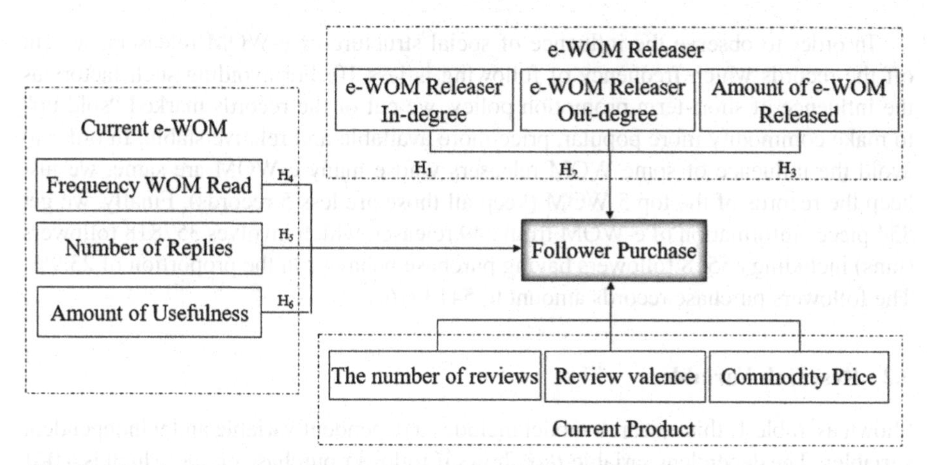

Fig. 2. Model of influencing factors of e-WOM on follower purchase in virtual communities

3 The Research Method

3.1 The Empirical Data

This research takes the data from jumei.com website to do empirical research. The main reason has three aspects: the first is the representativeness of data. Jumei.com is the first website of China's cosmetic industry. Its self-built center of e-WOM, koubei.jumei.com, has worked from 2011. In the several years, the characteristic virtual community named "jumei WOM" has achieved a better effect of propagating brands and products. The second is richness. The website of koubei.jumei.com provides its community numbers with technical support for satisfaction of their demands. We can see the content of rich data such as name of e-WOM releaser (wise person), products, purchase date, content of e-WOM, date of e-WOM release, number of follows(fans), frequency of being followed, frequency of reading, number of replies, usefulness etc. on the page of its website. We also can achieve tracking link and obtain more related records of all the fans of e-WOM releaser. The third is availability of data. If you access koubei.jumei.com, according to the identification of corresponding ID address, by composing program you can get the data of fans and wise person, which provide possibility for analyzing follower purchase behavior.

These empirical data are taking through Web Crawler. First, we randomly select 10 fans as initial entrances from about 200 wise persons, and through snowballing method we can get 1931 clients' IDs from which we randomly select 300 ID entrances. We download "client name, number of fans, frequency of following, commodity ID of e-WOM, frequency of e-WOM being browsed, number of replies, usefulness, date of release, fans' IDs and fans' "names of purchased products, price, and purchase date". Then we delete the incomplete records (such as the incomplete product information) and the clients who have duplicate fans. Finally, we get 280 first level clients whom are called wise persons. The records amount to 18298. All the fans of corresponding wise person make up the second level clients. Download date is March 2016. The final data begin with May 1, 2011 and end up February 29, 2016.

In order to observe the influence of social structure of e-WOM releasers, we cut off the records which frequency of following is less 10. For avoiding such factors as the influence of short-term promotion policy, we cut of the records marked "sold up" to make commodity more popular, price more available and relative stable. In order to avoid the influence of some WOM releasers whose many e-WOM are same, we just keep the records of the top 5 WOM (keep all those are less 5 records). Finally, we get 834 pieces information of e-WOM from 249 releaser, which involves 357818 followers (fans) including 85518 followers having purchase behavior in the proportion of 23.9%. The followers purchase records amount to 5413916.

3.2 Research Variable

Shown as Table 1, this research model includes a dependent variable and 9 independent variables. The dependent variable *Buy* shows if follower purchase or not, which is a 0–1 variable, namely "if followers purchased the commodities described by the corresponding e-WOM." It reflects if the followers whom an e-WOM releaser (wise person) belong to purchase the very commodity after the date when the e-WOM was released.

Table 1. Model variables and description

Name	Meaning	Description
Indegree	Frequency of e-WOM releaser being followed	Number of community members followed the e-WOM releaser, that is "number of fans"
Outdegree	Frequency of e-WOM releaser following	Number of community members being followed by the releaser of e-WOM
A'WOMs	Amount of released e-WOM	Amount of released e-WOM about the purchased commodities himself
Read	Frequency of e-WOM being read	Number of times of reading a certain e-WOM
Reply	Number of replies to e-WOM	Number of corresponding replies to a certain e-WOM
Useful	Amount of usefulness	Amount of a certain e-WOM being evaluated as usefulness
P'WOMs	Amount of commodity e-WOM	Total amount of released e-WOM on a certain commodity
Star	Score or star of a certain commodity	Number of scores or grades of a certain commodity
Price	Commodity price	Current price of the corresponding commodity
Buy	Follower purchase	If followers purchased the commodity described by the corresponding WOM

Considering independent variable, *Indegree* is called inside centripetal as well, which means the number of virtual community members who follow WOM releaser. *Outdegree* is called outside centripetal too, which means the number of virtual community members who followed by WOM releasers. *A'WOMs* means the total amount of e-WOM which are released by the e-WOM releasers after they experienced the concrete purchased commodities. The three variables are mainly used to measure the influence of e-WOM on follower purchase.

Except the network structure of virtual community, the other major feature of virtual community is WOM interaction. *Reply* means the replies made to the certain e-WOM. *Useful* means the accumulated amount of the corresponding e-WOM. Otherwise, *Read* means the number of times of the current e-WOM being read.

Lastly, *P'WOMs* means the total amount of e-WOM that is produced by the current commodity. *Star* means the overall evaluation of commodity which perfect score is 5 (5 stars). *Price* means the actual price of commodity.

3.3 Building Model

We take *Buy* as dependent variable of the model, which is virtual variable. $Buy = 1$ represents there are followers who made purchase. $Buy = 0$ represents there are not followers who made purchase. All the dependent variables are continuous variables. We process data with the form of logarithmic conversion. Mainly, through using logarithmic conversion to convert the nonlinear relation among latent variables to a linear relation, the results of regression will be more stable. And more, after logarithmic conversion, the discrepancy of dimension among variables will be reduced as a result the influence of the latent outliers will be reduced. In addition, because 0 value exists among the frequency of e-WOM releaser following namely *Outdegree, Reply, Useful*, when we take logarithm, we add 1 based on every original value.

Building Logit regression model

$$LogitP(Buy) = \ln \frac{P(Buy)}{1 - p(Buy)} = \beta_0 + \beta \ln X$$

Obviously, $P(Buy) \in [-1, 1]$, and $\frac{P(Buy)}{1-p(Buy)}$ represents the chance ratio of followers' purchasing to followers' not purchasing. X represents a row of independents variables, $\ln X$ is logarithmic conversion of corresponding variable, β is corresponding coefficient, thus

$$P(Buy) = \frac{\exp(\beta_0 + \beta \ln X)}{1 + \exp(\beta_0 + \beta \ln X)}$$

The parameters of Logit can be got by Maximum Likelihood Estimation.

4 Empirical Result and Discussion

4.1 Statistical Description

By SPSS21.0 statistically describe the above variables before taking logarithm and the results are listed in Table 2. There are 834 data samples in which the number of followers who didn't purchase is 580 which accounts for 69.5% of the total amount of samples meanwhile the number of followers made purchase is 254 which accounts for 30.5% of the total amount of samples. The discrepancy of frequency of e-WOM releaser being followed (indegree) is relatively big. The minimum is 10 and the maximum is greater than 1000. The discrepancy of Read and P'WOMs is relatively big as well.

Table 2. Descriptive statistical quantity

Valuable	Number(N)	Min	Max	Mean	Std. deviation
Buy	Buy = 0, N = 580 Buy = 1, N = 254	0	1	–	–
Indegree	834	10	11937	1581.362	2872.010
Outdegree	834	0	353	32.996	38.574
A'WOMs	834	2	310	67.444	60.350
Read	834	19	400278	3415.130	15822.539
Reply	834	0	190	9.510	22.623
Useful	834	0	432	3.480	16.935
P'WOMs	834	1	10325	995.310	1420.513
Star	834	2.8	5	4.619	0.170
Price	834	9	960	158.220	153.248

4.2 Logit Regression

(1) Result of regression
As for the built binary Logit regression, we analyze it with Enter method. The result shown as Table 3.

Table 3. The result of regress analysis.

Variable	Coe. estimator B	Std. deviation	Wald	p	EXP(B)
Indegree	0.474	0.064	53.984***	0.000	1.606
Outdegree	0.015	0.111	0.018	0.646	1.015
A'WOMs	0.994	0.170	34.268***	0.000	2.703
Read	0.635	0.123	26.563**	0.000	1.887

(continued)

Table 3. (*continued*)

Variable	Coe. estimator B	Std. deviation	Wald	p	EXP(B)
Reply	−0.101	0.099	1.043	0.307	0.904
Useful	−0.074	0.132	0.312	0.576	0.929
P'WOMs	0.533	0.080	44.229***	0.000	1.704
Star	16.174	3.783	18.275***	0.000	105756
Price	−0.436	0.130	11.211***	0.001	0.646
Overall adaptability of model	Omnibus test of model: $\chi^2 = 453.132, p = 0.000$ Hosmer–Lemeshow test value = 5.382, $p = 0.716$				
Association strength	Cox & Snell, R Square = 0.419; Nagelkerke, R Square = 0.592				

In the analysis of regression model, the most ideal state is the chi-square value reaches significance level, meanwhile, Hosmer–Lemeshow test do not reach signicance level. The Omnibus test result of this model indicates that the chi-square value got from the overall adaptability test for the regression model built by several variables is 453.132, $p = 0.000$, reaches significance level, meanwhile, the result of testing overall adaptability of regression model by the Hosmer&Lemeshow test method is: Hosmer–Lemeshow test chi-square value equals 5.382, $p = 0.716$, which doesn't reach significance level. It shows that the adaptability of model overall regression is better, that is to say, overall the dependent variables can effectively predict the follower purchase.

Considering association strength coefficient, the Cox-Snell association strength value is 0.416, and Nagelkerke association strength indicial value is 0.592. It indicates that there is a higher degree association.

In the test of the significance of individual parameter of Logit regression model, there are three independent variables which are not significant, namely, the frequency of e-WOM releaser following (*Outdegree*), number of replies to WOM (*Reply*) and amount of usefulness (*Useful*), and the same time the Wald tested values of other variables are bigger, meanwhile the p value is less than 1%, it has a very good statistical significance which shows that except the above three independent variables the other independent variables influence follower purchase significantly.

The Logit regression predication shown as Table 4. The correctness ratio of predication of follower's not purchase is 88.6%, the correctness ratio of predication of follower purchase is 74.0%, and the overall correctness ratio is 84.2%.

(2) Conclusion

The above analysis shows that the three hypotheses within the six hypotheses hold water and the others do not hold water. Among all the control variables the influence of 3 variables are significant. The conclusion summed up in Table 5.

Table 4. Predication classification table

Observation\Prediction	Prediction Buy = 0	Prediction Buy = 1	Percent
Observation Buy = 0	514	66	88.6%
Observation Buy = 1	66	188	74.0%
Overall percent			84.2%

Note: the cut value is 0.500

Table 5. The specific conclusion of Logit regression

Research hypotheses and control variables conclusion		Verified result
Hypothesis 1: between the frequency of e-WOM releaser's being followed and follower purchase there is a positive correlation		Support
Hypothesis 2: there is a positive correlation between the frequency of words-of mouth releaser's following (out-degree) and follower purchase		Not support
Hypothesis 3: there is a positive correlation between the amount of e-WOM released by e-WOM releaser and follower purchase		Support
Hypothesis 4: there is a positive correlation between the frequency of e-WOM being read and follower purchase		Support
Hypothesis 5: there is a positive correlation between the number of replies and follower purchase		Not support
Hypothesis 6: there is a positive correlation between the degree of helpfulness of e-WOM and follower purchase		Not support
Conclusion of control variable	Between the total amount of commodity e-WOM and followers' purchase there is a positive correlation	Significance
	Between the number of stars of commodity and followers' purchase there is a positive correlation	Significance
	Between the price of commodity and followers' purchase there is a negative	Significance

(3) Discussion

In terms of the network feature of e-WOM releaser, hypothesis 1 holds water and hypothesis 2 does not hold water. The conclusion of hypothesis 1 is more significant: necessarily, the more the number of e-WOM releasers who are followed is, the more possibly the followers make purchase. However, in terms of the number of people who are followed by e-WOM releaser of their own accord, although in some extent this data influences the structure of virtual community and the effect of interaction, especially when both of them follow each other, which shows even more that they have more similar fondness; the discrepancy in mean number of the frequency of e-WOM releaser's being followed and the frequency of e-WOM releaser's following is very big(1581 and 33), which shows that such situations are few. Therefore, as to a WOM releaser who possess a lot of people being followed, although he might read the e-WOM released by other e-WOM releasers, no matter if he displays of his own accord that he follows other people and pay attention to the displayed specific number of people to whom he pays attention, there has little influence on the purchase situation of people being followed. The analysis and empirical conclusion of Yin hold that both indegree and outdegree of network features influence usefulness of WOM and then can influence the purchase condition [27]. Besides, Yin's conclusion is opposed to the thought of Min et al. [28], which explains the different kinds influence on experimental products and search products. Compare with this research, both Yin and Min come to an approximately identical conclusion except some difference in the understanding of outdegree. The products on which this research makes are experimental products, but in the empirical test we draw a conclusion that the influence of outdegree is not significant. It needs to make a further empirical exploration through associating the network features of specific websites or getting more network data. To understand hypothesis 3 is more ease. Between the amount of the e-WOM which more releasers and their numerous people being followed are closely related, so its conclusion of positive correlation is nearly similar.

From the three dependent variables of e-WOM itself, we can see just the frequency of being read influences follower purchase significantly, and the conclusion of number of replies and amount of usefulness don't get empirical support. This may result from koubei.jumei.com having different way to set the values of *Read*, *Reply* and *Useful*. The data of *Read* don't need to register, whenever it is opened, a *Read* is added, and continuous hit result in continuous count, to a certain extent it relatively objectively reflects the times of this e-WOM being read. On the contrary, adding *Reply* and *Useful* need to register and log on personal account first, which is more capable of reflecting the real behavior of community member or follower, though it needs initiative participation consciousness, namely, the community members or followers show their attitude only through leaving their comments actively to express their real opinion or hitting hit the "useful" state. Statistics of *Reply* show that both 0 and 1 account for 51.4%, and statistics of *Useful* show that both 0 and 1 account for 68%, which indicates that the participation degree of all the member is on the low side. In addition, by taking *Useful* as dependent variable and using the ratio of "the useful" to "the non-useful" to measure, analysis and empirical probation of Yin avoid the current shortage that we use absolute amount to measure "*Reply*" and "*Useful*" [27]. In the comparison and appraisement of douban.com there is "useful" or "non-useful" option, by contrast, koubei.jumei.com just set "useful", which cannot avoid the shortage of the participation degree of all the members is on the low side. Therefore, koubei'jumei.com could adopt the comparison and appraisement method of douban.com and set "useful" or "non-useful" measure option in order to make followers directly get more explicit relative value contrast in the process of browsing e-WOM as well as make the useful e-WOM serve followers purchase more conveniently.

That all the three control variables of commodity are significant indicates the total amount of e-WOM (*A'WOMs*), grade (*Star*) and price (*Price*) influence follower purchase significantly. The higher the *A'WOMs* are, the higher the grade is; the lower the price is, the more possibly the followers make purchase.

5 Research Conclusion and Prospects

5.1 Research Conclusion

In the electronic commerce practice of recent years, numerous enterprises have been aware of the values of social network in extending and cultivating client relationship and stirring up client's creative behavior. The communication and interaction among the members of social network even produce great influence on business strategy and operation, in consequence, electronic commerce enterprises set about making electronic commerce marketing plans for blending into social communication. In social network perspective this paper builds influence factor model about network WOM influencing follower purchase and takes koubei.jumei.com as the empirical object to test the hypotheses mentioned, and the conclusion is shown as below:

In the perspective of e-WOM releaser, the frequency of e-WOM releaser's being followed(indegree) positively correlates with follower purchase. The correlation between the frequency of e-WOM releaser following(outdegree) and follower purchase is not significant. In the perspective of e-WOM itself, the frequency of e-WOM being read has positive influence on follower purchase and the influence of the amount of replies and the amount of usefulness on follower purchase is not significant. In commodity perspective, the total amount of e-WOM positively correlates with follower purchase, commodity grading positively correlates with follower purchase, and commodity price negatively correlates with follower purchase.

5.2 Theoretical Contribution and Practical Enlightenment

It is highly difficult and risky to determine the consumer who follows e-WOM purchase the corresponding commodity mainly affected by current e-WOM stimulating. Therefore, all the researchers always analyze influence factors of the usefulness of e-WOM or purchase intention from every perspective of network WOM. But the special sample data of koubei.jumei.com provide us with chances for further researching actual purchase behavior. So, the main contribution of this study lies in:

(1) Based on the unique data structure of e-WOM release and follower purchase from koubei.jumei.com, It is the first time to study the impact of online WOM on followers' purchase from an individual perspective. We build and empirically test the model for researching follower purchase and make a certain theoretical contribution for network WOM.

(2) As the electronic commerce blended into social communication, the network structure of "following and being followed" among members is a main feature. Because of having more similar interests and fondness, the follower and e-WOM releaser display more identification, and then the relation about network structure of "following and being followed" comes into existence, as a result the chances that purchase behavior arise from the influence of e-WOM display more significant. We could consider that if the follower purchase behavior arises after getting e-WOM mainly is decided by the influence of e-WOM. This Logit model empirical analysis provides a new way of thinking for further research of the influence of network WOM on purchase behavior from an individual perspective.

The main enlightenment of the research discounted in this paper on electronic commerce express in the following aspects:

(1) The social network feature of e-WOM releaser is the indispensable and important factor of e-WOM. The community platform should adopt specific measure for encouraging members to improve the amount of e-WOM, meanwhile, through some means such as sample display, comparing and appraising excellence etc. guide e-WOM releaser to improve the e-WOM quality for the purpose of drawing more members' attention and increasing the influence of e-WOM.

(2) In order to avoid the regret caused by the fact that the influence of absolute amount of usefulness on follower purchase is not significant, it can be considered to provide communication interaction platform and set "useful" or "non-useful" option for benefiting follower to form clear judgment and make purchase decision in comparison among relative values.

5.3 The Limitation of Research and Prospects

Though the research and its conclusion in this paper has a certain theoretical and practical meaning, there still are some shortages, for instance, in the analysis of the network feature of e-WOM releaser, the influence of the frequency of e-WOM releaser being followed (indegree) on purchase is similar to other researcher's conclusion on e-WOM usefulness, but in the empirical test of the frequency of e-WOM releaser following (outdegree), there is relative wider discrepancy in the research conclusion and explanation, which need to be further paid attention to and remain to be researched in depth from such aspects as commodity type and website feature.

Acknowledgments. This research was funded by Humanities and Social Science Fund of Ministry of Education of China, Grant No. 18YJA630071 and 19YJC630159; Soft science research project of Shanxi Province in China, Grant No. 2018041069–1; Business Management Advantage Discipline Climbing Project of Shanxi Province Higher Education, Grand No. 4[2018] (Shanxi Province Teaching Research) in China.

References

1. Park, M.S., Shin, J.K., Ju, Y.: Attachment styles and electronic word of mouth (e-WOM) adoption on social networking sites. J. Bus. Res. **99**(6), 398–404 (2019)
2. Han, H., Xu, H., Chen, H.: Social commerce: a systematic review and data synthesis. Electron. Commer. Res. Appl. **30**, 38–50 (2018)
3. Zhu, F., Zhang, X.: Impact of online consumer reviews on sales: the moderating role of product and consumer characteristics. J. Mark. **74**(2), 133–148 (2010)
4. Lee, J., Lee, J.N.: Understanding the product information inference process in electronic word-of-mouth: an objectivity–subjectivity dichotomy perspective. Inf. Manage. **46**(5), 302–311 (2009)
5. Chevalier, J.A., Mayzlin, D.: The effect of word of mouth on sales: online book reviews. J. Mark. Res. **43**(3), 345–354 (2006)
6. Gruen, T.W., Osmonbekov, T., Czaplewski, A.J.: eWOM: the impact of customer-to-customer online know-how exchange on customer value and loyalty. J. Bus. Res. **59**(4), 449–456 (2006)

7. Chen, X., Sheng, J., Wang, X.J., Deng, J.S.: Exploring determinants of attraction and helpfulness of online product review: A consumer behaviour perspective. Discrete Dyn. Nat. Soc. 1–19 (2016)
8. Jamil, R.A.: Consumer's reliance on word of mouse: influence on consumer's decision in an online information asymmetry context. Soc. Sci. Electron. Publishing **5**(5), 171–205 (2015)
9. Dholakia, U.M., Bagozzi, R.P., Pearo, L.K.: A social influence model of consumer participation in network- and small-group-based virtual communities. Int. J. Res. Mark. **21**(3), 0–263 (2004)
10. Ren, Y., Harper, F.M., Drenner, S.: Building member attachment in online communities: applying theories of group identity and interpersonal bonds. MIS Q. **36**(3), 841–864 (2012)
11. Xiong, Y., Cheng, Z.C., Liang, E.H., Wu, Y.B.: Accumulation mechanism of opinion leaders' social interaction ties in virtual communities: empirical evidence from China. Comput. Hum. Behav. **82**, 81–93 (2018)
12. Liang, R., Zhang, L., Guo, W.: Investigating active users' sustained participation in brand communities: effects of social capital. Kybernetes **48**(10), 2353–2363 (2019)
13. Cao, Q., Lu, Y., Dong, D.: The roles of bridging and bonding in social media communities. J. Am. Soc. Inf. Sci. **64**(8), 1671–1681 (2013)
14. Yang, J., Yecies, B., Zhong, P.Y.: Characteristics of Chinese online movie reviews and opinion leadership identification. Int. J. Hum. Comput. Interact. **25**, 1–6 (2019)
15. Susarla, A., Oh, J.H., Tan, Y.: Social networks and the diffusion of user-generated content: evidence from YouTube. Inf. Syst. Res. **23**(1), 23–41 (2012)
16. Ma, M., Agarwal, R.: Through a glass darkly: Information technology design, identity verification, and knowledge contribution in online communities. Inf. Syst. Res. **18**(1), 42–67 (2007)
17. Nambisan, S., Baron, R.A.: Virtual customer environments: testing a model of voluntary participation in value co-creation activities. J. Prod. Innov. Manag. **26**(4), 388–406 (2010)
18. Mudambi, S.M., Schuff, D.: What makes a helpful online review? a study of customer reviews on Amazon.com. MIS Q. **34**, 185–200 (2010)
19. Schlosser, A.E.: Can including pros and cons increase the helpfulness and persuasiveness of online reviews? the interactive effects of ratings and arguments. J. Consum. Psychol. **21**(3), 226–239 (2011)
20. Bansal, H.S., Voyer, P.A.: Word-of-mouth processes within a services purchase decision context. J. Serv. Res. **3**(2), 166–177 (2000)
21. Cheung, C.M.K., Lee, M.K.O., Rabjohn, N.: The impact of electronic word-of-mouth: the adoption of online opinions in online customer communities. Internet Res. **18**(3), 229–247 (2008)
22. Zhang, H., Takanashi, C., Si, S., Zhang, G., Wang, L.: How does multimedia word of mouth influence consumer trust, usefulness, dissemination and gender? Eur. J. Int. Manage. **13**(6), 785–810 (2019)
23. Adjei, M.T., Noble, S.M., Noble, C.H.: The influence of C2C communications in online brand communities on customer purchase behavior. J. Acad. Mark. Sci. **38**(5), 634–653 (2010)
24. Brodie, R.J., Ilic, A., Juric, B.: Consumer engagement in a virtual brand community: an exploratory analysis. J. Bus. Res. **66**(1), 105–114 (2013)
25. Sussman, S.W., Siegal, W.S.: Informational influence in organizations: an integrated approach to knowledge adoption. Inf. Syst. Res. **14**(1), 47–65 (2003)
26. Pan, Y., Zhang, J.Q.: Born unequal: a study of the helpfulness of user-generated product reviews. J. Retail. **87**(4), 598–612 (2011)
27. Yin, G.P.: What online reviews do consumers find more useful? – the effect of social factors. Manag. World **12**, 115–124 (2012)
28. Min, Q.F., Qin, L., Zhang, K.L.: Factors affecting the perceived usefulness of online reviews. Manag. Rev. **29**(10), 95–107 (2017)

Modeling Interactions Among Microservices Communicating via FIFO or Bag Buffers

Fei Dai[1], Jinmei Yang[1], Qi Mo[2(✉)], Hua Zhou[1], and Lianyong Qi[3]

[1] School of Big Data and Intelligent Engineering, Southwest Forestry University, Kunming, China
daifei@swfu.edu.cn

[2] School of Software, Yunnan University, Kunming, China

[3] School of Information Science and Engineering, Qufu Normal University, Jining, China

Abstract. Interactions among individual microservices communicating asynchronously via FIFO or bag buffers vary significantly even for the same buffer size. Different interactions among microservices will lead to different interaction behaviors, which can make microservices systems malfunction during their execution. However, these two asynchronous communication models with FIFO or bag buffers are seldom distinguished. In this paper, we present new results for the interaction differences between one asynchronous communication model with FIFO buffers and another asynchronous communication model with bag buffers. First, we propose a framework to uniformly define two asynchronous communication models. Second, we model interaction behaviors among microservices as sequences of send and receive message actions under these two asynchronous communication models. Finally, we compare these two asynchronous communication models using refinement checking to show their differences. Experimental results show that the asynchronous communication model with FIFO buffers is included in the asynchronous communication model with bag buffers.

Keywords: Interactions · Microservice · Asynchronous communication · FIFO buffers · Bag buffers

1 Introduction

The cloud computing paradigm drives many IT companies to build microservice systems [1, 2]. A microservices system consists of a set of individual microservices that interact with each other via exchanging messages [3]. Compared with traditional web services, these microservices are much more fine-grained and are independently developed and deployed. These characteristics of microservices drive many IT company's software systems to migrate from monolithic architecture to microservice architecture [4, 5].

In a microservices system, most of the interactions among microservices are asynchronous since synchronous interactions are considered to be harmful due to the multiplicative effect of downtime [4]. These interactions among individual microservices communicating asynchronously via FIFO or bag buffers vary significantly even for the

H. Song and D. Jiang (Eds.): SIMUtools 2020, LNICST 370, pp. 505–518, 2021.
https://doi.org/10.1007/978-3-030-72795-6_41

same buffer size. These different interactions will lead to different interaction behaviors, which can make microservices systems malfunction or deadlock during their execution. However, these two asynchronous communication models with FIFO or bag buffers are seldom distinguished [6].

Most of the existing works are restricted to asynchronous communication via FIFO buffers [7], e.g. [10–16]. However, asynchronous communication via bag buffers is meaningful [7–9] when the ordering of consuming messages does not matter. To ensure the sound execution of microservices systems, it is necessary to distinguish the differences between two asynchronous communication models with FIFO or bag buffers.

In this paper, we present new results for the interaction differences between one asynchronous communication model with FIFO buffers and another asynchronous communication model with bag buffers. Specifically, the contributions of this paper are as follows.

1) Define two asynchronous communication models with FIFO buffers and bag buffers uniformly.
2) Model interactions among microservices under these two different asynchronous communication models uniformly.
3) Compare these two asynchronous communication models using refinement checking to show their differences.
4) Conduct experiments to show the effectiveness of our proposed results.

The rest of this paper is organized as follows. Section 2 discusses related work. Section 3 presents formal definitions of two asynchronous communication models with FIFO or bag buffers. Section 4 formally model interaction behaviors among microservices under two asynchronous communication models. Section 5 compares asynchronous communication models with trace refinement. Section 6 discusses the implementation of our approach and experimental results. Section 7 concludes this paper.

2 Related Work

There have been many works on modeling interactions among peers communicating asynchronously via FIFO buffers. In [10, 11], Bultan et al. first proposed a framework to model interactions of e-service as a sequence of send messages, which is called a conversation. In their model, an e-service is composed of a set of peers where each peer is equipped with one FIFO queue for storing messages sent from other peers. Subsequently, Bultan et.al used to conversation protocol to model interactions of a composite Web service that consists of a set of peers under asynchronous communication with FIFO buffers [12]. In [13, 14], Basu et al. also used conservation protocols to model interactions of four distributed systems that consist of a set of peers communicating via FIFO queues. In [15, 16], Salaün et al. modeled interactions of asynchronously communicating systems with unbounded buffers as a sequence of sending messages via FIFO buffers.

Compared to these works above, we not only consider an asynchronous communication model with FIFO buffers but also consider an asynchronous communication model with bag buffers. Besides, we are particularly interested in modeling interactions among microservice as sequences of send and receive message actions rather than send message actions.

Recently, some research has begun to consider an asynchronous communication model with bag buffers. In [17], Clemente et al. studied the reachability problem for finite-state automata communicating via both FIFO and Bag buffers. In [7], Akroun et al. studied the verification problem for finite-state automata communicating via FIFO and bag buffers. However, our work focuses on the interaction differences between two asynchronous communication models with FIFO or bag buffers.

3 Asynchronous Communication Models

This section presents two asynchronous communication models with FIFO or bag buffers in a unified way.

In a microservices system, microservices interact with each other via messages.

Definition 1 (Message Set). A message set M is a tuple $(\Sigma, p, send, rec)$ where:

- Σ is a finite set of letters.
- $p \geq 1$ is a non-negative integer number which denotes the numbers of participating microservices.
- *src* and *dst* are functions that associate message $m \in \Sigma$ nonnegative integer numbers $send(m) \neq rec(m) \in \{1, 2, ..., p\}$.

We often write $m^{i \to j}$ for a message m such that $send(m) = i$ and $rec(m) = j$.

Definition 2 (Microservice). A microservice $MS_p = (S_p, s_p^0, F_p, A_p, \delta_p)$ is an LTS, where:

- $p \in \{1,2,\ldots,N\}$.
- S_p is the finite set of states.
- $s_p^0 \in S_p$is the initial state.
- $F_p \subseteq S_p$ is the finite set of final states.
- A_p is a set of actions
- $\delta_p \subseteq S_p \times (A_p \cup \{tau\}) \times S_p$ is the transition relation.

For a microservice p, an action over M is either send message action $!m^{i \to j}$ or receive message action $?m^{i \to j}$, with $m \in M_p$. The function $peer(a)$ of an action $a \in A_p$ is defined as $peer(!m^{i \to j}) = i$ and $peer(?m^{j \to i}) = i$.

In a microservice, a transition $t \in \delta_p$ can be one of the following three types:

(1) a send message transition $(s_1, !m^{1 \to 2}, s_2)$ denotes that the microservice MS_1 sends a message m to another microservice MS_2 where $m \in M_1$.
(2) a receive message transition $(s_1, ?m^{1 \to 2}, s_2)$ denotes that the microservice MS_1 consumes a message m from the microservice MS_2 where $m \in M_p$.
(3) an ε-transition (s_1, ε, s_2) denotes that the invisible action of MS_1.

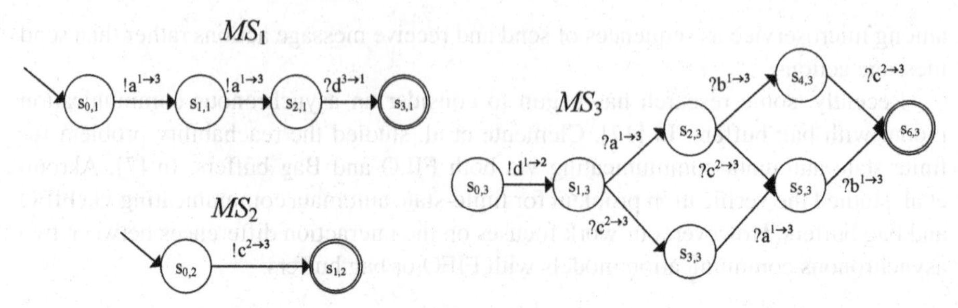

Fig. 1. Motivating example

We often write $s_m \xrightarrow{!m^{1\to 2}} s_k$ to denote that $(s_m, !m^{1\to 2}, s_k)$.

Figure 1 gives a motivating example where microservices are modeled LTS. The initial states of each microservice are subscripted with 0 and marked with an incoming half-arrow. The final states of each microservice are marked with double circles. Each transition of each microservice is labeled with send action (exclamation marks) or receive message action (question marks).

Definition 3 (Actions). Let a microservices system is composed of a set of microservices $MSs = (MS_1, MS_2, \ldots, MS_n)$, the set of actions is:

$$As \triangleq \{!m^{i\to j} \mid !m^{i\to j} \in MS.A \wedge MS \in MSs\} \cup \{?m^{j\to i} \mid ?m^{j\to i} \in MS.A \wedge MS \in MSs\}$$

Definition 4 (Execution Sequences). The execution sequences *ES* is set of all finite or infinite sequences of send and receive message actions over *As* such that a received message action is preceded by the send message action.

$$ES \triangleq \left\{ \begin{array}{c} \sigma \in As^* \mid \forall i, j \in dom(\sigma) : \sigma[i] = !m^{x\to y} \wedge \sigma[j] = !m \Rightarrow i = j \\ \wedge \forall i, j \in dom(\sigma) : \sigma[i] = ?m \wedge \sigma[j] = ?m \Rightarrow i = j \\ \wedge\ \forall j \in dom(\sigma) : \sigma[j] = ?m \Rightarrow \exists i \in dom(\sigma) : \sigma[i] = !m \wedge i < j \end{array} \right\}$$

If an action $a \in As$ occurs in an execution sequence σ, we write $a \in \sigma$ and have $\exists j \in dom(\sigma)$ such that $\sigma[j] = a$.

Concerning an execution sequence $\sigma \in ES$, we define the total order.

Definition 5 (Total Order).

$$a_1 \prec_t a_2 \triangleq \exists i, j \in dom(\sigma) : i \leq j \wedge \sigma[i] = a_1 \wedge \sigma[j] = a_2$$

In the following, we define two asynchronous communication models in terms of message ordering and buffer number, that are summed up in Table 1.

Table 1. Overview of two asynchronous communication models

	ACM_{FIFO}	ACM_{Bag}
Message ordering	$!m_1^{i\to j} \prec_t !m_2^{k\to j}$ $\wedge peer(?m_1^{i\to j}) = peer(?m_2^{k\to j})$ $\Rightarrow ?m_1^{i\to j} \prec_t ?m_2^{k\to j}$	$!m_1^{i\to j} \prec_t !m_2^{k\to j}$ $\wedge peer(?m_1^{i\to j}) = peer(?m_2^{k\to j})$ $\Rightarrow (?m_1^{i\to j} \prec_t ?m_2^{k\to j}) \vee (?m_2^{i\to j} \prec_t ?m_1^{k\to j})$
Buffer types	FIFO buffers	Bag buffers

3.1 Asynchronous Communication Model with FIFO Buffers

An asynchronous communication model with FIFO buffers that is called ACM_{FIFO} requires that the order of receiving message actions is the order of sending message actions and that each microservice has one buffer which is used to store all messages sent from other microservices in a FIFO fashion. The FIFO fashion means that the buffer can be viewed as a queue of messages. In other words, a send message action is to add a message at the tail of the buffer of the destination microservice while the corresponding receive message action is to consume a message at the head of the buffer of the receiver. This asynchronous communication model is used in [10–16].

Figure 2 (a) illustrates ACM_{FIFO}, where each microservice has one FIFO buffer. Figure 2 (b) illustrates an interaction scenario, where the microservice MS_1 sends a message a to the microservice MS_3 before sending a message b to MS_3 and then the microservice MS_3 consumes a from its $buffer_3$, and later b. Note that the messages a and b are stored in the buffer $buffer_3$ in the order they are sent.

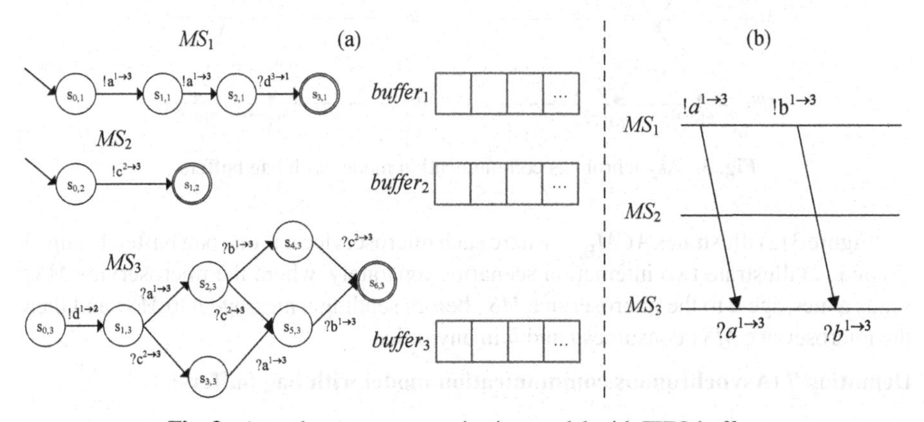

Fig. 2. Asynchronous communication model with FIFO buffers

Definition 6 (Asynchronous communication model with FIFO buffers).

$$ACM_{FIFO} \triangleq \left\{ \sigma \in A^* | \forall m_1, m_2 \in M,\ \exists i, j, k \in \mathrm{N}:\ !m_1^{i \to j} \prec_t\ !m_2^{k \to j} \wedge Buffer_j(m_1) \wedge Buffer_j(m_2) \Rightarrow ?m_1^{i \to j} \prec_t\ ?m_2^{k \to j} \right\}$$

3.2 Asynchronous Communication Model with Bag Buffers

An asynchronous communication model with bag buffers that is called ACM_{Bag} requires that the order of receiving message actions is the order of sending message actions and that all messages sent to one microservice from the other microservices are stored in a buffer in a bag fashion. The bag fashion means that the buffer can be viewed as a set. In other words, a receiver microservice can consume messages from its buffer in any order. This asynchronous communication model is used in [7, 17].

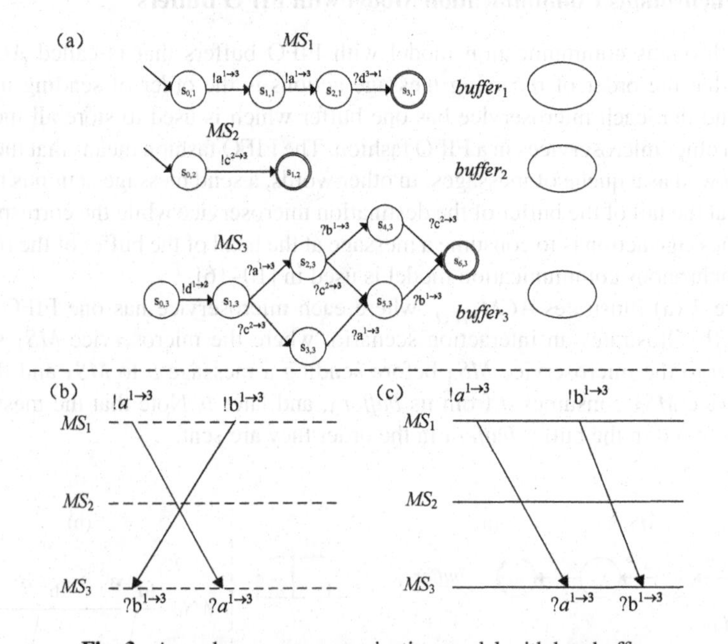

Fig. 3. Asynchronous communication model with bag buffers

Figure 3 (a) illustrates ACM_{Bag}, where each microservice has one bag buffer. Figure 3 (b) and (c) illustrate two interaction scenarios separately, where the microservice MS_1 sends a message a to the microservice MS_3 before sending a message b to MS_3 and then the microservice MS_3 consumes$a$ and b in any order.

Definition 7 (Asynchronous communication model with bag buffers).

$$ACM_{Bag} \triangleq \left\{ \begin{array}{l} \sigma \in A^* | \forall m_1, m_2 \in M,\ \exists i, j, k \in \mathrm{N}:\ !m_1^{i \to j} \prec_t\ !m_2^{k \to j} \wedge Buffer_j(m_1) \wedge Buffer_j(m_2) \\ \qquad \Rightarrow (?m_1^{i \to j} \prec_t\ ?m_2^{k \to j}) \vee (?m_2^{i \to j} \prec_t\ ?m_1^{k \to j}) \end{array} \right\}$$

4 Modeling Interactions

In this section, we model interaction behaviors among microservices as sequences of send and receive message actions under two asynchronous communication models defined in Sect. 3.

4.1 Modeling Interactions Among Microservices Under the Asynchronous Communication Model with FIFO Buffers

Definition 8 (FIFO buffer). A FIFO buffer is a queue of messages, where a send message action is to add a new message at the tail of the queue and a receive message action is to consume a message from the head of the queue.

Definition 9 (Interaction behavior under the asynchronous communication model with FIFO buffers). An interaction behavior among a set of microservices (MS_1, MS_2, …, MS_N) with $MS = (S_i, s_i^0, F_i, A_i, \delta_i)$ and Qi being its associated buffer under the asynchronous communication model with FIFO buffers is a labeled transition system $IB_{FIFO} = (S, s_0, F, A, \delta)$ where:

- $S \subseteq S_1 \times Q_1 \times S_2 \times Q_2 \times \ldots \times S_N \times Q_N$ where $\forall i \in \{1,2, \ldots, N\}$, $Q_i \in M^*$ with $M = MS_1.M_1 \cup MS_2.M_2 \cup \ldots \cup MS_N.M_N$.
- $s_0 \in S$ such that $s_0 = (s_1^0, \varepsilon, s_2^0, \varepsilon, \ldots, s_N^0, \varepsilon)$.
- $F \subseteq S$.
- $A = MS_1.A_1 \cup MS_2.A_2 \cup \ldots \cup MS_N.A_N$.
- $\delta_p \subseteq S \times (A \cup \{tau\}) \times S$ for $s = (s_1, Q_1, s_2, Q_2, \ldots, s_n, Q_n)$ and $s' = (s'_1, Q'_1, s'_2, Q'_2, \ldots, s'_n, Q'_n)$.

(a) send message action

$$s \xrightarrow{!m^{i \to j}} s' \in \delta \text{ if } \exists i, j \in \{1,2,\ldots,N\} \wedge m \in M:$$

(i) $send\ (m) = i \wedge rec(m) = j \wedge i \neq j$,
(ii) $s_i \xrightarrow{!m^{i \to j}} s'_i \in \delta_i$,
(iii) $Q'_j = Q_j m$,
(iv) $\forall k \in \{1,2, \ldots,N\}: k \neq i \Rightarrow s'_k = s_k$,
(v) $\forall k \in \{1,2, \ldots,N\}: k \neq i \Rightarrow Q'_k = Q_k$.

(b) receive message action

$$s \xrightarrow{?m^{i \to j}} s' \in \delta \text{ if } \exists i, j \in \{1,2,\ldots,N\} \wedge m \in M:$$

(i) $send\ (m) = i \wedge rec\ (m) = j \wedge i \neq j$,
(ii) $s_j \xrightarrow{?m^{i \to j}} s'_j \in \delta_j$,
(iii) $mQ'_j = Q_j$,
(iv) $\forall k \in \{1,2,\ldots,N\}: k \neq j \Rightarrow s'_k = s_k$,

(v) $\forall k \in \{1,2,\ldots,N\}: Q'_k = Q_k$.

(c) internal action

$$s \xrightarrow{\varepsilon} s' \in \delta \text{ if } \exists i, j \in \{1,2,\ldots,N\} \wedge m \in M:$$

(i) $s_i \xrightarrow{\varepsilon} s'_i \in \delta_i$,
(ii) $\forall k \in \{1,2,\ldots,N\}: k \neq i \Rightarrow s'_k = s_k$,
(iii) $\forall k \in \{1,2,\ldots,N\}: Q'_k = Q_k$.

According to Definition 9, there are three following interaction types among microservices communicating asynchronously via FIFO buffers:

(1) a send message action $s \xrightarrow{!m^{i \to j}} s'$ denotes that microservice MS_i sends a message m to another microservice MS_j where $m \in M_i$ (9a-i). After that, the state of the sender is changed (9a-ii), the message will be inserted to the tail of the *buffer*$_j$ of the receiver (9a-iii), the other microservices' states do not change (9a-iv), and the other buffers do not change (9a-v).

(2) a receive message action $s \xrightarrow{?m^{i \to j}} s'$ denotes that microservice MS_j consumes a message m sent from microservice MS_i where $m \in M_i$ (9b-i). After that, the state of the receiver is changed (9b-ii), the message at the head of the *buffer*$_j$ of the receiver will be consumed (9b-iii), the other microservices' states do not change (9b-iv), and the other buffers do not change (9b-v).

(3) an internal action $s \xrightarrow{\varepsilon} s'$ denotes that microservice MS_i executes an internal action (9c-i). After that, the other microservices' states do not change (9c-ii) and the other buffers also do not change (9c-iii).

We use $IB^k_{FIFO} = (S^k, s^k_0, F^k, A^k, \delta^k)$ to define the interaction behavior of a microservices system under the asynchronous communication model via FIFO buffers, where each buffer's bound is set to k. This IB^k_{FIFO} can be obtained from Definition 9 by allowing executing send message actions if the buffer of the receiver is not full (i.e., the buffer has less than k messages).

4.2 Modeling Interactions Among Microservices Under the Asynchronous Communication Model with Bag Buffers

Definition 10 (Bag buffer). A bag buffer is a multiset of messages.

Given a set messages{m1, m2, m3}, We often write B(m1) = 2, B(m2) = 2, and B(m3) = 1 to denote that a bag B = {m1, m1, m2, m2, m3}.

Definition 11 (Interaction behavior under the asynchronous communication model with bag buffers). An interaction behavior among a set of microservices (MS_1, MS_2, …, MS_N) with $MS_i = (S_i, s^0_i, F_i, A_i, \delta_i)$ and B_i being its associated buffer under the asynchronous communication model with bag buffers is a labeled transition system $IB^{Bag}_{Mailbox} = (S, s_0, F, A, \delta)$ where:

- $S \subseteq S_1 \times B_1 \times S_2 \times B_2 \times \ldots \times S_N \times B_N$ where $\forall i \in \{1,2,\ldots,N\}$, $B_i \subseteq M^*$ with $M = MS_1.M_1 \cup MS_2.M_2 \cup \ldots \cup MS_N.M_N$.
- $s_0 \in S$ such that $s_0 = (s_1^0, \varepsilon, s_2^0, \varepsilon, \ldots, s_N^0, \varepsilon)$.
- $F \subseteq S$.
- $A = MS_1.A_1 \cup MS_2.A_2 \cup \ldots \cup MS_N.A_N$.
- $\delta_p \subseteq S \times (A \cup \{tau\}) \times S$ for $s = (s_1, B_1, s_2, B_2, \ldots, s_n, B_n)$ and $s' = (s'_1, B'_1, s'_2, B'_2, \ldots, s'_n, B'_n)$.

(a) send message action

$$s \xrightarrow{!m^{i \to j}} s' \in \delta \text{ if } \exists i, j \in \{1, 2, \ldots, N\} \wedge m \in M:$$

(i)$send\ (m) = i \wedge rec\ (m) = j \wedge i \neq j$,
(ii) $s_i \xrightarrow{!m^{i \to j}} s'_i \in \delta_i$,
(iii)$B'_j = B_j \cup \{m\}$,
(iv) $\forall k \in \{1,2,\ldots,N\}: k \neq i \Rightarrow s'_k = s_k$,
(v) $\forall k \in \{1,2,\ldots,N\}: k \neq i \Rightarrow B'_k = B_k$.

(b) receive message action

$$s \xrightarrow{?m^{i \to j}} s' \in \delta \text{ if } \exists i, j \in \{1, 2, \ldots, N\} \wedge m \in M:$$

(i)$send\ (m) = i \wedge rec\ (m) = j \wedge i \neq j$,
(ii) $s_j \xrightarrow{?m^{i \to j}} s'_j \in \delta_j$,
(iii)$B'_j = B_j - \{m\}$,
(iv) $\forall k \in \{1,2,\ldots,N\}: k \neq j \Rightarrow s'_k = s_k$,
(v) $\forall k \in \{1,2,\ldots,N\}: B'_k = B_k$.

(c) internal action

$$s \xrightarrow{\varepsilon} s' \in \delta \text{ if } \exists i, j \in \{1, 2, \ldots, N\} \wedge m \in M:$$

(i) $s_i \xrightarrow{\varepsilon} s'_i \in \delta_i$,,
(ii) $\forall k \in \{1,2,\ldots,N\}: k \neq i \Rightarrow s'_k = s_k$,
(iii) $\forall k \in \{1,2,\ldots,N\}: B'_k = B_k$.

Note that the differences between Definition 9 and Definition 11 are the condition (11a-iii) and the condition (11b-iii).

Similarly, we use $IB^k_{Bag} = (S^k, s_0^k, F^k, A^k, \delta^k)$ to define the interaction behavior of a microservices system under the asynchronous communication model via bag buffers, where each buffer's bound is set to k.

5 Comparison of Two Asynchronous Communication Models

This section compares these two asynchronous communication models defined in Sect. 3 using refinement checking [18].

Definition 12 (Trace Refinement). Let ACM_i where $i \in \{1,2\}$ be two asynchronous communication models, for any a set of microservices (MS_1, MS_2, …, MS_n), ACM_1 is a trace refinement of ACM_2 if and only if *traces*(IB_1) ⊆ traces(IB_2), where IB_1 is the interaction behavior among microservices (MS_1, MS_2, …, MS_N) with under the ACM_1 and IB_2 is the interaction behavior among microservices (MS_1, MS_2, …, MS_N) with under the ACM_2:

$$ACM_1 \prec_r ACM_2 \triangleq \textit{for any a set of microservcies}\ (MS_1,\ MS_2, \ldots MS_n)\ traces(IB_1) \subseteq traces(IB_2)$$

Definition 13 (Model Inclusion). Let ACM_i where $i \in \{1,2\}$ be two asynchronous communication models. ACM_1 is included in ACM_2 iff ACM_1 is a trace refinement of ACM_2 and ACM_2 is not a trace refinement of ACM_1:

$$ACM_1 \subseteq ACM_2 \triangleq (ACM_1 \prec_r ACM_2) \wedge (ACM_2! \prec_r ACM_1)$$

Theorem 1 (Comparison of ACM_{FIFO} and ACM_{Bag} with regard to Refinement).

Proof. The proof follows directly from Definition 9 and 11. Since the FIFO buffers are the special case of bag buffers, for any a set of microservices (MS_1, MS_2, …, MS_n), *traces*(IB_1) ⊆ traces(IB_2) is always true, but not vice versa, where IB_1 is the interaction behavior among microservices (MS_1, MS_2, …, MS_N) with under the ACM_{FIFO} and IB_2 is the interaction behavior among microservices (MS_1, MS_2, …, MS_N) with under the ACM_{Bag}.

6 Implementation and Experiments

6.1 Implementation

We have implemented our approach under the support of the Process Analysis Toolkit (PAT) tool [19]. The snapshot of the comparison result of the running example using refine checking can be found in Fig. 4. This figure shows that the running example with FIFO buffers is included in that with bag buffers.

For the asynchronous communication model via FIFO buffers, we can use channel communication to implement it. More specifically, for a send message action ($s \xrightarrow{!m^{i \to j}} s'$), the operation of channel output can be written as $buffer_{ij}$!m, where $buffer_{ij}$ is a channel. For a receive message action ($s \xrightarrow{?m^{i \to j}} s'$), the operation of channel input can be written as $buffer_{ij}$?m.

For the asynchronous communication model via bag buffers, we define a new data type *Bag* using the C# library editor and compiler in the PAT to implement it. The *Bag* has two important operations to support message emission and reception, namely *add* and *remove.* For the *add* operation $buffer_{ij}.add(m)$ which is used to denote a send message action, the message *m* is stored in the *Bag* $buffer_{ij}$ if the $buffer_{ij}$ is not full yet. For the *remove* operation $buffer_{ij}.reomve(m)$ which is used to denote a receive message action, the message *m* in the *Bag* $buffer_{ij}$ is retrieved and if the buffer is not empty.

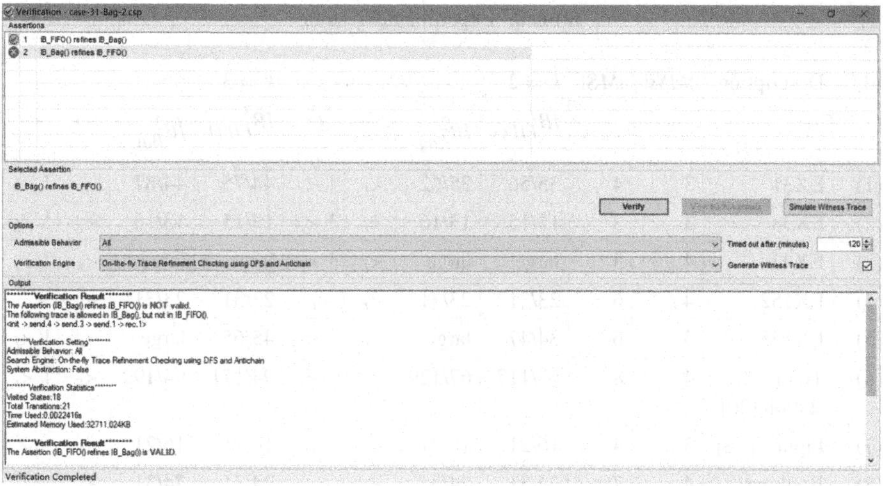

Fig. 4. The snapshot of comparison results of the running example

6.2 Experiments

For validating our approach, we use two datasets. The first dataset is obtained from [7], which contains more than 6 hundred examples. The second dataset contains more than 1 hundred examples, which are collecting from the literature on the close subject.

We conducted several modeling and comparing experiments on these two datasets. All the experiments were performed on the same Windows laptop running on a 2.5 GHz Intel Core i7 processor with 8 GB of memory.

Table 2 shows some of these examples, where we use FIFO buffers and bag buffers as asynchronous communication models and compare these two models using refinement checking in the PAT tool. The second column describes each case, where cases (1), (2), (3), (4) and (5) are from the first dataset and the other cases are from the second dataset. The third column shows the number of microservices in each case. The fourth column shows the number of messages exchanged among microservices. The fifth column shows the interaction behavior with buffers of size 2 under the ACM_{FIFO}(each tuple (X, Y) denotes the number of states X and transitions Y in the IB^2_{FIFO}), the interaction behavior with buffers of size 2 under the ACM_{Bag} (IB^2_{Bag}), and the comparisons of these two interaction behaviors, where " $\rightarrow$" denotes whether IB^2_{FIFO} is a trace refinement of IB^2_{Bag} and " $\leftarrow$" denotes whether IB^2_{Bag} is a trace refinement of IB^2_{FIFO}. The fifth column shows IB^3_{FIFO}, IB^3_{Bag}, and their comparisons. It should be noted that the word "large" in the fifth column means that we cannot build the state space of interaction behavior using PAT.

The experimental results show that IB^k_{FIFO} is a trace refinement IB^k_{Bag} where $k \in \{2,3\}$, but not vice versa (see, e.g., case (1), (2), (3), (5) and (6)). Furthermore, the experimental results also show that the ACM_{FIFO} with FIFO buffers is included in the ACM_{Bag} with bag buffers

Table 2. Experimental results

Id	Description	$\|MSs\|$	$\|MS\|$	$k = 2$				$k = 3$			
				IB^2_{FIFO}	IB^2_{Bag}	$\rightarrow$	$\leftarrow$	IB^3_{FIFO}	IB^3_{Bag}	$\rightarrow$	$\leftarrow$
(1)	EX31	3	4	35/56	35/62	$\prec_r$	$!\prec_r$	44/75	44/87	$\prec_r$	$!\prec_r$
(2)	EX38	3	3	13/15	13/16	$\prec_r$	$!\prec_r$	13/15	13/16	$\prec_r$	$!\prec_r$
(3)	EX43	4	5	large	large	$\prec_r$	$!\prec_r$	large	large	$\prec_r$	$!\prec_r$
(4)	EX152	4	6	23/31	23/31	$\prec_r$	$\prec_r$	23/31	23/31	$\prec_r$	$\prec_r$
(5)	EX155	3	6	34/47	large	$\prec_r$	$!\prec_r$	45/65	large	$\prec_r$	$!\prec_r$
(6)	Train station [20]	4	8	67/117	67/129	$\prec_r$	$!\prec_r$	94/171	94/192	$\prec_r$	$!\prec_r$
(7)	Figure 1 [8]	3	4	16/21	16/21	$\prec_r$	$\prec_r$	16/21	16/21	$\prec_r$	$\prec_r$
(8)	Booking system [21]	4	7	24/31	24/31	$\prec_r$	$\prec_r$	24/31	24/31	$\prec_r$	$\prec_r$
(9)	Online shopping [22]	3	6	13/13	13/13	$\prec_r$	$\prec_r$	13/13	13/13	$\prec_r$	$\prec_r$
(10)	Figure 8 [23]	3	3	16/20	16/20	$\prec_r$	$\prec_r$	16/20	16/20	$\prec_r$	$\prec_r$

7 Conclusions

In this paper, the interaction differrences between one asynchronous communication mod-el with FIFO buffers and another asynchronous communication model with bag buffers are discussed. First, we formally define two asynchronous communication models with FIFO or bag buffers in terms of message ordering and buffer number. Second, we model interactions among microservices as sequences of send and receive message actions under these two asynchronous communication models. Third, we compare these two asynchronous communication models using refinement checking. Our experiments show that the asynchronous communication model with FIFO buffers is included in the asynchronous communication model with bag buffers.

The future work is to study whether our proposed results are suitable for peer-to-peer communication rather than mailbox communication and compare more asynchronous communication models using refinement checking.

Acknowledgment. This work has been supported by the Project of National Natural Science Foundation of China under Grant No. 61702442 and 61862065, the Application Basic Research Project in Yunnan Province Grant No. 2018FB105, the Major Project of Science and Technology of Yunnan Province under Grant No. 202002AD080002 and No. 2019ZE005.

References

1. Qi, L., Chen, Y., Yuan, Y., Fu, S., Zhang, X., Xu, X.: A QoS-aware virtual machine scheduling method for energy conservation in cloud-based cyber-physical systems. World Wide Web **23**(2), 1275–1297 (2019). https://doi.org/10.1007/s11280-019-00684-y
2. Xiaolong, X., et al.: A computation offloading method over big data for IoT-enabled cloud-edge computing. Future Gener. Comput. Syst. **95**, 522–533 (2019). https://doi.org/10.1016/j.future.2018.12.055
3. Zhou, X., et al.: Poster: benchmarking microservice systems for software engineering research. In: Proceedings of the International Conference on Software Engineering: Companion Proceedings (ICSE 2018), pp. 323–324 (2018)
4. Zhou, X., et al.: Delta debugging microservice systems with parallel optimization. IEEE Trans. Serv. Comput. **99**, 1 (2019)
5. Zhou, X., et al.: Delta debugging microservice systems. In: Proceedings of the 33rd ACM/IEEE International Conference on Automated Software Engineering, pp. 802–807 (2018)
6. Chevrou, F., Hurau, A., Quéinnec, P.: On the diversity of asynchronous communication. Formal Aspects Comput. **28**(5), 847–879 (2016)
7. Akroun, L., Salaün, G.: Automated verification of automata communicating via FIFO and bag buffers. Formal Methods Syst. Des. **52**(3), 260–276 (2017). https://doi.org/10.1007/s10703-017-0285-8
8. Finkel, A., Lozes, É.: Synchronizability of communicating finite state machines is not decidable. In: Proceedings of the 44th International Colloquium on Automata, Languages, and Programming ICALP 2017, vol. 122, pp. 1–14 (2017)
9. Barbanera, F., van Bakel, S., de Liguoro, U.: Orchestrated session compliance. J. Logical Algebraic Method Program. **86**(1), 30–76 (2017)
10. Bultan, T., Fu, X., Hull, R., Su, J.: Conversation specification: a new approach to design and analysis of e-service composition. ACM 1–58113–680–3/03/0005, WWW, May 20–24, Budapest, Hungary (2003)
11. Xiang, F., Bultan, T., Jianwen, S.: Conversation protocols: a formalism for specification and verification of reactive electronic services. In: Ibarra, O.H., Dang, Z. (eds.) CIAA 2003. LNCS, vol. 2759, pp. 188–200. Springer, Heidelberg (2003). https://doi.org/10.1007/3-540-45089-0_18
12. Bultan, T., Su, J., Fu, X.: Analyzing conversations of web services. IEEE Internet Comput. **10**(1), 18–25 (2006)
13. Basu, S., Bultan, T., Quederni, M.: Deciding choreography realizability. ACM SIGPLAN Notices **47**(1), 191–202 (2012)
14. Basu, S., Bultan, T.: Automated choreography repair. In: Stevens, P., Wąsowski, A. (eds.) FASE 2016. LNCS, vol. 9633, pp. 13–30. Springer, Heidelberg (2016). https://doi.org/10.1007/978-3-662-49665-7_2
15. Ouederni, M., Salaün, G., Bultan, T.: Compatibility checking for asynchronously communicating software. In: Fiadeiro, J.L., Liu, Z., Xue, J. (eds.) FACS 2013. LNCS, vol. 8348, pp. 310–328. Springer, Cham (2014). https://doi.org/10.1007/978-3-319-07602-7_19
16. Akroun, L., Salaün, G., Ye, L.: Automated analysis of asynchronously communicating systems. In: Bošnački, D., Wijs, A. (eds.) SPIN 2016. LNCS, vol. 9641, pp. 1–18. Springer, Cham (2016). https://doi.org/10.1007/978-3-319-32582-8_1
17. Clemente, L., Herbreteau, F., Sutre, G.: Decidable topologies for communicating automata with FIFO and bag channels. In: Baldan, P., Gorla, D. (eds.) CONCUR 2014. LNCS, vol. 8704, pp. 281–296. Springer, Heidelberg (2014). https://doi.org/10.1007/978-3-662-44584-6_20

18. Yang, L., Chen, W., Liu, Y.A., Sun, J., Zhang, S., Less, J.: Verifying linearizability via optimized refinement checking. IEEE Trans. Softw. Eng. **39**(7), 1018–1039 (2013)
19. Sun, J., Liu, Y., Dong, J.S.: Model checking CSP revisited: introducing a process analysis toolkit. In: Margaria, T., Steffen, B. (eds.) ISoLA 2008. CCIS, vol. 17, pp. 307–322. Springer, Heidelberg (2008). https://doi.org/10.1007/978-3-540-88479-8_22
20. Salaün, G., Bultan, T., Roohi, N.: Realizability of choreographies using process algebra encodings. IEEE Trans. Serv. Comput. **5**(3), 290–302 (2012)
21. Poizat, P.: Checking the realizability of BPMN 2.0 choreographies. In: Proceedings of the 27th Annual ACM Symposium on Applied Computing, pp. 1927–1934 (2012)
22. Bultan, T.: Modeling interactions of web software. In: International Workshop on Automated Specification and Verification of Web Systems, pp. 45–52 IEEE (2006)
23. Bultan, T., Chris, F., Xiang, F.: A tool for choreography analysis using collaboration diagrams. In: Proceedings of the 7th IEEE International Conference on Web Services (ICWS 2009), Los Angeles, CA, July 6–10, pp. 856–863 (2009)

Multi-radio Relay Frequency Hopping Based on USRP Platforms

Luying Huang(✉), Yitao Xu, Xueqiang Chen, Dianxiong Liu, and Ximing Wang

College of Communications Engineering, Army Engineering University of PLA, Nanjing 210000, China
lgdxwxm@sina.com

Abstract. The wireless relay communication system can realize space diversity, expand the communication range, and increase capacity. The technology of frequency hopping communication helps the relay system randomly hop on multiple channels, which can improve system reliability. At present, most of the research related to relay communication and frequency hopping is based on theoretical analysis and numerical simulations, which can not precisely simulate the channel characteristics of the actual wireless communication environment. According to the decoding and forwarding (DF) relay transmission technology, this paper designs and implements the wireless multi-radio relay frequency hopping transmission system based on the software radio hardware platform USRP. The results show that the system realizes multimedia wireless data communication with a low packet loss rate through relay frequency hopping transmission.

Keywords: Relay communication · Frequency hopping · USRP · LabVIEW

1 Introduction

Complex wireless channel environment seriously affects the reliability and coverage of the signal transmission. Relay communication technologies can effectively increase the coverage, enhance reliability and availability of the communication system, which are widely used in wireless communication systems [1]. After Erlang put forward the basic principles of relay theory, relay technologies have been studied more and more extensively [2]. That is, the received wireless signal from the base station is processed through a series of relay stations and then forwarded. Microwave relay communication is the earliest communication system using relay technology. Its carrier wave is microwave, and long-distance radio communication is realized by relay on the ground. If node A and node B are far apart, to achieve communication between them, several relay stations need to be added. After receiving the signal, the relay node amplifies and forwards it to the

This work was supported by the National Natural Science Foundation of China under Grant No. 61771488, No. 61671473 and No. 61631020.

H. Song and D. Jiang (Eds.): SIMUtools 2020, LNICST 370, pp. 519–530, 2021.
https://doi.org/10.1007/978-3-030-72795-6_42

next relay, and finally realizes the communication between the two places [3]. The forwarding protocols of relay communication mainly include the amplify-and-forward (AF), compress-and-forward (CF) and decode-and-forward (DF) in [4]. The sending node sends the signal to the RN, and the RN simply amplifies the received signal and then forwards it, which is called amplified forwarding [5,6]. Differently, the RN in DF mode will decode and restore the original signal after receiving the signal from the previous node, and then encodes and modulates the signal and forwards it to the next node [7,8]. When the channel quality is poor, the RN with the AF protocol not only amplifies the received signal, but also accumulates noise, which ultimately affects the probability of correct decoding by the destination node. If DF protocol is used, the noise can be well controlled, ensuring the performance of the communication system. Therefore, the DF mode is employed in this paper.

The quality of information transmission is subject to harmful interference. To ensure the unimpeded transmission of information, the communication system needs to have a certain anti-interference ability. Frequency-hopping communication is a type of spread-spectrum communication [9], which is characterized by pseudo-random hopping of the carrier frequency at multiple frequencies and has a certain ability to resist interference.

Universal Software Radio Peripheral (USRP) [10–12] is a very nimble open source hardware device developed by Matt Ettus et al. The USRP works as a digital baseband, RF front-end, and digital intermediate frequency, and the remaining of the signal processing is done by software programming on the computer [13]. LabVIEW is a virtual instrument development software which can be used with USRP, with a flexible user interface and powerful interactivity. Combining USRP with LabVIEW can overcome the limitations of traditional communication experiment simulation methods, such as poor scalability of the curing test chamber and non-objective software simulation results.

In this paper, combining relay cooperative communication and frequency hopping communication, a wireless relay frequency hopping communication system based on USRP platform and LabVIEW software simulation is built. The main contribution of this paper is to achieve fixed sequence frequency hopping transmission from source node (SN) to relay node (RN) and relay node (RN) to destination node (DN), in addition, SN-RN link and R-D link do not interfere with each other. Through relay frequency hopping transmission, wireless communication can resist the attack of traditional interference, expand the scope of communication.

The remaining of the paper is arranged as follows. In Sect. 2, physical layer and data frame structure design are investigated. Section 3 is the system construction and the introduction of related functional modules. Section 4 introduces the design principles and system implementation of relay frequency hopping transmission. In Sect. 5, the experimental results are given. Finally, the paper is concluded in Sect. 6.

2 Related Works

The authors in [14] implemented a single relay wireless transmission system based on USRP and GNU Radio. The authors in [15] implemented a multi-relay wireless transmission system based on USRP and GNU Radio. However, the existing work [14,15] mainly focused on the construction of relay system, while communication protocol was not discussed, as well as the multi-channel model. In [18], a concept proof of LTE DF RN using two SDRs was proposed. In [19], a cooperative communication system experimental platform based on NI USRP2920 was proposed. The implemented experiments included voice communications and video streaming by GMSK modulation transmission to develop the cooperative advantages of multimedia communication. In [20], the authors used LabVIEW and USRP as the experimental platform. The three nodes cooperative communication based on amplifying and forwarding was simulated, and the system performance was analyzed by turning the system time slot settings by software.

Different from them, this paper combines wireless relay technology and frequency hopping technology to achieve picture transmission, based on the software radio platform USRP and software platform LabVIEW. The results show that the system can realize relay frequency hopping transmission with low packet loss rate.

3 General Design of Relay System

The designed wireless relay frequency hopping communication system is shown in Fig. 1. The system is mainly composed of one SN, one RN and one DN. The SN and DN are composed of a PC and a NI USRP2920, respectively. The RN is simulated by one PC, one switch and two NI USRP2920. The system consists of four transmission links, which are two data links and two ACK transmission links. Two data links include the data link f_{sr} from SN to RN, and the data link f_{rd} from RN to DN. Two ACK transmission links include an ACK link f_{ack}^{rs} from RN to SN and an ACK link f_{ack}^{dr} from DN to RN.

Before the experiment, the system framework of the experiment is introduced. The hardware of the experimental system in this paper mainly includes NI USRP 2920, supporting antenna, switch and computers. The software platform adopts LabVIEW visual simulation programming environment. LabVIEW is a virtual instrument development software that can be used with USRP, with flexible user interface and strong interactivity.

3.1 System Physical Layer Design

In the LabVIEW software platform, most modulation methods, such as FSK, BPSK, QAM, GMSK, etc., have corresponding modules and corresponding demodulation module. This experiment uses QPSK for modulation and demodulation to ensure transmission reliability.

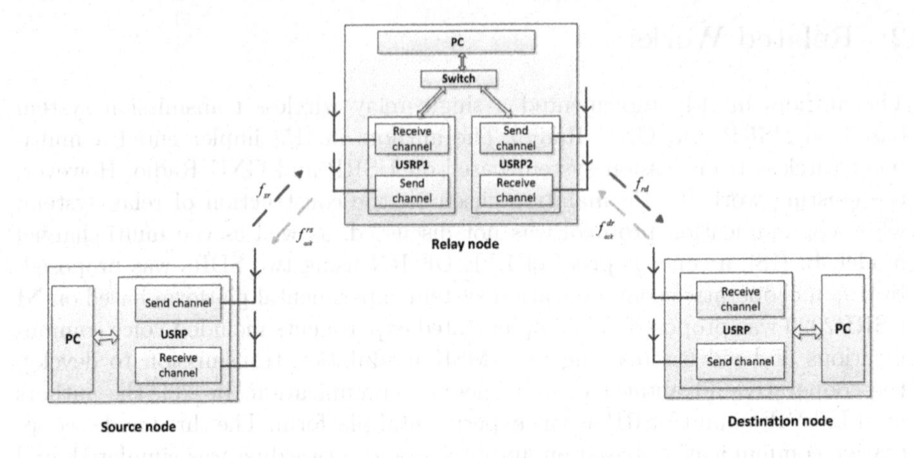

Fig. 1. USRP-based wireless relay frequency hopping system framework.

3.2 Data Frame Structure Design

In order to make sure the accuracy of the received data, we need to design the frame format of the data. In this system, the sender performs equal-length splitting, framing, and encapsulation of the data bit stream to form a data bit packet for modulation transmission. The encapsulation frame format is shown in Table 1. The definition of each data field function is shown in Table 2.

Frame synchronization is extremely important for the receiver correctly receiving data. Only when the sender and the receiver have achieved frame synchronization can they correctly distinguish the starting position of the data frame during reception and receive valid data. In a wireless communication system, data is inevitably subject to errors during transmission due to various adverse factors. Hence, data are detected and corrected, then the upper layers can transmit more reliably. Error detection codes do not correct transmission errors, but can increase transmission reliability and effectiveness. CRC (Cyclic Redundancy Check) is an error check code that is often used in the field of data communications to detect or verify data transmission. The verification method used in this experiment is CRC-16 [16].

Table 1. Encapsulated frame format.

Protection bit	Sync bit	Message type	Data frequency	ACK frequency	Package number	Data bit	Check bit	Padding

Table 2. Frame format function.

Data field	Number of bits	Function
Protection bit	30	Allows the automatic gain control module to reach a steady state quickly before processing useful data
Sync bit	30	Generate a 30-bit random sequence using a PN sequence generator for easy frame synchronization
Message type	8	0000 0000 represents data; 0000 0001 represents ACK
Data frequency	32	Data transmission channel
ACK frequency	32	ACK transmission channel
Package number	32	Received packet sequence number
Data bit	1024	Service data bit length in each frame, which can be set by the user
Check bit	16	CRC check the data
Padding	30	Used to eliminate filter effects

4 System Construction and Function Modules

4.1 System Construction

The NI USRP2920 has two antennas, TX/RX1 and RX2. TX/RX1 can be used to send and receive signals, and RX2 can only be used to receive signals. In the wireless relay frequency hopping transmission communication system, because the SN and the DN both need to send and receive information, the SN and the DN are composed of one PC and one NI USRP2920, respectively. The RN is relatively complicated which needs to receive the data information of the SN and forwards the data processing to the DN, at the same time initiates a frequency change request to the SN and receive the frequency change request initiated by the DN. It can be seen that the RN needs two transmissions and two receptions, so the RN needs to be composed of a PC, a switch, and two NI USRP2920. Figure 2 shows a schematic diagram of the experimental system in this paper.

4.2 Function Modules

The experimental system in this paper is composed of three nodes. In general, the design of RN is relatively complicated, and two receptions and two transmissions are required. To improve the development efficiency of the system, the SN and the DN can be selected to implement the corresponding functions of the RN.

The RN can be divided into two major blocks, the receiving module and the sending module. Among them, the receiving module is further divided into data receiving and ACK receiving, the sending module is further divided into sending data module and sending ACK module. The design of sending and receiving data and ACK are different except for the frame structure design, the other designs are the same, hence the processing flow of the transmitting module and receiving module of the node are introduced here.

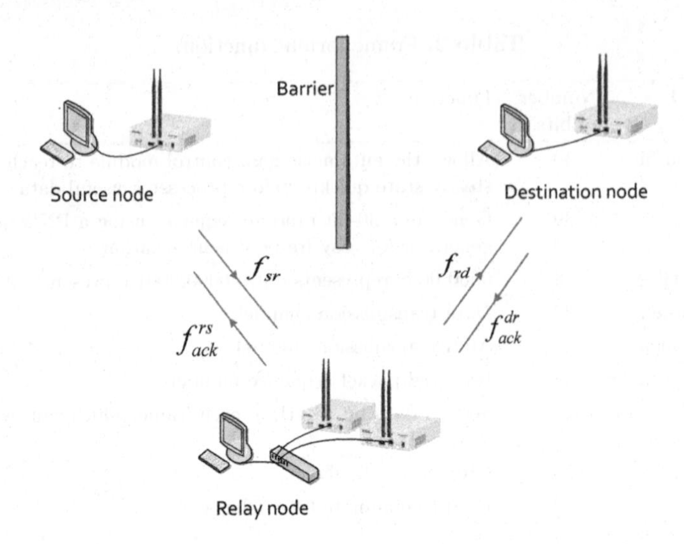

Fig. 2. Experimental system diagram.

At the transmitting end of the node, the service to be transmitted is first converted into a binary bit stream. According to the designed data frame format, the binary data bit stream is grouped into several data packets of the same length. The digital baseband signals are obtained through raised cosine filtering and QPSK modulation, and finally written into USRP through the Ethernet port and transmitted through the antenna.

On the receiving end of the node, the USRP obtains the baseband signal through the receiving antenna and reads it into the PC. The data processing process is completed with LabVIEW program. First, the received baseband signal is separated according to the packet (remove DC component, the corresponding position of each packet is detected by correlation operation, and the data packet is separated), then the extracted data packet is resampled and QPSK demodulated to obtain a binary bit stream. After the serial validity check (CRC check and frame synchronization check), the correctly received data packets are reconstructed and converted into a format to restore the original data information [17].

5 Principle and Implementation of Relay Frequency Hopping Transmission System

5.1 Design Principles of Relay Frequency Hopping

In actual wireless communication, comprehensive coverage is sought. For remote areas, the coverage of the cell is limited and severely affected by shadow fading. Relay communication technology can effectively conquer the fading of wireless channels and increase spectrum efficiency and communication range. The basic concept of wireless relay technology is to add a RN between a SN and a DN. The RN receives the signal from the SN and forwards it to the DN after a certain processing. Generally, in a wireless single-relay network, there are three communication links, namely SN-DN, SN-RN, and RN-DN. In this experimental system, the SN-DN link is not considered.

The handling of the received signal by the RN is diverse. Wireless repeater technology is divided into two categories, AF and DF. AF: After receiving the signal from the SN, the RN simply amplifies the signal and forwards it. DF: The RN decodes and demodulates the received signal to recover the original signal, then encodes and modulates the recovered signal to the DN. In this experiment, the RN uses DF.

The interaction procedure among SN, RN and DN in wireless relay frequency hopping transmission communication system is shown in the Fig. 3. The steps of wireless relay frequency hopping transmission are analyzed as follows:

Step 1: The SN sends data to the RN on the channel f_{sr} and receives ACK signal sent by the RN on the channel f_{ack}^{rs}, and reconfigures the transmission carrier frequency f_{sr} after receipting the ACK signal.

Step 2: The RN receives data on the channel f_{sr}, decodes and forwards it, and sends the data to the DN on the channel f_{rd}; after a time interval Tr, initiates a frequency change request, sends an ACK to the SN on the channel f_{ack}^{rs}, and reconfigures the RN receive the carrier frequency f_{sr}. At the same time receive the ACK signal sent by the DN on the channel f_{ack}^{dr}, and reconfigure the transmit carrier frequency f_{rd} after receiving the ACK signal.

Step 3: The DN receives data information on the channel f_{rd}. After a time interval Td, initiates a frequency change request, sends an ACK to the RN on the channel f_{ack}^{dr}, and reconfigure the receive carrier frequency f_{rd} of the DN.

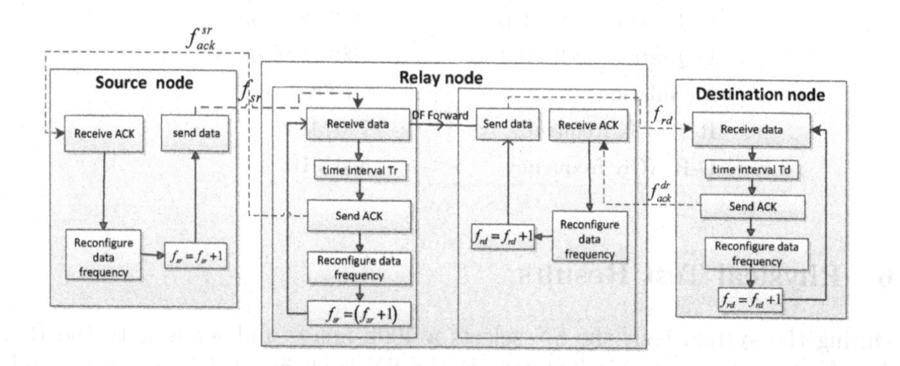

Fig. 3. Experimental flow chart.

5.2 System Implementation

Based on the above design analysis, a physical simulation system was built based on the USRP software radio platform and LabVIEW software. The system consists of 3 computers, 4 NI USRP2920, and 1 switch. The computer and USRP are connected through a Gigabit Ethernet port, which simulates the SN, RN, and DN.

The test steps of the system are as follows:

(1) Connect the USRP and the computer;
(2) Configure system related parameters (such as I/Q rate, carrier frequency, antenna, gain, modulation method, frequency hopping interval, frequency hopping range, frequency hopping time interval, etc.) on the front panel of the program, as shown in the Table 3;
(3) Power on the USRP and run the SN, RN, and DN programs at the same time for data transmission.

Table 3. Related parameter configuration.

Related parameters	Value
S-R frequency range	800 MHz–845 MHz
R-D frequency range	915 MHz–960 MHz
Frequency interval	5 MHz
S-R frequency hopping interval	3 s
R-D frequency hopping interval	3 s
Number of channels	10
Antenna gain	10 dB
S-R data I/Q rate	1M Sample/s
R-D data I/Q rate	800K Sample/s
I/Q rate of R-S ACK	400K Sample/s
I/Q rate of D-R ACK	400K Sample/s
Modulation	QPSK
R-S ACK frequency	750 MHz
D-R ACK frequency	500 MHz

6 Physical Test Results

During the system test, the SN selects a 400k image and sends it to the RN. The front panel operation interface of the SN is shown in Fig. 4, where module 1 is a modulation transmission constellation diagram, module 2 is a time domain waveform diagram of a transmission signal; module 3 is an access frequency point of each time slot, and module 4 is a picture to be transmitted. The access frequency sequence is obtained by the ACK of the RN. The set of available frequencies is {800 MHz, 805 MHz, 810 MHz, 815 MHz, 820 MHz, 825 MHz, 830 MHz, 835 MHz, 840 MHz, 845 MHz}. It can be seen from the Fig. 4. That the modulation method is QPSK.

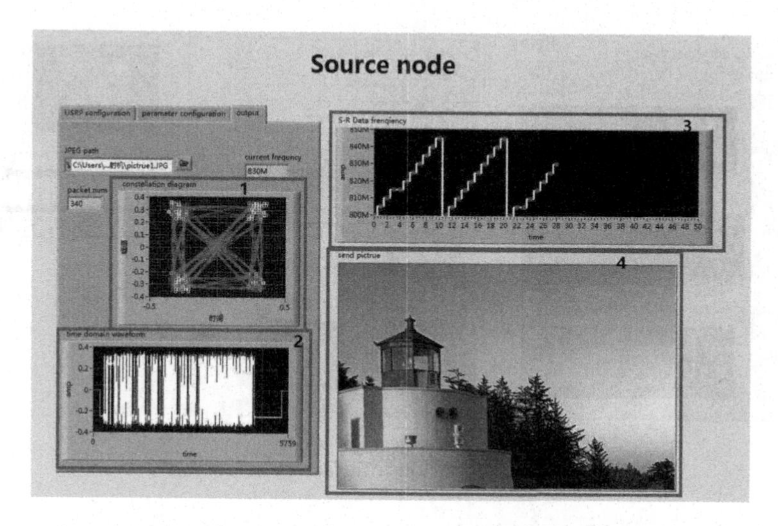

Fig. 4. User operation interface of source node.

The system RN uses the decoding and forwarding (DF) method to re-encode the received data information and send it to the DN. The RN is divided into four parts in terms of function realization, data reception and transmission, and ACK transmission and reception. The front panel operation interface of the RN is displayed in Fig. 5. In Fig. 5, module 1 is the frequency hopping sequence from the SN to the RN, module 2 is the packet loss rate of the RN receiving data, module 7 is the constellation diagram of the RN receiving data, module 8 is the time-domain waveform diagram of the RN receiving data, module 3 is the frequency hopping sequence from the RN to the DN, available frequency points are {915 MHz, 920 MHz, 925 MHz, 930 MHz, 935 MHz, 940 MHz, 945 MHz, 950 MHz, 955 MHz, 960 MHz}, module 5 is the constellation diagram and time domain waveform diagram of the RN sending data to the DN, module 6 is a constellation diagram for sending ACK to the SN and a time-domain waveform diagram for sending ACK information, module 4 is the original picture recovered by the RN. It can be seen from the picture that there is a packet loss phenomenon. There is a time lag between the SN and the RN during each frequency change, resulting in a small amount of packet loss during the frequency change, the packet loss rate floating around 0.1.

The DN is divided into two parts in terms of function implementation, which are divided into data reception and ACK transmission. As shown in the Fig. 6, module 3 is the DN to restore the received data information to the original image, module 1 is the access frequency point sequence, module 2 is the packet loss rate of the DN, the packet loss rate fluctuates around 0.2, module 4 is the constellation diagram of the DN receiving data, module 5 is the time domain waveform diagram of the received data.

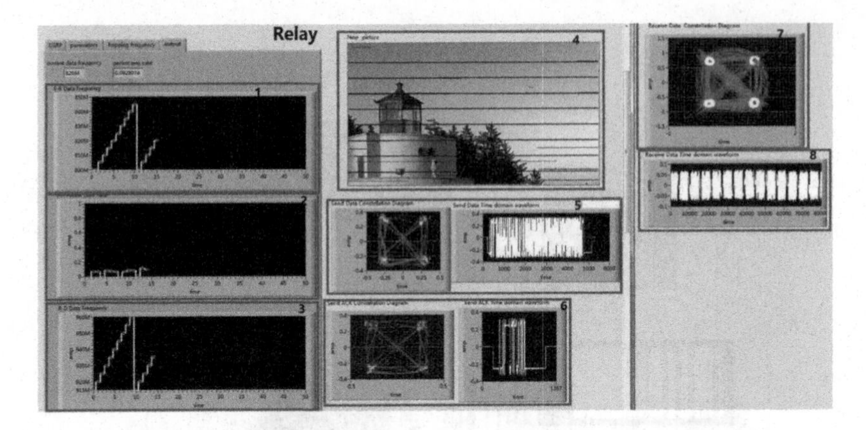

Fig. 5. User operation interface of relay node.

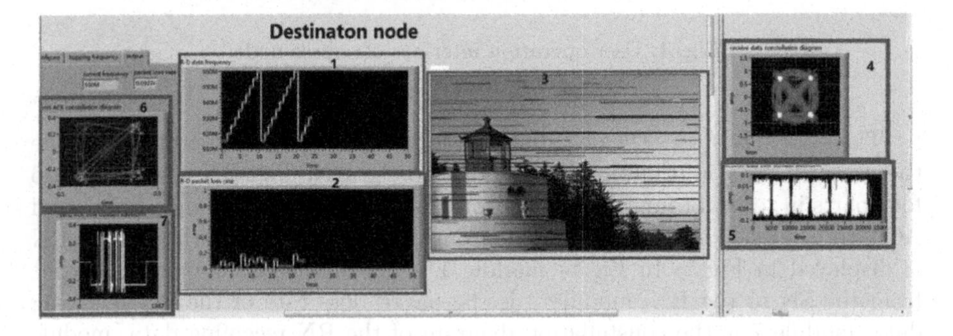

Fig. 6. User operation interface of destination node.

7 Conclusions

This paper makes full use of the reconfigurable, flexible and easy to operate characteristics of the NI USRP software radio platform to build a wireless relay frequency hopping communication transmission system. After many experiments, the system runs stably reliable, and packet loss rate is also within a certain range. On the basis of this article, the next step is to consider the existence of interference, and implement intelligent relay frequency hopping anti-interference combined with reinforcement learning.

References

1. Gouissem, A., Hasna, M.O., Hamila, R., et al.: Outage performance of OFDM adhoc routing with and without subcarrier grouping in multihop network. In: 2012 IEEE Vehicular Technology Conference (VTC Fall), pp. 1–5. IEEE (2012)
2. Laneman, J.N., Tse, D.N.C., Wornell, G.W.: Cooperative diversity in wireless networks: efficient protocols and outage behavior. IEEE Trans. Inf. Theory **50**(12), 3062–3080 (2004)

3. Chau, Y.A., Huang, K.Y.: Channel statistics and performance of cooperative selection diversity with dual-hop amplify-and-forward relay over Rayleigh fading channels. IEEE Trans. Wireless Commun. **7**(5), 1779–1785 (2008)
4. Lee, D., Kim, S.I., Lee, J., et al.: Performance of multihop decode-and-forward relaying assisted device-to-device communication underlaying cellular networks. In: 2012 International Symposium on Information Theory and its Applications, pp. 455–459. IEEE (2012)
5. Krikidis, I., Thompson, J., McLaughlin, S., et al.: Amplify-and-forward with partial relay selection. IEEE Commun. Lett. **12**(4), 235–237 (2008)
6. Hwang, K.S., Ko, Y.C., Alouini, M.S.: Performance analysis of two-way amplify and forward relaying with adaptive modulation over multiple relay network. IEEE Trans. Commun. **59**(2), 402–406 (2010)
7. Luo, J., Blum, R.S., Cimini, L.J., et al.: Decode-and-forward cooperative diversity with power allocation in wireless networks. IEEE Trans. Wireless Commun. **6**(3), 793–799 (2007)
8. Mills, D.G., Edelson, G.S., Egnor, D.E.: A multiple access differential frequency hopping system. In: IEEE Military Communications Conference, MILCOM 2003, vol. 2, pp. 1184–1189. IEEE (2003)
9. Sendonaris, A., Erkip, E., Aazhang, B.: Increasing uplink capacity via user cooperation diversity. In: Proceedings of 1998 IEEE International Symposium on Information Theory (Cat. No. 98CH36252), p. 156. IEEE (1998)
10. Abirami, M., Hariharan, V., Sruthi, M.B., et al.: Exploiting GNU radio and USRP: an economical test bed for real time communication systems. In: 2013 Fourth International Conference on Computing, Communications and Networking Technologies (ICCCNT), pp. 1–6. IEEE (2013)
11. ETTUS: Universal software radio peripheral (USRP). Ettus Research LLC (2008). http://www.ettus.com
12. Tong, Z., Arifianto, M.S., Liau, C.F.: Wireless transmission using universal software radio peripheral. In: 2009 International Conference on Space Science and Communication, pp. 19–23. IEEE (2009)
13. Xin, X., Hui, Z.: Design of experimental platform for wireless communication based on LabVIEW and USRP. Exp. Technol. Manag. **33** (2016)
14. Bo, F.: Experimental Study of Wireless Relay Transmission Based on USRP Platform
15. Yang, X.: Experimental Study of Wireless Multi-relay Transmission Based on USRP Platform
16. Keller, R.B.: Fast cyclic redundancy check (CRC) generation: U.S. Patent 6,701,479[P] (2004)
17. Lijun, K., Yuhua, X., Xueqiang, C., et al.: Design and implementation of data transmission system based on USRP and selective retransmission protocol. Commun. Technol. **51** (2018)
18. Marín-García, J.A., Romero-Franco, C., Alonso, J.I.: A software defined radio platform for decode and forward relay nodes implementation. In: 2019 IEEE Conference on Standards for Communications and Networking (CSCN), pp. 1–4 (2019). https://doi.org/10.1109/CSCN.2019.8931410.

19. Prince, A., Abdalla, A.E., Dahshan, H., Rohiem, A.E.: Multimedia SDR-based cooperative communication. In: 2018 13th International Conference on Computer Engineering and Systems (ICCES), Cairo, Egypt, pp. 381–386 (2018). https://doi.org/10.1109/ICCES.2018.8639312
20. Yu, Z., Luo, H., Li, L., Zhang, Y., Han, Z.: Research on the influence of system slot setting on the performance of three-node cooperative communication system. In: IEEE 3rd Information Technology, Networking, Electronic and Automation Control Conference (ITNEC), Chengdu, China, pp. 368–371 (2019). https://doi.org/10.1109/ITNEC.2019.8729297

Motion Simulation of Rocket Based on Simulink

Pengcheng Li[1](✉), Zhi Zhu[2], Haidong Chen[3], and Shipeng Li[1]

[1] Beijing Aerospace System Engineering Institute, Beijing 100076, China
[2] Department of Military Modeling and Simulation, College of System Engineering, Changsha 410073, China
[3] China Academy of Launch Vehicle Technology, Beijing 100076, China

Abstract. A system of carrier rockets tracing target points is modeled and simulated in Simulink platform of MATLAB, in order to reduce the workloads of designers in designing, analyzing, calculating and selecting the carrier rockets' motion parameters. Based on the dynamic formula of carrier rockets strictly deduced in this study for tracing targets, a three-degree free tracing system simulation, combined with model simulation techniques, is established, and the proper carrier rockets' design parameters are studied according to the results of the simulation. First, this paper introduces the background and stresses the importance of carrier rockets in national defense, and then analyzes the problems with system modeling and analysis to be solved. After designing some sub-system simulation modules and integrating them into Simulink platform, this research sets up an tracing simulation model. And finally, by executing the trascing system simulation under the condition of different guidance ratio values, this research obtains corresponding simulation results and compares the influences of various guidance ratio values on the tracing effects.

Keywords: Rockets · Simulation · Simulink · Proportion guidance

1 Introduction

Flying by the controlling law is an important task of carrier rockets for targeting to specific target point and fly on special trace. Traditionally, the research and design of carrier rockets' manufacture parameters rely on the relevant professional staff's deduction. The traditional way of completing the task is mainly to program the formula using computer or to calculate the parameters manually by professional staff [1]. There are problems such as large amount of calculation load, low efficiency to get accurate and suitable rocket parameters, tendentiousness of making mistakes and difficulty in programming due to formula constraints in scale and complexity [2]. As a rare method to obtain carrier rockets' actual flight parameters under guidance and control except for actual rocket flight test, the simulation technology is used to simulate the dynamic motion of carrier rockets working in the practical environment with the help of simulation software platforms such as MATLAB/Simulink, KD-HLA, LabVIEW, etc. [3], and modify the model simulation parameters to meet the design expectation, which has become the

H. Song and D. Jiang (Eds.): SIMUtools 2020, LNICST 370, pp. 531–544, 2021.
https://doi.org/10.1007/978-3-030-72795-6_43

essential method of carrier rockets' design and performance optimization. Between carrier rockets' design, test and finalization, the flight control design optimization, selection of guidance and control law, evaluation of rocket operational effectiveness and so on, almost depend on the trajectory simulation previously. Therefore, basic research on the trajectory simulation of carrier rockets is carried out, in the hope of developing a set of trajectory simulation specifications at some future date.

As one of the core components of MATLAB software, Simulink simulation platform provides users with an integrated environment of dynamic system modeling, simulation and comprehensive analysis [4]. By means of graphical interactive operation, dragging simulation modules, modifying simulation parameters, combining multiple modules as a whole and other humanized operations, complex simulation model system could be constructed effectively and humanly. Simulink provides carrier rockets' tracing system modeling with a wealth of useful library components. Simulink's commonly used blocks contains constant module, delay module in discrete system, integration module in continuous system, input port or output port required when creating tracing system's sub-system, etc., and continuous system simulation block includes modules used in continuous system, such as differential integration, state transfer function, delay, zero pole modules, while discrete system simulation block includes delay, difference, filter, signal control modules and math operation simulation part contains absolute, gain, deviation, division, exponent operation and lookup tables block, which realizes table mapping conveniently by inputting data into required dimensions blanks, includes various query table modules [5]. With the characteristics of wide application fields, clear and concise simulation structure and implementation, it has been widely used in the complex system simulation field of rocket control and guidance.

The existing trajectory simulation research of carrier rockets is almost based on MATLAB/Simulink [6]. For example, the department of automatic control engineering of naval aviation engineering university has established the anti-ship carrier rockets' trajectory simulation model with the help of Simulink, and verified the model; the air force engineering university has built carrier rockets' guidance and control dynamic simulation models in Simulink; based on MATLAB, Zhang Zhenzhong described the UAV's penetration simulation model under the missile interceptor [7]. It could be concluded that the Simulink/MATLAB simulation platform has become a trend in the development and design of carrier rockets. Given the characteristics of carrier rockets, this paper studies basic simulation models of the carrier rockets' trajectory control and guidance system when tracing the target point in Simulink, in order to develop a set of trajectory simulation specifications of the carrier rockets and assist the professional staff with high efficiency and quality.

There are two main practical application significances in this study. In the first place, repeatability and scalability. In actual rocket flight test, due to the random factors' dynamic influence of environment, it is difficult to ensure that the experiment results are obtained under the same condition [8]; however, the simulation method could control the random factors' impact by simulation environment parameters adjustment, which means strong repeatability. At the same time, each practical flight test needs to carry out a complete experiment process, while the trajectory simulation could start from any state among simulation, without needing to restart from the system's zero point, which is flexible and scalable.

The second is high test efficiency with low cost. Among the overall design process of the guidance and control of carrier rockets, it is necessary to carry out the trajectory calculation and optimization analysis for different guidance and control schemes to select the proper one to meet the trajectory index; generally, multiple trajectory flight tests are carried out to determine the rocket performance with high cost and long test time, while trajectory simulation of the carrier rockets as accurate as possible improves the test efficiency of different schemes and reduce the cost.

In this research, Simulink is used to execute the designed simulation model so as to improve the efficiency and quality of carrier rocket development. Section 2 gives a simple problem analysis of carrier rocket system, including the case study description and a collection of parameters. In Sect. 3, the modeling and simulation methodology is detailed. Simulation results are presented in Sect. 4 which follows conclusions in Sect. 5.

2 Problem Analysis

Case applied for this research's simulation is required to model and simulate the tracing system of the carrier rocket guided by the proportional guidance method and target point [9], as shown in Fig. 1. Assuming that an target point is flying horizontally at a constant speed along axis x in negative direction in the vertical plane, while at the same time, the defense side discovers the target and launches an carrier rocket in the same vertical plane in time to tracing the target point. The carrier rocket's flight stage is preset to be divided into two main working stages, of which the first stage is pushed by a high thrust rocket engine so as to be far away from the launching position as fast as possible and reach a high speed value in a short period of time to be comfortable for tracing in the next stage. The second stage engine's thrust is lower than that of the first stage, while the engine's working time of the second stage engine is longer than that of the first stage engine. In the second guidance stage, the rocket is controlled and guided by the proportional guidance method, which means that the rocket's elevation angle change will be corrected according to the change in the line of sight angle of between the rocket and target [10], so that the tracing rocket continuously approaches the target point to implement the task of tracing.

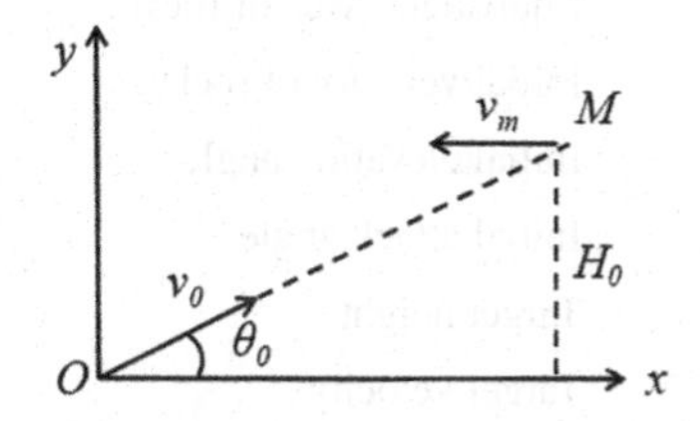

Fig. 1. The proportional guidance methodology of target tracing

The thrust and mass changes curves corresponding to two working stage of the primary and secondary engines in the rocket's flight phase are shown in Fig. 2. Time period 0 ~ t_1 corresponds to the working phase of the primary engine in the first stage, and time period t_1 ~ t_2 corresponds to the working phase of the secondary engine in the second stage.

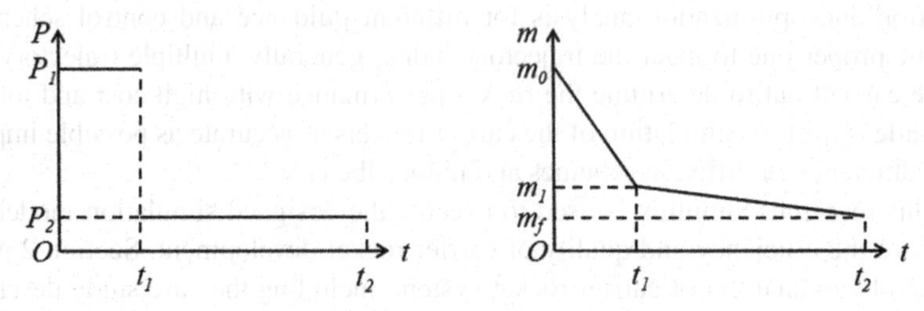

Fig. 2. Thrust and mass changes curves

The parameter symbol representation, initial values and physical definition of rocket design parameters, including dynamic characteristic parameters, motion parameters and target motion parameters [11] are shown in Table 1. The relationship between resistance coefficient caused by atmospheric acting force, Mach number of rocket's velocity and attack angle of rocket is shown in Table 2. The relationship between curve slope of lift coefficient and Mach number of its velocity is shown in Table 3.

Table 1. Simulation parameters

Parameter	Physical meaning
P_1	First stage thrust
P_2	Second stage thrust
t_1	First stage shutdown time
t_2	Second stage shutdown time
m_0	Initial mass
m_1	First stage mass when shut down
m_f	Second stage mass when shut down
S	Equivalent Area of rocket
v_0	Initial velocity of rocket
θ_0	Initial elevation angle
α_0	Initial attack angle
$H0$	Target height
v_m	Target velocity
r_0	Initial line of sight distance

Table 2. Resistance coefficient map

$M\backslash C_x\backslash\alpha$	0	2	4	6	8
1.5	C_{x11}	C_{x12}	C_{x13}	C_{x14}	C_{x15}
2	C_{x21}	C_{x22}	C_{x23}	C_{x24}	C_{x25}
2.5	C_{x31}	C_{x32}	C_{x33}	C_{x34}	C_{x35}
3	C_{x41}	C_{x42}	C_{x43}	C_{x44}	C_{x45}
3.5	C_{x51}	C_{x52}	C_{x53}	C_{x54}	C_{x55}
4	C_{x61}	C_{x62}	C_{x63}	C_{x64}	C_{x65}

Table 3. Lift coefficient curve slope map

M	$c_y{}^{\alpha}$
1.5	$c_y{}^{\alpha 1}$
2	$c_y{}^{\alpha 2}$
2.5	$c_y{}^{\alpha 3}$
3	$c_y{}^{\alpha 4}$
3.5	$c_y{}^{\alpha 5}$
4	$c_y{}^{\alpha 6}$
4.5	$c_y{}^{\alpha 7}$

3 Modeling and Simulation Methodology

3.1 Tracing System Modeling

Given the requirements of simulation case, it is necessary to establish dynamic system models for the carrier rocket and target point respectively [12]. Considering that the guidance model of the carrier rocket depends on the target point's dynamic models, hierarchical modeling method is applied to tracing system simulation at different model levels of granularity, as shown in Fig. 3.

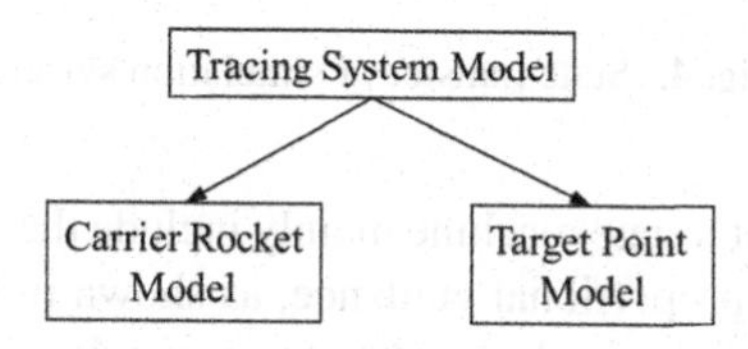

Fig. 3. Thrust and mass changes curves

The carrier rocket model and the target point model are at the same level in total system, and aggregated into the whole tracing system model. Conversely, the tracing system model could be divided into two sub-models: the carrier rocket model and the target point model. Through the multi-level modeling method of combination modeling, different granularity is refined based on the system model's structure [13], which is convenient for modeling and simulation.

The general form of FSM is described as Eq. 1 [13], where Q is the finite and non empty state set of simulation model system, and parameter I is the finite and non empty input set to change the system state, and parameter O is the finite and non empty output set to reflect the influences of input set I, and parameter δ is the transfer function of the system from original state Q to a new state Q with the input set I, representing the mapping relationship between states in a state transfer, and finally parameter λ is the function to describe relation between the input set I and output set O, which represents the mapping relationship between them.

$$FSM = < Q, I, O, \delta, \lambda > \tag{1}$$

The specific process of tracing target point is modeled and simulated by state modeling method of description modeling. The state transfer flow-process diagram is shown in Fig. 4 and Fig. 5, which respectively represents the internal state transfer of the whole tracing system and the inner state transfer of the carrier rocket. Figure 4 shows that the rocket starts from the initial launch state, and enters the first stage after the ignition of the primary engine, and transfers to the second state after the ignition of the secondary engine with the primary engine having shutdown, and finally ends the state transfer when successfully tracing the target point [14]. Relatively the target point starts from initial flight state after initialization, and continuously keeps the normal flight state as long as it's not traced by carrier rocket, or else it transfers to the end state.

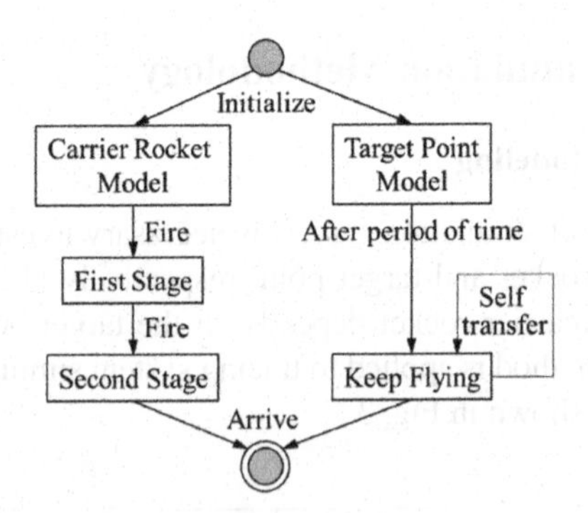

Fig. 4. State transfer of simulation system

The states of the rocket's state machine mainly include the no-guidance state and the guidance state guided by proportional guidance, as shown in Fig. 5. After the primary engine ignites, it enters the no-guidance state to accumulate speed as fast as possible, and then enters the guidance state when igniting the secondary engine. The proportional guidance state is further refined to two sub-states, of which the mutual conversion is realized when the change rate of the rocket's elevation angle or the line of sight angle varies following the proportional guidance law [15].

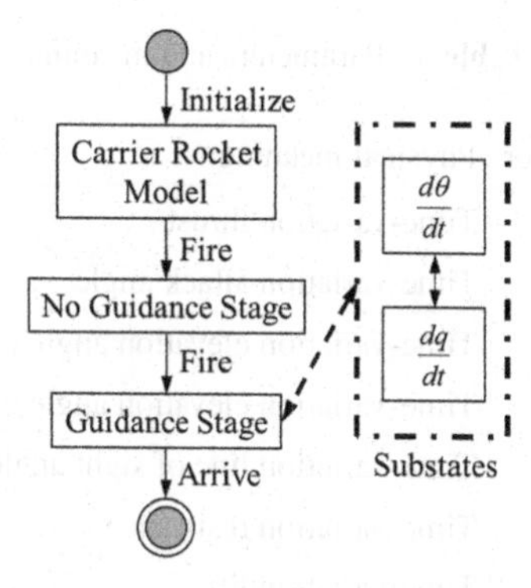

Fig. 5. State transfer of rocket model

The function modeling method could refine the system model for the composition of multiple distinguishable objects and describe the material signal flow with clear direction between different objects. The simulation objects and the material signal flow associated between the objects are designed depending on the carrier rocket model and the target point model.

The dynamic formulas of the carrier rocket model and the target point model are established with the constraint modeling of the function-based functional modeling [16]. The parameters' symbol representation involved and the corresponding physical meanings are listed in Table 4.

According to motion decomposition [17], the thrust of the rocket engine, gravity, lift and resistance are orthogonal decomposed along the speed direction and vertical direction. The resultant force along the speed direction provides the change of the speed value during the rocket flying and the resultant force along the vertical direction provides the change of the speed direction during the rocket flying, as shown in formula 2 and 3.

$$\frac{P(t)\cos\alpha(t) - m(t)g\sin\theta(t) - X(t)}{m(t)} = \frac{dv(t)}{dt} \tag{2}$$

$$\frac{P(t)\sin\alpha(t) - m(t)g\cos\theta(t) + Y(t)}{m(t)v(t)} = \frac{d\theta(t)}{dt} \tag{3}$$

The change rate equations of the rocket's velocity value and elevation angle value indicate the direction of signal flow during simulation modeling [18], that is, the output signal on the right side of the equation is calculated from the input on the left side, and the left side of the equation could be transformed into a corresponding sub-system in Simulink, in which the input ports represent the parameters on the equation's left side, and the output ports represent the right side of the equation.

According to the commonly calculating formula 4 of atmospheric density [19] and the calculation method of lift or resistance of rocket flying in the atmosphere [20] shown in formula 5 and 6.

$$\rho(t) = 1.293e^{-0.00015\,y(t)} \tag{4}$$

$$q_s(t) = \frac{1}{2}\rho(t)v(t)^2 \tag{5}$$

Table 4. Parameters and meanings

Parameter	Physical meaning
$P(t)$	Time-variation thrust
$\alpha(t)$	Time-variation attack angle
$\theta(t)$	Time-variation elevation angle of rocket
$\theta_m(t)$	Time-variation elevation angle of target
$q(t)$	Time-variation line of sight angle
$X(t)$	Time-variation resistance
$Y(t)$	Time-variation lift
$x(t)$	Time-variation location in x of rocket
$y(t)$	Time-variation location in y of rocket
$x_m(t)$	Time-variation location in x of target
$y_m(t)$	Time-variation location in y of target
$m(t)$	Time-variation mass
$v(t)$	Time-variation velocity
$r(t)$	Time-variation line of sight distance
$\rho(t)$	Time-variation atmospheric density
C_x	Resistance coefficient
C_y	Lift coefficient

$$\begin{cases} X(t) = C_x q_s(t) S \\ Y(t) = C_y q_s(t) S \end{cases} \tag{6}$$

On the basis of the proportional guidance method [21], the rocket's elevation angle change rate to time is adjusted depending on the line of sight angle change rate of the rocket and the target point. The parameter, the guidance ratio k, represents the ratio of the rocket's speed elevation angle's rotation angle speed to the target's line of sight angle rotation speed, as shown in the formula 7. Similarly, the constraint relation of the attack angle is shown as formula 8. When the angle between the line of sight and the horizontal plane increases, the rocket's flight elevation increases correspondingly in accordance with the guidance ratio set, in order that the rocket's flight direction is continuously corrected to aim until finally the tracing task is completed.

$$\frac{d\theta(t)}{dt} = k\frac{dq(t)}{dt} \tag{7}$$

$$\frac{\frac{d\theta(t)}{dt} m(t)v(t) + m(t)g\cos\theta(t) - Y(t)}{P(t)} = \alpha(t) \tag{8}$$

Decompose the two objects' velocity parameter along the line of sight between the rocket and the target point and the perpendicular direction to the line of sight. The velocity along the line of sight results in shortening the length of it, that is, the proximity of two objects, while the velocity vertical to the direction causes the effect of rotation centered on the other object.

$$\frac{-v(t)\sin(\theta(t)-q(t))-v_m\sin(q(t)-\theta_m)}{r(t)}=\frac{dq(t)}{dt} \tag{9}$$

$$-v(t)\cos(\theta(t)-q(t))+v_m\cos(q(t)-\theta_m)=\frac{dr(t)}{dt} \tag{10}$$

The line of sight angle's change rate between the rocket and target point and the length of the line's change rate to the time are separated into output signals in Simulink, and are integrated to get the solution results of the original time-variance angle [22]. The geometric relationship between rocket and target point is expressed as the following formula 11 and 12, which respectively represents the constraint relationship between horizontal position and vertical position of two objects.

$$x_m(t)-r(t)\cos q(t)=x(t) \tag{11}$$

$$y_m(t)-r(t)\sin q(t)=y(t) \tag{12}$$

The normal flight state of the target point is modeled based on the Eqs. 13 and 14, respectively according to the time-variance constraints of the horizontal position and the vertical position.

$$x_m(t)=r_0\cos\theta_0-v_m t \tag{13}$$

$$y_m(t)=r_0\sin\theta_0 \tag{14}$$

3.2 Simulation Methodology

In the Simulink environment, the constraint relations that function modeling depends on are represented using relative simulation modules. According to the parameter being left side or right side of the equation, set equation's left side the module's input ports, and right side the output ports of the simulation module. Overlapping ports are related to solve an equation in Simulink simulation process, and output ports' outputs are written to MATLAB variable workspace as results of the tracing simulation [23]. Each separate equation corresponds to sub-module at sub-system level and united sub-modules corresponds to the total system at system level in Simulink.

Tracing system simulation model are separated into 10 sub-modules of Simulink, which correspond 10 output ports, respectively representing the velocity, elevation angle and attack angle, x position and y position, cornering overload of rocket, x position and y position of target, line of sight length and angle between line and horizontal plane. According to the input signal from input ports and the output signal from the output ports in each sub-system, the related signal is connected through the signal flow direction line, and the total system simulation in Simulink is shown in the Fig. 6.

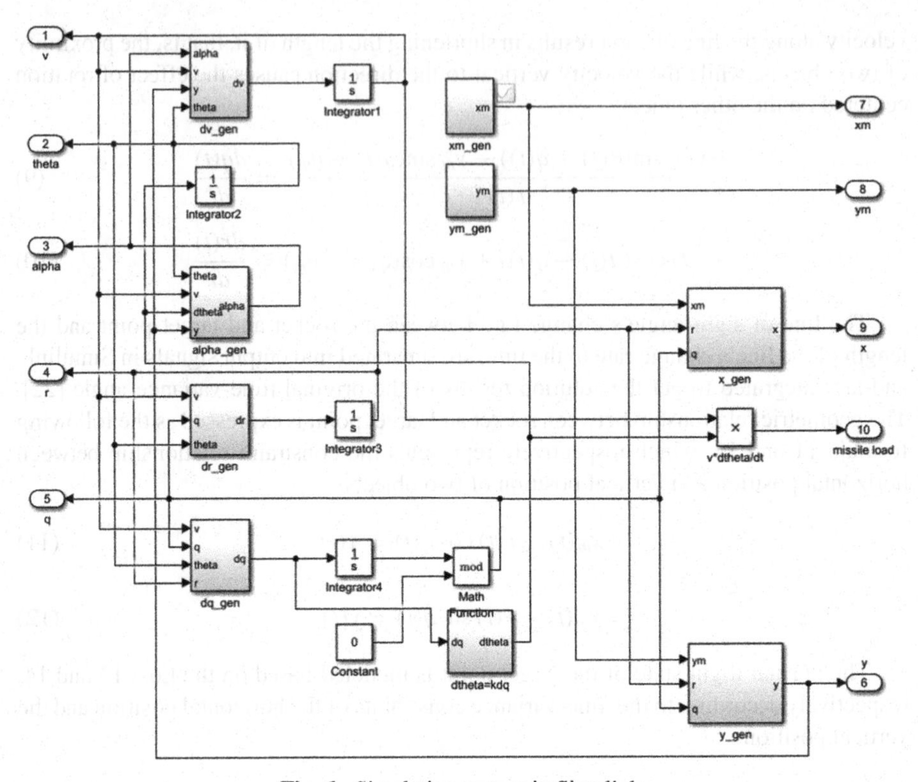

Fig. 6. Simulation system in Simulink

After the tracing system simulation model being built in Simulink, simulation execution options in the solver page and the data import or export page are configured. The total tracing phase continues nearly 45 unit time in this case, and thus set simulation time as the same time as the duration of total tracing phase with fixed time-step being set 0.025 with ode4 Runge-Kutta solver, meaning that 40 simulation points are sampled in 1 s and output the simulation execution results, time matrix *tout* and outputs matrix *yout*, into MATLAB variable workspace for further analysis. The specific simulation execution algorithm of Simulink is shown as following 4 steps:

1) Variable definition and initialization;
2) Loop each iteration in every minimum time step;
3) Each iteration, update the values of simulation time, *rtY.v, rtY.theta, rtY.alpha, rtY.r, rtY.q, rtY.y, rtY.xm, rtY.x* in turn, where *rtY* is the structural variable representing the output port of the simulation system;
4) Judge whether the simulation time is reached. If yes, the simulation ends, otherwise skip to step 2) for execution.

Simulink outputs the simulation time matrix *tout* and outputs matrix *yout* after performing the tracing system model simulation. In order to analyze the influences of different parameters on the carrier rocket's motion state and tracing effects, miss distance is selected as the main evaluation index, which describes the minimum distance deviation between the rocket and the target point [24] under the discrete-time sampling condition, and get the rocket's corresponding motion state under the circumstances of different guidance ratio values.

4 Simulation Results

Different tracing effects are listed in Table 5 under the circumstances of different guidance ratio k values ranging from 2.1 to 2.5. From the simulation results table, when the guidance ratio value equals 2.3, the corresponding miss distance reach the minimum distance as 8.20 unit distances. According to the common size of the target aircrafts, 20 unit distances is select as the maximum miss distance threshold of mission success and it means that the rocket's radius of explosion could reach to 20 unit distances at most. Therefore, to obtain the proper range of the guidance ratio value meeting the condition of tracing mission success, the curve of relationship between guidance ratio and miss distance is calculated and shown in Fig. 7.

Table 5. Simulation results

Guidance ratio	Miss distance
2.1	39.13
2.2	31.83
2.3	8.20
2.4	31.21
2.5	36.78

From the curve in Fig. 7, it is concluded that the guidance ratio k's proper value ranges from 2.25 to 2.35, and k greater than or less than proper range may cause the addition of miss distance, meaning the failure of tracing mission.

Under the circumstances of guarding tracing mission successfully with different k's value, the corresponding rocket's motion state changes are shown in Fig. 8. The relationship description curve between the guidance ratio k and carrier rocket's final velocity is shown in Fig. 8(a) when it just reaches the miss distance. The final velocity reaches higher than 700 unit velocity. In the meanwhile, the relationship description curve between the guidance ratio k and carrier rocket's elevation angle is shown in Fig. 8(b). With the increase of k, the final elevation angle decreases to less than 1 unit angle. Similarly, the relationship description curve between the guidance ratio k and carrier rocket's attack angle is shown in Fig. 8(c). With the increase of k, the final attack angle decreases and levels off to zero. The space position change curve of the rocket and target is shown in Fig. 8(d) in tracing process. Motion curve of the rocket is divided into two phase, the straight line phase corresponding to the first stage engine working and the

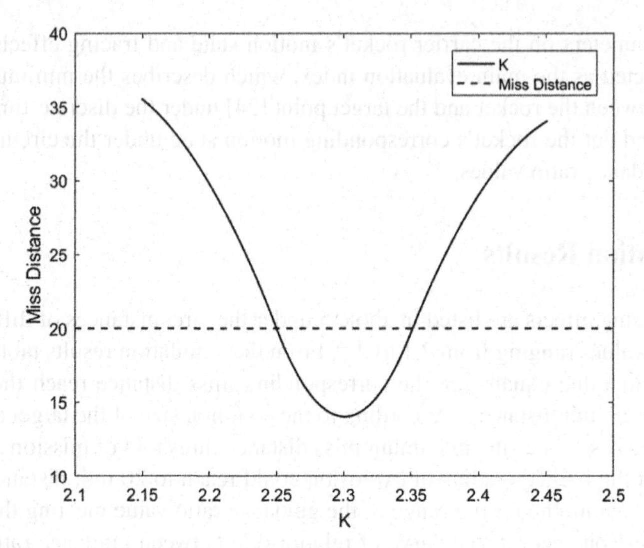

Fig. 7. *k*-Miss distance curve

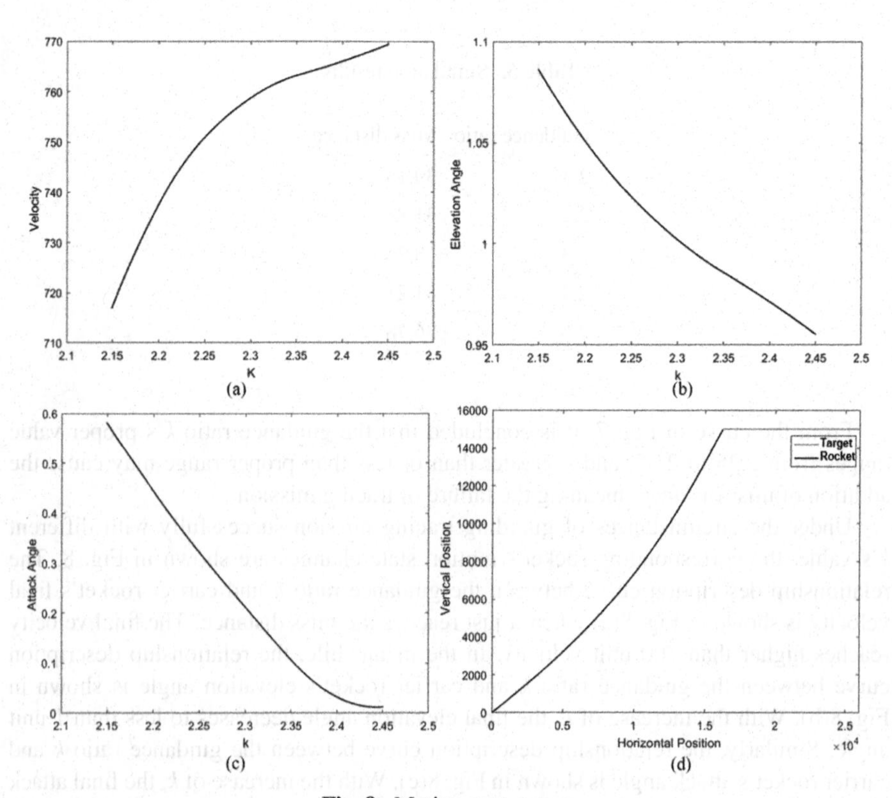

Fig. 8. Motion state curves

winding curve phase corresponding to the second stage engine working with guidance. Ultimately, the rocket arrives the target point in cross area of two motion curves.

Different guidance rate k represents the rotation speed of rocket velocity vector relative to line of sight, which could adapt to a variety of targets with different maneuvering characteristics. From the analysis of tracing effect, the appropriate k value is between 2.25 and 2.35, which meets the requirements of tracing mission. The higher the proportional guidance coefficient k is, the greater the final tracing speed of the rocket is, the smaller the final elevation angle and attack angle are. Within the tracing range, the interception time decreases with the increase of k, and the trajectory curvature also decreases, that is, the trajectory tends to be straight. Due to the limitation of missile structure and technology, the available overload could not be increased unlimited. Therefore, the selection of k value should not be too large, which means to select the appropriate value by the design indices.

5 Conclusions

Based on the Simulink platform and the trajectory characteristics, the tracing system model of the case that the carrier rocket aims to tracing the target point is developed on the dynamic differential equations' constraint conditions. And the simulation is carried out by Simulink simulation execution algorithm on condition of different guidance ratio k values to obtain proper guidance ratio and its corresponding influences on the rocket's motion state.

Finally, the proper range of guidance ratio meeting the tracing requirement and its influences on the carrier rockets' final motion state parameters, such as velocity, elevation angle, attack angle and motion position, are analyzed in detail, which aims at combining the method of simulation modeling with guidance and control laws of carrier rockets. With trajectory simulation models continuously developing, the simulation specifications of carrier rockets may be finally issued to promote design efficiency and quality.

References

1. Wang, Y., Chen, Q., Chang, S.: Visual simulation of six-degree free trajectory in multiple perspectives using Simulink and VRML. J. Weapon Equipment Eng. **40**(01), 143–147 (2019)
2. He, J.: Simulation of missile guidance and control using Simulink. Sci. Educ. Guide (08), 32–33 (2017)
3. Gao, S.: Design of missile six-degree free trajectory simulation system based on MATLAB/Simulink. Sci. Technol. Eng. (1), 29–33 (2011)
4. Cai, H., Xiong, J., Zheng, G., Liu, H.: Simulation system of missile terminal guidance based on MATLAB. Inf. Syst. Eng. (03), 36–37 (2014)
5. Gao, H., Zhu, D., Zhang, H.: Research on missile simulation system based on Simulink. Sci. Technol. Innov. Appl. (19), 16–17 (2016)
6. Yang, Y., Tang, S.: Total trajectory simulation of sub missile based on Simulink. J. Syst. Simul. (06), 1442–1444+1449 (2006)
7. Cheng, Y., Jiang, D., Wang, R.: Dynamic simulation based on matlab/simulink of SAM guidance and control system. Control Technol. Tactical Missile (3), 9–13 (2003)

8. Zhang, K.: Six-degree free modeling and simulation of missile based on simulink. Shipboard Electron. Countermeasure **34**(4), 72–76 (2011)
9. Geng, J., Chen, H., Cui, L.: Study on the calculation model of firing variables in the simulation modeling of surface to air missile. Naval Electron. Eng. **36**(10), 87–89+114 (2016)
10. Xia, F., Zhao, Y.: Design of simulation system of surface to air missile based on HLA. J. Syst. Simul. (02), 296–299 (2007)
11. Wang, H., Wan, X., Yang, J., Zhao, J., Liu, C.: Research on visual simulation of ground to air missile tracing air target. J. Missiles Guidance **36**(03), 22–24 (2016)
12. Liu, W., Li, X., Guan, Z., Shu, C.: Design and implementation of a general trajectory simulation system for surface to air missile. J. Projectiles Rockets Missiles Guidance **38**(2), 35–38 (2018)
13. Li, Q.: Simulation Model Design and Implementation. Electronic Industry Press, Beijing (2010)
14. Lan, F.: Research and application of ground to air missile training simulation system. Practical Electron. (6), 39–39 (2015)
15. Zhang, D., Zhao, Y., Lei, H., Wu, Y.: Modeling and simulating on kinematic trajectory of ground-to-air missile controled by three-point method. Flight Dyn. **30**(1), 57–60, 65 (2012)
16. Chen, J., Zhang, X., Zhang, G.: Research on launch dynamics modeling and simulating for surface-to-air missile. J. Projectiles Rockets Missiles Guidance **30**(1), 65–67, 71 (2010)
17. Zhou, L., Zhao, J., Lou, S.: Research into the trajectory simulation model of ground to air missile. Fire Control Command Control **31**(10), 69–72 (2006)
18. Ou, J., Li, G., Gao, Z.: Three-dimensional trajectory simulation of air defense missile based on matlab. Fire Control Command Control **35**(2), 166–168 (2010)
19. Wang, J., Zhou, L., Lei, H.: Forecast model and arithmetic on hit point of ground-to-air missile and aerial target. J. Syst. Simul. **21**(1), 80–83 (2009)
20. Gao, H., Zhang, H., Kong, F.: Research of missile system. Defense Manuf. Technol. Simul. Simulink (2), 67–72 (2016)
21. Zhang, Z., Tong, Y., Zhang, W.: Research on the proportional guidance trajectory simulation. Tactical Missile Technol. (2), 56–59 (2005)
22. Qiang, S., Yi, D., Tu, X.:Simulation research on tracing of tactical trajectory missile. J. Trajectorys (03), 34–37 (2007)
23. Wang, Y., Xu, C., Zheng, X., Sun, G., Jian, Y.: Initial proportional guidance method with fall angle constraint and overload constraint. J. Projectile Guidance **39**(04), 29–32+36 (2019)
24. Wang, P., Liu, H., Zou, H.: Research on simulation about the miss distance of the portable surface-to-air missile. Ship Electron. Eng. **31**(6), 108–110 (2011)

Fault Diagnosis of Vertical Pumping Unit Based on Characteristic Recalibration Residual Convolutional Neural Network

Youxiang Duan[1], Cheng Chang[1(✉)], Qifeng Sun[1], Chengze Du[1], and Zhengyang Li[2]

[1] School of Computer Science and Technology, China University of Petroleum (East China), Qingdao 266580, China

[2] School of Precision Instruments and Optoelectronics Engineering, Tianjin University, Tianjin 300072, China

Abstract. Rod pumping unit is the main equipment of oil exploitation. The automatic and intelligent of the management, control, running of pumping units are important goals for the construction of the smart oil field. The continuous development of network and deep learning technology provides strong support for the realization of intelligent pumping unit systems. The use of real-time collected pumping unit operating data for working condition supervision and intelligent analysis and decision-making has become an important part of the new generation of vertical pumping unit systems. This article is based on the collected operating data such as the dynamometer diagram of the pumping unit, using the deep learning technology, the intelligent working condition analysis, and the fault diagnosis model of the new generation vertical pumping unit are established. The training time of the model is short while the accuracy of identification and classification is high, which meets the practical application requirements well.

Keywords: Vertical pumping unit · Deep learning · Dynamometer card · Fault analysis

1 Introduction

With the rapid development and widespread application of technologies such as network information and artificial intelligence [1], intelligent oilfields are replacing conventional oilfields as the main target for the construction of next-generation oilfields [2]. Intelligent supervision of oil production is a top priority for future oilfield construction [3].

Oil is mainly lifted from the underground to the ground by a rod pumping unit, which is divided into a beam type and a beamless type. Due to the shortcomings of beam pumping units such as low efficiency and high consumption, and difficult to balance during long strokes [4]. The new generation of none-beam pumping units is gradually replacing beam pumping units. The vertical pumping unit is a tower pumping unit, with high efficiency and low consumption, safe and stable [5], and has been popularized and applied in domestic and foreign oil fields.

H. Song and D. Jiang (Eds.): SIMUtools 2020, LNICST 370, pp. 545–556, 2021.
https://doi.org/10.1007/978-3-030-72795-6_44

The pumping unit usually works in harsh and remote areas, and its operation is affected by many factors such as wind, snow, rain, and other severe weather, gearbox gear wear, and motor over-time operation. If failures are not found and diagnosed in time, it will seriously affect oil production and even cause production accidents.

Therefore, it is of great significance to intelligently analyze and diagnose the operating conditions of pumping units [6]. For many years, the intelligent analysis and diagnosis of pumping unit conditions using new computer technology have been an important part of oilfield application research. Artificial intelligence technologies such as neural networks and support vector machines have been used in the intelligent diagnosis of pumping unit failures [7, 8]. But the application effect did not meet people's expectations. The development of artificial intelligence technology, especially the emergence of deep learning technology, has provided new solutions for intelligent analysis and diagnosis of pumping unit failures [9].

The dynamometer method is the most effective fault analysis and diagnosis method for pumping units today [10]. The dynamometer diagram is a closed curve of the change of the suspension point load with the displacement of the suspension point in a pumping cycle. The shape characteristics of the dynamometer diagram can reflect the operating conditions of the pumping unit [11]. Although the vertical pumping unit and the beam pumping unit have different structures, the principle of oil extraction is the same. Therefore, the dynamometer method is used to analyze the operating conditions of the vertical pumping unit, and the fault diagnosis of the vertical pumping unit is also effective. However, different mechanical structures of pumping units have different shape characteristics of the dynamometer diagram.

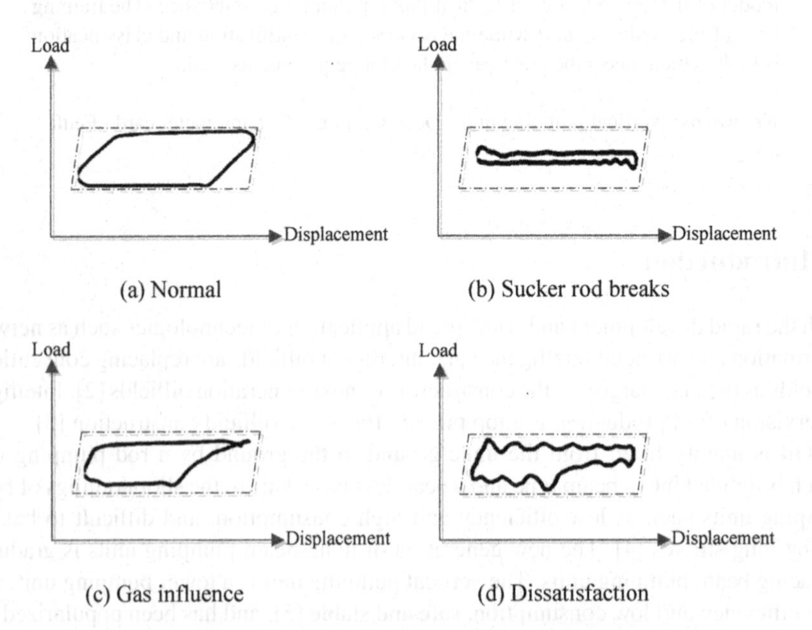

Fig. 1. Vertical pumping unit power diagram

Figure 1 is a typical dynamometer diagram of a vertical pumping unit collected and summarized. Figure 1 (a) is a dynamometer diagram of normal operation, and Fig. 1 (b) is a dynamometer diagram of a broken sucker rod. Figure 1 (c) is the dynamometer diagram of gas influence, and Fig. 1 (d) is the dynamometer diagram of dissatisfaction.

The experimental method of pumping unit fault diagnosis is as follows: First, collect the pumping unit dynamometer diagram data and use the data to draw the dynamometer diagram. Then, mark the dynamometer diagram according to the working condition of the pumping unit represented by the dynamometer diagram. Finally, perform a classification and recognition experiment on the dynamometer diagram, divide the dynamometer diagram data into a training set, a validation set, and a test set, use the training set to train the model, use the validation set to determine the model hyperparameters, select the optimal model, and use the test set perform performance evaluation on the trained model.

The significance of applying deep learning model to the analysis of pumping unit operating conditions is to use the deep learning model to analyze the dynamometer diagram instead of manual analysis by a large number of experienced professional staff. The analysis results are no longer affected by the staff's experience and knowledge, and the personal workload is greatly reduced Small, able to meet the needs of intelligent and automated oilfield construction.

The deep learning model can not only be applied to the fault diagnosis of pumping units but also be applied to the data processing of pumping units, for example, the power diagram converse to the dynamometer diagram. It has useful and broad application prospects.

2 Fault Diagnosis Model of Vertical Pumping Unit Based on Feature Recalibration Residual Convolutional Neural Network

2.1 Model Structure

Convolutional neural network (CNN) is a kind of feedforward neural network with convolution operation and deep structure. It consists of a convolutional layer, pooling layer, fully connected layer, etc. [12]. Different from other automatic image recognition and classification methods, CNN can automatically extract and filter features, without the need to manually extract features. Through the use of convolutional continuous extraction, more abstract features can be obtained. Using the obtained abstract features can complete a variety of tasks and perform well in the field of image recognition and classification [13]. The dynamometer diagram can be regarded as a special image, so it is feasible to analyze the dynamometer diagram with a CNN [14].

According to fault diagnosis needs, draw on the deep learning model with good image recognition and classification effect, such as, draw on the residual ideas and modularization in ResNet and DenseNet, draw on the SE substructure in SENet, etc. [15–17], and improve based on these, this paper proposes a feature recalibration residual CNN model based on dynamometer diagram data, which aims to enhance model performance and shorten training time.

Feature recalibration residual CNN model has 14 layers, and its module structure is shown in Fig. 2, where the convolutional layer contains 16 3×3 convolution kernels. 5 SE-residual modules (The residual module is embedded in the Squeeze-and-Excitation substructure) contains 2 convolutional layers, the convolution kernel sizes are 1×1 and 3×3, and the number of convolution kernels is 32, 64, 64, 128, 128. The convolution layer sets the convolution step size to 1, adding L2 regularization, and LeakyReLu as the activation function, the pooling filter size is 2×2, and the pooling step size is set to 2, the neurons of the fully connected layer The numbers are 1024, 512, and 4, and the activation functions are LeakyReLu, LeakyReLu, and softmax. The input of the model is the dynamometer diagram, and the output is the type of the dynamometer diagram (fault type).

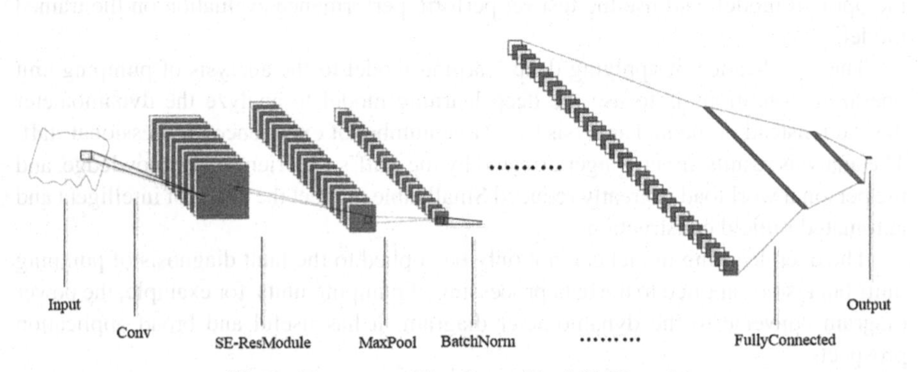

Fig. 2. Feature recalibration residual CNN model

The feature recalibration residual CNN model strengthens the effect of effective feature maps by embedding substructures, and the performance is significantly improved. Compared with the classical deep learning model to improve the performance by increasing the depth, the feature recalibration residual CNN model embeds the substructure adds little calculation and the modular structure is easy to modify to meet the application requirements.

2.2 Residual Module

The residual module designed in this paper consists of 1×1 convolutional layer, 3×3 convolutional layer, intermediate operations (BN is batch normalization, LeakyReLu is activation operation, Ave_pool is average pooling), and 1 The identity mapping is composed as shown in Fig. 3. Among them.

The module input is x. The module output is $H(x) = F(x) + x$. The formula changes to $F(x) = H(x) - x$.

$F(x)$ represents the residual, the module fits between the input and output is the residual $F(x)$.

Unlike most residual networks, this structure consists of 1 × 1 convolutional layer, 3 × 3 convolutional layer, and intermediate operations. It not only reduces the dimensions of extracted features, reduces the number of parameters, but also increases non-linear factors, and accelerates the model convergence. To play an important role in simplifying the model and enhancing the ability to express the model.

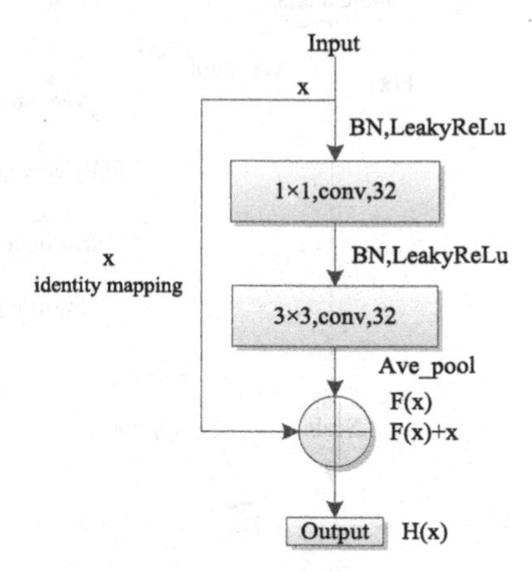

Fig. 3. Residual module

2.3 SE-Residual Module

After the residual module is built, the SE substructure is embedded in it, forming the main part of the feature recalibration residual convolution neural network model, namely the SE-residual module, as shown in Fig. 4 (in the Fig, c is the number of feature channels, H is the feature map height and w is the feature map width).

In the SE substructure, the first step is Squeeze operation, which uses average pooling to compress the features. Then, the fully connected layer is used to reduce the feature's dimension to 1/16 (16 is the setting parameter value, which can be modified). After the second fully connected layer, the feature dimensions are restored. Compared with only one fully connected layer, it can add more non-linearity, enhance the relationship between the channels, and reduce the number of parameters. The second step is Excitation operation, which uses the Sigmod function to calculate the weight for each feature channel, and the weight represents the importance of each feature channel. The third step is Scale operation, which uses multiplication to weigh each channel feature to complete the re-calibration of the original feature.

Squeeze operation expression (1).

$$Z_c = F_{sq}(U_c) = \frac{1}{H \times W} \sum\nolimits_{i=1}^{H} \sum\nolimits_{j=1}^{W} U_c(i,j) \tag{1}$$

F_{sq} is the Squeeze function, that is, average pooling and U_c is a feature map with height and width H × W. The input of H × W × C is converted into the output of 1 × 1 × C by the Squeeze function.

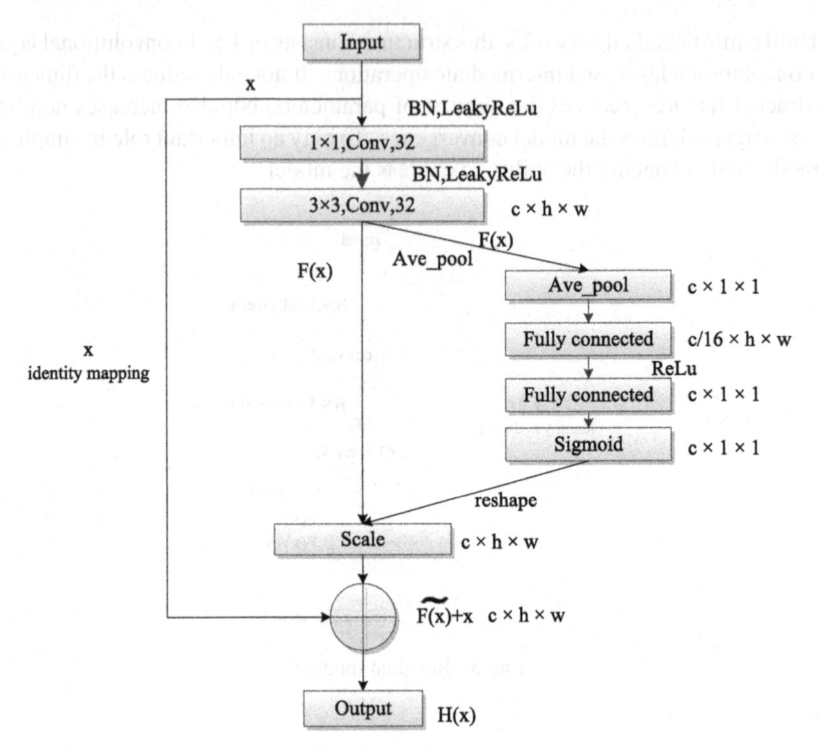

Fig. 4. SE-residual module

Excitation operation expression (2).

$$s = F_{ex}(z, W) = \sigma(W_2\delta(W_1 z)) \tag{2}$$

Fex is the Excitation function, z in formula (2) is Z_c in formula (1), and W_{1z} is the first fully connected calculation operation. The dimension of W_1 is C/16 × C. Taking 16 in this paper is to reduce the number of channels to the original 1/16. After the ReLu activation operation and the second fully connected layer are multiplied with W_2, the dimension of W_2 is C × C/16, and the output dimension is C × 1 × 1. Finally, s is obtained after passing the sigmoid function.

Scale operation expression (3).

$$F_{scale}(U_c, S_c) = S_c \cdot U_c \tag{3}$$

F_{scale} is the Scale function representing U_c multiplied by S_c, U_c is a two-dimensional matrix, and Sc is the weight.

Compared with the residual module, the performance of the SE residual module is improved. The added parameters only exist in the 2 fully connected layers, and the increased amount of computers is almost negligible. At the same time, the structure of the SE residual module is very simple and easy to implement, without introducing new functions or layers.

2.4 Model Features

Using the idea of residual to build the model changes the way of forward and backward information transmission. During model training, if the feature represented by x is already very mature, that is, an increase or decrease in x will increase the loss of the model. At this time, F(x) will tend to 0, x will continue to transmit information from the identity mapping path. This is conducive to the training of deep networks and solves the gradient descent problem to a certain extent. As the depth of the model deepens, its expression ability becomes stronger and the classification accuracy of the test set is higher.

The feature recalibration through the SE substructure is mainly based on the model's continuous learning process through the loss value. The feature weight calibration increases the weight of valid features, reduces the weight of invalid features, and trains a model with stronger performance. The SE substructure is simple to implement and suitable for embedding various deep learning models.

Adding a batch normalization operation between the convolutional layer can reduce the number of model parameters, improve the accuracy of model training, and reduce the problem of gradient descent.

The main steps of batch normalization.

1. Input.

$$x : \beta = \{x_1, \ldots, x_m\} \tag{4}$$

2. Output.

$$\{y_i = BN_{\gamma,\beta}(x_i)\} \tag{5}$$

3. Calculating the mean of batch data.

$$\mu_\beta = \frac{1}{m}\sum\nolimits_{i=1}^{m} x_i \tag{6}$$

4. Calculate the batch data variance.

$$\sigma_\beta^2 = \frac{1}{m}\sum\nolimits_{i=1}^{m} (x_i - \mu_\beta)^2 \tag{7}$$

5. Normalize.

$$\hat{x}_i = \frac{(x_i - \mu_\beta)}{\sqrt{\sigma_\beta^2 + \varepsilon}} \tag{8}$$

6. Scale change and migration processing.

$$y_i = y\hat{x}_i + \beta = BN_{\gamma,\beta}(x_i) \tag{9}$$

7. Return learning parameters γ and β.

The model uses the LeakyReLu function as the activation function. Compared with the ReLu function, the LeakyReLu function can modify the data distribution and update the network parameters when the input is negative [18]. LeakyReLu function formula such as (10).

$$y_i = \begin{cases} x_i \ \text{if} \ x_i \geq 0 \\ \frac{x_i}{a_i} \ \text{if} \ x_i < 0 \end{cases} \tag{10}$$

a_i is a fixed parameter in the interval (1,+∞).

The model uses Adam as the optimization function. Compared with SGD, Momentum, AdaGrad, and other optimization functions, the Adam optimization function is easy to implement, the range of hyperparameter adjustment is small, the update step can be limited to a certain range, and the learning rate does not need to be manually adjusted. Learning model [19], parameter description of Adam's optimization function.

α is the Step size.
β_1 is the first-order moment attenuation coefficient, initialized to 0.9.
β_2 is the second-order moment attenuation coefficient, initialized to 0.999.
f(θ) is the objective function. θ is the parameter to be optimized.
t is the number of update steps, initialized to 0.
g_t is Gradient for f (θ) derived from θ.
m_t is the first moment of g_t, which is the expectation of g_t.
v_t is the second moment of g_t, which is the expectation of g_t^2.
m_t is the offset correction of m_t, because m_t is initialized to 0, which will cause b to be offset to 0.

Similarly, $\widehat{v_t}$ is the offset correction for v_t.
The main update steps of the Adam optimization function.
Update steps.

$$\mathrm{t} = \mathrm{t} + 1 \tag{11}$$

Calculate the gradient of f (θ) to the parameter θ.

$$g_t = \nabla_\theta f_t(\theta_{t-1}) \tag{12}$$

Calculate the first moment of the gradient.

$$m_t = \beta_1 \cdot m_{t-1} + (1 - \beta_1) \cdot g_t \tag{13}$$

Calculate the second moment of the gradient.

$$v_t = \beta_2 \cdot v_{t-1} + (1 - \beta_2) \cdot g_t^2 \tag{14}$$

Correct the first moment.

$$m_t = m_t / (1 - B_1^t) \tag{15}$$

Correct the second-order moment.

$$v_t = v_t/(1 - B_2^t) \tag{16}$$

Update parameters.

$$\theta_t = \theta_{t-1} - \alpha \cdot m_t/(\sqrt{v_t} + \varepsilon) \tag{17}$$

The model uses the variance scaling metshod to initialize the weights, which has a stronger generalization ability than the conventional initialization method. The model is finally 3 fully connected layers, which not only plays a role in classification but also reduces the impact of feature distribution on classification [20].

3 Experiment and Analysis

3.1 Data Source

The dynamometer data set used in this article is from a domestic oil field using a new vertical intelligent pumping unit. The data collection period is from 12:00 on January 15, 2019, to 21:00 on August 20, 2019. After the manual screening, The classification and data enhancement totaled 19,000 dynamometer maps, of which 4,084 were "working normally", 2,697 were "filled with dissatisfaction", 8083 were "gas impact", and 4,136 were "pump rod disconnection".

The data set is divided using a random extraction method. The training set accounts for 90% and the test set accounts for 10% for the model.

3.2 Model Construction

This experiment uses NVIDIA Tesla P100 GPU graphics card and uses tflearn and sklearn as platforms to develop, and build logistic regression model, random forest model, XGBoost model, logistic regression, random forest, XGBoost integrated model, 10-layer ResNet model, 53-layer The DenseNet model and the feature recalibration residual CNN model proposed in this paper.

The last fully connected layer is classified using the Softmax function, the loss function is cross-entropy, and the optimization function is Adam.

3.3 Experiment and Analysis

Use the prepared data set to train the logistic regression model, random forest model, XGBoost model, integrated model, ResNet model, DenseNet model, and feature re-calibration residual CNN model. Use the trained model to show power on the test set Graph classification experiment. To reduce the error of the experiment, the average value obtained from multiple experiments is shown in Fig. 5, Fig. 6, Table 1, and Table 2.

From Fig. 5, Fig. 6, Table 1, and Table 2, the average accuracy, average accuracy, average recall, and average f1 scores of the feature recalibration residual CNN model on the test set compared with the rest of the models are the highest. The loss rate is the lowest, the average training time is very short, and the model obtained has a strong generalization ability, which can be applied to the application requirements of vertical pumping unit fault monitoring and analysis.

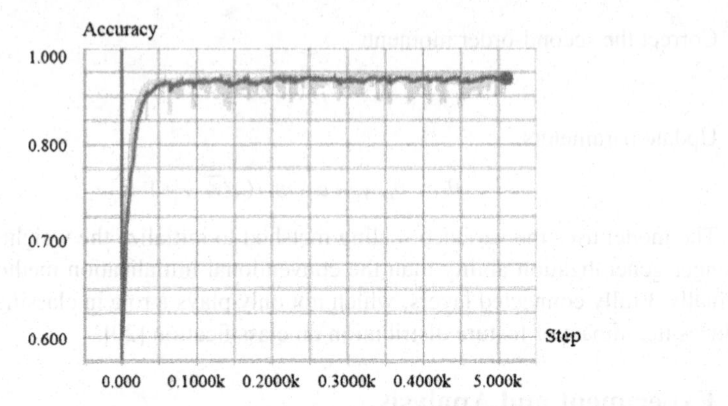

Fig. 5. Accuracy of the model test set in this paper

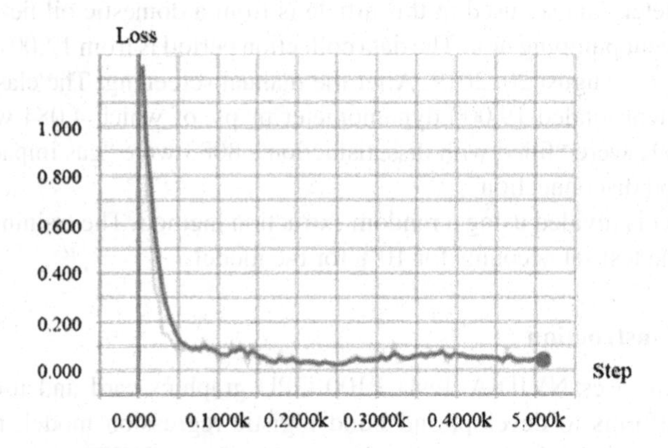

Fig. 6. Loss rate of the model test set in this paper

Table 1. Model accuracy, recall rate, f1 score of this paper

Class	Precision	Recall	F1-score
1	0.98	0.97	0.98
2	0.98	0.98	0.98
3	0.99	0.98	0.98
4	0.99	0.99	0.99
Total	0.98.5	0.98	0.98

Table 2. Comparison of experimental results. The first five columns of data in the table are the average of the model test set and the last column is the average training time of the model.

Model	Accuracy	Loss rate	Precision	Recall	f1 score	Time
Logistic regression	80%	10.9%	80%	80%	80%	27 min
Random forest	92%	10%	93%	93%	93%	23 min
XGBoost	70%	15%	73%	69%	66%	45 min
Integration	87%	7%	88%	87%	88%	1 h 35 min
ResNet	96%	13%	96%	96%	96%	2 h
DenseNet	97.5%	6.8%	97.5%	97.5%	98%	3 h
this article	98.4%	5.5%	98.5%	98%	98%	30 min

4 Total Knots

In oilfields at home and abroad, especially some tight and low-permeability special oil and gas reservoirs, vertical intelligent pumping units are increasingly used. It is of great significance to optimize the production decision by monitoring the oil and gas production process by analyzing the vertical pumping unit dynamometer diagram. In this paper, the new deep learning technology is used to analyze the working conditions of vertical pumping units, and a feature recalibrated residual convolution neural network model is designed to realize the automatic recognition and classification of the dynamometer diagram of vertical intelligent pumping units. In this paper, the residual idea is added to the model to effectively avoid vanishing gradient problem. The batch normalization layer, activation layer, and average pooling layer in the residual module can not only improve the training speed but also enhance the nonlinear ability of the model. The SE substructure is embedded in the residual module, which makes the weight of the effective feature graph increase and the weight of the invalid feature graph decreases continuously, which enhances the learning ability of the model. Experimental results show that compared with other models, The average classification accuracy, average accuracy, average recall, and average f1 score of the model on the test set are the highest, the average loss rate is the lowest, and the average training time is short, which meets the needs of applications such as accuracy and real-time performance. It has made useful explorations for the application of deep learning new technologies and the construction of smart oilfields.

References

1. Jufu, F., Jian, Y.: Foreword of the frontier progress of artificial intelligence. Comput. Res. Dev. **56**(08), 1604 (2019)
2. Yaowei, Z., Li, Z.: Application and development of intelligent oil field and intelligent drilling and production technology. Low Carbon World (02), 71–72 (2019)
3. Changshan, L.: Intelligent management mode in petrochemical production process. Zhonghua Constr. **35**(08), 90–91 (2019)

4. Liuzhu, Z.: Analysis of energy-saving and consumption-reducing measures at the oil production site. Cleaning World **35**(05), 31–32 (2019)
5. Chunming, X.: Intelligent control system of vertical energy-saving pumping unit. Electromech. Eng. **31**(04), 486–489 (2014)
6. Zhigang, T.: Research on fault dynamometer diagnosis method based on convolutional neural network. Xi'an University of Science and Technology, Xi'an (2019)
7. Shuguang, L., Chenhui, J., Yamin, T., et al.: An ART2-based pumping unit fault diagnosis. In: Chinese Control and Decision Conference 2015, Qingdao, pp. 2955–2958 (2015)
8. Zhidan, Z., Haojie, F., Penghui, L.: Application of CNN-SVM model in fault diagnosis of pumping unit wells. J. Henan Polytech. Univ. (04), 112–117 (2018)
9. Zhou, X., Zhao, C., Liu, X.: Application of cnn deep learning to well pump troubleshooting via power cards. In: Abu Dhabi International Petroleum Exhibition & Conference 2019, Abu Dhabi (2019)
10. Zhidan, Z., Penghui, L., Miaomiao, G.: Research on fault diagnosis of rod pumping unit in petroleum production. Comput. Simul. (02), 443–447 (2016)
11. Zhou, W., Li, X., Yi, J., et al.: A novel UKF-RBF method based on adaptive noise factor for fault diagnosis in pumping unit. IEEE Trans. Ind. Inform. **15**(03), 1415–1424 (2018)
12. Szegedy, C., Liu, W., Jia, Y., et al.: Going deeper with convolutions. In: Computer Vision and Pattern Recognition 2015, Boston, Massachusetts, pp. 1–9 (2015)
13. Simonyan, K., Zisserman, A.: Very deep convolutional networks for large-scale image recognition. In: Computer Vision and Pattern Recognition 2014, Columbus, Ohio (2014)
14. Tong, X.: Research on oilfield intelligent fault diagnosis system based on optimized RBF network. Xi'an Shiyou University, Xi'an (2019)
15. He, K., Zhang, X., Ren, S., et al.: Deep residual learning for image recognition. In: Computer Vision and Pattern Recognition 2016,Las Vegas, Nevada, pp. 770–778 (2016)
16. Huang, G., Liu, Z., Der Maaten, L.V., et al.: Densely connected convolutional networks. In: Computer Vision and Pattern Recognition 2017, Honolulu, Hawaii, pp. 2261–2269 (2017)
17. Hu J, Shen L, Albanie S, et al.: Squeeze-and-excitation networks. IEEE Trans. Pattern Anal. Mach. Intell. 7132–7141 (2018)
18. Gongpeng, W., Meng, D., Changyong, N.: Stochastic gradient descent algorithm based on convolutional neural network. Comput. Eng. Des. **39**(02), 441–445 (2018)
19. Hui, Z.: Research and improvement of optimization algorithms in deep learning. Beijing University of Posts and Telecommunications, Beijing (2018)
20. Fuyu, H., Jingzhong, W., Linhao, Z.: Public opinion classification method of heterogeneous data based on CNN and LSTM. J. Comput. Syst. **28**(06), 141–147 (2019)

Queue Regret Analysis Under Fixed Arrival Rate and Fixed Service Rates

Ping Cui, Lei Chen(✉), Yi Shi, Kailiang Zhang, and Yuan An

Jiangsu Province Key Laboratory of Intelligent Industry Control Technology,
Xuzhou University of Technology, Xuzhou 221018, China
chenlei@xzit.edu.cn

Abstract. In wireless communication, transmitter often need choose one channel from several available ones. Since the instantaneous channel rate is time-varying with unknown statistics, the channel selection is based on observation. Evaluating the lost of scheduling based on observation is an important for design scheduling policy. By adopting the concept of queue regret fact, we carry out simulation under different arrival rate and channel service rate. As arrival rate is approaching the service rate of the best channel, the queue regret has a shape increase in our simulation. However, even if the arrival rate is higher than best service rate, the transmitter have still chance to find the best channel, and the queue regret will converge. The relationship between arrival rate, service rate, queue length and queue regret is analyzed in the simulation.

Keywords: Queue regret · Scheduling policy · Wireless communication

1 Introduction

Recently, some new approaches are proposed to improve network routing and measurement [1–3]. Based on effective user behavior and traffic analysis methods [4–7], new scheduling strategies are designed to raise resources utilization [8–11] and energy-efficiency [12–14]. To test these new scheduling strategies, traffic reconstruction is important [15–17]. Because the network traffic changes very fast in mobile communication scenario [18, 19], some intelligent approaches are proposed for traffic reconstruction [3, 20, 21]. The traffic deeply affects the end user's experience [22–25]. Therefore, the resource management must be based on the network traffic model [26–29]. For air interface, a main issue of resource management is channel selection [30–33]. The channel selection approaches are usually based on the channel estimation in short term [34–37]. And most researches only concern about the short term performance index.

The system's regret index is defined to measure the system stability [38]. The regret compares the backlog of real learning controller which select policy based on statistics and the backlog under a controller that knows the best policy. As a stochastic multi-armed bandit problem, the regret bound is drawn in [39]. The practical policies have been studied for a long time to deal these problems [40]. To project a perfect service

H. Song and D. Jiang (Eds.): SIMUtools 2020, LNICST 370, pp. 557–567, 2021.
https://doi.org/10.1007/978-3-030-72795-6_45

rate observation in busy time under fixed arrival rate, the boundary of regret is obtained [41].

In some wireless communication situations, the channel state is stable in short term. Therefore, transmitter could evaluate the candidate channels during idle period. By the channel estimation information, transmitter is able to select optimal channel. The effect of channel estimations is affected by the length of idle period. In this paper, we design a stochastic estimation algorithm to select the optimal channel. In the simulation, the capacity of channel state is treated as service rate. Based on the record of queue length and queue regret, the relationship between arrival rate, service rate, queue length and queue regret is analyzed.

The paper is divided as six sections. In the second section, we give the related work about queue regret problem. The system model is given in the third section. Based on system model an analysis on queue regret facts is presented in the fourth section. The algorithm and simulation results analysis is given in the fifth section. We conclude in the sixth section.

2 Related Work

As a multi-armed bandit problem in which the controller need to allocate limited resource to alternative choices for optimal gain, the stochastic properties of choice must be described [40]. However, the choice's stochastic properties are unknown at the time of allocation. We must make the allocation under observation. The longer observations lead to that we can better understand the choices. In this classic reinforcement learning problem the controller meets a tradeoff dilemma of stochastic scheduling.

During the idle time periods, the offered service is unused, and therefore we can select a candidate service to observe. However, most researches only focus on the relationship between queue regret and the length of passed time [38]. The main issue in these research is that the necessary of take an observation in busy time. Someone argue that observation in busy time would increase the regret if the optimal service has been selected. Others argue that observing only in idle time would lose the chance to obtain a better choice when the busy time is too long.

3 System Model

For our wireless communication scenario, the capacity of N candidate channels is referred to as service rate of N servers in a single queue system. In our model, we assume that a controller schedules the servers over discrete time slots $t = 0, 1, 2\ldots$ Packets arrive to the queue as a Bernoulli process, written as $A(t)$ with rate $\lambda \in (0, 1]$. The service rate is defined as the amount of packets that server $i \in [N]$ can provide follows a Bernoulli process $D^i(t)$ with rate μ_i. The arrival process and server processes are arbitrarily assumed to be independent. If $\mu_i > \lambda$ the system is referred to as stabilizing; otherwise, it is referred to as non-stabilizing.

The controller must select one channel to serve the queue from the N channels (servers) to provide service when the queue is non-empty. We denote the controller's choice at time t as $u(t) \in [N]$ and the service offered to the queue as $D(t)$ which is equal to $D^{u(t)}(t)$. The queue length $Q(t)$ can be written as [38]:

$$Q(t+1) = (Q(t) - D(t))^{+} + A(t), \text{ for } t = 0, 1, 2, \ldots \quad (1)$$

where $(x)^{+}$ is used to denote the maximum of x and 0. And $Q(0)$ is assumed as 0. We assume that the arrival rate and service rate are stable, however, the controller do not know the values of $D^{i}(t)$ prior to making its decision $u(t)$. To make optimal action, that maximizes expected service, controller should select the best server

$$i^{*} \triangleq \underset{i \in [N]}{\operatorname{argmax}} \mu_i \quad (2)$$

to provide service. In this work, the controller does not a priori know the values of μ_i and must therefore use observations of $D(t)$ to identify i^{*}. We assume that the controller can observe $D(t)$ at all times t, even when the queue is empty. Define $Q^{*}(t)$ to be the queue length under the controller that always schedules i^{*} and $Q^{\pi}(t)$ the backlog under a policy that must learn the service rates. The performance of policy π is measured by queue length regret [38]:

$$R^{\pi}(T) \triangleq E\left[\sum_{t=0}^{T-1} Q^{\pi}(t) - \sum_{t=0}^{T-1} Q^{*}(t)\right]. \quad (3)$$

The π is scheduling policy, the assumption implies that $R^{\pi}(T)$ is monotonically.

4 Queue Regret Analysis

In the stabilizing scenarios, the controller has enough idle periods to observe the service rate of each channel. Therefore, the queue regret is expected to stop increasing after initial phase. In the non-stabilizing scenarios, the queue length would increase sharply and the queue will keep a backlogged situation for long time.

To investigate the possibility that the controller change a bad decision, we assume that the controller selects a channel with service rate μ_i and another channel have service rate μ_j, $\mu_j > \mu_i$. With a long busy period, the observation value of μ_i will approach the real value. The controller only keep use ith channel while the observation value of μ_j is less than μ_i. This possibility can be written as:

$$P\{X \leq n \cdot \mu_i\} = \sum_{k=0}^{n \cdot \mu_i} \binom{n}{k} {\mu_j}^{k} {\mu_j}^{n-k}, \quad (4)$$

where the n is the number of controller observing the jth channel in initial phase. We know that the Eq. (4) increase fast when $n \cdot \mu_i$ approaching to μ_j. Therefore, controller has higher possibility of changing channel as the value of $\mu_j - \mu_i$ increase. This implies the queue regret has chance to keep a low value within a long time busy period.

5 Simulation and Results

5.1 Algorithm

The algorithm we adopt in simulation is shown in Fig. 1. The algorithm uses idle period to observe the service rates of channels. In the busy period, the service rate of selected channel is updated at each time slot. The algorithm is presented as followed:

```
Step1    While (simulation time > 0)
Step2      If (the queue is empty)
Step3        Select a random channel
Step4        Observe the selected channel
Step5        Update the service rate of observed channel
Step6      else
Step7        Select randomly a channel from the set of servers with highest rate
Step8        Decide whether transmit the packet according to real service rate
Step9        Update service rate of the selected server
Step10     End if
Step11     Simulation time --
Step12   End while
```

Fig. 1. Simulation algorithm.

5.2 Simulation Parameters

We make two simulations, the simulation parameters are listed in Table 1.

Table 1. Simulation parameters.

Parameter name	Value of parameter in simulation 1	Value of parameter in simulation 2
Number of slots	15000 for each test	15000 for each test
Service rates	0.3, 0.325, 0.35, 0.375, 0.4	0.2, 0.225, 0.25, 0.275, 0.3
Arrival rates	0.2–0.5, increase 0.01 each test	0.2–0.5, increase 0.01 each test

In these two simulations, we have 5 candidate channels with a service rate range from 0.3 to 0.4 and from 0.2 to 0.3 respectively. Both the arrival rates increase 0.01 at each test from 0.2 to 0.5. Therefore, there are 31 tests in each simulation. And each test last 15000 time slots under the fixed arrival rates.

5.3 Simulation Results

The simulation results are shown in Fig. 2 and Fig. 3.

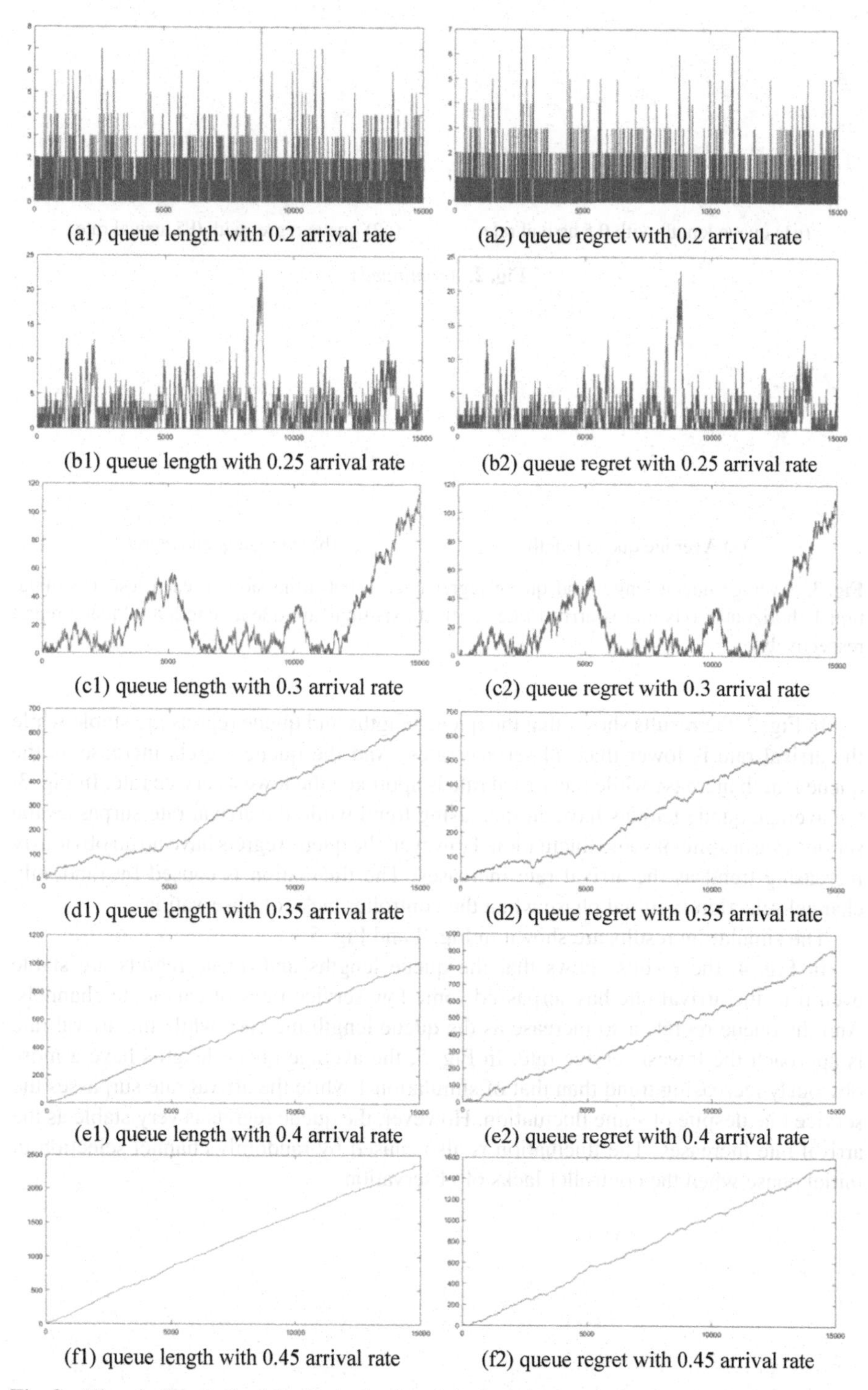

(a1) queue length with 0.2 arrival rate
(a2) queue regret with 0.2 arrival rate
(b1) queue length with 0.25 arrival rate
(b2) queue regret with 0.25 arrival rate
(c1) queue length with 0.3 arrival rate
(c2) queue regret with 0.3 arrival rate
(d1) queue length with 0.35 arrival rate
(d2) queue regret with 0.35 arrival rate
(e1) queue length with 0.4 arrival rate
(e2) queue regret with 0.4 arrival rate
(f1) queue length with 0.45 arrival rate
(f2) queue regret with 0.45 arrival rate

Fig. 2. Slices in Simulation1 (horizontal axis unit is time slot, vertical axis units are queue length and queue regret respectively)

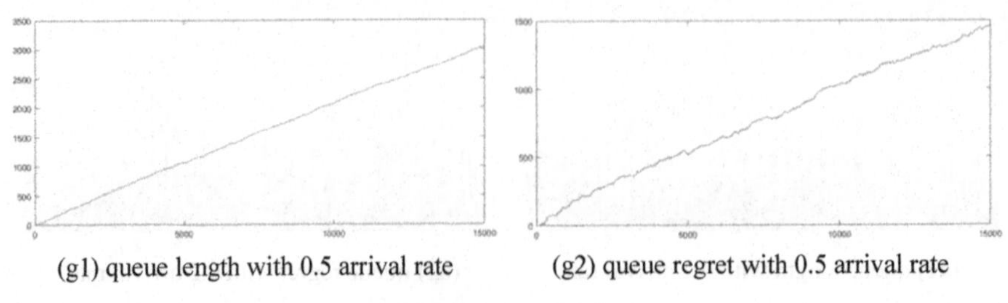

(g1) queue length with 0.5 arrival rate (g2) queue regret with 0.5 arrival rate

Fig. 2. (*continued*)

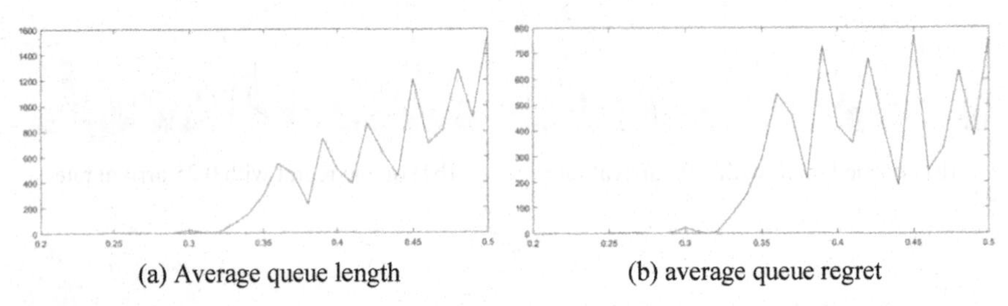

(a) Average queue length (b) average queue regret

Fig. 3. Average queue length and queue regret over 15000 time slots at each test in simulation 1 (horizontal axis unit is arrival rate, vertical axis units are queue length and queue regret respectively).

In Fig. 2, the results shows that the queue lengths and queue regrets are stable while the arrival rate is lower than all service rates. And the queue regrets increase as the queue length increase while the arrival rate is approach the lowest service rate. In Fig. 3, the average queue lengths have an increasing trend while the arrival rate surpasses the service rate despite of some fluctuation. However, the queue regrets have no an obviously increasing trend as the arrival rate increases. The fluctuation is caused by randomly channel selection in initial phase when the controller lacks of observation.

The simulation results are shown in Fig. 4 and Fig. 5.

In Fig. 4, the results shows that the queue lengths and queue regrets are stable even if as the arrival rate has surpassed some low service rates of candidate channels. And the queue regrets also increase as the queue length increase while the arrival rate is approach the lowest service rate. In Fig. 5, the average queue lengths have a more obviously increasing trend than that of simulation 1 while the arrival rate surpasses the service rate despite of some fluctuation. However, the queue regrets is very stable as the arrival rate increases. The fluctuation is also caused by randomly channel selection in initial phase when the controller lacks of observation.

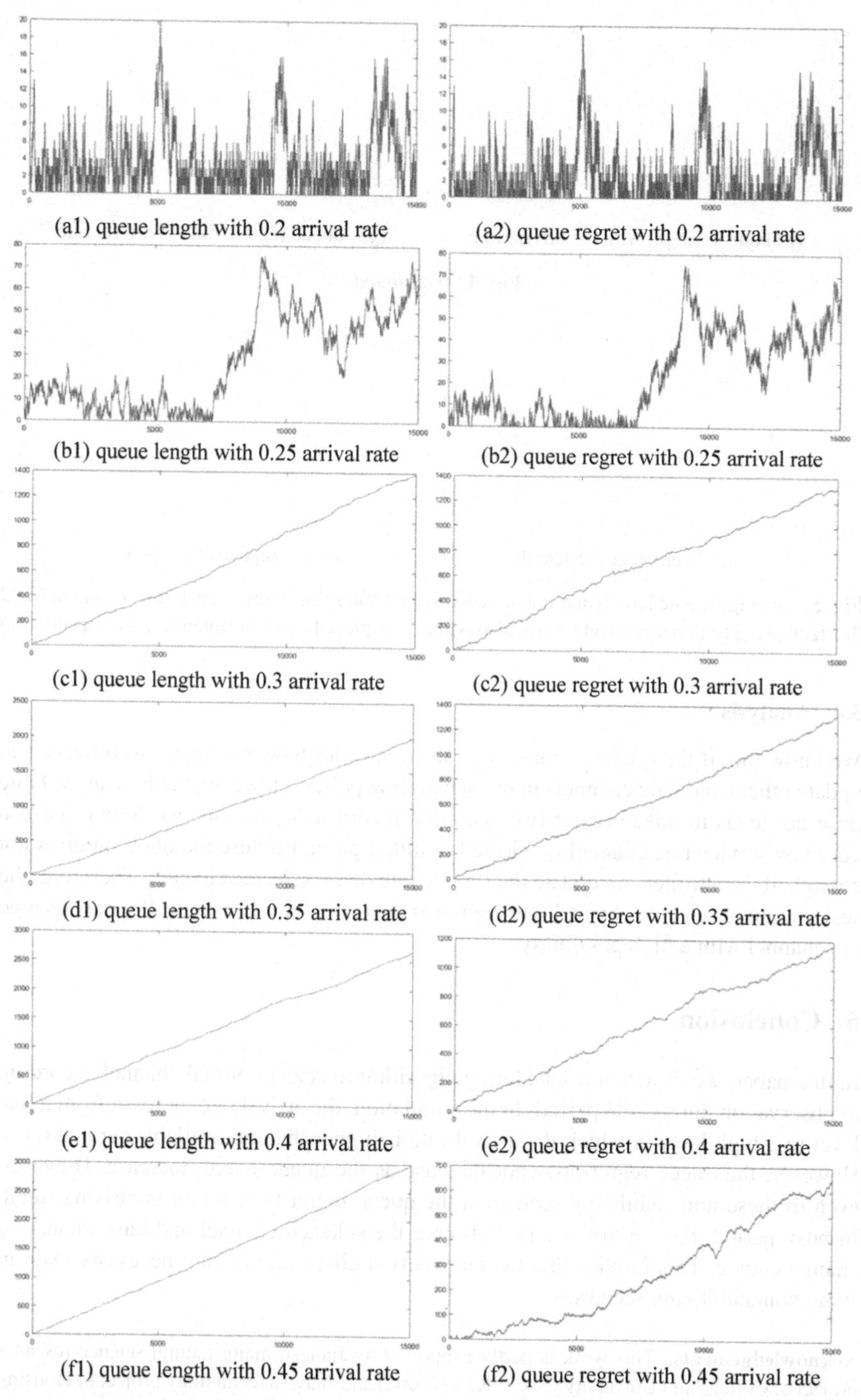

Fig. 4. Slices in Simulation2 (horizontal axis unit is time slot, vertical axis units are queue length and queue regret respectively).

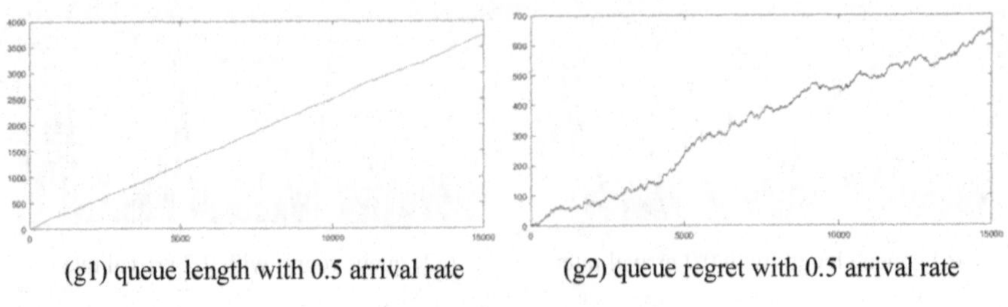

(g1) queue length with 0.5 arrival rate (g2) queue regret with 0.5 arrival rate

Fig. 4. (*continued*)

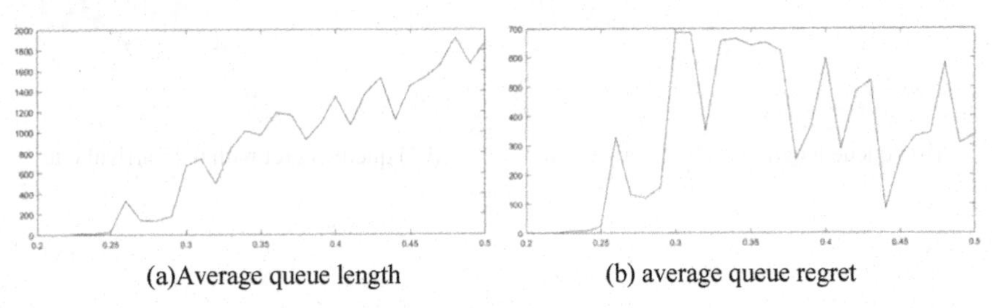

(a)Average queue length (b) average queue regret

Fig. 5. Average queue length and queue regret over 15000 time slots at each test in simulation 2 (horizontal axis unit is arrival rate, vertical axis units are queue length and queue regret respectively)

5.4 Analysis

We know that if the queue is non-empty the controller have no chance to observe and update other candidate channels in our scheduling policy. However, the high arrival rate dose not leads to queue regret rise. Through record data analysis, we found that if a very low service rate channel is selected in initial phase because the observation is not enough, the controller can update the service rate of the selected channel. Therefore, the service rate of selected channel will approach the real value. The controller may change the channel with a high possibility.

6 Conclusion

In this paper, we designed a scheduling algorithm to select optimal channel according to observation during idle period. In the simulation, the arrival rate increase from a low level to a high level in which the arrival rate is higher than all available service rates. However, the queue regret does not increase as the queue length increase. However, even in these non-stabilizing scenarios, the queue regret have no an increasing trend. In busy period, the controller could observe the selected channel and have chance to change choice. This implies that the busy period observation is not necessary even in some non-stabilizing scenarios.

Acknowledgements. This work is partly supported by Jiangsu major natural science research project of College and University (No. 19KJA470002) and Jiangsu technology project of Housing and Urban-Rural Development (No. 2019ZD041).

References

1. Zhang, K., Chen, L., An, Y., Cui, P.: A QoE test system for vehicular voice cloud services. Mobile Netw. Appl. **1**, 6 (2019). https://doi.org/10.1007/s11036-019-01415-3
2. Wang, F., Jiang, D., Qi, S.: An adaptive routing algorithm for integrated information networks. China Commun. **7**(1), 196–207 (2019)
3. Huo, L., Jiang, D., Lv, Z., et al.: An intelligent optimization-based traffic information acquirement approach to software-defined networking. Comput. Intell. **36**, 151–171 (2019)
4. Chen, L., Jiang, D., Bao, R., Xiong, J., Liu, F., Bei, L.: MIMO Scheduling effectiveness analysis for bursty data service from view of QoE. Chinese J. Electron. **26**(5), 1079–1085 (2017)
5. Jiang, D., Wang, Y., Lv, Z., et al.: Big data analysis-based network behavior insight of cellular networks for industry 4.0 applications. IEEE Trans. Ind. Inf. **16**(2), 1310–1320 (2020)
6. Jiang, D., Huo, L., Song, H.: Rethinking behaviors and activities of base stations in mobile cellular networks based on big data analysis. IEEE Trans. Netw. Sci. Eng. **1**(1), 1–12 (2018)
7. Chen, L., et al.: A lightweight end-side user experience data collection system for quality evaluation of multimedia communications. IEEE Access **6**(1), 15408–15419 (2018)
8. Chen, L., Zhang, L.: Spectral efficiency analysis for massive MIMO system under QoS constraint: an effective capacity perspective. Mobile Netw. Appl. **1**, 9 (2020). https://doi.org/10.1007/s11036-019-01414-4
9. Wang, F., Jiang, D., Qi, S., et al.: A dynamic resource scheduling scheme in edge computing satellite networks. Mobile Netw. Appl. (2019)
10. Jiang, D., Huo, L., Lv, Z., et al.: A joint multi-criteria utility-based network selection approach for vehicle-to-infrastructure networking. IEEE Trans. Intell. Transp. Syst. **19**(10), 3305–3319 (2018)
11. Jiang, D., Zhang, P., Lv, Z., et al.: Energy-efficient multi-constraint routing algorithm with load balancing for smart city applications. IEEE Internet Things J. **3**(6), 1437–1447 (2016)
12. Jiang, D., Li, W., Lv, H.: An energy-efficient cooperative multicast routing in multi-hop wireless networks for smart medical applications. Neurocomputing **220**(2017), 160–169 (2017)
13. Jiang, D., Wang, Y., Lv, Z., et al.: Intelligent optimization-based reliable energy-efficient networking in cloud services for IIoT networks. IEEE J. Sel. Areas Commun. (2019)
14. Dataesatu, A., Boonsrimuang, P., Mori, K., Boonsrimuang, P.: Energy efficiency enhancement in 5G heterogeneous cellular networks using system throughput based sleep control scheme. In: 2020 22nd International Conference on Advanced Communication Technology (ICACT), Phoenix Park, PyeongChang, Korea (South), pp. 549–553 (2020)
15. Jiang, D., Wang, W., Shi, L., et al.: A compressive sensing-based approach to end-to-end network traffic reconstruction. IEEE Trans. Netw. Sci. Eng. **5**(3), 1–2 (2018)
16. Jiang, D., Huo, L., Li, Y.: Fine-granularity inference and estimations to network traffic for SDN. PLoS ONE **13**(5), 1–23 (2018)
17. Wang, Y., Jiang, D., Huo, L., et al.: A new traffic prediction algorithm to software defined networking. Mobile Netw. Appl. (2019)
18. Derrmann, T., Frank, R., Viti, F., Engel, T.: How road and mobile networks correlate: estimating urban traffic using handovers. IEEE Trans. Intell. Transp. Syst. **21**(2), 521–530 (2020)
19. Kamath, S., Singh, S., Kumar, M.S.: Multiclass queueing network modeling and traffic flow analysis for SDN-enabled mobile core networks with network slicing. IEEE Access **8**, 417–430 (2020)
20. Qi, S., Jiang, D., Huo, L.: A prediction approach to end-to-end traffic in space information networks. Mobile Netw. Appl. (2019)

21. Huo, L., Jiang, D., Qi, S., et al.: An AI-based adaptive cognitive modeling and measurement method of network traffic for EIS. Mobile Netw. Appl. (2019)
22. Marí-Altozano, M.L., Luna-Ramírez, S., Toril, M., Gijón, C.: A QoE-driven traffic steering algorithm for LTE networks. IEEE Trans. Veh. Technol. **68**(11), 11271–11282 (2019)
23. Zhong, Y., Wang, G., Han, T., Wu, M., Ge, X.: QoE and cost for wireless networks with mobility under spatio-temporal traffic. IEEE Access **7**, 47206–47220 (2019)
24. Tian, F., Yu, Y., Li, D., Cui, J., Dong, Y.: QoE optimization for traffic offloading from LTE to WiFi. In: 2019 IEEE 8th Global Conference on Consumer Electronics (GCCE), Osaka, Japan, pp. 115–116 (2019)
25. Oliveira, T., Sargento, S.: QoE-based load balancing of OTT video content in SDN networks. 2019 IEEE Symposium on Computers and Communications (ISCC), Barcelona, Spain, pp. 1–6 (2019)
26. Seufert, M., Wassermann, S., Casas, P.: Considering user behavior in the quality of experience cycle: towards proactive QoE-aware traffic management. IEEE Commun. Lett. **23**(7), 1145–1148 (2019)
27. Ge, M., Chen, W., Zeng, Y.: A SDN-based QoE-aware routing algorithm on video. In: 2019 2nd International Conference on Information Systems and Computer Aided Education (ICISCAE), Dalian, China, pp. 147–152 (2019)
28. Kimura, T., Kimura, T., Matsumoto, A., Okamoto, J.: BANQUET: balancing quality of experience and traffic volume in adaptive video streaming. In: 2019 15th International Conference on Network and Service Management (CNSM), Halifax, NS, Canada, pp. 1–7 (2019)
29. Trakas, P., Adelantado, F., Verikoukis, C.: QoE-aware resource allocation for profit maximization under user satisfaction guarantees in HetNets with differentiated services. IEEE Syst. J. **13**(3), 2664–2675 (2019)
30. Saliba, D., Imad, R., Houcke, S.: Wifi channel selection based on load criteria. In: 2017 20th International Symposium on Wireless Personal Multimedia Communications (WPMC), Bali, pp. 332–336 (2017)
31. Ozduran, V.: Leakage rate based hybrid untrustworthy relay selection with channel estimation error. In: 2017 25th Telecommunication Forum (TELFOR), Belgrade, pp. 1–4 (2017)
32. Odeyemi, K.O., Owolawi, P.A.: Performance analysis of cooperative NOMA with partial relay selection under outdated channel estimate. In: 2019 IEEE 2nd Wireless Africa Conference (WAC), Pretoria, South Africa, pp. 1–5 (2019)
33. Hasegawa, S., Kim, S., Shoji, Y., Hasegawa, M.: Performance evaluation of machine learning based channel selection algorithm implemented on IoT sensor devices in coexisting IoT networks. In: 2020 IEEE 17th Annual Consumer Communications & Networking Conference (CCNC), Las Vegas, NV, USA, pp. 1–5 (2020)
34. Hussein, H.S., Hussein, S., Mohamed, E.M.: Efficient channel estimation techniques for MIMO systems with 1-bit ADC. China Commun. **17**(5), 50–64 (2020)
35. Zhao, Y., Zhao, W., Wang, G., Ai, B., Putra, H.H., Juliyanto, B.: AoA-based channel estimation for massive MIMO OFDM communication systems on high speed rails. China Commun. **17**(3), 90–100 (2020)
36. Miao, J., Chen, Y., Mai, Z.: A novel millimeter wave channel estimation algorithm based on IC-ELM. In: 2019 28th Wireless and Optical Communications Conference (WOCC), Beijing, China, pp. 1–5 (2019)
37. Rasheed, O.K., Surabhi, G.D., Chockalingam, A.: Sparse delay-doppler channel estimation in rapidly time-varying channels for multiuser OTFS on the uplink. In: 2020 IEEE 91st Vehicular Technology Conference (VTC2020-Spring), Antwerp, Belgium,pp. 1–5 (2020)
38. Stahlbuhk, T., Shrader, B., Modiano, E.: Learning algorithms for mining queue length regret. In: 2018 IEEE International Symposium on Information (2018)
39. Bubeck, S., Cesa-Bianchi, N.: Regret analysis of stochastic and nonstochastic multi-armed bandit problems. Found. Trends Mach. Learn. **5**(1), 1–122 (2012)

40. Auer, P., Cesa-Bianchi, N., Fischer, P.: Finite-time analysis of the multiarmed bandit problem. Mach. Learn. **47**(2–3), 235–256 (2002)
41. Krishnasamy, S., et al.: Regret of queueing bandits. In: Proceedings of Neural Information Processing Systems, pp. 1669–1677 (2016)

Empirical Analysis of Dynamic Relationship Between Green Economy and Green Finance by VAR Model

Xiaolin Li[1], You Li[1](✉), Jiali Cai[1], Yunzhong Cao[2], and Liangqiang Li[1]

[1] Business School, Sichuan Agricultural University, Chengdu 611830, China
[2] College of Architecture and Urban-Rural Planning, Sichuan Agricultural University, Chengdu 611830, China

Abstract. With the development of China's economy, some problems such as ecological imbalance, serious pollution of the environment, low utilization of natural resources and so on, are followed. Green economy and green finance are very necessary to alleviate these problems. Based on the annual data of China's green finance and green economy from 2004 to 2017, we build a VAR model to empirically analyze the dynamic relationship between green economy and green finance. The empirical results indicate that increasing investment in green finance positively affects the development of green economy; simultaneously, green economy also has a certain degree of impact on green finance, which has a restraining effect in the early stage, but with the growth of time, green economy is also important to green finance. This study explores the interaction mechanism between green finance and green economy, which is important to promote ecological civilization construction, green economic development and sustainable development.

Keywords: Green finance · Green economy · VAR model

1 Introduction

China has been exploring various new ways of development for decades to solve the problems of ecological imbalance and environmental pollution to ensure sustainable and stable economic growth on the premise of building a good ecological environment. The measures of sustainable development can make the natural resources be used reasonably and the waste of resources can be reduced obviously. After decades of continuous exploration and perfection of the way to solve the problems of ecology, environment and resources in China, although a good system has not yet been formed, the development direction of green finance and green economy in China is becoming more and more clear. Green finance contributes to the growth of green economy and green economy can also promote the upgrading of green finance. Therefore, the research on the relevance between green economy and green finance is very necessary to promote the sustainable development of economy and society and ecological environment protection, and has theoretical and practical guiding values.

H. Song and D. Jiang (Eds.): SIMUtools 2020, LNICST 370, pp. 568–579, 2021.
https://doi.org/10.1007/978-3-030-72795-6_46

This paper constructs a VAR model, uses impulse function analysis and variance analysis to empirically analyze the relationship between green economy and green finance. The innovation of this paper lies in the analysis of the interaction between green finance and green economy by estimating the VAR model; the dynamic impact characteristics of green economy and green finance are depicted by using impulse corresponding function; and the contribution of green economy and green finance to each other is analyzed by variance decomposition.

2 Literature Review

2.1 Green Economy

The concept of a "green economy" originated from the Green Economy Blueprint, a 1989 book by David W. Pearce, a British environmental economist. In the book, the author believed that economic development should be within the acceptable limits of the natural environment and human beings and should not lead to ecological environmental crisis and social and economic collapse due to the blind pursuit of economic growth [1, 2]. Bohong Huang (2020) defined green economy as an economic model that uses high and new technologies to reduce resource waste, realize recycling, and then promote the coordinated development of ecological benefits, economic benefits and social benefits [3]. He et al. (2019) believed that green economy is essentially a sustainable development integrating ecology and economy, which is conducive to improving high energy consumption, adjusting economic structure and promoting stable economic growth [4].

2.2 Green Finance

On August 31, 2016, the Guiding Opinions on Building a Green Financial System officially defined green finance for the first time as economic activities supporting environmental protection and improving resource utilization or financial services provided to projects in this related field [5]. Zhang et al. (2019), in view of existing studies, understood green finance as investment financing conducive to environment and sustainable development, mainly including climate finance and carbon finance [6]. Wang et al. (2016) believed that green finance is an innovative financial model combining economic interests with environmental protection and pays more attention to ecological environmental benefits and green environmental protection industry [7]. Different from traditional finance, green finance pays close attention to the green issues in economic development and tends to lead the capital chain to the green industry.

2.3 Green Economy and Green Finance

Domestic and foreign scholars have deeply discussed green economy and green finance from two aspects of theoretical research and empirical research. In terms of theoretical research, Jinqiang Yan and Xiaoyong Yang (2018) considered that the key to the construction of green economy system lies in the innovation of green technology, and green finance runs through the whole process of green technology innovation. Therefore, they believed that it was necessary to adhere to the green finance as the center, strengthen and coordinate the green financial system to promote the development of green economy [8]. Jianhua Zhu et al. (2019) pointed out that the environmental protection principle emphasized by green finance is consistent with the sustainable development followed by circular economy. The two complement each other and develop harmoniously. In other words, the sustainable growth of circular economy depends on the positive support of green finance. In turn, the continuous growth of circular economy promotes the creation and upgrading of green finance [9]. Owen et al. (2018) focused on the contribution of the public sector in solving the financing gap and proposed the need to establish a financial ecosystem to ensure financing support for investment in the context of considering low-carbon and environmental protection [10]. Taghizadeh-Hesary (2019) advocated the establishment of green credit guarantee scheme (GCGS) to improve the tax revenue effect generated by green energy supply, reduce the uncertainty of green finance, and promote more green investment, so as to promote the realization of sustainable development goals [11].

In empirical research, Zhang et al. (2020) used the sample data of Non-financial private enterprises in China from 2012 to 2017 to prove that green innovation can effectively broaden the limits of corporate financing considering the interests of all parties [12]. Based on the financial data from 2000 to 2018, Shao et al. (2020) constructed the green finance development index, and used vector error correction model (VECM) empirical analysis to conclude that green finance can reduce carbon intensity [13]. Xiaohong Dong and Yong Fu (2018) empirically analyzed the spatial dynamic evolution process of the coupling development of green finance and green economy by using the coupling degree model and the data of 2008–2016 in China. The research results showed that the coupling development of green economy and green finance is highly coordinated [14].

Through the research and analysis of domestic and foreign theoretical and empirical research results on green economy and green finance, it is shown that scholars have conducted in-depth research on the link between green finance and green economy. However, there is little research on the dynamic correlation within green finance and green economy, and have not yet described the dynamic impact characteristics between them. Therefore, this paper uses VAR model, impulse response function and variance decomposition analysis to empirically analyze the dynamic relationship between green economy and green finance, and reveal the dynamic characteristics of their interaction.

3 Sample Selection and Model Construction

3.1 Variable Selection

1. Environmental protection investment (INV). Based on the previous literature, environmental protection investment is used to measure the level of green investment, and green investment reflects the degree of financial support for green development, and then to measure the development level of green finance. Therefore, on the basis of reference to relevant literature, this paper chooses environmental protection investment (INV) as the representative of green financial indicators [15].

2. Green GDP (GGDP). On the basis of referring to relevant literatures, this paper chooses green GDP as the measurement index of green economy development [16,17]. Green GDP is defined as:

$$GGDP = GDP - \text{Natural Resource Loss} - \text{Pollution Loss} \quad (1)$$

Where pollution loss = three wastes treatment cost + pollution direct economic loss.

The data of green GDP and green financial indicators are processed logarithmically in order to eliminate possible heteroscedasticity. Variable definitions are presented in Table 1.

Table 1. Variable definition table.

Variable name	Variable definition	Variable name	Variable definition
GGDP	Green GDP	Log (GGDP)	Green GDP after taking logarithm
INV	Investment of Environmental Protection Funds in China	Log (INV)	Investment of Environmental Protection Funds after taking logarithm

3.2 Constructing VAR Model

Christopher Sims proposed Vector Auto Regressive model (VAR model) in 1980. VAR model breaks through the limitation of causality and action direction between fixed variables in traditional economic model, and can effectively analyze the interaction among multiple variables [18]. Therefore, in order to explore the dynamic correlation between green economic growth and green finance, this paper constructs a VAR model. The VAR model constructed is shown in Eq. (2):

$$Y_t = C + \phi_1 Y_{t-1} + \phi_2 Y_{t-2} + \varepsilon_t \quad (2)$$

Where $Y = \begin{bmatrix} GGDP \\ INV \end{bmatrix}$, C represents 2×1 constant vector, $\phi_i(\mathrm{i} = 1, 2)$ represents 2×2 Autoregressive Coefficient Matrix, ε_t represents 2×1 vector.

3.3 Data Sources

The time series data of green economy index and green finance index are selected from the annual data of China from 2004 to 2017. Data are collected from China Statistical Yearbook (2005–2018).

4 Empirical Analysis of the Dynamic Relationship Between Green Economy and Green Finance

4.1 Unit Root Test

For preventing the phenomenon of "pseudo-regression" in the two sets of time series data of Green Economy (GGDP) and Green Finance (INV), unit root test was carried out to determine that the two sets of time series data are stationary time series.

Eviews software was used to test the unit root of GGDP and Log (GGDP). We present the test results in Table 2. GGDP is a non-stationary time series and Log (GGDP) is a stationary time series.

Table 2. The unit root test results of GGDP and Log (GGDP).

GGDP		t-Statistic	Prob.*
Augmented Dickey-Fuller test statistic		−0.116547	0.9187
Test critical values:	1% level	−4.420595	
	5% level	−3.259808	
	10% level	−2.771129	
Log (GGDP)		t-Statistic	Prob.*
Augmented Dickey-Fuller test statistic		−3.684851	0.0357
Test critical values:	1% level	−4.803492	
	5% level	−3.403313	
	10% level	−2.841819	

*MacKinnon (1996) one-sided p-values

The unit root test of INV and Log (INV) was performed by Eviews. The test results as shown in Table 3 can be obtained as follows: INV is a non-stationary time series and Log (INV) is a stationary time series.

4.2 VAR Model Estimation Results

Eviews 8.0 is used to estimate the VAR model. The estimated results are presented in Table 4. As we can see: $R^2 = 0.992423$, and $\overline{R}^2 = 0.984847$, which shows that the established VAR model is very effective. The first and second lag stages of Log (GGDP) have a positive influence on Log (INV), and the first and second lag stages of Log (INV) also have significant positive effects on Log (GGDP).

4.3 Pulse Response Analysis Based on VAR Model

4.3.1 Stability Test of VAR Model

Referring to the previous stability test of VAR model, this study uses AR root to verify the stability of VAR model. According to the meaning of AR root test proposed in the

Table 3. INV and Log (INV) unit root test results.

INV		t-Statistic	Prob.*
Augmented Dickey-Fuller test statistic		0.001758	0.9306
Test critical values:	1% level	−4.582648	
	5% level	−3.320969	
	10% level	−2.801384	
Log (INV)		t-Statistic	Prob.*
Augmented Dickey-Fuller test statistic		−3.917012	0.0272
Test critical values:	1% level	−4.803492	
	5% level	−3.403313	
	10% level	−2.841819	

*MacKinnon (1996) one-sided p-values

Table 4. Estimated results of VAR model.

	LOG (GGDP)	LOG (INV)
LOG (GGDP(-1))	0.688183	1.169054
	(0.52431)	(0.73007)
	[2.31255]	[1.69929]
LOG (GGDP(-2))	0.149816	2.380886
	(0.47314)	(0.65881)
	[0.31664]	[3.61391]
LOG (INV(-1))	0.113643	0.040907
	(0.17201)	(0.23952)
	[2.66066]	[2.17079]
LOG (INV(-2))	0.033802	0.099681
	(0.16276)	(0.22663)
	[2.20768]	[0.43984]
C	0.918681	−4.589766
	(1.00471)	(1.39900)
	[0.91437]	[−3.28075]
R-squared	0.992423	0.988187
Adj. R-squared	0.984847	0.976373
Sum sq. Resids	0.013869	0.026890

(continued)

Table 4. (*continued*)

	LOG (GGDP)	LOG (INV)
S.E. equation	0.058884	0.081991
F-statistic	130.9830	83.64956
Log likelihood	16.36848	13.38900
Akaike AIC	−2.526329	−1.864222
Schwarz SC	−2.416760	−1.754653
Mean dependent	8.395943	4.372907
S.D. dependent	0.478341	0.533416

Note: Each column in the table represents a regression equation in the corresponding VAR model. Each coefficient can be tested by t statistics for the significance level of a single coefficient. The first row of each column in the table represents the regression coefficient, the values in parentheses represent the standard error, and the brackets represent the t value.

existing literature, if the reciprocal of all the root modules of VAR model are less than 1, that is, all the root modules are in the unit circle, then the model is stable; if the reciprocal of all the root modules of VAR model are greater than 1, that is, all the root modules are outside the unit circle, then the model is unstable [18]. Figure 1 can be obtained by analyzing data with Eviews 8.0 software. As can be seen from Fig. 1, the reciprocal of the four root modules lies in the circle with radius 1. Therefore, the VAR model constructed in this study is stable. This shows that the two variables selected in this paper have long-term stability, and impulse response function can be further applied to these two variables.

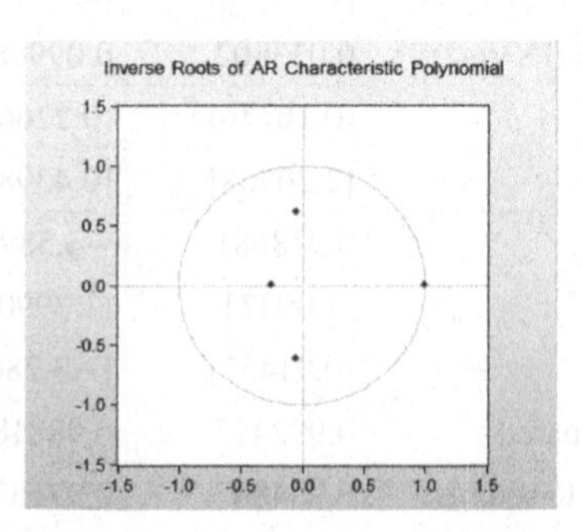

Fig. 1. AR root display of VAR model.

4.3.2 Impulse Response Analysis of Greèn Economy to Green Finance

The impulse response of green economy to green finance is shown in Fig. 2. If a standard deviation of the green financial indicators is given a positive impact, the green economic growth shows a fluctuating trend from top to bottom. When it reaches the second stage, the green economy growth reaches the maximum positive value, then gradually begins to decline and converge, and when it reaches the sixth stage, it begins to stabilize. This shows that green finance lags behind the growth of green economy. Green finance can drive the long-term growth of green economy. The positive effect is the greatest in the second period, and the driving effect will not gradually weaken with the increase of time. Generally speaking, increasing investment in green finance has a positive driving impact on green economy growth, that is, green finance supports the green economy.

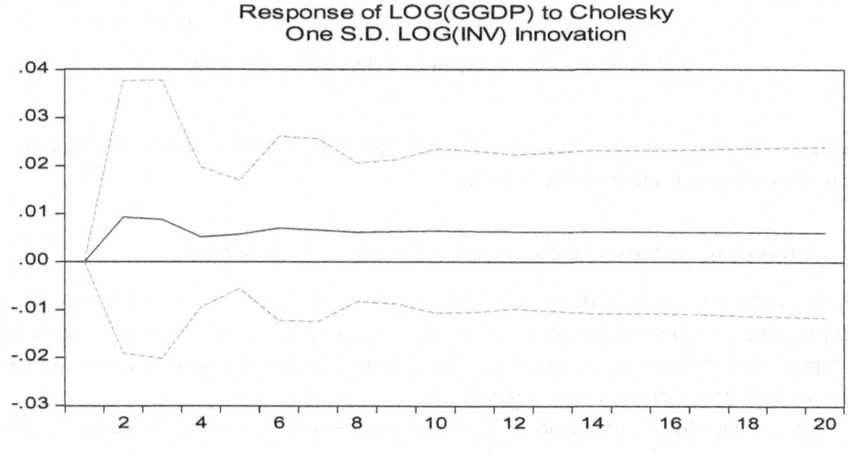

Fig. 2. Impulse response of Log (GGDP) to Log (INV).

4.3.3 Impulse Response Analysis of Green Finance to Green Economy

The impulse response of green finance to green economy is shown in Fig. 3. When a standard deviation is given to the green economy, green finance fluctuates from bottom to top, reaching the maximum negative response in stage 2, then fluctuates upward, reaching the maximum positive value in stage 3, and then converges gradually with the change of time. Finally, it tends to be stable in stage 10. This shows that in the first two stages, green economic growth has a negative influence on green finance, but with the passage of time, the impact of green economy on green finance begins to increase gradually. After reaching the positive peak, it begins to decline and tends to stabilize. Therefore, the impact of green economic growth on green finance changes greatly and lasts for a long time.

Generally speaking, the impulse response function analysis results manifest that increasing the input of green finance has a positive influence on green economy; and green economy also has a certain degree of impact on green finance, which will have a

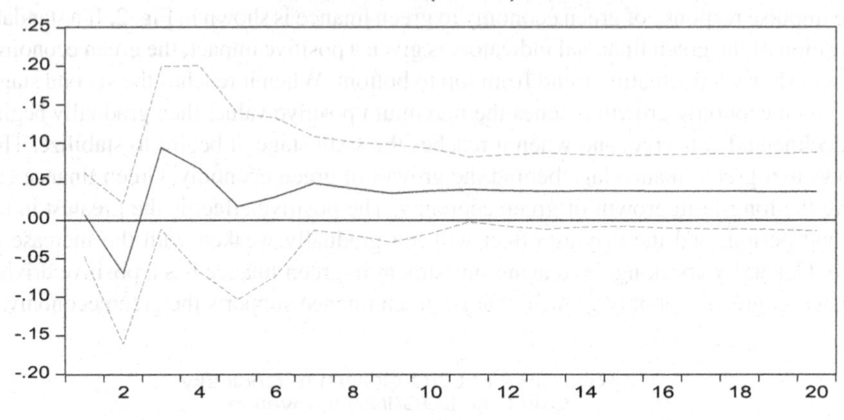

Fig. 3. Impulse response of Log (INV) to Log (GGDP).

negative impact in the early stage, but with the growth of time, green economy plays a positive role in promoting green finance.

4.4 Analysis of Variance Decomposition Based on VAR Model

Impulse response function describes the dynamic influence of one variable on another variable, and variance decomposition can decompose the variance of one variable into each perturbation term, so variance decomposition describes the relative impact level of each perturbation factor on each variable.

The variance decomposition of the green economy is shown in Table 5. From Table 5, we can see that in the contribution rate of green GDP, green GDP contributes the most to itself in the first stage, and shows a decreasing trend. In the sixth stage, it falls to 97.3%, and then shows a stable trend. This shows that the impact of green GDP on expectations decreases over time. However, the contribution rate of environmental protection investment to green GDP gradually increases with time, reaching 2.6% in the sixth stage, and then shows a steady trend. Therefore, the results of variance decomposition analysis show that the investment of environmental protection funds has a positive driving impact on green GDP, but the contribution rate of green GDP to itself is always far greater than the contribution rate of environmental protection funds to green GDP.

The variance decomposition of green finance is presented in Table 6. In the first stage, the contribution rate of environmental protection investment to itself is the greatest, and appears a decreasing trend. By the seventh stage, it falls to 96.9%, and then shows a stable trend. This shows that the impact of environmental protection funds on expectations decreases over time. However, the contribution rate of green GDP to the environmental protection investment increases gradually with time, reaching 3.0% in the seventh stage, and then shows a steady trend. Therefore, the results of variance decomposition analysis indicate that green GDP is positively driving environmental protection investment, but the contribution rate of environmental protection investment to itself is always far greater than the contribution rate of green GDP to environmental protection investment.

Table 5. Variance decomposition table of green economy.

Period	S.E	LOG (GGDP)	LOG (INV)
1	0.081991	100.0000	0.000000
2	0.107298	98.36272	1.637278
3	0.143879	97.37798	2.622015
4	0.159157	97.45830	2.541700
5	0.160949	97.54066	2.459335
6	0.163748	97.39960	2.600401
7	0.170621	97.30680	2.693196
8	0.175694	97.29287	2.707125
9	0.179284	97.27306	2.726942
10	0.183320	97.23863	2.761366

Table 6. The variance decomposition of green finance.

Period	S.E	LOG (INV)	LOG (GGDP)
1	0.0952	99.8621	0.1379
2	0.1184	98.0442	1.9558
3	0.1335	97.0594	2.9406
4	0.1488	97.1397	2.8603
5	0.1500	97.2221	2.7779
6	0.1534	97.0810	2.9190
7	0.1602	96.9882	3.0118
8	0.1653	96.9743	3.0257
9	0.1689	96.9545	3.0455
10	0.1730	96.9201	3.0799

5 Conclusion and Policy Recommendation

This paper chooses the annual data from 2004 to 2017 in China, uses VAR model, constructs impulse response function, and uses econometric analysis methods such as variance decomposition analysis to empirically analyze the dynamic link between green finance and green economy growth. The results make known that the first and second lag stages of Log (GGDP) have remarkable positive effects on Log (INV), and the first and second lag stages of Log (INV) also have prominent active effects on Log (GGDP). The results of impulse response function analysis show that green finance can promote the growth of green economy; concurrently, the influence of green economy on green finance has gradually developed from the early inhibition to the promotion over time, showing a positive impact on the whole. From the results of variance decomposition analysis, green finance is positively influencing green economy, but the impact of green economy on itself is far greater than that of green finance. Green economy also actively promotes green finance, but the contribution rate of green finance to itself is always far greater than that of green economy.

Based on the empirical results obtained in this paper, for developing China's green economy and green finance, we put forward the following policy suggestions:

Firstly, increase green investment and realize green transformation. We should support the development of green economy with green finance. Therefore, increasing investment in green industry can alleviate some problems, which are caused by environmental pollution and destruction, and bring about sustained economic growth and green transformation development.

Secondly, we should improve tax policies conducive to green economy. There are few laws and regulations on green economy and green finance in China compared with western developed countries. The establishment and improvement of related policies are the foundation and guarantee for the development of green economy. Therefore, based on existing laws and regulations, government departments at all levels in our country can timely adjust the relevant fiscal policies, such as reducing the enterprise tax rate of environmental protection industry, green industry and other environmentally friendly industries, raising the tax rate of consumption-oriented enterprises, and giving some preferential policies to those enterprises that have successfully realized the green transformation.

Thirdly, develop new green financing channels. Vigorously publicize the importance of environmental protection, so that the main polluting enterprises in our country can enhance their awareness of environmental protection, and then provide these enterprises with green financing channels to deal with environmental pollution. China's financial institutions should develop other new financial instruments on the basis of existing financial instruments. We should increase investment in green credit and innovate financing channels for energy conservation, emission reduction and environmental pollution control technologies.

References

1. Jia, H., Shi, X.: Measurement of green economy development level from provincial perspective – based on empirical analysis of 14 prefectures and cities in Gansu province . J. Lanzhou Univ. (Soc. Sci. Edn.) **48**(03), 113–121 (2020)
2. Pearce, D., Markandya, A., Barbier, E.B.: Blueprint for a Green Economy: A Report. Earths Can Publications Ltd., London (1989)
3. Huang, B.: Measurement and Countermeasures of the development level of Green Economy in Inner Mongolia. Northwest University for Nationalities (2020)
4. He, L., Zhang, L., Zhong, Z., Wang, D., Wang, F.: Green credit, renewable energy investment and green economy development: empirical analysis based on 150 listed companies of China. J. Clean. Prod. **208**, 363–372 (2019)

5. Mao, Y., Xu, W.: Research on the construction path of green financial system from the perspective of financial supply-side structural reform – a case study of green financial reform and innovation pilot zone. Credit Ref. **37**(12), 79–84 (2019)
6. Zhang, D., Zhang, Z., Managi, S.: A bibliometric analysis on green finance: current status, development, and future directions. Financ. Res. Lett. **29**, 425–430 (2019)
7. Wang, Y., Zhi, Q.: The role of green finance in environmental protection: two aspects of market mechanism and policies. Energy Procedia **104**, 311–316 (2016)
8. Yan, J., Yang, X.: Construction of green technology innovation system promoted by green finance. Fujian Forum (Humanit. Soc. Sci.) **03**, 41–47 (2018)
9. Zhu, J., Wang, H., Zheng, P.: Research on the coupling and coordinated development of circular economy and green finance in Guizhou Province. Econ. Geogr. **39**(12), 119–128 (2019)
10. Owen, R., Brennan, G., Lyon, F.: Enabling investment for the transition to a low carbon economy: government policy to finance early stage green innovation. Curr. Opin. Environ. Sustain. **31**, 137–145 (2018)
11. Taghizadeh-Hesary, F., Yoshino, N.: The way to induce private participation in green finance and investment. Financ. Res. Lett. **31**, 98–103 (2019)
12. Zhang, Y., Xing, C., Wang, Y.: Does green innovation mitigate financing constraints? Evidence from China's private enterprises. J. Clean. Prod. **264**, 121698 (2020)
13. Shao, Q., Zhong, R., Ren, X.: Nexus between green finance, non-fossil energy use, and carbon intensity: empirical evidence from China based on a vector error correction model. J. Clean. Prod. 122844 (2020)
14. Dong, X., Yong, F.: Spatial dynamic evolution analysis of the coordination of green finance and green economy. Ind. Technol. Econ. **12**, 94–101 (2018)
15. Wu, X.: Research on the support of green finance to green economy in China. Nanjing University (2019)
16. Wang, F., Li, J., Zhang, F., Yanjie, W.: Can financial agglomeration promote the development of green economy? – An empirical analysis based on 30 provinces in China . Financ. Forum **22**(09), 39–47 (2017)
17. Zhu, H.: Construction of green economy evaluation index system . Stat. Decis. Making **05**, 27–30 (2017)
18. Hou, Y., Wang, F.: Dynamic relationship between green innovation and economic growth: an empirical analysis based on VAR model. Ecol. Econ. **36**(05), 44–49 (2020)

An IoT Network Emulator for Analyzing the Influence of Varying Network Quality

Stefan Herrnleben[1(✉)], Rudy Ailabouni[2], Johannes Grohmann[1], Thomas Prantl[1], Christian Krupitzer[1], and Samuel Kounev[1]

[1] University of Würzburg, Würzburg, Germany
{stefan.herrnleben,johannes.grohmann,thomas.prantl, christian.krupitzer,samuel.kounev}@uni-wuerzburg.de
[2] Bosch Engineering GmbH, Stuttgart, Germany
rudy.ailabouni@de.bosch.com

Abstract. IoT devices often communicate over wireless or cellular networks with varying connection quality. These fluctuations are caused, among others, by the free-space path loss (FSPL), buildings, topological obstacles, weather, and mobility of the receiver. Varying signal quality affects bandwidth, transmission delays, packet loss, and jitter. Mobile IoT applications exposed to varying connection characteristics have to handle such variations and take them into account during development and testing. However, tests in real mobile networks are complex and challenging to reproduce. Therefore, network emulators can simulate the behavior of real-world networks by adding artificial disturbance. However, existing network emulators often require a lot of technical knowledge and a complex setup. Integrating such emulators into automated software testing pipelines could be a challenging task. In this paper, we propose a framework for emulating IoT networks with varying quality characteristics. An existing emulator is used as a base and integrated into our framework enabling the user to utilize it without extensive network expertise and configuration effort. The evaluation proves that our framework can simulate a variety of network quality characteristics and emulate real-world network traces.

Keywords: Mobile networks · Network quality · Network emulation · Application development

1 Introduction

In recent years, the number of IoT devices has increased rapidly [13]. These devices can be found in many different areas, such as smart homes, health care, smart buildings, vehicles, agriculture, smart cities, and industrial automation [8,12,17]. Since many IoT devices are mobile, communication over wireless networks is often necessary. For this, WiFi, mobile networks, as well as proprietary radio standards are used [15]. At the application level, IoT protocols such as MQTT or CoAP are often employed [3]. However, wireless networks are

H. Song and D. Jiang (Eds.): SIMUtools 2020, LNICST 370, pp. 580–599, 2021.
https://doi.org/10.1007/978-3-030-72795-6_47

exposed to significantly more interference factors than wired networks. Buildings, the velocity with which the device moves, weather, and the distance to the mobile radio station influence the connection quality [10,16,19]. These influencing factors can lead to packet delays, bandwidth restrictions, packet loss, or jitter.

IoT devices often require a stable connection to internet services. Temporary performance limitations could lead to undesired application behavior and dissatisfied users. Application developers should, therefore, test their applications in networks with different quality characteristics. However, tests in real-world networks are time-consuming, expensive, and not suitable for repetition. Network emulations can simulate the quality characteristics of networks and add artificial interference. The impact of varying network quality on the application can be simulated and tested by the developer through emulated networks. Unfortunately, network emulators often require time-consuming setup, complex configuration, and technical knowledge of networks. Furthermore, network emulators are often not designed for integration into software testing environments.

In this paper, we present a framework to emulate the characteristics of IoT networks. An existing network emulator is embedded in our framework to provide artificial interference at the network interface layer. By this, our emulator supports varying the maximum bandwidth, the packet delay, the packet loss rate, the packet duplication rate, and the jitter value. Besides the disturbance of the link, connection failures can be emulated as well. A scheduler adjusts the connection quality at predefined times to reconstruct a temporal behavior. The definition of time events enables the framework to simulate the movement of devices, changes in the environment, or a variation of disruptive factors. Besides manually specifying the quality characteristic changes and times, network measurement traces can also be imported into the event queue. Our work provides the following core contributions. First, the developed framework can be easily instantiated for efficient use by application developers. Second, measurements from real-world networks can be imported into the scheduler to reconstruct realistic scenarios. Third, a provided REST API for configuring and operating the emulator makes it ideal for integration into automated software test pipelines.

To ensure that the quality characteristics are correctly emulated, the first part of the evaluation focuses on these individual factors. The timing of the scheduler's events is also analyzed within this part of the evaluation. In the second part of the evaluation, a real-world network with Message Queuing Telemetry Transport (MQTT) traffic is measured and emulated. The evaluation will show that real network traces can be imported and emulated realistically.

The remainder of this work is structured as follows. Section 2 introduces existing emulators and their differences. Next, Section 3 presents our developed framework for the emulation of IoT networks and discusses the modules and features. Section 4 evaluates the emulator. The first part of the evaluation examines the precise appliance of limited quality characteristics on a link; the second part presents the test results of a real-world network's emulation. Section 5 concludes the work, identifies some limitations and states future work.

2 Related Work

Network emulators are already widely used to artificially limit networks in certain characteristics and execute measurements and benchmarks on the restricted networks. Most network emulators have been developed for specific purposes and fulfill different requirements. In this section, we introduce the open-source network emulators well-known in science. Generally, network emulators can be divided into the following two categories [11]: *Network Link Emulators (NLEs)* and *Virtual Network Emulators (VNEs)*.

2.1 Network Link Emulators (NLEs)

NLEs are lightweight and simple to use network emulators. They are usually limited to the link layer and are used to manipulate packets in such a way as to emulate specific network conditions. Most NLEs can influence the bandwidth, latency, packet error rate, and jitter of a network link but are limited to the machines they run on.

NISTnet [4] is a Linux based network emulation package that was originally developed and released by the American *National Institute of Standards and Technology (NIST)* in 2003. NISTnet can be set up on a single Linux computer and allows this computer to run as a virtual router. This virtual router can then be used to emulate variable network conditions by influencing different network parameters, such as packet delay distribution, congestion, loss, bandwidth limitation, as well as packet reordering and duplication. NISTnet is implemented as a Linux kernel extension and offers an X-Window System-based user interface, which can be used to monitor the network traffic and manage the emulation. Furthermore, NISTnet supports replaying real-world network traces. NISTnet is nowadays no longer maintained and actively developed. Most of its functionality has been incorporated into NetEM and the Linux *iproute2* kit.

NetEm [7] is currently one of the most popular NLEs and is included in the Linux kernel. Stephen Hemminger developed it at the Linux Foundation, and has incorporated many of the functionalities of NISTnet. NetEm is part of the Linux iproute2 kit, a collection of tools and utilities used to control TCP/IP networks and traffic in Linux. NetEm is an enhancement of the Linux *Traffic Control (TC)* system. It allows manipulating incoming and outgoing packets to a specific network interface by configuring delays, packet loss rates, and duplications rates, and packet reordering. Users of NetEm can either add discrete artificial delays or add delay distribution models. Furthermore, NetEm allows interfering traffic by adding random noise to a specific percentage of packets. Although NetEm is a powerful NLE, it has some limitations. One of the main issues with NetEm is related to Linux's timer granularity, since Linux does not run in real-time [7].

Similarly to Linux's TC, the Free-BSD based *Dummynet* [14] network emulator can also manipulate the kernel's IP queue to emulate variable network quality parameters. Dummynet allows emulating finite and bounded-size queues, bandwidth limitations, delays, and packet losses.

For individual use cases, NLEs are powerful network emulators, especially when multiple ones are combined, as is the case with the iproute2 of Linux. However, one of the main shortcomings of NLEs is their limitation to a single computer. This makes it difficult to emulate larger network topologies. Since most NLEs manipulate the packets coming into and leaving a specific network interface, one would need to use some virtualization technology to scale up the emulation. This makes the tools more complex and more challenging to use.

2.2 Virtual Network Emulators (VNEs)

VNEs aim to emulate entire network topologies. In VNEs, users can create all kinds of network topologies and can practically control every aspect of a network stack. However, this complexity often results in them being complicated and difficult to use.

EMULAB [18] is a network emulation platform that is run and hosted by the Flux Research Group, which is part of the School of Computing at the University of Utah in the United States. EMULAB can be used for research, development, and educational purposes. However, it is a time- and space-shared platform, which means not everyone can use it. EMULAB has multiple installations around the world and requires those who want to use it to apply in advance, stating the number of nodes they require and for how long. Multiple criteria are considered when granting access to EMULAB, such as being associated with an academic institution, having a clear deadline for a conference paper submission, and needing to perform experiments as part of a publication that will be peer-reviewed.

The *Integrated Multiprotocol Network Emulator/Simulator (IMUNES)* [20] is an open-source and free general-purpose IP network emulation tool that was developed in 2004 by the University of Zagreb in Croatia. It is available for both the FreeBSD and Linux operating systems and aims to offer realistic network topology emulation and simulation. It uses kernel-level virtualization techniques to instantiate multiple lightweight virtual nodes, which can be connected using kernel-level links and bridges to instantiate complex network topologies. Networks emulated using IMUNES can have a large number of virtual nodes all running on a single physical machine. IMUNES comes bundled with a console to configure the emulated network.

Mininet [6,9] is one of the best well known open-source network emulators and enables the creation of a virtual network containing virtual hosts, switches, and links on a single machine. Hosts are virtual machines running standard Linux network software. Mininet can create custom network testbeds that enable sophisticated topology testing without having to connect to physical networks. As Mininet can run real code, any network applications developed and tested in Mininet, specifically network control and routing software, can be migrated to a real system with minimal changes.

The *Common Open Research Emulator (CORE)* [1] is an open-source network emulator that originally bases on IMUNES. It has forked from IMUNES in

2004 and further expanded the emulator to add new functionality, such as support for wireless networks, mobility scripting, distributed emulation over multiple computers, and more. The Boeing company originally released it in 2008, and it is still further maintained and developed by a community of developers to this day on GitHub[1]. Like IMUNES, CORE uses the operating systems' virtualization technology to instantiate large network topologies. CORE allows for instantiating lightweight virtual networks and virtual nodes, e.g., Ethernet ports, routers, and PCs connected using the host operating system's kernel. These virtual machines are created using the Linux network namespace virtualization technology and are connected using Linux Ethernet bridging. This allows each virtual node to share resources such as the memory, CPU time, and the file system with the host system, but have their own network stack and process environment. Real network interfaces can also be connected to the virtual nodes to connect real and virtual worlds, which allows for even more powerful network emulation [2]. Users can launch a terminal from each of the emulated nodes, which in turn permits running any command and launching any application over the emulated network.

More recently, the NetSec Research Group from the Department of Engineering and Architecture at the University of Parma in Italy has developed *NEMO* [5]. NEMO is written in Java and aims to be a highly flexible and portable network emulator. It seeks to be platform-independent by taking advantage of Java's virtual machine and runs completely at the user-space. While NEMO can be used to develop and test new communication protocols, it also comes bundled with an extensive collection of standard protocols. It includes an implementation of the TCP/IP stack that is entirely independent of the underlying operating system. NEMO allows connecting real networks with the virtual network and can run emulations that are distributed among multiple physical machines.

Although many of the VNEs partly offer graphical tools for configuration, the creation of emulated networks is often complex. A lot of configuration effort is also required when connecting virtual and physical networks. During software testing, influencing quality parameters usually requires knowledge of the emulator and may require in-band reconfiguration of the emulators.

3 Approach

This section describes our framework for network emulation focusing on IoT communication. Our network emulator aims to test mobile IoT applications under varying quality characteristics at the development stage. The emulator artificially influences the quality of network connections so that the application can be tested under degraded quality characteristics.

Most emulators introduced in Sect. 2 offer great features and can influence networks in several aspects. However, using these emulators for application developers, especially in automated test environments, often requires much technical

[1] https://github.com/coreemu/core.

knowledge and manual intervention. We define the following requirements for a network emulator focusing on IoT application development:

1. Simple and easy to use: Application developers do not have to be network experts to use the emulator. Already defined disturbance scenarios can be reused to generate reproducible results.
2. Emulation of the temporal behavior of real-world networks: The quality characteristics should be changeable during a running emulation. This enables, e.g., simulating a road trip by a car passing different networks (e.g., 3G, LTE, or 5G).
3. Connection with real-world physical networks and devices: Both virtual and physical networks and devices should be able to be integrated into the emulator. For example, an IoT system benefits from including a real MQTT Message Broker.
4. Remote control: The network and connection characteristics can be controlled via API calls. This allows using the emulator in automated test environments.
5. Running programs and applications over the emulated network: The emulator can start applications and execute commands on the devices.

As indicated in Sect. 2, a large number of network emulators already exist, which in particular already supports many of the network characteristics to be controlled. However, to the best of our knowledge, there is currently no network emulator that meets the previously defined requirements. This is especially true for the easy integration of the emulator into software development pipelines. In this paper, we extend an existing emulator with these features.

Our research indicates that the *CORE* network emulator best suits as a foundation for our implementation. It partly fulfills the first requirement since it is a lightweight network emulation platform. Judging the ease of use is subjective as it is specified by many factors, such as the users' skills in writing Python scripts and their experience in network engineering. CORE can connect emulated virtual networks to real-world physical ones. However, this requires some complex configuration effort. Currently, CORE does not automatically support the emulation of the temporal behavior of networks and the ability to control the emulator remotely.

Figure 1 shows the architecture of our network emulator, including the components of the CORE network emulator. The emulator consists of four modules. The central part of our network emulator, denoted by *Network Emulator* in Fig. 1, abstracts the technical details from the users and combines all of the other modules into a working network emulator. On top of that, our emulator contains the modules *TopologyConfigurator*, *Scheduler*, and a *REST API*, which we describe in the following. Our emulation framework is available as open source and can be downloaded from our Git repository[2].

The *TopologyConfigurator* module is responsible for instantiating and managing the topology. It contains all the logic and methods required to manage

[2] https://gitlab2.informatik.uni-wuerzburg.de/descartes/iot-and-cps/iot-network-emulator.

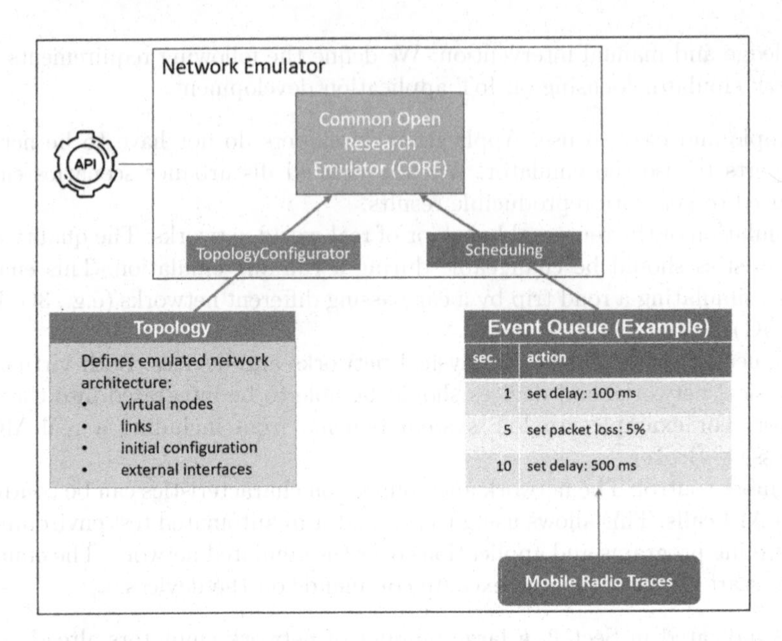

Fig. 1. The architecture overview of the network emulator.

the node and link entities. The module furthermore provides interfaces to add, modify, and remove these nodes and links. The hosts, connected to the emulator, can be virtual nodes, switches, hubs, or physical nodes, connected to Ethernet interfaces. The virtual nodes are deployed running an integrated SSH server by default to manage them and execute scripts. Starting traffic generators like *Iperf* or executing measurement scripts are some examples of using the SSH interface. If physical nodes connect to the emulated network, it is sufficient only to provide the network interface name. Our emulation framework instantiates all required virtual bridges between the physical and virtual networks. This is a great strength of our emulator compared to the original version of CORE, where these tasks have to be done manually. The module further provides the functionality to update the connection quality characteristics of links. It can add artificial network delay, jitter, packet loss rate, packet duplication rate, and limit the bandwidth of a link. The *TopologyConfigurator* module abstracts the technical complexity of network emulation from the user to easily integrating our emulator. At starting the emulator, users have the option to define a topology configuration as a YAML file. The topology configuration file includes fundamental information and parameters regarding the emulation. It defines the logical network addresses and the virtual nodes that will run in the emulation and the links between them. API calls can generate the emulated network entirely if no topology configuration is provided.

The *Scheduler* module emulates the temporal behavior of networks by scheduling the execution of certain events at specific times. Supported events are adding, removing, or modifying connection quality characteristics. This allows, for example, to add a delay later or to change the maximum supported bandwidth multiple times. Furthermore, the events can execute commands on virtual nodes, to, e.g., start programs or execute additional configuration tasks on the nodes. The events can be defined as an event queue in a configuration file. Each event is specified by the event type, the time when it should be executed, and additional optional execution arguments. One of the greatest strengths of the scheduler is the emulation of the temporal behavior of real-world networks. For this, the scheduler can import network measurement trace files. An event with the new value is created when a change in quality characteristics occurs in the measurement from the real-world network. A possible use-case for this is testing software for vehicular applications. The quality characteristics such as maximum bandwidth, delay, and packet loss can be varied along the route depending on network coverage and the mobile network used. Developers of such software can measure the internet connection quality on a certain route in advance and then use our emulator to reconstruct the entire route's network conditions. Developers can then perform experiments and investigate their software's behavior under these connection quality conditions, without having to drive the route again.

The *REST API* enables interaction with the network emulator at runtime. Upon starting our network emulator, a web server listening to HTTP requests is started in the background. Instead of configuration files, users can configure the topology and the events via REST API calls. Through REST API calls, users can further modify the emulation, i.e., adding or removing nodes and links. Also, the connection quality of specific links can be influenced (e.g., by adding artificial delays, bandwidth limitations, or increasing the packet loss and duplication rates) as well as the link jitter value. Further, users can retrieve information on the emulator's status via the REST API, including the current topology (i.e., nodes, their attributes such as IP addresses and interfaces), and the links between the nodes. Users can also get detailed information about a specific node, such as its neighbors and connections. Also, information about the links, including their current quality characteristic, can be queried. The REST interface is a powerful feature to integrate the network emulator into a software test and build pipelines, as both the instantiation and the execution of the emulator can be fully automated.

4 Evaluation

This section evaluates our emulator in two dimensions. Section 4.1 measures and validates the influence of quality characteristics on network traffic. Also, the temporal behavior of the scheduler is evaluated in this section. In Sect. 4.2, the quality characteristics of a real-world network are captured, and the communication behavior is then emulated in a virtual environment. Section 4.3 briefly discusses the validity of the measurements.

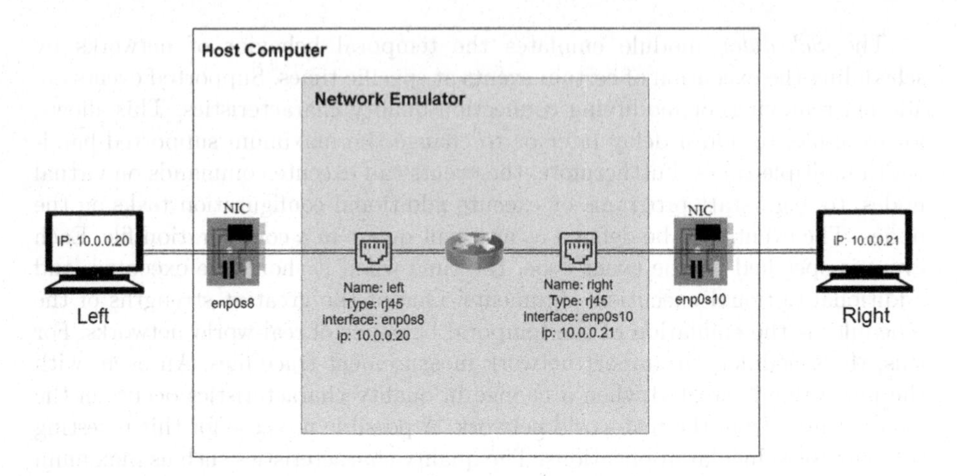

Fig. 2. Testbed setup for connection quality characteristic evaluation.

4.1 Connection Quality Characteristics

Our network emulation framework can manipulate different network quality characteristics: packet delay, bandwidth, packet loss rate, packet duplication rate, and jitter. In this evaluation, we verify that our network emulator correctly manipulates the quality of network links based on those five parameters. For each of these parameters, we perform an individual test using a common testbed.

Testbed Setup. All of the following experiments are performed on a notebook with an Intel Core i7-8550U CPU running at 1.80 GHz with a maximum turbo frequency of 4.0 GHz with 16 GB RAM. The network emulator is executed on a Linux virtual machine within the device and has access to four out of eight processor cores and 6 GB of RAM. The virtual machine runs Ubuntu version 18.04.2 LTS. For the hypervisor, we use version 5.2.22 of VirtualBox.

Figure 2 illustrates the experiment setup. Our topology consists of two *RJ45* interfaces connected to a virtual switch within the emulator. The two physical interfaces are connected to two virtual machines, simulating the connectivity to two other physical nodes. Each of the external virtual machines also runs Ubuntu version 18.04.2 LTS and have access to 2 GB of RAM, and two of the CPU cores.

Delay. The delay measurements verify that (i) the emulator applies the set delay correctly, and (ii) the scheduler executes the change at the correct time. To evaluate the network delay, we use an echo request/reply (ping) test. In this test, a packet is sent from one node to another, and upon receipt, a reply is sent back to the sending computer. The time it takes for the packet to be delivered and returned is called *Round Trip Time (RTT)*.

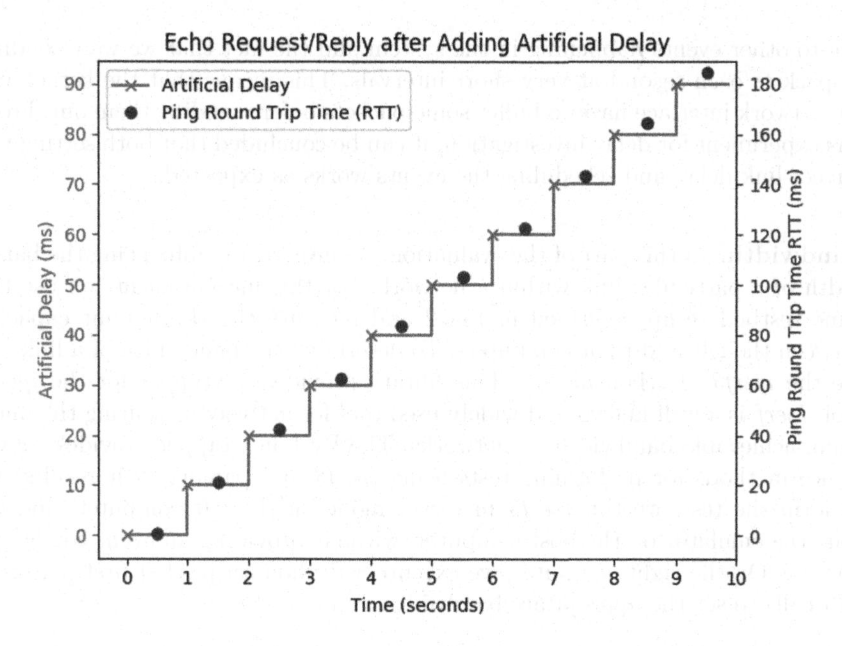

Fig. 3. The results of the delay measurements.

For the measurement, we send 50 echo requests sequentially per second from the right computer (ref. Fig. 2) to the left one. We run the test for a total of ten seconds. To do this, we schedule an echo request test every second.

To evaluate our emulator's scheduler module and the ability to variate the network delay, we defined an event queue containing ten link update events. Each of these events occurs every second and increases the delay value on the link between the left and right computers by ten milliseconds, starting by zero milliseconds at second zero.

The measurement results are depicted in Fig. 3. The x-axis of the graph shows the elapsed time of the experiment in seconds. The left y-axis shows the added artificial delay values in milliseconds. Likewise, the red step plot also denotes the added artificial delay values. More specifically, the red 'x' marks the time when the delay was set. Each delay value is stable for one second until the next delay value is set. The right y-axis in blue represents the RTT and has a linear relationship with the left y-axis. The RTT of a ping refers to the total time it takes for the echo request to be sent from one computer to another and the time for the echo reply back. Therefore, the RTT is expected to be twice the network delay/latency plus any required processing time. The y-value of the blue dots represents the average RTT of the 50 echo requests sent in that second. The x-value refers to the time interval of when those 50 echo requests were sent.

Optimally, each blue dot lies slightly above each step, on the red line. That means that the RTT is precisely double the delay plus the required processing time. Some of the RTT values are slightly higher than others, which could be due to other events happening in the system and the fact that we were sending 50 packets each second at very short intervals. This means that the kernel and the network interface have to buffer some packets before sending them out. From this experiment for delay investigation, it can be concluded that both setting the correct link delay and scheduling the events works as expected.

Bandwidth. In this part of the evaluation, we investigate influencing the bandwidth of a particular link within a network. For this measurement, we use the same testbed setup described in Fig. 2 and measure the throughput capacity between the left and right computers. To determine the bandwidth of a link, we use the *iperf3*[3] Python library. This library provides a wrapper for the *iperf3* tool. *Iperf* is a well known and widely used tool for actively measuring the maximum achievable bandwidth on networks. The Python wrapper provides convenience methods for performing tests using *iperf3* and parsing their results. To perform the test, we run *iperf3* in server mode on the left computer, and we start the emulator on the host computer, without providing an event configuration file. On the right computer, we execute a Python script that first performs API calls to set the appropriate bandwidth limits.

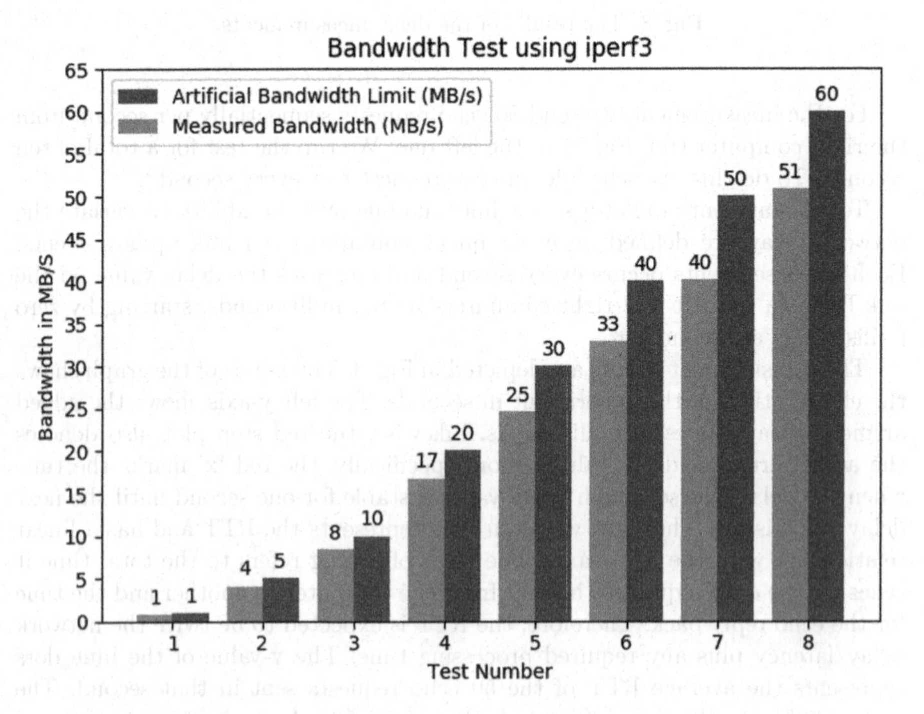

Fig. 4. Results of the bandwidth measurements. (Color figure online)

[3] https://pypi.org/project/iperf3/.

We run eight different measurements, each with an artificial limited emulated bandwidth of 1 MB/s, 5 MB/s, 10 MB/s, 20 MB/s, 30 MB/s, 40 MB/s, 50 MB/s, and 60 MB/s. For each bandwidth limit, the throughput between the left and right computer is measured for a total of 60 s. This measurement is repeated a total of eight times for each bandwidth limit. All of the throughput values from the eight measurements for each bandwidth limit are averaged.

The results of the measurement are depicted in Fig. 4. The x-axis denotes the measurement number. The y-axis shows bandwidth value in Megabytes per second. The blue bars refer to the bandwidth limit set using the emulator and denote the expected results. The orange bars represent the measured throughput itself, specifically the average of all eight measurements performed for each bandwidth limit. The numbers above the bars represent their values and are rounded down to the nearest integer.

From this investigation, it can be concluded that the actual throughput never exceeds the maximum defined bandwidth. Furthermore, it can be seen that by increasing bandwidth, a difference between the artificial limit and the measured bandwidth occurs. We assume that TCP's window size causes this difference. However, we asserted that our emulator appropriately modified the QDISC using NetEm.

Packet Loss Rate. This evaluation investigates the influence of our emulator on the packet loss rate of a specific link. The packet loss rate refers to the number of lost packets divided by the total number of sent packets. We use the testbed setup shown in Fig. 2, and evaluate the packet loss rate between the left and right computers by using echo requests and replies.

For the test, we perform five measurements, each with different packet loss rates of 20%, 40%, 60%, 80%, and 100%. Each test sends 1000 echo requests and is repeated 30 times. The packet loss rate is set via API call within the measurement script.

The results of our measurements can be seen in Fig. 5. The x-axis in the graph denotes the measurement number. The y-axis represents the packet loss rates in percent. The blue bars refer to the packet loss rate that was applied by the emulator and represent the expected results. The orange bars depict the average packet loss rate of all 30 measurements for each of the packet loss rates that we applied using the emulator. The numbers above the bars represent the bars' height, i.e., the expected values, respectively, the measured values and are rounded two decimal places.

It can be seen that the average packet loss rate of the measurements is very close to the rate that was applied by the emulator. From this, we can conclude that our network emulator is appropriately applying the packet loss rate parameter.

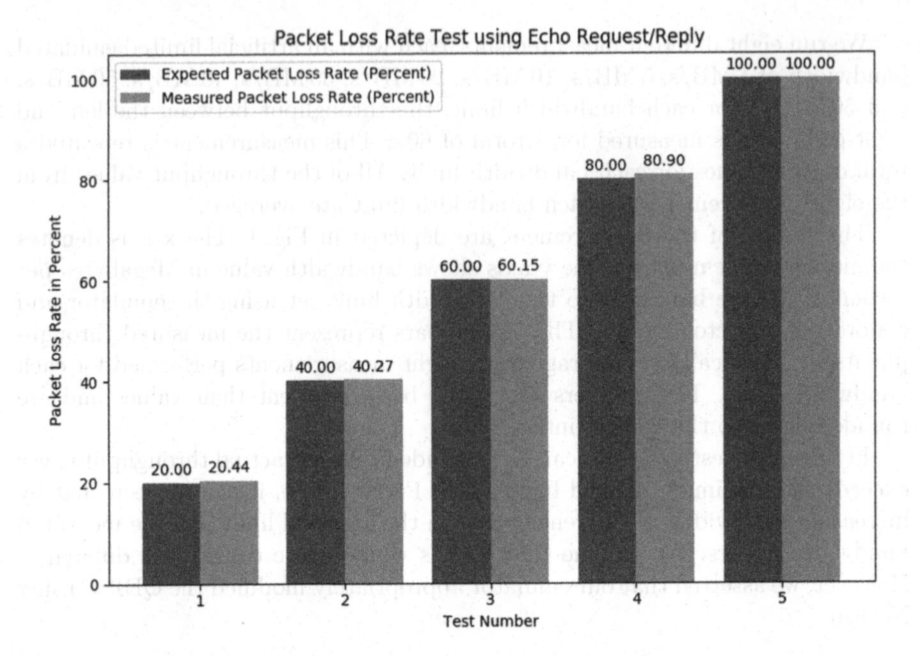

Fig. 5. The results of the packet loss rate measurements.

Packet Duplication Rate. This measurement verifies whether the set packet loss rate is correctly adopted. For this, the measurement setup from the previous section, including the echo reply/request measurement method, is used. The unique sequence number of the ICMP message identifies duplicate packets in the response log. We apply the packet duplication rates of 20%, 40%, 60%, 80%, and 100%. Each measurement sends 1000 echo requests and is repeated 30 times, before averaging the packet duplication rates.

The results of our measurements can be seen in Fig. 6. Like in the previous experiment, the X-axis represents the number of the measurement, and the y-axis represents the packet duplication rates in percent. The blue bars denote the packet duplication rate that we applied using our emulator and the results we expect to see. The orange bars represent the actual measurements, specifically the average of each of the 30 measurements for each of the five packet duplication rates. The numbers at the top of the bars are rounded to two decimal places.

It can be seen that the measured packet duplication rate of the emulated network matches the set value. From this, we can conclude that our network emulator is appropriately adjusting the packet duplication rate of network links accurately.

Jitter. This measurement evaluates the setting of an artificial jitter on a link. Jitter is defined as the variance in the time that data packets take to traverse a network path. For this measurement, we use the setup from Fig. 2 and measure the connection quality between the left and right computer using echo

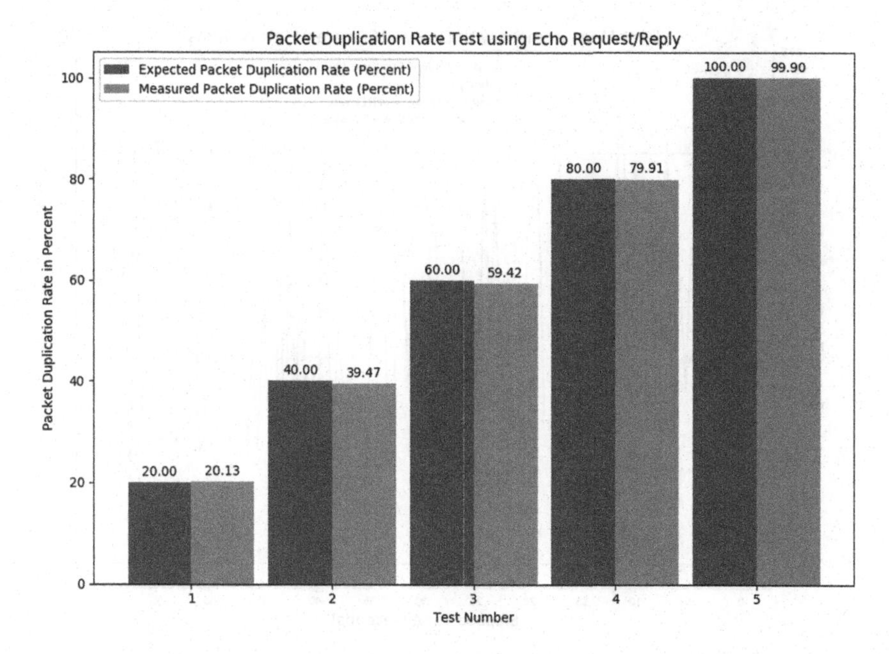

Fig. 6. The results of the packet duplication rate measurements.

requests/replies. As jitter is related to the delay value in a way, we first have to use the emulator to apply an artificial delay to the link between the right and left computers. Explicitly, we set the delay value to 100 ms. We then used the emulator to apply a jitter value of 50 ms to the link between the right computer and the switch.

For the measurement, we send echo requests for 60 s, with an interval of 500 ms between each request. From the individual timestamps of the ICMP replies, we calculate the standard deviation of the RTT. Since we set a link delay value of 100 ms, provided there is no network jitter, the average RTT of the link between the left and right computers should be approximately 200 ms plus some processing time.

The results of the measurement can be seen in Fig. 7. The x-axis denotes the elapsed time of the experiment. The y-axis represents the RTT values in milliseconds. The blue plot represents the variations between the RTT values. The orange line in the middle represents the theoretical average RTT if the jitter value was 0. The green line represents the maximum theoretical RTT value we should expect to see, i.e., the average theoretical RTT value plus the jitter value. Likewise, the red line at the bottom represents the minimum expected value, i.e., the average RTT minus the jitter.

In total, 120 echo requests were transmitted, 114 replies were sent back, and six were lost caused by a timeout of 60 s for the whole test. Figure 7 shows that most measurement results reflect our expectations, i.e., RTTs between 150 ms and 250 ms. The expected interval must be set to twice the set jitter value because

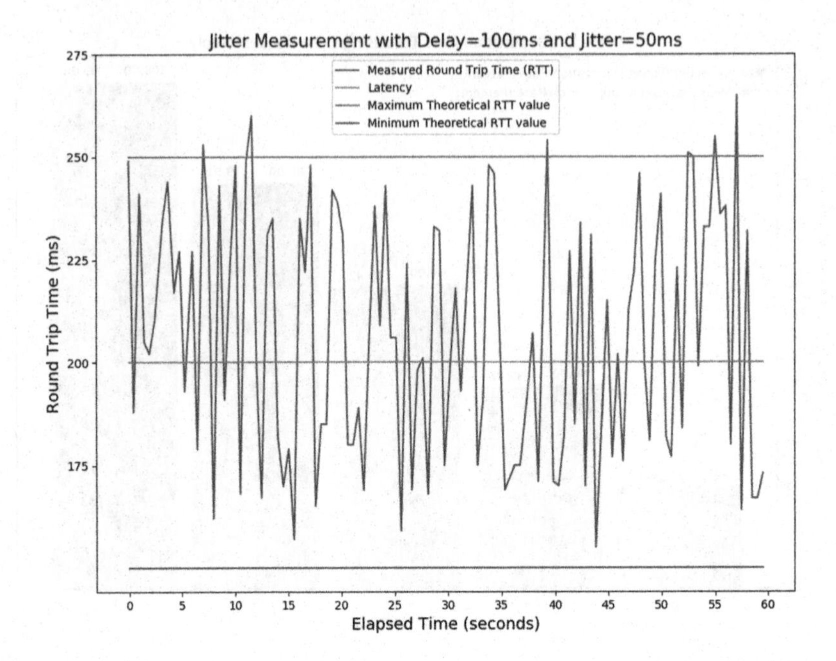

Fig. 7. The results of the jitter measurements. (Color figure online)

the packets pass through the link twice during an echo request and reply. The six values slightly above 250 ms can be explained by the processing time, which is not subtracted from the measured transmission time.

4.2 Evaluation of Temporal Behavior Emulation

This section evaluates the temporal behavior of the emulation framework within a real-world IoT use-case. For this, we employ a publish/subscribe communication scenario using the *Message Queuing Telemetry Transport (MQTT)* protocol for communication between two clients and one message broker. To emulate a real-world network's temporal behavior, we first need to capture the characteristics of such a network over a certain period. The capturing is done in a first experiment, using a public MQTT message broker. In a second experiment, we emulate the network to the public broker by the captured network trace and replace the broker by a local one.

Measuring the Temporal Behavior of a Real-World Network. In this measurement, we capture the temporal behavior of a real-world network in terms of network latency. The measurement setup is depicted in Fig. 8. Two virtual machines running an MQTT client, a subscriber and a publisher, communicate with each other over a public MQTT message broker[4].

[4] https://test.mosquitto.org/.

Fig. 8. The setup used to gather data of the network quality between the MQTT clients and the public MQTT Mosquitto broker.

For the publisher and subscriber clients, we developed an MQTT application using the *Eclipse Paho* library. The publisher application connects to the MQTT broker and publishes the system's current time along with a message sequence to a topic called "time" every second. With each published message, the sequence is incremented. We can control the time interval between each publication and the total number of messages that the application should publish. The application running on the subscriber client subscribes to the topic "time" and receives all messages related to this topic from the broker. Upon receiving a message, the subscriber subtracts the time in the message from the current time. By doing this, we can approximate the time it took for the message to be sent by the publisher, processed by the broker, forwarded to the subscriber, and processed by the subscriber. To obtain accurate results regarding each message's transport duration, we synchronized the clocks of the two MQTT clients directly before starting the experiment. Since in our setup, the publisher is publishing a message each second, we can use each message's sequence number to derive the sending time of the message. Upon calculating the transport duration of the various messages at the subscriber end, a trace file is generated, including the sequence number of each message and its measured transport duration. The total running time for the experiment is 60 s.

Emulating the Temporal Behavior of a Real-World Network. In this part of the evaluation, we emulate the network with the characteristic captured in the previous part. For the experiment, we replace the public MQTT message broker by a local broker, running within our the emulated network. The clients now connect to the local broker. The setup is shown in Fig. 9.

We use the trace file from the real-world measurement and import it into the event configuration so that the fluctuations are represented as events. For the evaluation, one client again publishes the current time and a sequence number, which is subscribed by the other client. As in the first part of the experiment, the subscriber measures the transport duration of each message and stores the duration along with each messages sequence number.

The data, captured from the real-word network and the measured data from the emulated network, are depicted in Fig. 10. The x-axis denotes the elapsed

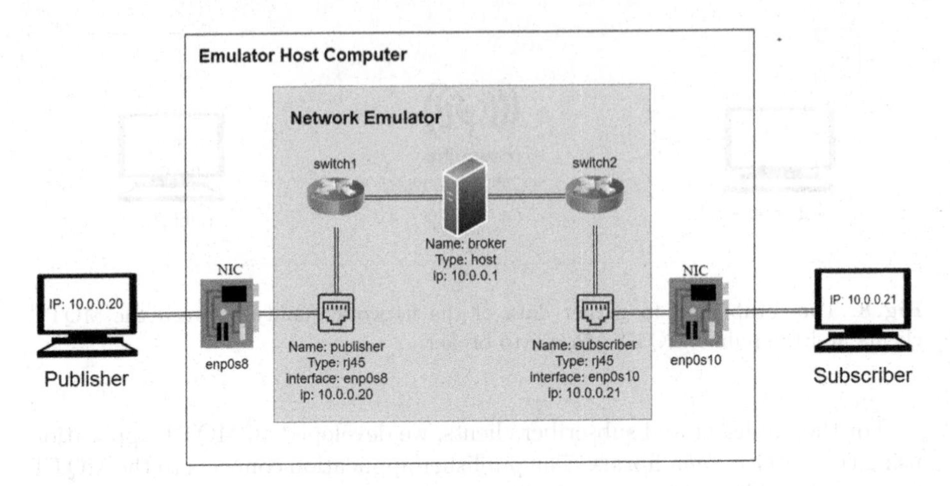

Fig. 9. Testbed setup for the MQTT scenario.

time in seconds. The y-axis indicates the transport duration of each message in milliseconds. The orange 'x' markers represent the transport duration of each message in the real-world scenario, i.e., when communicating with the public *Mosquitto* broker. The blue plot represents the transport duration of the messages in the emulated scenario.

It can be seen that our emulation framework can reconstruct the behavior of the real-work network accurately. The value measured from the emulation is always very close to the value of the real-world network. The minimal deviations can be explained by the delays in the local message broker. From this measurement, we can conclude that our emulation framework can accurately emulate the behavior of a real-world network.

4.3 Threats to Validity

Although the measurements of the evaluation were performed with great care, they might contain some inaccuracies. In measurements where time behavior such as RTT or jitter was measured, the processing time for the response was not included. Therefore, the reported time corresponds to the set delay plus the processing time. This explains the slightly discrepancy between expected value and measured value in Fig. 3 as well as Fig. 7. However, since the processing time of an echo request compared to the set delay is very short, this can be neglected. A similar consideration applies to the measured interference in the emulated network. In both virtual and physical networks, the connection quality is influenced by, e.g., packet queueing. Since our measurements for accuracy were performed on a single notebook and exclusively in a virtual network, this can also be neglected. Finally, in the real-world scenario, the transmission time was measured as the time for transport and processing in the MQTT message broker and emulated as an artificial delay. The local MQTT message broker also

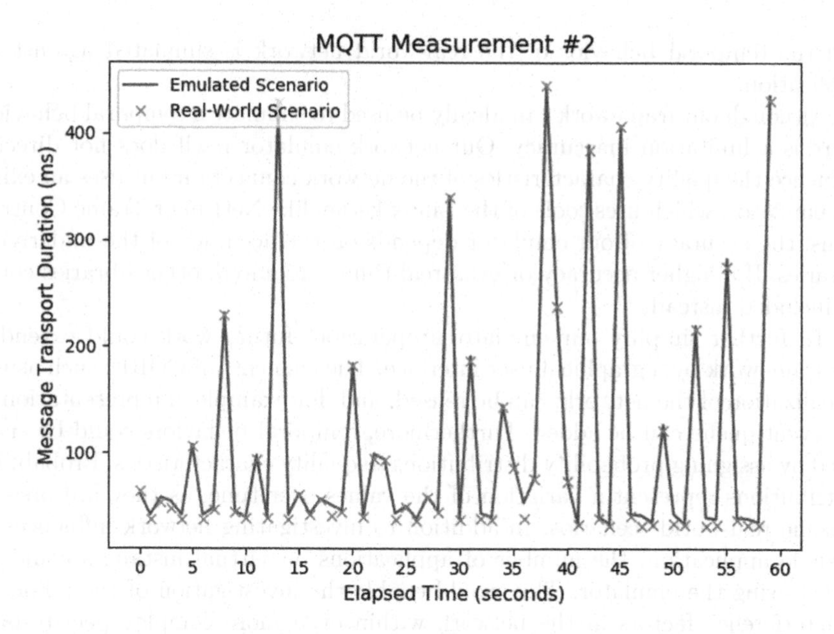

Fig. 10. Comparison of the connection quality with the public *Mosquitto* broker and the emulated connection over a period of 60 s. (Color figure online)

introduces a processing time, and this time is added to the measured transmission time, which explains the small offset in Fig. 10. Since the values for the real network's delay were very high, and the broker's processing time was very short, this aspect was not considered in depth.

5 Conclusion

Network emulators can artificially influence the quality characteristics of network connections. They can be powerful tools in the development of mobile IoT applications to test the application's behavior even if the connection quality of wireless networks varies. However, existing emulators are often complex to configure, require technical knowledge about networks, and are challenging to integrate into automated test environments.

In this paper, we presented a novel network emulation framework that can be easily integrated into software test pipelines. Our emulator does not require extensive expertise so that developers of IoT applications can focus on application development. The emulator supports changes in the interference parameters over time so that a mobile network's temporal behavior, e.g., during a car ride, could be simulated. In the evaluation, the correct influence of the quality characteristics delay, bandwidth, packet loss rate, packet duplication rate, and jitter value, as well as their temporal adjustment, is validated. Furthermore, a measurement of a real-world network with MQTT traffic is imported into the emulator, and the temporal behavior of the real-world network is simulated against an application.

Although our framework can ideally be used to simulate a temporal behavior, there is a limitation inaccuracy. Our network emulator itself does not directly influence the quality characteristics of the network connection but uses an existing emulator, which uses tools of the Linux kernel like NetEm or Traffic Control. Thus, the accuracy of our emulator depends on the accuracy of the underlying libraries. If a higher accuracy or even real-time is required, other libraries could be included instead.

To further simplify our emulator's operation, future work could extended our framework by a graphical user interface. The elements of CORE, such as the visualization of the network, can be reused, and, for example, a representation of the event queue can be added. Furthermore, temporal behaviors could be emulated by assigning probability distributions to quality characteristics. Probability distributions represent a variation of the values over time, as they are present in some real-world scenarios. In addition to investigating network influences of a single application, the number of applications on virtual instances could be scaled using the emulator. This would enable the investigation of the influence of interference factors in the network within even more complex peer-to-peer networks with a large number of participants.

Acknowledgements. This work was funded by the German Research Foundation (DFG) under grant No. (KO 3445/18-1).

References

1. Ahrenholz, J., Danilov, C., Henderson, T.R., Kim, J.H.: CORE: a real-time network emulator. In: MILCOM 2008–2008 IEEE Military Communications Conference, pp. 1–7, November 2008. https://doi.org/10.1109/MILCOM.2008.4753614
2. Ahrenholz, J.: Comparison of CORE network emulation platforms. In: 2010-Milcom 2010 Military Communications Conference, pp. 166–171. IEEE (2010)
3. Bormann, C., Castellani, A.P., Shelby, Z.: CoAP: an application protocol for billions of tiny internet nodes. IEEE Internet Comput. **2**, 62–67 (2012)
4. Carson, M., Santay, D.: NIST net: a linux-based network emulation tool. ACM SIGCOMM Comput. Commun. Rev. **33**(3), 111–126 (2003)
5. Davoli, L., Protskaya, Y., Veltri, L.: NEMO: a flexible java-based network emulator. In: 2018 26th International Conference on Software, Telecommunications and Computer Networks (SoftCOM), pp. 1–6, September 2018. https://doi.org/10.23919/SOFTCOM.2018.8555769
6. De Oliveira, R.L.S., Schweitzer, C.M., Shinoda, A.A., Prete, L.R.: Using Mininet for emulation and prototyping software-defined networks. In: 2014 IEEE Colombian Conference on Communications and Computing (COLCOM), pp. 1–6. IEEE (2014)
7. Hemminger, S., et al.: Network emulation with NetEm. In: Linux Conf AU, pp. 18–23 (2005)
8. Herrnleben, S., et al.: Towards adaptive car-to-cloud communication. In: Proceedings of the IEEE International Conference on Pervasive Computing and Communications Workshops (PerCom Workshops), Kyoto, Japan (2019)

9. Lantz, B., Heller, B., McKeown, N.: A network in a laptop: rapid prototyping for software-defined networks. In: Proceedings of the 9th ACM SIGCOMM Workshop on Hot Topics in Networks, p. 19. ACM (2010)
10. Luomala, J., Hakala, I.: Effects of temperature and humidity on radio signal strength in outdoor wireless sensor networks. In: 2015 Federated Conference on Computer Science and Information Systems (FedCSIS). IEEE (2015)
11. Nussbaum, L., Richard, O.: A comparative study of network link emulators. In: Proceedings of the 2009 Spring Simulation Multiconference, San Diego, CA, USA, pp. 85:1–85:8. SpringSim 2009. Society for Computer Simulation International (2009). http://dl.acm.org/citation.cfm?id=1639809.1639898
12. Porter, M.E., Heppelmann, J.E.: How smart, connected products are transforming competition. Harv. Bus. Rev. **92**(11), 64–88 (2014)
13. Ray, P.P.: A survey on Internet of Things architectures. J. King Saud Univ. Comput. Inf. Sci. **30**(3), 291–319 (2018)
14. Rizzo, L.: Dummynet: a simple approach to the evaluation of network protocols. ACM SIGCOMM Comput. Commun. Rev. **27**(1), 31–41 (1997)
15. Roux, J., Alata, E., Auriol, G., Nicomette, V., Kaâniche, M.: Toward an intrusion detection approach for IoT based on radio communications profiling. In: 2017 13th European Dependable Computing Conference (EDCC), pp. 147–150. IEEE (2017)
16. Sabu, S., Renimol, S., Abhiram, D., Premlet, B.: Effect of rainfall on cellular signal strength: a study on the variation of RSSI at user end of smartphone during rainfall. In: 2017 IEEE Region 10 Symposium (TENSYMP), pp. 1–4. IEEE (2017)
17. Shahid, N., Aneja, S.: Internet of things: vision, application areas and research challenges. In: 2017 International Conference on I-SMAC (IoT in Social, Mobile, Analytics and Cloud) (I-SMAC), pp. 583–587. IEEE (2017)
18. Stoller, M.H.R.R.L., Duerig, J., Guruprasad, S., Stack, T., Webb, K., Lepreau, J.: Large-scale virtualization in the emulab network testbed. In: USENIX Annual Technical Conference, Boston, MA, pp. 255–270 (2008)
19. Ukhurebor, K., Abiodun, C.: Assessment of Building Penetration Loss of Cellular Network Signals at 900 MHz Frequency Bands in Otuoke, Bayelsa State, Nigeria 119 (06 2018)
20. Zec, M., Mikuc, M.: Operating system support for integrated network emulation in imunes. In: Workshop on Operating System and Architectural Support for the on demand IT Infrastructure (1; 2004) (2004)

Detection of Moving Infrared Small Target Based on Fusion of Multi-gradient Filter and Vibe Algorithm

Guofeng Zhang[1,2] and Askar Hamdulla[1(✉)]

[1] Institute of Information Science and Engineering, Xinjiang University, Urumqi 830046, China
askar@xju.edu.cn

[2] Xinjiang Changji Vocational and Technical College, Changji 831100, China

Abstract. Aiming at the problem that dim small targets are submerged to complicated background in infrared images, it is difficult to complete extraction from background and noise clutter. An improved vibe algorithm is proposed for small target detection and tracking. First, target areas are extracted and stored by using vibe algorithm in every frame of video, meanwhile local multi-gradient filter is used to detect and store prominent edge information in each same frame of video. Then, fusion image is obtained through vibe algorithm and multi-gradient filter. Finally, a threshold separation technique is used to further eliminate background clutter and extract small targets. The experimental results show that proposed algorithm can quickly eliminate ghosts and is effective for detecting moving small targets. Compare to other background difference method, gaussian mixture model, experimental evaluation results show that our method outperforms vibe, background difference method and gaussian mixture model methods in terms of both tracking accuracy and computation speed for detection infrared small targets.

Keywords: Vibe algorithm · Ghost · Gradient filter · Image fusion

1 Introduction

High efficiency and high precision target detection and tracking technology have been widely applied in military and civil fields. For long-range infrared imaging with a distance of 13800 km, imaging target occupies only a few pixels and no shape, texture or other characteristics. Average brightness is low, motion law is indeterminate, the size of target is less than 9×9 pixels.

At the same time, due to jitter of infrared imaging equipment and effect of noise from imaging equipment, weak and small targets are submerged in the background. These problems make it more difficult to search and detect small targets in infrared sequence images. Traditional algorithms such as background difference method (BDM), is not friendly to adapt environmental change. It takes more time to build background model. Gaussian mixture model (GMM) is often used to detect moving small targets in complex background, However, it needs to extract a lot of frame of sequence images to

H. Song and D. Jiang (Eds.): SIMUtools 2020, LNICST 370, pp. 600–612, 2021.
https://doi.org/10.1007/978-3-030-72795-6_48

complete background model initialization, and real-time performance of the algorithm is not guaranteed. Vibe algorithm is mainly used for tracking of large targets, such as vehicle detection and tracking, face recognition, pedestrian detection, significance detection, etc. which is very effective [1], but there is no existing reference for infrared small target detection. There are two disadvantages in the application of vibe algorithm to detect small targets: one is that moving targets are regarded as background modeling in the first frame, which leads to generation of ghost in infrared images. The other is that ghost phenomenon caused by sudden movement targets which are stationary for a long time in previous several frames. Based on two problems, this paper proposes vibe algorithm of multi-gradient filter (GFVibe) to eliminate ghost and realize real-time target detection. The designing flow chart is shown in Fig. 1. Firstly, foreground feature map of moving target is obtained by using vibe algorithm, but detection result may generate ghost and background clutter. Therefore edge significant information of small targets is obtained by multi-gradient filter operator in the same time, this is target feature map. Then, fusion of foreground feature map and target feature map generates more accurate small targets. Finally, threshold separation technique is applied to eliminate residual background clutter and get real target area. Compared with traditional vibe algorithm, GFVibe algorithm proposed can compensate the disadvantage that moving targets generated ghost region, and also overcome flicker pixel point problem which is difficult to completely eliminate various noises when using vibe algorithm to detect small targets.

2 Feature Analysis for Point Target Image

The image of point target under complex conditions can be described as three parts [2, 3]:

$$I(x, y, t) = B(x, y, t) + T(x, y, t) + N(x, y, t) \tag{1}$$

where $I(x, y, t)$ represents infrared image contained point target, $B(x, y, t)$ represents background areas, $T(x, y, t)$ represents dim point targets, $N(x, y, t)$ represents random noise clutter, (x, y) represents current pixel location, t denotes different intervals of frames. Fig. 1 represents the detection flow chart of proposed algorithm.

3 The Vibe Algorithm [3, 4]

Vibe algorithm is a background modeling algorithm based on pixel level, which is mainly used for target detection. It has some advantages that other algorithms do not have, such as fast operation speed and small memory usage. The algorithm adopts random neighborhood pixel value to establish background modeling. Firstly, it establishes a background model sample set for each pixel. The elements of sample set come from random neighborhood pixel. Then, new pixel value at the next frame is compared with a set of values taken in the past at the same location or in the neighborhood. Euclidean distance is used to determine whether new pixel value belongs to the background, vibe algorithm adapts the model by choosing randomly which values to substitute from the background model. That is classification process of background pixel. The algorithm mainly consists of three core parts: background model initialization, extraction of foreground feature map and background model updating mechanism.

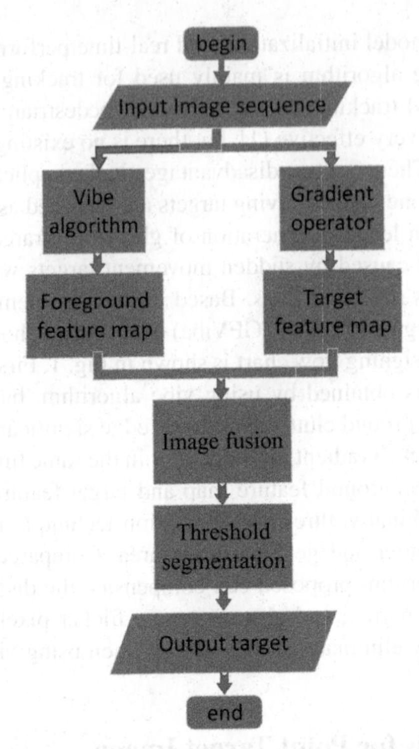

Fig. 1. The detection flow chart of proposed algorithm.

3.1 Background Model Initialization

In vibe algorithm, the essence of background model initialization is the process of establishing a sample set for each pixel. Since pixels in the image cannot contain spatio-temporal information, vibe algorithm uses image sequence and neighborhood pixel information to construct spatio-temporal distribution characteristics. Specific implementation method is as follows: each pixel in the image randomly selects gray value of its neighboring pixel as its sample values. In order to ensure that background model conforms to statistical laws, the range of neighborhood should be large enough. Figure 2 is a sample set of a center pixel point N(x, y) filled with the values of eight neighborhood pixels from background model.

$N(2)$	$N(2)$	$N(3)$
$N(4)$	$N(x, y)$	$N(5)$
$N(6)$	$N(7)$	$N(8)$

Fig. 2. Eight neighbor pixels of N(x, y).

3.2 Extraction of Foreground Feature Map

No matter whether there are targets in the image, the first frame is considered as background by default for vibe algorithm, and foreground feature map is detected from the second frame image. Moving target detection consists of following two steps:

The first step is target candidate, and the second step is target confirmation. In the first step, difference value $I_t(x, y)$ of grayscale is calculated and stored between the current pixel value $N_t(x, y)$ and all sample values N(k) that have been stored in the past in this neighborhood. Compared with the pre-setting Euclidean distance R. If the difference value $I_t(x, y)$ is greater than the Euclidean distance R, it is considered candidate target. The second step is to count the number of NUM which is more than Euclidean distance R in the first-step storage. If NUM is less than pre-set #min, it is confirmed as final target. The flow chart is shown in Fig. 3.

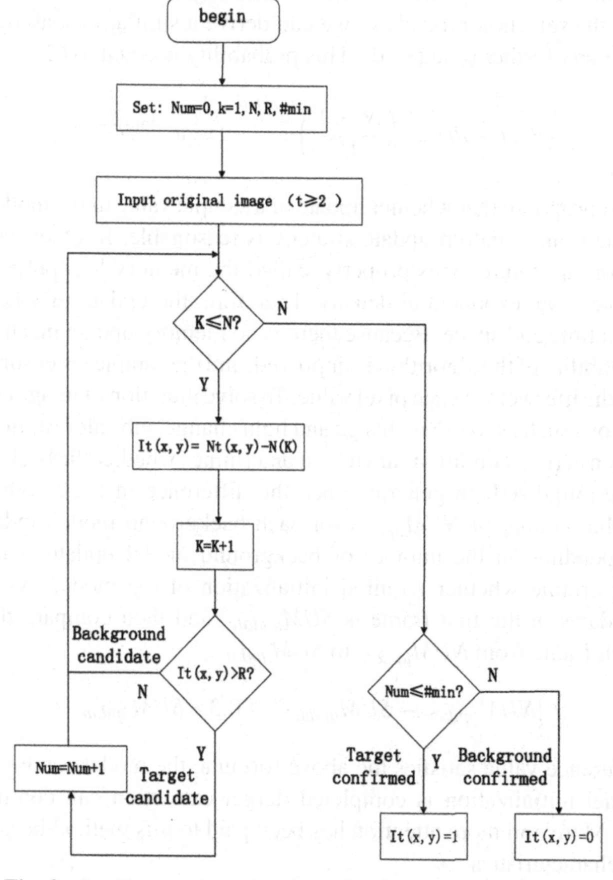

Fig. 3. Small target detection flow chart based on vibe algorithm.

3.3 Improvement of Model Update

In order to adapt to background environment of possible mutations, vibe establishes the time-sampling updating strategy [5] and the spatial neighboring updating mechanism [6]. Update strategy of time-sampling depends on time factor Ø, what can do is not need to update processing when process data of each frame, but according to a specific probability to update the background model. When a pixel is classified as background, it has a probability of 1/Ø to update background model samples or background samples of a neighboring pixel. Spatial neighboring update strategy: When background samples need to be updated, it is updated by randomly selecting a pixel value in the neighboring domain and replacing selected background model with newly randomly selected pixel. It is worth noting that when updating N sample values in a pixel sample's set, it is random to choose one from N samples to update. Therefore, the probability of the occurrence of sample value in the model decays exponentially, the (N − 1)/N probability of a sample present in the model is not updated at time t. Assuming time continuity and the absence of memory in the selection procedure, we can derive a similar probability, denoted as P(t, t + dt), for any further time t + dt. This probability is equal to (2).

$$p(t, t+dt) = \left(\frac{N-1}{N}\right)^{(t+dt)-t} = e^{-\ln(\frac{N}{N-1})dt} \tag{2}$$

Narayanan first proposed that whether update of a sample value in the model has nothing to do with time t, and random update strategy is reasonable. In other words, the past has no effect on the future. This property, called the memory less property, is known to be applicable to an exponential density. Therefore, the update of vibe algorithm is random in both time and space. Because there is no memory update mechanism [6], the dynamic adaptability of the algorithm is improved, and the sample set ensures the smooth attenuation of the life cycle of each pixel value. To solve mutation of image content caused by special reasons such as weather change and light change, vibe algorithm sets a specific threshold. When certain conditions are met, it determines whether the background model needs to be re-initialized. In general, since the difference in background changes is not obvious, the number of NUM_{update} for each background model update is similar. Therefore, depending on the number of background model updates each time, it is possible to determine whether required initialization of the model. Assume that the number of updates in the first frame is SUM_{update}, and then compare the number of updates in each frame from NUM_{update} to SUM_{update}.

$$\left|NUM_{update} - SUM_{update}\right| > 0.3 * SUM_{update} \tag{3}$$

When the difference value satisfies the above formula, the model can be re-initialized. After the model initialization is completed, target extraction can continue from the second frame. More and more attention has been paid to this method because of its fast and effective characteristics.

4 Multi-gradient Filter

Four gradient operators H1, H2, H3 and H4 (45°, 135°, 225°, 315°) are defined in this paper. Templates are respectively used to carry out convolution operation for each frame on the video image to obtain gradient convolution filter graph in four directions. Formula (4) below represents the convolution operation.

$$H_1 = \begin{bmatrix} 0 & 0 & 1 \\ 0 & -1 & 0 \\ 0 & 0 & 0 \end{bmatrix} \quad H_2 = \begin{bmatrix} 1 & 0 & 0 \\ 0 & -1 & 0 \\ 0 & 0 & 0 \end{bmatrix} \quad H_3 = \begin{bmatrix} 0 & 0 & 0 \\ 0 & -1 & 0 \\ 1 & 0 & 0 \end{bmatrix} \quad H_4 = \begin{bmatrix} 0 & 0 & 0 \\ 0 & -1 & 0 \\ 0 & 0 & 1 \end{bmatrix}$$

$$GF(x, y, t) = f(x, y, t) * H \tag{4}$$

Where f(x, y, t) represents sequence images in video at time t, GF(x, y, t) represents gradient convolution filter graph for corresponding sequence image in video at time t. H represents gradient operators. The size of image block is defined as W × W, and each gradient convolution filter graph $G_k (1 \le k \le 4)$ is traversed from left to right and from top to bottom. Each image block is divided into four sub-image blocks $\psi_k (1 \le k \le 4)$ on average, the gradient mean square value is respectively calculated in the corresponding direction of each sub-image block under the gradient convolution filter graph, it is denoted as g_{ψ_k} $(1 \le k \le 4)$, and calculate the maximum(Gmax) and minimum(Gmin) of the gradient mean square value as follows:

$$G_k = \frac{1}{N_j} \sum_{j=1}^{N_j} \left\| g^j_{\psi_k} \right\| \tag{5}$$

$$G_{\max} = \max_{1 \le k \le 4} G_k \tag{6}$$

$$G_{\min} = \min_{1 \le k \le 4} G_k \tag{7}$$

Where N_j represents the number of pixels in the sub-image block. After the gradient operation is realized, the threshold is set as η. According to Eq. (8), gradient filter graph (GF) can be obtained.

$$G = \begin{cases} \sum_{i=1}^{4} G_i & \frac{G_{\min}}{G_{\max}} > \eta \\ 0 & \text{其它} \end{cases} \tag{8}$$

5 Image Fusion

The image obtained by vibe algorithm and GF image are fused for each frame in the video to obtain a new fusion image. This effectively suppresses presence of ghost and problem of flicker pixels point. The blue rectangle represents the sliding window and the red rectangle represents small target. The calculation process of fusion is shown in Eq. (9) (Fig. 4).

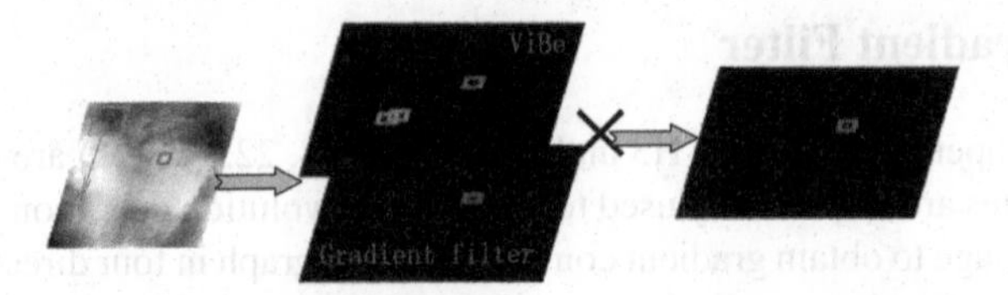

Fig. 4. The process of image fusion. (Color figure online)

$$GFVibe(x, y, t) = GF(x, y, t) \times Vibe(x, y, t) \tag{9}$$

Vibe algorithm can quickly and effectively extract moving small targets, and the gradient vector of small targets roughly points to the target center, but background clutter does not have such characteristics, so gradient operator filtering has a good performance of detecting edges, while fusion technology can effectively remove ghost or suppress isolated noise points to enhance the target areas.

6 Threshold Segmentation Method

After fusion image is obtained, in order to extract the small target, the segmentation method is used in fusion image, which is defined as Eq. 10.

$$T = \mu + k \times \sigma \tag{10}$$

Where μ and σ is mean and standard deviation of the final salient GFVibe map, k is a adjustment parameter. Our experiments show that the optimal range of k is from 3 to 10. If the pixel value in the GFVibe map is greater than T, we affirm that it is a target, otherwise, it should be considered as background.

7 Experimental Evaluation

7.1 Selection of Parameters

Setting of some parameters in this experiment consults the reference [6, 7]. The size of background sample (N equals 20) and the minimum cardinality (#min equals 2) are both selected as the default. The initial Euclidean distance is improved to $R = 0.5 \times \sigma_m$ and is set in the integer range [20,40], in order to better adapt to the changes for background clutter. The random update factor is set to 4, the updating probability of background model sample is 25%. As the detected target tends to be a point target, the image to be tested in the experiment is two sets of sequence images with a size of 240 × 320 video under different backgrounds of 200 frames. Each frame of video image 1 contains two small targets and video image 2 contains one small target. Figure 5 and Fig. 6 show tracking results of GMM, background difference method, vibe and proposed algorithm (GFVibe) at two different backgrounds. Extraction results of 20, 60, 120 and 180 frames are shown in Fig. 5 and Fig. 6.

In this paper, neighborhood pixels is set to 24, the gradient operator is improved on the basis of Robert operator [7–9]: the center coefficient is negative and the sum of all

(a) Original infrared image

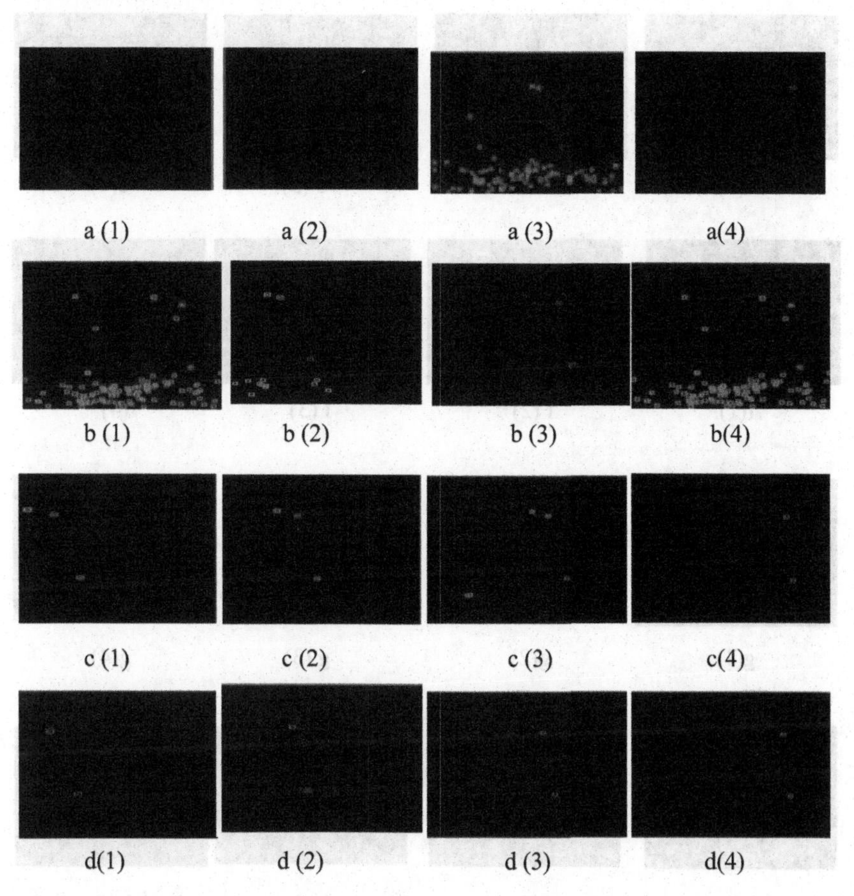

Fig. 5. Tracking results of different algorithm in video image 1.

coefficients is 0. The center of four gradient operators is set as the coefficient -1, and four diagonal edges (45°, 135°, 225°, 315°) are set as 1. The gradient is estimated according to the difference in any mutually perpendicular direction. Compared with horizontal and vertical edge detection, diagonal edge detection is more suitable for those edges with strong undulations whose boundary is not obvious. Gradient operator template matches gradient vector of small target region and is a template with a center, so the size of template must be odd. In order to achieve higher efficiency, the template in this paper adopts 3×3 for convolution filtering. According to literature [10–12], when the size of

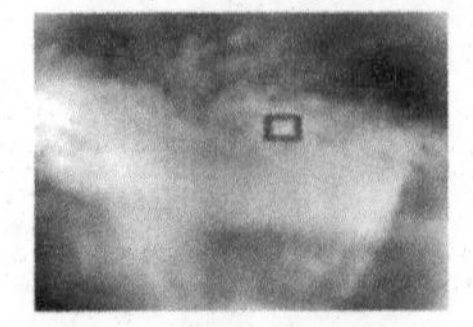

(b) Original infrared image

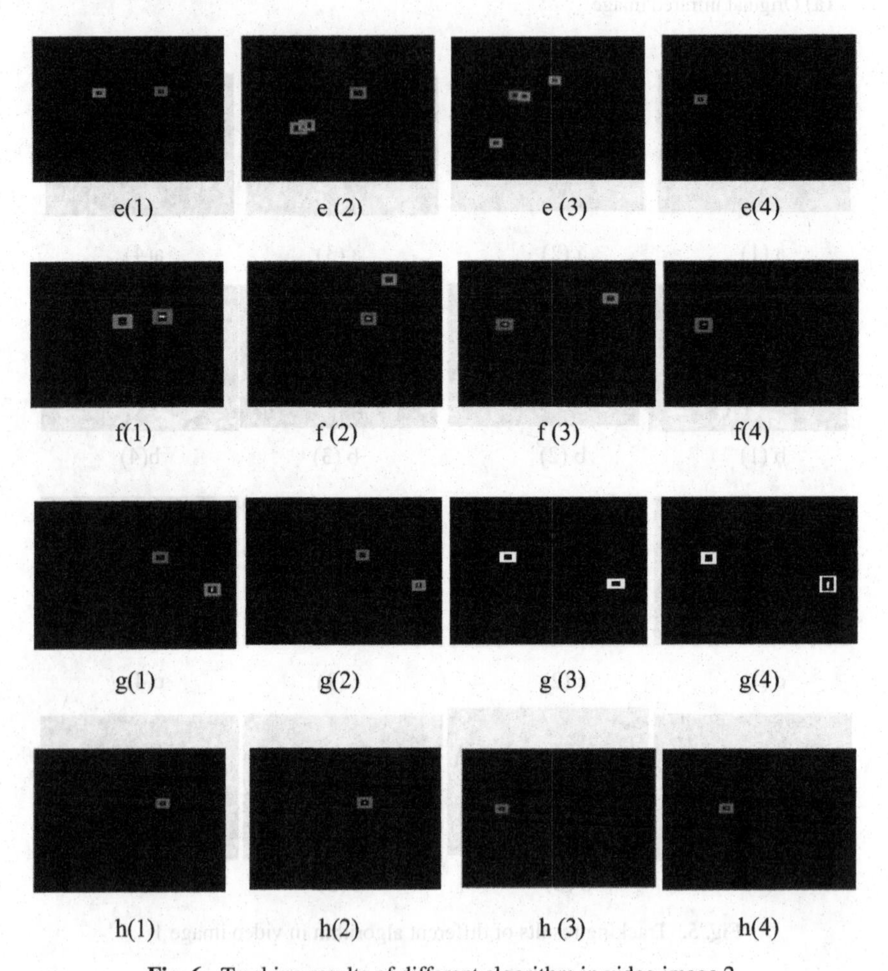

Fig. 6. Tracking results of different algorithm in video image 2.

image block increases beyond small target size, the new gradient data does not change significantly, so the size of local gradient sliding window only needs to be larger or closer to the target size. In order to evenly distribute 4 sub-image blocks, W is taken as an even number, that is 10×10, so that the sizes of 4 sub-image blocks are all set to 5×5, which can also satisfy the asymptotic change of gradient. After experimental data analysis, if it is target region, the maximum and minimum values of the gradient in the four directions have little difference. The edge area is quite different. Setting enough big is easy to miss targets, too small setting results in high false alarm rate. Experimental results show that η equal to 0.2 is the best effect. In the experimental image, the green rectangular represents the false targets and red rectangular represents real targets. The trajectory of small target is a curve, video image 1 and video image 2 are the tracking results of GMM (a1−a4, e1−e4), Background difference method (b1−b4, f1−f4),Vibe(c1−c4, g1−g4), GFVibe (d1−d4, h1−h4) in Fig. 5 and Fig. 6, respectively.

Vibe also achieved good tracking effect, but in video image 1, there is still a false alarm point in the 20th, 60th and frame 120th frame, until the false alarm point disappears in the 180th frame. It can be seen that successful background modeling of vibe requires a certain amount of frame accumulation to eliminate ghost and flicker pixels. The false alarm rate of GMM is very high in the complex sea background, and false negative rate is very serious, small targets are lost in 20th, 60th and 180 frames, respectively. Background difference method can correctly track small target in video image 1, but the false alarm rate is also very high in the 60th and 180th frame.The proposed algorithm GFVibe can correctly track small moving targets in the complex sea background. In video image 2, background difference method, GMM and vibe demonstrate good tracking results, The false alarm rate of GMM is slightly tall, ghost problems of vibe doesn't disappear for the same place in video image 2, the background difference method is slightly better than the vibe, to some extent, the background difference method can eliminate the ghosts, but it cannot eliminate flicker pixels, GFVibe not only can better eliminates ghost phenomenon existed, but also can restrains flicker pixel point, and achieve a good tracking effect.This good inhibition ability of GFVibe is due to good edge detection ability of gradient operator. Thus it can be seen that the proposed algorithm can correctly track moving small target under two different backgrounds. When there is a small target in the first frame. Ghost and flicker pixel removal for clutter noise is effective, robust, and excellent.

7.2 Quantitative Evaluation

Experimental results verify the comparison of runtime for GMM, background difference method, Vibe, GFVibe algorithm in two groups of video images, as shown in Table 1 and Table 2. T1, T2, T3 and T4 are the results of runtime of 200 frame sequence images under two different algorithms when the SNR (Signal-to-Noise Ratio) is 4, 3, 2, 1 dB. T(means) represents the average runtime of the same algorithm for 200 frame image sequences.

Table 1. Comparison of the mean runtime of different methods under video image 1 (unit: s).

Time	T1	T2	T3	T4	T(means)
GMM	76.368	267.739	268.142	269.153	220.351
BDM	77.587	269.766	270.586	269.616	221.888
ViBe	59.338	60.390	58.3390	59.300	59.342
GFVibe	**57.235**	**56.817**	**58.022**	**56.579**	**57.163**

Table 2. Comparison of the mean runtime of different methods under video image 2 (unit: s).

Time	T1	T2	T3	T4	T(means)
GMM	73.386	117.43	156.41	192.025	127.266
BDM	75.410	119.421	105.186	131.16	107.794
ViBe	53.783	50.190	51.431	54.012	52.354
GFVibe	**56.708**	**50.017**	**51.022**	**46.675**	**51.106**

Compared with GMM, BDM, Vibe algorithm consumes less time in video image 1 and in video image 2. It takes time about 0.297 s and 0.262 s to process a frame in video 1 and in video 2, and real-time is better. The runtime of GFViBe algorithm is basically stable under in two different background. It is shown in bold for Table 1 and Table 2. In Table 1 that average processing time of one frame is about 0.286 s and 0.256 s in video 1 and in video 2, which is about 4% and 2.3% shorter in video 1 and in video 2. Compared with the other three algorithms, the shorter runtime of the proposed method justifies its use in real time applications.

In addition, we also choose two common evaluation indicators for comparison, which are signal-to-clutter ratio gain (SCRG) and background suppression factor (BSF). The indicators are defined as follows:

$$SCRG = \frac{SCR_{out}}{SCR_{in}}, \quad BSF = \frac{C_{in}}{C_{out}}$$

where SCR_{in} and SCR_{out} represent the SCR values of the raw image and GFViBe map, respectively, and σ_{in} and σ_{out} represent the standard deviation of the raw image and GFViBe map, respectively. To a certain extent, SCRG can measure clutter suppression and target preservation, and BSF is metric which reflects the performance for clutter removal without the target information. Generally, the higher the SCRG is, the stronger ability of clutter suppression and target enhancement the method has.

Table 3. The average results of different detection methods for each sequence.

Seq.	Indicators	GMM	BDM	ViBe	GFVibe
1	$\overline{SCRG}$	14.57	24.06	49.42	**55.92**
	$\overline{BSF}$	6.632	8.54	23.65	**28. 87**
2	$\overline{SCRG}$	20.50	23.48	38.49	**42.31**
	$\overline{BSF}$	17.62	19.15	27.36	**38.47**

Table 3 presents the average results of different methods for each sequence. SCRG and BSF represent the average results. It can be seen that vibe achieve a better performance than the other baseline methods, because they have the steps of backgrounds prediction or preservation, and the predicted backgrounds are subtracted from the original image. Thus, they can achieve large SCRG and BSF. However, **GMM** and **BDM** do not specialize in the background suppression, thus their performance is not superior to others. The proposed approach achieves the best performance to enhance the target areas. On the two metrics. SCRG and BSF of our method are large extremely, because GFVibe fuses gradient operator on the basis of vibe, which has extremely capability of edge detection to remove ghost or flicker pixel points.

8 Conclusion

This paper focuses on the key technology of small target tracking, which involves the detection and tracking of moving small targets. From experimental evaluation, it can be seen that the proposed algorithm in this paper has certain practical significance in eliminating ghost and suppressing flicker pixel points. GFVibe has strong ability of background modeling and performance of good edge detection duo to introduction of gradient operator. Fusion image has obvious advantages in eliminating ghost and isolated noise. The proposed algorithm can be relatively stable and correctly track moving small target. This algorithm is the least time-consuming. In this paper, we only extract the simple small targets in the static background. Whether small targets can be accurately extracted in complex background will be explored in future work. Compared with GMM, BDM, Vibe tracking detection algorithms, GFVibe has better stability and shorter runtime.

Acknowledgments. This work has been supported by the National Natural Science Foundation of China (under grant number of 61563049).

References

1. Li, J., Li, S.-J., Zhao, Y.-J., Ma, J.-N., Huang, H.: Background suppression for infrared dim and small target detection using local gradient weighted filtering. In: International Conference on Electrical Engineering and Automation (2016)
2. Marvasti, F.S., Mosavi, M.R., Nasiri, M.: Flying small target detection in IR images based on adaptive toggle operator. IET Comput. Vis. **12**(4), 527–534 (2018)
3. Zhang, H., Zhang, L., Yuan, D., et al.: Infrared small target detection based on local intensity and gradient properties. Infrared Phys. Technol. **89**, 88–96 (2017)
4. Barnich, O., Van Droogenbroeck, M.: ViBe: a universal background subtraction algorithm for video sequences. IEEE Trans. Image Process. **20**(6), 1709–1724 (2011)
5. Henriques, J.F., Caseiro, R., Martins, P., Batista, J.: High-speed tracking with kernelized correlation filters. IEEE Trans. Pattern Anal. Mach. Intell. **37**(3), 583–596 (2015)
6. Bai,K., Wang, Y., Song, Q.: Patch similarity based edge-preserving background estimation for single frame infrared small target detection. In: IEEE International Conference on Image Processing, pp. 181–185 (2016)
7. Zhang, L.Y., Du, Y.X., Li, B.: Research on threshold segmentation algorithm and its application on infrared small target detection algorithm. In: International Conference on Signal Processing, pp. 678–682 (2015)
8. Wei, H., Tan, Y., Lin, J.: Robust infrared small target detection via temporal low-rank and sparse representation. In: International Conference on Information Science, pp. 583–587 (2015)

9. Yao, Y., Hao, Y.: Small infrared target detection based on spatio-temporal fusion saliency. In: 2017 17th IEEE International Conference on Communication Technology
10. Zhang, H., Niu, Y., Zhang, H.: Small target detection based on difference accumulation and Gaussian curvature under complex conditions. Infrared Phys. Technol. **87**, 55–64 (2017)
11. Xie, K., Fu, K., Zhou, T., Yang, J., Wu, Q., He, X.: Small target detection using an optimization-based filter. In: International Conference on Acoustic, pp. 1583–1587 (2015)
12. Deng, L., Zhu, H., Zhou, Q., Li, Y.: Adaptive top-hat filter based on quantum genetic algorithm for infrared small target detection. Multimed. Tools. Appl. **77**, 10539–10551 (2018)

Handwritten Uyghur Character Recognition Using Convolutional Neural Networks

Wujiahemaiti Simayi, Mayire Ibrayim, and Askar Hamdulla(✉)

Institute of Information Science and Engineering, Xinjiang University, Urumqi, China
askar@xju.edu.cn

Abstract. Handwritten Uyghur character recognition researches up to date have been based on traditional pattern recognition techniques that highly relies on handcrafted features. The similarity between character forms has been hindering the extraction of robust features. This paper proposed the deep learning based self-learned features to recognize 128 handwritten Uyghur characters forms. The first-hand online handwritten trajectory is first preprocessed and converted to a centralized binary image as input to the implemented deep neural network model. In experiments, the convolutional neural network models with 4, 5 and 8 convolutional layers are studied to get higher recognition accuracy. All models are trained implementing the same dropout regularization. The models 8 convolutional layers on 48 * 48 converted character images produced as high as 94.65% average accuracy on a test set of 10,240 handwritten character samples.

Keywords: Handwritten character recognition · Uyghur character forms · Size adjusting · Convolutional neural network

1 Introduction

Handwritten character recognition is one of the most typical implementations of pattern recognition [1]. Handwriting recognition has two main branches including online and offline handwriting recognition [2]. Recognizing the traced pen-tip trajectory is called online handwriting recognition, whereas offline handwriting recognition categorizes the already formed handwritten shape images. Online handwritten samples contain exact pen-tip trajectory with spatial and temporal information. A same handwritten object is perhaps formed in various sizes, orders and angles. Violation to the standard writing rules occurs very often in the handwriting process. For example, a stroke at the beginning part of a shape probably will be written at last, or a stroke is perhaps started without completing the previous one etc. These kinds of randomness decrease the advantage of temporal information [3–5].

Handwritten Uyghur character recognition studies have been based on traditional pattern recognition methods. This paper presents the first application of convolutional neural networks to recognize 128 Uyghur character forms. Different from the traditional methods, features are automatically learned and extracted during the model training [6].

H. Song and D. Jiang (Eds.): SIMUtools 2020, LNICST 370, pp. 613–624, 2021.
https://doi.org/10.1007/978-3-030-72795-6_49

The remaining content of the paper is given in the following sections. Section 2 gives a brief introduction to Uyghur characters and some related studies. Preprocessing techniques applied on the raw online samples before feeding them to the model is specified in Sect. 3. Section 4 introduces the convolutional neural network architecture implemented in this paper. The model training process and the experiment results are provided and discussed in Sect. 5. Finally, Sect. 6 draws conclusion.

2 Related Work on Uyghur Character Recognition

2.1 Uyghur Characters

Table 1. Uyghur character forms

End	Mid	Begin	Single	Rep	No.	End	Mid	Begin	Single	Rep	No.
ـڭ	ـڭـ	ڭـ	ڭ	ڭ	20	ـئا			ئا	ا	1
ـل	ـلـ	لـ	ل	ل	21	ـا			ا		
ـم	ـمـ	مـ	م	م	22	ـئە			ئە	ە	2
ـن	ـنـ	نـ	ن	ن	23	ـە			ە		
ـھ	ـھـ	ھـ	ھ	ھ	24	ـب	ـبـ	بـ	ب	ب	3
ـئو			ئو	و	25	ـپ	ـپـ	پـ	پ	پ	4
ـو			و			ـت	ـتـ	تـ	ت	ت	5
ـئۇ			ئۇ	ۇ	26	ـج	ـجـ	جـ	ج	ج	6
ـۇ			ۇ			ـچ	ـچـ	چـ	چ	چ	7
ـئۆ			ئۆ	ۆ	27	ـخ	ـخـ	خـ	خ	خ	8
ـۆ			ۆ			ـد			د	د	9
ـئۈ			ئۈ	ۈ	28	ـر			ر	ر	10
ـۈ			ۈ			ـز			ز	ز	11
ـۋ			ۋ	ۋ	29	ـژ			ژ	ژ	12
ـئې	ـئېـ	ئېـ	ئې	ې	30	ـس	ـسـ	سـ	س	س	13
ـې	ـېـ	ېـ	ې			ـش	ـشـ	شـ	ش	ش	14
ـئى	ـئىـ	ئىـ	ئى	ى	31	ـغ	ـغـ	غـ	غ	غ	15
ـى	ـىـ	ىـ	ى			ـف	ـفـ	فـ	ف	ف	16
ـي	ـيـ	يـ	ي	ي	32	ـق	ـقـ	قـ	ق	ق	17
	ـئـ	ئـ		ئ	33	ـك	ـكـ	كـ	ك	ك	18
ـلا			لا	لا	34	ـگ	ـگـ	گـ	گ	گ	19

Modern Uyghur script inherits alphabetic writing style that regular combination of characters generates words and further sentences according to the morphological and lexical rules. It is written from right to left and from top to down orientation. Table 1 provides the total Uyghur character forms [9].

As given in Table 1, there are 32 basic Uyghur characters containing 8 vowels (char.No.1–2, char.No.25–28, char.No.30–31) and 24 consonants. There are total 126 writing forms of 32 basic Uyghur characters. Each character has different writing forms according to character positions within a word or sub-word, such as initial, intermediate, ending and isolated forms. Furthermore, A special component character (char.No.33) and a compound character (char.No.34) are frequently used in both printing and handwriting. Each of them has two writing forms according to the position appeared in a word. To sum up, we have to handle with 130-character forms for handwritten Uyghur character recognition.

However, the isolated and ending forms of char.No.24 are very much similar to its beginning and intermediate forms respectively, not only in handwriting but also in printing fonts. Therefore, only the isolated and ending forms ﮬand ﮭare considered during character data collection for char24. The neglected character forms are labeled red in Table 1. Therefore, actually collected handwritten character forms are 128 in this paper.

2.2 Previous Studies

Preliminary researches on Uyghur handwritten character recognition investigated the handwriting characteristics of Uyghur characters and proposed different recognition methods. Different feature extraction methods have been presented according to the structural and statistical property of handwritten character forms [6]. Due to the high similarity between the Uyghur character forms which are hardly differentiated without context information, some studies were conducted only on the isolated character forms. However, recognizing the total 128 character forms has great significance in implementations. In this paper, we only review the character recognition studies on the total 128 character forms.

Coded 8-directional features from divided 16 grids of handwritten character forms were applied to recognize 128 character forms in [7]. Template matching based on Euclidean distance was adopted using minimum distance classifier. 51200 online character samples were split into train and test sets with 1:1 ratio in the conducted experiments. Experiments recorded modest recognition result of 60% average accuracy.

Center distance feature proposed by [8] produced character recognition accuracy to 80% using minimum distance classifier. The train and test sets are arranged with 7:3 ratio, using dataset size of 51200 samples. Later on, average recognition accuracy was improved to 87% by modifying feature extraction method.

MQDF classifier on NCFE features made even better recognition results on 128 character forms. 89.08% average recognition accuracy were recorded using combination of different features [9]. The same dataset used in [7] and [8] was used in experiments and samples are split into train and test set with ratio of 7:3, too. Using the same dataset with the same splitting into train and test sets, the experiment results in [9] fairly demonstrates the superiority of the MQDF with NCFE feature for Uyghur handwritten character recognition.

Some comparative character recognition experiments using minimum distance classifier, MQDF and BP neural networks were conducted on the online character samples from 115 writers [10]. Online character samples from 60 writers were used for training while the remained 55 writers are arranged for testing. Directional chain codes were extracted as extracted feature for each of the Minimum distance classifier, MQDF and BP models. 83.36%, 87.77% and 84.73% average recognition accuracies are obtained for the compared classifiers, respectively. Combination of the classifiers gave further improved results of 90.88% accuracy, later.

Time division directional feature with weighted Naïve Bayes algorithm has received 93.15% average recognition accuracy [11]. The dataset used in experiments were collected from 102 writers who contributed 13056 online character samples. Among them, samples from 60 writers were arranged for training, samples from 42 writers were used for testing.

3 Preprocessing

Besides unique handwriting styles of each writer, the first-hand collected handwritten samples always vary in size, orientation as well as the location on the handwritten tablet screen. Preprocessing improves the performance of the recognition system if well manipulated [12]. Some preprocessing operations including point insertion and size adjustment are incorporated to make the binary images to be the model input.

3.1 Point Insertion

In order to avoid generating extra points between strokes, this paper applies stroke-level point-insertion, which means the point insertion operation is applied only within a stroke and not between the strokes. Each neighbor point couples are checked to know if they need point insertion, by calculating Euclidian distance between them. Insertion is made between the neighbor points which have larger distance than a threshold value which is set 1 in this paper.

$$dist = \sqrt{(x_1 - x_2)^2 + (y_1 - y_2)^2} \tag{1}$$

$$x_i = x_1 + \frac{\Delta x}{N},\ y_i = y_1 + \frac{\Delta y}{N} \tag{2}$$

$P_1(x_1, y_1)$ and $P_2(x_2, y_2)$ are the neighbor points that need point insertion between them. Δx and Δy are the horizontal and vertical distances calculated between the neighbor points. Euclidian distance between them is obtained using Eq. (1). The integer value of the distance is used as the number of points to be inserted, denoted as N. Coordinates of each insertion point is determined according to Eq. (2) where (x_i, y_i) are the coordinate values of the i th insertion point.

3.2 Coordinate Range Normalization

The original coordinate range refers to the size of screen window that handwriting is recorded. The ranges of horizontal and vertical axis coordinates of the collected raw online samples are [1, 255] in this paper. The original coordinate range is always much larger than needed in most cases. Normalizing original sample to smaller coordinate range will save time and storage for later operations. Coordinate values of points are adapted to new coordinate range using simple linear normalization as in Eq. (3).

$$x = \frac{w}{W} * X, \; y = \frac{h}{H} * Y \tag{3}$$

where (X, Y) are the original coordinate values in the original interval or screen size (W, H), while (x, y) are the normalized coordinate values in new interval or the normalized widow size (x, y).

3.3 Size Adjusting

There are many quite similar characters forms in Uyghur. Direct implementation of the common linear normalization methods easily loses overall shape information of original samples and influences the classification accuracy. To keep original handwritten character forms unchanged in the generated images, the handwritten trajectories are adjusted to the normalized coordinate ranges by the following steps:

a. Calculate the width ratio and height ratio between the handwritten shape and the normalized window (coordinate ranges) by Eqs. (4)–(5).

$$W_{ratio} = \frac{w}{\max(XX) - \min(XX)} \tag{4}$$

$$H_{ratio} = \frac{h}{\max(YY) - \min(YY)} \tag{5}$$

w and h are the width and height of the imaginary window of the normalized coordinate ranges.

b. Adjust sample shape to the window by linear size normalization using adjust ratio obtained by Eq. (6), Sample shape is first moved to the coordinate origin prior to multiplying the adjust ratio, so that adjusted shape will still be in the normalized coordinate range, as Eqs. (7)–(8).

$$adjust_{ratio} = \min(W_{ratio}, H_{ratio}) \tag{6}$$

$$x = (X - \min(XX)) * adjust_{ratio} \tag{7}$$

$$y = (y - \min(YY)) * adjust_{ratio} \tag{8}$$

Where (X, Y) and (x, y) are the point coordinate values before and after size adjusting. Thus, the original sample shape will be kept as before after size adjusting and make good use the space in the normalized window size (coordinate ranges).

c. Centralize the sample trajectory to the normalized coordinate ranges using Eqs. (9)–(10).

$$x = X - \left(\frac{\max(XX) - \min(XX)}{2} - \frac{w}{2}\right) \tag{9}$$

$$y = Y - \left(\frac{\max(YY) - \min(YY)}{2} - \frac{h}{2}\right) \tag{10}$$

where $\left(\frac{\max(XX) - \min(XX)}{2}, \frac{\max(XX) - \min(XX)}{2}\right)$ is the center of the sample shape, $\left(\frac{w}{2}, \frac{h}{2}\right)$ the window center of the normalized coordinate ranges.

d. Convert the online trajectory to image. Since each point coordinate is clear to us, it is convenient to make binary images just marking black or white pixels at the corresponding positions in a matrix or window. Concerning the visibility for observation, background of the character sample images is set black while foreground is set white. The character forms shown in Fig. 1(a) are ones which easily lose their original shapes by careless use of linear size normalization. Implementing above described size adjusting kept the original shape of the samples, as shown in Fig. 1(b).

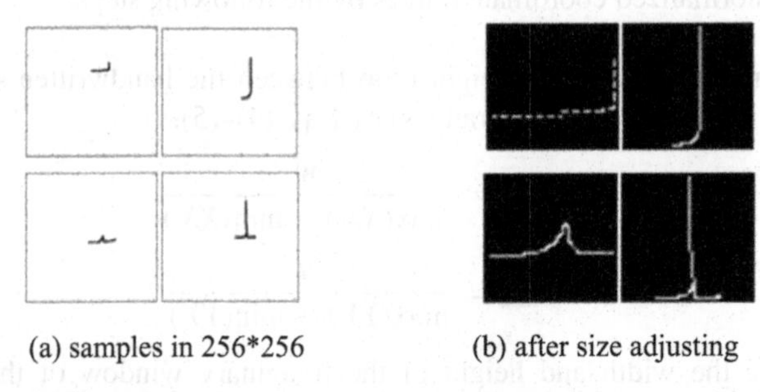

(a) samples in 256*256 (b) after size adjusting

Fig. 1. Binary images of character forms after preprocessing

4 Convolutional Neural Network Model

4.1 Convolutional Neural Networks-CNN

Convolutional Neural Networks-CNN is one the most popular deep neural network architectures and has been tremendously successful in practical applications [13]. CNN is given the first priority to tackle the pattern recognition problems with fixed number of classes. CNN has been providing very high accuracies for many computer vision tasks including handwriting recognition [14]. The main architecture of CNN consists convolution layer, pooling layer, fully connected layer and output layer for classification.

4.2 Model Architecture and Classification

We conducted some preliminary experiments using CNN models with 4, 5 and 8 layers to get higher recognition accuracy. Figure 2 shows the architecture of the convolutional networks with 8 convolutional layers, which produced highest recognition accuracy in this paper. It is consisted of 8 convolution layers and 4 pooling layers and 2 fully connected layer. Softmax layer with 128 neurons is fed with the flattened feature vector to perform classification.

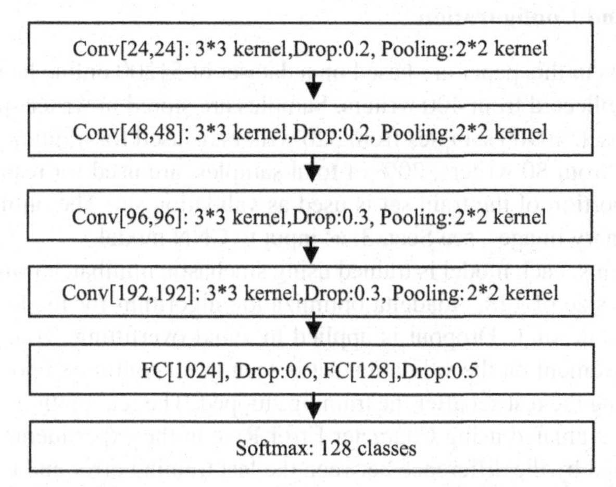

Fig. 2. The implemented CNN network structures with 8 conv layers (The number of neuron in a layer is given in brackets, for examples [24,24] means two successive layers with 24 and 24 neurouns, respectively)

All convolutional layers armed with ReLU and filter kernel size of 3 * 3 in our experiments. Lower layers of network are assumed to extract lower degree of feature-local feature, and higher layers correspond for higher level feature or global features. Keeping the image size same after convolution operation is an admirable effect of zero padding. Image or feature map size is kept same using zero padding at each convolution layer. Image size is decreased to its half via 2 * 2 max pooling, with stride of 2. All models implemented a few Fully connected layers to form more generalized features before softmax classification.

Softmax is one of the privileged classifiers with deep neural networks. It can be directly integrated to previous layers and produce a very good explainable classification results using Eq. (11).

$$P(y = i/x) = \frac{e^{x^T \cdot w_i}}{\sum_{k=1}^{K} e^{x^T \cdot w_i}} \tag{11}$$

With network parameters W, the normalized probability of the an input x to be class i in total K classes, the result at each output node reflects the normalized probability $P(y = i/x)$ of the input to be the corresponding classes. The number of output nodes is set by the number of classes and the node with the largest belonging probability is chosen to be the classification result y.

5 Experiments and Discussion

5.1 Dataset and Configuration

The experiments in this paper are based on a dataset of 51200 online handwritten character samples collected from 400 writers. Samples are stored in writer specific manner. 80% of the dataset, 40960 samples from 320 writers, is used for training. The remained 10240 samples from 80 writers, 20% of total samples, are used for testing. Randomly selected 10% portion of the train set is used as validation set. The online samples are converted to binary images, see Sect. 3, as input to CNN model.

In experiments, each model is trained using stochastic minibatch training algorithm with minibatch size of 128. Adadelta optimization algorithm for model training, with initial learning rate of 1. Dropout is applied to avoid overfitting. Training is stopped when no improvement on the validation set is seen in 10 continues epochs. The model is evaluated using the test set after the training stopped. The recognition performance of the models are evaluated using Character Error Rate in the experiments. Generality of model is reflected by the difference between the last training error and the test error.

5.2 Experiment Results

Table 2 gives the experiment records on the 48 * 48 character images with centralized 44 * 44 character forms. CNN model architectures with 4, 5 and 8 convolutional layers are compared.

Table 2. CNN training on 48 * 48 images

Num of Conv layers	Network architecture	Model size	#Epoch	Last train error	Test error
4 Conv	[C24-P2-d0.2]-[C48-P2-d0.2]-[C96-p2-D0.3]-[C192-P2-D0.3]-[FC256-D0.4]-[FC256-D0.4-FC128]-Softmax128	8.99M	45	4.28%	5.86%
5 Conv	[C24-P2-d0.2]-[C48-P2-d0.2]-[C96-p2-D0.3]-[C192-P2-D0.3]-[C384-p2-D0.4]-[FC256-D0.5]-[FC256-D0.5-FC128]-Softmax128	12.26M	68	4.36%	5.79%

(continued)

Table 2. (*continued*)

Num of Conv layers	Network architecture	Model size	#Epoch	Last train error	Test error
8 conv	[C24-C24-P2-d0.2]-[C48-C48-P2-d0.2]-[C96-C96-p2-D0.3]-[C192-C192-P2-D0.5]-[FC256-D0.5]-[FC256-D0.5-FC128]-Softmax128	14.1M	44	2.92%	5.35%

FC[24,24] means two successive convolutional layers with 24 and 24 neurouns, respectively. The implemented drop rate is noted D with a specific figure. P2 means the max pooling operation with kernel size of 2 * 2 and stride 2.

Recognition results of the CNN training experiments conducted in this paper are impressive that 4 and 5 conv layered CNN on 48 * 48 images reached as low as 5.86%, 5.79% average recognition error on 10240 test samples, which are equivalent to 94.14%, and 94.21% character recognition accuracy, respectively. Nevertheless, the training was not quite well, because the model performance on train set was not good enough due to the high drop rates implemented on layers.

Then, a deeper network which has 8 convolutional layers showed better training on train set than the other models in Table 2. The test error also reached to 5.35% which equals to 94.65% average accuracy. It is known that training neural networks requires many times of attempts and careful tuning. It is believed that the models used in this paper can be trained even better and get higher recognition accuracy.

5.3 Error Analysis

Most of the characters are recognized in high precision except only a few character forms. Some Uyghur character forms are quite simple in structure. However, the structural simplicity introduces higher similarity between different character forms and hinders to train a good classifier. The typical character forms of low precision are examined in more detail to find the most confused ones, as shown in Table 3 and Table 4.

The character samples illustrated in Table 4 are very hard to make true differentiation from a confused couple or some other similar character forms. Human eye perhaps cannot come up with true answer to most of the samples shown in Table 4, such as in No. 1–4. Any careless handwriting makes character shapes which does not belong to the expected character class. For example, most of the character samples in No. 5 of Table are already moved to another character class (the predicted one) just by a tiny difference or ambiguity at the stroke end.

The handwritten Uyghur character recognition research has been conducting on self-collected datasets which are different in total volume, number of writers and sample qualification. It is unfair to compare the experiment results from different databases. Therefore, we only refer the results reported using the same dataset to compare, as shown in Table 5. We would like to share out dataset for research activities and will make it public, soon.

Table 3. Easily confused character forms

Char (label)	char	Total misclassified	precision	Most confused char(label)	char	Ratio in misclassied
51	ﻏ	34	57%	55	ﻓ	0.85
55	ﻓ	26	68%	51	ﻏ	0.79
65	ﮓ	19	76%	66	ﮒ	1
106	ې	20	75%	111	ﯥ	0.91
111	ﯥ	27	66%	106	ې	0.95
119	ﻰ	22	72%	114	ى	0.83
124	ﻲ	36	55%	121	ي	1

Table 4. Some misclassified character samples

No.	Truth	Prediction	Some misclassified character samples
1	ﻏ	ﻓ	
2	ﻓ	ﻏ	
3	ﯥ	ې	
4	ى	ﻰ	
5	ﮒ	ﮓ	

For fair comparison, the dataset is rearranged into train and test set with ratio of 7:3, like to the works in [12, 13]. The network with 8 convolutional layers is trained and tested on the new split dataset. Table 5 compares the recognition results on 128 Uyghur character forms using same dataset. The convolutional networks implemented in this paper achieved higher recognition accuracy than the previous reported results on this dataset.

Table 5. Comparison of the recognition results using the same data

Ref.	Ratio	Train set	Test set	Feature	Classifier	Accuracy
[12, 20]	7:3	280 writers, 280 * 128 samples	120 writers, 120 * 128 samples	Center distance feature+other features	Nearest neigbor	80%
[12, 20]	7:3	280 writers, 280 * 128 samples	120 writers, 120 * 128 samples	Modified center distance feature+other features	Nearest neigbor	87%
[13]	7:3	280 writers, 280 * 128 samples	120 writers, 120 * 128 samples	NCFE-8 direction +other features	MQDF	89.08%
This paper	7:3	280 writers, 280 * 128 samples	120 writers, 120 * 128 samples	Learned features	CNN-Softmax	**94.03%**
This paper	4:1	320 writers, 320 * 128 samples	80 writers, 80 * 128 samples	Learned features	CNN-Softmax	**94.65%**

6 Conclusion

This paper studied the Implementation of Deep learning on handwritten Uyghur character recognition for the first time. One of the most welcoming deep neural network architecture-CNN is applied using converted binary images of online handwritten samples as input. CNN models with 4, 5 and 8 convolutional layers on image size 48 * 48 gave 94.17%, 94.21% and 94.65% average recognition accuracy on the test set of 10240 samples. For fair comparison with the previous reported results, the dataset is re-split into train and test sets with the same splitting used in literature. The best performed CNN model in our paper achieved higher recognition accuracy than other reported results on the dataset. The character forms with low recognition precision are investigated in more detail. Further improving the handwritten Uyghur character recognition accuracy will be the main content in next work.

Acknowledgment. This work is supported by National Key Research and Development Plan of China (2017YFC0820603), Natural Science Youth Foundation of Universities in Autonomous Region (No. XJEDU2019Y007) and Natural Science Foundation of Xinjiang (No. 2020D01C045). The first author is very much grateful to the National Laboratory of Pattern Recognition of CASIA for providing the experimental environment.

References

1. Gao, Y., Jin, L., He, C., Zhou, G.: Handwriting character recognition as a service: a new handwriting recognition system based on cloud computing. In: 2011 International Conference on Document Analysis and Recognition, Beijing, pp. 885–889 (2011)
2. Liu, C.L., Yin, F., Wang, D.H., Wang, Q.F.: Online and offline handwritten Chinese character recognition: benchmarking on new databases. Pattern Recognit. **46**(1), 155–162 (2013)
3. Okamoto, M., Yamamoto, K.: Online handwriting character recognition method using directional and direction-change features. Int. J. Pattern Recognit. Artif. Intell. **13**(7), 1041–1059 (1999)

4. Alom, M.Z., Sidike, P., Hasan, M., et al.: Handwritten Bangla character recognition using the state-of-art deep convolutional neural networks. Comput. Intell. Neurosci. **2018**, 1–3 (2017)
5. Kurban, A., Mamat, H.: Beida FangZheng Uighur text to unicode text code codeconversion. J. Xinjiang Univ. (Nat. Sci. Ed.) **23**(3), 343–347 (2006)
6. Simayi, W., Ibrayim, M., Tursun, D., et al.: Survey on the features for recognition of on-line handwritten Uyghur characters. Int. J. Signal Process. Image Process. Pattern Recognit. **8**(3), 850–853 (2015)
7. Dawut, R.: Research on the key technologies of online handwritten Uyghur word recognition, M.S. thesis, Xinjiang University (2011)
8. Simayi, W.: Online Uyghur handwritten character recognition technology based on center distance feature, MS thesis, Xinjiang University (2014)
9. Ibrayim, M.: Key technologies for recognition of on-line handwritten Uyghur characters and words. Ph.D. thesis, Wuhan University (2013)
10. Dai, X.L.: Online handwritten Uyghur character and word recognition based on the mobile platform. MS thesis, Xidian University (2012)
11. Xu, Y.M.: A study of key techniques for Uighur handwriting recognition. Ph.D. dissertation, Xidian University (2014)
12. Liu, C.L.: Normalization-cooperated gradient feature extraction for handwritten character recognition. IEEE Trans. Pattern Anal. Mach. Intell. **29**(8), 1465 (2007)
13. Lecun, Y., Bengio, Y., Hinton, G.: Deep learning. Nature **521**(7553), 436 (2015)
14. Krizhevsky, A., Sutskever, I., Hinton, G.E.: ImageNet classification with deep convolutional neural networks. In: International Conference on Neural Information Processing Systems, pp. 1097–1105. Curran Associates Inc. (2012)

Gravity Business Model Affection of One Belt – One Road Initiative

Marja-Liisa Tenhunen, Jun Chen(✉), Shiyan Xu, and Chenyang Zhao

SILC Business School, Shanghai University, Shanghai, China
marja-lisa.tenhunen@anvianet.fi, robert_chenj@aliyun.com, material@shu.edu.cn

Abstract. Since China launched the "The Belt and Road Initiative" (B&R) project in 2013, more than 70 countries have joined the project and signed a trade agreement with China. The rapid development of international trade between China and the rapid development of international trade between China and countries along the "Belt and Road" is a huge potential market, but at the macro-level, the development trend of the international economic and political situation is not ideal. The European Union and China have also broadened and deepened their relations under a complex architecture of now more than 60 political, economic, sector and people-to-people dialogue formats. The background of this paper comes from the status of trade and future development prospects.

This paper focuses on the research on the trade potential of China along The Belt and Road Initiative countries. At first, it explains the researching background, selects the data of 100 trading countries and China's import and export trade in 2008–2017 and explains the regression model, the meaning of each explanatory variable and the prediction of possible regression results. Then the data gathered in each database is substituted into the gravity model, and the stochastic frontier gravitation model is derived by reference to the stochastic frontier method. Based on this, the panel data are processed in the STATA-software by using the stochastic frontal gravitation model. The influence of variables on trade has yielded significant results, and the positive correlations of other factors except trade are obtained. Through the comparison of import and export, it proves that China is an export-oriented trading country. Export has more advantages and yet import also plays key role in the domestic economic economy development. The authors made the questionnaire research to the target students in the universities in Asia and European countries and made the analysis of the critical factors affect the perceptions of "One Belt and One Road Initiatives". The contribution of this article has set up the idea that education and E-commerce have paved the way for international economic growth.

Keywords: "One belt and One Road initiative" · International trade · Tangible and intangible service trade · E-commerce

National social science fund of China (16BJY057) project related.

H. Song and D. Jiang (Eds.): SIMUtools 2020, LNICST 370, pp. 625–634, 2021.
https://doi.org/10.1007/978-3-030-72795-6_50

1 Introduction: Challenges for European Economy and Higher Education

One Belt – One Road is based on the old Silk Road from 1500-century and represents a development strategy proposed by China's President Xi Jinping in 2013. One Belt – One Road, OBOR-Initiative focuses on connectivity and cooperation between Eurasian countries, primarily China, the land-based Silk Road Economic Belt and the oceangoing Maritime Silk Road. When Chinese leader Xi Jinping visited Kazakhstan, he raised the initiative of jointly building the Silk Road Economic Belt and the 21st-Century Maritime Silk Road. Essentially, the "belt" includes countries situated on the original Silk Road through Central Asia, West Asia, the Middle East, and Europe. The initiative calls for the integration of the region into a cohesive economic area through building infrastructure, infrastructure investment, construction materials, railway and highway, automobile, increasing cultural exchanges, and broadening trade. The strategy underlines China's push to take a larger role in global affairs by using global network. The vision is: The Belt and Road Initiatives are geographically structured roads along six (6) corridors, including the maritime road. With the Sino-American Trade dispute, the cooperation between China and European Union has been strengthening. That is the practical application of The Belt and Road Initiatives.

2 Research Objectives

The area of the initiative is primarily Asia and Europe, encompassing around 60 countries. Oceania and East Africa are also included in the project. The Belt and Road Initiative is expected to bridge the "infrastructure gap" and thus enhance economic growth across the Asia Pacific area and Central and Eastern Europe. The projects include both tangible and intangible investments. Education industry is actually a promising service industry for the project. The overseas education industry plays an important role in the culture communication and paves the way for the success of The Belt and Road Initiatives.

One Belt – One Road offers a huge amount of opportunities to apply new innovations, establish and expand new industry in Europe. For universities, the Belt and Road new research projects, which are important for co-operation between Western universities and Chinese universities. Publishing activities require a collaborative effort by the scientific community and the opportunity to participate in international scientific conferences. The most important opportunities, it offers to universities, are to take part in a wide range of domestic and international cooperation projects. One of the most important opportunities is the promotion of entrepreneurship and cooperation with existing companies and organizations. Another important opportunity is that Chinese universities can participate in various international projects and develop fruitful cooperation with international universities. These are special business economics students. They seem to know very little about the potential of such initiative, especially in Europe. Therefore, it is necessary to study the baseline.

3 Research Questions

According to the traditional International trading theory, we find that we can use Gravity trading model to explain the relationship among the different regions of The Belt and Road Initiative countries. We need to find the critical factors which affect the relationship. We limited the scope of our research to the European Union, Africa, Asia, and South American countries, and asked the following questions:

What are the critical factors to affect the relationship among the different regions of One Belt-One Road initiatives countries?

4 Literature Review

4.1 Gravity Model Application

The stochastic frontal gravitational model is widely used in the measurement of bilateral trade potential and trade efficiency. In order to make up for the deficiency of the traditional gravity model, this paper uses the stochastic frontier analysis method to study the gravity model. According to the stochastic frontier gravity model, the export scale Tin of the cross-border e-commerce of country n or region can be expressed as:

$$T_{in} = f(x_{in'})\exp(v_{in})$$

X_{in} is the core factor affecting trade volume in the trade gravity model, including population size, economic scale, geographical distance, etc.; v_{in} is a random interference term.

Model Building

This article is based on the gravity model and conducts empirical analysis based on national data. For the convenience of quantitative research, our commonly used method is to logarithm the gravity model into a bilateral logarithmic form. In actual operation, take the natural logarithm on both sides of the basic expression of the gravity model at the same time, and linearize it. According to the characteristics of cross-border e-commerce trade, based on the stochastic frontier gravity model, a model for measuring the export scale of my Country's cross-border e-commerce is established. The specific formula is as follows:

$$\begin{aligned} \ln EX_i = {} & \beta_0 + \beta_1 \ln GDP_i + \beta_2 \ln GNI_i + \beta_3 \ln POP_i + \beta_4 \ln Dis_i + \beta_5 \ln Tariff_i + \beta_6 \ln Net_i \\ & + \beta_7 \ln AC_i + \beta_8 \ln EAS_i + \beta_9 \ln LSC_i + \beta_{10} \ln ESS_i + \beta_{11} \ln Attitude_i + \beta_{12} \ln FTA_i + u_i \end{aligned}$$

Among them:

(1) EX represents China's exports to various trading partners.
(2) GDP represents the per capita gross domestic product (GDP) of major trading partner countries based on purchasing power parity (PPP).
(3) GNI represents the per capita GNI of the major trading partner countries based on purchasing power parity (PPP). The higher the per capita GNI, the higher the purchasing power level.

(4) POP indicates the total population of the major trading partner countries, which can indicate to some extent the market size and consumption potential of the exporting country.
(5) Di s represents the geographical distance between China and its major trading partners.
(6) Tariff expressed as the tariff level of major trading partners.
(7) Net represents the Internet penetration rate of major trading partner countries, that is, the proportion of netizens' population to the total population.
(8) AC (Airport connectivity) indicates the airport connectivity index of major trading partner countries. The larger the index, the closer the exporting country is to the global air transport network.
(9) EAS (Efficiency of air-transport service) indicates the air transport procedure burden index of major trading partner countries. The index is obtained by scoring the efficiency of processing the import and export trade at airports. The higher the index, the more efficient it is.
(10) LSC (Liner Shipping Connectivity Index) indicates the liner shipping connectivity index of major trading partner countries. The larger the index, the closer the exporting country is to the global shipping network.
(11) ESS (Efficiency of seaport service) indicates the customs procedure burden index of the major trading partner countries. The index is obtained by scoring the efficiency of customs processing of import and export trade by countries. The higher the index, the higher is the efficiency.
(12) Attitude (Attitudes toward entrepreneurial risk) indicates the trade risk attitude index of major trading partner countries. The higher the index, the more people think that the greater the trade risk, the less willing to engage in trade activities.
(13) The FTA expresses a free trade agreement, and the trading partner countries have free trade agreement with China remarks 1. Otherwise is 0 (zero).

4.2 Omni Channel Theory

Omni Channel Theory holds that in order to meet consumers' purchase demand in any way, enterprises should integrate various channels to provide services for customers. Ansari et al. (2008) found that customers use retailers' multiple channels to buy more products, thus improving customer satisfaction. The rise of Cross border e-commerce is providing a new channel for the sales of wigs industry. The existing literature on international trade mostly focuses on traditional mode. Shan (2016) found that potential factors such as product structure, export market and industrial specialization had a significant impact on the sales of the tangible products.

For the research of Cross border e-commerce construction, qualitative analysis methods are mostly used. Setal (2014) used the grey correlation entropy method to calculate the grey correlation entropy coefficients between different factors and Cross border e-commerce, founding that the logistics efficiency significantly affected the sales. Chang (2018) analyzed the sales data of the express platform and the development status of export Cross border e-commerce, and found the impact of logistics channels on Cross border e-commerce marketing. Zhang (2019) adopted multiple regression and found that the level of logistics development affects one of the important factors of Cross border e-commerce marketing.

Through our research, we aim to build up the bridge between the herein theories and the current achievements in economics and e-commerce in China. We discovered that we can extend Porter's theory in the new era of e-commerce. Alibaba and Jindong are branded independent distributors adopting the innovative channel and they are the extension of Michael Porter's 5 C-model. Furthermore, the service industry sets up another innovative model for e-commerce education as part of international economics.

4.3 Innovation

The existing literature mainly measures the trade potential between countries from the macro level, and there is less research based on the industry and product perspective. The previous literature focused on the research of the ten ASEAN countries, and the research objects were not comprehensive enough. This article studies the 100 trading countries along the "Belt and Road" as a whole, and the research objects are more comprehensive. This is consistent with China's trade policy, no matter how big or small a country is, it actively develops foreign trade with other countries to diversify foreign trade risks that may be brought about by changes in the situation of each country. Therefore, this article collects data on China's exports to 100 trading countries along the "Belt and Road" from 2008 to 2017. Taking 100 trading countries along the "Belt and Road" as the research object, this article intends to solve two problems: First, analysis based on gravity model The factors that affect China's export trade to countries along the "Belt and Road" provide a realistic basis for China to formulate trade policies related to countries along the "Belt and Road"; second, use the model to measure China's Export trade potential, to empirically analyze which countries China has "under-trade" and which countries have "excessive trade" in the export field in order to provide differentiated countermeasures.

5 Research Methods

Research methods include questionnaire survey and quantitative analysis. This paper uses the method of questionnaire survey, based on the current research on the factors affecting The Belt and Road Initiatives. The research tools involve different scales and types and try to make the research samples representative.

The survey was made in 2018 and 2019 in four universities in the following countries which have relationships with Shanghai University, such as ESIC (Spain), Bremen University (Germany), King University (Thailand) Christian University-Bucharest (Romania). The target group includes 20 master students major in business at each university. This questionnaire was similar to the one used for Shanghai University students in 2017 as the purpose is to make a clear comparison between them. According to the hypothesis, they should have a good understanding of the economy and foreign cooperation, including European wide project launched by China, based on their studies and life experience.

The questionnaire was conducted to the target group using a structured questionnaire in the classroom at the same time. The questionnaire included five questions and an opportunity to write their own ideas in the open box at the end. The questions are formulated into five main headings in the report, making sure that the title matches the content of the question. The method used to analyze qualitative research data is typing (classifying), where data are grouped according to similar types. The answers are accurately analyzed, searching for similarities in the answers for each classified question. The data are listed in the report by types. The result is the classification of the factors matching the answers with each question.

5.1 Sample and Data Source

This paper selects 10 major trading partners of China (United States, Russia, France, Great Britain, Brazil, Canada, Germany, Japan, South Korea, India) as research samples in 2018, and analyzes the data based on stochastic frontal gravity model, 10 data samples in all.

China's export to various trading partners come from the United Nations database. In 2018, data on per capita GDP, per capita national income, and total population of each country are derived from the World Bank's official figures of the development database (World Development Indicators); geographical distance is between the economic centers of capitals or regions and Beijing; distance (KM), data from Distance Calculator.

The World Economic Forum's 2017–2018 Global Competitiveness Report provides the following figures: the average tariff level, Internet penetration rate, airport connectivity index, air transport procedure burden index, liner shipping connectivity index, customs procedure burden index and entrepreneurial risk attitude index (Tables 1 and 2).

Table 1. Variable data 1

	EX	GDP	GNI	POP	Dis
America	1.2425E+13	62,641,000.00	63,390,000	327,167,430,000,000.00	8710
Brazil	3.976E+12	16,068,000.00	15,820,000	209,469,330,000,000.00	17598
Canada	3.195E+12	48,106,900.00	47,490,000	37,058,860,000,000.00	10615
France	9.372E+12	45,342,400.00	46,360,000	66,987,240,000,000.00	8237
Germany	2.627E+12	53,735,200.00	54,890,000	82,927,920,000,000.00	7377
India	1.704E+12	7,761,600.00	7,680,000	1,352,617,330,000,000.00	4759
Japan	2.414E+12	4,279,400.00	44,420,000	126,529,100,000,000.00	2099
Korea	1.775E+12	40,111,800.00	40,090,000	51,635,260,000,000.00	956
Russia	8.023E+12	27,147,300.00	26,470,000	144,478,050,000,000.00	5807
UK	5.964E+12	45,489,100.00	44,930,000	6,648,899,000,000.00	8161

Table 2. Variable data 2

	Tariff	NET	AC	EAS	LSC	ESS	ATER	FTA
America	76.7	64.3	33.2	56.6	NA	27.1	54.3	0
Brazil	17	60.9	89.7	57.3	35.6	34.3	51.6	0
Canada	82.3	91.2	96.3	73.8	45.4	68.4	56.7	0
France	92.4	79.3	95.8	75	72.2	66.3	46.2	0
Germany	92.4	89.6	100	77	85.9	72.4	67.5	0
India	0.8	29.5	100	64.1	52.9	60.4	62	0
Japan	86.4	93.2	100	85	66.4	77.3	53.6	0
Korea	39.7	92.8	91.7	80.6	100	72.8	47.5	1
Russia	69.5	73.1	89.2	65.3	32.2	59.7	54.8	0
UK	92.4	94.8	100	77.8	82.8	72.6	68.5	0

5.2 Model Regression Results

The least squares estimation of the constructed model was performed using the STATA-software.

(1) The coefficient of GDP per capita and GNI are both positive and highly significant. This means that the higher per capita GDP, the higher per capita gross national income of trading partner countries, that is, the more developed the economy and the stronger the purchasing power, the greater the scale of China's exports to its Cross-border e-commerce.

(2) The population (POP) coefficient is positively significant. This shows that the more people as trading partners, the greater the consumption potential, the greater the scale of China's exports to its Cross-border e-commerce.

(3) The geographical distance (Di s) has a positive coefficient, the geographical distance from trading partners has increased by 1%, and the export volume of Cross-border e-commerce in China has increased by about 7.43%. This shows that the farther the geographical distance of trading partners, the larger the export transactions of Cross-border e-commerce in China is. It is worth noting that under the traditional trade gravity model, geographical distance is an important factor hindering international trade.

(4) The coefficient of tariff level (Tariff) is negative, which indicates that the tariff level is an important factor hindering the scale of Cross-border e-commerce transactions. The higher the tariff level of trading partner countries, the smaller the scale of China's exports is to its Cross-border e-commerce.

(5) The coefficient of the Airport Connectivity Index (AC) and the liner shipping Unicom Index (LSC) is positive, which means that the proximity of the trading partners' global air and sea transportation network will promote the export scale of China's Cross-border e-commerce.

(6) The coefficient of the Air Transport Procedure Burden Index (EAS) and the Customs Procedure Burden Index (ESS) is positive indicating that the customs and airports of trading partners are more efficient in handling import and export trade. Therefore, China's exports to its Cross-border e-commerce has a bigger scale.

(7) The coefficient of the venture risk attitude index (Attitude) is negative, which indicates that the higher the risk assessment of trade in the trading partner countries, the smaller the scale of China's export transactions to its Cross-border e-commerce.
(8) The coefficient of the Free Trade Agreement (FTA) is positive, which is the same as expected. The conclusion of a free trade agreement has greatly promoted the export of Cross-border e-commerce in China.

6 Summary of the Students' Answers

The biggest risks of One Belt – One Road Initiatives are enumerated below and they should be carefully taken into account.

Environment degradation and project failure due to extremely high costs are the main risks. Espionage is a risk as well such as China spying other countries via this program. High political decisions of different governments and agreements involving China are also highly risky. Policy barriers impose strict rules. Infrastructure projects are always risky concerning quality. Huge major infrastructure projects always entail risks. The increasing amount of debt and loan in companies and countries mean risks. Also, the feeling that we are "selling our country" might be a risk too. Some countries' government will not be able to respect these rules.

Political barriers include economic risks. Cooperation among different nations and even division of benefits between participants has some risks. A correct priority of activities must be considered. European Union legislation includes some risks concerning BRI. Many of the markets may have a history of political turbulence. Policy barriers create strong borders. It would be harder for a contractor to win future projects. The price to bring goods and services from Asia to Europe is too high. The enlarging terrorism and movements trigger security risks. One solution to minimize risks is to offer timely implementation in order to get benefits. It is necessary to identify the risks regarding the transport, the environment and the amount of money. Overall the final effect will be excellent for all the involved partners. This is a win-win process. Some countries can benefit of their good political relations with China.

RBI might be successful in a more stable political environment and applicable to various countries. I think in Romania there would be some projects implemented in order to get further economic benefits.

This project can develop new opportunities for China to cooperate with various countries along the roads even though many of them are developing countries.

7 Conclusion

This paper establishes a stochastic frontal gravitational model, using the scale of China's exports to Cross-border e-commerce by 10 major trading partners in 2018 and the economic size, population size, tariff level, Internet penetration rate, liner shipping and air transport connectivity index of various trading countries as well as data indicators of the Trade Risk Attitude Index. The analysis quantifies the impact of distance factors on the scale of China's Cross-border e-commerce exports, and identifies other important factors affecting China's Cross-border e-commerce export scale and the direction and size of these impacts.

This paper finds that the traditional trade gravity model cannot explain the export scale of China's Cross-border e-commerce well because the geographical distance has a positive impact on China's Cross-border e-commerce scale. The geographical distance between China and its trading partners increased by 1%, and the export volume of Cross-border e-commerce increased by about 7.43%.

When formulating key areas for Cross-border e-commerce development, government departments should assess the economic development level, purchasing power level, consumption potential, the degree of connectivity and efficiency of sea and air transportation, and trade risk attitudes. Instead of putting the distance factor in the first place, choose to prioritize the development of neighboring national markets. At the same time, we must actively promote the establishment of a free trade zone between China's other trading partners. This will not only create a good external environment for China's Cross-border e-commerce exports, but also better promote the circulation of talents and information resources in the region in order to promote the common prosperity of all countries. Moreover, China's trading partners should work together to improve the efficiency of shipping and air transport by reducing import clearance time and simplifying procedures. Achieve a significant increase in China's export efficiency to major trading partners, thereby increasing the development of Cross-border e-commerce in China.

The survey clearly shows that the students embrace the idea of the collaboration of One Belt – One Road project, being a little bit cautious as for the eventual risks which may occur. However, if they are provided with more accurate and updated information, they will certainly be more aware of the importance and scope of this project and in the long run they might get involved in this worldwide process.

After all, this survey revealed something interesting to us - that the students in Southern American and African countries are willing to have their countries as part of One Belt-One Road Initiatives since they consider it necessary to stop relying as much on the U.S. and on the old-crippling Europe for development, and they know it is a competitive advantage that should not be disregarded. They think that it is indeed an exciting and ambitious project worth being part of. They think it works great in China for cultural and historical reasons. Anyhow, this article provides a perspective for many countries' higher education institutions to start and promote research, development and innovation (RDI)-projects regarding One Belt – One Road Initiative.

References

1. Geetha, R., Marie, M.S.: The gravity model and trade efficiency: a stochastic frontier analysis of eastern European countries' potential trade. World Econ. **37**(5), 690–704 (2014)
2. Geomina, T., Gomez-Herrera, E.: The drivers and impediments for cross-border e-commerce in the EU. Inf. Econ. Policy **28**, 83–96 (2014)
3. Martin, C., Frazer, L., Cumbo, E., et al.: Paved with Good Intentions: The Way Ahead for Voluntary, Community and Social Enterprise Sector Organisations the Voluntary Sector and Criminal Justice. Palgrave Macmillan, London (2016)
4. Tai, W.L., Katherine, H.T., Wen, X.L.: China's One Belt One Road Initiative (2016)

5. Trampe, D., Konu, U., Verhoef, P.C.: Customer responses to channel migration strategies toward the e-channel. J. Interact. Mark. **28**(4), 257–270 (2013)
6. Ussahawanitchakit, P.: Influences of knowledge acquisition and information richness on firm performance via technology acceptance as a moderator: evidence from Thai e-commerce businesses. J. Acad. Bus. Econ. **12**(1), 33–42 (2012)

Moderating Effects of Agricultural Product Category Characteristics on Consumers' Online Shopping Intention

Jiazhen Huo[1], Ying Xu[1], Jun Chen[2(✉)], Shiyan Xu[2], and Chenyang Zhao[2]

[1] School of Economic and Management, Tongji University, Shanghai 200092, China
[2] SILC Business School, Shanghai University, Shanghai 218000, China
robert_chenj@aliyun.com

Abstract. Based on the Technology Acceptance Model, this paper explores the relationship between consumers' online shopping perceived usefulness and online shopping intention, as well as the Tripartite for the five agricultural product characteristics, including price perception, purchase frequency, familiarity, thinking time and involvement. Experiment with two scenarios of high and low online shopping perceived usefulness is designed to test the hypothesis. Nine hundred sixty sample subjects' data are collected. The structural equation model is used to empirically test the variable regulation effect involving three groups of comparisons. The study finds that consumers' online shopping perceived usefulness has a positive effect on online shopping intention. Five agricultural product characteristics play a moderating effect in regulating this relationship. The tripartite of agricultural product characteristics have statistically significant differences as a moderator variable.

Keywords: Agricultural product · Moderating effect · Structural equation model · Experiment

1 Introduction

With the popularity of the Internet and terminal equipment in China, the Internet retail industry has made significant progress as a supplement to the physical retail industry. According to the data of China Internet Network Information Center, as of June 2019, the number of online shopping users in China was 639 million, with a half-year growth rate of 4.7%, accounting for 74.8% of the total Internet users [1]. From 2011 to 2018, the scale of online retail transactions in China continued to expand. However, the year-on-year growth rate of the scale of online retail transactions has shown a trend of floating decline and stabilization [2]. Simultaneously, the distribution of categories of online retail products differs from the distribution of categories of social consumer products. According to the classified sales and share data of China's retail sales of social consumer goods in 2018 [2], the top three categories of goods in China's total retail sales

National social science fund of China (16BJY057) project related.

H. Song and D. Jiang (Eds.): SIMUtools 2020, LNICST 370, pp. 635–648, 2021.
https://doi.org/10.1007/978-3-030-72795-6_51

of social consumer goods are automotive, petroleum and products, and grain, oil, food, and beverages, with the ratios at 11.51%, 5.78%, and 4.07%. In the same year, the top three product categories with the most significant online sales in China were clothing, shoes, and hats and textile products, daily necessities, and home appliances and audio-visual equipment, accounting for 25.2%, 14.4%, and 10.6% respectively, with only a small proportion of agricultural products [1]. It can be seen that some of the Chinese consumer goods retail categories, such as automobiles, petroleum and products, Chinese and Western medicine, tobacco and alcohol, construction and decoration materials, have not been sold online. At the same time, there is a difference between the categories with a higher proportion of retail value in China's total retail sales and those with a higher proportion of online shopping categories.

In this context, relevant literature in the theoretical world has pointed out to varying degrees the impact of commodity categories on consumer online shopping tendencies. Lee et al. [3] research on the online market of medical devices in China and Taiwan shows that product attributes and characteristics have a significant impact on consumers' willingness to buy online. Rudi and Çakır [4] studied the online and offline channels of healthy food and junk food when consumers buy food products, and found that consumers are more likely to buy junk food offline, while they are relatively restrained online. Yoo et al. [5] pointed out in the research on how the self-reinforcing mechanism of accessible information affects consumer purchases. Even within the same product category, there are many different categories within them. These different characteristics of products are essential and have a significant impact on consumers' online shopping decisions. These research results indicate that when exploring consumer online shopping trends, the regulatory role of commodity categories must be considered. Even in the same category of products, consumers' online shopping intentions will differ for products with different product characteristics.

Agricultural product retailing is also gradually realizing the influence of commodity categories on consumers' online shopping intentions. In the early days of the rise of online retail, e-commerce platforms pursued all categories, and a large number of brick-and-mortar retailers quickly set up online retail platforms to adapt to the new situation and list all the products in physical channels to online platforms with the reputation of "infinite shelves". E-commerce platforms and brick-and-mortar retail stores have gradually become more rational in channel category management. Brick-and-mortar retailers are more cautious about opening online businesses, and more successful emerging offline stores are paying more attention to their ability to select goods rather than hoping to increase e-commerce channels. Successful pure e-commerce companies are also more aggressive in opening physical stores. Alibaba Group has opened unmanned supermarkets since 2017; the e-commerce company Amazon has already established five offline store businesses, including bookstores, whole food supermarkets, full-time convenience stores, and Amazon Go in 2018. The sixth offline store model is equivalent to the "four-star Amazon" of a boutique [6]. These situations indicate that, whether in the theoretical world or the practice of the retail industry, there is a specific understanding that consumers' online shopping intentions will be affected by commodity categories. Research on how agricultural product categories affect consumers' online shopping intention is of great significance for expanding online sales of agricultural products, retailing theory, and practice.

When analyzing how consumers' online shopping intentions are affected by product categories, the academic tends to compare the respective strengths and weaknesses of physical and online retail channels to explain consumers' differences in online shopping intentions for different products. The original literature focused on merely analyzing the relative strengths and weaknesses of the two channels. In contrast, the recent literature analyzes channel selection decisions related to consumers, product categories, and channel adaptation from the perspective of the characteristics of the channels themselves and the advantages and disadvantages of other channels and fit analysis. For example, John et al. [7] discussed the channel design that retailers could carry out in a multi-channel environment using transaction cost methods corresponding to physical and online channels. Banerjee et al. [8] used the analysis of ideas based on online channels' advantages and disadvantages to evaluate the consumer happiness attributes of online grocery shopping sites in India. Harris et al. [9] investigated the corresponding perceived strengths and weaknesses of online shopping and traditional supermarket shopping of 871 British consumers who purchased daily necessities online and in physical stores based on behavior theory, thereby explaining different daily necessities characteristics of the channel.

In terms of the classification of product categories in the internet environment, most of the literature still uses information economics, regarding the methods, properties, and ease of searching for consumers to collect product information before purchasing different products to classify products into Taxonomy of search, experience and trust products. Chocarro et al. [10] discussed the different information sources of service goods with different characteristics such as search, experience, and trust in the process of consumer information search. They found that consumers have different information acquisition channels for different categories of service products. Other literature, from the perspective of consumers' emotions, is classified according to utilitarian goods and hedonistic goods with emotion and subjective experience. Zheng et al. [11] studied the impact of consumer browsing of hedonic and utilitarian products on mobile channels on their impulse shopping, and found that mobile channels increase consumers' impulse shopping.

It can be seen that previous research has the following three limitations. First of all, most of the literature that analyzes consumers' purchase intention for different products in online shopping channels is analyzed from the perspective of the advantages and disadvantages of online shopping channels compared to traditional physical channels, or transaction costs, which is hard to be exhaustive, and will change with the evolution of the times and technological progress; Second, in the research of online shopping channels and product categories, most of the research is temporarily limited to correlation analysis, rather than causal analysis and verification. The primary reason is mainly that in the analysis of categories and channels, the decision-making process of consumers cannot be studied as the main body of the model; Third, for the classification of commodity categories, the previous literature still uses dichotomy method in either information economics or hedonistic economics, resulting in category classification that simply considers the consumer's information search process rather than the consumer's decision-making process, making the study fail to reflect the impact of commodity categories on consumers' online shopping intention. This classification method is not suitable for agricultural products.

This article believes that under the circumstances that online channel sales are slowing down and the proportion of agricultural products online is still small, it is of great theoretical and practical significance to study the impact of agricultural product categories on consumers' online shopping intentions. This article will present hypotheses based on the literature review and use experimental methods to verify hypotheses. This article's structure is as follows: the second part is literature review; the third part builds models, puts forward hypotheses, and introduces experimental design; the fourth part analyzes data, and the fifth part deals with management discussions and deficiencies.

2 Literature Review

Technology Acceptance Model (TAM) was initially developed by Davis [12] and is considered a powerful and concise model. The TAM evolves from the psychological ABC Attitude Model, Rational Behavior Model, and Planned Behavior Model, assuming that a person's acceptance of a technology is determined by his willingness to use the technology voluntarily. This willingness, in turn, is determined by his attitude towards technology and its usefulness. An individual's attitude is formed from two beliefs in his technology use: the first belief is called perceived usefulness, that is, the use of an application system will increase the subjective possibility of his work performance; the second belief is that perception is easy to use, that is, the degree to which the user expects the target system to be used without much experience. The TAM was mainly used to detect the acceptance of new technologies by users and was used to predict the intention of online shopping after the rise of online shopping channels. If the online shopping channel is regarded as a new technology of consumer shopping, the TAM can be evolved into the consumer online shopping channel acceptance model, as shown in Fig. 1.

Valencia et al. [13] studied the online shopping intention of Colombian College students using TAM. This study validates that the TAM has a reliable and suitable effect in the interpretation and prediction of consumers' online shopping trends, and further confirms the consumer attitude, online shopping ease of use, online shopping usefulness perception, and trust in explaining online shopping trends. Ha and Nguyen [14] used the TAM to study the online shopping intention of Vietnamese consumers. Studies have found that perceived usefulness, ease of use, attitude, subjective norms, and trust positively affect consumers' online shopping intentions. Guzzo et al. [15] used social influence factors and demographic factors as the antecedents of TAM use in online channel shopping situations and studied how demographic factors affect consumers' online shopping intentions. These studies have proved that the TAM is useful in explaining online shopping behavior. However, none of the existing researches considers the target products of online shopping into the model or selects only individual products as scenarios for online shopping. Therefore, it fails to explain and verify the impact of agricultural product characteristics on consumers' online shopping decisions.

The purpose of this article is to explore the impact of agricultural product categories on consumers' online shopping propensities. The following literature illustrates five

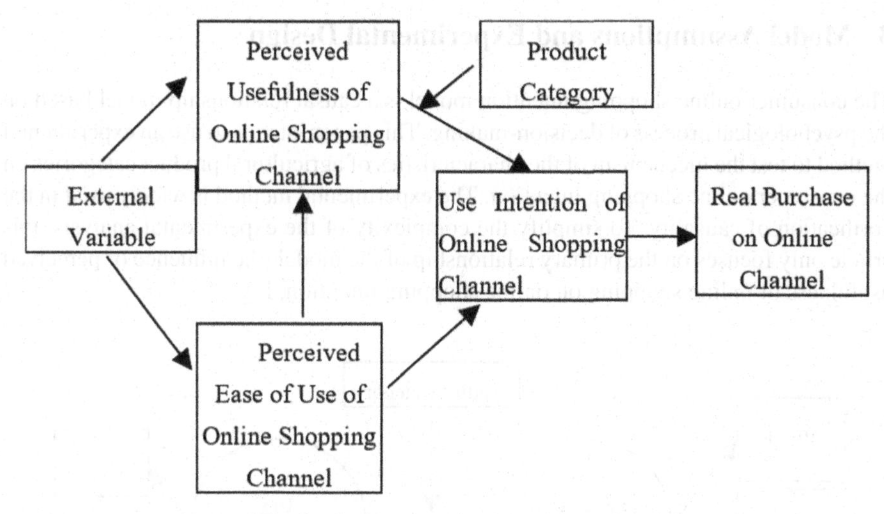

Fig. 1. Online shopping intention model

types of agricultural product categories that may affect consumers' online shopping propensities. Muhammad and Permana [16] studied the factors that affect consumers' online shopping intentions and found that there is a positive impact on product prices and consumer satisfaction on consumer online shopping tendencies. Liu and Li [17] studied the factors influencing consumers' willingness to purchase online on the mobile end. The study concluded that consumers' product familiarity has an impact on their online purchase intentions. Wen et al. [18] studied the factors that affect consumers' intention to use online shopping in China, including purchase frequency and purchase motivation, and found that as consumers' purchase frequency increases, their online shopping intention rise. Peng et al. [19] studied the impact of the time urgency of online shopping on the perceived value of online shopping and the intention of using online shopping on social platforms. The relationship played a moderating role. Ahmad et al. [20] studied the impact of consumer involvement in product purchases, lifestyles, and Internet self-efficacy on their intentions to use online shopping. They also found that consumer involvement in goods positively affects their online purchase intentions.

In summary, previous research on consumer online shopping intention using TAM has focused on the impact of demographic factors and social variables on channel use. In the literature concerning commodity categories or commodity category characteristics, the category and its characteristics are used as independent variables to study the correlation between them, thus failing to reveal the mechanism and mode of influence of commodity category characteristics on consumers' online shopping intention, and the lack of agricultural products the study. This article takes five objective category characteristics, including consumer price perception, purchase frequency, familiarity, time to think before purchase, and involvement, as examples. This article explicitly studies the mechanism and effect of commodity category characteristics on consumers' online shopping intentions.

3 Model Assumptions and Experimental Design

The consumer online shopping intention model is a causal relationship model based on the psychological process of decision-making. This paper intends to use an experimental method to test the mechanism of the characteristics of agricultural product categories on the consumer online shopping intention. The experimental method is widely used in the verification of causality. To simplify the complexity of the experimental analysis, this article only focuses on the primary relationship of the model: the influence of perceived usefulness of online shopping on online shopping intention, Fig. 2.

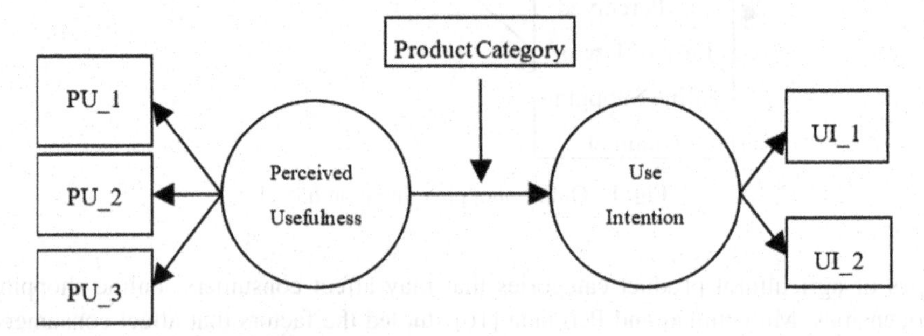

Fig. 2. Simplified model for experiment

The explanation of external factors will be analyzed in a separate paper. Ignoring the perceived ease of use variable differs from the new technology and new software fields in which TAM is widely used. The online shopping technology studied in this article is a relatively mature and straightforward application. Empirical evidence also shows that perceived ease of use has a weaker effect on behavioral intentions. The actual use of online shopping variables is not in the scope of this article, because the actual use of online shopping will be subject to objective conditions, such as the existing agricultural product e-commerce websites that can be improved but not improved, and the intensity of consumer willingness to buy [21]. This leads to Hypothesis 1.

Hypothesis 1: Consumers' perceived usefulness of online shopping has a positive effect on their online shopping intention.

According to the literature, this article selects fifteen specific agricultural products from grain and oil, dairy, vegetables and fruits, animal meat, and aquatic products, corresponding to consumers' price perception, purchase frequency, familiarity, time to think, and involvement. The five categories of characteristics must be low, medium, and high. This article argues that the causal relationship between consumers' perceived usefulness of online shopping for agricultural products and their online shopping intention is affected by the moderating effect of category characteristics. Specifically, when the perceived usefulness of online shopping is the same, the intention of shopping online affected by the characteristics of agricultural products is lower than shopping online that is not affected by the category characteristics. This leads to Hypothesis 2.

Hypothesis 2: Consumers' online shopping intention affected by specific agricultural product categories is lower than that of not affected by specific product categories.

The impact of involvement characteristics of agricultural products with different levels on consumers' online shopping intention is analyzed as an example below. According to the consumer's involvement in agricultural products at the time of purchase, agricultural products are classified into low-involved agricultural products, medium-involved agricultural products, and high-involved agricultural products, and specific commodities are selected as representatives from aquatic agricultural products. For the daily aquatic product purchases that must be completed, struggling to search will not be the choice of most consumers, so consumers are expected to have a lower tendency to purchase such agricultural products through online channels. For aquatic products that require time to choose, such as cod, prawns, salmon, etc., consumers will generally be willing to spend more energy searching for products that they need and cost-effective compared to lower-involved aquatic products. This article predicts their online shopping intention would be high. For lobsters, cods, crabs, etc., which require high freshness, consumers may be willing to invest great interest in searching and enjoying the process of careful selection. Therefore, consumers are expected to have lower online shopping intentions. This leads to Hypothesis 3.

Assumption 3-a/b/c/d/e: subject to three different levels of price perception (a), purchase frequency (b), familiarity (c), time to think before buying (d), As well as involvement (e), there is a statistically significant difference in the path coefficient of the relationship between perceived usefulness of online shopping and online shopping intention.

To avoid the highly concentrated perceived usefulness of online shopping perceived by participants participating in the experiment, and to form high and low levels of perceived usefulness of online shopping in the data to facilitate data analysis, we designed two scenarios that can affect consumers' perceived usefulness of online shopping. Six new subjects were asked to test and evaluate. The test results show that the two experimental scenarios have different guidance on the perceived usefulness of online shopping of the experimental subjects.

We used the experimental questionnaire group that did not mention specific agricultural product categories as the control group. The group that asked about the online shopping propensity variable for 15 specific commodities in the questionnaire as the experimental group, a total of 16 groups. There were 60 participants in each group. Half of the random participants received a questionnaire guided by low online shopping perceived usefulness, and the other half received a questionnaire guided by great online shopping perceived usefulness. According to the PLS Handbook [22], this experimental model involves a total of 2 variables, and each group of 20 samples can support the analysis. Each experimental questionnaire has a total of 5 items. The first three items are a scale for the perceived usefulness of online shopping, and the last two items are a scale for online shopping intention. The perceived usefulness of online shopping uses Frasquet et al. [23] scale and online shopping use the Venkatesh et al. [24] scale. Both scales are reflection scales, using 7 Likert scales. To ensure the objectivity and non-interference of the experiment, each subject can only see one type of questionnaire. At the same time, the subjects are told that they are participating in the online shopping willingness questionnaire without any knowledge of the experiment's objective.

The experiment uses the WJX website, which is distributed and collected in Shanghai through WeChat within two weeks from September 12 to September 26, 2019. Because the experimental questionnaire itself is relatively simple, with only five items, and no personal information is collected, the experimental recovery rate reaches 100%. A total of 960 valid questionnaires have been recovered. There is no defect value in the data.

4 Data Analysis

The data collection place for this experiment is located in Shanghai. The demand, acceptance, and penetration rate of online shopping in this area is generally high. At the same time, this experimental questionnaire was transmitted via WeChat, so we expect that the usefulness of online shopping perception in experimental data will be generally high. The distribution of data will deviate from the normal distribution. Therefore, this study uses Partial Least Squares (PLS) software to analyze the structural equation model's data. The advantage of PLS is that it can appropriately relax the normal distribution requirements of the data, and is suitable for data analysis with a small amount of data in each group. This paper uses SmartPLS 2 software for data analysis.

Each agricultural product characteristic that needs to be compared in this study involves three levels of the high, medium, and low, that is, for each product characteristic, we need to perform three sets of comparisons to verify whether the differences in the three groups of path coefficients are statistically significant. However, the PLS software can only perform a comparison between two groups, so we use the Omnibus Test of Group differences (OTG) method proposed by Sarstedt et al. [25], Chin et al. [26] for multi-group comparison and the Double Bootstrap by Shi et al. [27] using Programming in R to verify the statistical significance of the hypothesis that agricultural product category characteristics are used as moderators.

4.1 Preliminary Data Analysis

From the recovered data, the perceived usefulness is generally consistent with the experimental contextual guidance expectations; that is, the low contextual guidance group is lower than the high contextual guidance consumer online shopping perceived usefulness. In the control group without corresponding variables for specific agricultural products, the average value of consumers' online shopping intention (6.02) and the average value of perceived usefulness of online shopping (6.04) is almost at the same level. Consumers' judgment of online shopping intention and perceived usefulness are almost identical. Data from five experimental groups of agricultural product characteristics show that, at three different levels, the mean value of the perceived usefulness is basically at the same level. However, after the corresponding variables are inquired about specific agricultural product categories, the average value of consumers' online shopping intention is significantly lower than the corresponding average value of perceived usefulness by about 1 to 2 points. The difference is enlarged. Compared with the control group, consumers in the experimental group affected by agricultural product categories tend to have lower online shopping intentions than those in the control group who are not affected by agricultural product categories. This change is due to the moderating effect of agricultural product categories. Hypothesis 2 is verified.

Judging from the slope formed by the average value of consumers' online shopping intention in each of the agricultural product feature groups in the low context guidance group and the high context guidance group, intuitively speaking,

Under the influence of three different levels of all five category characteristics, consumers in the case of high and low perceived usefulness of online shopping have reported differences in their online shopping intention. In general, consumers with higher online shopping perceived usefulness tend to gain more than those with low online shopping perceived usefulness. The correlation coefficient of the effect of perceived usefulness on online shopping intention, and whether there are statistically significant differences between online shopping intention at three different characteristic levels of each agricultural product characteristic, needs to be calculated by structural equation calculation using PLS software and R programming.

4.2 Econometric Model Evaluation

The results of SmartPLS 2 show that the external load coefficients of all groups' independent and dependent variables are greater than the minimum standard of 0.708, indicating that the indicators of all variables are reliable. The composite reliability of independent and dependent variables is higher than the standard 0.708, which means that all variables' internal consistency reliability is high. The Average Variance Extracted (AVE) index of independent and dependent variables of all groups is higher than the standard 0.50, which means that all variables have good convergence validity. According to the Fornell-Larcker standard, the square roots of the independent and dependent variable's AVE of all groups are greater than their correlation coefficients, indicating that all variables enjoy good discriminant validity.

4.3 Structural Model Evaluation

Since this experimental model has only one independent variable and one dependent variable, there is no collinearity problem. The determination coefficient R^2 of endogenous variables for online shopping in all groups is higher than the standard of R^2 value of 0.20 in consumer behavior research. It can be said that the model predicts endogenous variables with higher accuracy. The path coefficients affected by the independent variables of all groups are positive and greater than the standard value of 0.20, indicating that online shopping's perceived usefulness has a positive effect on online shopping intention. Hypothesis 1 is verified.

To test the statistical significance of the path coefficients, we performed 5000 times bootstrapping calculations on each group's experimental data. The results show that the T-test values of the calculation were higher than the two-sided test T critical value of 2.57 (Significance level of 1%), showing that the path coefficients of each group of the model are significant. To evaluate the predictive correlation of the model, we ran a Blindfolding program with a missing distance of 7, and the results showed that Q^2 values were all greater than 0, which meant that the corresponding variables of the model had predictive correlation. Table 1 shows the index values of the calculation results of each group.

Table 1. The PLS results for the model.

Group		Path coefficient	R^2	T	Q^2
Control Group		0.8085	0.6537	9.9896	0.5741
Price Perception	Low	0.4272	0.1825	3.6124	0.1592
	Medium	0.6452	0.4162	9.6754	0.3678
	High	0.3705	0.1373	4.9041	0.1225
Purchase Frequency	High	0.3772	0.1423	4.3453	0.0685
	Medium	0.6669	0.4448	8.2632	0.4105
	Low	0.3834	0.1470	3.4658	0.1233
Familiarity	High	0.6110	0.3733	6.3149	0.3462
	Medium	0.4681	0.2191	3.0397	0.1958
	Low	0.7518	0.5652	8.5393	0.5321
Thinking Time	Low	0.6096	0.3716	7.1812	0.3471
	Medium	0.7837	0.6142	11.6271	0.5661
	High	0.4308	0.1856	4.9253	0.1813
Involvemetn	Low	0.7778	0.6050	10.8696	0.5617
	Medium	0.8574	0.7352	23.9951	0.7109
	High	0.5155	0.2658	5.5403	0.2353

To determine whether the independent variable's path coefficients on the dependent variable of each category group at every level have statistically significant differences, we use a multiple-group comparison OTG method (B = 5000, U = 5000). The FR of the price feature group is 11998.47, the FR of the purchase frequency feature group is 13793.46, the FR of the familiarity feature group is 5806.89, the FR of the think time feature group is 22634.39, the FR of the involvement feature group is 32721.87, and the P values are all 1. It is shown that all the differences between the five feature groups are significant at $p \leq 0.01$. At least one set of path coefficients of each agricultural product characteristic of the relationship between the perceived usefulness and online shopping intention is different from the other two groups. Hypothesis 3-a/b/c/d/e is verified.

To determine which group is significantly different from the other groups in the three levels of each category feature, we use the non-parametric Double Bootstrap simplified calculation method (B1 = B2 = 500) to use R programming to calculate confidence intervals for each group of path coefficients obtained. Table 2 lists the confidence intervals for each category characteristic group's 95% correction rate and the corresponding multi-group comparison analysis results. If the path coefficients of one group (Table 1) do not fall within the confidence interval of the other group, and vice versa, then we judge that the path coefficients of the two groups have a statistically significant difference at this level of significance. Table 2 shows that there was a significant difference in the path coefficients for medium and high price product groups, high and medium frequency product groups, medium and low-frequency product groups, medium and low familiar product groups, each commodity group in think time and involvement (significance level 95%). The path coefficients of the other groups were not significantly different at a significant level of 95%.

Table 2. Rectifying Deviation at 95% Confidence Interval (Shi 1992 methodology) and Multi-group Comparison Results.

Group	Confidence Interval 95%			Comparison	Significant @ 5%
Control Group	[0.7230153,0.8754204]				
	Low PP	Medium PP	High PP		
Price Perception	[0.07981551, 0.75299025]	[0.4205368, 0.8200263]	[0.1571652, 0.6151002]	Low PP vs. Medium PP	Not Sig
				Low PP vs. High PP	Not Sig
				Medium PP vs. High PP	Sig
	High Fr	Medium Fr	Low Fr		
Frequency	[0.3366327, 0.4571384]	[0.5913473, 0.7318270]	[0.2737296, 0.4878295]	High Fr vs. Medium Fr	Sig
				High Fr vs. Low Fr	Not Sig
				Medium Fr vs. Low Fr	Sig
	High Fa	Medium Fa	Low Fa		
Familiarity	[0.5323346, 0.6992964]	[0.3861742, 0.7148848]	[0.6977097, 0.8564919]	High Fa vs. Medium Fa	Not Sig
				High Fa vs. Low Fa	Not Sig
				Medium Fa vs. Low Fa	Sig
	Low TT	Medium TT	High TT		
Thinking Time	[0.5420411, 0.6900313]	[0.7227880, 0.8329695]	[0.3538202, 0.5241846]	Low TT vs. Medium TT	Sig
				Low TT vs. High TT	Sig
				Medium TT vs. High TT	Sig
	Low I	Medium I	High I		
Involvement	[0.7305404, 0.8287949]	[0.8352369, 0.8914794]	[0.4472404, 0.5803213]	Low I vs. Medium I	Sig
				Low I vs. High I	Sig
				Medium I vs. High I	Sig

5 Conclusion and Outlook

5.1 Analysis of Conclusion

Through an experimental design, this paper uses 960 sample data to study the moderating effect of agricultural product category characteristics on consumers' online shopping intentions. The results show that: (1) The customer online shopping perceived usefulness has a positive effect on online shopping intention. (2) The online shopping intention affected by the characteristics of agricultural products is lower than that of not affected by the characteristics of agricultural products. (3) The moderating effect of agricultural product category characteristics on consumers' online shopping intention is statistically significant.

5.2 Management Inspiration

This paper validates the causal relationship between online shopping perceived usefulness and online shopping intention in the hypothetical model, and the moderating effect of the five category characteristics of agricultural products on the above causal relationship. The results of this article show that, in the context of increasingly mature online shopping channels and high perceptions of consumers' usefulness of online shopping, agricultural product retailers must fully consider the impact of agricultural product categories on consumer online shopping intention in the formulation of retail competition strategies, to take a comprehensive and effective analysis of features to avoid blind distribution.

At the same time, by verifying that there are statistically significant differences between the three levels of agricultural product category characteristics, the results of this paper also show that the trichotomy method of agricultural products has use value in analyzing consumers' intention to use online shopping channels. Compared with the dichotomy method, the trichotomy method for category characteristics more accurately describes the differences and changes in consumers' intention to purchase different types of agricultural products through online channels. The purchase frequency, familiarity, thinking time, and involvement characteristics of agricultural products show the highest online shopping intention for agricultural products with a medium level of characteristics compared with other characteristic levels. As for the characteristics of price perception, consumers tend to have a higher online purchase intention for agricultural products with medium and low price perception levels. Therefore, the retailer can make decisions on the sales channels according to the different category characteristics of the agricultural products sold. For example, if a retailer sells agricultural products with a medium or low price perception, or a medium purchase frequency, familiarity, thinking time, or level of involvement, then sales on the online channel will be better than on the physical channel. Conversely, if the agricultural products sold belong to high price perception, purchase frequency, familiarity, thinking time, involvement level, or low purchase frequency, familiarity, thinking time, and involvement level, the sales in physical channels will be more than that of the shop online.

Judging from the size of each group's path coefficients, the path coefficients of the medium-price perception, purchase frequency, thinking time, involvement, and low familiarity of the agricultural product group are larger than in other level groups with corresponding characteristics. This shows that for agricultural products with these characteristic levels, the online shopping perceived usefulness has a greater effect on online shopping intention than other levels of corresponding characteristics. If a retailer sells agricultural products that belong to the medium price perception, frequency of purchase, thinking time, involvement, and low familiarity group, and also choose to sell online. Efforts to improve the perceived usefulness will greatly promote consumers' online shopping intention for these agricultural products, and get more results with less effort. If the retailer sells agricultural products with other levels of category characteristics, then even if the company strives to increase the level of perceived usefulness, it will seem an extremely limited increase in online shopping intention. The retailer should consider changing other variables that affect the online shopping intention to promote sales in online channels.

5.3 Future Research

Future related research can be carried out in the following aspects. First, consumers' online shopping intentions for different types of agricultural products change over time. This is a long-term process. Due to the limited time of this study, this evolutionary process cannot be tracked. Future research will try to track this process. Second, the experimental data were collected in Shanghai. Shanghai is a more economically developed region in China and enjoys mature online shopping. From a representative perspective, consumers in Shanghai are challenging to represent the situation of the entire Chinese consumer. Future research can consider expanding the geographic scope of research. Third, this study only selects categories of agricultural products for experiments. It does not involve whether the research model is universal to other commodities, so the research's general applicability needs to be strengthened. Future research can try to expand the range of product selection. It can also use more rigorous methods, such as clustering analysis to obtain classifications based on consumer perception of product characteristics.

References

1. China Internet Network Information Center. 44th Statistical Report on Internet Development in China (2019)
2. Official Website of China Statistics Bureau. 2018 China Statistical Bulletin (2019)
3. Lee, W.I., Cheng, S.Y., Shih, Y.T.: Effects among product attributes, involvement, word-of-mouth, and purchase intention in online shopping. Asia Pacific Manag. Rev. S1029313216303931 (2017)
4. Rudi, J., Çakır, M.: Vice or virtue: how shopping frequency affects healthfulness of food choices. Food Policy **69**, 207–217 (2017)
5. Yoo, B., Jeon, S., Han, T.: An analysis of popularity information effects: field experiments in an online marketplace. Electron. Commer. Res. Appl. **1**(17), 87–98 (2016)
6. https://www.cnbc.com/2019/03/01/grocery-store-stocks-take-a-beating-on-report-that-amazon-will-launch-its-own-chain-of-supermarkets.html
7. John, G., Viswanathan, M., Ghosh, M.: A transaction cost approach to channel design with application to multichannels settings. In: Handbook of Research on Distribution Channels 2019 22 February. Edward Elgar Publishing
8. Banerjee, T., Banerjee, A.: Web content analysis of online grocery shopping web sites in India. Int. J. Bus. Anal. **5**(4), 61–73 (2018)
9. Harris, P., Riley, F.D., Riley, D., Hand, C.: Online and store patronage: a typology of grocery shoppers. Int. J. Retail Distrib. Manag. **45**(4), 419–445 (2017)
10. Chocarro, R., Cortinas, M., Villanueva, M.L.: Different channels for different services: information sources for services with search, experience and credence attributes. Serv. Ind. J. **9**, 1–24 (2018)
11. Zheng, X., Men, J., Yang, F., Gong, X.: Understanding Impulsive Buying in mobile commerce: an investigation into hedonic and utilitarian browsing. Int. J. Inf. Manag. **48**, 151–160 (2019)
12. Davis, F.D.: Perceived usefulness, perceived ease of use, and user acceptance of information technology. MIS Q. **13**(3), 319–340 (1989)
13. Valencia, D.C., Alejandro, V.A., Bran, L., Benjumea, M., Valencia, J.: Analysis of e-commerce acceptance using the technology acceptance model. Scientific papers of the University of Pardubice. Series D, Faculty of Economics and Administration. 45/2019 (2019)
14. Ha, N., Nguyen, T.: The effect of trust on consumers' online purchase intention: an integration of TAM and TPB. Manag. Sci. Lett. **9**(9), 1451–1460 (2019)
15. Guzzo, T., Ferri, F., Grifoni, P.: A model of e-commerce adoption (MOCA): consumer's perceptions and behaviours. Behav. Inf. Technol. **35**(3), 196–209 (2016)
16. Muhammad, F., Permana, H.: Influence of information technology, quality of service, trust, price, customer satisfaction on e-commerce. Qual. Serv. Trust Price Cust. Satis. E-Commerce (2019)
17. Liu, D., Li, M.: Exploring new factors affecting purchase intention of mobile commerce: trust and social benefit as mediators. Int. J. Mobile Commun. **17**(1), 108–125 (2019)
18. Wen, X., Li, Y., Yin, C.: Factors influencing purchase intention on mobile shopping web site in China and South Korea: an empirical study. Tehnički vjesnik. **26**(2), 495–502 (2019)

19. Peng, L., Zhang, W., Wang, X., Liang, S.: Moderating effects of time pressure on the relationship between perceived value and purchase intention in social E-commerce sales promotion: considering the impact of product involvement. Inf. Manag. **56**(2), 317–328 (2019)
20. Ahmad, W., Attiq, S., Ahmad, A., Ilyas, A., Kulsoom, K.: Investigating the impact of consumer's involvement, risk-taking personality, internet self-efficacy, life style and privacy concern on online purchase intention and shopping adoption. Pakistan Bus. Rev. **20**(3), 582–599 (2019)
21. Weyden, W.: Introduction to Psychology (9th edition of the original book) (US). Mechanical Industry Press (2016)
22. Handbook of Partial Least Squares, pp. 171–193. Springer, Heidelberg (2010). https://doi.org/10.1007/978-3-540-32827-8
23. Frasquet, M., Mollá, A., Ruiz, E.: Identifying patterns in channel usage across the search, purchase and post-sales stages of shopping. Electron. Commer. Res. Appl. **14**(6), 654–665 (2015)
24. Venkatesh, V., Davis, F.D.: A theoretical extension of the technology acceptance model: four longitudinal field studies. Manage. Sci. **46**(2), 186–204 (2000)
25. Sarstedt, M., Henseler, J., Ringle, C.M.: Multigroup analysis in partial least squares (PLS) path modeling: alternative methods and empirical results. Soc. Sci. Electron. Publish. **22**, 195–218 (2011)
26. Chin, W.W., Dibbern, J.: An introduction to a permutation based procedure for multi-group PLS analysis: results of tests of differences on simulated data and a cross cultural analysis of the sourcing of information system services between Germany and the USA. In: Esposito Vinzi, V., Chin, W., Henseler, J., Wang, H. (eds.) Handbook of Partial Least Squares. Springer Handbooks of Computational Statistics, pp.171–193. Springer, Heidelberg (2019). https://doi.org/10.1007/978-3-540-32827-8_8
27. Shi, S.G.: Accurate and efficient double-bootstrap confidence limit method. Comput. Stat. Data Anal. **13**(1), 21–32 (1992)

A Novel Luby Transform Code with Improved Ripple Size

Dai Weina[1] and Zhao Yuli[2(✉)]

[1] College of Medicine and Biological Information Engineering, Northeastern University, Shenyang, China
20175565@stu.neu.edu.cn

[2] Software College, Northeastern University, Shenyang, China
zhaoyl@swc.neu.edu.cn

Abstract. Degree distribution of the encoded symbols is a critical factor that affects the performance of LT codes. Inspired by the SF-LT and the decreasing ripple size LT codes, we propose an SF-DRS degree distribution to construct LT codes, i.e., SF-DRS LT codes. Theoretical analysis proves that our proposed LT codes possess the property that the ripple size decreases as the decoding process continues. Moreover, with an overhead factor of slighter larger than one, the ripple size of SF-DRS LT codes remains larger than one in the entire decoding process. The performance of our proposed LT codes is compared to SF-LT codes and decreasing ripple size over a perfect channel. Simulation results reveal that the proposed SF-DRS LT codes outperform the decreasing ripple size LT codes with respect to the probability of successful decoding, the average overhead factor with a relatively large number of input symbols. Moreover, in contrast to the SF-LT codes, the SF-DRS LT codes achieve a better probability of successful decoding.

Keywords: Luby transform codes · Fountain codes · Degree distribution

1 Introduction

It has been proven that fountain codes, including Luby transform codes [1, 2], raptor codes [3], online codes [4, 5], could achieve excellent performance over time-varying channels and highly impaired channels [6]. Fountain codes could potentially be applied in the field of scalable video transmission [7], deep communications [8], and wireless relay networks [9, 10].

Luby transform codes, LT code for short, is a typical type of fountain codes. Given K input symbols, LT codes could generate an unlimited number of encoded symbols. When slighted larger than K encoded symbols are received, the decoder could successfully recover the original input symbols [6]. The degree distribution and the short-cycle structure in the Tanner graph are two factors that affect the performance of LT codes the most. The LT codes using ideal soliton degree distribution theoretically emerge one degree-1 encoded symbol at each iteration in the decoding process [1]. However, due to

H. Song and D. Jiang (Eds.): SIMUtools 2020, LNICST 370, pp. 649–658, 2021.
https://doi.org/10.1007/978-3-030-72795-6_52

the randomness, any small fluctuation of the number of degree-1 encoded symbols will lead to the failure of the whole decoding process. Therefore, LT codes based on ideal soliton distribution are not available in practice [3]. The LT codes using robust solution degree distribution are the de facto codes discussed in recent researches.

Based on the robust soliton distribution, by adjusting the proportion of degree-1, degree-2 and the maximum degree, Yen et al. derive a modified robust soliton distribution [11]. The LT codes using modified robust soliton distribution is aimed at increasing the mean of the expected ripple size to improve the ripple evolution. In [12], Zhu et al. theoretically analyze the LT BP (belief propagation) decoding process and propose an optimal degree distribution algorithm. Based on this for simplicity, a "suboptimal" degree distribution is proposed to improve the coding and decoding performance. Zhao et al. analyze the LT decoding process with the perspective of complex networks and propose a scale-free Luby transform code, SF-LT code for short [2]. This code is based on the fact that the iteration process between input symbols and encoded symbols in the Tanner graph can be seen as a message traveling process in a complex network. Complex networks with scale-free degree distribution have the shortest path length compared with other similar scale-size complex networks. In [13], Savchenko et al. propose using deep reinforcement to learn an optimal degree distribution of LT based code. In [14, 15], Jesper et al. have proved that the ripple size of a well-performed LT code should decrease with the decoding iteration process. The degree distribution proposed by the authors could further reduce the number of encoded symbols required for decoding and significantly improving the performance.

Other schemes combing improved degree distribution and the selection of input symbols when generating an encoded symbol are also proposed. In [16], Yen et al. show that in LT code, most of the decoding termination is caused by no ripple in the early stage. Hence, the authors modify the robust Soliton distribution and introduce a non-repetitive encoding scheme to avoid repeated degree-1 encoded symbols. This scheme requires a smaller number of encoded symbols to get a larger decoding probability. In [17], the authors modify the encoding process to maximize the minimum degree of the input symbols. This scheme improves the robustness of the LT codes over erasure channels. However, the encoding complexity increases with the enlarged average degree of the encoded symbols.

Inspired by scale-free LT codes proposed in [2] and LT codes with decreasing ripple size, we combine the modified scale-free distribution and the degree distribution proposed in [14] to construct a novel LT code with balanced decreasing ripple size. The proposed method could improve the decoding performance including average overhead factor and probability of successful decoding. Moreover, the complexity also outperforms the codes in [14]. The rest of this paper is organized as follows. Section 2 reviews the principle of LT codes. Section 3 provides the degree distribution of the encoded symbols we proposed and its theoretical analysis. Simulation results are shown in Sect. 4. Finally, Sect. 5 concludes the paper.

2 Review of the LT Codes

Given the set of input symbols $\{s_1, s_2, \ldots, s_{K-1}, s_K\}$ and the degree distribution of the encoded symbols $\Omega(x)$, the ith encoded symbol c_i is generated as follows.

1) Select a degree according to the degree distribution and denote it as d_i.
2) uniformly select d_i input symbols $\{s_{i1}, s_{i2}, s_{i3}, \ldots s_{id_i}\}$;
3) the encoded symbol c_i is assigned with $\sum_{j=1}^{d_i} s_{ij}$.

Repeat step 1)–3), encoded symbols can be continually generated. Then, taking the input symbols as one node set and the encoded symbols as the other node set, a tanner graph can be constructed according to the encoding process. Only if input symbol s_i is taken as one selected symbol for generating encoded symbol c_j, there exists an edge between node s_i and c_j in the Tanner graph. Figure 1 shows a Tanner graph with 5 input symbols and 6 encoded symbols.

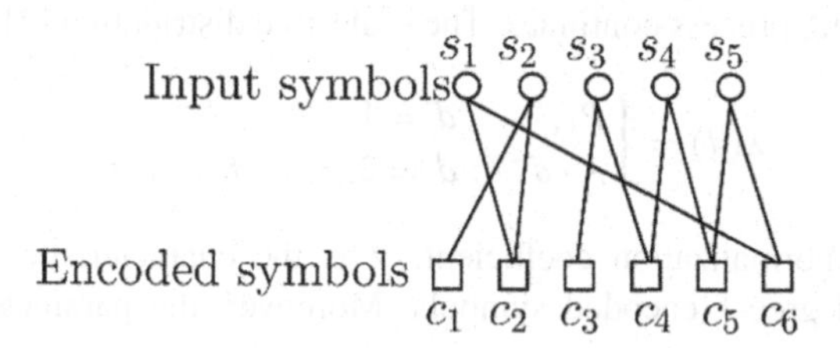

Fig. 1. A Tanner graph with 5 input symbols and 6 encoded symbols.

The BP decoding process over the erasure channel could be considered as a message-passing process. When N encoded symbols have been received (N is slightly larger than K), the decoder starts to recover the input symbols. A particular set that contains all degree-1 encoded symbols is defined as a ripple. The BP decoding process could be described as follows.

1) find all degree-1 encoded symbols and put them in the ripple set;
2) randomly select an encoded symbol c_j from the ripple set.
3) the neighbor input symbol (s_i) of the selected encoded symbol is recovered immediately by copying the value of c_j to the input symbol. The edge that connects these two symbols is removed from the tanner graph.
4) updating the value of the encoded symbols that connect with the newly recovered input symbol by obtaining the XOR value of the encoded symbol and the input symbol; remove the edges connecting the newly recovered input symbol and its neighboring encoded symbols.
5) repeat 1)–4), until all input symbols have been recovered. Then, the decoder informs the successful decoding event to the sender.

3 Proposed SF-DRS LT Code and Its Theoretical Analysis

3.1 SF-DRS Degree Distribution

In [14], Jesper et al. proved that an encoded symbol with a higher degree is more likely to be redundant than lower degree ones. Early in the decoding process, the low degree encoded symbols decrease their degree to one. At this stage, an input symbol has a relatively low probability of being repeatedly added to the ripple set. However, when high degree encoded symbols are becoming degree-1 ones late in the decoding process, the input symbol neighbored with the degree-1 encoded symbol is more likely existing in the ripple set already. Thus, the ripple size should be decreasing as the decoding process continues. Moreover, it has been proven that in contrast to other same scale networks, scale-free networks possess the lowest short-length path property. But, the ripple size of LT codes using modified scale-free degree distribution does not decrease with the iteration of the decoding process.

In this section, we add the degree distribution proposed in [14] to the scale-free degree distribution in [2] and obtain an SF-DRS distribution. The objection of the addiction is to get a distribution both possessing scale-free property and having a decreasing ripple size when decoding process continues. The scale-free distribution [2] is given by

$$\lambda(d) = \begin{cases} P_1, & d = 1, \\ A \cdot d^{-\gamma}, & d = 2, 3, \cdots K-1, K. \end{cases} \tag{1}$$

Where A is the normalization coefficient, γ is the characteristic index, and P_1 is the fraction of degree-1 encoded symbols. Moreover, the parameters in (1) satisfy $\sum_{d=1}^{K} \lambda(d) = 1$.

The degree distribution of LT codes proposed by Jesper in [14] is given by

$$\begin{cases} \theta(1) = \frac{R}{n}, \\ \theta(2) = \frac{K(K-1)}{2n(K-R)}, \\ \theta(i) = \frac{i-1}{i}\theta(i-1),\ i < \frac{K}{3}, \\ \theta(i) = \theta(i-1),\ \frac{K}{3} \le i < \frac{2K}{3}, \\ \theta(i) = \frac{K-i+1}{K-i}\theta(i-1),\ \frac{2K}{3} \le i \le K-R+1. \end{cases} \tag{2}$$

Where the parameters n and R should satisfy $\sum_{d=1}^{K} \theta(d) = 1$.

Our proposed SF-DRS distribution is obtained by normalizing (1) and (2), and expressed as

$$\varphi(d) = \frac{\lambda(d) + \theta(d)}{\sum_{t=1}^{K} \lambda(t) + \theta(t)}, \quad d = 1, 2, 3, \ldots, K. \tag{3}$$

The LT codes which degree distribution of encoded symbols follows the SF-DFS distribution is named as SF-DFS LT codes. Figure 2 shows the degree distribution curve of encoded symbols in an SF-DFS LT codes when $P_1 = 0.1$, $\gamma = 1.9$, $n = 1049$, $R = 20$, $K = 1000$. It can be seen that a large number of encoded symbols are low degree nodes. Hence, the decoder could start decoding and iteratively recover the input symbols. With some high degree encoded symbols, we can guarantee that all the input symbols participate in the generation process of encoded symbols.

3.2 Ripple Set Analysis

Define overhead factor α as the ratio of the number of encoded symbols and the number of input symbols. Assume N encoded symbols are received, the overhead factor is calculated

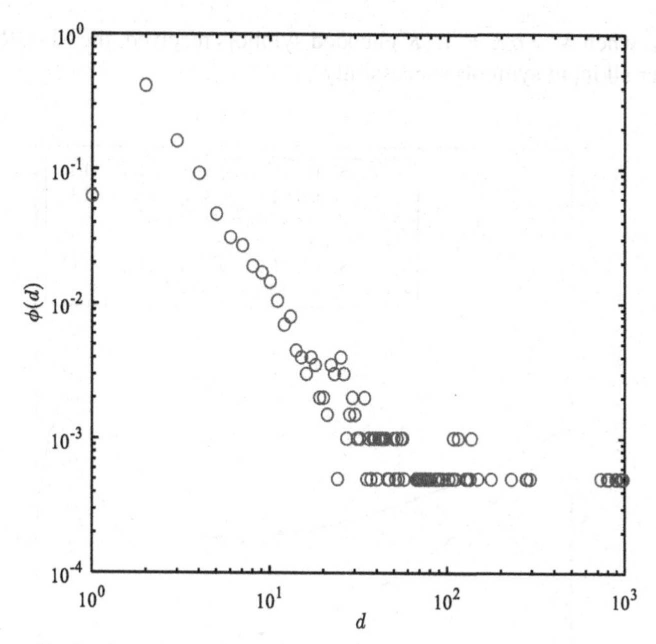

Fig. 2. Degree distribution of encoded symbols in SF-DFS LT codes. $P_1 = 0.1$, $\gamma = 1.9$, $n = 1048$, $R = 20$, $K = 1000$.

by $\alpha = N/K$. Given the degree distribution of the encoded symbols, the evolution of the ripple size [14] can be calculated by

$$\begin{cases} \eta^K(i) = \alpha K \varphi(i), i = 1, 2, \ldots, K; \\ \eta^{L-1}(1) = \eta^L(1) - 1 + \frac{2(L-\eta^{(L)}(1))}{L(L-1)} \eta^{(L)}(2) \\ \eta^{L-1}(i) = \eta^L(i) - \frac{i}{L}\eta^L(i) + \frac{i+1}{L}\eta^L(i+1), i = 2, 3, \ldots, L-1; \\ \eta^{L-1}(L) = 0. \end{cases} \tag{4}$$

Where $\phi(i)$ is the degree distribution of encoded symbols; $\eta^L(i)$ is the number of degree-i encoded symbols left when L input symbols are not recovered. Equation (4) is based on the assumption that each input symbol only has one opportunity to be added into the ripple set. However, due to the randomness of the connection between encoded symbols and the input symbols, it could not guarantee that the input symbols were put into the ripple set without repetition. Theoretically, the codes with ripple size larger than 1 from $L = 1$ to $L = K$ could recover the input symbols successfully. Hence, setting $\alpha = 1.1$, we substitute (3) to (4), and get the theoretical ripple evolution process. Figure 3 illustrates the theoretical ripple evolution of SF-DRS LT codes and LT codes based on robust soliton distribution. It can be seen that both SF-DRS LT codes ($P_1 = 0.1$, $\gamma = 2.0$) and SF-DRS LT codes ($P_1 = 0.1$, $\gamma = 1.9$) achieve ripple size of larger than 1. Moreover, compared with LT codes using robust soliton degree distribution, these two codes achieve a larger ripple size at the beginning phase of the decoding process. The ripple size of the SF-DRS LT codes ($P_1 = 0.1$, $\gamma = 2.0$) decreases as the decoding process continues. Thus, theoretically, when $N = \alpha K = 1.1K$ encoded symbols received, the SF-DRS LT codes could recover all input symbols successfully.

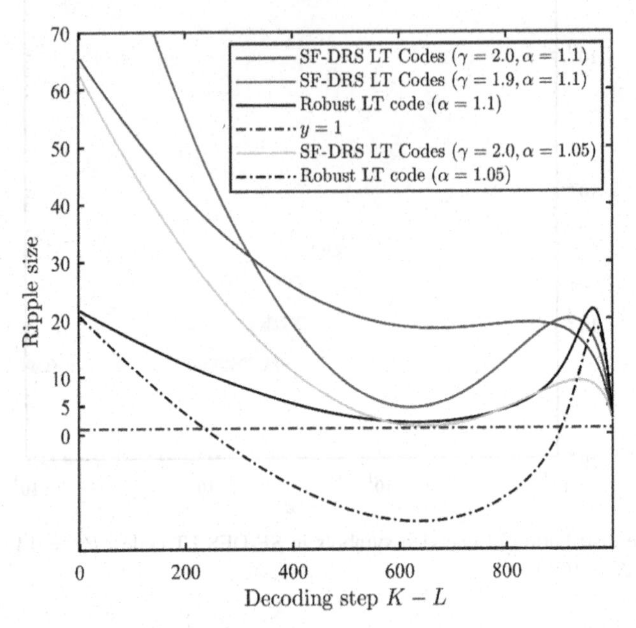

Fig. 3. The ripple evolution of the proposed LT codes. $K = 1000$.

Moreover, when the overhead factor decreases to 1.05, the ripple size of robust LT codes becomes smaller than one at some points. It indicates that using robust LT codes, the decoder could not recover all input symbols when 1050 encoded symbols are received. Inversely, the SF-DRS LT codes ($P_1 = 0.1$, $\gamma = 2.0$) always achieve a ripple size of larger than one at each iteration. Hence, when the receiver gets 1050 encoded symbols, the SF-DRS LT codes ($P_1 = 0.1$, $\gamma = 2.0$) could decode all the 1000 input symbols successfully.

4 Simulation Result

In this section, we assume each input symbol contains $l = 1$ bit, and randomly generate $K = 1000$ ($K = 2000$) input symbols. Then, we use the following codes to encode each group of input symbols.

(i) SF-DRS1 LT code: our proposed SF-DRS LT code with $P_1 = 0.1$, $\gamma = 2.0$, $n = 1049$, $R = 20$;
(ii) SF-DRS2 LT code: our proposed SF-DRS LT code with $P_1 = 0.1$, $\gamma = 1.9$, $n = 1049$, $R = 20$;
(iii) Decreasing ripple size LT code: LT code proposed in [14] with $n = 1049$, $R = 20$ (when $K = 1000$); when $K = 2000$, parameters are $n = 2056$, $R = 24$;

(iv) SF-LT1 code: the LT code proposed in [2] using parameters $P_1 = 0.1$, $\gamma = 2.0$;
(v) SF-LT2 code: the LT code proposed in [2] using parameters $P_1 = 0.1$, $\gamma = 1.9$.

As we consider the effect of the number of received encoded symbols, we ignore the erasure probability of the erasure channel, and assume the generated encoded symbols are transmitted over a perfect channel, i.e. the erasure probability is zero. For each LT code, as the receiver continuously obtain encoded symbols and recover parts of the input symbols, we record the number of input symbols being recovered. Figure 4 shows the probability of successful decoding versus overhead factor when $K = 1000$.

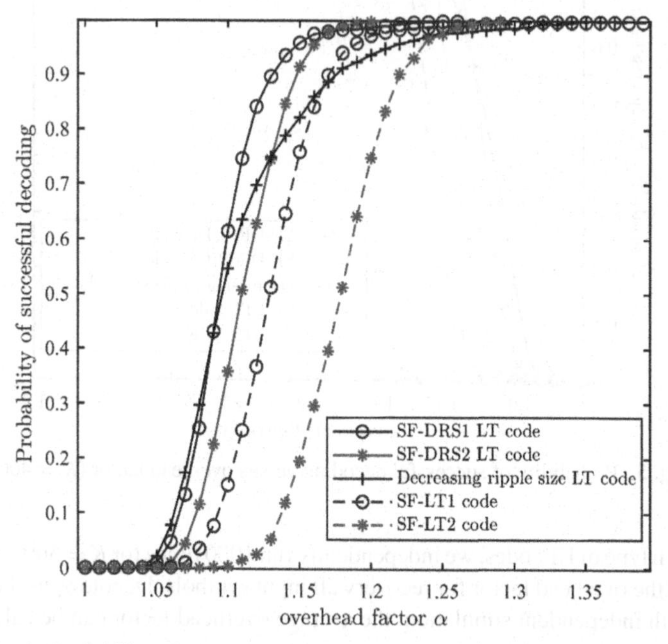

Fig. 4. Probability of successful decoding versus overhead factor ($K = 1000$)

It can be seen from Fig. 4 that when enough encoded symbols have been received, the receiver begins to recover the original input symbols. Initially, the decreasing ripple size LT codes start to decode some input symbols prior to other LT codes. As encoded symbols continuously being delivered to the receiver, SF-DRS1 LT codes outperform decreasing ripple size LT codes with respect to the immediate probability of successful decoding when the overhead factor is larger than 1.09. Moreover, in contrast to the SF-LT codes, the SF-DRS LT codes with the same parameters achieve a larger probability of successful decoding.

Figure 5 shows the probability of successful decoding versus overhead factor when $K = 2000$. When the overhead factor increases to 1.08 (1.12), the performance of SF-DRS1 (SF-DRS2) LT codes surpasses that of the decreasing ripple size LT codes. Moreover, when the overhead factor approaches 1.16, the SF-DRS LT codes and SF-LT1 codes recover 99% input symbols. However, even when the overhead factor increases to 1.24, the decreasing ripple size LT codes could recover 98.5% input symbols. With the same parameters, SF-DRS LT codes achieve a much better probability of successful decoding than SF-LT codes.

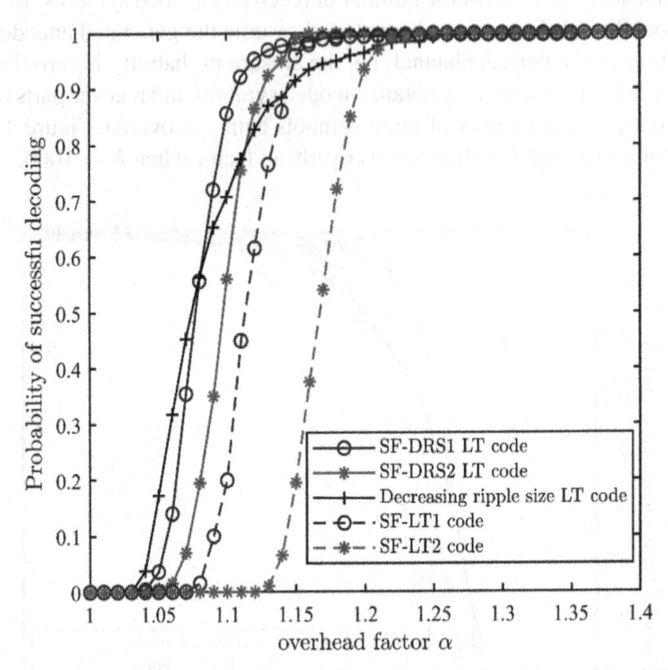

Fig. 5. Probability of successful decoding versus overhead factor ($K = 4000$)

For each type of LT codes, we independently run 1000 times for $K = 500, 1000, 2000$, and record the overhead factor for recovery all input symbols. Denote α_i as the overhead factor of i-th independent simulation, the average overhead factor can be calculated by

$$\overline{\alpha} = \sum\nolimits_{t=1}^{1000} \alpha_i \quad (5)$$

Figure 6 illustrates the average overhead factor of SF-DRS1 LT codes, SF-DRS2 LT codes and decreasing ripple size LT codes when $K = 500, 1000, 2000$. It can be seen that for the three LT codes, the average overhead factor decreases with the increase of K. Moreover, SF-DRS1 LT code achieves the minimum average overhead factor compared with SF-DRS2 LT code and decreasing ripple size LT code. When $K = 500$, the decreasing ripple size LT code requires less number of encoded symbols to recover the input symbols than the SF-DRS2 LT code. However, when K increases to 1000 and 2000, the SF-DRS2 LT code outperforms the decreasing ripple size LT code in terms of the average overhead factor.

For further evaluating the efficiency of our proposed LT codes, we calculate the average degree of the encoded symbols using (6).

$$\overline{d} = \sum\nolimits_{d=1}^{K} d \cdot \varphi(d) \quad (6)$$

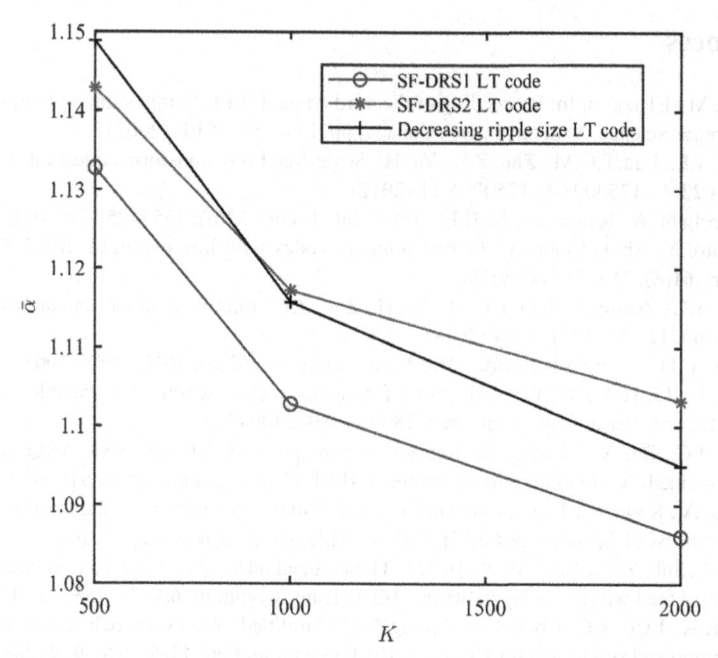

Fig. 6. Average overhead factor of SF-DRS1 LT code, SF-DRS2 LT code and Decreasing ripple size LT code. $K = 500, 1000, 2000$.

The average degrees of SF-DRS1 LT codes are 13.84 ($K = 1000$) and 15.41 ($K = 2000$), which are smaller than the average degrees of the decreasing ripple LT codes (15.32, when $K = 1000$; 17.56, when $K = 2000$). Thus, the coding efficiency of SF-DRS1 LT codes is superior to that of the decreasing ripple size LT codes.

5 Conclusion

In this paper, we combine the advantage of SF-LT codes and decreasing ripple size LT codes, and propose an LT code using SF-DRS degree distribution. Using the ripple evolution formula, we theoretically analyze the overhead factor of the SF-DRS LT codes. By selecting appropriate parameters, the ripple size of the SF-DRS LT codes continuously decreases with the decoding process. Simulation results further show that the SF-DRS1 LT codes obtain better performance than SF-DRS2 codes, SF-LT codes, and decreasing ripple size LT codes.

Acknowledgment. This research was supported by the National Natural Science Foundation of China (Grant Nos. 61603082, 61977014, 61902056 and 61902057), and the Fundamental Research Funds for the Central Universities (Grant Nos. N2017016).

References

1. Luby, M.: LT codes. In: Proceeding of the 43rd Annual IEEE Symposium on Foundations of Computer Science, Vancouver, BC, Canada, pp. 271–280. IEEE (2002)
2. Zhao, Y.L., Lau, F.C.M., Zhu, Z.L., Yu, H.: Scale-free Luby transform codes. Int. J. Bifurcat. Chaos **22**(4), 1250094-1–1250094-11 (2012)
3. Shokrollahi, A.: Raptor codes. IEEE Trans. Inf. Theory **52**(6), 2551–2567 (2006)
4. Cassuto, Y., Shokrollahi, A.: Online fountain codes with low overhead. IEEE Trans. Inf. Theory **61**(6), 3137–3149 (2015)
5. Zhao, Y.L., Zhang, Y., Lau, F.C.M., Yu, H., Zhu, Z.L.: Improved online fountain codes. IET Commun. **12**(18), 2297–2304 (2018)
6. MacKay, D.J.C.: Fountain codes. IEE Proc. Commun. **152**(6), 1062–1068 (2005)
7. Yuan, L., Li, H.A., Yi, W.: A novel UEP fountain coding scheme for scalable multimedia transmission. IEEE Trans. Multimed. **18**(7), 1389–1400 (2016)
8. Fang, J.C., Bu, X.Y., Yang, K.: Retransmission spurts of deferred NAK ARQ in fountain coding aided CCSDS file-delivery protocol. IEEE Commun. Lett. **20**(4), 816–819 (2016)
9. Nessa, A., Kadoch, M.: Joint network channel fountain schemes for machine-type communications over LTE-advanced. IEEE Internet Things J. **3**(3), 418–427 (2016)
10. Baik, J., Suh, Y., Rahnavard, N., Heo, J.: Generalized unequal error protection rateless codes for distributed wireless relay networks. IEEE Trans. Commun. **63**(12), 4639–4650 (2015)
11. Yen, K.K., Liao, Y.C., Chen, C.L., Chang, H.C.: Modified robust soliton distribution (MRSD) with improved ripple size for LT codes. IEEE Commun. Lett. **17**(5), 976–979 (2013)
12. Zhu, H.P., Zhang, G.X., Li, G.X.: A novel degree distribution algorithm of LT code. In: Proceeding of the 11th IEEE International Conference on Communication Technology, Hangzhou, China, pp. 221–224. IEEE (2008)
13. Savchenko, Y., Liu, Y.: Optimizing degree distributions of LT-based codes with deep reinforcement learning. In: IEEE Conference on Computer Communications Workshops, IEEE INFOCOM 2019, Paris, France, pp. 228–233. IEEE (2019)
14. Sørensen, J.H., Popovski, P., Ostergaard, J.: On LT codes with decreasing ripple size. In: Information Theory and Applications Workshop (ITA), La Jolla, USA (2011)
15. Sørensen, J.H., Popovski, P., Ostergaard, J.: Design and analysis of LT codes with decreasing ripple size. IEEE Trans. Commun. **60**(11), 3191–3197 (2012)
16. Yen, K.K., Liao, Y.C., Chen, C.L., Zao, J.K., Chang, H.C.: Integrating non-repetitive LT encoders with modified distribution to achieve unequal erasure protection. IEEE Trans. Multimed. **15**(8), 655–658 (2013)
17. Hussain, I., Xiao, M., Rasmussen, L.K.: Design of LT codes with equal and unequal erasure protection over binary erasure channels. IEEE Commun. Lett. **17**(2), 261–264 (2013)

Cross-border E-commerce Competitive Strategy Based on the Survey of Product Differentiation

Jun Chen, Shiyan Xu(✉), Chenyang Zhao, and Hui Chen

SILC Business School, Shanghai University, Shanghai 218000, China

Abstract. With the development of international trade and information technology, cross-border e-commerce is booming globally. It involves the movement of goods between domestic and foreign markets, domestic warehouses, and consumers. However, as cross-border e-commerce products often need long-distance transport, the commodities suitable for sale through cross-border e-commerce are limited. Therefore, merchants are increasingly facing fiercer competition. In this paper, a game model was built to analyze how differentiation of products affects the competitive strategy making in the cross-border dual-channel supply chain. The optimal pricing and service level under centralized and decentralized decisions were formulated. The relationship between multiple variables and service levels and profits were interpreted in combination with spatial geography. The equilibrium solution revealed that a centralized supply chain can generate more profits. And merchants should increase the degree of product differentiation and strive to improve service levels.

Keywords: Cross-border e-commerce · Product differentiation · Supply chain coordination · Dual-channel

1 Introduction

Cross-border E-commerce (CBE) is an electronic and information zed form of international trade. Buyers and sellers are in different countries or regions; they realize online transactions through information technology such as the Internet and complete the transportation of goods through international logistics (Wang 2014, Liu et al. 2015). Today, it is equivalent to cross-border e-retail, a business to consumer (B2C) model (iResearch 2014).

Advanced technology, growing demand and favorable policies have led to the explosive growth of global cross-border e-commerce. In 2014, the global cross-border e-commerce market exceeded $230 billion and is expected to grow further to $1 trillion by 2020 (Alizila 2015). In the same year, nearly one billion people worldwide will engage in cross-border online shopping, and their transaction volume will account for one-third of global B2C transactions (AliResearch 2015).

In the short term, due to the superior infrastructure and advanced technical conditions, the development of CBE will be concentrated in developed countries. In Germany,

H. Song and D. Jiang (Eds.): SIMUtools 2020, LNICST 370, pp. 659–678, 2021.
https://doi.org/10.1007/978-3-030-72795-6_53

for example, more than half of online merchants are selling their products to multiple foreign markets (Research and Markets 2015). However, in the long run, developing countries expect a leap-forward development after overcoming some of the limitations of information technology. China is expected to become the largest market for CBE by 2020 (iResearch 2016).

Since July 2014, including Announcement on Regulating Issues Concerning Import and Export Goods and Articles of Cross-border Trade E-Commerce, Announcement on Adding Customs Supervision Method Codes (i.e. Announcement No. 56 and No. 57) of the General Administration of Customs, various favorable policies have been introduced continuously, involving customs, commodity inspection, logistics, and payment, which have stimulated the development of cross-border e-commerce.

So far, cross-border e-commerce companies have emerged and gradually entered the track of normal development. They have various business models, mainly platform-based or self-operated, vertical or integrated.

Platform-based cross-border e-commerce companies develop and operate third-party e-commerce websites. They earn profit from providing related services and get commissions from settled businesses. Some of them directly their products. Self-operated cross-border e-commerce companies develop and operate e-commerce websites by themselves, and they are also responsible for a series of services such as procurement, warehousing, transportation and sales.

Integrated cross-border e-commerce has a wide range of goods while vertical cross-border e-commerce focuses on specific markets. According to this, it can be subdivided into four types.

Representative enterprises of integrated platform-based cross-border e-commerce include Jingdong Global, Tmall.HK, Taobao Global, ymatou.com and others. Representative enterprises of integrated self-operated cross-border e-commerce include Amazon Global, Wal-Mart Global e-shop, Net Ease Koala, Red, Lightinthebox and any others.

There are fewer vertical platform-based cross-border e-commerce businesses, with the focus of clothing, beauty and other products, such as Meilishuo, haimi.com. Vertical self-operated cross-border e-commerce is also rare. Representative companies include VIPSHOP, womai.com, mia.com, and JUMEI.COM.

However, no matter which kind of business model the company adopts, cross-border products often need long-distance transportation; the types of products suitable for cross-border e-commerce are limited. Cross-border e-commerce has to face increasingly severe channel competition, and how to set up the price to ensure maximum revenue is the current focus of cross-border supply chain competition strategy.

There has been a lot of academic research on the issue of channel competition. Among them, the channel coordination problem in the supply chain under the e-commerce environment is particularly prominent.

Huang et al. (2012) built a dual-channel supply-chain model consisting of a single manufacturer and retailer, discussing how to adjust prices and production plans in the event of an outage to achieve the maximum potential profit. They found that whether centralized or decentralized dual-channel supply chain, the optimal pricing is affected by the customer's channel preferences and market size changes (Huang et al. 2012).

Based on the distribution efficiency of different channels and the online channel acceptance of products, Liu Hanging et al. (2015) explored the effects of different pricing strategies on channels introduced by manufacturers when introducing online channels. The results show that the introduction of online channels is beneficial to manufacturers and the overall channel, and whether it is beneficial to retailers depends on the distribution efficiency of online channels relative to traditional channels (Hostelling 1929).

Deng Mingrong et al. (2016) constructed a signal game model between retailers and manufacturers under the condition of asymmetric demand information and analyzed the influence of market share and channel substitution coefficient on the dual-channel supply chain. It was found that the retailer would send the wrong market demand information to the manufacturer according to the wholesale price, and the manufacturer's wholesale price decision was related to the sales cost of the direct sales channel (Chamberlin 1965).

Compared with the traditional e-commerce market, cross-border e-commerce has a more serious homogenization competition trend. A simplistic and specialized business strategy will not bring excess profits to the company; in contrast, it will reduce the company's living space. Therefore, to reduce the degree of homogenization competition in the market, the implementation of the differentiation strategy is inevitable, which can help enterprises gain advantages and improve sales performance in the cross-border e-commerce market (Porter 1980).

This paper study the problem of how differentiation of product affects the competitive strategy making in the cross-border dual-channel supply chain which is significantly important in service levels. With the development of international trade and information technology, cross-border e-commerce is booming globally. However, as cross-border e-commerce products often need long-distance transport, the commodities suitable for sale through cross-border e-commerce are limited. Therefore, this paper formulates the optimal pricing and service level under the centralized and decentralized decision. The relationship between multiple variables and service levels and profits were interpreted in combination with spatial geography. Finally, the equilibrium solution revealed that a centralized supply chain can generate more profits.

2 Summary of Related Research

2.1 Differentiation Strategy

Differentiation has received extensive attention from scholars in early research, especially product differentiation.

The research by Hotelling (2015) and Chamberlin (2014) reveals how companies choose their position in the product space to cushion the impact of direct price competition.

Michael Porter (Xizheng et al. 2016) makes the concept of product differentiation more and more widely accepted by scholars. He believes there are two forms of competitive advantage: one is cost advantage and the other is differentiated operation. The company's generic competitive strategy includes Cost leadership, Differentiation, and Focus.

The three competitive strategies are not isolated. For example, Cost leadership can also realize the differentiation of the company's product performance, price, brand, and any aspects in the competition; and differentiation can also control and reduce the cost to achieve the cost by providing different levels of products or services. Focus is the specific implementation of these two strategies. Therefore, in a sense, Cost leadership and Focus can also be regarded as a broader Differentiation.

Diao Xinjun et al. (2014) studied Bertrand price competition or Cournot quantity competition strategy for vertically differentiated products with asymmetric network externalities. The results show that the market profits and social welfare of the two products in the Cournot quantity competition are greater than that in the Bertrand price competition.

In the Bertrand price competition and Cournot quantity competition, when the network externality of low-quality products is large and certain conditions are met, low-quality products can also obtain large market profits. When high-quality products have large network externalities, or although the network externalities are small but meet certain conditions, the network externalities are equal or the products do not have network externalities, high-quality products gain greater market profit.

Xiao Di et al. (2013) studied the coordinated operation strategy of supply chain members under the influence of quality competition and price competition in a supply chain system consisting of two suppliers and one manufacturer. Furthermore, the effects of price competition and quality competition on the equilibrium solution of the supply chain in different scenarios are discussed separately.

Research shows that supplier cooperation can help improve the quality efforts of suppliers, but it will lead to a decline in the overall profit of the supply chain. The more intense the quality competition, the higher the quality of the suppliers in most scenarios. However, the intensity of price competition has a limited impact on the degree of supplier effort and can even be ignored in the context of supplier cooperation (Di et al. 2013).

Zhang Xizheng et al. (2016) extended the model of Salop's circular city, studied the dual-channel supply chain pricing strategy of alternative products in the e-commerce environment, and proposed an improved revenue-sharing contract coordination strategy, which can implement Pareto improvement of supply chain members to alleviate channel conflicts.

The study finds that the degree of product substitution has a positive impact on the pricing of various channels of alternative products and the demand for retail channels, while the impact on the demand and pricing of existing products is negative. The efficiency of the supply chain increases with the degree of product substitution while the coordination effect becomes better with the degree of product substitution first and then worse.

2.2 Consumer Utility

The market demand function in the above studies is derived by the manufacturer or retailer. The difference between products and services is essentially the difference in the eyes of consumers. When consumers perceive different prices or services from other competitive products, the differentiation strategy can be realized. Therefore, differentiation should be expressed as a function of consumer preferences, and research on differentiation can be carried out at the level of consumer utility. With the utility function, the demand function can be derived from it, so the utility function is the key to model building.

The product brings benefits to consumers because the specific needs of consumers can be met through the product. However, consumers' demand for products is not limited to the products themselves but also includes the services provided by the merchants. The products sold in the market are broadly a binary combination containing the products themselves and merchant services. Providing perfect services can add value to customers and also bring a good reputation to the company, thereby enhancing the brand and market competitiveness of the company.

The research literature on services has long shown that consumers are often willing to pay for the promotion of services such that consumers who willing to pay a higher price do not have to wait anymore. As one of the product portfolios, the service level will inevitably become an important way for enterprises to obtain differentiated competitive advantages. Especially in the case that the quality of similar products is close and relatively fixed, the slight changes in the provision of services sometimes leads to a higher willingness to pay. For example, JD Logistics can achieve next-day delivery, and consumers usually choose JD as a shopping platform to meet the demand for timeliness of goods regardless of the higher price (Meng and Wang 2015).

Consumers make purchase decisions based on the principle of maximum utility when purchasing products. Whether consumers have a willingness to buy depends on the sum of the positive effects of products and services. Thus, the consumer will only be willing to purchase the product if and only if the total utility of the merchant's products and services is greater than the negative utility of the consumer's payment.

However, providing high-quality products or high-level services to consumers requires additional costs, that is, the higher the product quality or service quality, the higher the production cost of the merchants.

Under the restriction of market price competition, the excessive pursuit of high-quality products or services will bring the risk that the final profit will not rise and fall. So in real business activities, the merchants must balance the creation of added value and the corresponding cost increase. When designing a product portfolio, they must weigh the product differentiation competition and the differentiated competition strategy. The utility of the same product or service to different consumers is different, so there should be a parameter representing the type of consumer in the utility function.

Besides, products or services are different, different products or services have different effects for the same consumer, so there should be a parameter in the utility function that reflects the degree of differentiation of products and services.

To study the situation that the electronic direct selling channel and the traditional retail channel separately sell heterogeneous products in the dual-channel supply chain, Wang Yao et al. (2014) constructed a consumer utility function that characterizes product differences and service spillover effects and established a demand model and a profit model for a dual-channel supply chain.

The research results show that the differentiation strategy is beneficial to both manufacturers and retailers. Under decentralized decision-making, although the service has a negative spillover effect, within the appropriate scope, the improvement of service level is beneficial to the manufacturer. Therefore, manufacturers have the incentive to motivate retailers to work together to improve both sides' benefits (Diyun and Yaozhong 2015). But the assumption that manufacturers dominate in the game model does not match the facts.

Wang Chunping et al. (2016) introduced the differentiated network effect into the Hotelling model under linear cost. Based on the pricing strategy selection of basic information products and additional services, a game model of two companies' bundled sales and separate sales strategies in the duopoly market was established. They analyzed the impact of heterogeneous consumers on the duopoly market and the maximization of pricing strategies (Wujun et al. 2019). However, the products and services considered in the study are homogeneous.

Previous studies have achieved many results in channel competition or differentiation strategies or consumer utility. However, the cross-border e-commerce competition strategy is the result of comprehensive integration of the above aspects. So far, no relevant research has systematically combined them. Therefore, this paper aims to fill this gap and study the cross-border e-commerce supply chain coordination problem based on the differentiation strategy and consumer utility theory. It provides guidelines for merchants to develop cross-border e-commerce competition strategies.

3 Competitive Strategy Based on Product Differentiation

3.1 Model Parameters and Assumptions

Suppose there are one overseas supplier and two cross-border e-commerce in the cross-border supply chain. Merchant 1 is platform-based e-commerce, and merchant 2 is self-operated e-commerce. The products that the supplier provides to the merchants with a certain degree of difference, that is, the products sold through the two merchants, have a certain degree of substitutability.

There are two types of consumers in the market: price-sensitive and service-sensitive. The preferences of different types of consumers are reflected in the perception of product prices and service levels. For the price, price-sensitive consumers are more sensitive than service-sensitive consumers, and product price changes have a greater impact on their utility. On the contrary, for services, service-sensitive consumers are more sensitive than price-sensitive consumers, and service level changes have a greater impact on their utility.

Consumers are rational; they consider the benefits they get when purchasing products, so they will buy products from merchants with greater perceived utility. In order not to lose the generality, it is assumed that the demand faced by the two merchants is the unit market demand, and only the cost required to provide the logistics service is considered, the other costs are assumed to be zero.

The variables used in the model and their definitions are as follows (Table 1):

Table 1. Product differentiation model parameters

Parameters	Explanation
p_i	The price of the product, p_1 is the price of merchant 1, p_2 is the price of merchant 2
v	Valuation of the product, $v \in [0, 1]$
s_i	The level of logistics services provided by the merchant.s_1 is the level of logistics services provided by merchant 1, s_2 is the level of logistics services provided by merchant 2
$C(s_i)$	The cost functions of logistics services. Refer to the definition of the cost function in the *Principle of economics*, $C(s_i) = (\eta/2)s_i^2$, C_1 is the cost of logistics services of merchant 1, C_2 is the cost of logistics services of merchant 2
θ	The degree of difference between the products sold by merchant 2 and merchant 1. $\theta \in [0, 1]$, When $\theta = 0$, it means the product sold by merchant 2 is completely different from merchant 1 and there is no substitutability. When $\theta = 1$, it means there is no difference between the product sold by Merchant 2 and Merchant 1 and they are completely homogeneous
α	Consumer sensitivity to product prices, $\alpha \in [0, 1]$
β	Consumer sensitivity to service levels, $\beta \in [0, 1]$
D_i	Market demand function. D_1 is market demand faced by merchant 1; D_2 is market demand faced by merchant 2
π_i	Profit function π_1 is profit for merchant 1; π_2 is profit for merchant 2
U_i	The utility consumers get when they purchase the product.U_1 is the utility when they purchase the product from merchant 1, U_2 is the utility when they purchase the product from merchant 2

The cross-border e-commerce supply chain system relationship at this time is shown in Fig. 1:

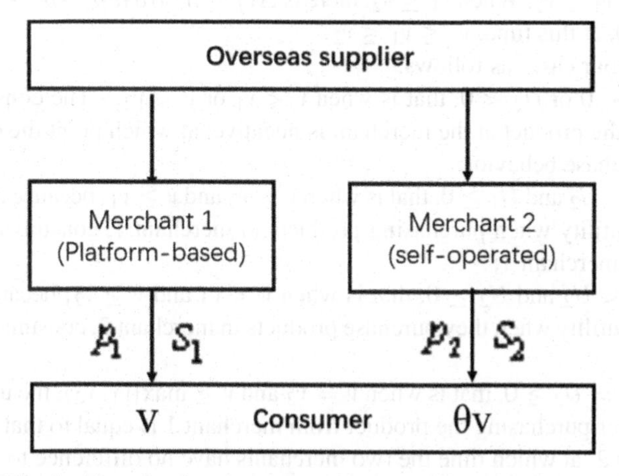

Fig. 1. Cross-border e-commerce product differentiation model

For the supply chain of cross-border e-commerce, the overseas suppliers transport the commodities to the self-operated or third-party platform for the next step of the sales according to the information of the consumers.

Referring to the model of Chen and Yang (Xizheng et al. 2016), the utility of the consumer to purchase items from merchant 1 is:

$$U_1 = v - \alpha p_1 + \beta s_1 \tag{1}$$

The utility of the consumer to purchase items from merchant 2 is:

$$U_2 = \theta v - \alpha p_2 + \beta s_2 \tag{2}$$

Consumers are rational. When the utility obtained is greater than or equal to zero, they will choose to purchase. When faced with two different channel choices, they will choose a channel with a large utility to purchase products, that is, consumers choose to follow max $(U_1, U_2, 0)$.

We define product valuations v_1, v_2, v_3 under three critical conditions $U_1 = 0$, $U_2 = 0$, $U_1 = U_2$; we can get:

$$\begin{aligned} v_1 &= \alpha p_1 - \beta s_1 \\ v_2 &= \frac{\alpha p_2 - \beta s_2}{\theta} \\ v_3 &= \frac{\alpha(p_1 - p_2) - \beta(s_1 - s_2)}{1 - \theta} \end{aligned} \tag{3}$$

The difference between the critical values is defined as:

$$\begin{aligned} \Delta v_1 = v_1 - v_2 = \frac{\alpha(\theta p_1 - p_2) - \beta(\theta s_1 - s_2)}{\theta} \\ \Delta v_2 = v_3 - v_1 = \frac{\alpha(\theta p_1 - p_2) - \beta(\theta s_1 - s_2)}{1 - \theta} \end{aligned} \tag{4}$$

When $v_1 \geq v_2$, there is $\Delta v_1 \geq 0$, so $\alpha(\theta p_1 - p_2) - \beta(\theta s_1 - s_2) \geq 0$ and $\Delta v_2 \geq 0$, at this time, $v_3 \geq v_1 \geq v_2$; When $v_1 \leq v_2$, there is $\Delta v_1 \leq 0$, so $\alpha(\theta p_1 - p_2) - \beta(\theta s_1 - s_2) \leq 0$ and $\Delta v_2 \leq 0$, at this time, $v_3 \leq v_1 \leq v_2$.

There are four cases as follows:

When $U_1 < 0$ or $U_2 < 0$, that is when $v < v_1$ or $v < v_2$, - The consumer's utility in purchasing the product at the merchant is negative, at which point the consumer will not have a purchase behavior;

When $U_1 > U_2$ and $U_1 \geq 0$, that is when $v > v_3$ and $v \geq v_1$; because consumers can obtain higher utility when purchasing products in merchant 1, consumers are inclined to purchase at merchant 1;

When $U_1 < U_2$ and $U_2 \geq 0$, that is when $v < v_3$ and $v \geq v_2$; because consumers can get higher utility when they purchase products in merchant 2, consumers tend to buy at merchant 2;

- When $U_1 = U_2 \geq 0$, that is when $v = v_3$ and $v \geq \max\{v_1, v_2\}$, the utility obtained by the consumer purchasing the product from merchant 1 is equal to that of purchasing from merchant 2, at which time the two merchants have no difference to the consumer.

This allows different product valuations in different product price ranges:

If $v_1 \leq v_2$, then $v_3 \leq v_1$, at this time, no one is buying in merchant 2, that is, the demand of merchant 2 is $D_2 = 0$. When $v \in [v_1, 1]$, the consumer purchases the product in merchant 1, the demand of merchant 1 is $D_1 = 1 - v_1$. When $v \in [0, v_1]$, consumers will not have purchase behavior. The price of this interval satisfies the conditions:

$$p_1 \leq \frac{\alpha p_2 + \beta(\theta s_1 - s_2)}{\alpha\theta} \tag{5}$$

If $v_1 \geq v_2$ and $v_3 \geq 1$, no one is buying in merchant 1, that is, the demand of merchant 1 is $D_1 = 0$. When $v \in [v_2, 1]$, the consumer purchases the product in merchant 2, the demand of merchant 2 is $D_2 = 1 - v_2$. When $v \in [0, v_2]$, consumers will not have purchase behavior. The price of this interval satisfies the conditions:

$$p_1 \geq p_2 + \frac{1 - \theta + \beta(s_1 - s_2)}{\alpha} \tag{6}$$

If $v_1 \geq v_2$ and $v_3 \leq 1$, when $v \in [v_3, 1]$, the consumer purchases the product in merchant 1, the demand of merchant 1 is $D_1 = 1 - v_3$; when $v \in [v_2, v_3]$, the consumer purchases the product in merchant 2, the demand of merchant 2 is $D_2 = v_3 - v_2$. The price of this interval satisfies the conditions:

$$\frac{\alpha p_2 + \beta(\theta s_1 - s_2)}{\alpha\theta} \leq p_1 \leq p_2 + \frac{1 - \theta + \beta(s_1 - s_2)}{\alpha} \tag{7}$$

This study aims to analyze the cross-border e-commerce competition strategy under differentiation. Since the first two cases are single-channel issues, they are not included in the following discussion. In summary, in the third case, the demand functions of merchant 1 and merchant 2 are, respectively:

$$\begin{aligned} D_1 &= 1 - \frac{\alpha(p_1 - p_2) - \beta(s_1 - s_2)}{1 - \theta} \\ D_2 &= \frac{\alpha(\theta p_1 - p_2) - \beta(\theta s_1 - s_2)}{\theta(1 - \theta)} \end{aligned} \tag{8}$$

To explore the impact of product differentiation on competitive strategies more clearly, assume that merchant 1 and merchant 2 provide the same level of logistics services, including delivery time, reliability, flexibility, after-sales service and others, that is $s_1 = s_2 = s$, at this time, the utility obtained by the consumer in merchant 1 and merchant 2 is:

$$U_{1s} = v - \alpha p_1 + \beta s \quad , \quad U_{2s} = \theta v - \alpha p_2 + \beta s \tag{9}$$

The demand functions of merchant 1 and merchant 2 are:

$$D_{1s} = 1 - \frac{\alpha(p_1 - p_2)}{1 - \theta} \quad , \quad D_{2s} = \frac{\alpha(\theta p_1 - p_2) + \beta s(1 - \theta)}{\theta(1 - \theta)} \tag{10}$$

The profits obtained are:

$$\begin{aligned} \pi_{1s} &= p_1 D_{1s} - C_1 = p_1 - \frac{\alpha p_1(p_1 - p_2)}{1-\theta} - \frac{1}{2}\eta s^2 \\ \pi_{2s} &= p_2 D_{2s} - C_2 = \frac{\alpha p_1(\theta p_1 - p_2) + \beta p_1 s(1-\theta)}{\theta(1-\theta)} - \frac{1}{2}\eta s^2 \end{aligned} \tag{11}$$

The total profit of cross-border e-commerce is:

$$\pi_s = \pi_{1s} + \pi_{2s} = \frac{p_1(1-\theta-\alpha p_1 + \alpha p_2)}{1-\theta} + \frac{\alpha p_1(\theta p_1 - p_2) + \beta p_1 s(1-\theta)}{\theta(1-\theta)} - \eta s^2 \tag{12}$$

3.2 Optimal Strategy Under the Centralized Model

Under the centralized decision-making, cross-border e-commerce closely cooperates to develop product prices and sales strategies jointly, and both parties aim at maximizing the total profit of cross-border e-commerce π_s.

$$\begin{aligned} \frac{\partial \pi_s}{\partial p_1} &= 1 - \frac{2\alpha(p_1 - p_2)}{1-\theta} \\ \frac{\partial \pi_s}{\partial p_2} &= \frac{2\alpha(\theta p_1 - p_2)}{\theta(1-\theta)} + \frac{\beta s}{\theta} \end{aligned} \tag{13}$$

The Hessian matrix of the optimization problem according to its second-order condition is:

$$H = \begin{pmatrix} \frac{-2\alpha}{1-\theta} & \frac{2\alpha}{1-\theta} \\ \frac{2\alpha}{1-\theta} & \frac{-2\alpha}{\theta(1-\theta)} \end{pmatrix} \tag{14}$$

$$|H_1| = \frac{-2\alpha}{1-\theta}\ , \quad |H_2| = \frac{4\alpha^2}{\theta(1-\theta)} \tag{15}$$

Because $\theta \in [0, 1]$, we can get $|H_1| < 0$, $|H_1| > 0$; the Heather matrix is negative, so there is an optimal pricing strategy (p_1^* , p_2^*). It can maximize the total profit of cross-border e-commerce and can be calculated simultaneously $\partial \pi_s / \partial p_1 = 0$, $\partial \pi_s / \partial p_2 = 0$.

In the case of providing homogeneous services, the optimal pricing strategies for merchant 1 and merchant 2 are:

$$\begin{aligned} p_1^* &= \frac{1+\beta s}{2\alpha} \\ p_2^* &= \frac{\theta + \beta s}{2\alpha} \end{aligned} \tag{16}$$

Bring p_1^* , p_2^* into π_{1s}, π_{2s}, π_s, we can get under the homogenous service concentration strategy, the profit of merchant 1, merchant 2 and the total profit of cross-border e-commerce are:

$$\pi_{1s}^{c} = \frac{1 + \beta s - 2\alpha\eta s^2}{4\alpha}$$
$$\pi_{2s}^{c} = \frac{\theta\beta s + (\beta^2 - 2\alpha\eta\theta)s^2}{4\alpha\theta} \tag{17}$$
$$\pi_{s}^{c} = \frac{\theta(1 + 2\beta s) + (\beta^2 - 4\alpha\eta\theta)s^2}{4\alpha\theta}$$

Let $\partial\pi_s^c / \partial s = 0$, the service level provided by the merchants under the homogeneous service concentration strategy is:

$$s^c = \frac{\beta\theta}{4\alpha\eta\theta - \beta^2} \tag{18}$$

3.3 Optimal Strategy Under the Decentralized Model

In the decentralized decision-making strategy, the merchants each make decisions based on their profit maximization principle. This is a dynamic oligopolistic game model, namely the Stackelberg game.

Because the platform-based merchants have more abundant resources, it is assumed that the platform-based merchant 1 in the distributed model is the leader of the game, and the self-operated merchant 2 is the follower.

The decision order is that merchant 1 first formulates the product price to maximize its profit; merchant 2 then sets its income level according to the pricing of merchant 1. Merchant 1 knows that merchant 2 will make such a decision, so merchant 1 will utilize the response function of merchant 2 to itself as the basis for the final price decision. We use the inverse inductive method to calculate the optimal solution of the model.

Merchant 2 sets the price to maximize its profits, that is $\partial\pi_{2s} / \partial p_2 = 0$, the price of the product of merchant 2 can be determined as:

$$p_2 = \frac{\alpha\theta p_1 + \beta s(1 - \theta)}{2\alpha} \tag{19}$$

Bringing p_2 into the profit function of merchant 1, we can get:

$$\pi_{1s} = \frac{\alpha p_1^2(\theta - 2) + p_1(2 + \beta s)(1 - \theta)}{2(1 - \theta)} - \frac{1}{2}\eta s^2 \tag{20}$$

Let $\partial\pi_{1s} / \partial p_1 = 0$, get the best product pricing p_1^* that maximizes the profit of merchant 1:

$$p_1^* = \frac{(2 + \beta s)(1 - \theta)}{2\alpha(2 - \theta)} \tag{21}$$

Bring p_1^* into p_2^*, get the best product pricing p_2^* that maximizes the profit of merchant 2:

$$p_2^* = \frac{[\beta s(4 - \theta) + 2\theta](1 - \theta)}{4\alpha(2 - \theta)} \tag{22}$$

Bring p_1^*, p_2^* into π_{1s}, π_{2s}, π_s, under the homogenous service decentralization strategy, the profits of merchant 1, merchant 2 and the total profits of cross-border e-commerce are:

$$\begin{aligned}\pi_{1s}^d &= \frac{8\theta(1+s\beta)A_1 + s^2(2\beta\theta A_1 - \alpha\eta A_5)]}{2\alpha A_5}\\ \pi_{2s}^d &= \frac{4\theta^2(1-\theta) + 4s\beta\theta A_2 + s^2[\beta^2(4-\theta)A_2 - \alpha\eta A_5]}{2\alpha A_5}\\ \pi_s^d &= \frac{4\theta A_2 + 4s\beta\theta A_3 + s^2[\beta^2 A_4 - 2\eta\alpha A_5]}{2\alpha A_5}\end{aligned} \tag{23}$$

Among them:

$$\begin{aligned}A_1 &= (1-\theta)(2-\theta)\\ A_2 &= (1-\theta)(4-\theta)\\ A_3 &= (1-\theta)(8-3\theta)\\ A_4 &= (1-\theta)(16-4\theta-\theta^2)\\ A_5 &= 8\theta(2-\theta)^2\end{aligned} \tag{24}$$

Because $\theta \in [0, 1]$, A_1–A_5 are all positive.

Let $\partial\pi_s^d / \partial s = 0$, the level of service provided by merchants under the homogeneous service decentralization strategy is:

$$s^d = \frac{2\beta\theta A_3}{2\alpha\eta A_5 - \beta^2 A_4} \tag{25}$$

3.4 Comparative Analysis

3.4.1 Impact on Service Levels

First, we compare the level of service under centralized decision-making and decentralized decision-making. When merchants provide homogeneous services, the difference between their service level in centralized and decentralized decision-making is:

$$\Delta s = s^c - s^d = \frac{\beta\theta^2[8\alpha\eta\theta(3-\theta) - \beta^2 A_1]}{(4\alpha\eta\theta - \beta^2)(2\alpha\eta A_5 - \beta^2 A_4)} \tag{26}$$

Because $\theta \in [0, 1]$, $s^c > 0$, $s^d > 0$, $4\alpha\eta\theta - \beta^2 > 0$, $2\alpha\eta A_5 - \beta^2 A_4 > 0$, and weather Δs is negative or positive is depends on $8\alpha\eta\theta(3-\theta) - \beta^2 A_1$. When $8\alpha\eta\theta(3-\theta) - \beta^2 A_1 > 0$, $s^c > s^d$, the level of service under centralized decision is greater than that under decentralized decision-making, and vice versa.

That is, when merchants in a cross-border e-commerce supply chain provide the same service, the service level is affected by consumer behavior (α, β), operating costs (η), and product attributes (θ). Therefore, merchants should consider these influencing factors when formulating service strategies.

Second, we analyze in detail the impact of variables $\alpha,\ \beta,\ \eta,\ \theta$ on service levels. Under centralized decision:

$$\begin{aligned} \frac{\partial s^c}{\partial \alpha} &= \frac{-4\beta\eta\theta^2}{(4\alpha\eta\theta-\beta^2)^2} \\ \frac{\partial s^c}{\partial \beta} &= \frac{\theta(4\alpha\eta\theta+\beta^2)}{(4\alpha\eta\theta-\beta^2)^2} \\ \frac{\partial s^c}{\partial \eta} &= \frac{-4\beta\eta\theta^2}{(4\alpha\eta\theta-\beta^2)^2} \\ \frac{\partial s^c}{\partial \theta} &= \frac{-\beta^3}{(4\alpha\eta\theta-\beta^2)^2} \end{aligned} \tag{27}$$

Under decentralized decision:

$$\begin{aligned} \frac{\partial s^d}{\partial \alpha} &= \frac{-4\beta\eta\theta A_3 A_5}{(2\alpha\eta A_5-\beta^2 A_4)^2} \\ \frac{\partial s^d}{\partial \beta} &= \frac{2\theta A_3(\beta^2 A_4+2\alpha\eta A_5)}{(2\alpha\eta A_5-\beta^2 A_4)^2} \\ \frac{\partial s^d}{\partial \eta} &= \frac{-4\alpha\beta\theta A_3 A_5}{(2\alpha\eta A_5-\beta^2 A_4)^2} \\ \frac{\partial s^d}{\partial \theta} &= \frac{-8\beta(4\alpha\eta\theta^2 A_6+\beta^2 A_7)}{(2\alpha\eta A_5-\beta^2 A_4)^2} \end{aligned} \tag{28}$$

Among them:

$$\begin{aligned} A_6 &= (2-\theta)(6-\theta) \\ A_7 &= (1-\theta)^2(32-24\theta+5\theta^2) \end{aligned} \tag{29}$$

Because $\theta \in [0, 1]$, so $A_6 > 0,\ A_7 > 0$ are all positive.
Conclusion as below:

(1) $\partial s^c / \partial \alpha < 0,\ \partial s^d / \partial \alpha < 0$

Whether centralized or decentralized, cross-border e-commerce service levels s will decrease as consumers become more sensitive to product prices. This is because, when consumers are more sensitive to product prices, they will be attracted by lower prices. At this time, both platform-based and self-operated companies tend to adopt low-cost competition strategies to attract more consumers. Merchants often transfer logistics costs to consumers, as they are reflected in the final price of the product. In the case of selling homogeneous products, the reduction in prices will inevitably lead to a reduction in logistics investment, which will eventually lead to a decline in service levels.

(2) $\partial s^c / \partial \beta > 0,\ \partial s^d / \partial \beta > 0$

Whether it is centralized or decentralized, the level of cross-border e-commerce services s will increase as consumers become more sensitive to service levels.

This is because, when consumers are more sensitive to service levels, they will pay more attention to the utility of services. At this time, both platform-based and self-operated enterprises tend to adopt competitive strategies to improve service levels and then attract more consumers.

(3) $\partial s^c / \partial \eta < 0, \; \partial s^d / \partial \eta < 0$

Whether it is centralized or decentralized, the level of cross-border e-commerce services s will decrease as the service cost factor η increases.

This is because when the service cost coefficient rises, or when the efficiency of service cost input and output is low, providing the same level of service requires more cost. The merchant is rational, and the pursuit of profit maximization is bound to the reduction of service costs, and ultimately leads to a decline in service levels.

(1) $\partial s^c / \partial \theta < 0, \; \partial s^d / \partial \theta < 0$

Whether it is centralized or decentralized, the level of cross-border e-commerce services s will decrease as the degree of product differentiation decreases.

When the products sold by different merchants tend to be the same, the needs of consumers are roughly the same. At this time, the warehouses will be placed close to the suppliers (because consumers are widely distributed), and the service level would then reduce. (The warehouse is far from the consumer, which means that the delivery time increases, the waiting time of consumers increases, and may not be punctual. If there are after-sales problems, the return and exchange problem may be difficult to solve, which affects reliability; long-distance transportation adopts a single form and is not flexible.) At the same time, the warehouse is far from consumers, which means that some consumers may not be in the delivery area, which is the low service level. Also, the reduction in the diversity of commodities means that there may be no more strict requirements for special logistics and distribution of different types of goods, the overall level of logistics will decline.

3.4.2 Impact on Cross-border E-commerce Total Profit

First, look at the comparison of cross-border e-commerce total profit under centralized- and decentralized decision-making. When the merchants provide homogeneous services, the difference between the total profit of cross-border e-commerce in centralized- and decentralized decision-making is:

$$\Delta\pi_s = \pi_s^c - \pi_s^d = \frac{4\theta + 4s\beta\theta(3-\theta) + s^2\beta^2(4+\theta-\theta^2)}{16\alpha(2-\theta)^2} \tag{30}$$

Because $\theta \in [0, 1]$, $\Delta\pi_s > 0$. This shows that the total profit of cross-border e-commerce under centralized decision-making is larger than that under decentralized decision-making, which means that if the merchants work closely together, conclude alliances, and formulate product prices and sales strategies jointly, they can bring greater benefits to the entire cross-border e-commerce.

Next, analyze in detail the impact of variables $\alpha, \; \beta, \; \eta, \; \theta$ on total profit.

Under centralized decision:

$$\begin{aligned}\frac{\partial \pi_s^c}{\partial \alpha} &= \frac{-(s^2\beta^2 + 2s\beta\theta + \theta)}{4\alpha^2\theta} \\ \frac{\partial \pi_s^c}{\partial \beta} &= \frac{s^2\beta + s\theta}{2\alpha\theta} \\ \frac{\partial \pi_s^c}{\partial \eta} &= -s^2 \\ \frac{\partial \pi_s^c}{\partial \theta} &= \frac{-s^2\beta^2}{4\alpha\theta^2}\end{aligned} \tag{31}$$

Under decentralized decision:

$$\begin{aligned}\frac{\partial \pi_s^d}{\partial \alpha} &= \frac{-(4\theta A_2 + 4s\beta\theta A_3 + s^2\beta^2 A_4)}{2\alpha^2 A_5} \\ \frac{\partial \pi_s^d}{\partial \beta} &= \frac{s(2\theta A_3 + s\beta A_4)}{\alpha A_5} \\ \frac{\partial \pi_s^d}{\partial \eta} &= -s^2 \\ \frac{\partial \pi_s^d}{\partial \theta} &= \frac{-4(4\theta^2 s\beta A_6 + 4\theta^2 A_8 + s^2\beta^2 A_9)}{\alpha A_5^2}\end{aligned} \tag{32}$$

Among them:

$$\begin{aligned}A_8 &= (2-\theta)(2+\theta) \\ A_9 &= (2-\theta)(32 - 48\theta + 34\theta^2 - 7\theta^3)\end{aligned} \tag{33}$$

Because $\theta \in [0, 1]$, $A_8 > 0$, $A_9 > 0$ are all positive.
Conclusion as below:

(1) $\partial \pi_s^c / \partial \alpha < 0, \ \partial \pi_s^d / \partial \alpha < 0$

Whether it is centralized or decentralized, the total profit of cross-border e-commerce π_s will decrease as consumers become more sensitive to product prices.

This is because, when consumers are more sensitive to product prices, they will be attracted by lower prices. At this time, both platform-based and self-operated companies tend to adopt a competitive strategy of lowering product prices to attract more consumers. This will result in a decline in the total profit of cross-border e-commerce. Therefore, price competition is not beneficial to the overall supply chain. Both enterprises and governments should actively guide consumers to pay more attention to the quality of products and services.

(2) $\partial \pi_s^c / \partial \beta > 0, \ \partial \pi_s^d / \partial \beta > 0$

Whether it is centralized or decentralized, the total profit of cross-border e-commerce π_s will increase as consumers become more sensitive to service levels.

This is because, when consumers are more sensitive to service levels, they will pay more attention to the utility of services. At this time, both platform-based and self-operated enterprises tend to adopt competitive strategies to improve service levels to attract more consumers, this will increase the total profit of cross-border e-commerce.

(3) $\partial \pi_s^c / \partial \eta = \partial \pi_s^d / \partial \eta < 0$

Whether it is centralized or decentralized, the total profit of cross-border e-commerce π_s will decrease as the service cost factor η increases.

This is because when the service cost coefficient rises, or when the efficiency of service cost input and output is low, providing the same level of service requires more cost, which will inevitably lead to a decline in the total profit of cross-border e-commerce. At this time, the government should give enterprises certain subsidies, improve the corresponding cross-border e-commerce infrastructure construction, reduce the service cost of enterprises, and promote the healthy development of cross-border e-commerce.

$\partial \pi_s^c / \partial \theta < 0$, $\partial \pi_s^d / \partial \theta < 0$

Whether it is centralized or decentralized, the total profit of cross-border e-commerce π_s will increase as the degree of difference in products sold by merchant's θ decreases. This means that to increase profits, companies should widen the differences between products, including differences in category and quality.

Under the circumstance of commodity convergence, large-scale platforms have certain advantages in the logistics service level because of their own logistics system. Small enterprises have to reduce the funds for maintaining the logistics service level due to the loss of consumers, and then their profits may be affected. At the same time, the profits of large enterprises will also be affected because of the decrease in service level caused by commodity convergence.

4 Numerical Analysis

To visualize the impact of the differentiation strategy on the cross-border e-commerce supply chain, this section does some simulation analysis with specific values.

Let $\alpha = 0.5$, $\beta = 0.5$, $\eta = 0.6$, $s = 1$, the impact of product differentiation on the total profit of cross-border e-commerce supply chain under centralized decision-making and decentralized decision-making is shown in Fig. 2. The red line refers to the centralized decision while the blue dotted line refers to the decentralized decision.

It can be seen from the figure that as the degree of product differentiation increases, the profit of cross-border e-commerce supply chain is decreasing under both centralized and decentralized strategies, and the profit attenuation under the centralized strategy is smaller than that under the decentralized strategy. When the product differentiation between cross-border e-commerce is certain, the supply chain that implements the centralized strategy can obtain more profits.

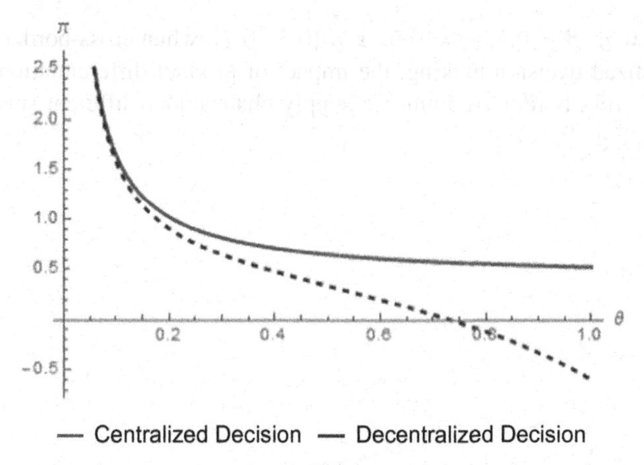

Fig. 2. Profit of cross-border e-commerce supply chain under centralized decision-making and decentralized decision-making (Color figure online)

(1) Let $\alpha = 0.5$, $\beta = 0.5$, $\eta = 0.6$, $s = \{0.3, 0.7\}$, when cross-border e-commerce centralized decision-making, the impact of product differentiation on the total profit of cross-border e-commerce supply chain under different service levels is shown in Fig. 3:

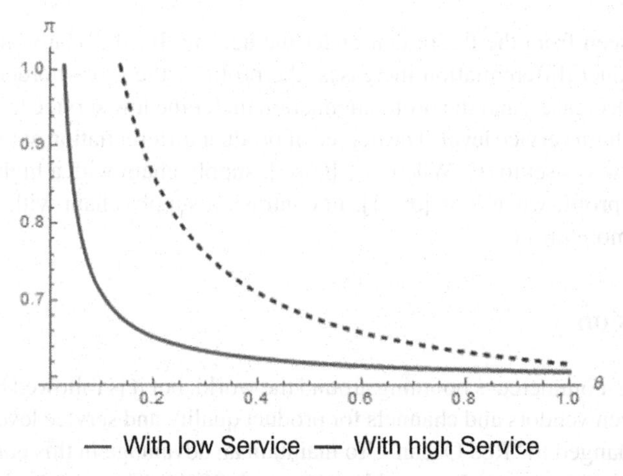

Fig. 3. Profit of cross-border e-commerce supply chain under different service level centralized decision (Color figure online)

It can be seen from the figure that, under the centralized decision, as the degree of product differentiation increases, the profit of the cross-border e-commerce supply chain decreases, and the profit attenuation under the low service level is less than that under the high service level. When the product differentiation between cross-border e-commerce is certain, the supply chain with a high service level can obtain more profits.

(2) Let $\alpha = 0.5,\ \beta = 0.5,\ \eta = 0.6,\ s = \{0.3,\ 0.7\}$, when cross-border e-commerce decentralized decision-making, the impact of product differentiation on the total profit of cross-border e-commerce supply chain under different service levels is shown in Fig. 4:

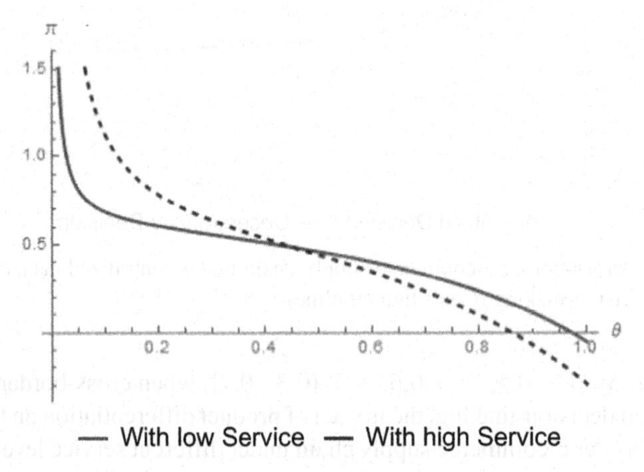

Fig. 4. Profit of cross-border e-commerce supply chain under different service level decentralized decision-making (Color figure online)

It can be seen from the figure that under the decentralized decision-making, as the degree of product differentiation increases, the profit of the cross-border e-commerce supply chain decreases, and the profit attenuation under the low service level is less than that under the high service level. The degree of product differentiation at the intersection of the two curves is set to θ^*. When $\theta \in [0, \theta^*]$, supply chain with a high service level can get more profit; when $\theta \in [\theta^*, 1]$, in contrast, a supply chain with a low service level can get more profit.

5 Conclusion

Cross-border e-commerce is booming around the world, but it is followed by fierce competition between vendors and channels for product quality and service levels. The "Blue Ocean" has changed to "Red Ocean". To maintain an advantage in this competition, the differentiation strategy is indispensable (Cao et al. 2019). This paper adopts the theoretical method of game theory analysis and the empirical research method of designing questionnaire interviews.

This paper considers a dual-channel cross-border supply chain that includes overseas suppliers, platform-based e-commerce, and self-operated e-commerce. The game model is used to analyze the impact of product differentiation on the competitive strategy of the merchants, and the optimal pricing and optimal service levels under centralized- and decentralized decision-making. According to the collection of effective questionnaires and interview information, the correctness of the product differentiation analysis strategy model was verified.

From model equilibrium solutions and numerical analysis, the following practical guidance can be derived:

(1) As the degree of product differentiation increases, the profit of cross-border e-commerce supply chain increases. To increase profits, companies should actively expand the differences in sales products, including differences in quality and quality. Provide consumers with more choices and better services.

(2) When the product differentiation between cross-border e-commerce is certain, the cross-border supply chain that implements the centralized strategy can obtain more profits. For example, Wal-Mart's global store in JD.com has proven this (Zhang 2016). Wal-Mart has increased its customers and secured domestic supply chain support through the JD platform. JD.com has expanded its product offerings through Wal-Mart and gained stronger international supply chain support.

(3) When the products of the merchants tend to be the same, the role of service differentiation is highlighted, and enterprises that can provide better logistics service levels can gain a competitive advantage. Price competition is not beneficial to the overall supply chain. Both enterprises and governments should actively guide consumers to pay more attention to the quality of products and services.

Cross-border e-commerce companies focused on the construction of overseas warehouses is evidence of the importance of service differentiation. Cross-border e-commerce has experienced modes such as direct mail to overseas warehouse collection and domestic bonded area collection, aiming to improve the satisfaction of cross-border e-commerce customers and improve competitiveness. At this time, the government should give enterprises certain subsidies, improve the corresponding cross-border e-commerce infrastructure construction, and reduce the service cost of enterprises, which can promote the healthy development of cross-border e-commerce.

(4) Promote the transformation and upgrading of cross-border enterprises, accelerate the process of enterprise branding, foster international Internet brands, enhance the technological innovation capability and international competitiveness of enterprises, drive the development of high-end service industries, and foster new service formats. Enhance export competitive advantage and profit margin, master the international market to expand import scale and industry coverage and develop local service industry to drive domestic enterprises to participate in the global network division of labor and competition.

This paper is only the initial step in the study of the cross-border e-commerce operation strategy. It has a positive guiding role in the practical operation of modern Chinese cross-border e-commerce. Since cross-border e-commerce includes international supply chain and domestic supply chain, and the operation strategy is complex and extensive, we will further improve the competitive strategy in future research work, such as research on the risk of cross-border e-commerce operation mode, cross-border e-commerce warehousing, international trade transportation, and last-mile delivery management.

References

Wang, J.: Opportunities and challenges of international e-commerce in the pilot areas of China. Int. J. Mark. Stud. **6**(6), 141–149 (2014)

Liu, X.J., Chen, D.Y., Cai, J.S.: The Operation of the Cross-border e-commerce Logistics in China. Int. J. Intell. Inf. Syst. **4**(2), 15–18 (2015)

iResearch. 2014 China Cross-border E-commerce Report (Brief Edition) [EB/OL]. https://californiacenter.us/wp-content/uploads/2015/02/2014-China-Cross-border-E-commerce-Report-Brief-Edition.pdf. Accessed 05 Jan 2015

Alizila. Cross-border E-Commerce to Reach $1 Trillion in 2020 [EB/OL]. https://www.alizila.com/cross-border-e-commerce-to-reach-1-trillion-in-2020/. Accessed 11 Jun 2015

AliResearch. The prospect for the global cross-border B2C e-commerce market. Beijing, AliResearch (2015)

Research and Markets: Europe Cross-border B2C E-Commerce 2015. Business Wire, Dublin (2015)

iResearch. 2016 China Cross-border Online Shoppers Report [EB/OL]. https://californiacenter.us/wp-content/uploads/2015/01/2016-Chinas-Cross-border-Ecommerce-Report.pdf. Accessed 21 Mar 2016

Huang, S., Yang, C., Zhang, X.: Pricing and production decisions in dual-channel supply chains with demand disruptions. Comput. Ind. Eng. **62**, 70–83 (2012)

Hanjin, L., Xiaojun, F., Hongmin, C.: Research on dual channel pricing strategy under retailer price leadership structure. CMS **23**(6), 91–98 (2015)

Deng Mingrong, L., Xiujuan. : Research on dual channel supply chain based on signal game under demand information asymmetry. Oper. Res. Manage. **25**(4), 125–133 (2016)

Porter, M.E.: Competitive strategy: techniques for analyzing industries and competitors. Soc. Sci. Electron. Publishing **2**, 86–87 (1980)

Xinjun, D., Deli, Y., Bin, T.: Service cooperation policy in a dual-channel supply chain under service differentiation. Oper. Manage (2014)

Xiao, D., Yuan, J.X., Bao, X.: Analysis of supply chain coordination strategy under the double competition of quality and price. Chin. Manage. Sci (2013)

Zhang, X., Liu, C., Zhang, R.: Pricing and coordination of dual-channel supply chain based on alternative product competition. Soft Sci. **30**(3) (2016)

Diyun, M., Yaozhong, W.: Research on the sources of competitive advantages of online retail enterprises from the perspective of value networks-taking Jingdong as an example. Search **278**(10), 45–49 (2015)

Yao, W., Bin, D., Can, L.: Heterogeneous product dual channel supply chain improvement strategy with negative spillover effect. J. Manage (2014)

Cunping, W., Guofang, N., Minqiang, L.: Optimal pricing strategy for oligopolistic information products and services. J. Manage. Sci. (2016)

Chen, J., Yang, Y.: Service cooperation policy in a dual-channel supply chain under service differentiation. Am. J. Ind. Bus. Manag. **4**(6), 284–294 (2014)

Wujun, C., Mengna, Y., Chaogai, X.: Construction of a logistics enterprise-led cross-border e-commerce ecosystem-multi-case study. Sci. Technol. Manage. Res. **16** (2019)

Tao, Z.: Wal-Mart's JD alliance, the retail industry is brewing huge changes. Chin. Bus. **7**, 78–79 (2016)

Image Denoising Method Based on Curvelet Transform in Telemedicine

Yang Yu[1], Dan Li[1](✉), Likai Wang[2], Weiwei Liu[2], Kailiang Zhang[1], and Yuan An[1]

[1] Xuzhou University of Technology, Xuzhou 221000, Jiangsu, China
[2] Traffic Police Detachment of Xuzhou Public Security Bureau, Xuzhou 221000, Jiangsu, China

Abstract. To resolve the problems that the traditional image denoising methods are easy to lose details such as edges and textures, a new method of image denoising was proposed. It based on the Curvelet denoising algorithm, using polynomial interpolation threshold method, combining with Wrapping and Cycle spinning techniques to determine the adaptive threshold of each Curvelet coefficient for denoising the medical images. Simulation experiments confirm that the new method reduces the pseudo Gibbs phenomenon, retains the details and texture of the image better, and obtains better visual effects and higher PSNR values.

Keywords: Curvelet transform · Cyclic translation · Image denoising · Telemedicine

1 Introduction

Images are noisy after acquisition and remote network transmission, which affects the visual and application effects [1–3]. Noise in images has greatly affected people's extraction of information from images, such as medical images, remote sensing images, and computer vision images. In many fields such as medicine and scientific, the requirements for image quality have become higher and higher [4]. Therefore, we need to explore from various angles to improve the quality of the image.

In the field of image denoising, people continue to improve the use of wavelet transform for image denoising [5–7]. The theory of wavelet denoising is still developing. Research is made on transform methods by selecting different basis functions or using frames to transform non-decimating wavelet transforms or selecting the optimal basis to transform the wavelet packets. Wavelet can not express Linear Singularity sparsely and can only be applied to isotropic singular objects. It can't fully study the geometric characteristics of objects, such as the edge, contour and other main features of two-dimensional image, which are the most interesting places for people.

E.J.Candes and D.L.Donoho established a multi-scale method particularly suitable for representing anisotropic singularities-Ridgelet transform based on wavelet theory [8, 9]. It has a strong ability of direction selection and discrimination, but for image denoising of curve singularity, Ridgelet transformation needs to be carried out in blocks. In order to further express the more general curve singularity of multi-dimensional

H. Song and D. Jiang (Eds.): SIMUtools 2020, LNICST 370, pp. 679–690, 2021.
https://doi.org/10.1007/978-3-030-72795-6_54

signals, on the basis of Ridgelet transformation, the local Ridgelet transformation and Curvelet curve transformation methods ware proposed to solve the curve singularity, local straight lines with multiple scales are used to approximate the entire curve.

The first generation of Curvelet is a multi-scale pyramid, which has many direction and location elements in each scale. Its construction idea is to treat the curve as a straight line in each block through a small enough block, and then use the local analysis of its characteristics. E.J.Candes and D.L.Donoho proposed the second-generation Curvelet transform theory, which is completely different in structure from the first generation Curvelet transform [10, 11]. The second generation of Curvelet is a new method based on frequency domain. It no longer uses different scales to decompose objects, no longer analyzes them in blocks, directly analyzes them in frequency domain. It also has high anisotropic characteristics. It can accurately and effectively express important information such as image edges with less non-zero coefficients, and it also has high approximation accuracy for images and better sparse expression ability, which makes the implementation simpler, faster, easier to understand, and the redundancy is greatly reduced.

At present, telemedicine provides a broader development space for the application of modern medicine [12–15]. Telemedicine technology has developed from the initial TV monitoring and telephone remote diagnosis to the comprehensive transmission of digital, image and voice using high-speed network, and realized the communication of real-time voice and high-definition image [16–18]. In this paper, the second-generation Curvelet transform is applied to image denoising. Based on this, existing algorithms are improved to make the image signal-to-noise ratio higher, and the image display effect is more real and clear.

2 Fast Discrete Curvelet Transform

Applying the Curvelet transform to image processing requires its discrete form. Compared with other traditional multi-scale transform (such as wavelet transform) [19–21], the discrete Curvelet transform is a new multi-scale analysis method [22–26], which has obvious advantages in the expression of image edge direction features, and the anisotropic features can better describe the edge and detail of the image [27, 28].

In the Cartesian coordinate system, $f[t_1, t_2]$, '$0 \leq t_1,t_2 \leq n$ is the input, and the discrete form of the Curvelet transform can be expressed as:

$$c^D(j, l, k) := \sum_{0 \leq t_1, t_2 < n} f[t_1, t_2]\overline{\varphi^D_{j,l,k}[t_1, t_2]} \tag{1}$$

Where j, k, l represent the scale, direction, and position respectively, using a bandpass function:

$$\Psi(\omega_1) = \sqrt{\phi(\omega_1/2)^2 - \phi(\omega_1)^2} \tag{2}$$

Definition $\Psi_j(\omega_1) = \Psi(2^{-j}\omega_1)$, using this function to achieve multi-scale segmentation. For each $\omega = (\omega_1, \omega_2), \omega_1 > 0$, corner window is:

$$V_j(S_{\theta_1}\omega) = V(2^{\lfloor j/2 \rfloor}\frac{\omega_2}{\omega_1} - l) \tag{3}$$

$S_{\theta l}$ is a shear matrix.

$$S_{\theta l} := \begin{pmatrix} 1 & 0 \\ -\tan\theta_l & 1 \end{pmatrix} \tag{4}$$

θ_1 is not equally spaced, but the slope is equally spaced. Delimiting $\widetilde{U}_j(\omega) := \psi_j(\omega_1)V_j(\omega)$, for each $\theta_1 \in [-\pi/4, \pi/4)$, it has

$$\widetilde{U}_{j,1}(\omega) := \psi_j(\omega_1)V_j(S_{\theta_1}\omega) = \widetilde{U}_{j,1}(S_{\theta_1}\omega) \tag{5}$$

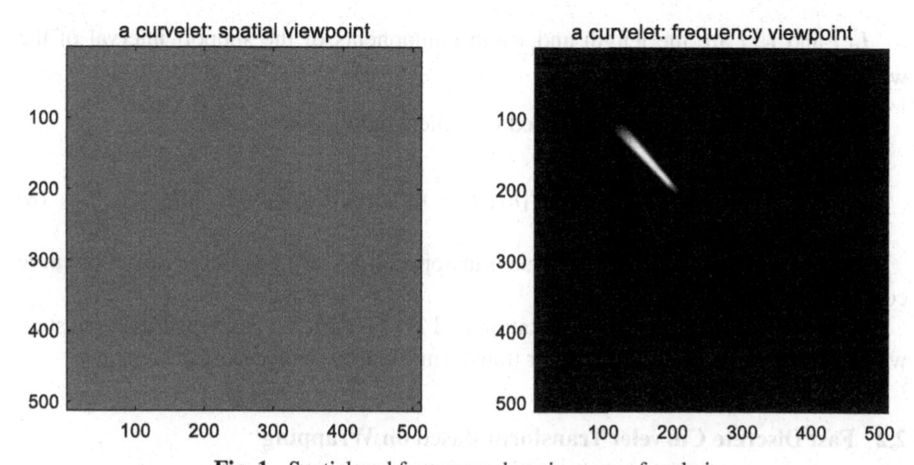

Fig. 1. Spatial and frequency domain maps of scale j

The discrete Curvelet transform method first transforms into the frequency domain and then localizes it in the frequency domain. After localization, the inverse two-dimensional fast Fourier transform is used to obtain the curve coefficient. Figure 1 is a spatial and frequency domain image where the scale is *j*. There are two discrete Curvelet transform algorithms. The first is the discrete Curvelet transform algorithm USFFT (Unequally-Space Fast Fourier Transform) based on fast Fourier transform in non-equivalent space. The second is the fast discrete Curvelet transform algorithm based on wrapping.

2.1 Fast Discrete Curvelet Transform Based on USFFT

The fast discrete Curvelet transform method based on USFFT first transforms the frequency domain [29–32], then localizes in the frequency domain, and then uses two-dimensional Fourier transform after localization. The implementation processes of USFFT are as follows:

Step1: For a given two-dimensional function $f[t_1, t_2]$, $0 \le t_1, t_2 \le \omega$ in Cartesian coordinates, performing 2DFFT(Two-Dimensional Fast Fourier Transform), to get a two-dimensional frequency domain representation:

$$\hat{f}[n_1, n_2],\ -n/2 \le n_1,\ n_2 \le n/2 \tag{6}$$

Step2: In the frequency domain, for each pair (j, l), j is scale, l is angle, re-sampling $\hat{f}[n_1, n_2]$, getting sampled value $\hat{f}[n_1, n_2 - n_1 \tan\theta_1]$,$(n_1, n_2) \in P_j$, among them,

$$P_j = \{(n_1, n_2) : n_{1,0} \le n_1 < n_{1,0} + L_{1,j}, n_{2,0} \le n_2 < n_{2,0} + L_{2,j}\} \tag{7}$$

$L_{1,j}$ and $L_{2,j}$ are the length and width components of the support interval of the window function $\tilde{U}_j[n_1, n_2]$.

Step3: Multiplying the interpolated $\hat{f}$ by the window function $\tilde{U}_j$:

$$\tilde{U}_{j,1}[n_1, n_2] = \hat{f}[n_1, n_2 - n_1 \tan\theta_1]\tilde{U}_j[n_1, n_2] \tag{8}$$

Step4: Performing 2DIFFT inverse transform on $\tilde{f}_{j,1}$, we get a discrete set of curve coefficients $C^D(j, l, k)$.

Mathematically speaking, localization and 2D FFT can be combined into one step, which is to multiply the local Fourier transform basis by the localization window.

2.2 Fast Discrete Curvelet Transform Based on Wrapping

The core idea of fast discrete Curvelet transform based on Wrapping [33–35] is to wrap around the origin, that is to say, in the specific implementation, any area is mapped to the affine area of the origin one by one through periodic techniques. Curvelet wrapping wave is shown in Fig. 2.

The algorithm processes are as the follows:

Step1: Fourier transform of two-dimensional image data in a Cartesian coordinate, to get a frequency domain representation $\hat{f}[n_1, n_2]$,$-n/2 \le n_1$,$n_2 \le n/2$.

Step2: For each scale and direction parameter (j, l), interpolation method is used to obtain $\hat{f}[n_1, n_2 - n_1 \tan\theta_1]$ for $\hat{f}[n_1, n_2]$.

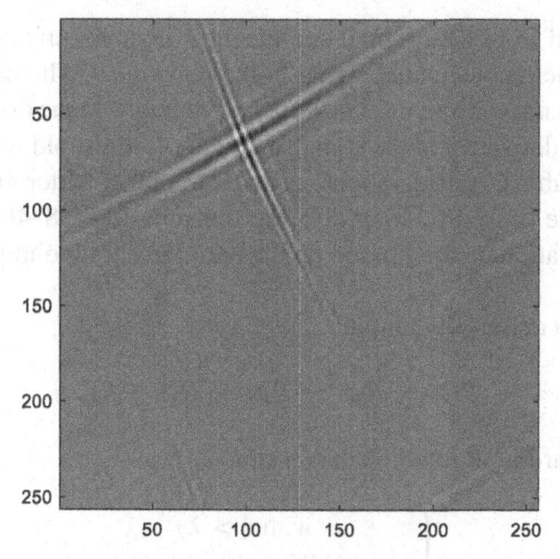

Fig. 2. Curvelet wrapping_wave

Step3: Multiply $\hat{f}$ with the fitting window $\hat{U}_j$ to get $\tilde{f}_{j,l}$.

$$\hat{f}_{j,l}[n_1, n_2] = \hat{f}[n_1, n_2 - n_1 \tan\theta_1]\tilde{U}_j[n_1, n_2] \tag{9}$$

Step4: Wrapping around the origin, localizing $\hat{f}$.

Step5: Inversing $\tilde{f}_{j,l}$ with 2DFFT, this gives the discrete Curvelet coefficient set $C^D(j, l, k)$, j, l, k represent the scale, direction, and matrix coordinates of each direction.

These two fast discrete Curvelet transform algorithms have the same output, but the latter is faster and more efficient. When the edge contour direction is consistent with the direction of the Curvelet wave, there will be a large Curvelet coefficient. Otherwise, the Curvelet coefficient is close to zero. When the speckle noise or debris that is much smaller than the target object which appears in the image, applying the Curvelet transform not only makes it easy to filter it out, but also does not lose edge details, which is helpful for accurate edge extraction.

3 Improved Image Denoising Algorithm

According to the Curvelet transform theory, a larger Curvelet coefficient corresponds to a stronger edge, and a noise corresponds to a smaller coefficient. Although the threshold method can be used to obtain a good denoising effect and local features such as image edge can be retained well, it will cause visual distortion such as ringing, pseudo Gibbs effect [36, 37]. So the image display effect is not ideal. The hard threshold function will lose a lot of detailed information in the denoised image, because the hard threshold is discontinuous at the threshold point, which will overkill the transform coefficients. Although the soft threshold function has good overall continuity, it is an amount that subtracts a threshold from the original coefficient, which directly affects the degree of approximation of the reconstructed signal to the real signal. The derivative of the soft threshold function is not continuous, but we need to operate on the first derivative, second derivative or higher derivative of the signal, and the soft threshold is not suitable for this case. Based on the above considerations, in order to obtain better medical visual effect and reduce the noise caused by long-distance transmission [38–40], this paper uses a polynomial interpolation method based on the hard threshold to improve the denoising algorithm.

Define p (x) as a cubic polynomial:

$$P(x) = R_0x^3 + R_1x^2 + R_2x + R_3 \tag{10}$$

The improved threshold function expression is:

$$\hat{w} = \begin{cases} w,\ |w| > \lambda_2 \\ sign(w)P(w),\ \lambda_1 \leq |w| \leq \lambda_2 \\ 0,\ |w| < \lambda_1 \end{cases} \tag{11}$$

From the continuous and differentiable properties, the following relationship is obtained:

$$P(\lambda_1) = 0, P(\lambda_2) = \lambda_2,\ P'(\lambda_1) = 0, P'(\lambda_2) = 1 \tag{12}$$

The cubic function has high-order differentiability.

$$P(x) = \frac{\lambda_1+\lambda_2}{(\lambda_1 - \lambda_2)^3}x^3 - \frac{2\lambda_1^2+2\lambda_1\lambda_2+2\lambda_2^2}{(\lambda_1 - \lambda_2)^3}x^2 + \frac{\lambda_1^3+\lambda_1^2\lambda_2+4\lambda_1\lambda_2^2}{(\lambda_1 - \lambda_2)^3}x - \frac{2\lambda_1^2\lambda_2^2}{(\lambda_1 - \lambda_2)^3} \tag{13}$$

Where λ_1, λ_2 meet the conditions: $0 < \lambda_1 < \lambda < \lambda_2$. Interpolating with a cubic polynomial in the interval $\lambda_1 \leq |w| \leq \lambda_2$, the improved threshold function is continuous and differentiable at λ_1, λ_2. The threshold function is continuous as a whole and has high-order derivability. λ_1,λ_2 can be selected according to actual needs. When the values of λ_1 and λ_2 are almost equal, the threshold function is the hard threshold function. In addition, adjusting λ_2 when it is greater than the threshold can keep the details well, which overcomes the defect of soft threshold to some extent, so it has the incomparable flexibility of soft and hard threshold function.

In order to suppress the pseudo Gibbs phenomenon caused by the lack of invariance of the transformation during the threshold denoising process, this paper uses a translation image to change the position of the discontinuous points. The Cycle spinning method performs cyclic translation on the signal. This method first performs cyclic translation on the noisy signal, then performs threshold denoising, and finally performs reverse cyclic translation.

Suppose the translation is $(i, j), i \in (0, M), j \in (0, N)$, M and N are image width and height. After each translation denoising, we will get a result $Y_{i,j}$, the two-dimensional image can translate the row and column at the same time. The linear average of the denoising results can suppress the pseudo Gibbs phenomenon and get the denoising image $\hat{Y}(i, j)$. The specific process is shown in the following formula:

$$\hat{Y}(i,j) = \frac{1}{K_1 K_2} \sum_{i=1,j=1}^{K_1 K_2} Y_{-i,-j}(F^{-1}(T[F(Y_{i,j})])) \tag{14}$$

In the above formula, K_1, K_2 are the maximum translation of rows and columns. Full translation is based on the entire signal length. T is the threshold, F and F^{-1} are wavelet transform and inverse transform, $Y_{i,j}$ represents panning the image. $Y_{-i,-j}$ represents the opposite translation operation after denoising. This method can well suppress the pseudo Gibbs phenomenon. The translation invariant wavelet denoising method can also reduce the RMSE(Root Mean Square Error) between the original signal and the estimated signal.

In this paper, based on the Curvelet transform, combining the Wrapping method and Cycle spinning, the pseudo Gibbs effect is suppressed, and the improved polynomial interpolation threshold function method is used to denoise. The specific methods are as follows:

Step1: The noisy image is cyclically translated with a translation amount of 1. The translation method is as described above.

Step2: The Wrapping-based Curvelet transform is performed on the translated image to obtain the discrete Curvelet coefficient set $C^D(j, l, k)$ at all scales and directions.

Step3: The improved polynomial interpolation thresholding function is used to denoise the Curvelet coefficients at different scales and directions.

Step4: Performing Curvelet inverse transform, and translate to reconstruct the image to get the denoised image.

Step5: Reverse cycle translation of denoised image, repeat the above steps and average the results after iteration, and the final denoising result $x(i, j)$ is obtained.

This method can not only effectively remove the pseudo Gibbs phenomenon and show better surface quality, but also can obtain smaller RMSE and improve the signal-to-noise ratio.

4 Simulation Experiment and Performance Analysis

In order to test the superiority of the new denoising algorithm, we selected a human lung medical image with a size of 256 × 256 for experiments in the Matelab2018 environment. The original image is shown in Fig. 3. Figure 4 is a comparison of Wrap Curvelet and the new algorithm in different noise standard deviations of 5, 10, 15, 20, and 30. Table 1 is the results of PSNR under different noise standard deviations. Test methods include: mean filtering denoising, wavelet threshold denoising, Wrap Curvelet algorithm denoising and denoising algorithm of this paper.

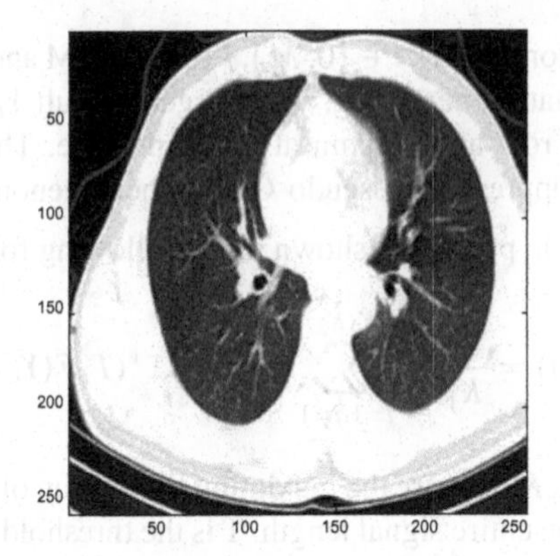

Fig. 3. The original image

As shown in Fig. 4, compared with Wrap Curvelet, the algorithm of this paper has a clearer image of different noise standard deviation. The new algorithm can better restore the texture in the image, and retains the edge information better with different noise standard deviation.

As shown in Table 1, from the perspective of PSNR, among the four denoising methods, wavelet threshold based method is better than mean filtering, and Wrap Curvelet based method is better than wavelet threshold method. Combined with other algorithms, the new algorithm has the highest PSNR value, better reduces the pseudo Gibbs phenomenon and ringing effect in the image. It also reduces the RMSE of the signal, better preserves the high frequency details and texture of the image.

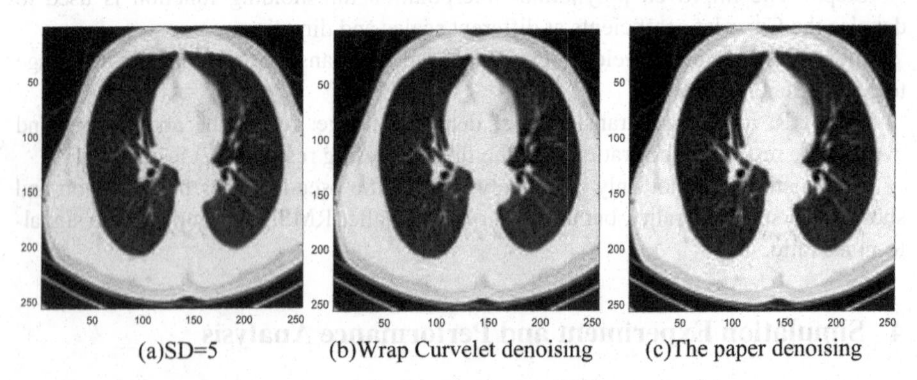

(a)SD=5 (b)Wrap Curvelet denoising (c)The paper denoising

Fig. 4. Comparison of denoising results of different algorithms

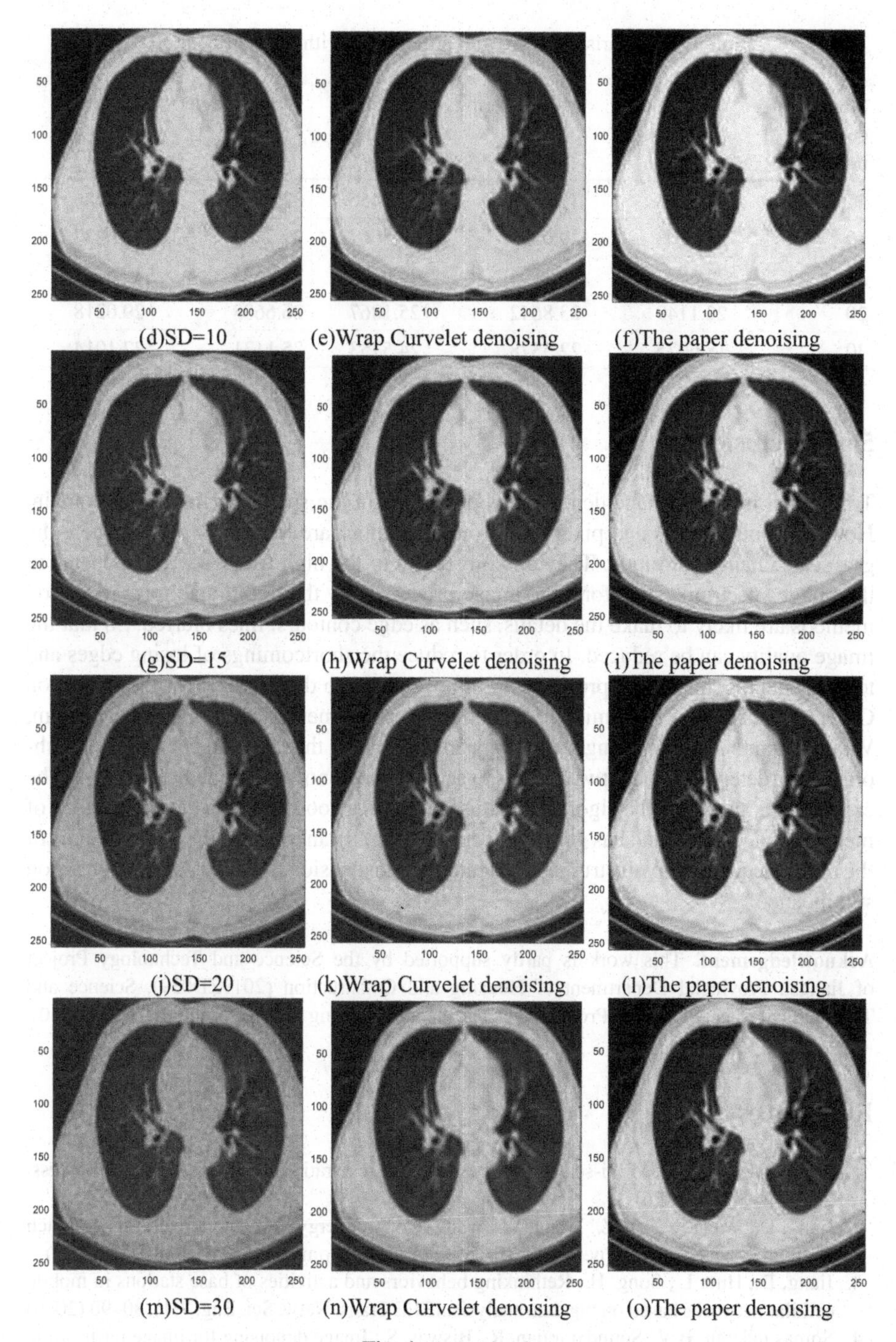

Fig. 4. (*continued*)

Table 1. Comparison PSNR with different algorithms in different SD

Standard deviation of noise	Before noise reduction	Mean filtering	Wavelet threshold	Wrap curvelet	This paper
5	31.1235	31.4453	31.8456	32.3592	36.4286
10	28.1703	29.0435	29.5667	29.7805	32.6237
25	24.587	25.1864	26.6482	27.9592	30.5172
20	22.1149	23.8642	25.1867	26.6603	29.0218
30	18.6075	22.5378	24.5753	25.1131	27.1014

5 Conclusion

Traditional medical image denoising is processed in the spatial or frequency domain. However, noise, edges and pixel points on the contour are located in places where the grayscale changes abruptly. They all correspond to the high frequency components in the image spectrum. Therefore, the image processed by the traditional noise reduction methods are likely to make the details, such as edge contours, lines blurred. So that the image quality can be reduced. In order to reduce the shortcomings of losing edges and textures details, this paper proposes an improved image denoising algorithm based on Curvelet transform. Polynomial interpolation threshold method is used in this algorithm, Wrapping and Cycle spinning technology are combined, the Curvelet coefficient threshold can be determined adaptively, so as to achieve the denoising of medical images. The experiments show that the algorithm in this paper has a good effect on the suppression of medical image noise, and has good subjective vision. It can improve the shortcomings of the image acquisition by instrument and remote transmission, it has practical application significance.

Acknowledgement. This work is partly supported by the Science and Technology Project of Jiangsu Provincial Department of Housing and Construction (2019ZD039), Science and Technology Project of Jiangsu Provincial Department of Housing and Construction (2019ZD040).

References

1. Papyan, V., Elad, V.: Multi-scale patch-based image restoration. IEEE Trans. Image Process. **25**(1), 249–261 (2016)
2. Jiang, D., Wang, Y., Lv, Z., Wang, W., Wang, H.: An energy-efficient networking approach in cloud services for IIoT networks. IEEE J. Sel. Areas Commu-n. **38**(5), 928–941 (2020)
3. Jiang, D., Huo, L., Song, H.: Rethinking behaviors and activities of base stations in mobile cellular networks based on big data analysis. IEEE Trans. Netw. Sci. Eng. **7**(1), 80–90 (2020)
4. Somasundaran, B.V., Soundararajan, R., Biswas, S.: Image denoising for image retrieval by casscading a deep quality assessment network. In: 25th IEEE International Conference on Image Processing (ICIP), pp. 525–529. IEEE (2017) https://doi.org/10.1109/ICIP.2018.8451132

5. Ding, Y., Selesnick, I.W.: Artifact-free wavelet denoising: non-convex sparse regularization, Convex Optimization. IEEE Trans. Signal Process. **22**(9), 1364–1368 (2015)
6. Wu, Y., Gao, G., Cui, C.: Improved wavelet denoising by non-convex sparse regularization under double wavelet domains. IEEE Access **7**, 30659–30671 (2019)
7. Li, D., Xiao, L.Q., Tian, J., Sun, J.P.: Mine image stitching based on invariant feature and lifting wavelet. J. Chin. Comput. Syst. **35**(07), 1671–1675 (2014)
8. Candes, E.J.: Ridgelet: theory and applications. Department of Statistics, Stanford University, USA (1998)
9. Mangaiyarkarasi, P., Arulselvi, S.: A new digital image watermarking based on Finite Ridgelet Transform and extraction using ICA. In: International Conference on Emerging Trends in Electrical and Computer Technology, pp. 837–841 (2011)
10. Mahdinejad, N., Mota, H.O., Silva, E.J., Adriano, R.: Improvement of system quality in a generalized finite-element method using the discrete curvelet transform. IEEE Trans. Magn. **53**(6), 1–4 (2017). https://doi.org/10.1109/TMAG.2017.2659652
11. Guo, J.M., Prasetyo, H., Farfoura, M.E., Lee, H.: Vehicle verification using features from curvelet transform and generalized gaussian distribution modeling. IEEE Trans. Intell. Transp. Syst. **16**(4), 1989–1998 (2015)
12. Jeng-Miller, K.W., Yonekawa, Y.: Telemedicine and pediatric retinal disease. Int. Ophthalmol. Clin. **60**(1), 47–56 (2020)
13. Ray, K.N., Mehrotra, A., Yabes, J.G., Kahn, J.M.: Telemedicine and outpatient subspecialty visits among pediatric medicaid beneficiaries. Acad. Pediatr. **20**(5), 642–651 (2020)
14. Huo, L., et al.: An intelligent optimization-based traffic information acquirement approach to software-defined networking. Comput. Intell. **36**(1), 151–171 (2019)
15. Hernando-Requejo, V., Huertas-González, N., Lapeña-Motilva, J., Ogando-Durán, G.: The epilepsy unit during the covid-19 epidemic: the role of telemedicine and the effects of confinement on patients with epilepsy. Neurología (English Edition) **35**(4), 274–276 (2020)
16. Wang, F., Jiang, D., Qi, S.: An adaptive routing algorithm for integrated information networks. China Commun. **7**(1), 196–207 (2019)
17. Huo, L., et al.: An AI-based adaptive cognitive modeling and measurement method of network traffic for EIS. Mob. Netw. Appl. pp. 1–11 (2019)
18. Jiang, D., et al.: Big data analysis based network behavior insight of cellular networks for industry 4.0 applications. IEEE Trans. Ind. Inform. **16**(2), 1310–1320 (2020)
19. Liu, Y., Liu, S., Wang, Z.: A general framework for image fusion based on multi-scale transform and sparse representation. Inf. Fusion. **24**, 147–164 (2015)
20. Jiang, D., Huo, L., Li, Y.: Fine-granularity inference and estimations to network traffic for SDN. PLoS ONE **13**(5), 1–23 (2018)
21. Hardalac, F., Yaşar, H., Akyel, A., Kutbay, U.: A novel comparative study using multi-resolution transforms and convolutional neural network (cnn) for contactless palm print verification and identification. Multimedia Tool Appl. **79**, 22929–22963 (2020)
22. Ja'Afar, N.H.: Implementation of fast discrete curvelet transform using field-programmable gate array. Int. J. Adv. Trends Comput. Sci. Eng. **9**(1.2), 167–173 (2020)
23. Vyas, R., Kanumuri, T., Sheoran, G., Dubey, P.: Efficient iris recognition through curvelet transform and polynomial fitting. Optik **185**, 859–867 (2019)
24. Jiang, D., Zhang, P., Lv, Z., et al.: Energy-efficient multi-constraint ligent optimization-brouting algorithm with load balancing for smart city applications. IEEE Internet Things J. **3**(6), 1437–1447 (2016)
25. Jiang, D., Li, W., Lv, H.: An energy-efficient cooperative multicast routing in multi-hop wireless networks for smart medical applications. Neurocomputing **220**, 160–169 (2017)
26. Khadilkar, S.P., Das, S.R., Assaf, M.H., Biswas, S.N.: Face identification based on discrete wavelet transform and neural networks. Int. J. Image Graph. **19**(04), 634–654 (2019)

27. Mahdinejad, N., Mota, H.O., Silva, E.J., Adriano, R.: Improvement of system quality in a generalized finite-element method using the discrete curvelet transform. IEEE Trans. Magn. **53**(6) 1–4 (2017)
28. Yang, Y., Tong, S., Huang, S.Y., Lin, P., Fang, Y.M.: A hybrid method for multi-focus image fusion based on fast discrete curvelet transform. IEEE Access **5**, 14898–14913 (2017)
29. Ahmed, R., Riaz, M.M., Ghafoor, A.: Attack resistant watermarking technique based on fast curvelet transform and robust principal component analysis. Multimedia Tool Appl. **77**(8), 9443–9453 (2018)
30. Wang, K., Yang, X., Tian, Z., Du, T.: The finger vein recognition based on curvelet. In: Proceedings of the 33rd Chinese Control Conference, pp. 4706–4711. IEEE (2014)
31. Wang, Y., Jiang, D., Huo, L., Zhao, Y.: A new traffic prediction algorithm to software defined networking. Mob. Netw. Appl. pp. 1–10 (2019)
32. Chaki, J., Parekh, R., Bhattacharya, S.: Plant leaf recognition using texture and shape features with neural classifiers. Pattern Recogn. Lett. **58**, 61–68 (2015)
33. Agrawal, D., Karar, V.: Generation of enhanced information image using curvelet-transform-based image fusion for improving situation awareness of observer during surveillance. Int. J. Image Data Fusion **10**(1), 45–57 (2019)
34. Qi, S., Jiang, D., Huo, L.: A prediction approach to end-to-end traffic in space information networks. Mob. Netw. Appl. 1–10 (2019)
35. Elnemr, H., Elnemr, H.A.: Color histogram with curvelet and cedd for content-based image retrieval. Int. J. Comput. Inf. Secur. **15**(12) (2018)
36. Sharif, B., Dharmakumar, R., Labounty, T., Arsanjani, R., Shufelt, C., Thomson, L., et al.: Towards elimination of the dark-rim artifact in first-pass myocardial perfusion mri: removing gibbs ringing effects using optimized radial imaging. Magn. Reson. Med. **72**(1), 124–136 (2014)
37. Veraart, J., Fieremans, E., Jelescu, I.O., Knoll, F., Novikov, D.S.: Gibbs ringing in diffusion MRI. Magn. Reson. Med. **76**(1), 301–314 (2016)
38. Jiang, D., Huo, L., Lv, Z., Song, H., Qin, W.: A joint multi-criteria utility-based network selection approach for vehicle-to-infrastructure networking. IEEE Trans. Intell. Transp. Syst. **19**(10), 3305–3319 (2018)
39. Jiang, D., Wang, W., Shi, L., Song, H.: A compressive sensing-based approach to end-to-end network traffic reconstruction. IEEE Trans. Netw. Sci. Eng. **7**(1), 507–519 (2020)
40. Bitunguhari, L., Manzi, O., Walker, T., Mukiza, J., Clerinx, J.: Pathological features seen on medical imaging in hospitalized patients treated for tuberculosis in a reference hospital in rwanda. Rwanda Med. J. **76**(4), 1–9 (2020)

Research on Image Segmentation of Complex Environment Based on Variational Level Set

Hang Li, Dan Li(✉), Kailiang Zhang, and Chuangeng Tian

Xuzhou University of Technology, Xuzhou 221000, Jiangsu, China

Abstract. An improved image segmentation model was established to achieve accurate detection of target contours under high noise, low resolution, and uneven illumination environments. The new model is based on the variational level set algorithm, which improves the C-V (Chan and Vese) model, fuses the contour and area models to segment the image information, and solves the problem of optimal solution of the energy model by finding the steady-state solution of the partial differential equation. It can improve the calculation accuracy, topological structure adaptability, anti-noise ability, and reduce the light sensitivity effectively. Experiment shows that the new model has good robustness, high real-time performance, and it can effectively improve detection accuracy.

Keywords: Variational level set · Contour model · Image segmentation · C-V model

1 Introduction

In recent years, active contour models have received widespread attention in the fields of machine vision and image segmentation [1–3]. Active contour models include parametric active contours and geometric active contour models [4–6]. The parametric Snake active contour model is a commonly used active image segmentation contour model, but the Snake active contour model is sensitive to noise [7–9], it cannot adaptively change the evolution curve topology structure, and requires the segmentation object to be a closed curve, which will not break during segmentation [10–13]. Therefore, the Snake model is not applicable when detecting scattered targets. With the attenuation of the signal in the long-distance network transmission, the image noise will be increased, the resolution will be reduced, and the accurate detection of the target contour will be affected [14, 15]. Therefore, a robust image segmentation algorithm that is not easily disturbed by noise is needed [16–19]. The variational level set geometric active contour model method is a hot topic for scholars due to its topological self-adaptation ability and model integration ability. The method minimizes the energy function, and obtains the PDE (Partial Differential Equation) [20, 21] partial differential equation of level set evolution [22, 23]. By adding constraint information in the energy function, the image has good topological adaptability during segmentation. Li and his team proposed a level set-based ICP model, which used gray-scale clustering to segment images [24], but there will be

H. Song and D. Jiang (Eds.): SIMUtools 2020, LNICST 370, pp. 691–701, 2021.
https://doi.org/10.1007/978-3-030-72795-6_55

over-segmentation problems for weak-boundary images. The C-V model based on region information proposed by Chan and his team is a non-linear image processing method, which can segment images without obvious texture features, discrete distribution, blurred borders, but it has poor anti-light sensitivity [25].

High noise, uneven illumination, image signal attenuation and low resolution are widely existed in complex environment [26–30]. It is not ideal to use the existing model directly. In this paper, geometric active contour model was studied, the C-V model based on the variational level set was improved, and a new model which fused edge and area information was proposed. The new model combines contour and area models to segment image information. By finding steady-state solutions to partial differential equations, it can better solve the problems of obtaining optimal solutions for energy models, have certain topological structure adaptability and anti-light sensitivity.

2 Variational Level Set Method

The active contour model expresses the deformation of curves and surfaces in the form of parametric curves and surfaces [31, 32]. The active contour model expresses the deformation of curves and surfaces in the form of parametric curves and surfaces [33, 34]. Its parameterization is as follows: $\mathbf{v}(s) = (x(s), y(s)), s \in [0, 1]$, where s is the curve parameter, x, y is the coordinate of the contour point. The expression of total energy of dynamic contour is as follows:

$$E_{snake} = \int_0^1 E_{snake}(\mathrm{v}(s))ds = \int_0^1 E_{int}(\mathrm{v}(s)) + E_{ext}(\mathrm{v}(s))ds \tag{1}$$

E_{int} represents the internal energy generated by curve bending, which makes the model smooth and continuous. E_{ext} represents the external energy of the image, which comes from external constraints or image features and attracts the contour to the image feature location. Under the joint action of internal and external energy, the curve converges to the target boundary and has the minimum energy.

Since Snake's active contour model is an edge-based algorithm, it requires the target to be a closed curve, so no segmentation problems will occur when using it for segmentation. In order to detect multiple targets, the level set method was used in this paper.

2.1 Level Set Method

A closed plane curve can be defined as the level set u(x,y) of a two-dimensional function: $\mathrm{C} = \{(\mathrm{x,y}), u(\mathrm{x,y}) = c\}$.

If C changes, then it can be considered that the u(x,y) changes. A closed curve over time can be expressed as a level set changing with time.

$$\mathrm{C(t)} := \{(\mathrm{x,y}), u(\mathrm{x,y,t}) = c\} \tag{2}$$

When the curve C(t) is evolving, the evolution of the embedded function u(x,y,t) follows the following rules:

Total derivative $\frac{du}{dt} = \frac{\partial u}{\partial t} + \nabla u \cdot \frac{\partial(x,y)}{\partial t} = 0$, due to $\frac{\partial(x,y)}{\partial t} = \frac{\partial C}{\partial t} = V$, so

$$\frac{\partial u}{\partial t} = -\nabla u \cdot V = -|\nabla u| \frac{\nabla u}{|\nabla u|} \cdot V = |\nabla u| N \cdot V = \beta |\nabla u| \tag{3}$$

Where $\beta = V \cdot N$ represents the motion velocity normal vector. The above formula is the basic evolution equation of horizontal set curve.

If u (x,y) > c, (x,y) is outside the closed curve C.

If u (x,y) < c, (x,y) is inside closed curve C.

If u(x,y) = c, (x,y) is on closed curve C.

For convenience, c = 0 is often taken, which is the zero level set of the curve. The evolution of the closed curve C is the evolution of the embedded function which is given the initial value $u_0(x,y)$. As long as the level set of u(x,y) = 0 is obtained at time t, the curve C(t) can be determined.

The evolution process is a curve-oriented evolution of a two-dimensional function in space u(x,y,t).The level set method is a no-argument method, the partial differential equations of which are given in a fixed coordinate system. In the process of curve evolution, there is no need to track topological changes because the changes in topology will be automatically embedded in the numerical changes of u(x,y,t).

2.2 C-V Variational Level Set Model

T.Chan and L.Vese proposed the C-V model, it was also known as the geodesic active area model [35, 36], which can be distinguished by the average gray level of the inner and outer regions of the image [37, 38]. The following energy function was proposed:

$$E(c_1, c_2, C) = \mu \oint_C ds + \lambda_1 \iint_{\Omega_1} (I - c_1)^2 dxdy + \lambda_2 \iint_{\Omega_2} (I - c_2)^2 dxdy \tag{4}$$

In the above formula, c_1 and c_2 are scalars, C represents the curve, the first term represents the full arc length of the curve C, the second term represents the square error between the gray value of the internal area and c_1, and the third term represents the square error between the gray value of the external area and c_2. Only when C is in the correct position, two and three terms can reach the minimum at the same time.

The Heaviside function is introduced in the above formula, and the variational level set method is used to modify the functional of the embedded function u:

$$E(c_1, c_2, u) = \mu \iint_{\Omega} \delta(u)|\nabla u| dxdy + \lambda_1 \iint_{\Omega} (I - c_1)^2 H(u) dxdy + \lambda_2 \iint_{\Omega} (I - c_2)^2 (1 - H(u)) dxdy \tag{5}$$

If u is fixed, the above formula can be minimized relative to c_1 and c_2, and we can get the follow formula:

$$c_j = \frac{\iint_{\Omega_j} I dxdy}{\iint_{\Omega_j} dxdy} \quad j = 1, 2 \tag{6}$$

c_1 is the average value of the input image I inside the curve, and c_2 is the average value of the image I outside the curve. When c_1 and c_2 are fixed, the relative minimization can be obtained as the following formula. A steady state solution of the segmentation results can be obtained by simultaneous.

$$\frac{\partial u}{\partial t} = \delta_\varepsilon \left[\mu div(\frac{\nabla u}{|\nabla u|}) - \lambda_1 (I - c_1)^2 + \lambda_2 (I - c_2)^2 \right] \tag{7}$$

Figure 1 is a gray image with size of 100 * 100 pixels which iterated under the common and the lighting conditions, respectively. In Fig. 1-a, from the left to the right, the images were iterated 0, 30, 40 and 50 times. The convergence speed is fast and the effect is good. The round holes inside the tool can also be accurately detected. In Fig. 1-b, from the left to the right, the images were iterated 0, 50, 100 and 200 times. Due to the effects of light, the C-V model was affected by light, it converged slowly, and finally failed the segmentation.

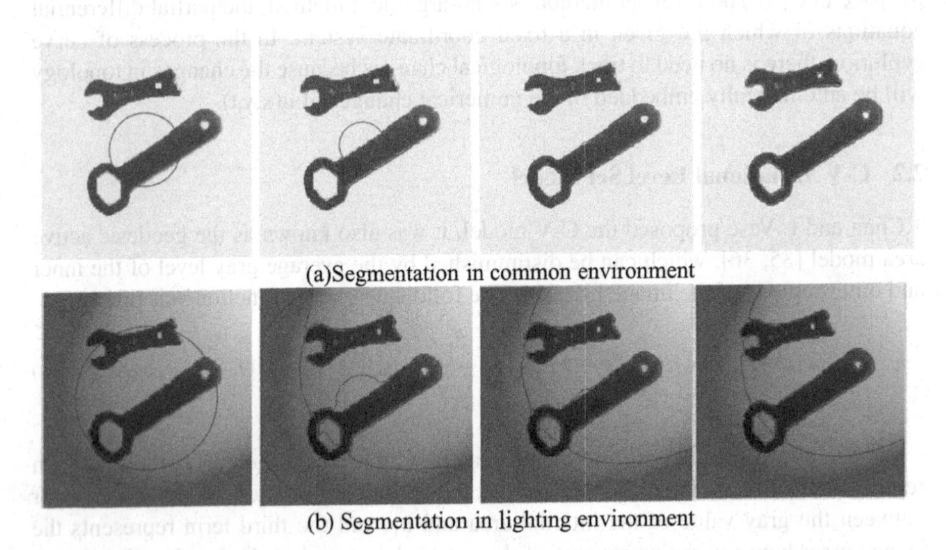

(a)Segmentation in common environment

(b) Segmentation in lighting environment

Fig. 1. Segmentation of C-V model in common and uneven lighting environment

Figure 2 is a gray image with size of 152 * 152 pixels. The C-V model segmentation with different initial values was performed separately. The results are as follows:

As shown in Fig. 2, image is segmented under different initial contour conditions for the same noise. The initial contour in the first line is small, and the numbers of iterations are 0, 40, 60, and 80. There is severe noise interference in the background, but it still converges quickly to the edges. The initial contour of the second image is large, and the numbers of iterations are 0, 50, 100 and 300. Although the number of iterations is higher and the speed is slower than the first line, the final segmentation effect is still very accurate. At 300 frames, it converges completely with the boundary.

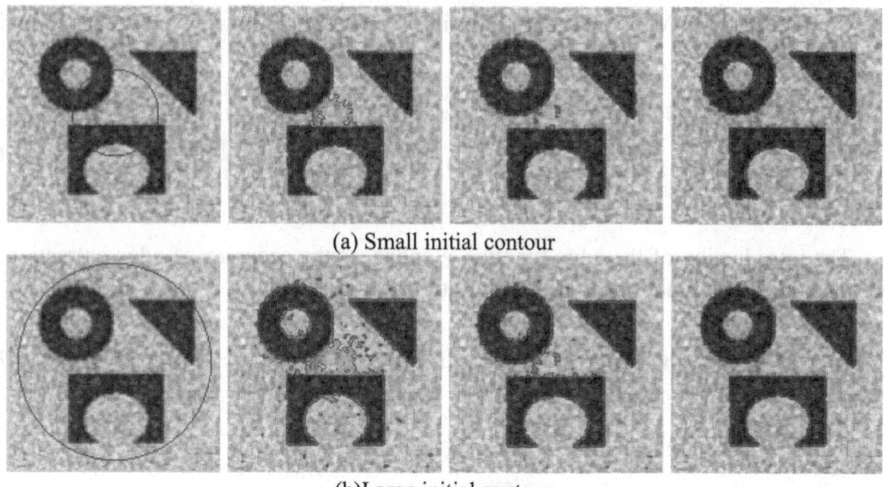

(a) Small initial contour

(b)Large initial contour

Fig. 2. Segmentation of C-V model in noisy environment

Fig. 3. Segmentation of C-V model in noisy and uneven lighting environments

Figure 3 is C-V model segmentation under noisy and uneven lighting environment, The algorithm is very easily interfered by illuminance. The figure shows the results of iterations of 0, 10, 50, and 300 times. It is easy to fall into local extremes and cannot effectively segment the target.

3 Improved Image Segmentation Model

The above experiments show that the C-V model uses the global information of the homogeneous region of the image to effectively improve the ability of adaptive adjustment of the curve topology [39, 40]. In the segmentation process, the image boundary is not required to be clear, and it has good noise immunity. However, it does not consider the inherent characteristics of the level set function, and does not use the target boundary information. Therefore, the edge positioning is not accurate, and it is extremely sensitive to light. Based on the research, this paper proposes a new method to improve the C-V model for the environment with noisy and uneven illumination. The new model combines the boundary-based segmentation method with the region-based segmentation method, which can be relatively complemented.

Combining the C-V model and the edge information, the following energy functional is obtained:

$$E(c_1, c_2, C) = E_{GAC}(C) + E_{C-V}(c_1, c_2, C)$$
$$= \mu \oint_C gds + \lambda_1 \iint_{\Omega_1} (I - c_1)^2 dxdy + \lambda_2 \iint_{\Omega_2} (I - c_2)^2 dxdy \quad (8)$$

The introduction of the edge function g enhances the accuracy of edge extraction. Rewrite the above functional with the variational level set:

$$E(c_1, c_2, u) = \mu \iint_{\Omega} \delta(u) g |\nabla u| dxdy + \lambda_1 \iint_{\Omega} (I - c_1)^2 H(u) dxdy + \lambda_2 \iint_{\Omega} (I - c_2)^2 (1 - H(u)) dxdy \quad (9)$$

For the formula (10):

$$E(u) = \iint_{\Omega} g(x, y) \delta(u) |\nabla u| dxdy \quad (10)$$

The paper improves the above formula. Adding a diffusion term in the energy functional to keep the embedded function as a distance function, it can be transformed into:

$$E(u) = \mu \iint_{\Omega} \frac{1}{2} (|\nabla u| - 1)^2 dxdy + \iint_{\Omega} g(x, y) \delta(u) |\nabla u| dxdy \quad (11)$$

Where u is the constant. The improved variational level set model can make the embedding function $u_0(x, y)$ approximate to the distance function, the work of initializing the embedding function can be greatly simplified, and re-initialization can be completely avoided.

When u is fixed and minimized relative to c_1, c_2, and the following partial differential equation is obtained:

$$\frac{\partial u}{\partial t} = \mu_1 \left[\Delta u - div\left(\frac{\nabla u}{|\nabla u|}\right) \right] + \delta_\varepsilon \left[\mu_2 div\left(g \frac{\nabla u}{|\nabla u|}\right) - \sum_{i=1}^{m} \lambda_{1i} (I - c_{1i})^2 + \sum_{i=1}^{m} \lambda_{2i} (I - c_{2i})^2 \right] \quad (12)$$

The discretization of Δu adopts 4-neighbor difference scheme.

$$(\Delta u)_{i,j} = u_{i+1,j} + u_{i-1,j} + u_{i,j+1} + u_{i,j-1} - 4u_{i,j} \quad (13)$$

The magnitude of the effect of the forced term is controlled by μ. In the case of $\tau\mu \leq 0.25$, the gradient descent flow explicit scheme is stable, and $\tau \approx 0.1$, $\mu \approx 2$ can be taken. Discrete operators need to be discretized. The "half-point discretization" scheme can be used.

$$div\left(\frac{\nabla u}{|\nabla u|}\right) = \frac{\partial}{\partial x}\left(g \frac{u_x}{|\nabla u|}\right) + \frac{\partial}{\partial y}\left(g \frac{u_y}{|\nabla u|}\right) \quad (14)$$

From the above formula, the following formula can be obtained.

$$div(g\frac{u}{|\nabla u|}) \approx g_{i,j+1/2}(\frac{u_x}{|\nabla u|})_{i,j+1/2} - g_{i,j-1/2}(\frac{u_x}{|\nabla u|})_{i,j-1/2} + g_{i+1/2,j}(\frac{u_y}{|\nabla u|})_{i+1/2,j} - g_{i-1/2,j}(\frac{u_y}{|\nabla u|})_{i-1/2,j} \tag{15}$$

Express each term in the above formula with the values of u and g at the Integer value. For example, in the first item of formula (15):

$$(\nabla u)_{i,j+1/2} = ((u_x)_{i,j+1/2}, (u_y)_{i,j+1/2}) \tag{16}$$

$$(u_x)_{i,j+1/2} = u_{i,j+1} - u_{i,j} \tag{17}$$

$$(u_y)_{i,j+1/2} = (u_{i+1,j+1/2} - u_{i-1,j-1/2})/2 = (u_{i+1,j+1} + u_{i+1,j} - u_{i-1,j+1} - u_{i-1,j})/4 \tag{18}$$

Therefore, the following formula can be got.

$$\begin{aligned}(\frac{u_x}{|\nabla u|})_{i,j+1/2} &= (u_x)_{i,j+1/2}/\sqrt{(\nabla u)^2_{i,j+1/2} + (u_y)_{i,j+1/2}} \\ &= (u_{i,j+1} - u_{i,j})/[(u_{i,j+1} - u_{i,j})^2 + (u_{i+1,j+1} + u_{i+1,j} - u_{i-1,j+1} - u_{i-1,j})^2/16]^{1/2}\end{aligned} \tag{19}$$

$$\begin{aligned}&= C_{1,i,j}(u_{i,j+1}, u_{i,j}) \\ &C_{1,i,j} = 1/(u_{i,j+1} - u_{i,j})^2 + (u_{i+1,j+1} + u_{i+1,j} - u_{i-1,j+1} - u_{i-1,j})^2/16]^{1/2}\end{aligned} \tag{20}$$

The value of g at half point can be approximated by the average of two adjacent points:

$$g_{i+1/2,j} = (g_{i+1,j} + g_{i,j})/2 \tag{21}$$

The other three items are also processed like above formulas to get the formula (15). Then the upwind scheme can be used for calculations.

$$\begin{aligned}div(g\frac{\nabla u}{|\nabla u|})_{i,j} \approx\ & C_{1,i,j}g_{1,i,j}(u_{i,j+1} - u_{i,j}) - C_{2,i,j}g_{2,i,j}(u_{i,j} - u_{i,j-1}) \\ & + C_{3,i,j}g_{3,i,j}(u_{i+1,j} - u_{i,j}) - C_{4,i,j}g_{4,i,j}(u_{i,j} - u_{i-1,j})\end{aligned} \tag{22}$$

4 Experiment and Analysis

The experimental environment includes Windows 10 operating system, 8G memory, 3.6 GHZ CPU, and MATLAB 2018. In Fig. 4, (b) is the result of the improved model in the common environment for coin image segmentation. The numbers of iterations are 0, 15, 30 and 50 times respectively. (b) is the result of the improved model in the

(a)Segmentation of coins in common environment

(b)Segmentation of part in low noisy environment

Fig. 4. Segmentation of improved model in common and low noisy environments

Table 1. Number and time of iterations in improved models

Images	Number of iterations	Time (seconds)
(a)	15	0.419718
	30	0.749150
	50	1.170998
(b)	20	0.736783
	40	1.339194
	60	1.727235

low noise environment. The numbers of iterations are 0, 20, 40, 60 times respectively. The starting position in figure (b) is different from figure(a). The improved algorithm has better shrinkage effect no matter it shrinks inward or expands outward, and it can effectively detect the empty area in figure (b). The time consumption in Fig. 4 is shown in Table 1.

The high-noise image and the uneven illumination image are segmented by the new method in Fig. 5.

In Fig. 5, it can be seen that the new model implemented by improving the variational level set has better convergence results for the problems that the above C-V model cannot solve. It has fast convergence speed and good robustness in complex environments such as low contrast, noisy, and blur. During the shrinking process, not only the outer boundary of the target is detected clearly, but also the inner boundary, whether it is concave or hollow, can be detected well. The improved new model has a good effect on the contour

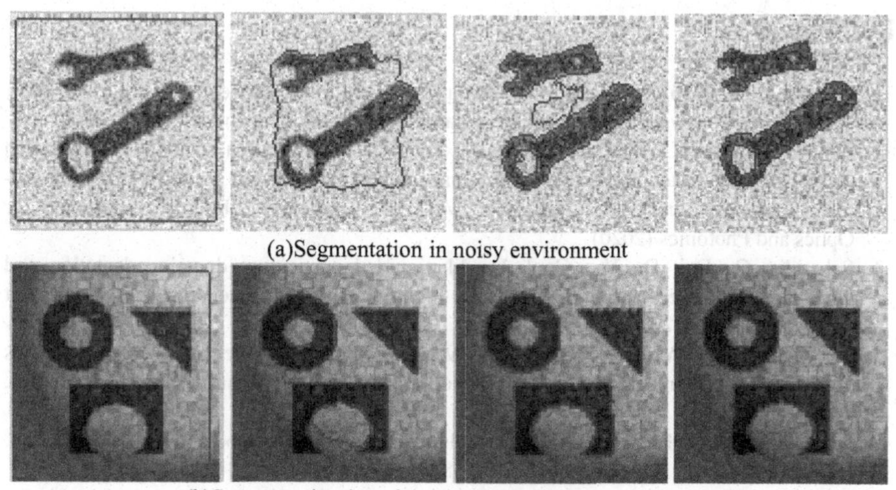

(a)Segmentation in noisy environment

(b)Segmentation in noisy and uneven lighting environment

Fig. 5. Segmentation of improved model in noisy and uneven lighting environments

detection of multiple targets. The results show that the new model proposed in this paper is more accurate than C-V model segmentation and has better ability to resist complex environments.

5 Conclusion

This paper analyzes and studies the image segmentation model based on variational level set. Aiming at the complex environment of uneven illumination, blurred video image and noisy environment, this paper proposes an improved model that combines edge and area information. State solution, the optimal solution of the energy model is obtained, and the numerical calculation is performed by using a half-point discretization difference scheme, thereby improving the calculation accuracy. Experiments show that compared with other models, the complex environment segmentation model established in this paper has more accurate segmentation and high real-time performance. It solves the problems of illumination, noise sensitivity and inaccurate edge positioning of the original model.

Acknowledgement. This work is partly supported by the Science and Technology Project of Jiangsu Provincial Department of Housing and Construction (2019ZD039), Major Projects of Natural Science Research in Universities of Jiangsu Province (19KJA470002).

References

1. Zhou, S., Kan, P., Silbernagel, J., Jin, J.: Application of image segmentation in surface water extraction of freshwater lakes using radar data. ISPRS Int. J. Geo-Inf. **9**(7), 424 (2020)
2. Zhang, Y., Chen, P., Hong, H., Huang, Z., Zhou, C.: The research of image segmentation methods for interested area extraction in image matching guidance. In: MIPPR 2019: Automatic Target Recognition and Navigation, Vol. 11429, p. 114290R International Society for Optics and Photonics (2020)
3. Sakaridis, C., Dai, D., Van Gool, L.: Map-guided curriculum domain adaptation and uncertainty-aware evaluation for semantic nighttime image segmentation. arXiv preprint arXiv:2005.14553 (2020)
4. Xia, G.S., Liu, G., Yang, W., Zhang, L.P.: Meaningful object segmentation from sar images via a multiscale nonlocal active contour model. IEEE Trans. Geosci. Remote Sens. **54**(3), 1860–1873 (2016)
5. Li, H., Gong, M.G., Liu, J.: A local statistical fuzzy active contour model for change detection. IEEE Trans. Geosci. Remote Sens. **12**(3), 582–586 (2015)
6. Jiang, D., Huo, L., Li, Y.: Fine-granularity inference and estimations to network traffic for SDN. PLoS ONE **13**(5), 1–23 (2018)
7. Jiang, D., Zhang, P., Lv, Z., et al.: Energy-efficient multi-constraint routing algorithm with load balancing for smart city applications. IEEE Internet Things J. **3**(6), 1437–1447 (2016)
8. Yu, S., Lu, Y., Molloy, D.: A dynamic-shape-prior guided snake model with application in visually tracking dense cell populations. IEEE Trans. Image Process. **28**(3), 1513–1527 (2019)
9. Jiang, D., Li, W., Lv, H.: An energy-efficient cooperative multicast routing in multi-hop wireless networks for smart medical applications. Neurocomputing **220**(12), 160–169 (2017)
10. Ren, H., Su, Z.B., Lv, C.H., Zou, F.J.: An improved algorithm for active contour extraction based on greedy snake. In: IEEE/ACIS 14th International Conference on Computer and Information Science (ICIS), pp. 589-592 (2015) https://doi.org/10.1109/ICIS.2015.7166662
11. Celestine, A., Peter, J.D.: Investigations on adaptive connectivity and shape prior based fuzzy graph-cut colour image segmentation. Expert Syst. **37**(5), e12554 (2020)
12. Feng, C., Yang, J., Lou, C., Li, W., Zhao, D.: A global inhomogeneous intensity clustering-(GINC-) based active contour model for image segmentation and bias correction. Comput. Math. Methods Med. **2020**(5), 1–8 (2020)
13. Wang, Y., Jiang, D., Huo, L., Zhao, Y.: A new traffic prediction algorithm to software defined networking. Mob. Netw. Appl. pp. 1-10 (2019)
14. Jiang, D., Wang, Y., Lv, Z., Wang, W., Wang, H.: An energy-efficient networking approach in cloud services for IIoT networks. IEEE J. Sel. Areas Commun. **38**(5), 928–941 (2020)
15. Huo, L., et al.: An intelligent optimization-based traffic information acquirement approach to software-defined networking. Comput. Intell. **36**(1), 151-171 (2019)
16. Arbeláez, P., Maire, M., Fowlkes, C., Malik, J.: Contour detection and hierarchical image segmentation. IEEE Trans. Pattern Anal. Mach. Intell. **33**(5), 898–916 (2011)
17. Mariano, R., Oscar, D., Washington, M., Alonso, R.M.: Spatial sampling for image segmentation. Comput. J. **55**(3), 313–324 (2018)
18. Jiang, D., Wang, Y., Lv, Z., Qi, S., Singh, S.: Big data analysis based network behavior insight of cellular networks for industry 4.0 applications. IEEE Trans. Ind. Inform. **16**(2), 1310–1320 (2020)
19. Huo, L., et al.: An AI-based adaptive cognitive modeling and measure-ment method of network traffic for EIS. Mob. Netw. Appl. 1-11 (2019)
20. Avalos, G., Geredeli, P.G.: Exponential stability of a non-dissipative, compressible flow–structure PDE model. J. Evol. Eqn. **20**(1), 1–38 (2020)

21. Xia, M., Greenman, C.D., Chou, T.: PDE models of adder mechanisms in cellular proliferation. SIAM J. Appl. Math. **80**(3), 1307–1335 (2020)
22. Kolářová, E., Brančík, L.: Noise influenced transmission line model via partial stochastic differential equations. In: 2019 42nd International Conference on Telecommunications and Signal Processing (TSP), pp. 492-495. IEEE (2019). https://doi.org/10.1109/TSP.2019.8769101
23. Pels, A., Gyselinck, J., Sabariego, R.V., Schops, S.: Solving nonlinear circuits with pulsed excitation by multirate partial differential equations. IEEE Trans. Magn. **54**(3), 1–4 (2017)
24. Li, C., Huang, R., Ding, Z., Gatenby, J.C.: A level set method for image segmentation in the presence of intensity inhomogeneities with application to MRI. IEEE Trans. Image Process. **20**(7), 2007–2015 (2011)
25. Chan, T.F., Vese, L.A.: Active contours without edges. IEEE Trans. Image Process. **10**(2), 266–277 (2001)
26. Jiang, D., Huo, L., Song, H.: Rethinking behaviors and activities of base stations in mobile cellular networks based on big data analysis. IEEE Trans. Netw. Sci. Eng. **7**(1), 80–90 (2020)
27. Qi, S., Jiang, D., Huo, L.: A prediction approach to end-to-end traffic in space information networks, Mob. Netw. Appl. pp. 1-10 (2019)
28. Jiang, D., Wang, W., Shi, L., Song, H.: A compressive sensing-based approach to end-to-end network traffic reconstruction. IEEE Trans. Netw. Sci. Eng. **7**(1), 507–519 (2020)
29. Jiang, D., Huo, L., Lv, Z., Song, H., Qin, W.: A joint multi-criteria utility-based network selection approach for vehicle-to-infrastructure networking. IEEE Trans. Intell. Transp. Syst. **19**(10), 3305–3319 (2018)
30. Wang, F., Jiang, D., Qi, S.: An adaptive routing algorithm for integrated information networks. China Commun. **7**(1), 196–207 (2019)
31. Li, D., Tian, J., Xiao, L.Q., Sun, J.P., Cheng, D.Q.: Target tracking method based on active contour models combined camshift algorithm. Video Eng. **39**(19), 101–104 (2015)
32. Liu, G., Dong, Y., Deng, M., Liu, Y.: Magnetostatic active contour model with classification method of sparse representation. J. Electr. Comput. Eng. **2020**(9), 1–10 (2020)
33. Zhang, H., Wang, G., Li, Y., Wang, H.: Faster r-cnn, fourth-order partial differential equation and global-local active contour model (FPDE-GLACM) for plaque segmentation in IV-OCT image. SIViP **14**(3), 509–517 (2020)
34. Ali, H., Sher, A., Saeed, M., Rada, L.: Active contour image segmentation model with de-hazing constraints. IET Image Proc. **14**(5), 921–928 (2020)
35. Xiao, J.S., et al.: The improvement of C-V level set method for image segmentation. In: International Conference on Computer Science and Software Engineering, pp. 1106–1109 (2008)
36. Tan, H.Q., et al.: C-V level set based cell image segmentation using color filter and morphology. In: International Conference on Information Science, Electronics and Electrical Engineering, Vol. 2, pp. 1073-1077. IEEE (2014). https://doi.org/10.1109/InfoSEEE.2014.6947834
37. Yu, S., Yiquan, W.: A morphological approach to piecewise constant active contour model incorporated with the geodesic edge term. Mach. Vis. Appl. **31**(4), 1–25 (2020). https://doi.org/10.1007/s00138-020-01083-4
38. Sarotte, C., Marzat, J., Piet-Lahanier, H., Ordonneau, G., Galeotta, M.: Model-based active fault-tolerant control for a cryogenic combustion test bench. Acta Astronautica **177**, 457-477 (2020)
39. Kai, L.I., Jianhua, Z., Shuqing, H., Fantao, K., Jianzhai, W.U.: Target extraction of cotton disease leaf images based on improved C-V model. J. China Agric. Univ. (2019)
40. Lakra, M., Kumar, S.: A CNN-based computational algorithm for nonlinear image diffusion problem. Multimedia Tool Appl. **79**(33), 23887-23908 (2020)

Research on Recognition Method of Test Answer Sheet Based on Machine Vision

Ping Cui[1], Dan Li[1(✉)], Kailiang Zhang[1], Likai Wang[2], and Weiwei Liu[2]

[1] Xuzhou University of Technology, Xuzhou 221000, Jiangsu, China
[2] Traffic Police Detachment of Xuzhou Public Security Bureau, Xuzhou 221000, Jiangsu, China

Abstract. When using the cursor reading technology to mark the answer card, the cursor machine can only be used for special card, which is expensive and difficult to popularize. A new method of answer sheet recognition based on machine vision and image processing was proposed. Firstly, the improved curvelet algorithm was used to preprocess the image to solve the problem of low resolution and high noise caused by different acquisition methods. Secondly, Hough transform was used to detect lines and correct deformation of binary image. Finally, the answer area was segmented, and the vertical and horizontal projections were used to detect the question and option interval, generate grid lines, mark the center of rectangle and judge the option results. Experiments show that this method is accurate, efficient and robust to low resolution, tilt and noise.

Keywords: Answer sheet recognition · Automatic marking · Curvelet algorithm · Hough transform

1 Introduction

With the rapid development of computer science, cursor reading technology [1, 2] has been widely used in large-scale examinations. It is very convenient for objective questions using the cursor reader to avoid the problem of low efficiency and easy to make mistakes compared with manual marking. However, there are many problems that are difficult to be solved in the process of using them: 1) the answer card must be unified, and it is a special-purpose machine for special cards; 2) it transforms the light signal of filling position into electrical signal. When the color is light, the effect is poor; 3) the paper of the answer card has strict requirements in the printing process, the paper must be smooth, thin and thick, uniform, foldable and defaced, and 2B pencil must be used.

Aiming at the problems of low efficiency of traditional marking and special card for cursor reader, which are expensive and difficult to popularize, this paper combines machine vision [3, 4]and digital image processing technology [5–7]to study the preprocessing, feature extraction and recognition judgment technology. In order to make the automatic marking system more flexible and convenient for network transmission [8–10], it should be adaptive to the complex environment such as angle tilt, different style, low resolution, noise and so on.

H. Song and D. Jiang (Eds.): SIMUtools 2020, LNICST 370, pp. 702–714, 2021.
https://doi.org/10.1007/978-3-030-72795-6_56

2 Image Processing Based on Curvelet

After image acquisition, it is easy to have noise, which greatly affects the subsequent image recognition [11–13]. In the preprocessing, the denoising algorithm can improve the image recognition degree and facilitate the later image recognition. E. J. Candes and D. L. Donoho first proposed the second generation Curvelet transform theory based on frequency domain [14–18], which has high approximation accuracy and sparse expression [19–21] ability. The noise and image edge information can be separated well, and the implementation is simple and the redundancy is reduced. It can not only keep the edge, but also suppress the noise well.

In this paper, the improved curvelct transform [22, 23] is used to denoise the answer card image, improve the signal-to-noise ratio and transmission rate [24–26] of the image, and obtain a clearer display effect. Applying curvelet transform to image processing requires its discrete form. Let $f[t_1, t_2], 0 \le t_1, t_2 \le n$ in Cartesian coordinate system be input, its discrete form as is shown in formula (1), and the information of scale, direction and position are respectively represented by j, k and l.

$$c^D(j, l, k) := \sum_{0 \le t_1, t_2 < n} f[t_1, t_2] \overline{\varphi^D_{j,l,k}[t_1, t_2]} \tag{1}$$

$$\Psi(\omega_1) = \sqrt{\phi(\omega_1/2)^2 - \phi(\omega_1)^2} \tag{2}$$

The above formula uses functions $\Psi(\omega_1)$ to realize multi-scale segmentation, which is defined as follows:

$$\Psi_j(\omega_1) = \Psi(2^{-j}\omega_1) \tag{3}$$

For each, $\omega = (\omega_1, \omega_2), \omega_1 > 0$, the corner window is as follows:

$$V_j(S_{\theta_1}\omega) = V(2^{j/2}\frac{\omega_2}{\omega_1} - l) \tag{4}$$

$$S_{\theta l} := \begin{pmatrix} 1 & 0 \\ -\tan\theta_l & 1 \end{pmatrix} \tag{5}$$

In formula (5), where $S_{\theta l}$ is a shear matrix. Defining $\widetilde{U}_j(\omega) := \psi_j(\omega_1)V_j(\omega)$, for each $\theta_1 \in [-\pi/4, \pi/4)$, it has $\widetilde{U}_{j,1}(\omega) := \psi_j(\omega_1)V_j(S_{\theta_1}\omega) = \widetilde{U}_{j,1}(S_{\theta_1}\omega)$.

Figure 1 below shows the spatial and frequency domain map of scale j.

In this paper, the wrapping curve transformation method [27, 28] combined with cycle spinning [29–31] is adopted. The specific steps are as follows:

Step 1: The image with noise is cyclically translated, assuming that the amount of translation is $(i, j), i \in (0, M), j \in (0, N)$, the width and height of the image are M and N respectively. The row and column are translated at the same time, and the denoising results are linearly averaged to get the image $\hat{Y}(i, j)$, as shown in Eq. (6) below.

$$\hat{Y}(i, j) = \frac{1}{K_1 K_2} \sum_{i=1, j=1}^{K_1 K_2} Y_{-i,-j}(F^{-1}(T[F(Y_{i,j})])) \tag{6}$$

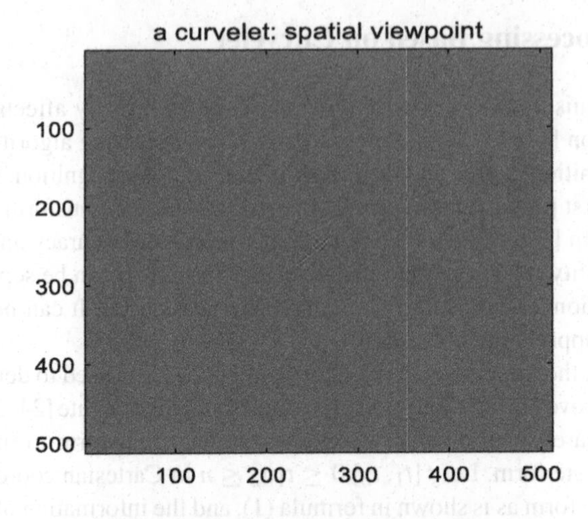

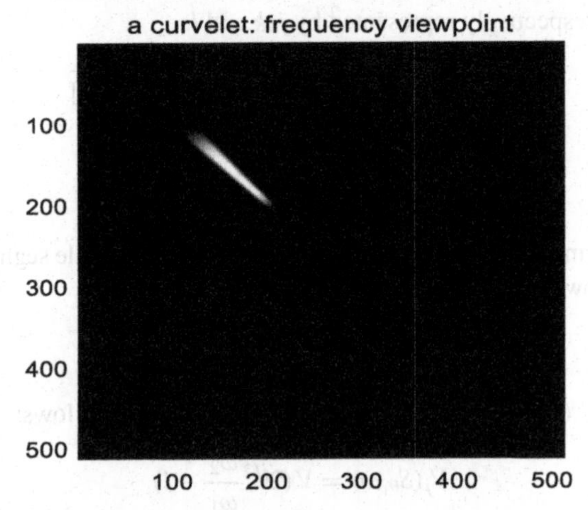

Fig. 1. Spatial and frequency domain map

F represents wavelet transform, F^{-1} represents inverse transformation, T is the threshold, K_1, K_2 are the maximum translations of row and column respectively, $Y_{i,j}$ is the panning image, and $Y_{-i,-j}$ is the reverse panning operation after image denoising.

Step 2: The curved transform based on wrapping is used to process the translated image, and the set $C^D(j, l, k)$ of discrete curved coefficients of each scale and direction is obtained.

Step 3: The polynomial interpolation threshold function is used to denoise the curvelet coefficients of different scale and direction subbands.

Step 4: The curvelet algorithm performs translation operation after inverse transformation, reconstructs the image, and then obtains the denoised image.

Step 5: Repeat the above steps for the image after inverse cycle translation denoising. Finally, average the results of multiple iterations to obtain the final denoising result image $I(i, j)$.

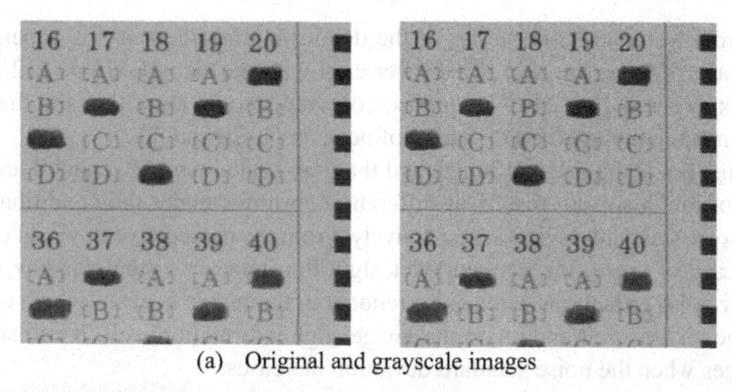

(a) Original and grayscale images

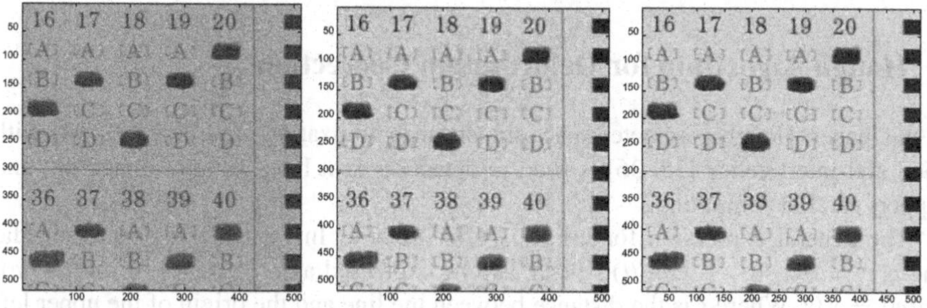

(b) Standard deviation 10, wrapcurvelet and this paper

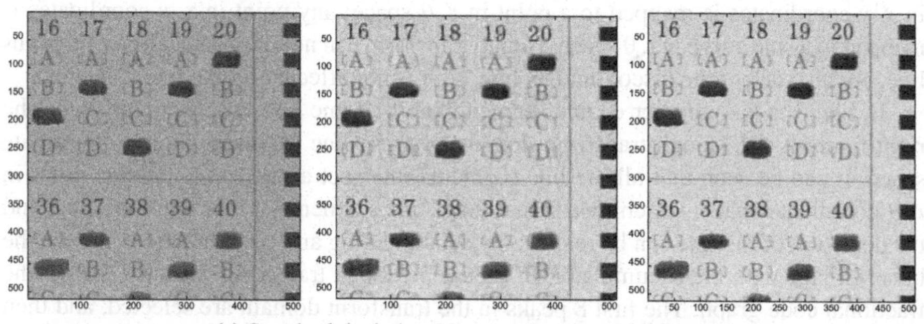

(c) Standard deviation 20, wrapcurvelet and this paper

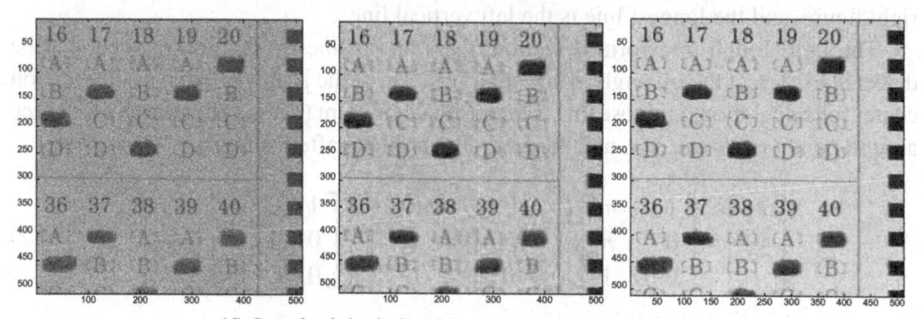

(d) Standard deviation 30, wrapcurvelet and this paper

Fig. 2. Comparison of image denoising algorithms of answer card

In order to test the superiority of the denoising algorithm in this paper, the gray-scale image of some areas of the answer card with the size of 512×512 is selected for the experiment. Fig. 2 is a comparison of wrapcurvelet algorithm and the improved algorithm when the standard deviation of noise is 10, 20 and 30.

In Fig. 2, a is the original image and the gray-scale image, b, c and d are the comparison of the denoising results of different algorithms under the condition of 10, 20 and 30 noise standard deviation respectively. From the perspective of visual effect, with the increase of noise, the wrapcurvelet algorithm appears virtual shadow and image blur after noise reduction. After noise reduction, the image is clear, which can recover the image texture well and retain the image edge information well. It still shows good advantages when the noise standard deviation increases.

3 Hough Algorithm for Deformation Correction

In this paper, Hough parameter space transform optimization algorithm [32–34] with good anti-interference [35, 36] is used to detect straight lines, correct image tilt, and support parallel computing.

The general expression for a line is $y = ax + b$. In Hough transformation, the line is $r = x\cos(\theta) + ysin(\theta)$, It can also be expressed as:$r = \sqrt{x^2 + y^2}sin(\theta + \varphi)$, $\tan(\varphi) = x/y$. Where r is the distance between the line and the origin of the upper left corner of the image, and the angle between the line and the perpendicular is θ. A line in x, y coordinates is mapped to a point in r, θ space; any point in x, y coordinates is mapped to a sine curve in r, θ space. Hough transform can not only detect the continuous line, but also detect the discontinuous line, so it is not affected by the linear fracture.

In Fig. 3, a is a number of line segments on the same line in x, y coordinates. The right figure is the Hough transform domain of these line segments in r, θ coordinate space. It can be seen that all the line segments intersect at the same extreme point in r, θ coordinate space, which means that these line segments are on the same line, and the detection of the line can be realized by accumulating and counting the points on the binary image with the accumulator. In Figure b, Hough transform is used to detect the fractional code graph. The first 8 peaks in the transform domain are selected, and then 8 extreme points are detected in the transform domain. The first 8 lines are shown in the right figure, and the longest line is the left vertical line.

Through Hough transform detection, the straight line in the answer sheet can be detected to correct the tilt angle. The center position of the image is taken as the rotation center for calculation. Suppose that the coordinate of point (x_0, y_0) after turning θ degree around point (s,t) is (x_1, y_1), and the central coordinate after rotation is (m, n), then

$$\begin{bmatrix} x_1 \\ y_1 \\ 1 \end{bmatrix} = \begin{bmatrix} 1 & 0 & m \\ 0 & -1 & n \\ 0 & 0 & 1 \end{bmatrix} \begin{bmatrix} \cos(\theta) & \sin(\theta) & 0 \\ -\sin(\theta) & \cos(\theta) & 0 \\ 0 & 0 & 1 \end{bmatrix} \begin{bmatrix} 1 & 0 & -s \\ 0 & -1 & t \\ 0 & 0 & 1 \end{bmatrix} \begin{bmatrix} x_0 \\ y_0 \\ 1 \end{bmatrix} \tag{7}$$

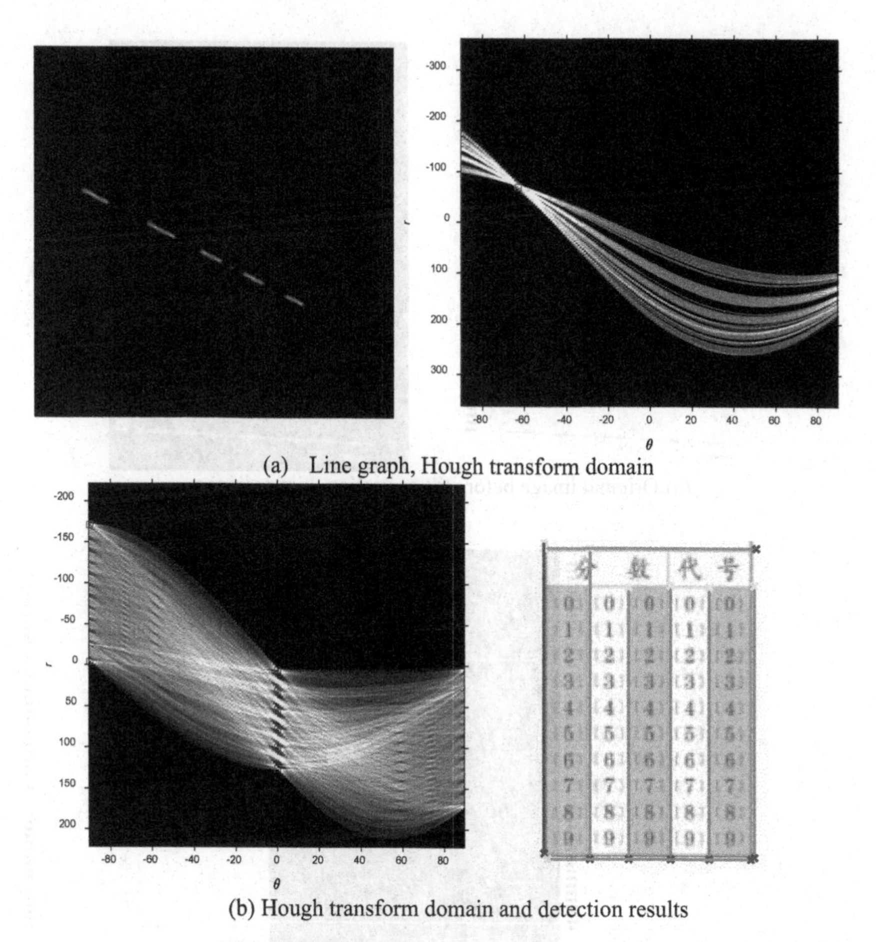

(a) Line graph, Hough transform domain

(b) Hough transform domain and detection results

Fig. 3. Hough transform domain line detection

In Fig. 4, a is the original image and the binary image before the tilt correction, b is the corrected image and the binary image. Using Hough transform, when the number of extreme points in the transform domain is 3, the straight line detection result is obtained. In the detection image, the vertical line on the right is the longest, which can be used to adjust the tilt angle. The other two lines are the horizontal lines that separate the information area and the objective question area of the examinee. You can use the binary image line length of Hough detection to arrange in descending order. The two lines with the longest horizontal direction to get the area segmentation position, so as to realize the segmentation of the examinee's information area and the objective question area.

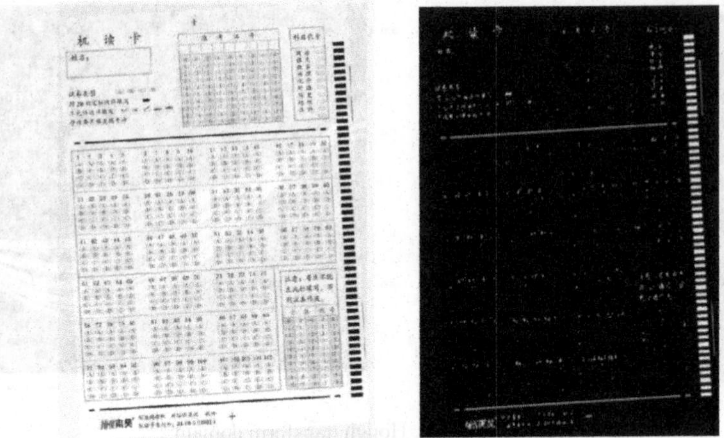

(a) Original image before tilt correction, binary image

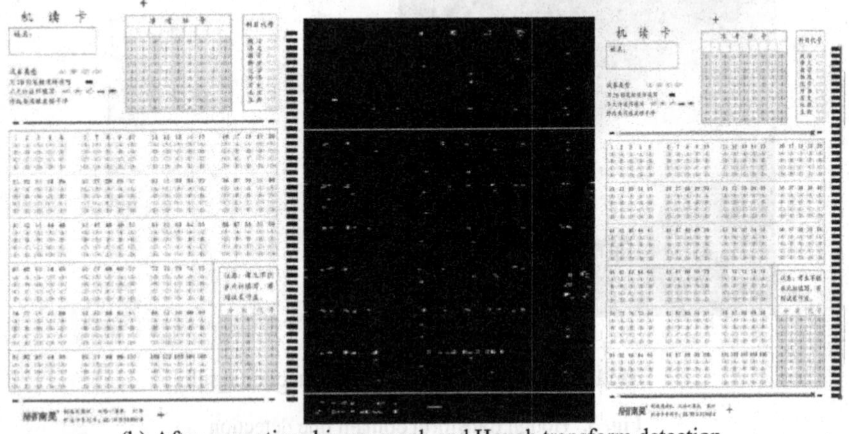

(b) After correction, binary graph and Hough transform detection

Fig. 4. Image tilt correction of answer sheet

4 Option Identification and Judgment

After the correction of the answer card image, it is necessary to locate and segment different areas, complete the detection of the smear area of the answer card options and the identification and judgment of the answers.

4.1 Impurity Removal

The connected area of smear information in the binary image of the answer card needs to calculate the number of pixels in each area. The small noise points contain fewer pixels. Set the threshold T. If the connected area is smaller than T, discard it. The pixels larger than the threshold T are effective pixels. In addition, the burr on the edge of the filling area is further removed by morphological corrosion operation [37–41], and then the edge and internal holes are filled by expansion operation to smooth the filling area.

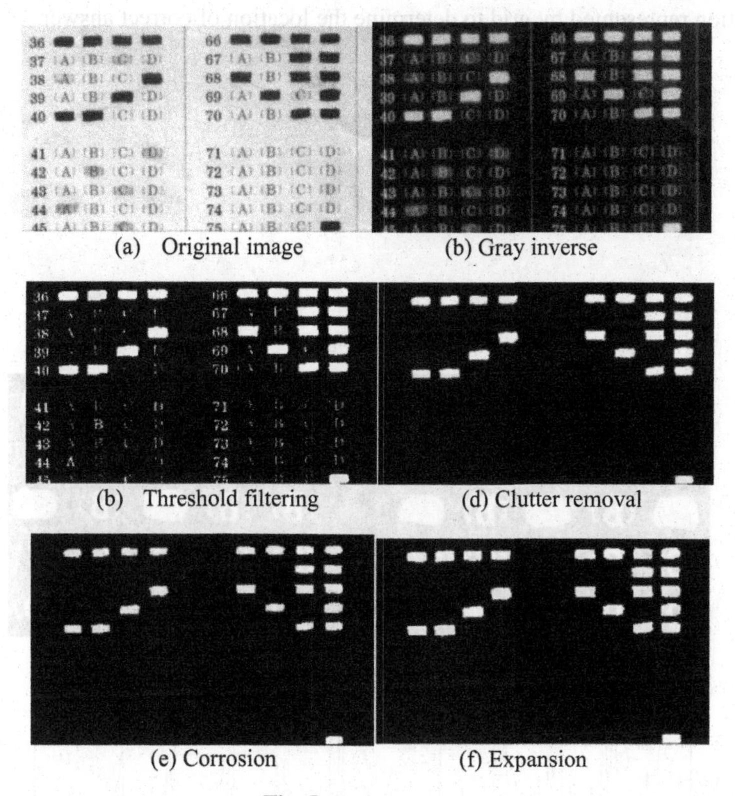

(a) Original image (b) Gray inverse

(b) Threshold filtering (d) Clutter removal

(e) Corrosion (f) Expansion

Fig. 5. Noise removal

In Fig. 5, a and b are the original image and gray-scale inverse image of part of the answer area. Through threshold filtering, the background information, separation vertical line, question number and option information with lighter gray level are removed, as shown in c. If the threshold value is increased, the information such as question number can be further removed, but the pencil smear area will reduce the area and affect the recognition. Therefore, through the above method of removing small area speckles, the connected area with small pixel area is removed while the complete smear information is retained, as shown in d. After further corrosion and expansion, as shown in e, f, the edge of the coating area is smoother.

4.2 Option Positioning and Marking

Generally, the options of the answer card are regular and have the same interval, so the answer area of the answer card can use the horizontal or vertical projection curve of the gray image to determine the distance between the option areas of each row and column of the image. Through the distance between each question and each option, the objective question area of the answer card is divided into grids, each grid corresponds to a question number, or an option. Store and record the question number or option information represented by grid to determine the location of correct answer.

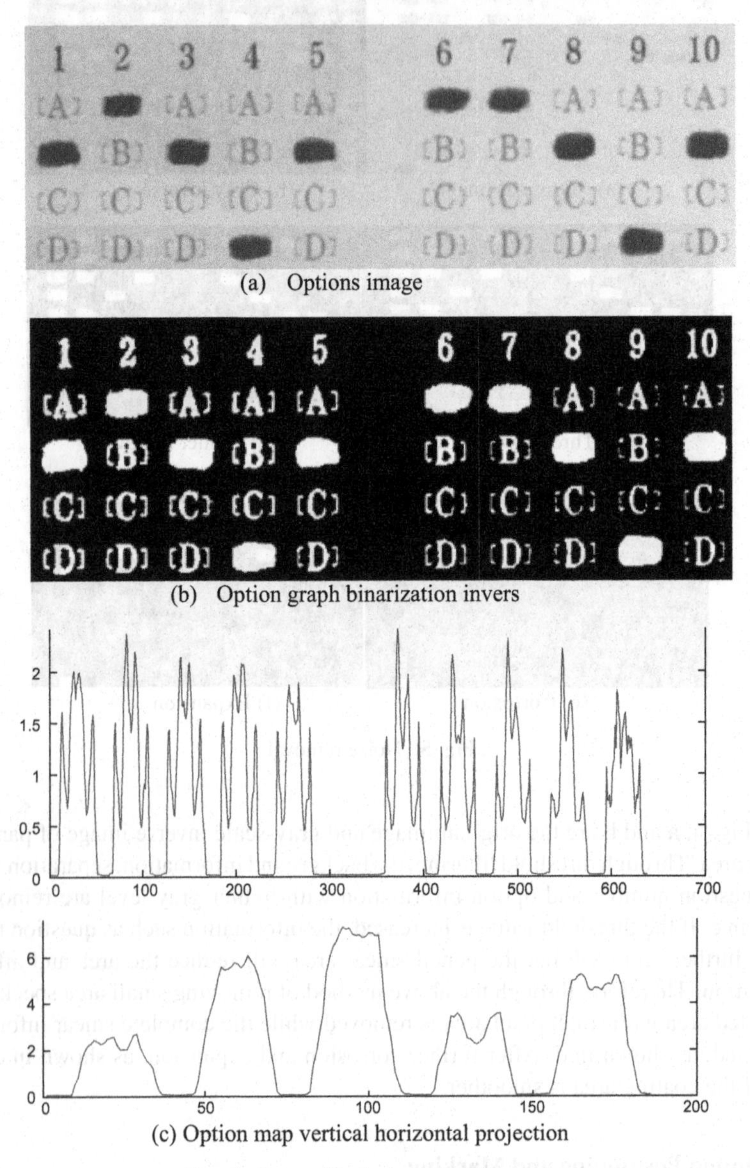

Fig. 6. Interval judgment of option area

As shown in Fig. 6, a is a partial option area image, and b is a binary inverse image of a. According to the projection of b in the vertical and horizontal directions, the horizontal and vertical coordinates in the c represent the pixel width of the option image in the column and row directions, the change of the sum of the gray values of the questions, A, B, C and D, and the valley value represents the binary value in the image. The blank areas between the original pictures of 5 and 6 are the largest, so the valley value is the longest in the vertical projection. In the horizontal projection, the peak value of about 100 pixels in the middle is the largest, because option B is the most selected and the accumulated gray value is the largest. In this paper, the middle point whose valley value is zero in each segment is taken as the grid line to divide horizontally and vertically, so as to realize the grid line division between the questions and between the options.

Locate the connected area of the objective question after filtering out the impurity points and invalid areas, extract the central position of the objective question filling area, and mark it. Compare the location of the answer sheet options according to the central position, and determine which one of A, B, C and D options corresponds to the location, so as to realize the location and marking of the options (Fig. 7).

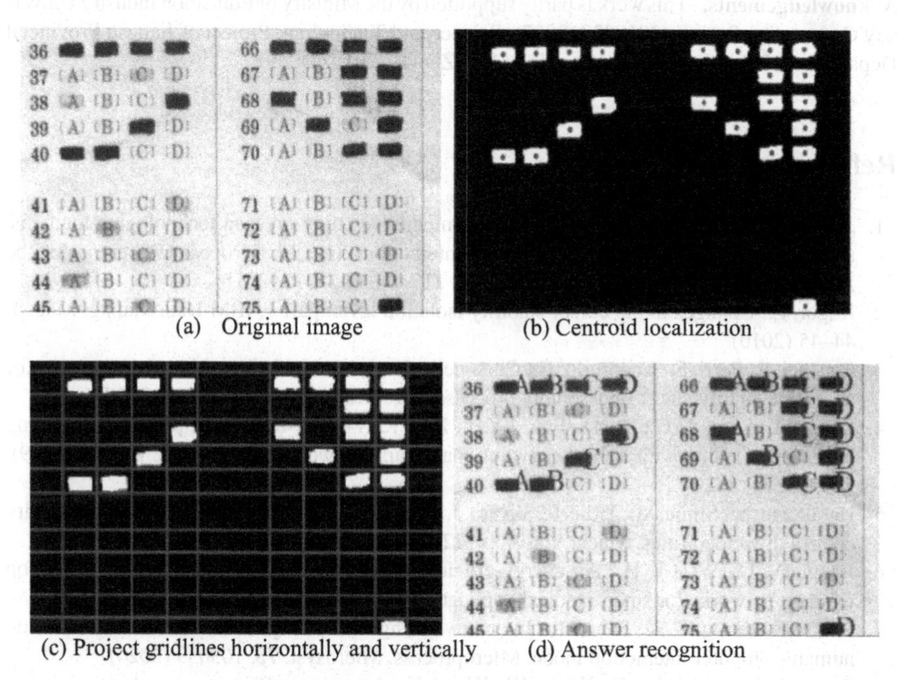

(a) Original image (b) Centroid localization

(c) Project gridlines horizontally and vertically (d) Answer recognition

Fig. 7. Results of filling location and option identification

The above figure shows the process of recognition of smear information in the objective question area of the answer card. After the image is preprocessed and corrected for binarization, the center of mass of the smear area can be accurately extracted, b marks the center of mass of the filled rectangle in a. C is the grid line divided by the horizontal projection and vertical projection of the answer area image. If the center of b is located in the cell corresponding to an option in the grid, then mark the corresponding option value next to the filled area in d. According to the test result of the answer card and the standard answer, we can judge whether the filling result is correct.

5 Conclusion

In this paper, image preprocessing, segmentation, feature extraction and automatic recognition are applied in the marking of answer card. In this method, the image is preprocessed based on curvelet to improve the SNR and image quality. The Hough transform is used to detect the line effectively in polar coordinate space, and the candidates' information area and answer area are segmented effectively while the tilt angle is corrected. Using the horizontal and vertical projection segmentation option, realizes marking the filling position in the answer sheet and results recognition, improves the accuracy of marking papers.

Acknowledgements. This work is partly supported by the Ministry of Education Industry University Cooperation Project (201802312007), Science and Technology Project of Jiangsu Provincial Department of Housing and Construction (2019ZD041, 2019ZD039).

References

1. Jitngernmadan, P., Miesenberger, K.: A comparative study on Java technologies for focus and cursor handling in accessible dynamic interactions. In: 13th European Conference on the Advancement of Assistive Technology. Vol. 217, pp. 267–273 (2015)
2. Rogor, E.: Changes to the cursor stability isolation level: part 1. IBM Data Manag. Mag. **1**, 44–45 (2010)
3. Frustaci, F., Perri, S., Cocorullo, G., Corsonello, P.: An embedded machine vision system for an in-line quality check of assembly processes. Procedia Manuf. **42**, 211–218 (2020)
4. Zhang, Y., Soon, H.G., Ye, D., Fuh, J.Y.H., Zhu, K.: Powder-bed fusion process monitoring by machine vision with hybrid convolutional neural networks. IEEE Trans. Ind. Inf. **16**(9), 5769–5779 (2020)
5. Herakovic, N., Simic, M., Trdic, F., Skvarc, J.: A machine-vision system for automated quality control of welded rings. Mach. Vis. Appl. **22**(6), 967–981 (2011)
6. Jiang, D., Li, W., Lv, H.: An energy-efficient cooperative multicast routing in multi-hop wireless networks for smart medical applications. Neurocomputing **220**, 160–169 (2017)
7. Shu, Y.F., Xiong, C.W., Fan, S.L.: Interactive design of intelligent machine vision based on human–computer interaction mode. Microprocess. Microsyst. **75**, 103059 (2020)
8. Jiang, D., Wang, Y., Lv, Z., Wang, W., Wang, H.: An energy-efficient networking approach in cloud services for IIoT networks. IEEE J. Sel. Areas Commun. **38**(5), 928–941 (2020)
9. Wang, F., Jiang, D., Qi, S.: An adaptive routing algorithm for integrated information networks. China Commun. **7**(1), 196–207 (2019)
10. Huo, L., et al.: An intelligent optimization-based traffic information acquirement approach to software-defined networking. Comput. Intell. **36**(1), 151-171 (2019)
11. Jiang, D., Huo, L., Lv, Z., et al.: A joint multi-criteria utility-based network selection approach for vehicle-to-infrastructure networking. IEEE Trans. Intell. Transp. Syst. **19**(10), 3305–3319 (2018)
12. Huo, L., et al.: An AI-based adaptive cognitive modeling and measurement method of network traffic for EIS. Mob. Netw. Appl. (2019)

13. Jiang, D., Huo, L., Song, H.: Rethinking behaviors and activities of base stations in mobile cellular networks based on big data analysis. IEEE Trans. Netw. Sci. Eng. **7**(1), 80–90 (2020)
14. Hou, S.J., Wu, S.L.: Image denoising by the curvelet transform for doppler frequency extraction. Chin. J. Electron. **17**(1), 178–182 (2008)
15. Raju, C., Reddy, T.S., Sivasubramanyam, M.: Denoising of remotely sensed images via Curvelet transform and its relative assessment. Procedia Comput. Sci. **89**, 771–777 (2016)
16. Altan, A., Karasu, S.: Recognition of covid-19 disease from x-ray images by hybrid model consisting of 2d curvelet transform, chaotic salp swarm algorithm and deep learning technique. Chaos Solitons Fractals **140**, 110071 (2020)
17. Jiang, D., Wang, W., Shi, L., Song, H.: A compressive sensing-based approach to end-to-end network traffic reconstruction. IEEE Trans. Netw. Sci. Eng. **7**(1), 507–519 (2020)
18. He, T., Shang, H.: Direct-wave denoising of low-frequency ground-penetrating radar in open pits based on empirical curvelet transform. Near Surf. Geophys. **18**(3), 295-305 (2020)
19. Li, J., Wang, Y., Xiao, H., Xu, C.: Gene selection of rat hepatocyte proliferation using adaptive sparse group lasso with weighted gene co-expression network analysis. Comput. Biol. Chem. **80**, 364–373 (2019)
20. Wang, Y. Jiang, D., Huo, L., Zhao, Y.: A new traffic prediction algorithm to software defined networking. Mob. Netw. Appl. 1–10 (2019)
21. Kedar, A., Ligthart, L.P.: Wide scanning characteristics of sparse phased array antennas using an analytical expression for directivity. IEEE Trans. Antennas Propag. **67**(2), 905–914 (2019)
22. Jiang, D., Wang, Y., Lv, Z., Qi, S., Singh, S.: Big data analysis based network behavior insight of cellular networks for industry 4.0 applications. IEEE Trans. Ind. Inform. **16**(2), 1310–1320 (2020)
23. Mahdinejad, N., Mota, H.O., Silva, E.J., Adriano, R.: Improvement of system quality in a generalized finite-element method using the discrete Curvelet transform. IEEE Trans. Magn. **53**(6), 1–4 (2017)
24. Jiang, D., Huo, L., Li, Y.: Fine-granularity inference and estimations to network traffic for SDN. PLoS ONE **13**(5), 1–23 (2018)
25. Qi, S., Jiang, D., Huo, L.: A prediction approach to end-to-end traffic in space information networks. Mob. Netw. Appl. 1–10 (2019)
26. Jiang, D., Zhang, P., Lv, Z., et al.: Energy-efficient multi-constraint ligent optimization-brouting algorithm with load balancing for smart city applications. IEEE Internet Things J. **3**(6), 1437–1447 (2016)
27. Luo, J., Chai, S., Zhang, B., Xia, Y., Gao, J., Zeng, G.: A novel intrusion detection method based on threshold modification using receiver operating characteristic curve. Practice and Experience, Concurrency and Computation (2020)
28. Vuppala, A., Krämer, A., Braun, A., Lohmar, J., Hirt, G.: A new inverse explicit flow curve determination method for compression tests. Procedia Manufacturing **47**, 824–830 (2020). https://doi.org/10.1016/j.promfg.2020.04.257
29. Caraka, R.E., Chen, R.C., Toharudin, T., Pardamean, B., Yasin, H.: Ramadhan short-term electric load: a hybrid model of cycle spinning wavelet and group method data handling (CSWGMDH). IAENG Int. J. Comput. Sci. **46**, 670–676 (2020)
30. Luo, P., Yao, W., Susmel, L.: An improved critical-plane method and cycle counting method to assess damage under variable amplitude multiaxial fatigue loading. Fatigue Fract. Eng. Mater. Struct. **43**(9), 2024–2039 (2020)
31. Kozhemyakin, G.N., Kovalev, S.Y., Soklakova, O.N.: Fabrication of bismuth films by a melt spinning method and the influence of annealing on their microstructure. Inorg. Mater. Appl. Res. **11**(3), 727–730 (2020)
32. Fernández, A., Alonso, J.R., Ayubi, G.A., Osorio, M., Ferrari, J.A.: Optical implementation of the generalized Hough transform with totally incoherent light. Opt. Lett. **40**(16), 3901 (2015). https://doi.org/10.1364/OL.40.003901

33. Chandrasekar, L., Durga, G.: Implementation of hough transform for image processing applications. In: International Conference on Communication and Signal Processing, pp. 843–847 (2014)
34. Wei, W., Dong, X.Q., Shen, Y.Y.: Research on a two value generalized Hough transform method of identification. In: Proceedings of 2011 International Conference on Computer Science and Network Technology, ICCSNT 2011, Vol. 1, pp. 278–281 (2011)
35. Zhang, W., Zhu, P., Cheng, L., Zhu, H.: Improved centripetal force type-magnetic bearing with superior stiffness and anti-interference characteristics for flywheel battery system. Int. J. Precis. Eng. Manuf. Green Technol. **7**(3), 713–726 (2020)
36. Ma, J., Yang, Y., Li, H., Li, J.: FH-BOC: generalized low-ambiguity anti-interference spread spectrum modulation based on frequency-hopping binary offset carrier. GPS Solutions **24**(3), 1–16 (2020).
37. Mandhala, V.E.N., Bhattacharyya, D., Kim, T.H.: Face detection using image morphology - a review. Int. J. Secur. Appl. **10**(4), 89–94 (2016)
38. Nazre, B., Rama, C.: Fast detection of facial wrinkles based on Gabor features using image morphology and geometric constraints. Pattern Recogn. **48**(3), 642–658 (2015)
39. Sangpongsanont, Y., Chenvidhya, D., Chuangchote, S., Kirtikara, K.: Corrosion growth of solar cells in modules after 15 years of operation. Sol. Energy **205**, 409–431 (2020)
40. Ryu, H.S., Lee, H.S., Jalalzai, P., Kwon, S.J., Aslam, F.: Sodium phosphate post-treatment on al coating: morphological and corrosion study. J. Therm. Spray Technol. **7**(2), 1–21 (2019)
41. Jafari, A., et al.: Statistical, morphological, and corrosion behavior of pecvd derived cobalt oxide thin films. J. Mater. Sci. Mater. Electr. **30**(24), 21185–21198 (2019). https://doi.org/10.1007/s10854-019-02492-6

A Biomechanical Model and Simulation of Vertebral Support and Lumbar Spinal Discs in Student Backpacks

Meryl Liu(✉)

Spruce Creek High School, Port Orange, FL 32127, USA

Abstract. Nowadays, many students are having to carry excessive amounts of weight in their backpacks every single day. Thousands of children are treated for backpack-related injuries every year, and the continuous strain has been linked in many studies to chronic spinal injuries in later life. The purpose of this project was to develop and simulate a new potential backpack design utilizing a theoretical biomechanical model in order to analyze the efficiency of applying exterior support to alleviate strain on lumbar spinal discs L4/L5 and L5/S1. Using anthropometric data collected from CDC statistics of 16-year-old male and female high school students, a simulation was conducted based on the 10th, 50th, and 90th percentiles, using the 3DSSPP program. The compression and shear forces were compared between the new design and current backpack models, with results demonstrating a significant reduction of forces, (around 44–67% and 25–38% for compression and shear forces, respectively). Thus, it was concluded that the new design was able to help to prevent the discomfort and pain on back, even at a very heavy backpack weight. Results were discussed and future research directions were given to improve the applicability of the new design.

Keywords: Biomechanics · Ergonomics · 3DSSPP

1 Introduction

1.1 Background Information

Increased backpack carriage loads throughout childhood and adolescence have been generally attributed through research to the chronic degeneration of the intervertebral lumbar spinal discs due to the compression of the vertebrae and the stress on the spine, with disc height decreasing over time [1, 2]. As a disc progresses into prolapse and extrusion due to compression, the nucleus pulposus breaks through the anulus fibrosus, causing the final stage of disc herniation [3]. Moreover, there are many parts of the body affected, stemming from long-term wear of high carriage loads, including the appendicular joints, which can sustain 7 lb of weight for each pound of a load. Additionally, potential nerve damage can occur, with increased reports of numbing appendages resulting from improper usage of the student's backpack [4–8]. The cervical area is strained,

H. Song and D. Jiang (Eds.): SIMUtools 2020, LNICST 370, pp. 715–727, 2021.
https://doi.org/10.1007/978-3-030-72795-6_57

and the posterior shoulders can become rounded as the four natural curvatures of the body are disrupted, thus affecting the balance and standing weight distribution of the body as well. Therefore, it is imperative that a proper backpack wearing position is established and enforced [9–11].

Chronic lower back injuries are one of the most common health issues present for American people today. Eighty percent of adults in the United States suffer from some degree of lower back pain (LBP) during their lives, according to the United States National Institute of Health [12]. It was reported that back injuries were the cause of 10–30% of medical claims in the United States and Europe, with the rate of back injuries increasing from previous years. Yet, despite heightened awareness of these statistics and taking action to protect the adult workforce, the Centers for Disease Control and Prevention (CDC) reports that back injuries still account for up to 30% of medical compensations cases every year. Additionally, the American Chiropractic Association estimates that a staggering 3.1 million people are suffering from lower back pain in the United States and that one half of the entire American working community has complained of back pain related to their occupation[13]. The total cost of LBP-related injuries could reach more than $90 billion per year in the U.S [14]. Many of such back-pain problems find their root cause in early age and adolescent habits, as studies showed that more than up to 70% of the teens can have a lifetime prevalence of lower back pain, without treatment or preventive intervention/correction [15].

There are many factors that lead to lower back pain, including physical, bodily conditions and health status, sports, and daily, sustained activities such as lifting weights and carrying loads. Among these aforementioned factors, carrying heavy backpacks are one of the most prevalent factors [16], as it exerts a constant load that must be counteracted by the spine, and together with repetitive, dynamic body movement and improper body posture such as excessive slouching and lordosis, it can easily lead to long-term strain, that develops to LBP and disc herniation at an older age. Nowadays, high school students often are found to be carrying more objects and weight in their backpacks, partially due to the increasing level of study and activities involved in the everyday school, as well as disregard for carrying weights exceeding a tolerable threshold. In conjunction, back pain and shoulder/neck discomfort have become more a common problem reported in high school students as well. School backpacks have been found to be directly associated with back pain, especially the strain and forces exerted in the L4/L5 and L5/S1 lumbar discs of the spine: the lumbar and sacral intersection. Exposed to such heavy stress for a sustained period, this can lead to long term – even permanent – back problems and injuries in later adult life. Children under 18 carrying a heavy backpack have higher risks of musculoskeletal disorders, as they are still in a growing developmental phase; their body is more prone to the injuries of the heavy carriage in the back. The American Chiropractic Association recommends a backpack should not weigh more than 10% of a child's body weight; however [17], most high school students often carry backpacks exceeding recommended load.

Despite general safety guidelines having been published [18], the studies related to adolescence backpack issues are still very limited, partially due to the complexity and multifactorial nature of the lower back problem and large degree of variability in the human anthropometry and shapes/weights of the different backpack. Heavy backpacks mainly impact the human body by posing external forces and torque on the torso, interacting with internal factors and individual characteristics, thus generating greater internal loads on the individual, altering posture and even gaits to become unnatural. An accurate simulation of LBP should account for both external load and internal loads effects, including the individual human characteristics, spinal structure, spine load model, torso kinematics such as body posture, along with the consideration of human body anthropometric data (stature, body mass distribution, segmental dimensions, age, biological sex etc.) [19].

On the other hand, directly and empirically measuring the pressure exerted on the lower back, especially on the fibrocartilaginous lumbar spinal discs, is nearly impossible. Such physical measures could be surgically invasive, unfeasible, and pose risks and danger to the participants. As an alternative, biomechanical models and simulations provide an accurate, safe and effective way to investigate the impact of various external exertions on the human musculoskeletal system. Biomechanical simulation integrates findings and knowledge from the physical and engineering sciences with the principles of human physiology, biology and behavioral sciences, in order to suggest protection from injuries and disorders from these loads [20]. Biomechanical models have been previously utilized in modeling and predicting lower back pain [19].

As a unique population group, the understanding of impact wearing heavy backpacks on adolescents are still limited, most analyses are focused on the evaluation of current available commercial backpacks. There are few studies that take a further step and utilize biomechanical modeling to simulate the effectivity of a proposed, novel design to actively help mitigate forces exerted on the spinal discs and, thus, chronic LBP. In order to prevent LBP from an early age, a comprehensive analysis based on biomechanical modeling is in demand, improvement of current backpack designs should be based on the results derived from these models.

2 Method

2.1 Problem Description

The objective of this study is to utilize a Biomechanical simulation model to investigate the stress/strain on the vertebral column, specifically focusing on lumbar spinal discs L4-L5 and L5-S1, and the musculoskeletal systems caused by the weight of student backpacks and varying standing postures, to compare and analyze how exterior vertebral back support may alleviate possible damage and excessive stress in order to propose and model a conceptual external structure applied to current function student backpacks that provide sufficient support, reinforces a healthy, normal standing posture, and reduces and prevents possible musculoskeletal disorders, injuries to the spine, deformations, etc.

It is hypothesized that if stress on the lumbar spinal discs from heavy backpack wear disproportionate to human body weight is analyzed mathematically, along with external structures that possibly alleviate the amount of strain, then the proposed structure resulting from the elements derived from previous analyses, which provides additional back support, will be able to sustain a healthy standing posture and substantially minimize the amount of strain on the lumbar spinal discs, which would effectively prevent herniation, bulging, compression, and injury inflicted on the spinal area and nearby tissues from heavy backpacks, as the conceptual model would enforce a straight, perpendicular alignment, not leaning forward on the torso and pushes the waist forward to maintain optimal posture. Additionally, a pivot operated by hand from pushing down on handles extending to the front in order to reduce the amount of weight dragging down on the back may also mitigate the amount of stress caused by the heavy backpack compressing down on the spine.

Therefore, the goal of this project would ultimately be to develop and test a theoretical biomechanical model to compare the efficiency of the new design to previous conditions in students and meeting the design criteria of efficiency would require a substantial reduction in the compression and shear forces acting upon the L4-L5 and L5-S1 lumbar spinal discs.

2.2 Biomechanical Model

At the Backpack Level: Figure 1 illustrates the biomechanical model at the backpack level.

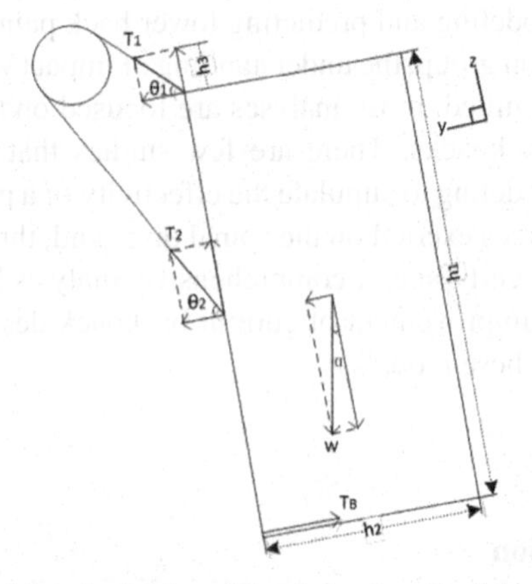

Fig. 1. Biomechanical model at the backpack level. h_1: length of the backpack, h_2: depth of the backpack; w: weight of the backpack; h_3: distance from the top of the backpack to the shoulder; T_1, T_2: forces on the backpack straps; T_B: forces acting on the touching point of the backpack to torso; $\theta_1, \theta_2, \alpha$: angles of the straps and backpack to the reference axis.

According to the Newton's law, at the backpack level, the condition of equilibrium requires

$$\sum \vec{F} = 0, \sum \vec{M} = 0 \tag{1}$$

The force is denoted by F and the torque is denoted by M.

So we have

$$\begin{cases} W \cdot \cos\alpha = T_1 \cdot \sin\theta_1 + T_2 \cdot \sin\theta_2 \\ W \cdot \sin\alpha + T_1 \cdot \cos\theta_1 + T_2 \cdot \cos\theta_2 = T_B \end{cases} \tag{2}$$

And

$$T_B \cdot (h_1 + h_3) = W \cdot \cos\alpha \cdot \frac{h_2}{2} + W \cdot \sin\alpha \cdot (\frac{h_1}{2} + h_3) \quad (3)$$

Giving the value of backpack weight and dimensions, we can derive the value of the forces on shoulders.

At the Shoulder Level: Figure 2 illustrates the biomechanical model at the shoulder level.

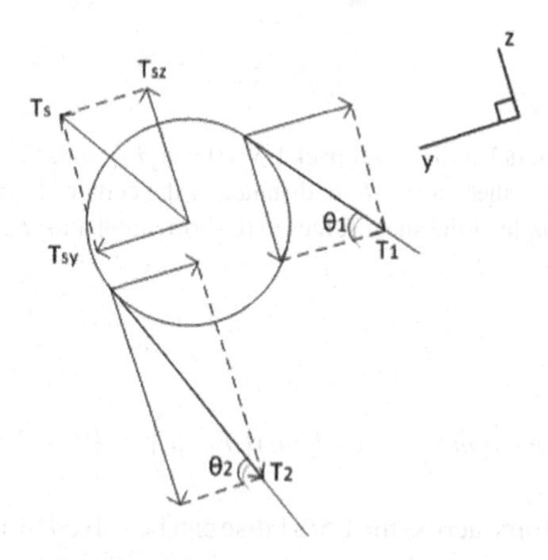

Fig. 2. Biomechanical model at the shoulder level. T_S: counter forces of the shoulders on the backpack straps; T_{SY}, T_{SZ}: forces components on the y and z axis; θ_1, θ_2: angles of the straps to the reference axis.

At the shoulder level, according to the Newton's law,

$$\begin{aligned} T_{SZ} &= T_1 \cdot \sin\theta_1 + T_2 \cdot \sin\theta_2 \\ T_{SY} &= T_1 \cdot \cos\theta_1 + T_2 \cdot \cos\theta_2 \end{aligned} \quad (4)$$

At the Lower Back Level: Figure 3 illustrates the biomechanical model at the L5/S1 level.

$$\sum \overrightarrow{M_{L5/S1}} = 0 \quad (5)$$

Or

$$b\left(\vec{mg}\right) + h\left(\vec{F_{load}}\right) + E(\vec{F_m}) = 0 \quad (6)$$

The forces acting parallel to the disc compression forces can be described as

$$\sum \vec{F_c} = 0 \quad (7)$$

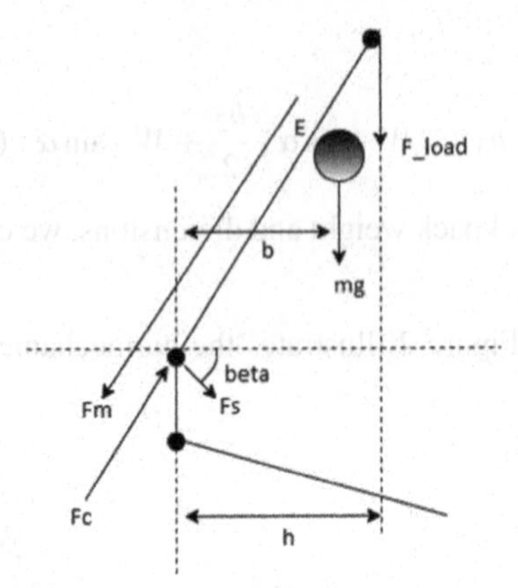

Fig. 3. Biomechanical model at the lower back L5/S1level. F_m: muscle forces on the lower back; F_c: compression force; F_s: sheer force; b, h: distance of the center of gravity and external forces to the lower back; beta: angle of the shear forces to the horizontal axis; E: distance of muscle force to the spine.

Or

$$\cos(beta)\left(\vec{mg}\right) + \cos(beta)\left(\vec{F_{load}}\right) + \vec{F_m} + \vec{F_c} = 0 \tag{8}$$

The reactive shear force across the L5/S1 disc can be solved by the similar equilibrium conditions.

$$\sum \vec{F_s} = 0 \tag{9}$$

Or

$$\sin(beta)\left(\vec{mg}\right) + \sin(beta)\left(\vec{F_{load}}\right) - \vec{F_s} = 0 \tag{10}$$

2.3 Biomechanical Simulation

With data on commercial backpack dimensions and anthropometrics (including weight and dimensions and angles of the torso), we can easily derive the physical forces acting on the shoulders. However, for the L4/L5 and L5/S1 lumbar spinal discs, the measures of compression and shear forces can only be estimated by considering the external forces on the shoulder based on the static model outlined above, and human anthropometric data. This data includes body stature and weight. In this case, a static strength simulation software, 3D SSPP was used to predict the compression force and shear forces operating at the L4/L5 and L5/S1 disc level. Developed by the Center for Ergonomics at the University of Michigan, 3D SSPP is aimed at investigating and analyze human material handling tasks. It utilizes principles and models in Biomechanics to derive the static strength exerted for tasks such as lifting, pushing and carrying loads (Umich.edu). Figure 4 shows a screenshot of the 3D SSPP software.

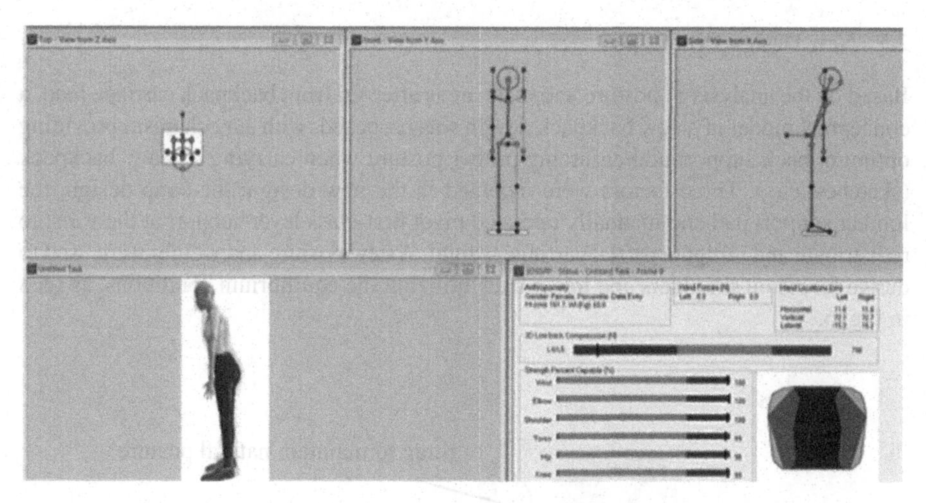

Fig. 4. Screenshot of 3D SSPP.

2.4 Data Collection

Human anthropometric data was retrieved from the Center of Disease Control and Prevention, more specifically, the 10th, 50th (mean) and 90th percentile value on body height and body weight for U.S female and male at the age of 16 (Fig. 5) [21]. The student backpack dimensions were collected from online retail. Using Eqs. (1)–(4), the external forces on the shoulder level and lower back were calculated, by using three levels of backpack weight (7 kg, 10.5 kg and 14 kg, respectively, simulating the backpack weight of 5%, 10% and 15% of the average male body weight).

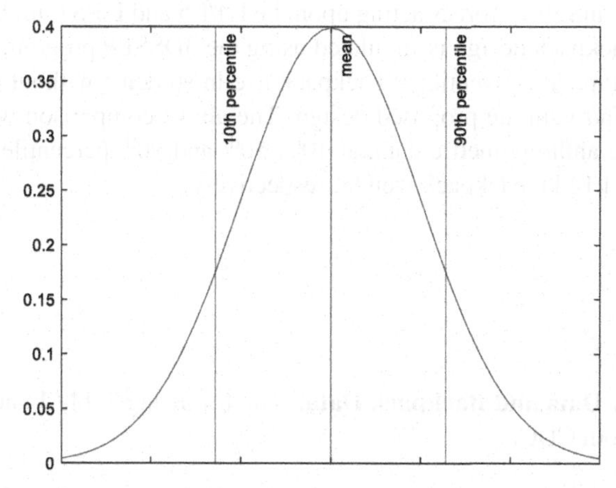

Fig. 5. Illustration of normal distribution percentile value

2.5 New Design Prototype

Based on the analysis of posture and main areas affected from backpack carriage load, a conceptual model of a new backpack design was proposed, with a mechanism providing optimum back support and enforcing proper posture when carrying a heavy backpack (sketches, etc.). Three factors were included in the new design: the strap design, the lumbar support pad and manually operated pivot first-class lever support at the waist to redistribute and mitigate partial carriage weight. A new posture and weight were used to derive forces on shoulders and lower back utilizing the equilibrium conditions, as seen in Fig. 6.

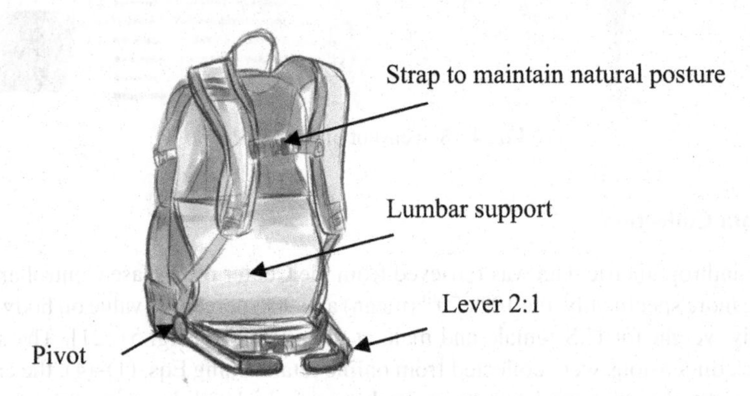

Fig. 6. Illustration of normal distribution percentile value

Therefore, the proposed design would ultimately lead to a substantial reduction in the compression and shear forces acting upon the L4/L5 and L5/S1 lumbar spinal discs. The proposed backpack design is simulated using the 3DSSPP program to compare the forces between typical conventional backpack use in students without enforcement of proper posture and with the proposed design. The above comparison was repeated for female and male anthropometric data, at 10th, 50th and 90th percentile values and for 7 kg, 10.5 kg and 14 kg backpack weight, respectively.

3 Results

3.1 Data

Anthropometry Data and Backpack Data. The following Table 1 shows the anthropometric data from CDC.

Table 1. Anthropometric Data

		10th	50th	90th
Female	Height (cm)	151.9	161.7	172.1
	Weight (kg)	46.9	60.0	95.3
Male	Height (cm)	166.0	174.5	182.3
	Weight (kg)	55.7	68.7	96.1

The backpack data are determined as follows

Length = 18 in. (h_1)
Width = 11 in.
Depth = 8 in. (h_2)
Shoulder distance to backpack = 3 in. (h_3)

3.2 Simulation Results

Based on the backpack weight of 7 kg, 10.5 kg and 14 kg, the lower back simulation results are summarized in Table 2 and 3, for female and male, respectively.

Table 2. Lower back simulation results for female

Weight (KG)	Forces (Newton)	Old design			New design		
		10%	50%	90%	10%	50%	90%
7	L4/L5 compression	644	798	1195	302	377	575
	L4/L5 Shear	140	165	233	82	106	169
	L5/S1 compression	704	872	1305	202	250	376
	L5/S1 Shear	192	230	331	123	158	255
10.5	L4/L5 compression	713	867	1261	403	482	687
	L4/L5 Shear	164	189	256	99	123	186
	L5/S1 compression	787	955	1386	265	316	447
	L5/S1 Shear	221	258	360	144	180	276
14	L4/L5 compression	782	936	1326	503	587	799
	L4/L5 Shear	187	212	279	116	140	204
	L5/S1 compression	869	1037	1466	328	382	517
	L5/S1 Shear	249	287	389	166	202	298

Table 3. Lower back simulation results for male

Weight (KG)	Forces (Newton)	Old design			New design		
		10%	50%	90%	10%	50%	90%
7	L4/L5 compression	773	948	1276	385	472	647
	L4/L5 Shear	167	208	269	104	129	183
	L5/S1 compression	870	1069	1440	261	317	431
	L5/S1 Shear	244	290	383	174	216	304
10.5	L4/L5 compression	842	1008	1333	480	572	752
	L4/L5 Shear	202	231	292	115	140	194
	L5/S1 compression	955	1145	1513	321	380	497
	L5/S1 Shear	275	319	412	196	237	325
14	L4/L5 compression	902	1067	1389	575	672	852
	L4/L5 Shear	225	254	315	127	152	203
	L5/S1 compression	1031	1220	1586	381	443	559
	L5/S1 Shear	303	347	440	217	259	344

To further summarize the results, the main differences between female and male, between the current design and proposed design are illustrated in Table 4.

Table 4. Mean Differences (in Newton/percentage) between two design for female and male

Mean force differences (N/percentage)	Female	Male
L4/L5 compression	423.00/44.87%	459.00/43.50%
L4/L5 shear	66.67/33.93%	90.67/38.11%
L5/S1 compression	699.78/67.00%	804.33/66.92%
L5/S1 shear	79.44/29.44%	82.33/25.10%

Results demonstrated that with the new design (reduced load by the waist/hand lever support) and improved body posture (curved lumbar support and new strap design), the forces of both L4/L5 and L5/S1 discs will be significantly reduced, as shown in Table 3. This implies that the new design will help to relieve the pressure forces at the lower back for adolescent populations of varying anthropometric profiles, thus preventing the lower back from injury as a result of carrying a heavy backpack load.

4 Discussion and Conclusions

Students are carrying heavy backpacks that exceed the recommended carriage limit; long term exposure to heavy loads leading to discomfort at the lower back level, especially back pain and injury in the L4/L5 and L5/S1 disc in the vertebrae. This lower back pain largely is due to the unnatural posture caused by the load and unbalanced weight distribution of the backpack on the torso. A new design of student school backpacks was proposed based on the principles of human biomechanics, more specifically to use proper design of straps and lumbar support to reinforce a more natural standing posture and alleviation, including distributing external forces on the shoulder and lower back by using a pivot lever at the waist level. A theoretical biomechanics model was developed and a simulation study was conducted based on the 10th, 50th and 90th percentile high school 16-year old student anthropometric data in the 3DSSPP program. The compression force and shear force were compared between the new designed back pack and current backpack model with varying weights, with results showing that with the reinforced natural posture, the reduction of the forces on the L4/L5 and L5/S1 are significant (with around 44%–67% for the compression force and about 25%–38% for the shear force, respectively), thus helping to prevent discomfort and back pain, even at the very heavy carriage level. The design was able to meet the goals and criteria originally established in that it was able to decrease significantly the strain on the lumbar vertebrae and discs, ultimately aiding in the prevention of backpack-related injuries.

The study utilized a theoretical biomechanics model to assess the effectiveness of the new design. In order to simplify the model, it was assumed that the center of gravity of the backpack was at the center of the backpack, although this could change with different sizes of the backpack and different load distribution; additionally, the friction and materials of the backpack were not considered, which could have possibly correlated with an effect on the comfort of carrying the backpack.

For the future direction of this study, a physical prototype shall be developed to conduct human testing, including the pressure measurement at different locations of the body to obtain a more precise assessment. Participants should include a variety of anthropometric profile, such as different sex, different body sizes/weight and a broader range of weight to be carried in the backpack. These tests will accommodate the variability of the backpack carriage and present a more realistic assessment on the effectiveness of the new design.

According to the U.S. Consumer Product Safety Commission, at least 14,000 children are treated for backpack-related injuries every year. Quite often, students' backpacks are too heavy in relationship to their own body weight, as the American Academy of Orthopedic Surgeons recommends that a backpack should be no more than 10–15% of a student's bodyweight, while many children are carrying backpacks as heavy as forty, fifty, and sixty pounds due to the amount of textbooks and classwork required in school every day. Furthermore, many students develop consistent, poor habits, wearing backpacks incorrectly on one shoulder or hanging far too low, or having a poor, hunched posture, therefore increasing the compression on the spine and increasing the risk for injury. This research and investigation are important scientifically in that it analyzes and tests a possible conceptual way to provide back support on a conventional backpack that efficiently reduces strain and reinforces a healthy posture, which is crucial to students, especially in middle and high school as coursework becomes more demanding, tiring, and rigorous. Many will suffer from chronic back problems and pain in the future and irreversible damage to the spine, surrounding muscles, ligaments, and tissues stemming from a constant compression from a heavy backpack are a key factor of many of the back problems resulting later on in elderly life as well. If the design of the model can be put to use and commercialized in the market, it can considerably resolve the preventable health issues of the youth today.

References

1. Negrini, S., Carabalona, R.: Backpacks on! schoolchildren's perceptions of load, associations with back pain and factors determining the load. Spine **27**(2), 187–195 (2002)
2. Sheir-Neiss, G.I., Kruse, R.W., Rahman, T., Jacobson, L.P., Peili, J.A.: The association of backpack use and back pain in adolescents. Spine **28**(9), 922–930 (2003)
3. Abitol, J., Haid, Jr. R.W.: Lumbar herniated disc: risk factors, diagnosis, treatment (2019). https://www.spineuniverse.com/conditions/herniated-disc/lumbar-herniated-disc. Accessed 30 Mar 2020
4. Rai, A., Agarawal, S.: Back problems due to heavy backpacks in school children. IOSR J. Hum. Soc. Sci. **10**(6), 22–26 (2013)
5. Sayler, M.H., Shamie, A.N.: The Encyclopedia of the Back and Spine Systems and Disorders, 1st edn. Facts On File, New York (2007)
6. Skoffer, B.: Lower back pain in 15- to 16-year old children in relation to school furniture and carrying of the school bag. Spine **32**(24), 713–717 (2007)
7. Udesky, L.: Back pain in children and teens. https://consumer.healthday.com/encyclopedia/back-care-6/backache-news-53/back-pain-in-children-and-teens-645950.html. Accessed 30 Mar 2020
8. Walicka-Currys, K., Skalska-Izdebska, R., Rachwa, M., Truszczynska, A.: Influence of the weight of a school backpack on spinal curvature in the sagittal plane of seven-year-old children. Biomed. Res. Int. **2015**, 1–6 (2015)
9. Viry, P., Creveuil, C., Marcelli, C.: Nonspecific back pain in children. A search for associated factors in 14-year-old school children. Rev. Rhum. Engl. Ed. **66**(79), 381–388 (1999)
10. American Friends of Tel Aviv University: Heavy backpacks may damage nerves, muscles and skeleton, study suggests. ScienceDaily (2013). https://www.sciencedaily.com/releases/2013/02/130221141604.htm. Accessed 30 Mar 2020
11. Pauza, K.: How backpacks affect the adolescent spine. Dr. Kevin Pauza Biologic Disc Treatment. N.p. (2016). https://drkevinpauza.com/how-backpacks-affect-the-adolescent-spine/Accessed 30 Mar 2020
12. Freburger, J.K., et al.: The rising prevalence of chronic low back pain. Arch. Internal Med. **169**(3), 251–258 (2009). https://doi.org/10.1001/archinternmed.2008.543
13. American Chiropractic Association: Back pain facts and statistics. https://www.acatoday.org/Patients/What-is-Chiropractic/Back-Pain-Facts-and-Statistics. Accessed 04 May 2020
14. Luo, X., Pietrobon, R., Sun, S.X., Liu, G.G., Hey, L.: Estimates and patterns of direct health care expenditures among individuals with back pain in the United States. Spine **29**, 79–86 (2004)
15. Balagué, F., Troussier, B., Salminen, J.: Non-specific low back pain in children and adolescents: risk factors. Eur. Spine J. **8**(6), 429–438 (1999)
16. Skaggs, D.L., Early, S.D., D'Ambra, P., Tolo, V.T., Kay, R.M.: Back pain and backpacks in school children. J. Pediatr. Orthop. **26**, 358–363 (2006)
17. American Chiropratic Association: Backpack safety. https://www.acatoday.org/Patients/Health-Wellness-Information/Backpack-Safety. Accessed 04 Jan 2020
18. Academy of Pediatrics: Backpack safety. http://www.aap.org/advocacy/backpack_safety.pdf. Accessed 04 Jan 2020

19. Davis, K.G., Jorgensen, M.: Biomenchanical modeling for understanding of lower back injuries: a systematic review. Occup. Ergon. **5**(1), 57–76 (2005)
20. Chaffin, D.B., Andersson, G.B.J., Martin, B.J.: Occupational Biomechanics. 4th edn. Wiley, Hoboken (2006)
21. Fryer, C.D., Gu, Q., Ogden, C.L., Flegal, K.M.: Anthropometric reference data for children and adults: United States, 2011-2014. Vital Health Stat **3**(39), 1–46 (2016)

The Influence of Enterprise Financing Structure in China on Business Performance—Simulation Analysis Based on the System Dynamics

Chenggang Li[1,2](✉) and Jun Wan[1]

[1] College of Big Data Application and Economics, Guizhou University of Finance and Economics, Guiyang 550025, China
[2] Guizhou Key Laboratory of Big Data Statistics Analysis, Guizhou University of Finance and Economics, Guiyang, Guizhou, People's Republic of China

Abstract. Enterprise financing structure has an impact on the profitability and solvency of the enterprise to some extent, which is also called the influence on business performance. Enterprise managers and scholars have always focused on the influence of enterprise financing structure on business performance. This paper selects the data of the listed companies from 2010 to 2018, uses the method of the system dynamics, constructs the system dynamics model of enterprise financing structure, and simulates and analyzes the influence of equity financing and debt financing on enterprise business performance. The simulation shows: (1) Equity financing has an active impact on the rate of total assets and scale, while a negative impact on asset liability ratio. (2) Debt financing has an active impact on the rate of total assets, liability assets and scale. (3) Between the two external financing, the influence of equity financing on the business performance is better than the debt financing's.

Keywords: Enterprise financing structure · Business performance · System dynamics · Simulation analysis

1 Introduction

Since 21st century, the financing market in China has been developed fast and the enterprise financing structure has changed. Enterprise business performance is not only the core problem of corporate financing decision-making, but also affects the business performance of enterprises. The influence of enterprise financing structure on its business performance is becoming more and more significant factor in the process of enterprise development. Therefore, the study on the impact of enterprise business performance on business performance is conducive to providing scientific financing decisions and promoting the fast development of the enterprises.

Different financing methods have different effects on the enterprise operation performance and enterprise value. Equity financing has the best operating performance, followed by debt financing (Hu and Li 2019) [1]. Equity financing has a positive impact

H. Song and D. Jiang (Eds.): SIMUtools 2020, LNICST 370, pp. 728–746, 2021.
https://doi.org/10.1007/978-3-030-72795-6_58

on enterprise value, while debt financing has a negative impact on enterprise value (Wang and Yang 2018) [2]. Scholars' research on the influence of enterprise business performance and business performance mainly focuses on two aspects: One is research on the optional styles of enterprise financing structure from Gu (2014) [3], Wang and Song (2017) [4]. The other is research on the influence of enterprise financing structure on the business performance, such as Zhao (2012) [5], Ding (2013) [6] and Huang et al. (2018) [7]. However, these studies fail to reflect the future trend of the influence of enterprise financing structure on the business performance.

The purpose of this paper is to study the influence of enterprise business performance on business performance. Based on the research method of system dynamics adopted by Lu et al. (2017) [8], Li et al. (2018) [9], Yao et al. (2018) [10], Gao and Wang et al. (2019) [11], this paper simulates the trend of the influence of different financing structures on business performance in the future.

The contribution of this paper is that based on the theory of system dynamics, this paper uses the system dynamics model to simulate the evolution trend of the influence of enterprise business performance on business performance in the future period.

The structure of this paper is organized as following: The second part is the literature review, which summarizes the research achievements of the existing scholars. The third part is the introduction, boundary determination, and variable declaration of system dynamics, draws the causality diagram and flow diagram, and constructs the system dynamic model. The fourth part tests the simulation model and simulates the future influence of equity financing and debt financing on the enterprise total assets, asset liability ratio and production scale. The last part is the conclusion of the simulation experiment and policy recommendation.

2 Literature Review

Research on the choice of enterprise financing, under the complete condition of market assumption, Li et al. (2013) [12] indicated that internal and external financing had an active relevant with company innovation investment, and the external financing has better promotion effect on innovation investment than internal financing. Wen (2013) [13] used Ordered-Logistic to range the financing preference of listed company in China. He thought that listed companies most preferred the short term lending, next is internal financing. Gu (2014) [3] analyzed the financing preference of listed company and optimization of capital structure. The result showed that listed companies should further optimize the capital structure in order to make enterprise financing more rationalized. Rao and Wu (2017) [14] discussed the influence of financing structure and R&D input on the industrial structure. The result showed that direct financing and indirect financing has an active influence on the upgrading of industrial structure. Wang and Song (2017) [4] used enterprise financing structure theory, combined with the special property of enterprise innovative financing, then proposed suggestions about enterprise financing in China. Zhao and Chen (2018) [15] discussed the influence of financing structure on the choice of financing mode. The research showed that financing structure has a significant influence on the choice of financing mode, and the enterprise prefer to equity financing. Pan and Hu (2019) [16] chose the data from A-share listed company between 2011 and

2016. They empirical analyzed the relationship between financing structure and innovation level. The results showed that there is interaction between enterprise financing structure and innovation level. What's more, there is a negative influence of enterprise innovation level on the financing structure.

Research on the influence of Enterprise business performance on business performance, Ling and Hu (2012) [17] discussed about the influence of strategic emerging industries on the business performance. The research showed that equity capital has improved the business performance of strategic emerging industrial. Zhao (2012) [5] made a research about financing structure of listed companies which is based on the different life-cycle theory. The results showed that there are different features of financing behavior during different stages of enterprise life cycle. Ding (2013) [6] chose data from 128 technological small and medium-sized enterprises (SEMs), analyzed the influence of financing structure in technological SMEs on the business performance. The result showed that there is a negative relationship between the asset liability ratio of technological SMEs and improving business performance. Pan (2015) [18] used factors analysis and analyzed the relationship between financing structure and business performance in 25 listed companies. The results showed that internal financing and equity financing has an active influence on business performance, while debt financing has a negative influence on business performance. Yu and Xu (2016) [19] analyzed the influence of equity financing of emerging industrial on the business performance in listed companies. The results showed that enlarge financing scale and increase the equity financing ratio can improve the company's profitability. Huang et al. (2018) [7] made a research on the relationship between the equity structure and business performance. The results showed that at the micro macro level, there is different influence of equity concentration on different business performance. Wang and Zhang (2018) [20] studied the impact of financing cost on enterprise productivity and found that the financing cost had a negative and significant impact on productivity. The lower the financing cost, the higher the enterprise performance. Gu and Nian (2018) [21] selected the data of listed companies from 2000 to 2016, conducted an empirical study on the influence of liquidity generality on corporate financing, and found that the liquidity of individual stocks had a negative effect on the external equity financing and debt financing of enterprises. Wang and Yang (2018) [2] selected the data of A-share enterprises from 2009 to 2016, and studied the influence of financing methods on enterprise value relationship. The research results showed that different financing methods had different effects on enterprise value. Debt financing had a negative and significant impact on enterprise value, while equity financing had a positive and significant impact on enterprise value. Ma (2019) [22] used the panel data of 45 A-share listed sports companies and empirical researched the influence of financing structure on the business performance and risk. The results showed that the financing efficiency of these sports company was a low. What's more, it had an inverse relationship between business performance and risk. Wu and Luo (2019) [23] selected data of Chinese listed enterprises from 2011 to 2017, and researched the impact of One Belt And One Road on TFP of Chinese listed enterprises. The research results showed that One Belt And One Road initiative promoted the TFP of listed enterprises in the provinces along the routes of China. Qian et al. (2019) [24] selected the data of listed companies in 2015 and 2017 to study the impact of endogenous financing, debt financing and equity financing on innovation efficiency of enterprises. The research results showed that endogenous financing and debt financing had a positive and significant impact on innovation efficiency of enterprises, while equity financing had a negative and significant impact on innovation efficiency. Hu and Li (2019) [1] empirically analyzed the influence of different financing methods on the enterprises operating performance. The analysis results showed that the optimal method is the equity financing. Li and Gao (2019) [25] simulation analyzed the economic mechanism of structural monetary policy. The research showed that under the direct financing mode, the economic effect of structural monetary policy had a more significant impact on SMEs. Lin and Zhang (2019) [26] selected relevant data from 2012 and 2016 to analyze the impact of corporate income tax exemption on enterprise business performance. The results showed that the corporate income tax exemption increased endogenous financing and extruded current debt financing, and improved total factor productivity.

As for the research on the system dynamics method for simulation analysis, Lu et al. (2017) [8] adopted the system dynamics method to simulate the tourism development model and analyzed the growth trend of tourism on economic development. The research results showed that industrial economy, social culture and resource environment had a positive and significant impact on the tourism development. Li et al. (2018) [9] constructed a system dynamics model to conduct a simulation analysis of the development of trust industry by monetary policy and trust industry policies. The results showed that monetary policy has a positive and significant impact on the scale of trust issuance. The change of net capital ratio had a significant impact on the development of trust industry. Yao (2018) [10] constructed a system dynamics model, and simulation analyzed the impact of traditional policies and carbon tax policies on rural logistics. The results showed that traditional policies could promote the transformation of rural logistics to low-carbon economy, while carbon tax policies had high sensitivity. Gao (2019) [11] selected the panel data of Guizhou province from 2007 to 2016, constructed the system dynamics model, and conducted a simulation analysis on the GDP policy of the region with the industrial structure policy, and the logistics self-growth coefficient. The research results showed that there was a positive correlation between the GDP, logistics demand, logistics supply capacity and the logistics self-growth coefficient.

Although we have achieved rich conclusions, the existing research remains some deficiency, which gives this paper the space to expand: (1) A part of literature research only considered about the choose mode of enterprise financing structure, such as adopting econometric model to study the ranking of financing preference of listed enterprises (Wen 2013) [13], but don't study the influence of enterprise financing structure on the business performance. (2) A part of literature study the relationship between the enterprises financing structure of single industry type and business performance. For example, they studied the relationship between the business performance and the enterprise structure in technology-based SMEs and agricultural listed companies (Ding 2013 [6] and Pan 2015 [18]). These research object is too single and the results is not universal. (3) A part of research only reached the conclusion that there is an influence of the equity financing and debt financing on business performance in a single industry. For example, there is an inverse relationship between the financing structures of sports companies and business performance (Ma 2019) [22]. However, it is not possible to carry out simulation analysis on the influence of enterprise financing structure on business performance in the future period. Therefore, this paper try to innovate in the following aspects aiming at the above questions: (1) Selecting the sample data of listed enterprises from 2010 to 2018, we study the choice of financing methods for enterprises in the whole industry and analyzes the influence of enterprise equity financing and debt financing on their business performance. (2) This paper adopts the system dynamics methods to construct the simulation model of enterprise financing structure and predicts the influence of enterprise financing structure on its business performance during the time in the future.

3 Introduction of System Dynamics Method and Model Construction

3.1 Introduction of System Dynamics Method

The system dynamics method is a subject which mixes with system science and management science. It was founded by Forrester of the Massachusetts Institute of Technology in 1956. This subject combines the system theory with computer simulation technique tightly and studies the complex system feedback structure and behavior (Zhong 2009) [27]. The system dynamics model is the complex laboratory of society, economic and ecology. When solving questions, it usually combines qualitative and quantitative together. It is a common tools for learning and policy analysis. Generally, key steps such as system boundary determination, system structure analysis, model construction, model test, simulation and evaluation are required.

3.2 Model Construction

3.2.1 Introduction' of System Dynamics Method

By drawing from the research on economics structure and economic development relationship by Lin (2006) [28], the exploration of optimal financial structure theory during economics development by Lin (2009) [29], research on the dynamics model of financial efficiency system from the perspective of capital operation by Liu (2011) [30], research on financing structure by Dong (2016) [31], research on enterprise scale and different bank financing characteristics by Zhang (2019) [32], this paper takes the enterprise financing structure as the research object, constructs the system dynamics model and researches the influence of enterprise financing on the business performance by exploring the relationship between enterprise financing and economic growth, stock market development and social savings. This paper adopts with the software called Vensim to draw the causal diagram of the enterprise financing structure in China, shown in Fig. 1. From Fig. 1, we can see that the model includes GDP, economic growth, social saving, bank deposit and loan rates, securities market development and other key variables. There are 18 feedback loops in total. The symbols "+" or "−" in front of the variables represent the tendency to change in the same direction or in the opposite direction which is caused by change of the former variable. The specific loop is shown as below:

Feedback loop 1: Enterprise financing structure → + enterprise production scale → + enterprise foreign exchange earning → + enterprise internal financing → + enterprise financing structure.

Feedback loop 2: Enterprise financing structure → + GDP → + economic growth → + social saving → + enterprise equity financing → + enterprise financing structure.

Feedback loop 3: Enterprise financing structure → + enterprise production scale → + enterprise foreign exchange earning → + enterprise internal financing → - financing cost → + enterprise financing structure.

Feedback loop 4: Enterprise financing structure → + enterprise production scale → + enterprise foreign exchange earning → + social saving → + enterprise equity financing → + enterprise financing structure.

Feedback loop 5: Enterprise financing structure → + GDP → + economic growth → + financial needs → + enterprise equity financing → + enterprise financing structure.

Feedback loop 6: Enterprise financing structure → + GDP → + economic growth → + social saving → + enterprise equity financing → + financing cost → + enterprise financing structure.

Feedback loop 7: Enterprise financing structure → + GDP → + economic growth → + financial needs → + enterprise equity financing → + financing cost → + enterprise financing structure.

Feedback loop 8: Enterprise financing structure → + GDP → + economic growth → + financial needs → + securities market development → + enterprise equity financing → + enterprise financing structure.

Feedback loop 9: Enterprise financing structure → + enterprise production scale → + enterprise foreign exchange earning → + social saving → + enterprise equity financing → + financing cost → + enterprise financing structure.

Feedback loop 10: Enterprise financing structure → + GDP → + economic growth → + financial needs → + securities market development → + enterprise equity financing → + financing cost → + enterprise financing structure.

Feedback loop 11: Enterprise financing structure → + GDP → + national income → + disposable income → + household savings → + bank capital → + enterprise debt financing → + enterprise financing structure.

Feedback loop 12: Enterprise financing structure → + GDP → + national income → + disposable income → + household savings → + securities market development → + enterprise equity financing → + enterprise financing structure.

Feedback loop 13: Enterprise financing structure → + GDP → + national income → + disposable income → + household savings → + bank capital → + enterprise debt financing → + financing cost → + enterprise financing structure.

Feedback loop 14: Enterprise financing structure → + GDP → + national income → + disposable income → + household savings → + securities market development → + enterprise debt financing → + financing cost → + enterprise financing structure.

Feedback loop 15: Enterprise financing structure → + GDP → + national income → + disposable income → + consumption → + enterprise production scale → + enterprise foreign exchange earning → + enterprise internal financing → + enterprise financing structure.

Feedback loop 16: Enterprise financing structure → + GDP → + national income → + disposable income → + consumption → + enterprise production scale → + enterprise foreign exchange earning → + enterprise internal financing → + financing cost → + enterprise financing structure.

Feedback loop 17: Enterprise financing structure → + GDP → + national income → + disposable income → + consumption → + enterprise production scale → + enterprise foreign exchange earning → + social saving → + enterprise equity financing → + enterprise financing structure.

Feedback loop 18: Enterprise financing structure → + GDP → + national income → + disposable income → + consumption → + enterprise production scale → + enterprise foreign exchange earning → + social saving → + enterprise equity financing → + financing cost → + enterprise financing structure.

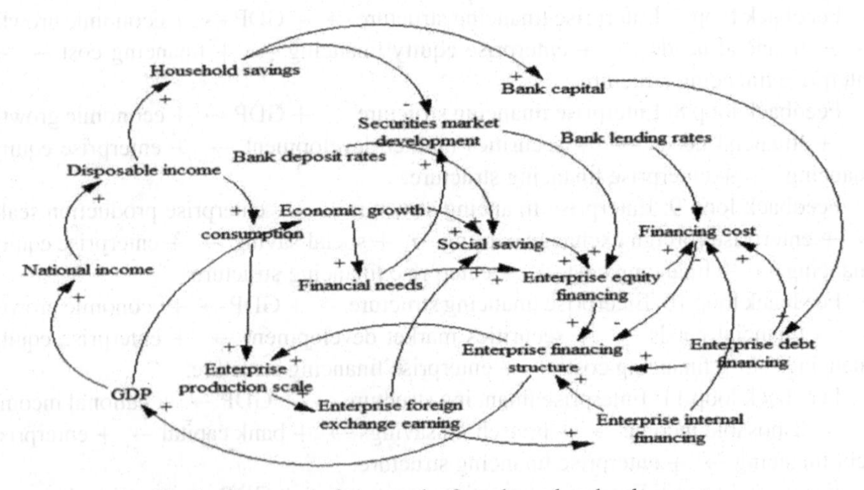

Fig. 1. Systematic framework of stock market development.

3.2.2 Variable Declaration of the Model

This paper selects 46 indicator variables to analyze the causal feedback relationship of enterprise financing structure system, and simulte the influence of the enterprise financing structure on the business performance. Considering the scientific and rationality of variables and the data availability, this paper selects 7 flow potential variables (L), 8 rate variables (R), 8 auxiliary variables (A) and 2 constant variables (C), as shown in Table 1.

Table 1. Main variables of system dynamics model of enterprise financing structure

Flow potential variables (L)	Rate variables (R)	Auxiliary variables (A)	Constant variables (C)
Bank capital	Added value of bank capital	The loan interest rate	Growth rate of capital demand
Household savings	Added value of household savings	Interest rates on deposits	The extent to which Banks satisfy listed companies
GDP	The decline in household savings	Contribution to total capital formation	
Enterprise deposits	The added value of GDP	Final consumption contribution	
Listed enterprises own capital	The increase in enterprise deposits	Amount of stock raised	
SME own capital	Satisfaction of capital needs	Enterprise bond issuance	
Market value of shares	Capital needs	The GDP growth rate	
	Stock market value added	Enterprise financing cost	

3.2.3 Construct Enterprise Financing Structure Flow Figure

According to the causal diagram of enterprise financing structure as shown in Fig. 1 and the own characteristic of enterprise financing structure, this paper draws the flow diagram of enterprise financing structure as shown in Fig. 2. In Fig. 2, the mathematical formulas for the variables of the system are the basis of assuring the correlations. Mathematical formulas can be established by regression fitting form. A part of variables are stated as

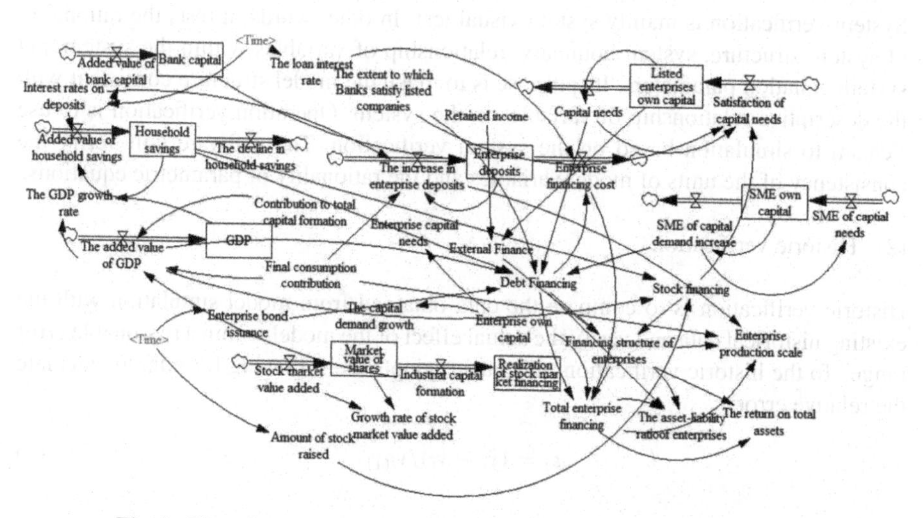

Fig. 2. Flow diagram of financing structure system of Chinese enterprises

a constant value by experience and historic data. Meanwhile, this paper adopts the table function in the system dynamics to simulate the future development of each variables by stating the table function.

4 Model Verification and Simulation

4.1 Data Source

This paper simulates the influence of the different proportions of equity financing and debt financing on the return of total assets, asset-liability ratio and enterprises production scale through the trend analysis of various variables in enterprise financing structure. Through the summary of literature, expert experience and other methods, this paper sets the bank loan satisfaction to the enterprises at 0.8, the capital demand growth at 5%. The initial value of bank loan interest rate, deposit interest rate, enterprise stock financing amount and enterprise debt financing amount is set from 2010. The simulation interval is between 2010 and 2015. All data are from Chinese National Bureau of Statistics, website of the People's Bank of China, China Securities Regulatory Commission, China Association of Listed Companies, Shanghai Stock Exchange, and Shenzhen Stock Exchange.

4.2 Model Verification

There are mainly two sides about system dynamics: one is system verification and operation verification, the other is fitness comparison between simulation results and historic data, which is called historic verification.

(1) System verification and operation verification

System verification is mainly system visual test. In other words, it tests the rationality of system structure, system boundary, relationship of variables within the system and system equation parameters. Its purpose is to make the model structure consistent with the descriptive relationship of variables in the system. Operation verification is to use Venism to simulation based on the system verification. The software will verify the consistency of the units of model variables and the rationality of parametric equations.

(2) Historic verification

Historic verification is to compare the data obtained from model simulation with the existing historical data and verify the actual effect of the model within a reasonable error range. To the historic verification, this paper adopts the following formula to calculate the relative error:

$$e_i = (\hat{y}_i - y_i)/y_{i(1)}. \tag{1}$$

In the Formula (1), y_i represents the true value of the indicators in i-th period. $\hat{y}_i$ represents the simulation value of the indicators in i-th period. The historic verification results are shown in Table 2. From Table 2, we can see that the average estimated errors of the return of total asset and asset-liability ratio of listed companies are −0.66% and −2.85%, respectively. There is some error between the simulation value and the actual value. However, this model is a reduction model of China's enterprise financing structure. Therefore, we believes that the simulation ability of this model is good. It can be used for subsequent simulation.

Table 2. Return of total asset and asset-liability ratio

Indicators	Return on total assets (%)			Asset-liability ratio (%)		
Year	Simulation value	True value	Relative error	Simulation value	True value	Relative error
2010	1.85	1.91	−3.14%	81.13	85.74	−5.37%
2011	1.76	1.86	−5.37%	83.79	85.72	−2.25%
2012	1.61	1.65	−2.42%	84.52	85.77	−1.45%
2013	1.79	1.78	0.56%	83.96	85.75	−2.08%
2014	1.64	1.71	−4.02%	83.08	85.25	−2.54%
2015	1.63	1.53	6.53%	82.80	84.65	−2.18%
2016	1.46	1.45	0.69%	79.09	84.59	−6.50%
2017	1.61	1.57	2.55%	76.64	80.23	−4.47%
2018	1.58	1.60	−1.25%	78.97	78.04	1.19%
Average estimated error	−0.66%			−2.85%		

4.3 Simulation

According to the historic test, the system dynamics model of enterprise financing structure can better reflect the impact of enterprise profitability and solvency. Dong (2016) simulated the impact of different financing structures on the economic subject in China. He found that when the society direct financing scale arrives to 30%–50%, a crucial period for economic and social development is coming. Therefore, this paper assumes that equity financing and debt financing in the enterprise financing structure increase by 30%, 40% and 50% respectively, and simulates the impact of these changes on the enterprise's return of total assets (ROA), asset liability ratio (LEA), and scale of production (SCALE), as well as the changing trend of various variables in a certain period in the future. In this paper, Vensim software is used for simulation analysis.

(1) The influence of changes in enterprise financing structure on the enterprise return of total assets.

The simulation results of the impact of changes in equity financing and debt financing on the return of total assets are shown in Fig. 3(a)–(e). As can be seen from Fig. 3(a) and (b), with the increase of the ratio of equity financing and debt financing, the return of total assets also increases gradually. This shows that the increase of the ratio of equity financing and debt financing has a positive impact on the profitability of enterprises.

It can be seen from Fig. 3(c), (d) and (e) that, with the same proportion of financing increased, the impact of equity financing on the return on total assets of enterprises is greater than that of debt financing.

(2) The influence of changes in enterprise financing structure on corporate asset-liability ratio.

The asset-liability ratio determines the solvency of an enterprise. The higher the asset-liability ratio, the weaker the solvency of the enterprise and the greater the financial risk. For enterprises, the asset-liability ratio should be controlled within a certain range. In China, the ideal debt ratio of enterprises is about 40%. According to the financial reports of listed companies, the asset-liability ratio of enterprises remains high. It reaches 75–85% from 2010 to 2018. This paper simulates the influence of equity financing and debt financing on the asset-liability ratio under different proportions. This paper analyzes which financing method has more significant influence on the asset-liability ratio. The simulation results are shown in Fig. 4(a)–(e).

As can be seen from Fig. 4(a), with the increase of equity financing, the asset-liability ratio of enterprises decreases. Without changing the amount of debt financing, the increase of the equity financing will increase the total financing amount and lead to the reduction of the asset-liability ratio. As can be seen from Fig. 4(b), with the increase of the debt financing, the asset-liability ratio of enterprises increases. The increase of debt financing further increases the financial risk of enterprises. It can be seen from Fig. 4(c), (d) and (e) that, under the same proportion of financing, the increase of enterprise debt financing will make the asset-liability ratio of enterprises higher than that of equity financing. The financial risk caused by the increase of debt financing is greater than that caused by the increase of equity financing. Enterprises should reasonably control the amount of debt financing.

(3) The influence of changes in the financing structure of enterprises on the production scale.

The production scale is affected by labor, technology and capital. By increasing the equity financing and debt financing, capital is increased to influence the production scale of the enterprise. The simulation results of the impact of changes in enterprise financing structure on production scale are shown in Fig. 5(a)–(e).

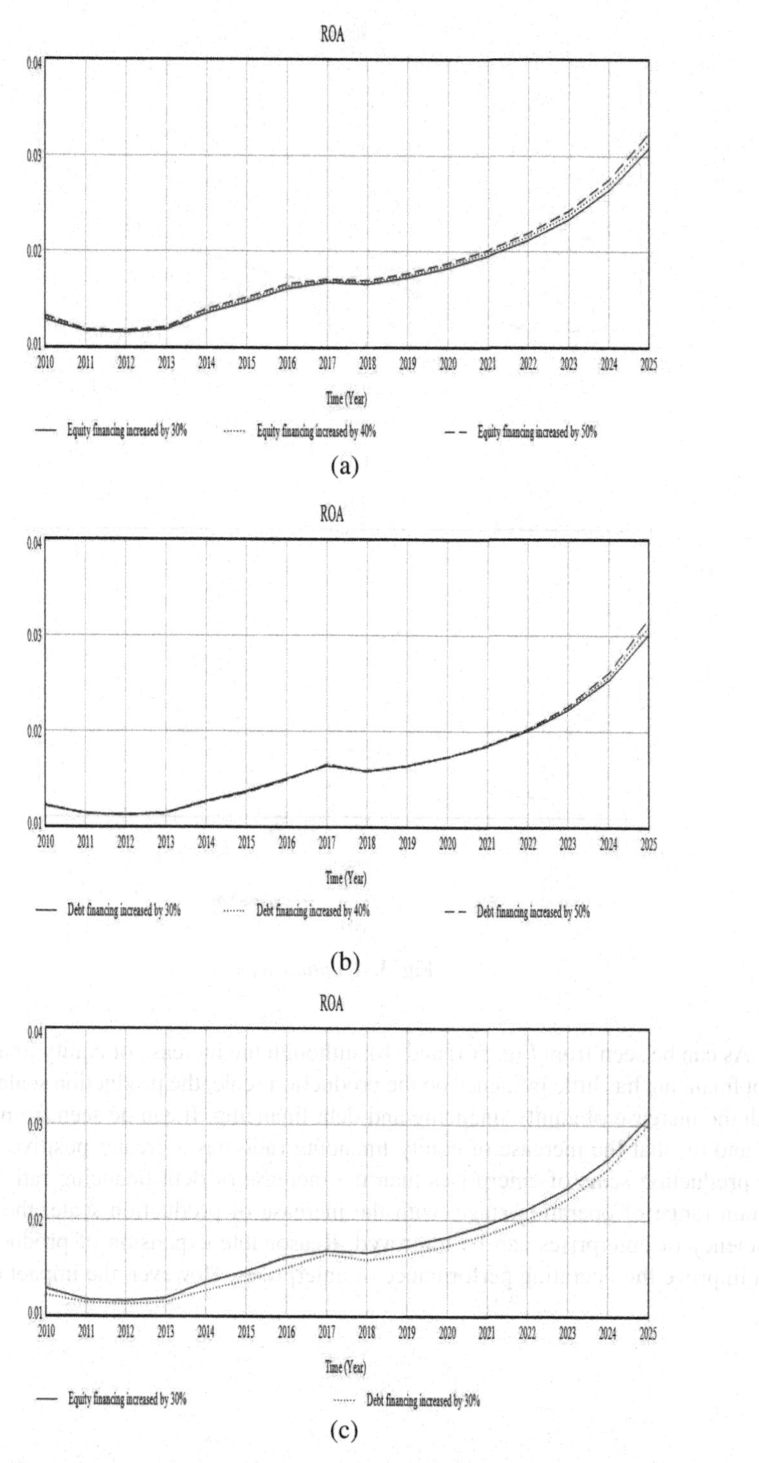

Fig. 3. The influence of equity financing and debt financing on the return of total assets

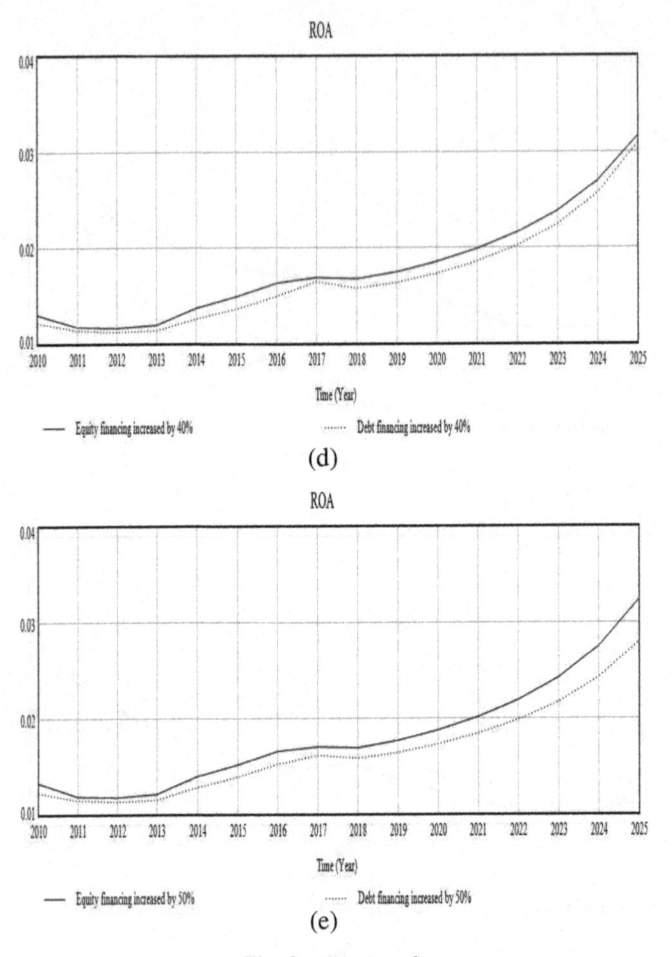

Fig. 3. (*continued*)

As can be seen from Fig. 5(a) and (b), although the increase of equity financing and debt financing has little influence on the production scale, the production scale increases with the increase of equity financing and debt financing. It can be seen from Fig. 5(c), (d) and (e) that the increase of equity financing ratio has a greater positive impact on the production scale of enterprises than the increase of debt financing ratio. Within a certain range of operation stage, with the increase of production scale, the operating efficiency of enterprises can be improved. Reasonable expansion of production scale can improve the operating performance of enterprises. However, the impact of the two

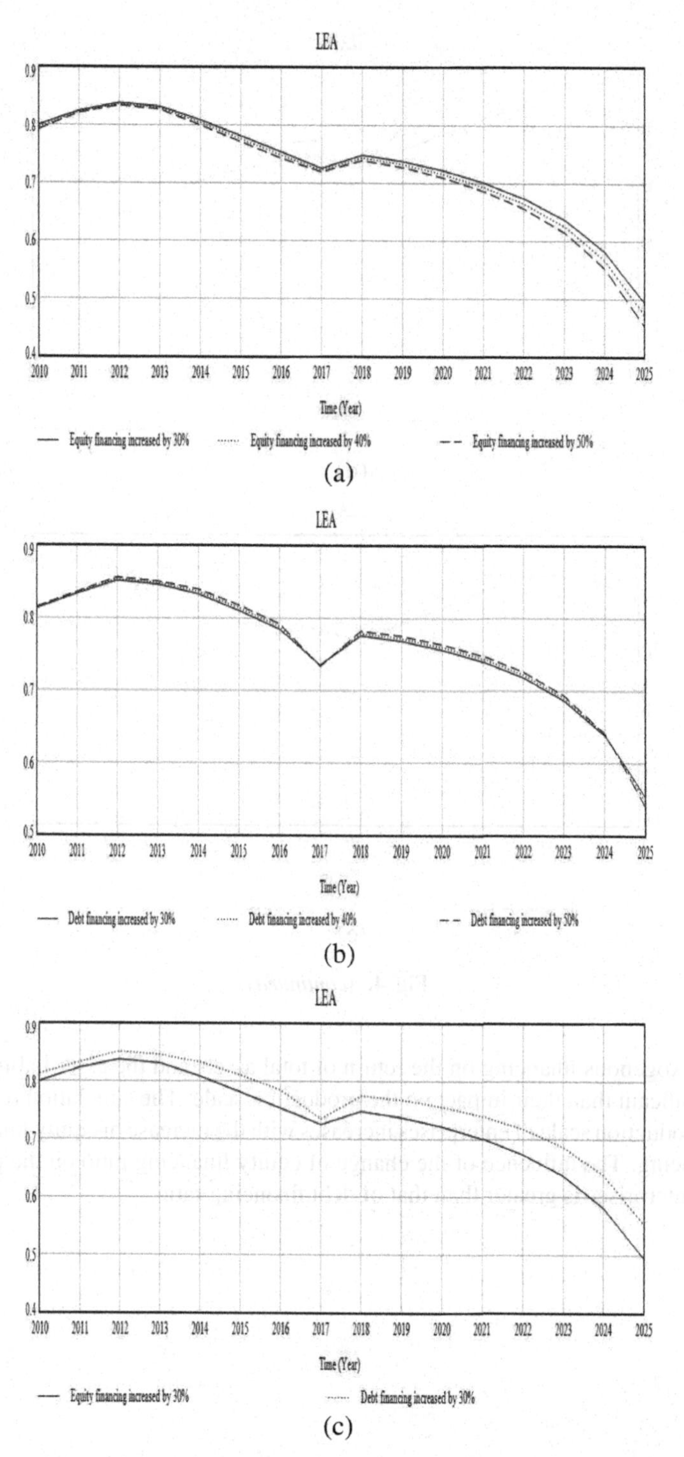

Fig. 4. The influence of equity financing and debt financing on the asset-liability ratio

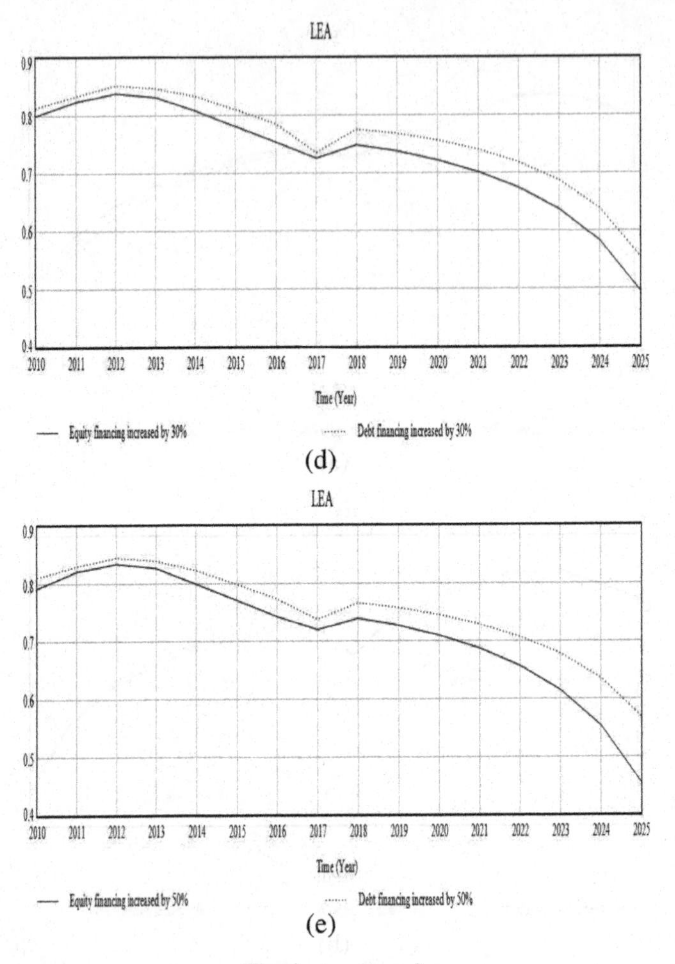

Fig. 4. (*continued*)

kinds of exogenous financing on the return of total assets and the asset-liability ratio is more significant than their impact on the production scale. The simulation results show that the production scale of enterprises increases with the increase of equity financing and debt financing. The influence of the change of equity financing ratio on the production scale of enterprises is greater than that of debt financing ratio.

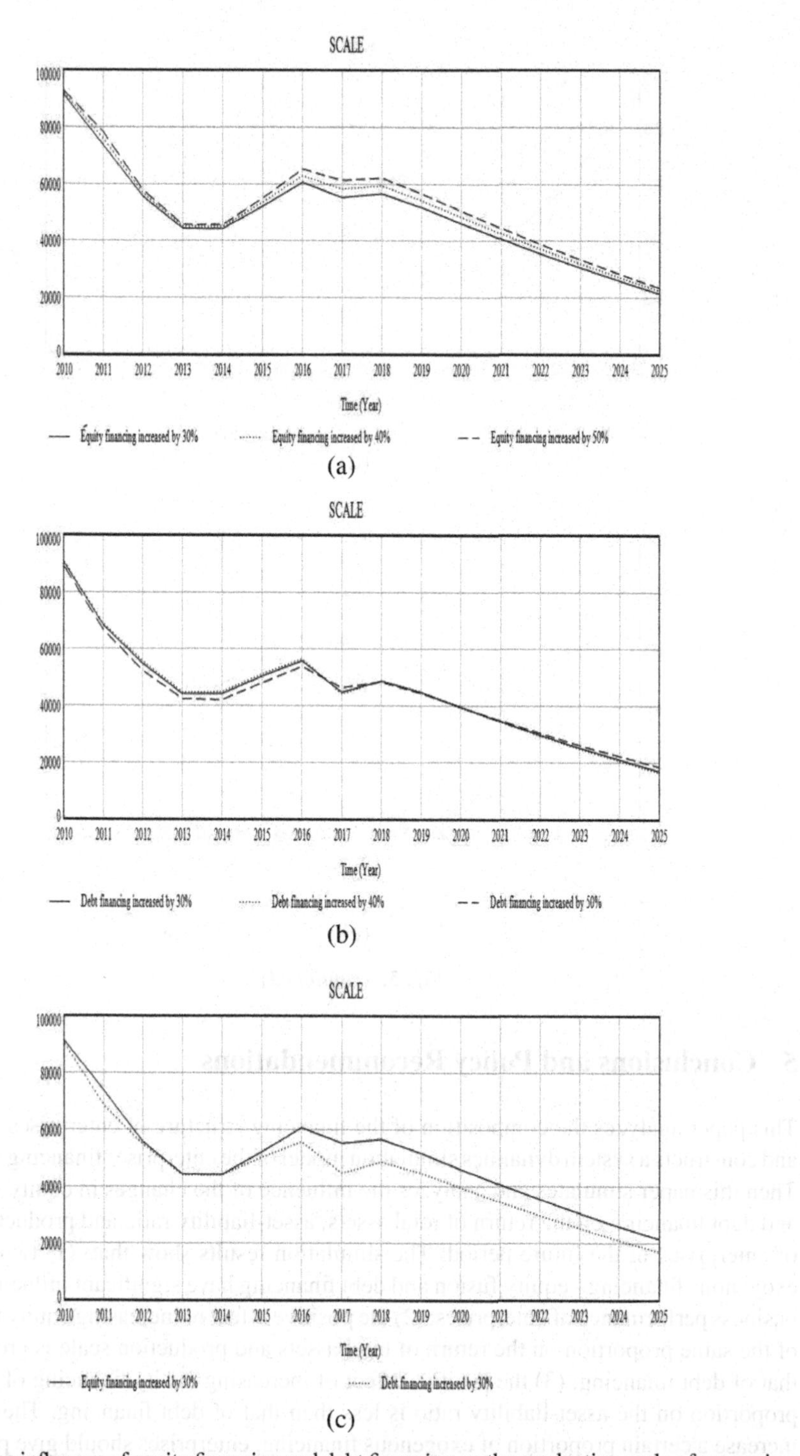

Fig. 5. The influence of equity financing and debt financing on the production scale of enterprises

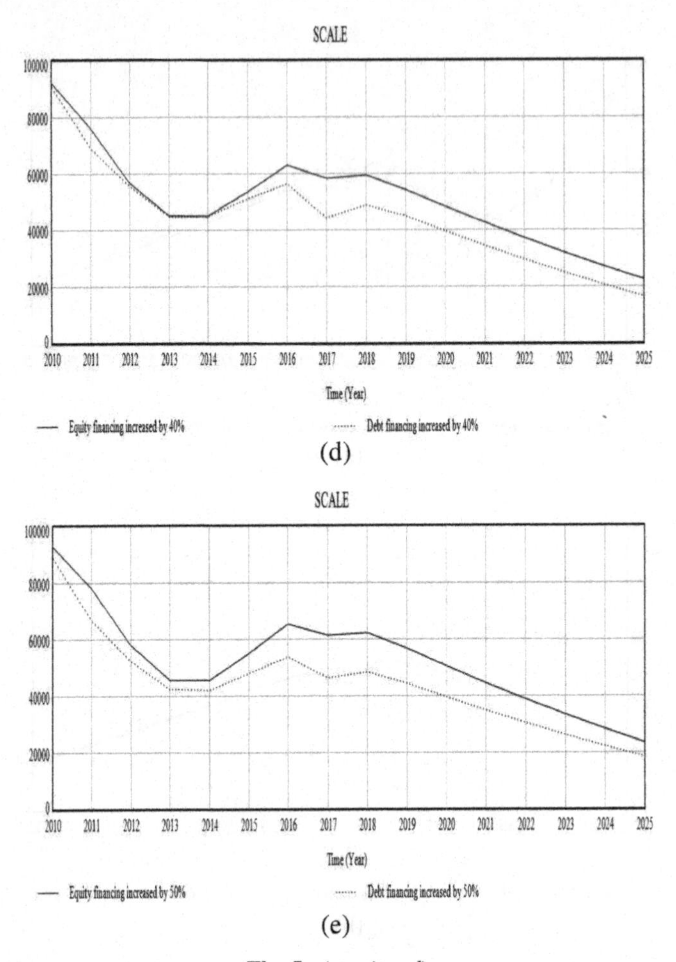

Fig. 5. (*continued*)

5 Conclusions and Policy Recommendations

This paper analyzes the composition of the financing structure of enterprises in China, and constructs a system dynamics simulation model of the enterprises financing structure. Then this paper simulates and analyzes the influence of the changes in equity financing and debt financing on the return of total assets, asset-liability ratio and production scale of enterprises in the future period. The simulation results show that: (1) two kinds of exogenous financing - equity fusion and debt financing have significant influence on the business performance of enterprises; (2) the positive effect of increasing equity financing of the same proportion on the return of total assets and production scale is greater than that of debt financing; (3) the positive effect of increasing equity financing of the same proportion on the asset-liability ratio is less than that of debt financing. Therefore, to increase a certain proportion of exogenous financing, enterprises should give priority to equity financing, followed by debt financing.

Based on the results of the simulation analysis, the following policy recommendations are proposed: (1) Enterprises should prefer to choose the equity financing. There are many methods for the enterprises to raise capital. Equity financing is conducive to expanding capital channels, increasing the scale of capital and promoting the development of enterprises. (2) Enterprises should strictly control debt financing. Corporate debt financing increases the asset-liability ratio. Enterprises need to control the debt ratio to a certain extent, otherwise it will increase the financial risk of enterprises. So from the perspective of avoiding financial risks, enterprises should use the equity financing, and then use the debt financing when choosing two external financing methods. (3) Enterprises should increase the proportion of equity financing to expand production scale. Production scale is a crucial factor to measure the operation capacity of enterprises. Especially for enterprises in the growth period, equity financing should be preferred when expanding production scale.

Of course, the actual environment which the enterprises confront is very complex. The system dynamics model constructed in this paper may fail to consider some aspects, such as the individual investment preference and the operation decision of SMEs. This will be further refined in future studies.

References

1. Hu, W., Li, Z.: A comparative study of M&A performance of science and technology enter prises with different financing approach– an empirical analysis based on factor analysis and Wilcoxon 's SignRank test. Shanghai J. Econ. **11**, 94–107 (2019)
2. Wang, Y., Yang, H.: Financing methods capitalized R&D choice and firm value. Forecasting **37**(02), 44–49 (2018)
3. Gu, J.: Financing preference and capital structure optimization of listed companies in China. Jiangxi Soc. Sci. **34**(10), 34–38 (2014)
4. Wang, S., Song, L.: The enlightenments on the study of enterprise R&D financing structure theory and experience. J. Liaoning University (Philosophy and Social Sciences Edition) **45**(4), 42–48 (2017)
5. Zhao, X.: Research on the financing structure of listed companies based on life cycle theory. Collected Essays Fin. Econ. **2**, 84–89 (2012)
6. Ding, N., Su, Y., Li, J.: The impact of technological SMES financing structure on enterprise performance. J. Xi'an Polytechn. Univ. **27**(6), 816–820 (2013)
7. Huang, F., Feng, D., Wang, Q., Yang, M., Zhu, X.: Equity structure and corporate performance based on A comparative study of investor protection environment in A and H share markets. Rev. Invest. Studie **37**(7), 131–157 (2018)
8. Lu, X., Sui, L., Guo, L., Wu, C.: Study on simulation of driving mechanism of rural Tourism: a case of Dalian. Syst. Eng. **35**(11), 121–129 (2017)
9. Li, S., Zhong, J., Liu, J., Dong, Z., Fang, Y., Dong, J.: A simulation study of the development of China's trust industry based on system dynamics. Manage. Rev. **30**(04), 3–11 (2018)
10. Yao, G., Bian, X., He, Y.: Construction and Simulation of Rural Logistics System Dynamic Model from the Perspective of Low Carbon Economy. Soft Sci. **32**(02), 60–66 (2018)
11. Gao, K., Wang, M.: A study on the system dynamics of the coordinated development of regional economy and logistics. Stat. Dec. **35**(08), 60–63 (2019)
12. Li, H., Tang, Y., Zuo, J.: Innovate with your own money or with other people's money? Based on the research of financing structure and innovation of Chinese listed companies. J. Fin. Res. **02**, 170–183 (2013)

13. Wen, S.: Financing preference of listed companies in China and Its Influencing factors. Chengdu: southwestern university of finance and economics (2013)
14. Rao, P., Wu, Q.: The influence of financing structure and R&D investment on industrial structure upgrading – from the perspective of social financing scale. Modernizat. Manage. **37**(6), 25–27 (2017)
15. Zhao, X., Chen, J.: Research on the impact of target capital structure on M&A financing decision. Reform Econ. Syst. (05):126–132 (2018)
16. Pan, H., Hu, Q.: A research on the interaction effect between enterprise financing structure and innovation level from the perspective of life cycle: based on the empirical evidence of A-share listed companies in strategic emerging industries. J. Nanjing Audit Univ. **16**(4), 81–92 (2019)
17. Ling, J., Hu, W.: Firm size, capital structure and operating performance comparative study on traditional industries and strategic emerging industries. Fin. Trade Econ. **12**, 71–77 (2012)
18. Pan, Y., Zang, D., Ji, Y.: A study on the influence of PLS regression based financing structure on business performance of agricultural listed companies. J. Agrotechn. Econ. **9**, 107–116 (2015)
19. Yu, J., Xu, Y.: The influence of equity financing on business performance is based on the research of listed companies in strategic emerging industries. J. Southeast Univ. (Philosophy Soc. Sci.), **18**(6), 88–94+147 (2016)
20. Wang, Y., Zhang, Y.: The impact of collaboration between internal governanceand external financing on enterprises' productivity. Ind. Econ. Rev. **9**(05), 101–111 (2018)
21. Gu, N., Nian, R.: Commonality in liquidity corporate financing behavior and capital structure. J. Manage. Sci. China **21**(08), 34–53 (2018)
22. Ma, Y., Su, K., Zhang, C.: Impact of current listed sports company's financing structure on corporate performance and risk: also on optimal asset-liability ratio. J. Shenyang Sport Univ. **38**(4),64–69+85
23. Wu, M., Lou, X.: "The Belt and Road Initiative", Financing structure and total factor productivity of listed companies in China. J. Ind. Technol. Econ. **38**(12), 119–126 (2019)
24. Qian, Y., Duan, S., Zhang, L.: Financing structure and innovation efficiency of science and technological enterprises: empirical evidence from listing corporations of GEM. Sci. Technol. Manage. Res. **39**(21), 53–60 (2019)
25. Li, J., Gao, H.: A comparative study of M&A performance of science and technology enter can structural monetary policy reduce financing constraints for smes? Analysis based on heterogeneous dynamic stochastic general equilibrium model. Econ. Sci. **06**, 17–29 (2019)
26. Lin, X., Zhang, K.: Corporate income tax deduction, financing structure and total factor productivity: an empirical study based on the data of national tax survey from 2012 to 2016. Contemp. Fin. Econ. **04**, 27–38 (2019)
27. Zhong, Y.: System Dynamics. Science Press, Beijing (2009)
28. Lin, Y., Jiang, Y.: The empirical analysis of the structure and economic development of banking industry based on the panel data of provinces. J. Fina. Res. **1**, 7–22 (2006)
29. Lin, Y., Sun, X., Jiang, Y.: Toward a theory of optimal financial structure in economic development. Econ. Res. **44**(8), 4–17 (2009)
30. Liu, W.: Construction of system dynamics model of financial efficiency from the perspective of capital operation. Ocean University of China, Qingdao (2011)
31. Dong, Z., Li, X., Dong, J.: Simulation of financing structure based on system dynamics. Syst. Eng.-Theory Pract. **36**(5), 1109–1117 (2016)
32. Zhang, Y., Lin, Y., Gong, Q.: Firm size bank size and optimal banking structure: from the perspective of new structural economics. Manage. World **35**(3), 31–47+206 (2019)

An Action of the Picard Group on Generalized Euler Classes

Bingjun Li(✉), Jianjun Jiao, and Dan Yang

School of Mathematics and Statistics, Guizhou University of Finance and Economics, Guiyang, China

Abstract. In this paper, we construct an action of Picard group on the generalized Euler classes defined in Chow-Witt groups $\widetilde{CH^d}(X, \mathscr{L})$ with twisted line bundle $\mathscr{L}$, which is the generalization of the Euler classes defined in Chow groups. This result gives a new method to compare the generalized Euler classes with different twisted bundle.

Keywords: Picard group · Euler classes · Chow groups

1 Introduction

Let X =Spec(A) be a smooth affine Noetherian scheme of dimension d over a field k, $\mathscr{E}$ be a vector bundle over X of rank d with determinant isomorphic to a line bundle $\mathscr{L}$. One could give an obstruction class in algebraic geometry, such that its vanishing is the only obstruction to $\mathscr{E}$ splitting as the direct sum of a rank $d-1$ vector bundle and a trivial line bundle. Or more precisely, to give an obstruction classes in algebraic geometry which is an analog of the Euler classes in algebraic topology. Historically there are mainly three different theories be constructed for this problem. One theory comes from the work of S. Bhatwadekar and R. Sridharan [10], which are based upon the ideal of M.V. Nori [16]. This theory defines the obstruction in a pure algebraically constructed group named generalized Euler class group $E^d(X; \mathscr{L})$. Second theory comes from the lecture of F. Morel [20], where one constructs an Euler class as the primary obstruction to existence of a non-vanishing section of $\mathscr{E}$. Let Gr_d denotes the infinite Grassmannian, and γ_d is the universal rank d vector bundle on Gr_d, the first non-trivial stage of the Moore-Postnikov factorization in A^1−homotopy theory of the map $Gr_{d-1} \to Gr_d$ gives rise to a canonical morphism

$$Gr_d \to K^{G_m}(\mathbf{K}_d^{MW}, d)$$

This gives a canonical cohomology class

$$o_d \in H^d_{Nis}(Gr_d, \mathbf{K}_d^{MW}(det(\gamma_g^{vee})))$$

This work was supported by Guizhou University of Finance and Economics (NO: 2018YJ19, 2019XYB21).

H. Song and D. Jiang (Eds.): SIMUtools 2020, LNICST 370, pp. 747–754, 2021.
https://doi.org/10.1007/978-3-030-72795-6_59

And further more, given any smooth scheme X and vector bundle $\mathscr{E}$ as above. The map $\zeta : X \to Gr_d$ which classifying $\mathscr{E}$ defines a class

$$e_m(\mathscr{E}) := \zeta^*(o_d) \in H^d_{Nis}(X, \mathbf{K}^{MW}_d(det(\mathscr{E}))) = \widetilde{CH^d(X, (\mathscr{E}))}$$

This definition of obstruction class is evidently more homotypical.

The third method is constructed by J. Barge and F. Morel in [9], which is a "cohomological" version of the obstruction. Or more precisely, they give an cohomological class

$$e_c(\mathscr{E}) \in H^d_{Nis}(X, \mathbf{K}^{MW}_d(det(\mathscr{E}))) = \widetilde{CH^d(X, (\mathscr{E}))}$$

for the vector bundle $\mathscr{E}$ using push-forward operation in cohomology theory of sheaf of Milnor-Witt groups. The associated obstruction theory is called by A. Asok and J. Fasel in paper [6] as the "cohomological obstruction theory".

By the works of A. Asok and J. Fasel in papers [6] and [4], and the work of Yang in paper [26], we know that all of the theories above are equivalent, and all of the Euler classes (obstruction classes) defined by the above theories are the same under these equivalence.

By the results in paper [11] of S. M. Bhatwadekar and Raja Sridharan, we know that for different line bundles $\mathscr{L}$ and $\mathscr{L}'$, the Euler class groups $E^d(X; \mathscr{L})$ and $E^d(X; \mathscr{L}')$, or equivalently Chow-witt groups $\widetilde{CH^d(X, \mathscr{L})}$ and $\widetilde{CH^d(X, \mathscr{L}')}$, are not isomorphic generally. Now for any pair of line bundles $\mathscr{L}$ and $\mathscr{L}'$, we could regard them as the elements in the Picard group $\mathrm{Pic}(X)$ of X, and there is an element $\alpha = \mathscr{L}^{-1}\mathscr{L}'$, such that $\alpha \cdot \mathscr{L} = \mathscr{L}'$. In this paper, we will give a construction for a lifting of this action to an action of Picard on the generalized Euler class groups. More precisely, for any element $\alpha \in \mathrm{Pic}(X)$ and element $\rho \in E^d(X; \mathscr{L})$ we could define an element $\alpha \cdot \rho \in E^d(X; \alpha \cdot \mathscr{L})$ which defines an action of group on sets.

Notations

In this paper, a base field k is always denoted a (infinite) perfect field. We write Sm_k for the category of separated, smooth and finite type schemes over $\mathrm{Spec}(k)$. $sPre(Sm_k)$ denotes the category of simplicial presheaves over Sm_k, objects of which will be called $k-$spaces. We equip the category $sPre(Sm_k)$ with its usual injective Nisnevich local model structure [13]. The associated homotopy category is denoted by $\mathscr{H}_{Nis}(k)$. If $\mathcal{X}$ and $\mathcal{Y}$ are spaces in $\mathscr{H}_{Nis}(k)$, then we write $[\mathcal{X}, \mathcal{Y}]_{Nis}$ for the set $Hom_{\mathscr{H}_{Nis}(k)}(\mathcal{X}, \mathcal{Y})$.

Using a left Bousfield localization of $\mathscr{H}_{Nis}(k)$, one could get the Morel-Voevodsky A^1-homotopy category $\mathscr{H}(k)$. In particular, there is an endofunctor $\mathscr{L}_{A^1}$ of the category $sPre(Sm_k)$, together with a natural transformation $\theta : Id \to \mathscr{L}_{A^1}$ such that for any space $\mathcal{X}$, the map $\mathcal{X} \to \mathscr{L}_{A^1}(\mathcal{X})$ is an trivial A^1-cofibration and $\mathscr{L}_{A^1}(\mathcal{X})$ is simplicially fibrant and A^1-local. We refer the reader to [1] for a convenient summary of properties of the A^1-localization functor and note in passing that $\mathscr{L}_{A^1}$ commutes with the formation of finite products.

We set $[\mathcal{X}, \mathcal{Y}]_{A^1} := Hom_{\mathscr{H}_k}(\mathcal{X}, \mathcal{Y})$. Beside these, we write $S^{i+j,j} = S^i \wedge \mathbf{G}_m^{\wedge j}$ for the bigrade spheres in $\mathscr{H}(k)$. Particular S^i denotes the constant presheave associated with the i−th simplicial sphere. Further, unless we mention otherwise, sheaf cohomology will always be taken with respect to Nisnevich topology.

Beside these, in this paper, we sometimes use the letters with an arrow hat $\boldsymbol{X}$ or $\boldsymbol{a}$ to denote the sequences such as $(x_1, x_2, \cdots, x_n) \in A^n$, and use $<x_1, x_2, \cdots, x_n>$ to denote the ideal of A generated by the elements

$$\{x_1, x_2, \cdots, x_n\} \subset A,$$

where A is a commutative ring with identity.

2 The Representation of $E^d(X; \mathscr{L})$

In this section we will recall the construction of the representation of group $E^d(X; \mathscr{L})$ in the paper [26], which is the base point of this paper.

First, Assume

$$Q_{2(n+d)} = Spec \frac{k[x, x', \cdots, x^d, x_2, \cdots, x_n, y, y', \cdots, y^d, y_2, \cdots, y_n, s]}{<\Sigma x^j y^j + \Sigma x_i y_i + s^2 = 1>}$$

which is the affine quadric hypersurface defined by A. Asok, D, Brent and J. Fasel in the paper [5]. Now let $\widehat{Q_{2(n+d)}}$ be the blow-up of $Q_{2(n+d)}$ at the subscheme defined by the ideal $< x, x', \cdots, x^d >$. So we have a canonical map

$$\phi : \widehat{Q_{2(n+d)}} \to P^d$$

where $P^d =$Proj$k[t_0, \cdots, t_d]$ is the projective space of dimension d. By the analysis in the paper [26] we know that this construction give a motivic sphere ($P^{\wedge n}$ or $S^n \wedge G_m^{\wedge n}$) bundle over P^d.

Let A be a regular Noetherian ring of dimension d, smooth over the field k. Let $X =$Spec(A) be the affine scheme, and $[X, \widehat{Q_{2(n+d)}}]_{A^1}$ be the homotopy mapping classes in $\mathscr{H}(k)$, i.e. $[X, \widehat{Q_{2(n+d)}}]_{A^1} = Hom_{\mathscr{H}(k)}(X, \widehat{Q_{2(n+d)}})$.

For any $\rho_{\mathscr{L}} \in Hom_{Sm_k}(X, P^d)$, define

$$E_{\rho_{\mathscr{L}}}(X) \subseteq Hom_{\mathscr{H}(k)}(X, \widehat{Q_{2(n+d)}})$$

to be the subset of $Hom_{\mathscr{H}(k)}(X, \widehat{Q_{2(n+d)}})$ consisting of elements $f \in Hom_{\mathscr{H}(k)}(X, \widehat{Q_{2(n+d)}})$ such that $f \circ \phi = \rho_{\mathscr{L}} \in Hom_{Sm_k}(X, P^d)$. In the paper [26], with the condition $d \leq 2n - 2$, we construction a monoid structure on $E_{\rho_{\mathscr{L}}}(X)$, and prove that there is an isomorphism between the generalized Euler class group and $E_{\rho_{\mathscr{L}}}(X)$. More precisely we have the following theorem

Theorem 1. *Let A be a regular Noetherian ring of dimension d, smooth over the field k, Spec$(A) = X$ be the associated affine scheme. Let $E_n(A, \mathscr{L})$ be the*

generalized Euler Class Group where $\mathscr{L}$ is an invertible ideal and $d \leq 2n-2$. Then the Segre Class Map

$$S(\ ,\ ):\ E_n(A,\mathscr{L}) \to E_{\rho_{\mathscr{L}}}(X)$$

is an isomorphism. Thus we get that the Abelian monoid $E_{\rho_{\mathscr{L}}}(X)$ is a group isomorphic to the Euler Class Group $E_n(A,\mathscr{L})$.

Furthermore, by the analysis in the first section of the paper [26], we know that for any affine scheme $\mathrm{Spec}(A) = X$, any element $f \in E_{\rho_{\mathscr{L}}}(X)$ could be defined by a concrete map $f \in Hom_{Sch}(X, \widehat{Q_{2(n+d)}})$. Thus we could use concrete map $f \in Hom_{Sch}(X, \widehat{Q_{2(n+d)}})$ to represent the Euler classes in $E_n(A,\mathscr{L})$.

3 Action of Picard Group on $E^d(X;\mathscr{L})$

Let $\mathrm{Pic}(X)$ be the Picard group of X. Now for any element $mathscrL' \in \mathrm{Pic}(X)$ there is a map in $Hom_{Sch}(X, P^d)$ represent it which is noted by α. The element α defines an action on $\mathrm{Pic}(X)$ by the group operation.

Now for any element $\rho \in E_n(A,\mathscr{L})$ by the analysis above we have a map $\rho \in Hom_{Sch}(X, \widehat{Q_{2(n+d)}})$ representing it. The composition $\phi \circ \rho:\ X \to P^d$ represent the line bundle $\mathscr{L}$. Now α acts on the element $\mathscr{L}$ to get the line bundle $\mathscr{L}\mathscr{L}'$. Or more precisely, we have the following diagram

$$\begin{array}{ccccc} X \xrightarrow{\rho} \widehat{Q_{2(n+d)}} & & \widehat{Q_{2(n+d)}} & & (*) \\ \downarrow \phi & & \downarrow \phi & & \\ P^d & \xrightarrow{\cdot\alpha} & P^d & & \end{array}$$

Let $Gr_{d+1}(\mathscr{O}^N)$ be the Grassmannian of subbundles of dimension $d+1$ in $\mathscr{O}^N$ and $\mathscr{M}_{N\times(d+1)}$ be the scheme over k defined by the $(N\times(d+1))$ matrixes over k of rank $d+1$. Of course we have a map

$$\tau:\ \mathscr{M}_{N\times d+1} \to Gr_{d+1}(\mathscr{O}^N).$$

By the analysis in paper [27], we get that this is an A^1−fibration.

Lemma 1. *The above action α could be lifted to an element in $\mathscr{M}_{N\times(d+1)}(X)$.*

Proof. We know that the action α comes from the Segre Embedding $S:\ P^d \times P^d \to P^{N-1}$. Or equivalently the action α could be defined by the composition

$$X \xrightarrow{(\phi\circ\rho,\alpha)} P^d \times P^d \xrightarrow{S} P^{N-1}$$

Now for any point $p \in P^d$, we could define the subbundle $\mathscr{V}_p \subset \mathscr{O}^N$ as the bundle generated by

$$\{S(e_1,p), S(e_2,p), \ldots, S(e_{d+1},p)\}$$

in $\mathscr{O}^N$, where e_i are the standard basis of the bundle $\mathscr{O}^{d+1}$. Thus we define a map

$$S:\ P^d \to Gr_{d+1}(\mathscr{O}^N).$$

Thus we have the following diagram

$$\begin{array}{ccccc} X & \xrightarrow{(\phi\circ\rho,\alpha)} & P^d \times P^d & \xrightarrow{S} & P^{N-1} \\ & & \Big\downarrow{\scriptstyle (Id_{P^d}\times S)} & & \Big\downarrow \\ & & P^d \times Gr_{d+1}(\mathscr{O}^N) & & \\ & & \Big\uparrow{\scriptstyle (Id_{P^d}\times \tau)} & & \\ & & P^d \times \mathscr{M}_{N\times(d+1)} & \longrightarrow & P^{N-1} \end{array} \qquad (**)$$

Now since the map τ is an A^1-fibration and is $d+1-$connected. By the obstruction arguments in the paper [20], the map $(Id_{P^d\times S})(\phi \circ \rho, \alpha)$ could be lifted uniquely to a map $(\phi \circ \rho \times \gamma) : X \to P^d \times \mathscr{M}_{N\times(d+1)}$.

The following lemma comes from the paper [27], which could be proved by the main theorem of A. Asok, M. Hoyois and M. Wendt about the localization of A^1-fibration sequences.

Lemma 2. *The following sequence is an A^1-fibration sequence*

$$SL_{N-(d+1)} \longrightarrow SL_N \longrightarrow \mathscr{M}_{N\times(d+1)}$$

Now the map $\gamma : X \to \mathscr{M}_{N\times(d+1)}$ could be lifted to a map $\Gamma : X \to SL_N$ if and only if the Euler Class defined by the map γ in $E_N(A)$ is zero. But since the dimension of A is d, so this element must be zero by the Bass Cancellation Theorem.

Now we could get the following theorem about the action of the Picard group on the Euler classes.

Theorem 2. *Let the notations be as above. By the analysis in Sect. 1, for the any element $\rho \in E_n(A, \mathscr{L})$, there is a map $\rho \in Hom_{Sch}(X, \widehat{Q_{2(n+d)}})$. Now for any element $\alpha \in$Pic(X), we have an action of α on $\phi \circ \rho$. Now combining the* *diagram (*) and (**) we have the following digram consisting of concrete arrows*

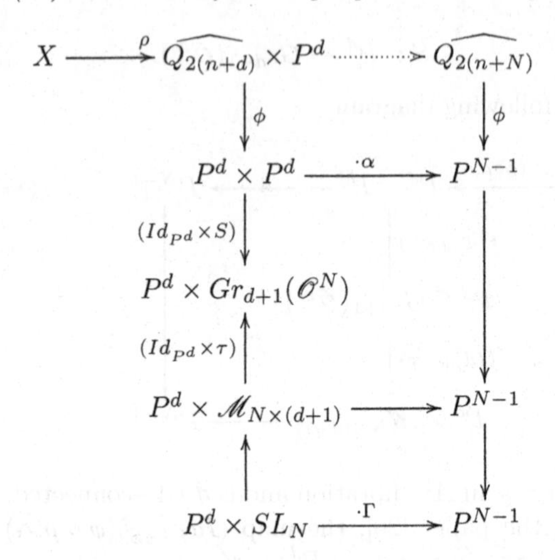

Now we could complete the diagram to a commutative diagram with a dot arrow. That is to say, we could define an action of $\alpha \in Pic(X)$ on the Euler class group $E_n(A, \mathcal{L})$.

Proof. On one hand, by the Lemma 2.1 and Lemma 2.2 the lifting of the map ρ is defined by an invertible matrix $SL_N(A)$. On another hand, since $\dim(A) = d$, any invertible matrix in $SL_N(A)$ must be a product of some elementary matrix. So in order to prove the theorem, we just need to prove that for elementary matrix δ, the action Γ could be lifted to the $\widehat{Q_{2(n+N)}}$. But this is the conclusion in the last section of the paper [12].

Corollary 1. *For any line bundle $\mathcal{L}$ and $\mathcal{L}'$ in Pic(X), there is a map $E_n(A, \mathcal{L}) \to E_n(A, \mathcal{L}')$ such that if the bundle $\mathcal{L}'$ is trivial the map is identity.*

Corollary 2. *For any line bundle $\mathcal{L}$ and $\mathcal{L}'$ in Pic(X), then there is an isomorphism $E_n(A, \mathcal{L}) \cong E_n(A, \mathcal{L}')$.*

Proof. For line bundles $\mathcal{L}$ and $\mathcal{L}'$ in Pic(X), we always have $\mathcal{L}^{-1}\mathcal{L}'$ and $\mathcal{L}\mathcal{L}'^{-1}$ in Pic(X). Using the action of $\mathcal{L}^{-1}\mathcal{L}'$ on $E_n(A, \mathcal{L})$ and action of $\mathcal{L}\mathcal{L}'^{-1}$ on $E_n(A, \mathcal{L}')$, we can construct the isomorphisms between $E_n(A, \mathcal{L})$ and $E_n(A, \mathcal{L}')$ by the above lemma.

Conclusion

From Theorem 1, we get that the Abelian monoid $E_{\rho_{\mathcal{L}}}(X)$ is a group isomorphic to the Euler Class Group $E_n(A, \mathcal{L})$. Thus, we could use a concrete map $f \in Hom_{Sch}(X, \widehat{Q_{2(n+d)}})$ to represent the Euler classes in $E_n(A, \mathcal{L})$. From Theorem 2, we could define an action of $\alpha \in$Pic(X) on the Euler class group $E_n(A, \mathcal{L})$. This result gives a new method to compare the generalized Euler classes with different twisted bundle.

References

1. Asok, A., Hoyois, M., Wendt, M.: Affine representability results in A^1−homotopy theory I: vector bundles. Duke Math. J. **166**(10), 1923–1953 (2017)
2. Asok, A., Hoyois, M., Wendt, M.: Affine representability results in A^1−homotopy theory II: principal bundles and homogeneous spaces. Geom. Top. **22**(2), 1181–1225 (2018)
3. Asok, A., Jean, F.: Splitting vector bundles outside the stable range and homotopy theory of punctured affine spaces. J. Amer. Math. Soc. **28**(4), 1031–1062 (2014)
4. Asok, A., Jean, F.: Euler class groups and motivic stable cohomotopy. ArXiv:1601.05723
5. Asok, A., Doran, B., Jean, F.: Smooth models of motivic spheres and the clutching construction. Int. Math. Res. Not. (6), 1890–1925 (2017)
6. Asok, A., Jean, F.: Comparing Euler classes. Q. J. Math **67**(4), 603–635 (2016)
7. Asok, A., Wickelgren, K., Williams, T.B.: The simplicial suspension sequence in A^1−homotopy. Geom. Top. **21**(4), 2093–2160 (2017)
8. Bass, H.: K-theory and stable algebra. Inst. Hautes tudes Sci. Publ. Math. **22**, 5–60 (1964)
9. et Fabien Morel, J.B.: Groupe de Chow des cycles orientes et classe d'Euler des fibres vectoriels, C.R. Acad. Sci. Paris, t. 330, Serie I, pp. 287–290 (2000)
10. Bhatwadekar, S.M., Sridharan, R.: Zero cycles and the Euler class groups of smooth real affine varieties. Inventiones mathematicae **136**(2), 287–322 (1999)
11. Bhatwadekar, S.M., Sridharan, R.: The Euler class group of a noetherian ring. Compositio Mathematica **122**(2), 183–222 (2000)
12. Fasel, J.: On the number of generators of ideals in polynomial rings. Ann. Math **184**(1), 315–331 (2016)
13. Jardine, J.F.: Local Homotopy Theory. Springer Monographs in Mathematics. Springer, New York (2015)
14. Jardine, J.F.: Documenta Mathematica **5**, 445–553 (2000)
15. Georgies, R., Strunk, F.: Documenta Mathematica **23**, 1757–1797 (2018)
16. Mandal, S.: On efficient generation of idealsProjective Modules and Complete Intersections. Inventiones mathematicae **75**(1), 59–67 (1984)
17. Mandal, S.: Projective Modules and Complete Intersections. LNM, vol. 1672. Springer, Heidelberg (1997). https://doi.org/10.1007/BFb0093560
18. Manda, S., Yang, Y.: Excision in algebraic obstruction theory. JPAA **216**, 2159–2169 (2012)
19. Mandal, S., Yang, Y.: Intersection theory of algebraic obstructions (with Yong Yang). JPAA **214**, 2279–2293 (2010)
20. Fabien Morel A^1-Algebraic Topology over a Field, LNM 2052, (2010)
21. Fabien Morel and Vladimir Voevodsky; A^1−homotopy theory of schemes. Publ. Math. IHES, (90), 45–143 (1999)
22. Mohan Kumar, N.: Complete intersections. J. Math. Kyoto Univ. **17**(3), 533–538 (1977)
23. Rezk, C.: Toposes and homotopy toposes (2005). http://www.math.uiuc.edu/?rezk/homotopy-topos-sketch.pdf
24. Strunk, F.: On motivic spherical bundles. Thesis Institut fur Mathematik Universit at Osnabruck (2012)

25. Suslin, A.A.: On the structure of the special linear group over polynomial rings. Math. USSR Izv. **11**, 221–238 (1977)
26. Yang, Y.: Generalized Euler Class Group and Chow-Witt Group with Twist, preprint
27. Schlichting, M., Tripathi, G.S.: Geometric models for higher Grothendieck–Witt groups in A^1-homotopy theory **362**, 1143–1167(2015)
28. Bergeron, N., Charollois, P., Garcia, L.E.: Transgressions of the Euler class and Eisenstein cohomology of GL N (Z). Jpn. J. Math. **15**(2), 311–379 (2020)
29. Naolekar, A.C., Singh, T.A.: Euler classes of vector bundles over iterated suspensions of real projective spaces. Mathematica Slovaca **68**(3), 677–684 (2018)

Dynamical Analysis of a Predator-Prey Economic Model with Impulsive Control Strategy

Airen Zhou(✉) and Jianjun Jiao

School of Mathematics and Statistics, Guizhou University of Finance and Economics, Guiyang 550025, People's Republic of China

Abstract. In the present work, a predator-prey economic model with impulsive stocking immature predator and impulsive harvesting mature predator is studied. First, the method of stroboscopic mapping of the discrete dynamics is used in our demonstrate. After that, it is proved that the relevant solution of the studied system is globally asymptotically stable. Next, based on the principle of comparison of pulsed differential equations, it is also proved that the studied system possesses its persistence. In the end, the numerical experiment is introduced to demonstrate our results. The obtained conclusions provide a methodological guidance for the actual biological economic managements.

Keywords: Control strategy · Extinction · Permanence

1 Introduction and Background

The ultimate goal of scientific management and rational development and application of renewable resources (fisheries, forestry resources, etc.) is to make them an inexhaustible resource. However, if unreasonable utilization of resources damages resources and disrupts their renewal cycle, it will cause resource depletion, which will not only cause economic losses, but also affect the living environment of human beings in severe cases. Therefore, in order to protect the natural environment on which humans depend, the development of renewable resources must be reasonable and appropriate. Under the premise of sustainable resources, it is a feasible way to analyze, evaluate and predict the development and utilization of renewable resources by means of mathematical models [1].

Supported by National Natural Science Foundation of China(11761009), Guizhou Science and Technology Platform Talents ([2017] 5736-019), Science and Technology Foundation of Guizhou Province ([2020]1Y001), and Guizhou Provincial Education Department's Project for the Growth of Young Science and Technology Talents (KY[2018]157), Guizhou Team of Scientific and Technological Innovation Talents (No.20175658), Guizhou University of Finance and Economics Project Funding (No. 2019XYB11), Guizhou University of Finance and Economics Introduced Talent Research Project (2017).

H. Song and D. Jiang (Eds.): SIMUtools 2020, LNICST 370, pp. 755–765, 2021.
https://doi.org/10.1007/978-3-030-72795-6_60

In recent years, Predator-prey model and Stage-structured model have been received extensively and deeply studied [2,10–15].

2 The Research Model

Once the immature predator population grows into mature predator population, it should be harvested [7]. In order to examine the rational development and sustainable development of resources, we build an economic model as follows:

$$\begin{cases} \left.\begin{aligned} \frac{dp(t)}{dt} &= rp(t)(1-\frac{p(t)}{K}) - \frac{\beta_1 p(t)}{1+c_1 p(t)}p_1(t) - \frac{\beta_2 p(t)}{1+c_2 p(t)}p_2(t), \\ \frac{dp_1(t)}{dt} &= \frac{l_1\beta_1 p(t)}{1+c_1 p(t)}p_1(t) - (d_1+\kappa)p_1(t), \\ \frac{dp_2(t)}{dt} &= \frac{l_2\beta_2 p(t)}{1+c_2 p(t)}p_2(t) + \kappa p_1(t) - d_2 p_2(t), \end{aligned}\right\} t \neq nU, \\ \left.\begin{aligned} &\Delta p(t) = 0, \\ &\Delta p_1(t) = \frac{\delta_1 p(t)}{1+\lambda_1 p(t)} + \mu_1, \\ &\Delta p_2(t) = -\mu_2 p_2(t), \end{aligned}\right\} t = nU, n \in Z^+, \end{cases} \tag{1}$$

where $p(t), p_1(t), p_2(t)$ are the population densities of prey, immature predator and mature predator, respectively. $\frac{\delta_1 p(t)}{1+\lambda_1 p(t)}$ is a nonlinear input term, $\delta_1 \geq 0$ is a proportional coefficient (capture rate), $\lambda_1 \geq 0$ is the half saturation constant for the prey due to stocking immature predator. $\Delta p_i(t) = p_i(t^+) - p_i(t)$. $\mu_1 (\mu_1 \geq 0)$ denotes the stocking quantity of immature predator at $t = nU, n \in Z^+$. $\mu_2 (1 > \mu_2 \geq 0)$ represents the gathering portion of mature predator population owing to behaviors of biological economic management at $t = nU, n \in Z^+$. The meaning of the other variables can be referred to [6,8,9].

3 Some Lemmas

Obviously, the positivity of solutions of (1) results from the positive initial value conditions. The following Lemma 1° is not difficult to be proved.

Lemma 1°. It is bounded for the solution of system (1). i.e., there exists a positive constant $L > 0$ satisfying $p(t) \leq L/l, p_1(t) \leq L, p_2(t) \leq L$, for each solution $(p(t), p_1(t), p_2(t))$ of (1) for t big enough.

Proof. Denote that $V(t) = lp(t) + p_1(t) + p_2(t)$, $l = \max\{l_1, l_2\}$, $H = \min\{d_1, d_2\}$ and $b = \frac{r}{K}$. When t is not equal to nU,

$$\begin{aligned} D^+V(t) + HV(t) &\leq l[(r+H)p(t) - bp^2(t)] - (d_1 - H)p_1(t) - (d_2 - H)p_2(t) \\ &= -bl\big(p(t) - \frac{r+H}{2b}\big)^2 + \frac{l(r+H)^2}{4b} \\ &\leq \frac{l(r+H)^2}{4b} = L_0. \end{aligned} \tag{2}$$

When t is equal to nU, we have $V(nU^+) < V(nU) + \frac{\delta_1}{\lambda_1} + \mu_1$. Thanks the Lemma in [4], as $t \in (nU, (n+1)U]$, holds

$$\begin{aligned} V(t) &\leq V(0)\exp(-Ht) + \frac{L_0}{H}(1 - \exp(-Ht)) \\ &+ \frac{(\frac{\delta_1}{\lambda_1} + \mu_1)\exp(-H(t-U))}{1 - \exp(HU)} + \frac{(\frac{\delta_1}{\lambda_1} + \mu_1)\exp(HU)}{\exp(HU) - 1} \\ &\longrightarrow \frac{L_0}{H} + \frac{(\frac{\delta_1}{\lambda_1} + \mu_1)\exp(HU)}{\exp(HU) - 1}, \quad t \to \infty, \end{aligned} \tag{3}$$

which implies the consistent boundedness of $V(t)$. Therefore, Lemma 1° holds in terms of the above definition.

Taking the subsystem of (1) into account:

$$\begin{cases} \left.\begin{aligned} \frac{dp_1(t)}{dt} &= -(\kappa + d_1)p_1(t), \\ \frac{dp_2(t)}{dt} &= \kappa p_1(t) - d_2 p_2(t), \end{aligned}\right\} t \neq nU, n \in Z^+, \\ \left.\begin{aligned} \Delta p_1(t) &= \frac{\delta_1 p(t)}{1 + \lambda_1 p(t)} + \mu_1, \\ \Delta p_2(t) &= -\mu_2 p_2(t), \end{aligned}\right\} t = nU, n \in Z^+. \end{cases} \tag{4}$$

Lemma 2° can also be obtained:

Lemma 2°. For system (4), there is only one positive U-periodic solution $(\widetilde{p_1(t)}, \widetilde{p_2(t)})$ with globally asymptotical stability, here

$$\begin{cases} \widetilde{p_1(t)} = \dfrac{\mu_1 e^{-(\kappa+d_1)(t-nU)}}{1 - e^{-(\kappa+d_1)U}}, nU < t \leq (n+1)U, n \in Z^+, \\ \widetilde{p_2(t)} = \dfrac{\mu_1 \kappa}{(\kappa + d_1 - d_2)(1 - e^{-(\kappa+d_1)U})} \left[\dfrac{1 - (1-\mu_2)e^{-(\kappa+d_1)U}}{1 - (1-\mu_2)e^{-d_2 U}} e^{-d_2(t-nU)} \right. \\ \qquad \left. - e^{-(\kappa+d_1)(t-nU)} \right], nU < t \leq (n+1)U, n \in Z^+. \end{cases} \tag{5}$$

4 Dynamical Analysis of (1)

Next, we are going to show an important conclusion in this paper, that is, the sufficient condition for the globally asymptotic stability of solution $(0, \widetilde{p_1(t)}, \widetilde{p_2(t)})$ of Eq. (1).

Theorem 1°. The solution $(0, \widetilde{p_1(t)}, \widetilde{p_2(t)})$ of (1) is globally asymptotically stable under the condition

$\mu_1 > rU$

$[\frac{\beta_1}{\kappa+d_1} + \frac{\kappa\beta_2}{(\kappa+d_1-d_2)(1-e^{-(\kappa+d_1)U})}(\frac{1-(1-\mu_2)e^{-(\kappa+d_1)U}(1-e^{-d_2U})}{d_2(1-(1-\mu_2)e^{-d_2U})} + \frac{e^{-(\kappa+d_1)U}-1}{\kappa+d_1})]^{-1}$

is true.

Proof. According to local stability and global attraction, we can derive global asymptotic stability [2,3]. The first step aims at the local stability. Let $p(t), p_1(t) - \widetilde{p_1(t)}, p_2(t) - \widetilde{p_2(t)}$ relabel as $p(t), q_1(t), q_2(t)$, respectively. We do the following linearization:

$$\begin{pmatrix} p'(t) \\ q_1'(t) \\ q_2'(t) \end{pmatrix} = \begin{pmatrix} r - \beta_1\widetilde{p_1(t)} - \beta_2\widetilde{p_2(t)} & 0 & 0 \\ l_1\beta_1\widetilde{p_1(t)} & -(\kappa+d_1) & 0 \\ l_2\beta_2\widetilde{p_2(t)} & \kappa & -d_2 \end{pmatrix} \begin{pmatrix} p(t) \\ q_1(t) \\ q_2(t) \end{pmatrix}.$$

Assume $\Phi(t)$ is the fundamental matrix above system, so $\Phi(t)$ should meet

$$\Phi(U) = \begin{pmatrix} \exp(\int_0^U (r - \beta_1\widetilde{p_1(s)} - \beta_2\widetilde{p_2(s)})ds) & 0 & 0 \\ \dagger & \exp(-(\kappa+d_1)U) & 0 \\ \star & * & \exp(-d_2U) \end{pmatrix},$$

it is not necessary to calculate the terms $\dagger, \star$ and $*$ with the exact forms for the rest of analysis. The linearization of the 4th, 5th and 6th equations of (1) turns to be

$$\begin{pmatrix} p(nU^+) \\ q_1(nU^+) \\ q_2(nU^+) \end{pmatrix} = \begin{pmatrix} 1 & 0 & 0 \\ 0 & 1 & 0 \\ 0 & 0 & 1-\mu_2 \end{pmatrix} \begin{pmatrix} p(nU) \\ q_1(nU) \\ q_2(nU) \end{pmatrix} = A(U) \begin{pmatrix} p(nU) \\ q_1(nU) \\ q_2(nU) \end{pmatrix}.$$

Thus, the absolute values of three eigenvalues of the following matrix will control the local stability of the solution $(0, \widetilde{p_1(t)}, \widetilde{p_2(t)})$ of equations (1)

$$\begin{aligned} M &= A(U)\Phi(U) \\ &= \begin{pmatrix} \exp[\int_0^U (r - \beta_1\widetilde{p_1(s)} - \beta_2\widetilde{p_2(s)})ds] & 0 & 0 \\ 0 & \exp(-(\kappa+d_1)U) & 0 \\ 0 & 0 & (1-\mu_2)\exp(-d_2U) \end{pmatrix}. \end{aligned}$$

The eigenvalues of matrix M satisfy

$$|\Lambda_1| = \exp[\int_0^U (r - \beta_1\widetilde{p_1(s)} - \beta_2\widetilde{p_2(s)})ds],$$

$$|\Lambda_2| = \exp(-(\kappa+d_1)U) < 1, |\Lambda_3| = (1-\mu_2)\exp(-d_2U) < 1.$$

This manifests that the solution $(0, \widetilde{p_1(t)}, \widetilde{p_2(t)})$ is locally stable if and only if

$$rU - \beta_1 \int_0^U \widetilde{p_1(s)} ds - \beta_2 \int_0^U \widetilde{p_2(s)} ds < 0.$$

Therefore, the local asymptotical stability of solution $(0, \widetilde{p_1(t)}, \widetilde{p_2(t)})$ of (1) is dominated by

$$\mu_1 > rU$$

$$[\frac{\beta_1}{\kappa + d_1} + \frac{\kappa\beta_2}{(\kappa + d_1 - d_2)(1 - e^{-(\kappa+d_1)U})}(\frac{1 - (1 - \mu_2)e^{-(\kappa+d_1)U}(1 - e^{-d_2U})}{d_2(1 - (1 - \mu_2)e^{-d_2U})} + \frac{e^{-(\kappa+d_1)U} - 1}{\kappa + d_1})]^{-1}.$$

The second step, we will go to prove the global attraction of the solution $(0, \widetilde{p_1(t)}, \widetilde{p_2(t)})$. Let $(0, p_1(t), p_2(t))$ be any prey-extinction boundary periodic solution of equaions (1), for this goal, we just have to prove $(0, p_1(t), p_2(t))$ converges to $(0, \widetilde{p_1(t)}, \widetilde{p_2(t)})$ when $t \to \infty$. To pick up a positive number ε, which is little enough, such that

$$\eta = \exp[\int_0^U (r - \beta_1(\widetilde{p_1(t)} - \varepsilon) - \beta_2(\widetilde{p_2(t)} - \varepsilon))dt] < 1.$$

From system (1), we have

$$\begin{cases} x_1'(t) = -(\kappa + d_1)x_1(t), t \neq nU, \\ x_2'(t) = \kappa x_1(t) - d_2x_2(t), t \neq nU, \\ \Delta x_1(t) = \dfrac{\delta_1 x(t)}{1 + \lambda_1 x(t)} + \mu_1, \quad t = nU, \\ \Delta x_2(t) = -\mu_2 x_2(t), \quad t = nU. \end{cases} \tag{6}$$

With the helps of Lemma 2° and the principle of comparison of differential equations [4], to get

$$\begin{cases} \widetilde{p_1(t)} - \varepsilon \leq x_1(t) \leq p_1(t), \\ \widetilde{p_2(t)} - \varepsilon \leq x_2(t) \leq p_2(t), \end{cases} \tag{7}$$

for t large enough. For simplicity and clarity, we make the admission that (7) be true for all $t \geq 0$. Using (1) and (7), yields

$$p'(t) \leq p(t)[r - \beta_1(\widetilde{p_1(t)} - \varepsilon) - \beta_2(\widetilde{p_2(t)} - \varepsilon)]. \tag{8}$$

Consequently $p(t) \leq p(0^+) \exp[\int_0^t (r - \beta_1(\widetilde{p_1(s)} - \varepsilon) - \beta_2(\widetilde{p_2(s)} - \varepsilon))ds]$, hence $p((n+1)U) \leq p(nU^+) \exp[\int_{nU}^{(n+1)U} (r - \beta_1(\widetilde{p_1(s)} - \varepsilon) - \beta_1(\widetilde{p_2(s)} - \varepsilon))ds]$. As a result, $p(nU) \leq p(0^+)\eta^n$ and $p(nU) \to 0$ as $n \to \infty$. In consequence, $p(t) \to 0$, when t goes to infinity.

Further, need to prove when t goes to ∞, $p_1(t) \to \widetilde{p_1(t)}$, $p_2(t) \to \widetilde{p_2(t)}$. Without loss of generality, making the following assumption: for all $t \geq 0$, $0 < p(t) < \varepsilon$. From the above assumption, for equations (1), yield

$$-(\kappa + d_1)p_1(t) \leq \frac{dp_1(t)}{dt} \leq [-(\kappa + d_1) + l_1\beta_1\varepsilon]p_1(t), \tag{9}$$

and

$$\kappa p_1(t) - d_2 p_2(t) \leq \frac{dp_2(t)}{dt} \leq \kappa p_1(t) - (d_2 - l_2\beta_2\varepsilon)p_2(t). \tag{10}$$

Therefore, we can take the following systems of equations into account:

$$\begin{cases} y_1'(t) = -(\kappa + d_1)y_1(t), & t \neq nU, \\ y_2'(t) = \kappa y_1(t) - d_2 y_2(t), & t \neq nU, \\ \Delta y_1(t) = \dfrac{\delta_1 y(t)}{1 + \lambda_1 y(t)} + \mu_1, & t = nU, \\ \Delta y_2(t) = -\mu_2 y_2(t), & t = nU, \end{cases} \tag{11}$$

and

$$\begin{cases} m_1'(t) = -[(\kappa + d_1) - l_1\beta_1\varepsilon]m_1(t), & t \neq nU, \\ m_2'(t) = \kappa m_1(t) - (d_2 - l_2\beta_2\varepsilon)m_2(t), & t \neq nU, \\ \Delta m_1(t) = \dfrac{\delta_1 m(t)}{1 + \lambda_1 m(t)} + \mu_1, & t = nU, \\ \Delta m_2(t) = -\mu_2 m_2(t), & t = nU, \end{cases} \tag{12}$$

according to the principle of comparison of pulsed differential equations, we get $m_1(t) \geq p_1(t) \geq y_1(t)$, $m_2(t) \geq p_2(t) \geq y_2(t)$, together with $y_1(t) \to \widetilde{p_1(t)}, y_2(t) \to \widetilde{p_2(t)}, m_1(t) \to \widetilde{m_1(t)}, m_2(t) \to \widetilde{m_2(t)}$ when t goes to ∞. Here

$$\begin{cases} \widetilde{m_1(t)} = \dfrac{\mu_1 e^{-(\kappa+d_1-l_1\beta_1\varepsilon)(t-nU)}}{1 - e^{-(\kappa+d_1-l_1\beta_1\varepsilon)U}}, \quad nU < t \leq (n+1)U, n \in Z^+, \\ \widetilde{m_2(t)} = \dfrac{\mu_1\kappa}{(\kappa + d_1 - l_1\beta_1\varepsilon - d_2)(1 - e^{-(\kappa+d_1-l_1\beta_1\varepsilon)U})}. \\ \left[\dfrac{1-(1-\mu_2)e^{-(\kappa+d_1-l_1\beta_1\varepsilon)U}}{1-(1-\mu_2)e^{-(d_2-l_2\beta_2\varepsilon)U}} e^{-(d_2-l_2\beta_2\varepsilon)(t-nU)} - e^{-(\kappa+d_1-l_1\beta_1\varepsilon)(t-nU)}\right], \\ \hfill nU < t \leq (n+1)U, n \in Z^+. \end{cases} \tag{13}$$

Consequently, for any sufficiently small positive number ε_1, there being a t_1 , such that $t \geq t_1 > 0$, holds

$$\widetilde{m_1(t)} + \varepsilon_1 > p_1(t) > \widetilde{p_1(t)} - \varepsilon_1,$$

and

$$\widetilde{m_2(t)} + \varepsilon_1 > p_2(t) > \widetilde{p_2(t)} - \varepsilon_1.$$

Here $\varepsilon \to 0$ should be allowed, generates

$$\widetilde{p_1(t)} + \varepsilon_1 > p_1(t) > \widetilde{p_1(t)} - \varepsilon_1,$$

and

$$\widetilde{p_2(t)} + \varepsilon_1 > p_2(t) > \widetilde{p_2(t)} - \varepsilon_1.$$

So, $p_1(t) \to \widetilde{p_1(t)}$ and $p_2(t) \to \widetilde{p_2(t)}$ when t goes to ∞. The proof is completed.

Next, we set out to study the persistence of equations (1).

Theorem 2°. Equations (1) is persistent under the condition

$$\mu_1 < rU$$

$$[\frac{\beta_1}{\kappa + d_1} + \frac{\kappa\beta_2}{(\kappa + d_1 - d_2)(1 - e^{-(\kappa+d_1)U})}(\frac{1 - (1 - \mu_2)e^{-(\kappa+d_1)U}(1 - e^{-d_2U})}{d_2(1 - (1 - \mu_2)e^{-d_2U})} + \frac{e^{-(\kappa+d_1)U} - 1}{\kappa + d_1})]^{-1}$$

is true.

Proof. From Lemma 1°, for convenience, it should be able to assume $p(t) \leq L/l, p_1(t) \leq L, p_2(t) \leq L$ and $L > \frac{r}{\beta_1}$ for all $t > 0$. System (7) indicates that $p_1(t) > \widetilde{p_1(t)} - \varepsilon_2$ and $p_2(t) > \widetilde{p_2(t)} - \varepsilon_2$ for sufficiently large t and some $\varepsilon_2 > 0$, so it is easy to deduce that $p_1(t) \geq \zeta_2$ and $p_2(t) \geq \zeta_2'$ for sufficiently large t. As a result, just need to find $\zeta_1 > 0$ such that $\zeta_1 \leq p(t)$ for sufficiently large t. We intend to take two steps to implement it.

The first step: Takeing sufficiently small positive numbers ζ_3 and ε_1, such that

$$\min\left\{\frac{\kappa + d_1}{l_1\beta_1}, \frac{d_2}{l_2\beta_2}\right\} > \zeta_3, \quad \delta = \max\{l_1\beta_1\zeta_3, l_2\beta_2\zeta_3\} < \min\{\kappa + d_1, d_2\}$$

and

$$\begin{aligned}\sigma = {} & rU - \frac{r}{K}\zeta_3 U - \beta_1\varepsilon_1 U - \beta_2\varepsilon_1 U - \frac{\beta_1\mu_1}{\kappa + d_1 - \delta} \\ & - \frac{\mu_1\kappa\beta_2(1 - (1 - \mu_2)e^{-(\kappa+d_1-\delta)U})(1 - e^{-(d_2-\delta)U})}{(\kappa + d_1 - d_2)(1 - e^{-(\kappa+d_1-\delta)U})(d_2 - \delta)(1 - (1 - \mu_2)e^{-(d_2-\delta)U})} \\ & + \frac{\beta_2\mu_1\kappa}{(c + d_1 - d_2)(\kappa + d_1 - \delta)} \\ > {} & 0.\end{aligned}$$

Presume for all $t \geq 0$, $\zeta_3 < p(t)$, yields

$$p_1'(t) \leq [-(\kappa + d_1) + \delta]p_1(t),$$

and

$$p_2'(t) \leq \kappa p_1(t) - (d_2 - \delta)p_2(t).$$

According to Lemmas 2°, we obtain $x_1(t) \geq p_1(t), x_2(t) \geq p_2(t), (x_1(t), x_2(t)) \rightarrow (\overline{x_1(t)}, \overline{x_2(t)})$, when t goes to infinity, here $(x_1(t), x_2(t))$ is the solution to

$$\begin{cases} \left.\begin{array}{l} x_1'(t) = -[(\kappa + d_1) - \delta]x_1(t), \\ x_2'(t) = \kappa x_1(t) - (d_2 - \delta)x_2(t), \end{array}\right\} t \neq nU, \quad n \in Z^+, \\ \left.\begin{array}{l} \Delta x_1(t) = \dfrac{\delta_1 x(t)}{1 + \lambda_1 x(t)} + \mu_1, \\ \Delta x_2(t) = -\mu_2 x_2(t), \end{array}\right\} t = nU, \quad n \in Z^+, \end{cases} \tag{14}$$

and

$$\begin{cases} \overline{x_1(t)} = \dfrac{\mu_1 e^{-(\kappa+d_1-\delta)(t-nU)}}{1 - e^{-(\kappa+d_1-\delta)U}}, nU < t \leq (n+1)U, n \in Z^+, \\ \overline{x_2(t)} = \dfrac{\mu_1 \kappa}{(\kappa + d_1 - d_2)(1 - e^{-(\kappa+d_1-\delta)U})} \Big[\dfrac{1 - (1 - \mu_2)e^{-(\kappa+d_1-\delta)U}}{1 - (1 - \mu_2)e^{-(d_2-\delta)U}} e^{-(d_2-\delta)(t-nU)} \\ \qquad -e^{-(\kappa+d_1-\delta)(t-nU)} \Big], nU < t \leq (n+1)U, n \in Z^+. \end{cases} \tag{15}$$

Thereupon, there being a positive number T_1, for all $t \geq T_1$, satisfies

$$\overline{x_1(t)} + \varepsilon_1 \geq x_1(t) \geq p_1(t), \quad \overline{x_2(t)} + \varepsilon_1 \geq x_2(t) \geq p_2(t),$$

and

$$p'(t) \geq p(t) \Big[r - bm_3 - \beta_1(\overline{x_1(t)} + \varepsilon_1) - \beta_2(\overline{x_2(t)} + \varepsilon_1) \Big]. \tag{16}$$

Suppose $N_1 \in Z^+$ and $N_1 U > T_1$, let's integrate (16) on both sides with $t \in (nU, (n+1)U), n \geq N_1$, yields

$$\begin{aligned} p((n+1)U) &\geq p(nU^+) \exp(\int_{nU}^{(n+1)U} [r - bm_3 - \beta_1(\overline{x_1(t)} + \varepsilon_1) - \beta_2(\overline{x_2(t)} + \varepsilon_1)]dt) \\ &= p(nU)e^{\sigma}, \end{aligned}$$

so $p((N_1 + k)U) \geq p(N_1 U^+)e^{k\sigma} \rightarrow \infty$, when k goes to ∞. Boundedness and unboundedness of $p(t)$ contradict each other. Therefore, it has got a positive number t_1, satisfying $p(t_1) \geq \zeta_3$.

The second step: The proof of the second step can be carried out by referring to [8,9], which is omitted here. So we've done the proof of the Theorem 2°.

5 Discussion

In the present section, we will do some numerical simulations. Taking a group of variables as shown in the table below:

$p(0)$	$p_1(0)$	$p_2(0)$	r	K	β_1	β_2	l_1	l_2	c_1	c_2	d_1	κ	d_2	U	μ_1	μ_2
1.1	1.1	1.1	1.1	2.1	2.1	3.1	0.7	0.9	2.1	2.1	0.4	2.1	0.4	1.1	2.1	0.8

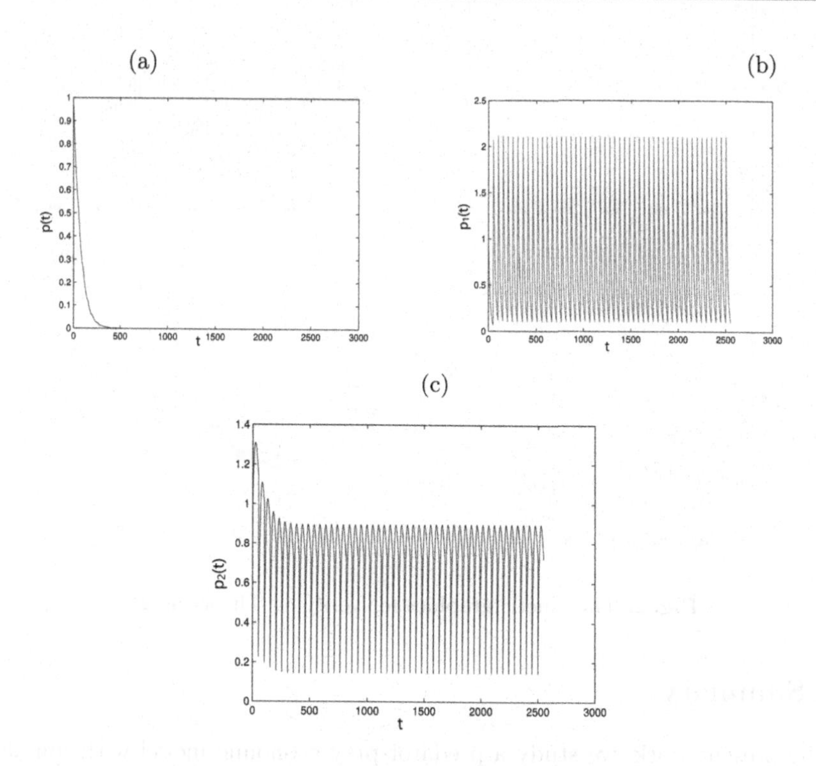

Fig. 1. The kinetic graphics for verifying Theorems 1°.

Clearly, the condition for Theorem 1° is satisfied, wherefore, the solution $(0, p_1(t), p_2(t))$ of (1) possesses global asymptotic stability (see Fig. 1.).
Taking another group of variables as shown in the table below:

$p(0)$	$p_1(0)$	$p_2(0)$	r	K	β_1	β_2	l_1	l_2	c_1	c_2	d_1	κ	d_2	U	μ_1	μ_2
0.9	0.9	0.9	0.9	1.9	1.9	2.9	0.5	0.8	1.9	1.9	0.3	1.9	0.3	0.9	0.9	0.8

Evidently, the condition for Theorem 2° is also satisfied, thereupon, system (1) can preserve its permanence (see Fig. 2.).
According to Theorem 1° and Theorem 2°, there must being a threshold $\Omega = rU[\frac{\beta_1}{\kappa+d_1} + \frac{\kappa\beta_2}{(\kappa+d_1-d_2)(1-e^{-(\kappa+d_1)U})}(\frac{1-(1-\mu_2)e^{-(\kappa+d_1)U}(1-e^{-d_2U})}{d_2(1-(1-\mu_2)e^{-d_2U})} + \frac{e^{-(\kappa+d_1)U}-1}{\kappa+d_1})]^{-1}$.
The solution $(0, p_1(t), p_2(t))$ of (1) possesses global asymptotic stability provided that $\mu_1 > \Omega$; On the other hand, system (1) can preserve its permanence provided that $\mu_1 < \Omega$. The results of numerical experiments verify the validity of our obtained theorems.

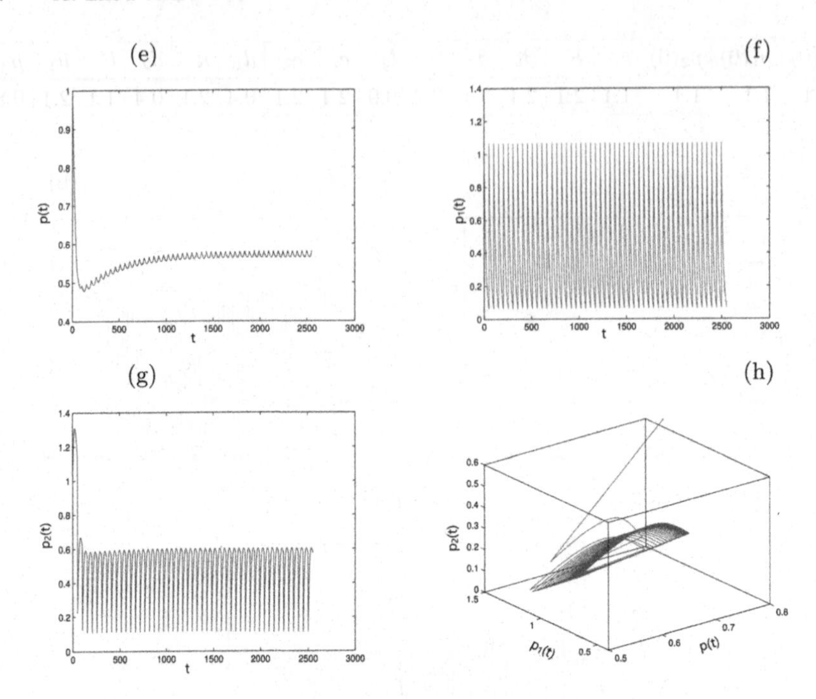

Fig. 2. The kinetic graphics for verifying Theorems 2°.

6 Summary

In the present work, we study a predator-prey economic model with impulsive control strategy. Some threshold conditions (Such as possessing global asymptotic stability, preserving its permanence) have been proved in terms of theory of dynamical system. Further, simulation experiments demonstrate our theorems. The obtained results provide a methodological guidance for the actual biological economic managements.

References

1. Tang, S., Xiao, Y., et al.: Mathematical Biology. IAM, vol. 17. Springer, New York (2002). https://doi.org/10.1007/978-0-387-22437-4_9
2. Allen, L.J.S.: An Introduction to Mathematical Biology. Prentice Hall, New Jersey (2007)
3. Li, C., Tang, S., Cheke, R.A.: Complex dynamics and coexistence of period-doubling and period-halving bifurcations in an integrated pest management model with nonlinear impulsive control. Adv. Difference Equations **2020**(1), 1–23 (2020). https://doi.org/10.1186/s13662-020-02971-9
4. Lakshmikantham, V., Bainov, D., Simeonov, P.: Theory of Impulsive Differential Equations. World scientific, Singapore (1989)
5. Klausmeier, C.: Floquet theory: a useful tool for understanding nonequilibrium dynamics. Theor. Ecol-Neth. **1**(3), 153–161 (2008)

6. Jiao, J., Cai, S., Chen, L.: Dynamical behaviors of a biological management model with impulsive stocking juvenile predators and continuous harvesting adult predators. J. Appl. Math. Comput. **35**(1–2), 483–495 (2011)
7. David, L., Alfred, E.: Predators and prey in fishes. In: David, N., David, L., Gene, H., Jack, W. (eds.) Proceedings of the 3rd Biennial Ecology of Fishes, 1983, pp. 55–67. Kluwer Academic Publishers Group, Kordrecht (1983)
8. Zhou, A., Sattayatham, P., Jiao, J.: Dynamics of an SIR epidemic model with stage structure and pulse vaccination. Adv. Diff. Eq. **2016**(1), 1–17 (2016). https://doi.org/10.1186/s13662-016-0853-z
9. Jiao, J., Cai, S., Li, L.:Dynamics of an oasis-vegetation degradation model with impulsive irrigation and diffusion in arid area. J. Appl. Math. Comput., **53**(1-2), 555–570 (2017)
10. Dalziel, B.D., Thomann, E., Medlock, J., et al.: Global analysis of a predator-prey model with variable predator search rate. J. Math. Biol. **81**, 159–183 (2020)
11. Hoang, M.T.: On the global asymptotic stability of a predator-prey model with Crowley-Martin function and stage structure for prey. J. Appl. Math. Comput. **2020**, 1–16 (2020)
12. Becker, O., Leopold-Wildburger, U.: Optimal dynamic control of predator-prey models. Cent. EUR. J. Oper. Res., **28**(2), 425–440 (2020)
13. Neverova, G.P., Zhdanova, O.L., Ghosh, B., et al.: Dynamics of a discrete-time stage-structured predator-prey system with Holling type II response function. Nonlinear Dyn. **98**(1), 427–446 (2019)
14. Dubey, B., Kumar, A.: Dynamics of prey-predator model with stage structure in prey including maturation and gestation delays. Nonlinear Dyn. **96**(4), 2653–2679 (2019)
15. Qiang, L., Wang, B., Zhao, X.: A stage-structured population model with time-dependent delay in an almost periodic environment. J. Dyn. Differ. Equ., **2020**, 1-24 (2020)

Ellipse Fitting Based on a Hybrid $l1l2$-Norm Algorithm

Lingling Luo[1(✉)], Wenqing Yang[2], Wei Ren[2], and Xiaojun Yu[3]

[1] School of Mathematics and Statistics, Guizhou University of Finance and Economics, Guizhou 550025, China

[2] Guangzhou Power Supply Bureau of Guangdong Power Grid Co., Ltd., Guangzhou, China

[3] School of Mathematics and Statistics, Guizhou University of Finance and Economics, Guizhou 550025, China

Abstract. This paper proposes a new efficient direct method noted the Hybrid $l1l2$-Norm Model for fitting ellipse to resist the Gaussian and Laplacian disturb simultaneously which will be solved by split Bregman iteration and shrink operator. The experimental results reveal that the proposed method not only works well with Gaussian Noise data but Laplacian Noise data and its mixed version. Several experimental examples are reported to demonstrate the robustness of the proposed approach.

Keywords: Ellipse fitting · $l1l2$-Norm · Split-Bregman iteration

1 Introduction

Ellipse fitting is much more difficult than circle fitting due to the reason that the curvature of the ellipse is not uniform and has more parameters. And Ellipse fitting is widely applied in the fields of computer vision. Thus, a large amount of ellipse fitting algorithms have been developed. Liang et al. [1] adopte the $l(p)$-norm with $p < 2$ in the direct least square fitting method to achieve outlier resistance, and also developed a robust ellipse fitting approach using the alternating direction method of multipliers. A robust and direct algorithm for the least-square fitting of ellipses to scattered data is proposed. Maini [2] propose a new algorithm to improve the robustness of direct least squares fitting under the premise of increasing the amount of computation. Zhang et al. [3] present an ellipse-fitting algorithm based on residual p-norm minimum is proposed to deal with the outliers of massive 3D point data. The experiments validate that the p-norm minimum is more robust than the least-squares algorithm, and the application of an adaptive threshold allows the algorithm to clearly distinguish the

Supported by Humanities and Social sciences project of Guizhou Provincial Department of Education, No.: 2019qn032, the National Natural Science Foundation of China, No.: 71761005 and 2017 Academic New Seedling Cultivation and Innovation Exploration Project of Guizhou University of Finance and Economics, No.: [2017]5736-002.

H. Song and D. Jiang (Eds.): SIMUtools 2020, LNICST 370, pp. 766–776, 2021.
https://doi.org/10.1007/978-3-030-72795-6_61

outliers. Zhang et al. [4] note that in order to achieve high measurement accuracy, a random phase retrieval approach based on difference map Gram-Schmidt othonormalization and Lissajous ellipse fitting method (DGS-LEF) is proposed. Shcherbakova et al. [5] demonstrate a feasibility of a novel, ellipse fitting approach, for simultaneous estimation of relaxation times T-1 and T-2 from a single 3D phase-cycled balanced steady-state free precession sequence. Liang et al. and Mulleti et al. [6,7] show a robust ellipse fitting method to alleviate the influence of outliers. The proposed algorithm solves ellipse parameters by linearly combining a subset of data points. Kurt et al. [8] propose an error measure which can be used both to measure the accuracy of any ellipse fitting method and to compare the accuracy of the ellipses fitted with different methods. By computing this measure it is possible to obtain the precision of the ellipse parameters with respect to the orthogonal distance residuals. Fu et al. [9] demonstrate a rapid and precise phase-retrieval method based on Lissajous ellipse fitting and ellipse standardization. After compensating phase-shift errors by ellipse standardization, the phase distribution is extracted with high precision. In fact, ellipse fitting is widely used in many fields. Liao et al. [10] use Ellipse fitting to judge the correct splitting point pair of the automatic segmentation for cell images. Mitchell et al. [11] develope Ellipse fitting to analyse electron diffraction patterns from polycrystalline materials, the result shown that the technique is robust and can be used to determine the pattern centre and the diameter of diffraction rings with a high degree of accuracy. Gabriel Mistelbauer et al. [12] show an elliptical blood vessel model can quantitatively gauge blood vessels in order to better understand their morphology. Augustyn et al. [13] apply the ellipse fitting algorithm in incoherent sampling measurements of complex ratio of AC voltages, the result show that it was possible to perform coherent and incoherent sampling measurements with controlled frequency deviation. Li [14] propose a class of multi ellipse fitting problems for densely connected contours and propose a evaluation framework of various composite data sets in practical application. Yatabe et al. [15] propose a method of avoiding such untrusted pixels within the estimation processes of HEFS, and used an experiment of measuring a sound field in air for real data.

Despite the ellipse fitting is importance and has many methods to solve it, however, there has been until now no computationally efficient ellipse specific fitting algorithm. In this case, this paper introduces a new fitting method called hybrid $l1l2$-Norm algorithm which absorb the advantages of $l1$-norm and $l2$-norm algorithm. By introducing weight w and discussing the ellipse fitting with different noise, we will show its robustness by compare with other convention approaches.

The remaining paper is organized as follows: In Sect. 2, we present a new model called the hybrid $l1l2$-Norm algorithm, and then we introduce the split Bregman iteration method to solve these parameters basing direct least squares fitting of ellipse. In Sect. 3, by analyzing different noises, we conducted some experiments to verity the robustness of the proposed method in ellipse fitting. In Sect. 4, it is the conclusion of the paper.

2 The Hybrid *l*1*l*2-Norm Algorithm

2.1 Problem Statement

For a two-dimensional plane coordinate system, a general conic can be expressed by an implicit second order polynomial as follows:

$$F(D,\delta) = D \cdot \delta = ax^2 + bxy + cy^2 + dx + ey + f = 0$$

where $D = (x^2, xy, y^2, x, y, 1)$ represents data points, $\delta = (a, b, c, d, e, f)^T$ represents parameter of the conic. (x_k, y_k), where $k = 1, 2, ..., n$, represents the coordinates of n data points, $F(D_k, \delta)$ is called the algebraic distance between the point (x_k, y_k) to the conic $F(D, \delta) = 0$. The fitting of a general conic may be approached by minimizing the sum of the squared algebraic distance of points, note

$$\sum_{k=1}^{n} F(D_k, \delta)^2$$

According to change the conic into an ellipse, we must add an ellipse-specific constraint $b^2 - 4ac < 0$ on the conic parameter [1].

The elliptic parameters are solved by minimizing the L2-norm of the algebraic distance, but when there are outliers in the data, the *l*2-norm will expand the influence of outliers, so that the fitting results are biased towards the parameters contained in the outliers. So *l*1-norm presented to estimate parameters can get a more robust. Thus, the parameters of the elliptic equation can be estimated by solving the following L1-norm minimization problems:

$$\begin{cases} \arg\min_\delta \|D\delta\|_1 \\ s.t. \quad b^2 - 4ac < 0 \end{cases} \tag{1}$$

In order to force the conic to be an ellipse, the conic parameter should satisfied the necessary condition $b^2 - 4ac < 0$, for convenience we take its subset $4ac - b^2 = 1$ instead, its matrix form can be presented as $\delta^T C \delta = 1$, where

$$\mathbf{C} = \begin{pmatrix} 0 & 0 & 2 & 0 & 0 & 0 \\ 0 & -1 & 0 & 0 & 0 & 0 \\ 2 & 0 & 0 & 0 & 0 & 0 \\ 0 & 0 & 0 & 0 & 0 & 0 \\ 0 & 0 & 0 & 0 & 0 & 0 \\ 0 & 0 & 0 & 0 & 0 & 0 \end{pmatrix}$$

By using Lagrange multiplier method, the constrained problem in Eq. (1) can be transformed into the following unconstrained problem:

$$\arg\min_\delta \{\|D\delta\|_1 + \lambda(\delta^T C \delta - 1)\} \tag{2}$$

where $\lambda > 0$ is Lagrange parameter.

The advantage of using $l1$-norm is that it can get a satisfied performance on Laplacian noise data, while $l2$-norm fits better for Gaussian noise data than $l1$-norm. In order to improve the accuracy of ellipse fitting and reduce the complexity of ellipse fitting, the hybrid $l1l2$-Norm Model is proposed to achieve data denoising. Unlike the $l1$-norm and $l2$-norm algorithm, the hybrid $l1l2$-Norm Model can impose different constraints on different data points using weight Matrix w. Obviously, when the selection of w is reasonable, the hybrid model can inherit the advantages of $l1$-norm and 12-norm algorithm. It has good robustness to Laplacian noise and Gaussian noise data.

2.2 Solution of the Hybrid $l1l2$-Norm Algorithm

According to the Eq. (2), the hybrid $l1l2$-Norm Algorithm can be shown as:

$$\arg\min_{\delta}\{\|w \circ D\delta\|_1 + \|(1-w) \circ D\delta\|_2 + \lambda(\delta^T C\delta - 1)\} \tag{3}$$

where λ is the Lagrange multiplier, w is the weigh. The simulation is a minimization problem with $l1$-norm and $l2$-norm at the same time. In order to get the accurate results, the split Bregman method [16] is adopted to solve the problem. For Eq. (3), it is a very difficult to solve the general $l1$-regularized optimization problem, so we introduce an auxiliary variable $s = D\delta$ in it, then the original problem is converted as

$$\begin{cases} \arg\min_{s}\{\|w \circ s\|_1 + \|(1-w) \circ s\|_2 + \lambda(\delta^T C\delta - 1)\} \\ \qquad\qquad\qquad\qquad\qquad\qquad s.t. \quad s = D\delta \end{cases} \tag{4}$$

By introducing the Bregman iteration variable t, the Eq. 4 can be transformed into the following unconstrained problems:

$$\arg\min_{\delta,s,t}\{\|w \circ s\|_1 + \|(1-w) \circ s\|_2 + \frac{\mu}{2}\|s - D\delta - t\|^2 + \lambda(\delta^T C\delta - 1)\} \tag{5}$$

Now, we solve these subproblems about δ, s, t one by one.

(1) the δ subproblem is

$$\arg\min_{\delta}\{\frac{\mu}{2}\|s - D\delta - t\|^2 + \lambda(\delta^T C\delta - 1)\} \tag{6}$$

It is a typical least square problem. Thus, the following closed solutions can be obtained as:

$$\delta^{k+1} = \frac{\mu D^T(s^k - t^k)}{\mu D^T D + 2\lambda C} \tag{7}$$

(2) the s subproblem is

$$\arg\min_{s}\{\|w \circ s\|_1 + \|(1-w) \circ s\|_2 + \frac{\mu}{2}\|s - D\delta - t\|^2\} \tag{8}$$

Set $F = \|w \circ s\|_1 + \|(1-w) \circ s\|_2 + \frac{\mu}{2}\|s - D\delta - t\|^2$. For Eq. (8), we first get the partial derivation of variable s, and then, we note the necessary conditions, i.e. $\frac{\partial F}{\partial s} = 0$, at the extreme value of the function. So we can obtain

$$\frac{\partial F}{\partial s} = w\frac{s}{|s|} + 2(1-w)s + \mu(s - D\delta - t)$$

where the symbol $|s|$ means to take absolute value for variable s. It can divide the issue into two circumstances.

1) set $s > 0$, get $w + 2(1-w)s + \mu(s - D\delta - t) = 0$
2) set $s < 0$, get $-w + 2(1-w)s + \mu(s - D\delta - t) = 0$

Since w and $1 - w$ are the coefficient matrix, each of these elements are between 0 and 1. According to this, the standard shrink operator [17] is used to solve these problems, we can get as:

$$s^{k+1} = \frac{shink(\mu(D\delta^{k+1+t^k}), w)}{2(1-w) + \mu} \tag{9}$$

where $shink(x, r) = \frac{x}{|x|}\max(|x| - r, 0)$

(3) the t subproblem is

$$t^{k+1} = t^k + D\delta^{k+1} - s^{k+1} \tag{10}$$

where t is Bregman variable, μ is the penalty parameter.

The complete algorithm of the minimization problem Eq. (3) with Bregman iteration is shown in Algorithm 1.

Algorithm 1. the Hybrid $l1l2$-Norm Algorithm based direct least squares fitting of ellipse with Bregman iteration

(1) set $w, \mu, \lambda, \delta^0, s^0, t^0, k = 0, \varepsilon = 10^{-6}.n_{maxiter} = 300$;
(2) For $k < n_{maxiter}$ and $\frac{\|D\delta^{k+1} - D\delta^k\|_2}{\|D\delta^{k+1}\|_2} > \epsilon$;
(3) Solve δ^{k+1} using Eq. (7);
(4) Solve s^{k+1} using Eq. (9);
(5) Solve t^{k+1} using Eq. (10);
(6) set $k = k + 1$;
(7) end.

when the ellipse parameter $\delta = (a, b, c, d, e, f)'$ is solved, according [18] and [19], we can normalize $f = 1$, so it gets $\delta = (a, b, c, d, e, 1)'$. Thus, ellipse's center (x_c, y_c), major semiaxis r_c, minor semiaxis r_c and angle θ have the following coordinates:

$$\begin{cases} x_c = \frac{2cd - be}{b^2 - 4ac} \\ y_c = \frac{2ae - bd}{b^2 - 4ac} \\ r_x = \sqrt{\frac{2(ax_c^2 + cy_c^2 + bx_cy_c - 1)}{a + c - \sqrt{(a-c)^2 + b^2}}} \\ r_y = \sqrt{\frac{2(ax_c^2 + cy_c^2 + bx_cy_c - 1)}{a + c + \sqrt{(a-c)^2 + b^2}}} \\ \theta = \frac{1}{2}\tan^{-1}\frac{b}{a-c} \end{cases} \tag{11}$$

3 Experimental Results

In this section, we conduct simulations and experiments to evaluate the performance of the proposed $l1l2$-Norm approach for ellipse fitting with simulated data. We main analysis our algorithm to compare with the $l2$-Norm based direct least-squares ellipse fitting (DLS). We note the scale normalization parameter $f_0 = 600$ when we calculate the center position of the circle and note the max iteration for convergence in hyper-renormalization 30. At the same time, we test the proposed algorithm with idea data from a set of ellipses points with parameters $x_c = 128, y_c = 256, r_x = 80, r_y = 64, \theta$ and the weight $w = 0.25$. Unless otherwise explained, all parameters are the same. In this paper, three kinds of experiments are discussed as follows.

Experiment 1: Containing Gaussian Noise in Ellipse Fitting
In this experiment, a serial of ellipse points polluted with gaussian noise with mean 0 and standard deviation 6 is simulated for averting the scatter matrix $D^T D$ to be a singular matrix. And the simulated ellipse parameters are set as $x_c = 128, y_c = 256, r_x = 80, r_y = 64, \theta$. If we keep the center and axis unchanged and rotation angle θ, in this paper, we will discuss the angle $\theta = 25, 45, 65, 85$, thus, 5 ellipses are acquired. For the DLS method and the proposed method, we can compute the parameters $x_c, y_c, r_x, r_y, \theta$, and then the result is represented in Table 1. We can get that no matter how you change the angle, all methods get a very precise results. But on the whole, the proposed method get higher accuracy than DLS method.

Experiment 2: Containing Laplacian Noise in Ellipse Fitting
In this experiment, a serial of ellipse points polluted with Laplacian noise with mean 9 and standard deviation 16 is simulated. The simulated ellipse parameters are set as $x_c = 128, y_c = 256, r_x = 80, r_y = 64, \theta = 45^0$. In this paper, we apply the DLS method and the proposed method to analyze different noise point, i.e. 5 noisy points, 15 noisy points, 35 noisy points and 65 noisy points. Comparison of their results (see Fig. 1), we can see that our approach gives a more precise than the DLS. While under the condition of laplacian noise, as the number of noise increases, our approach achieves a very competitive performance of the ellipses parameters estimate issue. The more detail are shown in Table 2.

Experiment 3: Containing Laplacian Noise and Gaussian Noise in Ellipse Fitting
In this experiment, we set a serial of ellipse points polluted with gaussian noise with zero-mean and standard deviation 2 and at the same time with Laplacian noise with mean 16 and standard deviation 32. The simulated ellipse parameters are set as $x_c = 128, y_c = 256, r_x = 80, r_y = 64, \theta = 45^0$. In this paper, when we set the weight to a fixed value $w = 0.25$, $l1l2$-Norm approach obtain the highest accuracy and far superior to $l2$-norm method (see Fig. 2). It also show that when the input data is polluted with a low density Laplacian Noise and Gaussian Noise, with the increasing of the mixed noise, the hybrid $l1l2$-Norm algorithm produces satisfactory fitting result. The ellipse fitting results of $x_c, y_c, r_x, r_y, \theta$ of the two methods are shown in Table 3. Now, let's talk about different weights w. In this case, the simulated ellipse parameter is set as $x_c = 128, y_c = 256$,

$r_x = 80, r_y = 64, \theta = 45°$ and we choose 25 noise points. It should be noted that other parameter are not change. When we change the weight w, we can estimate the ellipse parameters $x_c, y_c, r_x, r_y, \theta$. According to the Eq. (3), we can see that when our weight w is greater than 0.5, the proportion of $l1$-norm method is larger than that of $l2$-norm method. Otherwise, it's the opposite. The more detail are shown in Table 4. We obtain that it is not the bigger of the weight w the better and at the same it is not the smaller the weight w the better. $w = 0.25$ is more accurate than other data for estimating ellipse parameters.

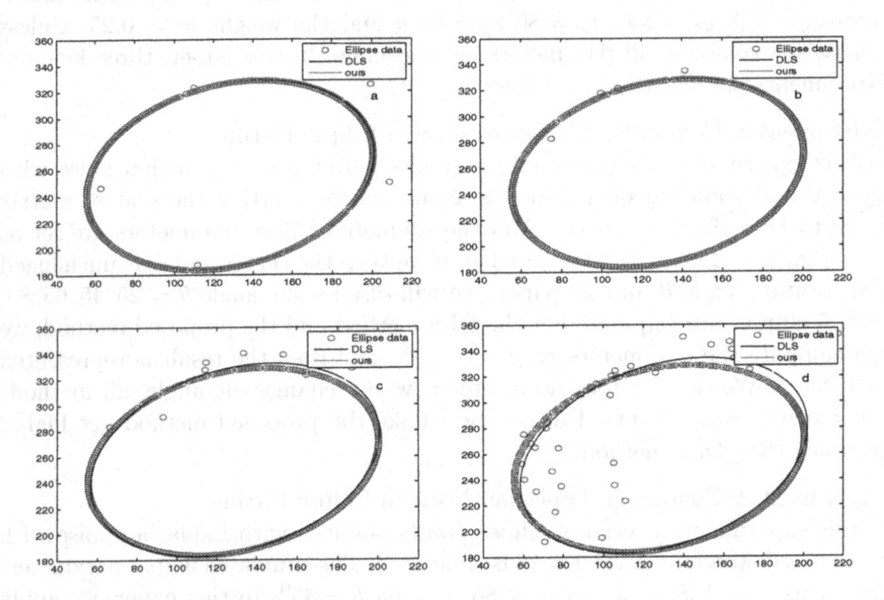

Fig. 1. Ellipse Fitting with Laplacian noise.(a) 5 noisy points.(b) 15 noisy points.(c) 35 noisy points.(d) 65 noisy points.

Table 1. Ellipse parameters fitting results with Gaussian noise.

Angle	Method	r_x	r_y	x_c	y_c	θ
25	DLS	79.7815	64.5943	127.6936	256.3401	23.6879
25	Ours	79.8369	64.6328	127.6527	256.3573	24.3919
45	DLS	80.6144	63.8754	127.8803	256.1238	45.1474
45	Ours	80.3639	63.9584	127.9081	256.0573	44.8488
65	DLS	80.2682	64.3844	128.0290	256.3719	65.9796
65	Ours	79.7223	64.8028	128.1065	256.5538	66.0449
85	DLS	80.6070	63.9293	128.2050	256.4321	84.4931
85	Ours	80.0320	64.1070	128.2758	256.3606	84.7569

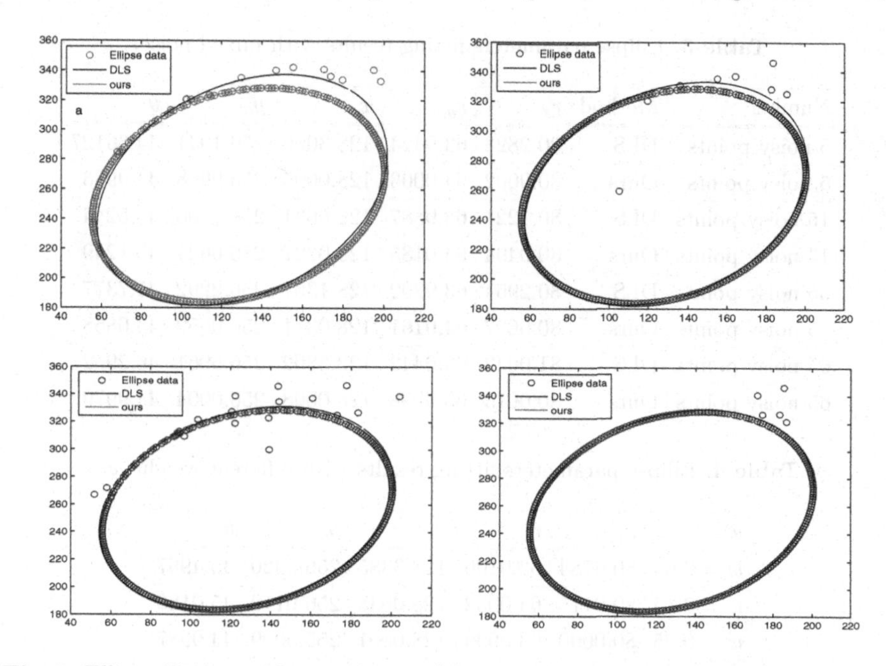

Fig. 2. Ellipse Fitting with mixed noise.(a) 5 noisy points.(b) 15 noisy points.(c) 35 noisy points.(d) 65 noisy points.

Table 2. Ellipse parameters fitting results with Laplacian noise.

Number	Method	r_x	r_y	x_c	y_c	θ
5 noisy points	DLS	80.3658	63.9807	128.0540	256.2458	44.1923
5 noisy points	Ours	79.9992	63.9995	127.9986	256.0000	45.0039
15 noisy points	DLS	81.2586	63.6584	128.5470	256.6195	47.0657
15 noisy points	Ours	79.9932	64.0006	127.9956	255.9967	45.0030
35 noisy points	DLS	84.1981	62.4486	127.5874	256.0333	45.5419
35 noisy points	Ours	80.0087	64.0040	128.0051	256.0035	45.0075
65 noisy points	DLS	88.7864	60.7384	131.8915	259.9820	46.0048
65 noisy points	Ours	79.8599	64.0552	128.0141	256.0038	44.9933

Table 3. Ellipse parameters fitting results with mixed noise.

Number	Method	r_x	r_y	x_c	y_c	θ
5 noisy points	DLS	80.2828	63.9424	128.3090	256.1941	44.66127
5 noisy points	Ours	80.0002	64.0003	128.0005	255.9998	44.9983
15 noisy points	DLS	80.1221	63.9887	128.0631	256.1300	45.5244
15 noisy points	Ours	80.0494	64.0185	128.0227	256.0621	45.1249
35 noisy points	DLS	80.2954	63.9292	128.1387	256.2292	45.7377
35 noisy points	Ours	80.0697	64.0161	128.0311	256.0652	45.0858
65 noisy points	DLS	81.0642	63.6442	128.3892	256.6061	46.2952
65 noisy points	Ours	80.0013	63.9998	128.0008	256.0004	44.9979

Table 4. Ellipse parameters fitting results with different weight w.

w	r_x	r_y	x_c	y_c	θ
$w = 0.05$	80.6784	63.9100	128.3385	256.5320	45.4967
$w = 0.15$	80.0115	64.0032	128.0038	256.0118	45.0186
$w = 0.25$	80.0000	63.9999	128.0001	255.9999	44.9987
$w = 0.5$	80.0019	63.9993	128.0008	256.0009	44.9996
$w = 0.65$	80.2000	64.0419	128.1200	256.0378	45.1254
$w = 0.75$	79.9596	64.0534	127.9523	256.0484	45.2212
$w = 0.85$	80.0193	63.9942	128.0033	256.0007	44.9756
$w = 0.95$	80.5697	63.7796	128.0124	256.1112	44.8990

4 Conclusions

This paper has presented a new method for direct $l1l2$-Norm algorithm based ellipse fitting. In this study, we describe three experiments: ellipse parameters fitting results with Gaussian noise, ellipse parameters fitting results with Laplacian noise and ellipse parameters fitting results with mixed noise.

For the experiment 1, under the condition that other parameters remain unchanged, the ellipse parameters with DLS model and the hybrid $l1l2$-Norm model are predicted by the rotation angle. At 45°, the prediction result of our proposed model is the most accurate. For the experiment 2, two models are used to estimate the ellipse parameters by introducing different amount of noise. No matter how the amount of noise changes, we can see that the prediction result of our proposed model is better than that of DLS model. For the experiment 3, we mainly discuss the influence of the change of noise quantity and weight on the prediction results of ellipse parameters for DLS model and the hybrid $l1l2$-Norm model. It is also concluded that the prediction of our proposed is better than that of DLS model no matter how the amount of noise changes. As for the change of weight w, $w = 0.25$ is the optimal value. In fact, we can also discuss that the weight is a matrix form, which will be discussed in the next topic.

It can see that the proposed model is good for Laplacian and all kinds of mixed noise without degradation of Gaussian performance. Comparison experiments suggest that the proposed method can bring great challenge for artificial intelligence regardless of the noise type.

References

1. Liang, J.L., Li, P.L., Zhou, D.Y.: Robust ellipse fitting via alternating direction method of multipliers. Signal Process. **164**, 30–40 (2019)
2. Maini, E.S.: Enhanced direct least square fitting of ellipses. Int. J. Pattern Recogn. Artif. Intell. **20**, 939–953 (2006)
3. Zhang, L., Cheng, X., Wang, L.Y.: Ellipse-fitting algorithm and adaptive threshold to eliminate outliers. Survey Rev. **51**, 250–256 (2019)
4. Zhang, L., Cheng, X., Wang, L.Y.: Random three-step phase retrieval approach based on difference map Gram-Schmidt orthonormalization and Lissajous ellipse fitting method. Optics and Lasers in Eng. **121**, 11–17 (2019)
5. Shcherbakova, Y., Van den Berg, C.A.T., Moonen, C.T.W., Bartels, L.M.: PLANET: An ellipse fitting approach for simultaneous T-1 and T-2 mapping using phase-cycled balanced steady-state free precession. Magnetic Resonnace Med. **79**, 711–722 (2018)
6. Liang, J.L., Zhang, M.H., Liu, D., Zeng, X.J.: Robust ellipse fitting based on sparse combination of data points. IEEE Trans. Image Process. **22**, 2207–2218 (2013)
7. Mulleti, S., Seelamantula, C.S.: Ellipse fitting using the finite rate of innovation sampling principle. IEEE Trans. Image Process. **25**, 1451–1464 (2015)
8. Kurt, O., Arslan, O.: A general accuracy measure for quality of elliptic sections fitting. Measurement **145**, 640–647 (2019)
9. Fu, Y., Wu, Q., Yao, Y., Gan, Y., Liu, C., Yang, Y., Tian, J., Xu, K.: Rapid and precise phase retrieval from two-frame tilt-shift based on Lissajous ellipse fitting and ellipse standardization. Optics Express **28**(3), 3952–3964 (2020)
10. Liao, M., Zhao, Y.Q., Li, X.H.: Automatic segmentation for cell images based on bottleneck detection and ellipse fitting. Neurocomputing **173**, 615–622 (2016)
11. Mitchell, D.R.G., Van den Berg, J.A.: Development of an ellipse fitting method with which to analyse selected area electron diffraction patterns. Ultramicroscopy **160**, 140–145 (2016)
12. Mistelbauer, G., Zettwitz, M., Schernthaner, R.: Visual assessment of vascular torsion using ellipse fitting. Proc. Eurographics Workshop on Visual Comput. Biol. Med. **12**, 129–133 (2018)
13. Augustyn, J., Kampik, M.: Application of ellipse fitting algorithm in incoherent sampling measurements of complex ratio of AC voltages. IEEE Trans. Instrument. Measurement **66**, 1117–1123 (2017)
14. Li, H.: Multiple ellipse fitting of densely connected contours. Inf. Sci. **502**, 330–345 (2019)
15. Yatabe, K., Ishikawa, K., Oikawa, Y.: Hyper ellipse fitting in subspace method for phase-shifting interferometry: practical implementation with automatic pixel selection. Optics Express **25**(23), 29401–29416 (2017)
16. Miao, C., Yu, H.Y.: A general-thresholding solution for l(p)(0<p<1) regularized CT reconstruction. IEEE Thans. Image Process. **24**, 5455–5468 (2015)

17. Goldstein, T., Osher, S.: A new wavelet threshold function and denoising application. SIAM J. Imag. Sci. **2**, 323–343 (2009)
18. Zhou, G., Zhong, K., Li, Z., Shi, Y.: Direct least absolute deviation fitting of ellipses. Math. Prob. Eng. **6**, 1–11 (2020)
19. Rosin, P.L.: Further five point fit ellipse fitting. Graph. Models Image Process. **61**(5), 245–259 (1999)

Author Index

s
sher Services